Molecular Immunology

A Textbook

edited by

M. Zouhair Atassi
Carel J. van Oss
Darryl R. Absolom

MARCEL DEKKER, INC. New York and Basel

Library of Congress Cataloging in Publication Data
Main entry under title:

Molecular immunology.

Includes indexes.
1. Immunochemistry. I. Atassi, M. Z. II. van Oss, Carel J. III. Absolom, Darryl R. [DNLM: 1. Immunochemistry. 2. Molecular biology. QW 504.5 M718]
QR183.6.M64 1984 616.07'9 83-23991
ISBN 0-8247-7045-5

MARCEL DEKKER, INC.
270 Madison Avenue, New York, New York 10016

Current printing (last digit):
10 9 8 7 6 5 4 3 2

PRINTED IN THE UNITED STATES OF AMERICA

Preface

It all started with a phone call from our publisher, Dr. Maurits Dekker, who had read a book review by one of us. The review contained an oblique allusion to the need for an up-to-date textbook of immunochemistry. Dr. Dekker asked, "Why don't you write one yourself?" The answer was brief and simple. "You can't do it; no single immunologist can, or indeed should, do it any more." Advances in recent years have made the field so vast that it is virtually impossible for any one individual to write an advanced textbook on the subject. Only a panel of experts could accomplish this goal.

Now the fact remains that although no single person can write such a textbook any more, for those of us who are in the habit of teaching a graduate course on the fundamentals of immunochemistry, a good up-to-date textbook would still be uncommonly handy. A few of us therefore combined forces and decided to organize and edit such a work. The advantage of being able to call upon more than three dozen experts in the various subdisciplines is that it reasonably well assured the expertise in each subject; the disadvantage may be a residual heterogeneity of style and organization with, however, only minor instances of factual overlap, not to mention rare items of actual contradiction.

Accordingly, the book *Molecular Immunology* was conceived as a textbook for advanced graduate students and postdoctoral fellows aiming at a research career in molecular immunology. The subject matter was designed to cover topics currently at the forefront of molecular immunology.

A disadvantage of aiming at a textbook (and not a handbook) is that the graduate student ought to be able to afford it. This imposes a limit on the overall size of the book

and hence on the number of pages per chapter. Accordingly the number of pages per chapter as well as the number of references have been kept to the minimum required for optimal coverage of the subject matter.

To allow ourselves a somewhat wider latitude in the subjects treated, we decided to adopt the title "Molecular Immunology" for the textbook. This was done with the kind blessing of the editor in chief of the journal *Molecular Immunology*.

M. Zouhair Atassi
Carel J. van Oss
Darryl R. Absolom

Contributors

C. John Abeyounis, Ph.D. Department of Microbiology, School of Medicine, State University of New York at Buffalo, Buffalo, New York

Darryl R. Absolom, Ph.D. Division of Surgical Research, Research Institute, The Hospital for Sick Children, and Department of Mechanical Engineering, University of Toronto, Toronto, Ontario, Canada

Boris Albini, M.D. Department of Microbiology, State University of New York at Buffalo, Buffalo, New York

Aftab Ahmed, Ph.D. Scientific Director, U.S. Naval Research Unit 3, FPO, New York

M. Zouhair Atassi, Ph.D., D.Sc.* Department of Immunology, Mayo Clinic, Rochester, Minnesota

Richard B. Bankert, V.M.D., Ph.D. Department of Molecular Immunology, Roswell Park Memorial Institute, Buffalo, New York

Constantin A. Bona, M.D., Ph.D. Department of Microbiology, The Mount Sinai Medical Center, New York, New York

Nickolas J. Calvanico, Ph.D. Department of Medicine, The Medical College of Wisconsin, and Research Service, Wood Veterans Administration Medical Center, Milwaukee, Wisconsin

Nicholas Catsimpoolas, Ph.D. Department of Biochemistry and Behavioral Sciences, Boston University School of Medicine, Boston, Massachusetts

*Current affiliation: Department of Biochemistry, Baylor College School of Medicine, Texas Medical Center, Houston, Texas.

Roger K. Cunningham, Ph.D. The Witebsky Center for Immunology and Department of Microbiology, School of Medicine, State University of New York at Buffalo, Buffalo, New York

Chella S. David, Ph.D. Department of Immunology, Mayo Medical School, Rochester, Minnesota

Keith J. Dorrington, Ph.D., D.Sc.* Department of Biochemistry, University of Toronto, Toronto, Ontario, Canada

Philip Early, Ph.D. Department of Physiology, School of Medicine, University of California, San Francisco, California

Roger B. Eaton, Ph.D. Department of Microbiology, School of Medicine, State University of New York at Buffalo, Buffalo, New York

A. M. Fagundus, Ph.D. Department of Microbiology, State University of New York at Buffalo, Buffalo, New York

Douglas T. Fearon, M.D. Department of Medicine, Harvard Medical School, Boston, Massachusetts

Matthew A. Gonda, Ph.D. Electron Microscopy Laboratory, NCI-Frederick Cancer Research Facility, Frederick, Maryland

Charles A. Janeway, Jr., M.D. Department of Pathology, Yale University School of Medicine, and Howard Hughes Medical Institute, New Haven, Connecticut

Yi-Her Jou, Ph.D. Department of Molecular Immunology, Roswell Park Memorial Institute, Buffalo, New York

Takeo Juji, M.D. Blood Transfusion Service, The Tokyo Women's Medical College, Shinjuku, Tokyo, Japan

Kyoichi Kano, M.D. Department of Immunology, The Institute of Medical Science, University of Tokyo, Minato-ku, Tokyo, Japan

Christopher J. Krco, Ph.D. Department of Immunology, Mayo Clinic, Rochester, Minnesota

W. G. Laver, Ph.D. Department of Microbiology, John Curtin School of Medical Research, Australian National University, Canberra City, Australia

Erik P. Lillehoj, Ph.D.† Division of Immunopathology, Department of Clinical Pathology, William Beaumont Hospital, Royal Oak, Michigan

Gary William Litman, Ph.D. Macromolecular Biochemistry Section, Walker Laboratory, Sloan-Kettering Institute, Rye, New York

Patrick M. Long, Ph.D. Department of Immunology, Mayo Clinic and Mayo Medical School, Rochester, Minnesota

George L. Mayers, Ph.D. Department of Molecular Immunology, Roswell Park Memorial Institute, Buffalo, New York

*Current affiliation: Connaught Research Institute, Willowdale, Canada.
†Current affiliation: Laboratory of Immunogenetics, NIAID, National Institutes of Health, Bethesda, Maryland.

Daniel J. McCormick, Ph.D. Department of Immunology, Mayo Graduate School of Medicine, Mayo Clinic, Rochester, Minnesota

Felix Milgrom, M.D. Department of Microbiology, School of Medicine, State University of New York at Buffalo, Buffalo, New York

Miercio E. A. Pereira, M.D., Ph.D. Division of Geographic Medicine, Tufts University School of Medicine, Boston, Massachusetts

Terry A. Potter, Ph.D. Albert Einstein College of Medicine, Bronx, New York

M. D. Poulik, M.D. Section of Immunopathology, Department of Clinical Pathology, William Beaumont Hospital, Royal Oak, Michigan

Milton R. J. Salton, Ph.D. Department of Microbiology, New York University School of Medicine, New York, New York

Konrad Schauenstein, Ph.D. Institute of General and Experimental Pathology, University of Innsbruck Medical School, Innsbruck, Austria

H. Edward Schmitz, Ph.D. Department of Immunology, Mayo Clinic, Rochester, Minnesota

Regina R. Skelly, Ph.D.* Department of Immunology, Merck Sharp and Dohme Research Laboratories, Rahway, New Jersey

Arthur G. Steinberg, Ph.D. Case Western Reserve University, Cleveland, Ohio

Carel J. van Oss, Ph.D. Immunochemistry Laboratory, Department of Microbiology, School of Medicine, and Department of Chemical Engineering, School of Engineering and Applied Sciences, State University of New York at Buffalo, Buffalo, New York

Adrian Vladutiu, M.D., Ph.D. Departments of Pathology, Microbiology, and Medicine, School of Medicine, State University of New York at Buffalo, Buffalo, New York

Kenneth W. Walls, Ph.D. Protozoal Diseases Branch, Division of Parasitic Diseases, Center for Infectious Diseases, Centers for Disease Control, Public Health Service, U.S. Department of Health and Human Services, Atlanta, Georgia

Georg Wick, M.D. Institute of General and Experimental Pathology, University of Innsbruck Medical School, Innsbruck, Austria

Takeshi Yoshida, M.D., D.Med.Sci. Department of Pathology, University of Connecticut Health Center, Farmington, Connecticut

Colin R. Young, Ph.D.† Department of Immunology, Mayo Clinic and Mayo Medical School, Rochester, Minnesota

Marek B. Zaleski, M.D., Ph.D. Department of Microbiology, School of Medicine, State University of New York at Buffalo, Buffalo, New York

*Current affiliation: Department of Immunology, U.S. Naval Medical Research Unit 3, FPO, New York.
†Current affiliation: Department of Pathology, Division of Immunology, University of Cambridge, Cambridge, England.

Contents

1

Structural Requirements for Immunogenicity and Antigenicity

Colin R. Young* Mayo Clinic and Mayo Medical School, Rochester, Minnesota

I. INTRODUCTION

A. Definitions

Some years ago (Sela, 1969) proposed the following definitions for the main immunological functions of an antigen:

1. Immunogenicity: the capacity to induce an immune response characterized by the formation of specific immunoglobulins and/or specifically committed lymphocytes
2. Antigenicity *sensu stricto*; the capacity of reacting with and binding to specific immunoglobulins and/or cellular receptors
3. Allergenicity: the capacity to elicit various types of allergic reactions and tissue lesions in sensitized animals having specific immunoglobulins and/or committed lymphocytes
4. Tolerogenicity: the capacity to induce specific immunological unresponsiveness, including antibody formation and/or cellular immunity

An understanding of the molecular basis of immunity is partly due to the increase in knowledge over the past 20 years of the number, size, distribution, chemical nature, and specificities of antigenic determinants. A large body of evidence suggests that the induction of humoral and/or cellular immunity depends in part on the molecular structure, and that the structural requirements for immunogenicity differ from those for antigenicity. For example, denatured proteins are often less immunogenic than the corresponding native protein. Also, bacterial proteins are generally better immunogens

*Current affiliation: University of Cambridge, Cambridge, England.

than serum proteins but show little difference in their behavior as antigens, whereas self-aggregation of a protein is usually associated with a negligible change in antigenicity but with a considerable increase in immunogenicity.

Knowledge of the sizes and structures of antigenic determinants of polysaccharides, nucleic acids, and synthetic polypeptides has been obtained from studies of the ability of fragments of antigen to inhibit the reaction of the whole molecule with its corresponding antibody. Synthetic antigens provide much information for delineating relationships between antigen structure and lymphocyte activation, since it is possible to synthesize tailor-made antigens for asking specific questions. The relationship between structure and antigenicity is more complex for proteins in that it depends to a large extent on the overall three-dimensional configuration of the molecule. Fibrous proteins, however, represent an antigenic situation that could be regarded as intermediate between linear polysaccharides and globular proteins, because of their repetition of amino acid sequence in the polypeptide chain. Studies on the antigenicity of globular proteins are more complex because of the absence of repeating sequences of amino acid residues.

The discussion of antigenicity and immunogenicity in this chapter is divided into three sections. The first section is basically a brief résumé of substances that can act as antigens: proteins, polysaccharides, synthetic polypeptides, nucleic acids, chemically modified antigens, and low-molecular-weight substances. The second section is devoted to studies on the immunogenicity of molecules: criteria for immunogenicity and the factors affecting immunogenicity. The third section deals with an analysis of the relationship between chemical structure, immunogenicity, and lymphocyte function: the hapten-carrier relationship, antigen structural requirements for cell cooperation and lymphocyte activation, and structural requirements for the induction of immune responsiveness and immune suppression. This chapter is not intended to be a detailed discussion encompassing all of the structural requirements for antigenicity and immunogenicity of proteins, polypeptides, polysaccharides, blood group antigens, bacterial antigens, viral antigens, or tumor antigens, since chapters dealing with each of these specifically appear in this book. Rather, this chapter is intended to be more of a general discussion of some of the concepts involved in antigenicity and immunogenicity.

II. ANTIGENS

We should now consider which substances may be antigenic, that is, will induce the formation of antibodies and will then react with these antibodies. Conventionally, a large amount of information has been derived from: first, investigations using rabbits as the laboratory animal of first choice; second, from the necessity for large-scale immunization of humans to prevent various infectious disease; third, from clinical observations of people having diseases, in whom antibody production was taking place.

A. Proteins

Early studies clearly established that proteins were good antigens when injected into an animal species other than the one from which they originated. For example, bovine serum albumin is a good antigen when used to immunize rabbits. When immunizing mammals with homologous proteins, foreignness of protein has conventionally been thought to be a requirement for a protein to be antigenic; that is, the antibody-forming system recognized its own proteins as being not foreign and hence did not react with them. However,

recent data with rabbit serum albumin and mouse myoglobin have demonstrated that these self-proteins, when used to immunize a host, can readily elicit an auto-antibody response; see Chap. 2.

B. Polysaccharides

These have proven to be very useful in immunochemical studies because they provide antigens of relatively simple structure by which many of the detailed structural aspects of antigenic determinants and antibody combining sites have been elucidated. Examples of polysaccharides often used as antigens are dextran and levan. Polysaccharide antigens are somewhat unusual in that in the purified form they may stimulate antibody formation in some animal species and not in others; see Chap. 3.

C. Synthetic Polypeptides

These peptides provide a series of antigens of desired specifications for studies on the structural requirements for antigenicity. Chemical methods now allow the preparation of the various types of poly-α-amino acids: first, homopolymers of a single amino acid; second, copolymers in which short peptides of known sequence are linked together; third, random copolymers of several amino acids; fourth, multichain copolymers; fifth, polymers with chains in which a peptide is repeated at intervals.

A very important concept discovered using synthetic polypeptides is that the antigenically important regions of the molecule must be accessible to the antibody-forming system. This was demonstrated as follows: two multichain copolymers were synthesized, each containing lysine, alanine, and a peptide of glutamic acid and tyrosine. Both had a poly-L-lysine backbone. One was designated (T,G)–A--L; in the other the order of addition was reversed (Sela, 1969). A schematic representation of these two synthetic polypeptides is shown in Figure 1.

Only the polymer with the glutamic acid-tyrosine residues on the outside was found to be antigenic in rabbits. If, however, the polypeptide backbone was made as a copolymer

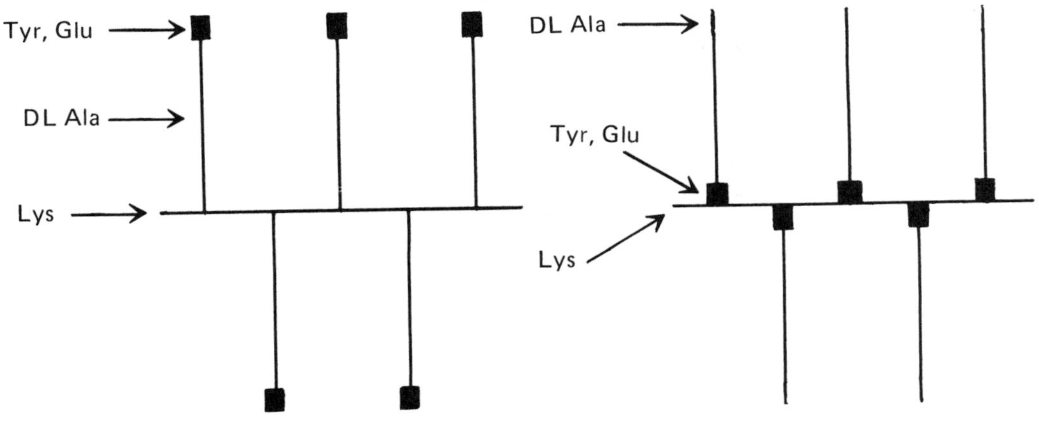

Figure 1 Synthetic diagram of two synthetic copolymers, p(Tyr,Glu)-pDL Ala-pLys and pDLAla-p(Tyr-Glu)-pLys.

of lysine and alanine, so that there was more space between the ϵ-amino groups of lysine to which the side chains were attached, then a polymer with glutamic acid-tyrosine attached directly to the lysines with alanine side chains at the outside was found to be a strong antigen. Thus accessibility does not require per se a position at the end of the chains.

D. Nucleic Acids

It is very difficult to induce antibody formation to purified nucleic acids. However, it is quite possible to prepare antibodies to DNA, using denatured DNA coupled to methylated bovine serum albumin and injected with adjuvant. Serum antibodies to nucleic acids occur naturally in humans suffering from lupus erythematosus and in NZB/NZW mice. The antibodies in some sera react with denatured DNA, and in others with native DNA, and some antibodies apparently react to antigenic determinants present in both native and denatured DNA. The mechanism by which these antibodies are produced in patients with lupus erythematosus is not as yet fully understood.

E. Chemically Modified Antigens

Much knowledge of the relationships of antigenic determinants came from experiments in which groups of known structure were attached by covalent bonds to protein antigens, after which antibodies were obtained with complementarity directed toward the groups that had been introduced. This work was pioneered by Landsteiner in the 1940s (Landsteiner, 1945), using proteins coupled to low-molecular-weight substances by means of a diazonium salt. The low-molecular-weight groups which are not antigenic in their free form are called *haptens*, and the groups attached to the protein are referred to as the *haptenic group*. The main reactions used routinely in laboratories for introducing such groups into proteins are: iodination, diazotization and coupling, reaction with isocyanates and isothiocyanates, dinitrophenyl (DNP) derivatives, the mixed anhydride reaction, reaction with carbodiimides, reactions with penicillin, and coupling of ribonucleosides and nucleotides to proteins. However, there are certain drawbacks to all of these techniques. First, antibodies may be produced to determinants on the protein which have not reacted with the reagent. Second and more important, the groups introduced may affect the conformation of the protein.

F. Low-Molecular-Weight Substances

Many recent studies have shown that low-molecular-weight substances can function as complete antigens, either in their free form (i.e., not coupled to a large protein) or when adsorbed to carbon particles. Examples of the former are very small synthetic peptides of natural molecules, such as myoglobin and influenza virus. Examples of the latter are hormones such as angiotensin II amide, an octapeptide Asn-Arg-Val-Tyr-Ile-His-Pro-Phe of molecular weight 1045. Also, antibiotics such as bacitracin, gramicidin S, and oxytetracycline are antigenic.

III. IMMUNOGENICITY

A. Criteria for Immunogenicity

Until recently the immunogenicity of an antigen was established by two criteria: the production of specific immunoglobulins and the establishment of cell-mediated immunity.

However, there are some low-molecular-weight antigens, such as DNP-oligolysines, that are nonimmunogenic on their own but are able to induce antibody formation, but not delayed hypersensitivity, when immunized with complete Freund's adjuvant. More recent data have shown that very small free synthetic peptides of natural molecules of 6 to 10 amino acids can induce the formation of antibody, with or without cell-mediated immunity. Since many factors obviously affect the immune response to an antigen, the characterization of a substance as immunogenic or nonimmunogenic can be assigned only in operational terms.

Antibody Formation

Cellular processes involved in antibody formation require the participation of macrophages, bone-marrow-derived lymphocytes (B cells), and thymus-derived lymphocytes (T cells). For many antigens T cells function as helper cells. However, for a small number of antigens, T-helper cells are not required to activate B lymphocytes. These antigens are termed *thymus-independent antigens*. Examples of some thymus-independent antigens are bacterial polysaccharides and some polymerized proteins. The mechanism by which thymus-independent antigens act is not fully understood, but the immune response to such antigens is different from the response to thymus-dependent antigens in that the antibody produced is predominantly of the IgM class and little or no immunological memory is acquired.

The normal course of events in the primary and secondary immune response to most defined antigens is as follows: The first contact with antigen induces an immunological memory and proliferation of antigen-sensitive cells, and a second contact with antigen causes massive differentiation of B cells into antibody-secreting plasma cells. This overall picture for the antibody response is not necessarily the case for all antigens, particularly multideterminant antigens. In multideterminant antigens, antibodies toward the various determinants do not appear all at the same time during the course of immunization (see Chap. 24).

Cellular Immunity

Cell-mediated immunity and delayed-type hypersensitivity is an essential function of T cells. Interaction of specific antigen with antigen-sensitive T cells induces them to proliferate and release soluble mediators responsible for the tissue lesions characteristic of delayed-type hypersensitivity. As yet it is unclear whether the structural determinants on an antigen responsible for lymphocyte proliferation are the same as or different from the structural determinants on the antigen responsible for the molecular mechanisms leading to the release of the lymphocyte mediators. It seems likely that these may well require different molecular properties from the antigen.

B. Factors Affecting Immunogenicity

Size

Traditionally, it has been held that the immunogenicity of antigens is directly related to their molecular size. Very large molecules such as horse spleen ferritin, keyhole limpet hemocyanin, or particulate antigens such as viruses and bacteria are highly immunogenic. On the other hand, polymers of glutamic acid, lysine, and tyrosine elicit the same antibody response even over a threefold variation of molecular weight. However, a

Table 1 Examples of Some Low-Molecular-Weight Immunogens

Immunogen	Molecular weight	Immune response
Artificial and synthetic peptides		
Amino acid copolymers such as GAT, GL, GLΦ	4000–5000	Antibody
Bifunctional haptens	600–1000	Antibody and delayed hypersensitivity
DNP-amino acids	⩾300	Antibody and delayed hypersensitivity
Synthetic encephalitogenic peptides	⩾700	Delayed hypersensitivity
Synthetic myoglobin peptides	900–3000	Antibody and lymphocyte proliferation
Native peptides		
Insulin and peptide chains	⩽6000	Antibody
Gastrin	2000	Antibody
Angiotensin II	1000	Antibody and delayed hypersensitivity
Oxytocin	1007	Antibody

parallel does exist between molecular size and the ability to induce antibody formation for polymerized antigens of various sizes having similar repeating structures. It has been shown that there is a progressive decrease in the immunogenicity and tolerogenicity of fractions of decreasing molecular weight of type III pneumococcal polysaccharide (Howard et al., 1971). However, recently it has been shown that very small peptides, of only 7 to 21 amino acids, of sperm whale myoglobin are very effective both as immunogens and tolerogens (Young and Atassi, 1982a). With levan, on the other hand, a decline in immunogenicity with reduction of molecular weight was not accompanied by loss of tolerogenicity. Generally, high molecular size favors immunogenicity; however there are examples in which large macromolecules are nonimmunogenic, for example high-molecular-weight fully substituted polylysines, polyglutamic acid, and various celluloses.

Generally, low-molecular-weight antigens are not the best immunogens. As shown in Table 1, several compounds below 5000 molecular weight have been demonstrated to be immunogenic. One important question to ask in this context is: What is the minimum molecular size for immunogenicity? Using a series of oligolysines of increasing sizes monosubstituted on the α-N-terminal lysine by a DNP group, Schlossman (Schlossman et al., 1965) showed very clearly that mono-α-DNP-oligolysines smaller than heptalysine were incapable of eliciting anti-DNP-antibody production or cellular hypersensitivity. However, exceptions to this heptapeptide rule are the p-azobenzenearsonate derivative of tyrosine and histidine and dinitrophenylated amino acids. Using free synthetic peptides of increasing size, it has recently been shown that a free antigenic site (i.e., uncoupled) of as few as six amino acids is effective in eliciting both humoral and cellular immunity (Young and Atassi, 1982b).

The experiments with DNP-oligolysines of varying length established that delayed-type hypersensitivity requires a molecule of at least seven amino acid residues, whereas smaller substituted oligolysines were capable of inducing antibody formation. However, this finding occurred only when the immunogens were injected together with complete Freund's adjuvant containing mycobacterium. However, without adjuvant a reverse phenomenon occurs; that is, the smaller the molecule, the more it tends to induce delayed-type hypersensitivity in the absence of circulating antibody. It is possible that these smaller molecules containing a single immunodeterminant may preferentially induce delayed-type hypersensitivity over humoral immunity. In general, weak immunogens favor the establishment of the delayed-hypersensitivity state.

Conformation

For large antigens the antigenic determinants are those portions which are superficially exposed or protruding parts of the antigenic molecule. For example, the antigenic sites of sperm whale myoglobin have been shown (Atassi, 1975) to reside at conformationally sensitive portions of the polypeptide chain, located on the surface of the molecule and sometimes referred to as the "elbow regions." For multichain amino acid copolymers it has been demonstrated that the immune response can be directed against conformational determinants produced by both the tertiary and quarternary structures. For example, the tripeptide Tyr-Ala-Glu attached to a branched polymer of DL-alanine and lysine induces antibodies inhibitable by the tripeptide, whereas a high-molecular-weight polymer of the tripeptide forms α helices and induces antibodies against conformational determinants not present in the tripeptide.

It has been assumed that an α-helical configuration favors immunogenicity, and that oligopeptides should only be immunogenic when large enough to assume the α-helical conformation. Oligopeptides derived from L-alanine and co-oligomers of L-alanine with ν-methyl-L-glutamate possess a random coil conformation in trifluoroacetic and dichloroacetic acids, whereas in trifluoroethanol the pentamer and larger oligomers are in helical form. The lowest polymeric size for an α helix in nonaqueous solvents is an octapeptide of γ-methyl glutamate nonalanine or undecapeptide of β-methylaspartate. However, studies on the immunogenicity of oligopeptides requires more knowledge not only of their primary structure but also of their stereochemical conformation.

Antigenic determinants of globular proteins have been designated to be either of a continuous or a discontinuous type. Continuous antigenic determinants are those formed by a linear sequence of amino acids of a polypeptide chain; for example, antigenic site 1 of sperm whale myoglobin occurs in amino acid sequence 15–21. Discontinuous antigenic determinants are those formed by amino acid residues which are contiguous but not part of a linear amino acid sequence, for example the antigenic sites of lysozyme. A much more detailed discussion of these continuous and discontinuous antigenic sites appears in Chapter 2.

Effect of alteration of conformation on antigenicity. It has been known since the 1940s that denaturation due to heating or to chemical modification results in a reduction of immunogenicity and an alteration in antigenicity. The structural requirements for antigenicity and immunogenicity have since been examined in much more detail using various well-characterized procedures.

Cleavage of all the intramolecular disulfide bonds in proteins, destroying non-covalent interactions and causing unfolding of polypeptide chains, has been shown to

result in a large reduction in activity of antibodies elicited to the native protein, for example ribonuclease, papain, trypsin, lysozyme, and bovine serum albumin. However, when only two of the four disulfide bonds of ribonuclease are cleaved, the partially reduced protein is immunologically indistinguishable from native ribonuclease. This result, taken together with results from other partially reduced proteins, suggests that disulfide bonds per se do not play a major role in determining antigenicity. The secondary role of disulfide bonds in determining antigenicity is probably due to the fact that disulfide bridges do not direct the folding of the polypeptide chain but stabilize the most thermodynamically stable conformation.

Because of the close relationship between conformation and antigenicity, it would be expected that a small conformational change would result in a change in antigenic reactivity. One of the best characterized systems demonstrating this is the removal of the heme group from sperm whale myoglobin. The loss of the heme group in apomyoglobin is associated with a decrease in helical content of the protein and an increase in asymmetry of the protein; that is, the folded polypeptide chain of apomyoglobin is more flexible. Differences between the antigenic reactivities of metmyoglobin (heme group present) and apomyoglobin have been clearly shown (Crumpton and Wilkinson, 1965). Figure 2 shows that metmyoglobin forms, at equivalence, larger amounts of precipitate than that formed by apomyoglobin, with antisera to metmyoglobin.

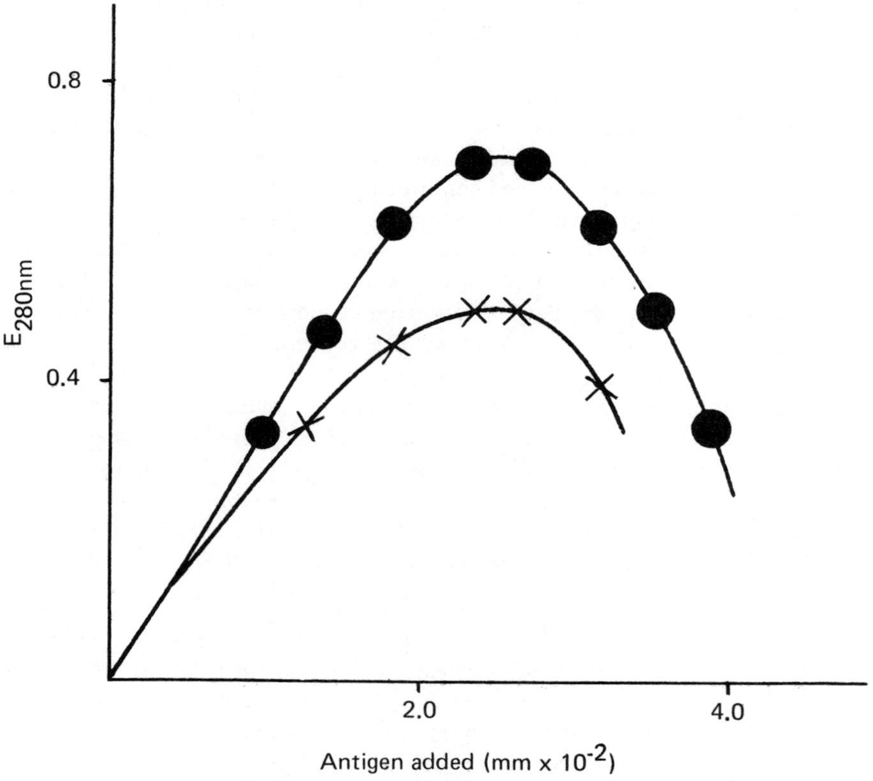

Figure 2 Amounts of precipitate formed by the addition of sperm whale metmyoglobin (●——●) and of apomyoglobin (X——X) to an antimetmyoglobin serum.

Electrical Charge

There are two aspects of electric charge which should be discussed. First, is an electrical charge on the immunogen required for eliciting an immune response? Second, does a net electrical charge on the immunogen affect the characteristics of immunoglobulins evoked by the immunogens?

A net electrical charge per se on the immunogen does not seem to be required, since nonionizable synthetic polypeptide antigens, soluble in water and nonionizable at neutral pH, are immunogenic. However, basic peptides and basic proteins with clusters of positively charged amino groups are very good immunogens, for example the basic protein of myelin. High electrical basic or acidic net charge may secondarily enhance immunogenicity by the formation of complexes or aggregates with the adjuvant used in immunization. Since lymphocyte and macrophage membranes are negatively charged, it could be argued that positively charged antigens are attracted nonspecifically to their membranes, thereby facilitating antigen presentation.

It has been clearly shown that there is an inverse relationship between the net electrical charges of immunogens and the immunoglobulins elicited by them. This has been demonstrated very clearly with antibodies to acidic diptheria toxoid and to negatively charged synthetic amino acid copolymers. Interestingly, antibodies produced to p-azobenzene-arsonate groups bound to basic poly-L-lysyl ribonuclease or to acidic rabbit serum albumin showed that the antibody charge was the opposite of the net charge of the molecule, rather than charges within limited areas of the azobenzenearsonate determinant. For large globular proteins it is possible that charged residues may contribute strongly to the specificity of antigenically reactive determinants (see also Chap. 16).

Adjuvants

Many substances, such as oil emulsions and aluminum hydroxide, have been described as enhancing immunogenicity. There are various effects of adjuvants that should be emphasized in this context: first, enhancing the immunogenicity of a substance that is immunogenic when immunized into an animal on its own; second, conferring immunogenicity to substances that are not immunogenic when immunized into animals on their own; third, modifying the type of immune response of an immunogen (e.g., humoral immunity or delayed-type hypersensitivity). The enhancing effect of adjuvants, such as Freund's complete or incomplete adjuvant, has been well demonstrated for macromolecular and polypeptide antigens. However, not all antigens are equally sensitive to the effects of adjuvants; for example, polysaccharide antigens show no enhanced immunogenicity when immunized in complete Freund's adjuvant. For macromolecular antigens, complete Freund's adjuvant and mycobacteria have been shown to shift the type of immune response elicited; for example, in guinea pigs complete Freund's adjuvant increased the production of γ_2 immunoglobulin and enhances delayed-type hypersensitivity.

Various adjuvants contain mycobacteria or *Bordetella pertussis*, which stimulate macrophage activity, and it is possible that the increased digestion of immunogens by macrophages could result in the formation of active peptides containing single antigenic sites and thereby activating antigen-sensitive lymphocytes. On the other hand, inert polystyrene latex particles of about the same size and hydrophobicity as mycobacteria also are excellent adjuvants, at least for the elicitation in rabbits of antibody formation to human IgG (van Oss et al., 1976). It thus may be the simple hydrophobicity of such particles that favors the adsorption of the antigen and allows its presentation in a regular array on the surface of insoluble particles.

Dose of Antigen and Route of Administration

Both of these factors can play a considerable role in the development of an immune response and the type of response obtained. These factors are of great importance for both high- and low-molecular-weight antigens, since the former can contain very low molar concentrations of single antigenic sites which can function as very low molecular weight antigens. One very good example to quote here is the response to the synthetic octapeptide angiotensin II: injection in complete Freund's adjuvant results in delayed-type hypersensitivity, whereas intradermal immunization evokes antibody production. When investigating the immunogenicity of high- or low-molecular-weight antigens, one should carry out an examination of the effects of different routes of immunization as well as complete dose-response curves for each route of immunization. Thus the definition of a substance as immunogenic or nonimmunogenic can be accepted only in operational terms.

Genetic Factors

Genetic factors in the immune response, specifically the immune response (Ir) genes, have been reviewed extensively (McDevitt and Landy, 1972). Information clearly defining an immune response gene to synthetic polypeptides came from the study of (T,G)-A--L (see Fig. 1), which bears a restricted number of antigenic determinants. The antigen (H,G)-A--L is a related polypeptide in which the tyrosine residue has been substituted by a histidine residue in the antigenic determinant. Studies with (T,G)-A--L and (H,G)-A--L have clearly emphasized the importance of genetic background of the animal species used in defining the immunogenicity of these substances. The CBA strain of mouse is a low responder to (T,G)-A--L and a high responder to (H,G)-A--L, whereas the C57 strain of mouse is a high responder to (T,G)-A--L and a low responder to (H,G)-A--L. It was shown that the immune response to each of these antigens is under separate Ir-gene control. Since this initial demonstration of the genetic basis of the immune response, many antigens have been shown to be under genetic control, not only in mice, but also in guinea pigs, rats, chickens, monkeys, and humans. Examples of some of the many antigens the response to which has been shown to be under genetic control are: lactate dehydrogenase, ovalbumin, ovomucoid, ribonuclease, staphylococcal nuclease, bovine gamma globulin, and myoglobin.

Studies on the genetic control of the immune response to complex antigens, such as sperm whale myoglobin, have revealed some very interesting features concerning the structural recognition of antigenic portions of the molecule. For example, it has been shown that each of the five individual antigenic sites in myoglobin are controlled by unique, distinct Ir genes. Furthermore, strains of mice which are high responders to myoglobin do not necessarily respond to the same antigenic sites of the molecule. Also, the contribution that the dose of antigen used in immunization plays in the genetic recognition of antigenic sites has been clearly demonstrated. For example, strains of mice which are nonresponders to myoglobin or any of its antigenic sites at an immunizing dose of 50 μg will respond well to myoglobin and all of its antigenic sites at higher immunizing doses (see also Chap. 25).

IV. ANALYSIS OF RELATIONSHIP BETWEEN CHEMICAL STRUCTURE, IMMUNOGENICITY, AND LYMPHOCYTE FUNCTION

A. Hapten-Carrier Relationship

Much of the understanding of the nature and specificity of antigen-antibody reactions results from the early pioneering work of Landsteiner, using small substances which are

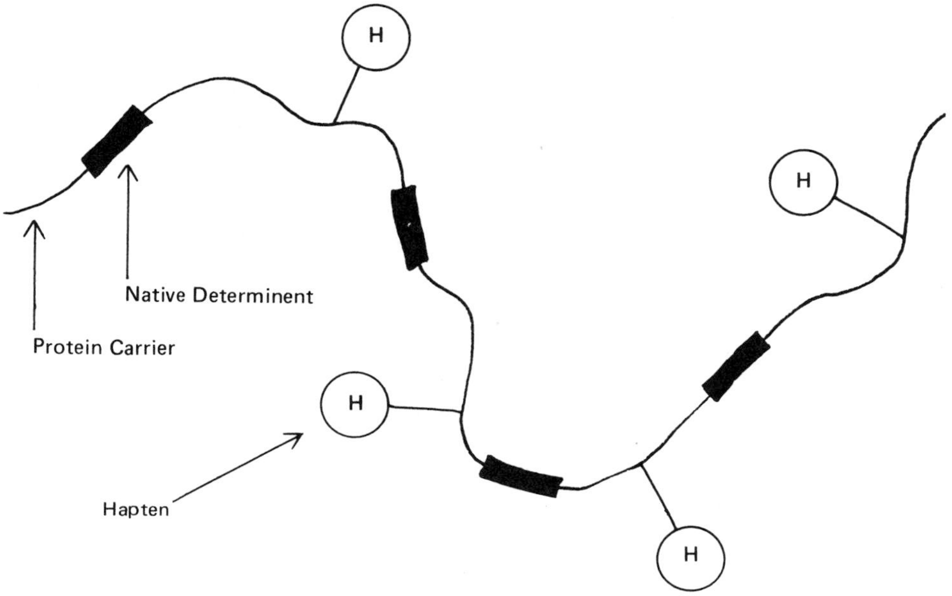

Figure 3 Illustration of hapten-protein conjugate.

not immunogenic but can react with antibodies of appropriate specificity. In order to elicit antibodies to the haptenic determinant, the hapten is first coupled to a protein. The conjugated proteins elicit antibodies specific for the hapten, as demonstrated by the capacity of the free hapten to bind antibody. The protein, called the carrier, has its own set of native determinants as well as the new determinants introduced by the conjugated hapten, as represented in Fig. 3.

It has been known for some time that the production of antibodies against the haptenic determinant involves a structural configuration of the carrier molecule. The importance of the carrier moiety is demonstrated by the finding that animals that cannot respond by antibody production to the hapten DNP when carried on a nonimmunogenic carrier can respond when the DNP is bound to an immunogenic carrier. It thus appeared that a recognition of antigenic determinants on the carrier molecule was a prerequisite for the induction of antibody as well as for recognition of the haptenic determinant. The work of Mitchison (1971) has clearly shown that different lymphoid cells respond to carrier and hapten determinants and that both responses are required for the production of antihapten antibody. Experiments in mice have clearly demonstrated that the carrier-sensitive cells are T lymphocytes (see Chap. 24).

The insertion of spacer molecules between the hapten and carrier determinants on the conjugate does not influence the immunogenicity of the conjugate, indicating that the hapten and carrier portions are not acting as part of a single determinant. The separation of the carrier determinant and haptenic determinant of bifunctional molecules with spacers of varying size has allowed an assessment of the spatial requirements between hapten and carrier for an antihapten response. Using the hapten DNP, the carrier L-tyrosine-p-azobenzenearsonate (RAT), and the spacer 6-aminocaproic acid it has been established that cooperation can be implemented by an antigen in which the hapten and carrier moieties were separated by less than 8 Å.

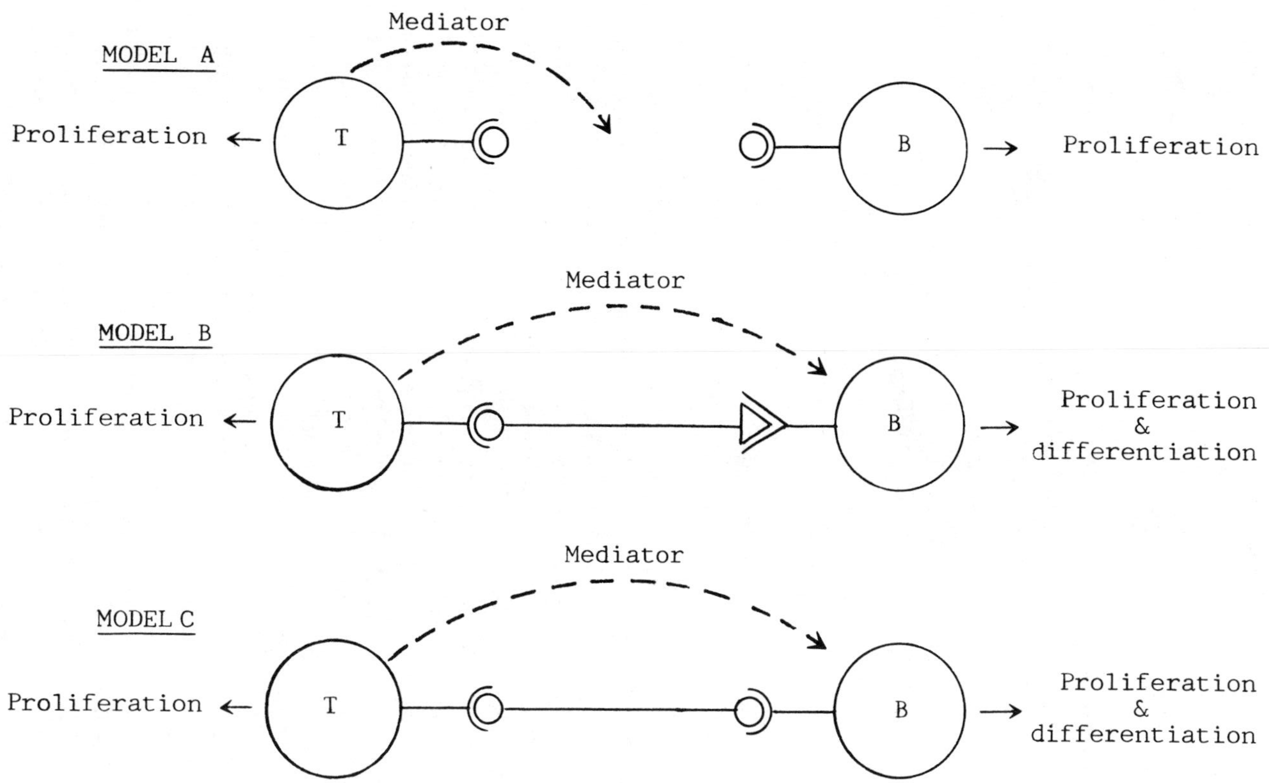

Figure 4 Various models of cooperation between T and B cells in response to: A, a monofunctional antigen containing one immunogenic determinant only; B, an asymmetric bifunctional antigen containing one immunogenic and one haptenic determinant only; C, symmetrical bifunctional antigen containing two immunogenic determinants. O, Immunogenic determinant; Δ, haptenic determinant.

However, there are certain exceptions to the hapten-carrier concept. For example, some polymeric antigens composed of identical repeating units such as endotoxin lipopolysaccharides, pneumococcal polysaccharides, and polymerized flagellin do not require the intervention of T cells and of carrier-specific cells. Also, recent work has shown that uncoupled peptides of macromolecules of only six amino acids and containing only one antigenic determinant can elicit antibody production when immunized with complete Freund's adjuvant. However, most multivalent hapten-protein conjugates show a strong carrier specificity in the anamnestic antibody response.

B. Structural Requirements of Antigens for Cell Cooperation and Lymphocyte Activation

Cell cooperation mediated by identical determinants on an antigen molecule has been investigated using a series of bifunctional antigens with different spacers (Goodman et al., 1974). Using symmetrical bifunctional antigens it was found that there was a good correlation between helper activity and the steric availability of determinants, thus strengthening the concept that at least two determinants, which may be identical, must be accessible to elicit antibody response. This finding also strengthens the thesis of antigen-bridging models of cell cooperation, which are summarized in Fig. 4.

This same series of bifunctional antigens was also used for comparing the efficiency of single-point and two-point antigen binding in triggering lymphocyte proliferation. Using these antigens it was demonstrated that rigidly spaced bifunctional antigens are poor stimulators of antigen-induced proliferation, whereas they served very well for cell cooperation. Thus the structural requirements for cell triggering and for cell cooperation differ markedly.

C. Structural Requirements for Induction of Immune Responsiveness and Immune Suppression

A particularly good example to demonstrate these structural requirements is the L-tyrosine-p-azobenzenearsonate (RAT) immunogenic system. Two series of RAT analogues were synthesized, one in which substitutions were made at the arsonate position and the other in which the side chain of tyrosine was modified. It was found that

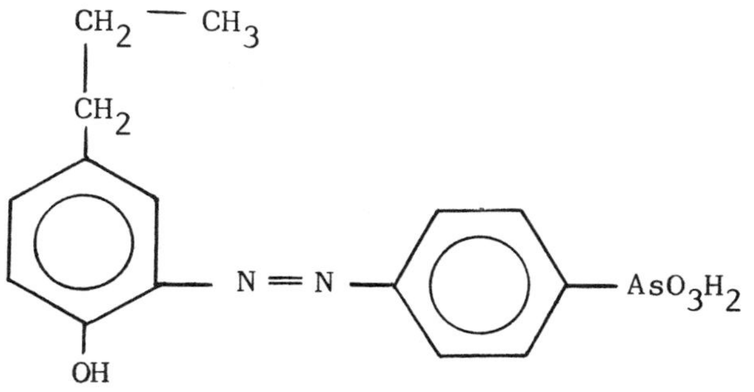

Figure 5 (p-Hydroxyphenyl)propane-azobenzene-p′-arsonate (RAN).

other charged moieties could substitute for arsonate without loss of immunogenicity. On the other hand, removal of both charged groups from the tyrosyl side chain completely abolished immunogenicity. This nonimmunogenic analogue of RAT with a propyl side chain was designated (p-hydroxyphenyl)–propane-azobenzene-p′-arsonate or RAN (see Fig. 5).

Since RAN is a very close analogue of RAT, and lacks perceptible immunogenicity, it has been used to look at the structural requirements for tolerogenicity. RAN, as well as RAT, induces specific unresponsiveness to RAT when immunized with incomplete but not with complete Freund's adjuvant, as measured by delayed hypersensitivity and antigen-induced DNA synthesis. Thus a compound without perceptible immunogenicity is capable of rendering T cells unresponsive. Although not clearly proven yet, it seems likely that this phenomenon is mediated by suppressor T cells.

REFERENCES

Atassi, M. Z. (1975). Immunochemistry *12*, 423.

Crumpton, M. J., and Wilkinson, J. M. (1965). Biochem. J. *94*, 545.

Goodman, J. W., Bellone, C. J., Hanes, D., and Nitecki, D. E. (1974). In *Progress in Immunology II*, Vol. 2 (L. Brent and J. Holborow, Eds.). North-Holland, Amsterdam, p. 27.

Howard, J. G., Zola, H., Christie, G. H., and Courtenay, B. M. (1971). Immunology *21*, 535.

Landsteiner, K. (1945). *The Specificity of Serological Reactions.* Harvard University Press, Cambridge, Mass.; Dover, New York, 1962.

McDevitt, H. O., and Landy, M. (1972). *Genetic Control of Immune Responses* Academic Press, New York.

Mitchison, N. A. (1971). Eur. J. Immunol. *1*, 18.

Schlossman, S. F., Yaron, A., Ben-Efraim, S., and Sober, H. A. (1965). Biochemistry *4*, 1638.

Sela, M. (1969). Science *66*, 1365.

van Oss, C. J., Singer, J. M., and Gillman, C. F. (1976). Immunol. Commun. *5*, 18.

Young, C. R., and Atassi, M. Z. (1982a). Adv. Exp. Med. Biol. *150*, 73.

Young, C. R., and Atassi, M. Z. (1982b). Immunol. Commun. *11*, 9.

2

Immune Recognition of Proteins

M. Zouhair Atassi* Mayo Clinic, Rochester, Minnesota

I. INTRODUCTION

The immune response is largely dependent on molecular interactions involving proteins. The recognition of antigen molecules, whether they are proteins or nonproteins, whether they are self or nonself, takes place at the molecular cellular interface through membrane receptor molecules that are proteins. The initial step of recognition activates a complex series of cellular events requiring some mechanism of cell-cell interactions and communications, eventually leading to antibody production. This biological cascade is controlled at several positions along its consecutive pathways by protein molecules, either in the free form or as receptors on membranes of cells committed to this activity. Thus proper understanding of the response by cells of the immune system will depend, to a great measure, on the definition of molecular events involving protein interactions. Obviously, cells work via molecules and molecules work via cells and, at this level of functional resolution, molecular immunology and cellular immunology will merge and will depend heavily on functional correlation with the structures of the participating proteins.

The great majority of antigens associated with immune disorders and invasive agents (viruses, bacteria, allergens, etc.) are attributed to protein molecules. Clearly, then, knowledge of the molecular features responsible for the antigenicity of native protein molecules lies at the basis of understanding in molecular terms the cellular events of the immune response. Therefore, defining the antigenic sites of protein antigens and charting a strategy for such undertakings will be critical for the molecular elucidation and the ultimate manipulation of these functions. Localization and synthesis of antigenic sites on invasive agents (e.g., viruses, bacteria, allergens, toxins, etc.) should afford valuable synthetic vaccines which will, in principle, be expected to have few or no side effects.

*Current affiliation: Baylor College School of Medicine, Texas Medical Center, Houston, Texas.

Progress in this field has been very slow, and the elucidation of the entire antigenic structure of a protein frustrated many attempts. A considerable number of chemical and technical factors were responsible for the slow progress in this field; these have been discussed elsewhere in detail (Atassi, 1975, 1977b). In the last few years great advances have been made in the investigation of the antigenic sites of proteins, which have resulted in the determination of the entire antigenic structures of the first two proteins to be thus completed, sperm whale myoglobin (Mb) (Atassi, 1975) and hen egg white lysozyme (Atassi, 1978). Also, all the antigenic sites of the α chain of human adult hemoglobin (Hb) (Kazim and Atassi, 1980a, 1982) have recently been localized and synthesized, as well as the major antigenic sites of serum albumin (Atassi et al., 1979; Sakata and Atassi, 1980a,b; Atassi, 1982). Determination of the antigenic structure of Mb and then that of lysozyme has answered many questions relating to the molecular immune recognition of native proteins with surprising accuracy (Atassi, 1975). Many of the observations and findings first made from the Mb work (Atassi, 1972, 1975) have since become established concepts that have been confirmed with a variety of other proteins. However, it is necessary to caution that this is a very complex field, and hence proper understanding of the immunochemistry of proteins requires the elucidation of the complete antigenic structures of a few native proteins (Atassi, 1972, 1975). Already, the precise determination of the entire antigenic structure of lysozyme (Atassi, 1978) has shown fascinating differences between these two proteins, as briefly outlined in this chapter.

This chapter focuses on the immunology of proteins in which either all (Mb and lysozyme) or most (serum albumin and hemoglobin) of the antigenic sites have been chemically localized and synthetically confirmed. The antigenic sites of other proteins are being investigated. One antigenic site has been well characterized and synthetically confirmed in tobacco mosaic virus protein (Benjamini, 1977; Benjamini et al., 1978). The immunogenic and disease-inducing sites on myelin basic protein have been studied and thus far, the site near the tryptophan residue 115 has been carefully characterized by synthesis (for a review, see Eylar, 1978).

II. CHEMICAL STRATEGY FOR THE DETERMINATION AND SYNTHESIS OF ANTIGENIC SITES

A. General Chemical Strategy

The antigenic structure of a protein cannot be determined by the exclusive application of a single approach. A strategy was, therefore, developed (Atassi, 1972) which relied on five approaches. This strategy enabled the precise determination of the entire antigenic structure of Mb (Atassi, 1975) and subsequently was equally effective in obtaining a similar result with lysozyme (Atassi, 1978). These approaches were: (1) to study the effect of conformational changes on the immunochemistry of the protein; (2) to study the immunochemistry and conformation of chemical derivatives of the protein, specifically modified at appropriate amino acid locations; (3) to isolate and characterize immunochemically reactive fragments that can account quantitatively for the total reaction of the native protein; (4) to study the effect of chemical modification at selected amino acid locations on the immunochemistry and conformation of immunochemically reactive peptides; (5) after narrowing down (using approaches 1 to 4) each of the antigenic sites to a region of conveniently small size, the final delineation would rely on studying the immunochemistry of synthetic peptides corresponding to many overlaps around this

region. It is critical to note that each of these chemical approaches has advantages as well as shortcomings. The application, usefulness, and shortcomings of these approaches in protein immunochemistry have been discussed in considerable detail (Atassi, 1975, 1977a,b). It is necessary to stress here that none of these approaches by itself is capable of yielding the full antigenic structure. We used the results from one approach to confirm and/or correct those from the others. The complete structure is a composite, logical coordination of all the information.

It should be noted that the strategy described above, although first employed in the delineation of protein antigenic sites, is applicable, with appropriate adaptations, to the precise delineation and chemical synthesis of other types of protein binding sites. The introduction of the concept of surface-simulation synthesis (Atassi et al., 1976b; Lee and Atassi, 1976) has provided a methodology by which in principle any type of protein binding site can be mimicked synthetically after careful chemical characterization.

B. Comprehensive Synthetic Approach for Continuous Antigenic Sites

The study of overlapping peptide fragments is a key approach in the delineation of protein antigenic structures (Atassi, 1972). This approach, when systematically employed, can yield valuable information on the number and locations of antigenic sites and their relative contributions to the total protein activity. In practice, however, this approach has severe shortcomings (for a review, see Atassi, 1977a). These include the relatively limited number of reproducible cleavage procedures for proteins; inadvertent scission within an antigenic site, usually affording inactive or only slightly active peptides; chemical modification, which frequently occurs at internal locations within a chemical-cleavage fragment; and any traces of contamination of an isolated peptide. Furthermore, the distribution of residues whose peptide bonds are potentially susceptible to cleavage in the protein does not permit the desired overlapping peptides to be obtained. Accordingly, the aforementioned strategy developed (Atassi, 1972, 1975) for the determination of protein antigenic structures employed five distinct chemical approaches.

In spite of the complexity of each chemical approach, however, the molecular details of the antigenic sites of Mb and lysozyme were derived by that strategy and have been valuable in furthering our comprehension of immune recognition processes. As will be seen in the following sections, the antigenic sites of Mb and lysozyme consist of discrete surface portions of their respective native molecules. Each site comprises five to seven amino acid residues that come into close proximity either through direct peptide linkage, such as the "continuous" antigenic sites of Mb, or in a more complex manner, through polypeptide-chain folding, such as the "discontinuous" sites of lysozyme. In view of the structural alternatives in their architecture (continuous or discontinuous sites; Atassi and Smith, 1978), the determination ab initio of the antigenic structures of other proteins should anticipate both alternatives. The logical progression, therefore, is first to localize and characterize the more readily identifiable continuous antigenic sites, with attention later paid to discontinuous sites.

Any approach that could facilitate the localization of continuous antigenic sites would certainly diminish the overall effort required for complete immunochemical characterization. Thus, with this goal in mind, and to circumvent the aforementioned problems and render the determination of protein antigenic structures more feasible within a reasonable time a novel and comprehensive synthetic approach was devised (Kazim and Atassi, 1980b), designed to yield the full antigenic profile of the protein under

study. This approach consists of the direct synthesis and examination of the immunochemical activities of a series of overlapping peptides that encompass the entire primary structure of the protein, from the beginning to the end. We have employed this approach to determine the antigenic sites on the α and β chains of human adult Hb. Its application to the α chain has been reported (Kazim and Atassi, 1980b, 1982). A more detailed discussion of this approach is given in these two references.

III. MAIN FEATURES OF THE ANTIGENIC STRUCTURES OF MYOGLOBIN, LYSOZYME HEMOGLOBIN, AND SERUM ALBUMIN

Derivation of the antigenic structures of Mb and lysozyme spanned numerous publications. A concise review of the determination of the antigenic structure of Mb has appeared (Atassi, 1975) and a very comprehensive account of the derivation has also been published (Atassi, 1977b). The climax of our studies on the antigenic structure of lysozyme can be found in the papers dealing with the synthesis of the three antigenic sites of lysozyme (Atassi et al., 1976b; Lee and Atassi, 1976, 1977a,b; Atassi and Lee, 1978a), which quantitatively accounted for its entire antigenic reactivity, and the work has recently been reviewed (Atassi, 1978). The main features of the antigenic structures of Mb and of lysozyme will be outlined below. Also, the antigenic sites of the α chain of Hb (Kazim and Atassi, 1982) and the six major antigenic sites of serum albumin (Atassi et al., 1979; Sakata and Atassi 1980a,b; Atassi, 1982) will be given. The emphasis in this chapter is on the implications and impact of the determination of these antigenic structures, rather than reviewing the step-by-step immunochemical derivations of these structures. A presentation of these derivations will not be given here because they have already been documented in detail in the aforementioned reviews and articles.

A. Special Features of the Antigenic Structure of Myoglobin

By application of the first four chemical approaches mentioned above, it was possible to achieve delineation of the antigenic sites in Mb to within regions of 8 to 10 residues. Having reached a chemical delineation down to such a conveniently small size, precise narrowing down of the sites was achieved by the organic synthesis and immunochemical studies of peptides corresponding to various parts of the regions carrying the sites.

This section gives a very brief outline of the first antigenic structure of a protein to be completed (Atassi, 1975). Readers desiring more detail may consult a very comprehensive recent review (Atassi, 1977b), while the initial description of that work (Atassi, 1975) offers a more concise account.

Five antigenic sites are present in the native protein and are situated as follows. Site 1: residues 16 to 21 + 1 or 0 residue on one side only of this segment, depending on the antiserum. This antigenic site exhibits a certain degree of "shift" or "displacement" and minor variability in size (limited to ±1 residue only) from one antiserum to the next. Its location in the three-dimensional structure is on the bend between helices A and B. Site 2: residues 56 to 62, on the bend between helices D and E. This antigenic site has exhibited no variability in size with the antisera so far studied. Site 3: residues 94 to 99 on the bend between helices F and G. Site 4: residues 113 to 119, on the end of helix G and only part of the bend GH. Site 5: residues 146 to 151 (+ lysine 145 with some antisera). This antigenic site is situated on the end of helix H and part of the randomly coiled C-terminal pentapeptide. The covalent structures of the five antigenic sites are

Site	Structure and Location	No. of Residues
1	15 16 21 22 (Ala)-Lys-Val-Glu-Ala-Asp-Val-(Ala)	6 (or 7)
2	56 62 Lys-Ala-Ser-Glu-Asp-Leu-Lys	7
3	94 99 Ala-Thr-Lys-His-Lys-Ile	6
4	113 119 His-Val-Leu-His-Ser-Arg-His	7
5	145 146 151 (Lys)-Tyr-Lys-Glu-Leu-Gly-Tyr	6 (or 7)

Figure 1 Covalent structures of the five antigenic sites of sperm whale Mb. Residues in parentheses are part of the antigenic site only with some antisera. Thus for site 1, the reactive region invariably comprises residues 16 to 21, and with some antisera Ala-15 is part of the region (which will then correspond to residues 15 to 21), while with other antisera Ala-22 is an essential part of the region (which will then correspond to residues 16 to 22). This site occupies either six or seven residues, depending on antiserum. For sites 2 to 4 no such displacement or shift to one side or the other has been observed (at least with the antisera studied so far). In the case of site 5, Lys-145 can be part of the antigenic region only with some antisera, and this site will therefore comprise six or seven residues, depending on the antiserum. (From Atassi, 1975).

shown in Fig. 1. The locations of the antigenic sites in the three-dimensional structure of native Mb are shown schematically in Fig. 2.

We have previously cautioned (Atassi, 1975) against the likely formulation of an erroneous conclusion that every bend constitutes an antigenic site. No such statement is made or implied here, and indeed, examination of Fig. 2 immediately reveals that bends B-C, C-D, and E-F do not carry antigenic sites. Also, site 4 (i.e., region 113–119) is located mostly on a helical portion.

The antigenic sites are surprisingly small (six to seven residues, or about 19 to 23 Å in their extended dimensions) and possess sharp boundaries. They may exhibit a limited variability in boundaries with various antisera, which, when it exists, has so far been ±1 residue. The size, surface locations, and shape of these antigenic sites make them quite accessible for binding with antibody combining sites.

The types of amino acids present in the antigenic sites are to be expected from the surface locations of the sites (Atassi, 1972, 1975). Lysine is present in four sites (Fig. 1), and arginine is present in the fifth. Three out of five antigenic sites contain asparatic acid or glutamic acid or both. Two sites contain histidine. From this and the demonstrated detrimental effect of appropriate modifications of these polar residues on the antigenic reactivity, it is concluded (Atassi, 1972, 1975) that interactions of the Mb antigenic sites with antibody must be predominantly polar in nature. Stabilizing effects are contributed by hydroxy and nonpolar amino acids through hydrogen bonding and hydrophobic interactions. The sequence and three-dimensional structure features that confer immunogenicity on these regions are not too clear.

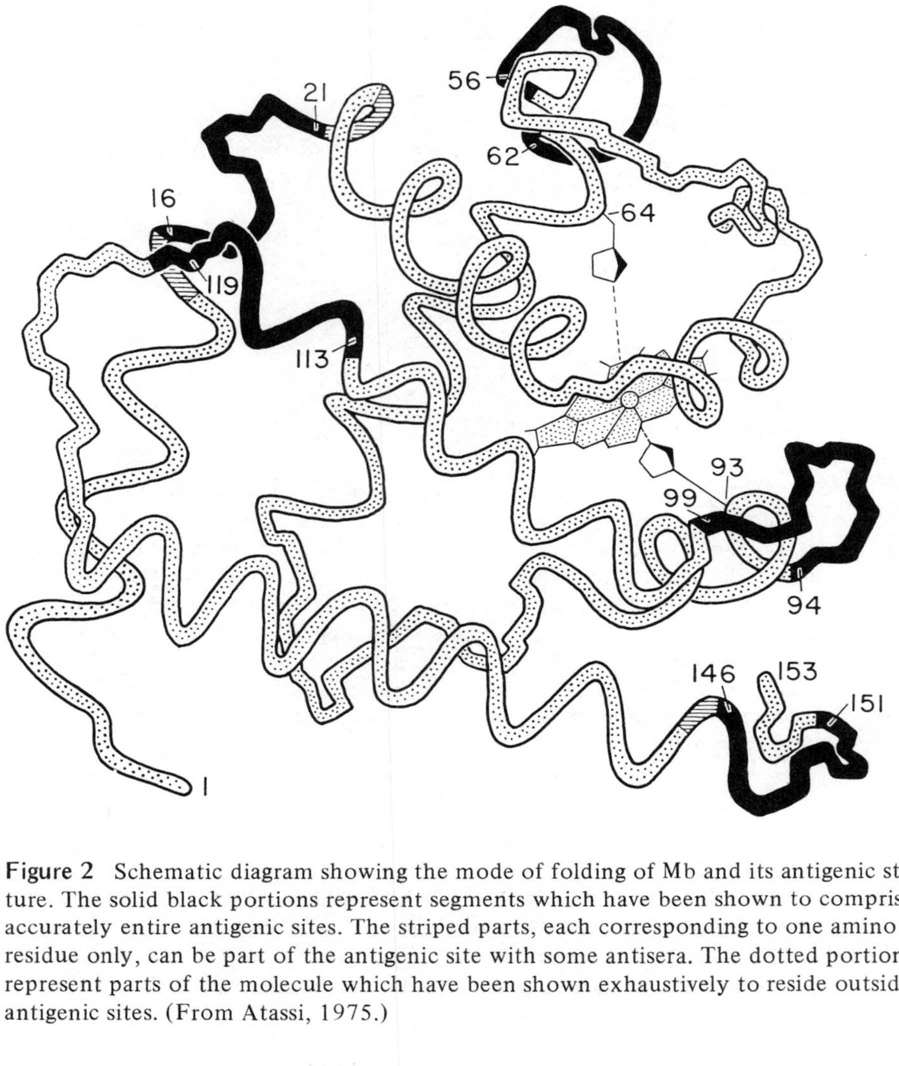

Figure 2 Schematic diagram showing the mode of folding of Mb and its antigenic structure. The solid black portions represent segments which have been shown to comprise accurately entire antigenic sites. The striped parts, each corresponding to one amino acid residue only, can be part of the antigenic site with some antisera. The dotted portions represent parts of the molecule which have been shown exhaustively to reside outside antigenic sites. (From Atassi, 1975.)

The sum of the reactivities of these sites accounts quantitatively for all (98 to 100%) of the antibody response (Tables 1 to 3) directed against native Mb (Twining and Atassi, 1979; Twining et al., 1981a). The affinity of an antigenic site and its share of the total reactivity of Mb may vary with the antiserum. However, with all the antisera studied thus far, an antigenic site is invariably an antigenic site, but its potency or immunodominance vary with the immunized animal. Initially, the antigenic structure was determined with early antisera elicited in rabbits and goats (Atassi, 1975). In more recent studies (discussed later in this chapter) these findings (obtained with rabbit and goat antisera) were extended to four other species and the time dependence of the response from 9 days up to a year after the initial immunization was examined. It was determined that the antigenic sites recognized are not dependent on the host species and that the recognition is essentially directed to the same antigenic sites over extended periods of immunization.

Table 3 Binding of Anti-Mb [^{125}I] Antibodies from Selected Congenic Mice with Immunoadsorbents of Mb and Its Synthetic Sites[a]

	Binding of [^{125}I] antibodies from congenic strains							
	ASW		BALB/c		DBA/2		B10.S	
Immunoadsorbent	cpm	Percent	cpm	Percent	cpm	Percent	cpm	Percent
Mb	23,910	100	27,630	100	19,750	100	20,090	100
Site 1	4,330	18.1	6,460	23.4	5,530	28.0	5,420	27.0
Site 2	6,410	26.8	7,430	26.9	3,770	19.1	6,230	31.0
Site 3	6,620	27.7	8,040	29.1	4,780	24.2	1,060	5.3
Site 4	597	2.5	980	3.5	630	3.2	920	4.6
Site 5	5,570	23.3	4,560	16.5	4,660	23.6	6,610	32.9
Control peptide 1-6	<30	–	<35	–	0	–	<30	–
Control peptide 121-127	0	–	0	–	<25	–	<25	–
Total for sites	23,527	98.4	27,470	99.4	19,370	98.1	20,240	100.8
Sites active in T-cell proliferation	1, 2, 3, 5		1, 2, 3, 5		1, 2, 3, 5		1, 2, 5	

[a]A fixed amount of ^{125}I-labeled antibodies against Mb from selected mouse strains was titrated with increasing amounts of Mb-Sepharose or peptide-Sepharose. For examples of such titrations, see Twining and Atassi (1979). The results represent the *maximum* (plateau) amounts of antibodies that were bound by a given adsorbent. Values have been corrected for nonspecific binding and are the average of six replicate analyses which varied ±1.2% or less.
Source: Twining et al. (1981a).

Site 1. This antigenic site is constructed (Atassi and Lee, 1978a) by the side chains of the spatially adjacent five surface residues: Arg-125, Arg-5, Glu-7, Arg-14, and Lys-13. The extended dimension of the site from Arg-125 to Lys-13 (C^{α} to C^{α}) is 30 Å. These residues, which fall in an imaginary line circumscribing part of the surface topography of the protein, bind with antibody as if they are in direct peptide linkage (Atassi et al., 1976b). In fact, the immunochemical reactivity of this site is fully satisfied by the surface-simulation-synthetic peptide Arg-Gly-Gly-Arg-Gly-Glu-GlyGly-Arg-Lys, which does not exist in native lysozyme but mimics the arrangement of residues in this surface region. Following the introduction of this concept (Atassi et al., 1976b), we proposed the term *surface-simulation synthesis* (Lee and Atassi, 1976) to describe it. The surface-simulation-synthetic site exhibits a directional preference (Arg-125 to Lys-13), which appears to be independent of the species of the immunized animal, at least with the rabbits and goats so far tested. The site has restricted conformational freedom, clearly evidenced by its sensitivity to variation of the spacing between its constituent residues in surface-simulation synthesis. The intactness of the disulfide bond 6-127 in native lysozyme is critical for the integrity of this site (Atassi et al., 1973a, 1976a).

Table 4 Binding of Radioiodinated Antibodies against Lysozyme to Immunoadsorbents Carrying the Three Surface-Simulation-Synthetic Sites[a]

	Antibody bound from G9		Antibody bound from G10	
Immunoadsorbent	cpm	Percent	cpm	Percent
Lysozyme[b]	50,180	100	73,330	100
Site 1[b]	15,760	31.4	20,335	27.7
Site 2[b]	18,015	35.9	27,600	37.6
Site 3[b]	14,950	29.8	22,300	30.0
Total of independent binding by three sites	48,425	97.1	70,235	95.8
Binding by passage through sites serially[c]	49,880	99.4	71,670	97.7

[a]The specific ^{125}I-labeled antibody fractions from antisera G9 and G10 were isolated on a lysozyme immunoadsorbent prior to use in these studies. Each value represents the average of four replicate analyses which varied ±1.3% or less. Results have been corrected for the amount of antibody bound in control experiments using glycine-Sepharose, histidine-Sepharose, and myoglobin-Sepharose. Another set of controls were employed using nonimmune goat ^{125}I-labeled IgG. The amount of nonspecific background binding in the various controls ranged from 1 to 3% of the total label applied.
[b]Results for independent binding were obtained by passage of an aliquot of the antibody solution on only one of the immunoadsorbents indicated.
[c]Results obtained by serial passage of the same antibody sample on the immunoadsorbent of site 1, then site 2, then site 3.
Source: Atassi (1979).

Site 2. This antigenic site also consists (Atassi et al., 1976b; Lee and Atassi, 1977b) of the spatially adjacent surface residues Trp-62, Lys-97, Lys-96, Asn-93, Thr-89 and Asp-87. The extended dimension of the site from Trp-62 to Asp-87 (C^{α}-to-C^{α} distance) is 27.3 Å. As with site 1, site 2 also describes a region that occupies part of the surface topography of the protein molecule. This region comprises the residues forming the site, which behave functionally as if they are directly linked by peptide bonds. Thus the surface-simulation-synthetic peptide Phe-Gly-Lys-Lys-Asn-Thr-Asp, which does not exist in lysozyme but simulates a surface region of it, carries the full reactivity of the site (Atassi et al., 1976b; Lee and Atassi, 1977b). With the antisera studied so far, the antigenic site exhibits in surface-simulation synthesis a preferred directionality (Trp-62 to Asp-87) as well as conformational constraints (indicated by requirements for appropriate spacing between the constitutent residues) for maximum binding with antibody (Lee and Atassi, 1977b). The intactness of the disulfide bonds 64-80 and 76-94 is essential to bring together the various constitutent residues of the site (Atassi et al., 1973a; Lee and Atassi, 1975). This antigenic site overlaps with the enzymic active site because they both share Trp-62 (Lee and Atassi, 1975, 1977b; Atassi et al., 1976b).

Site 3. Like sites 1 and 2, this site is constructed of spatially adjacent surface residues. The antigenic site comprises (Lee and Atassi, 1977a) the five residues Lys-116, Asn-113, Arg-114, Phe-34, and Lys-33, and in its extended form has a C^{α}-to-C^{α} dimension (from Lys-116 to Lys-33) of 21 Å. These residues describe a region that circumscribes part of the surface of the molecule and act functionally toward the antibody as if they

Table 1 Comparison of the Antibody Binding Capacity of the Synthetic Antigenic Sites with the Expected Reactivity of the Sites[a]

Adsorbent	Antiserum G3: Percent[b] antibody bound	Antiserum G3: Expected[c] reactivity of the site	Antiserum G4: Percent[b] antibody bound	Antiserum G4: Expected[c] reactivity of the site
Site 1	26.5	27.4	11.5	8.2
Site 2	20.6	22.2	27.8	33.6
Site 3	17.3	17.9	21.3	17.9
Site 4	18.5	17.8	24.6	23.6
Site 5	18.0	18.5	14.9	14.0
Total	100.9	103.8	100.1	97.3
Total by sequential binding to the five sites	99.6[d]		99.8[d]	

[a]The independent binding ability of each site-(Sepharose to ^{125}I-labeled goat anti-Mb antibodies) was determined by numerous immunoadsorbent titration studies (16 to 24 experiments, each in triplicate). The results for each site represent plateau values. The total binding values for the five sites were also obtained from the amount of label bound by serial adsorption of the same aliquot of [^{125}I] immune IgG onto plateau amounts of adsorbents of sites 1 to 5.
[b]Values represent the amount of ^{125}I-labeled antibody bound in the plateau relative to the label bound in the plateau by Mb-Sepharose as 100%. Results represent the average of 48 to 72 replicate analyses which varied by ±2.2% or less.
[c]Values are from Atassi (1977b).
[d]These values are the average of triplicate analyses.
Source: Twining and Atassi (1979).

The findings that purely conformational changes in Mb will influence its reaction with antisera to the native protein (Atassi, 1967b; Andres and Atassi, 1970), and the immunochemical results on numerous peptide fragments have made it possible to conclude (Atassi, 1967b, 1972, 1975) that the antibody response is directed against the native three-dimensional structures of proteins. These conclusions, initially derived with early course antisera, were shown in more recent studies (which are discussed subsequently in this chapter) to apply to antisera obtained up to a year after the initial immunization.

An intact antigenic site free of extraneous nonreactive residues would usually react with a lower binding energy than when isolated as part of a longer peptide (Atassi, 1972; Koketsu and Atassi, 1973, 1974a). The nonreactive parts may assist the achievement of the correct folding for binding of the antigenic site with an antibody combining site (Atassi and Saplin, 1968). On the other hand, evidence was also obtained that nonreactive parts composed of bulky residues linked to an antigenic site may exert unfavorable steric or conformational effects on the ability of the site to bind with antibodies (Koketsu and Atassi, 1974a,b; Atassi and Pai, 1975).

B. Summary of the Features of the Antigenic Structure of Lysozyme

Through the application or appropriate adaptations of the general chemical strategy outlined above, lysozyme was found to have three antigenic sites, each constituting spatially adjacent residues (that are otherwise distant in sequence) that occupy a discrete area on the surface topography of the protein (for a review, see Atassi, 1978). Although we

Table 2 Binding of Mouse (Outbred) ^{125}I-Labeled Antibodies to Immunoadsorbents of Mb and Its Synthetic Sites[a]

	Binding of [^{125}I] antibodies from outbred mice					
	Antiserum MS2		Antiserum MS3		Antiserum MS5	
Immunoadsorbent	cpm	Percent	cpm	Percent	cpm	Percent
Mb	29,650	100	31,750	100	21,840	100
Site 1	4,320	14.6	5,080	16.0	3.150	14.4
Site 2	5,910	19.9	5,680	17.9	3,060	14.0
Site 3	8,250	27.8	5,120	16.4	4,590	21.0
Site 4	7,690	25.9	8,700	27.4	5,790	26.5
Site 5	3,460	11.7	7,020	22.1	5,200	23.8
Control peptide 1-6	0	0	0	0	30	–
Control peptide 121-127	40	–	30	–	0	0
Total for sites 1-5	29,630	99.9	31,690	99.8	21,790	99.7

[a]A fixed amount of ^{125}I-labeled antibodies against Mb from individual outbred mice was titrated with increasing amount of Mb-Sepharose or peptide-Sepharose. For examples of such titrations, see Twining and Atassi (1979). The results represent the *maximum* (plateau) amounts of antibodies that were bound by a given adsorbent. Values have been corrected for nonspecific binding and are the average of six replicate analyses which varied ±1.2% or less.
Source: Twining et al. (1981a).

had proposed the existence of this type of protein antigenic site quite early (Atassi and Saplin, 1968), we subsequently found that they did not exist in Mb and it was in lysozyme that we identified the first such site (Atassi et al., 1976b). The need for chemical verification of such sites led to the introduction (Atassi et al., 1976b; Lee and Atassi, 1976) of the concept of surface simulation synthesis, by which the spatially adjacent surface residues constituting a protein binding site are linked directly via peptide bonds with appropriate spacing where necessary. The introduction of the concept of surface-simulation synthesis made it possible to delineate precisely and to synthesize chemically the three antigenic sites of lysozyme (for review, see Atassi, 1978). Surface-simulation synthesis afforded a new concept in protein molecular recognition, providing a novel and versatile chemical strategy for the synthesis of any type of protein binding site (Atassi et al., 1976b; Atassi, 1978; Twining and Atassi, 1978; Kazim and Atassi, 1980a).

For the sake of direct comparison, the highlights of the antigenic structure of lysozyme will be outlined. Native lysozyme has three antigenic sites. Table 4 gives an example of the relative contribution of the sites to the total antigenic reactivity of lysozyme with two goat antisera, and Fig. 3 shows an example of the time dependency of the response in serial mouse antilysozyme antisera up to a year after the initial immunization. The identities of the sites are shown in Fig. 4 and their locations in the three-dimensional structure of lysozyme can be seen in Fig. 5 and 6. The antigenic sites are briefly described below.

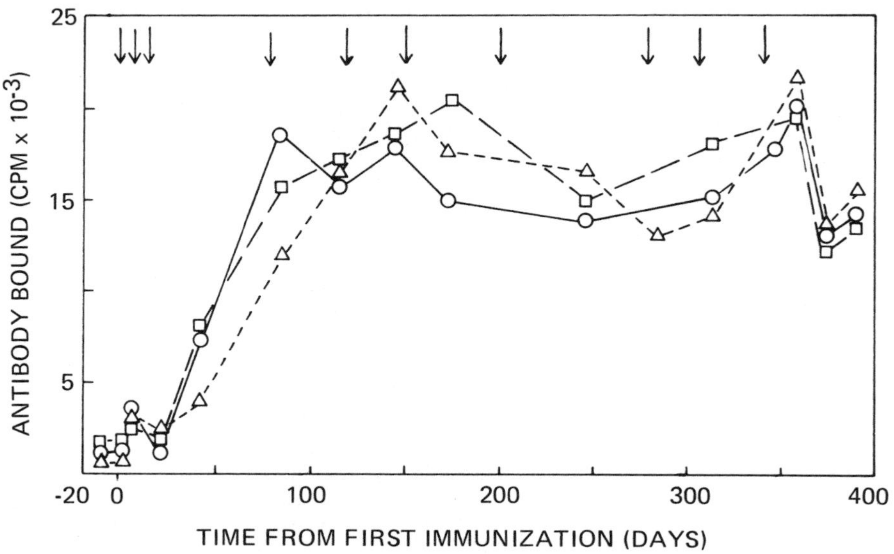

Figure 3 Time dependence of the antilysozyme antibodies' specificity profile toward the three antigenic sites of lysozyme. Antisera were raised in outbred mice against native lysozyme and the vertical arrows indicate points of immunization. Results were obtained by immunoadsorbent titrations using fixed amounts of ^{125}I-labeled immune IgG and varying amounts of adsorbents (for details of the technique, see Twining and Atassi, 1979). Each point represents the maximum (plateau) binding value of a given bleeding with a site adsorbent. (○), antibodies bound to site 1 adsorbent; (△), antibodies bound to site 2 adsorbent; (□), antibodies bound to site 3 adsorbent. The total amounts of antibodies bound by the three sites in any given bleeding over the entire period studied ranged from 95.6 to 98.8% relative to the antibodies bound to lysozyme. (From Yoshida and Atassi, 1983.)

are in direct peptide bond linkage. In fact, the surface-simulation-synthetic peptide having the structure Lys-Asn-Arg-Gly-Phe-Lys (which does not exist in lysozyme) carries the full immunochemical reactivity of the site (Lee and Atassi, 1977a). With the antisera studied so far, the synthetic antigenic site exhibited a preferred directionality (Lys-116 to Lys-33), since the reverse surface-simulation-synthetic sequence was immunochemically inefficient. The intactness of the disulfide bond 30-115 is critical for the integrity of this antigenic site in lysozyme (Atassi et al., 1973a). Antigenic site 3 overlaps with the hexasaccharide substrate binding site at the carbonyl group of Phe-34 and the side chain of Arg-114 (Lee and Atassi, 1977a).

C. Summary of the Antigenic Structure of Human Adult Hemoglobin

Two antigenic sites of hemoglobin (Hb), one on the α chain and one on the β chain, were initially predicted and confirmed by synthesis (Kazim and Atassi, 1977a) through extrapolation of the three-dimensional location of an antigenic site of sperm whale Mb (Atassi, 1975). The finding that antigenic sites on immunochemically unrelated members of the same protein family can exist in conformationally homologous regions gave

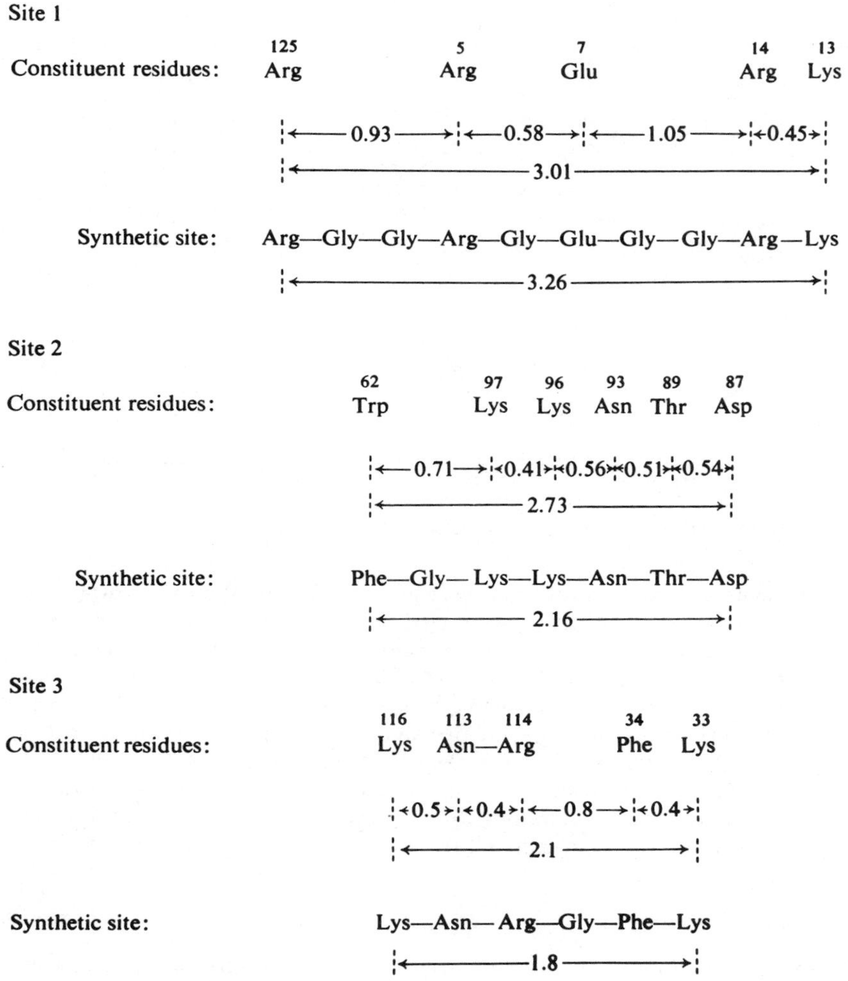

Figure 4 The three antigenic sites representing the entire antigenic structure of lysozyme. The diagram shows the spatially adjacent residues constituting each antigenic site and their numerical positions in the primary structure. The distances (in nm) separating the consecutive residues and the overall dimension of each site (in its extended form) are given, together with the dimension of each surface-simulation synthetic site. The latter assumes an ideal C-to-C distance of 0.362 nm. The precise boundary, conformational, and directional definitions of the sites are described in the text. The three sites account quantitatively for the entire (96 to 100%) antigenic reactivity of lysozyme. (From Atassi and Lee, 1978b.)

considerable support to the proposal of "structurally inherent antigenic sites" (Atassi and Kazim, 1978), namely that the antigenicity of protein antigenic sites is derived largely from their conformational locations in the respective protein chains (see later sections). Although the individual α and β subunits of Hb have similar overall tertiary structures to Mb, its tetrameric subunit structure endows Hb with unique conformational

Figure 5 Photograph of a lysozyme model showing the relative positions of the residues constituting antigenic sites 1 and 2. The side chains of the residues in the sites are outlined, those making up site 1 with horizontal bars and those constituting site 2 with vertical bars, to avoid confusion. The preferred direction of site 1 (at least by surface-simulation synthesis) is Arg-125 to Lys-13. Site 2 also had a preferred direction (Trp-62 to Asp-87). (From Atassi and Lee, 1978b, as modified from color in M. Z. Atassi, 1978, Adv. Biol. Med. *98*, 41.)

properties that contrast with those of the monomeric Mb molecule (Perutz et al., 1968). Therefore, the presence of structurally inherent antigenic sites on Hb and Mb does not preclude these proteins from having antigenic sites at dissimilar three-dimensional positions.

In order to identify other continuous (see Sec. IV for nomenclature) antigenic sites on Hb, a comprehensive synthetic approach was devised (Kazim and Atassi, 1980b) that would localize antigenic sites occurring throughout the α and β chains of the Hb molecule. This approach consists of studying the immunochemical activities of a series of overlapping synthetic peptides that encompass the entire primary structure of the protein

Figure 6 Photograph of a lysozyme model showing the position of antigenic site 3 on the molecule relative to sites 1 and 2. The side chains of the residues comprising the sites are outlined. The residues constituting site 3 have diagonal bars. This view is taken by rotating the model 125° anticlockwise on the vertical axis relative to the view shown in Fig. 5. From this perspective only parts of site 1 can be seen, which are the residues Lys-13, Arg-5, and Arg-125 (horizontal bars). Of site 2, only Trp-62 can be seen (vertical bars). Site 3 showed the same directional preference (Lys-116 to Lys-33) toward rabbit and goat antisera. (From Atassi and Lee, 1978b, as modified from color in M. Z. Atassi (1978), Adv. Exp. Biol. Med. *98*, 41.)

chain. The rationale for this approach was presented in detail by Kazim and Atassi (1980b, 1982).

This approach enabled the indentification of a full profile of immunochemically active α chain peptides and the localization of its major "continuous" antigenic sites. Antibodies to Hb raised in each of three different species (goat, rabbit, and mouse) recognize similar sites on the α chain (Kazim and Atassi, 1982). Accordingly, five antigenic sites on the α chain that are recognized by goat, rabbit, and mouse antisera to Hb were tentatively assigned to reside within, but not necessarily include all of, the general regions shown in Fig. 7. Further, except for site 5, the molecular locations of these sites coincide with α chain regions extrapolated from antigenic sites of the conformationally similar

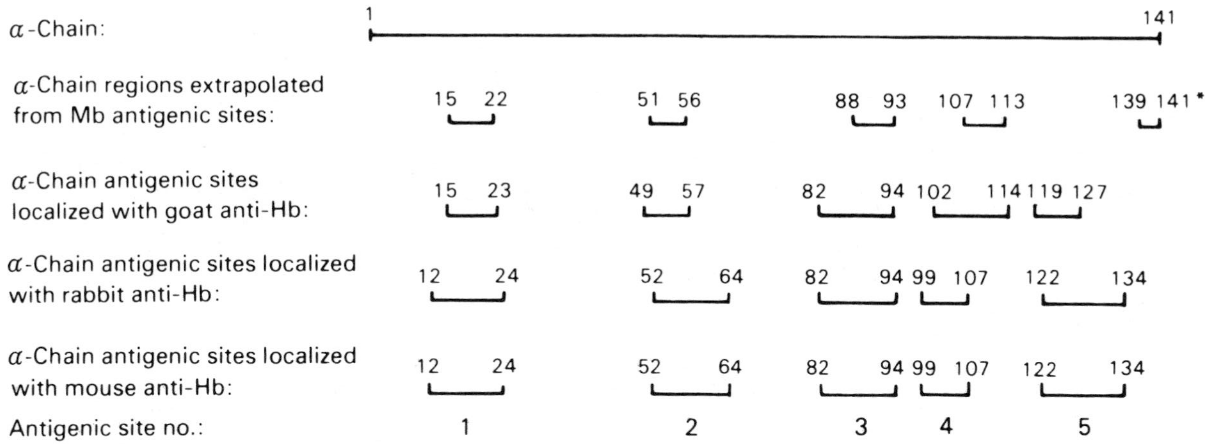

Figure 7 Schematic diagram of the α chain of Hb, α chain regions that have been extrapolated from the five antigenic sites of sperm whale Mb, and the antigenic sites of the α chain that were localized with goat, rabbit, and mouse antisera to Hb. The numbers refer to the locations of the residues in the α chain. The asterisk (*) indicates that, owing to differences in the sizes of the polypeptide chains of Mb and the α subunit, the antigenic site Mb 145-151 does not really have a full structural counterpart in the α chain. (From Kazim and Atassi, 1982.)

Mb molecule. Because the α-chain is shorter than Mb, site 5 of Mb has no true structural counterpart on the α-chain. These findings confirm the earlier proposed concept of *structurally inherent antigenic sites* (Atassi and Kazim, 1978), namely that antigenicity is conferred on certain surface regions of proteins by virtue of their three-dimensional locations. Thus the antigenic sites of conformationally related proteins are likely to have similar molecular locations. On the other hand, small conformational differences may cause some of the antigenic sites to occur at dissimilar locations.

It should be kept in mind that the precise boundaries of these sites have not yet been determined, and they were intentionally enlarged to increase the probability of a correct assignment (Kazim and Atassi, 1982). Further studies with synthetic peptides overlapping within the assigned regions are required to determine the boundaries. Upon such refinement of the data, however, small shifts in the boundaries of these sites should not be unexpected (Koketsu and Atassi, 1973, 1974b; Atassi, 1975).

Inspection of the three-dimensional structure of Hb (Perutz et al., 1968; Fermi, 1975) reveals that three of the five antigenic sites localized on the α chain contain residues that participate in α_1-β_1 (sites 4 and 5) or α_1-β_2; (site 3) subunit interactions, and are therefore potentially obstructed in the Hb dimer and tetramer. This indicates that the immune recognition of Hb occurs, at least in major part, at the level of its individual subunits. Studies on the genetic control in mice of the immune response to Hb (Krco et al., 1981a–c) have shown that the immune response to the α and β subunits in Hb are under separate genetic control. However, it has been reported (Tan-Wilson et al., 1976) that a minor subpopulation of antibodies can occur in goat antisera to Hb that binds only to Hb (probably the $\alpha\beta$ dimer) and not to either of the isolated subunits. It is likely therefore that, on immunization, Hb is recognized predominantly at the level of individual subunits and to a much smaller degree in higher states of subunit association.

At this level of delineation it cannot be excluded that residues outside the regions assigned to the antigenic sites, but in close spatial proximity to the sites in the native α chain, influence, to various degrees, their reactivities. In this context, antigenic sites were described (Atassi and Kazim, 1980; Kazim and Atassi, 1980c) as existing in steric and electronic equilibrium with neighboring residues, so that their reactivities are affected by changes or pertubations in these environmental residues (Twining et al., 1980).

D. Summary of the Antigenic Structure of Serum Albumin

The antigenic structure of serum albumin has been the subject of considerable interest for over two decades. Recent studies on bovine and human serum albumins have enabled the delineation of most of the antigenic structures of these two proteins. The location of six major antigenic sites (probably accounting for the bulk of the antibody response) have been determined (and confirmed by synthetic peptides) in bovine serum albumin (Atassi et al., 1979; Sakata and Atassi, 1980a; Atassi, 1982) and in human serum albumin (Sakata and Atassi, 1980b). The six antigenic sites that have been delineated in bovine and human serum albumins are shown in Fig. 8.

Because the three-dimensional structure of the albumin molecule is not known, the conformational locations of these antigenic sites cannot yet be analyzed. However, sites 1 to 3 in both albumin molecules comprise continuous parts of the polypeptide chain residing in the third, sixth, and ninth subdomains (disulfide double loops) far from the disulfide bonds. Antigenic sites 4 to 6 comprise the regions on both sides of disulfide bonds 166-175, 314-359, and 565-556, respectively, in bovine albumin and bonds

Site 1:

BSA: Tyr(137)-Leu-Tyr-Glu-Ile-Ala-Arg-Arg-His-Pro(146)

HSA: Tyr(138)-Leu-Tyr-Glu-Ile-Ala-Arg-Arg-His-Pro(147)

Site 2:

BSA: Phe(328)-Leu-Tyr-Glu-Tyr-Ser-Arg-Arg-His-Pro(337)

HSA: Phe(329)-Leu-Tyr-Glu-Tyr-Ala-Arg-Arg-His-Pro(338)

Site 3:

BSA: Ala(526)-Leu-Val-Glu-Leu-Leu-Lys-His-Lys-Pro(535)

HSA: Ala(527)-Leu-Val-Glu-Leu-Val-Lys-His-Lys-Pro(536)

Site 4:

BSA: Gln(168)-Ala-Glu-Asp-Lys-Gly-Ala-Cys-Leu-Leu-Pro-Lys(179)

HSA: Gln(170)-Ala-Ala-Asp-Lys-Ala-Ala-Cys-Leu-Leu-Pro-Lys(181)

Site 5:

BSA: Ala(308)-Glu-Asp-Lys-Asp-Val-Cys(314) Cys(359)-Ala-Lys-Asp(362), with S—S bond between Cys 314 and Cys 359

HSA: Val(309)-Glu-Ser-Lys-Asp-Val-Cys(315) Cys(360)-Ala-Ala-Asp(363), with S—S bond between Cys 315 and Cys 360

Site 6:

BSA: Ala(559)-Asp-Asp-Lys-Glu-Ala-Cys(565) Cys(556)-Lys-Asp-Val(553), with S—S bond between Cys 565 and Cys 556

HSA: Ala(560)-Asp-Asp-Lys-Glu-Thr-Cys(566) Cys(557)-Lys-Glu-Val(554), with S—S bond between Cys 566 and Cys 557

Figure 8 Structure and location of the six regions of bovine serum albumin (BSA) and of human serum albumin (HSA) that we have shown to carry antigenic sites. It is not implied that the antigenic sites comprise the entire size of the regions shown, but rather that they fall within these regions. (From Sakata and Atassi, 1980b; Atassi, 1982.)

168-177, 315-360, and 566-557, respectively, in the human albumin. Each of the six antigenic sites is rich in polar and hydrophilic amino acids, which is compatible with an exposed conformational location. It is noteworthy that the antigenic sites in the two albumins occupy analogous structural locations. Sites 1 to 3 in a given albumin show a similarity in binding function which increases with progressively later antisera after the first immunization. This functional similarity, which is a reflection of the structural similarity (see Fig. 8) of these sites, is expressed in an immunochemical cross-reaction of the sites, increasing in magnitude with progressively later antisera. Sites 4 to 6 also exhibit a functional similarity (in terms of binding with antibody) which improves progressively with the time at which time the antiserum is obtained after the initial immunization. Very recently, we have shown (Atassi et al., 1982a) that the cross-reactivity at the antibody level is also accompanied by extensive cross-reactions at the T-cell level.

IV. COMPARATIVE ANALYSIS OF PROTEIN ANTIGENIC STRUCTURES

Determination of the antigenic structures of Mb and lysozyme has permitted the definition in precise molecular terms of the antibody recognition of protein antigens. Many of the general conclusions relating to antigenic structures of proteins which have been derived from the precise definition of the antigenic structure of Mb (Atassi, 1972, 1975) are applicable to lysozyme as well (Atassi, 1978). These include the small size and sharp boundaries of the antigenic sites, their presence in only a limited number, their surface locations, their sensitivity to changes in the conformation and to changes in the environment of the site (e.g., amino acid substitution), the variation of the immunodominancy with the antiserum, and many other features. For a detailed discussion of these general conclusions, see Atassi (1975, 1977b).

The sizes of the lysozyme antigenic sites in their extended forms are 30, 27, and 21 Å, respectively (Atassi, 1978). These resemble the dimensions of the extended antigenic sites of Mb, which range between 19 and 23 Å (Atassi, 1975). Since the antigenic sites of a protein are not in the extended form, the actual dimensions of the sites in their folded shapes will be smaller than the values given. Nevertheless, the sizes of the antibody combining sites needed for binding with antigenic sites on either of the two proteins will have to be somewhat larger than the combining sites found for haptens by x-ray crystallography (Amzel et al., 1976; Padlan et al., 1976).

Significantly, the location of the antigenic sites on a given protein is independent of the immunized species. Thus rabbits, cats, goats, and mice make antibodies to native lysozyme with identical specificities for the same three antigenic sites (Yoshida and Atassi, 1983). Antibodies produced in rabbit, goat, pig, cat, mouse (outbred), and chicken against sperm whale Mb recognize the same antigenic sites on Mb (Twining et al., 1980). The same antigenic sites on serum albumin are recognized by rabbit and mouse antisera against this protein (Atassi et al., 1979; Sakata and Atassi, 1980a). The antigenic sites so far localized on Hb are similarly recognized by rabbit, goat, and mouse antibodies against Hb (Kazim and Atassi, 1982). The antigenic sites thus far localized and confirmed by synthesis on influenza virus hemagglutinin (Fig. 9) are similarly recognized by human, rabbit, goat, and mouse anti-influenza antibodies (Atassi and Webster, 1983, and unpublished results).

Examination of the antigenic sites of lysozyme reveals that they are very rich in basic amino acids. This was also seen in Mb, but we caution against premature generalizations. Obviously, interactions with antibody are predominantly polar in nature (as to be expected from the surface location of protein antigenic sites), with a considerable number of stabilizing effects being contributed by hydrophobic interactions and some hydrogen bonding (Atassi, 1975, 1978). However, it should be emphasized that the basicity of the lysozyme antigenic sites cannot be the only underlying factor for their antigenic expression.

Recently, based on the foregoing qualitative observations (Atassi, 1975, 1978, 1980), an empirical approach was suggested (Hopp and Woods, 1981) which, by linear analysis of the protein sequence in relation to a hydrophilicity index using the Chou and Fasman (1974) method, attempted to correlate antigenic sites with hydrophilicity maxima. However, such a relationship is not supported by what we know about protein antigenic structures. For example, in the classic case of Mb, analysis has revealed that only two (out of five) antigenic sites coincide with hydrophilicity maxima and, furthermore, that other hydrophilicity maxima are not antigenic. The antigenic structure of Mb was first to show that all antigenic sites are exposed, but *not* every exposed region constitutes an antigenic site. Thus exposure is not a sufficient criterion for antigenicity.

```
              201           205               210                       218
HA (201-218): Arg-Val-Thr-Val-Ser-Thr-Arg-Arg-Ser-Gln-Gln-Thr-Ile-Ile-Pro-Asn- Ile-Gly

              300               305               310               315
HA (300-315): Ile-Thr-Tyr-Gly-Ala-Gly-Pro-Lys-Tyr-Val-Lys-Gln-Asn-Thr-Leu-Lys

                             1                                       11
HA2 (1-11)  Type A:   Gly-Leu-Phe-Gly-Ala-Ile-Ala-Gly-Phe-Ile-Glu

            Type B:   Gly-Phe-Phe-Gly-Ala-Ile-Ala-Gly-Phe-Ile-Glu
```

Figure 9 Antigenic sites of influenza virus hemagglutinin that have been localized and their antigenic activity confirmed by synthesis and reactivity with rabbit, goat, and mouse antivirus and antihemagglutinin antibodies. Also, each of the peptides bound considerable amounts of anti-influenza antibodies from sera of human individuals after influenza attack. (From Atassi and Webster, 1983, and M. Z. Atassi and R. G. Webster, unpublished data, 1983.)

Furthermore, antigenic sites are not necessarily in highly hydrophilic locations. Hydrophobic interactions frequently provide major contributions to the binding energy. For example, in the case of the α chain of Hb, the two most immunodominant locations are not hydrophilic maxima. A strongly immunogenic location on influenza virus hemagglutinin has recently been reported (Atassi and Webster, 1983) which is entirely hydrophobic, and in fact the synthetic peptide comprising this region (HA2 1-11, Fig. 9) is so bydrophobic that it is insoluble in aqueous solvents. Two regions that may be "expected" to be antigenic on cytochrome *c* on the basis of hydrophilicity (regions 60--65 and 89–94) have been found to be nonantigenic by synthetic peptides around these regions (Atassi, 1981). Clearly, application of hydrophilicity analysis of the linear sequence for prediction of protein antigenic sites is not supported by the findings because it does not take into account the influences of the three-dimensional structure. It is very common, for example, for hydrophobic regions to become exposed by virtue of the three-dimensional folding of the protein. It is well to caution here that the sequence and three-dimensional features that confer immunogenicity on given parts or surface areas of a protein molecule are still not too clear. Undue speculation is inadvisable at this stage.

In the recent literature, there appears to have occurred a confusion as to what a protein antigenic site is. This term has always been used to describe a region of a protein antigen that is recognized (by antibodies and/or T cells) in the immune response to the *whole* protein. However, it has long been known that any peptide from a given protein, if immunized as a conjugate on an appropriate large carrier, will stimulate the formation of antipeptide antibodies. This even applies to peptides representing regions that are not antigenic when the whole protein is used as immunogen. As pointed out previously (Atassi and Habeeb, 1977), "antibodies obtained by immunization with a peptide (coupled or free) corresponding to a surface region of a protein can recognize that region in the protein, even though that region itself is not part of an antigenic site when the intact protein is used as immunogen." Thus it is necessary not to confound this fact with antigenic sites which define the regions that are immunogenic when the *whole* protein is the antigen.

In spite of the aforementioned points of resemblance between the antigenic structures of Mb and lysozyme, the antigenic sites of these two proteins are radically different in structural terms (Atassi, 1978). The five antigenic sites of Mb are each made up of residues that are directly linked to one another by peptide bonds (Atassi, 1975). In

contrast, in lysozyme each of the three antigenic sites constitutes spatially adjacent surface residues that are mostly distant in sequence, reacting with antibody as if they are in direct peptide linkage (Atassi, 1978). An important question here is: What factors determine the type of the site? Since the full antigenic structures of only two proteins are now determined, an unequivocal answer to this question cannot be formulated at this stage. Perhaps it can be tentatively concluded that an important factor in determining the type of the antigenic site may be dependent to a great extent on the stabilization or otherwise of the structure by internal disulfide cross-links (Lee and Atassi, 1976). A more definitive understanding of this subject must await knowledge of the antigenic structures of several proteins.

Even though it had been suggested earlier (Atassi and Saplin, 1968) that protein antigenic sites exist which are made up of spatially adjacent surface residues that are distant in sequence, their identification and precise definition in lysozyme is the first example in protein immunochemistry (Atassi et al., 1976b; Lee and Atassi, 1976, 1977a,b). A common feature of these two types of antigenic sites is that they occupy exposed regions on the surface topography of the respective protein, and this has invariably been the situation with other protein antigenic sites such as Hb (Kazim and Atassi, 1982), cytochrome *c* (Atassi, 1981), and influenza virus hemagglutinin (Atassi and Webster, 1983; M. Z. Atassi and R. G. Webster, unpublished data, 1983). It should be stressed that the antigenic sites of each of these proteins are sensitive to conformational changes in the respective protein, with those of lysozyme showing, as expected, a much higher sensitivity. Accordingly, it is inadequate to identify the antigenic sites of Mb and Hb by the terms "linear," "sequential," or "primary" or other such terms, while identifying the antigenic sites of lysozyme by the terms "spatial," "conformational," and so on. We have proposed that antigenic sites of the type seen in Mb and Hb be named *continuous* antigenic sites, which implies that they consist of conformationally distinct continuous surface portions of the polypeptide chain (Atassi, 1978; Atassi and Smith, 1978). For antigenic sites of the type seen in lysozyme, the term *discontinuous* antigenic sites will be most appropriate (Atassi and Smith, 1978). A discontinuous site is made up of conformationally (or spatially) contiguous surface residues that are totally or partially not in direct peptide bond linkage.

In Hb, the antigenic sites of the α chain and the one on the β chain that has so far been delineated and synthesized all reside in exposed parts of the individual subunit (Kazim and Atassi, 1977a, 1980b, 1982). This, as already mentioned, led to the conclusion that Hb is recognized predominantly at the level of its individual subunits, and that (at least for the α-chain) its antigenic sites occur mostly at molecular locations that coincide with the regions obtained from a structural extrapolation of the Mb antigenic sites to Hb (Kazim and Atassi, 1982). These findings afford further confirmation of our concept of structurally inherent antigenic sites (see Sec. VI. A).

The antigenic sites of serum albumin have not yet been narrowed down to their precise boundaries. Delineation of the peptide regions as shown in Fig. 8 is not intended to imply that the entire region constitutes an antigenic site, but rather that the site falls within that region. Nevertheless, it can be readily seen, even at this level of delineation, that the antigenic sites of serum albumin exhibit the expected characteristics of having discrete boundaries and being limited in number, sensitive to conformational and environmental changes, and variable in immunodominancy with the antiserum.

In albumin, the status of exposure of its antigenic sites is not known. Unfortunately, the three-dimensional structure of a serum albumin has not yet been determined.

Therefore, it is not possible to correlate the locations of the antigenic sites with the shape of the protein molecule. Three of the antigenic sites (sites 1 to 3 in Fig. 8) occupy continuous portions of the polypeptide chain of albumin and are, therefore, continuous antigenic sites. The other three antigenic sites (sites 4 to 6) are each localized around a disulfide bond and belong to the type termed discontinuous antigenic sites. However, by analogy with Mb, lysozyme, and Hb, it would be expected that the antigenic sites of serum albumin also occupy exposed structural locations. The predominance of polar and hydrophilic amino acids in the albumin antigenic sites (Fig. 8) tends to support this expectation.

V. REGIONS OUTSIDE THE ANTIGENIC SITES

The exhaustive strategy on which the immunochemical analyses were based included probing in detail every part of the molecule for its immunochemical reactivity and permitted the determination of the contribution of each part of a given protein molecule to the overall immune response against that protein. As already mentioned, the amounts of antibodies that can be bound by the synthetic antigenic sites of Mb (Tables 1 to 3) and the synthetic antigenic sites of lysozyme (Table 4 and Fig. 4) account for the bulk (usually over 98%) of the total antibodies against the respective protein in a given antiserum. But how much antibody is directed against regions outside these antigenic sites?

In studies on Mb (Atassi, 1975) and lysozyme (Atassi, 1978) a "general background response" to regions between their respective antigenic sites was found to account for 0.1 to 1.1% of the total antibodies to these proteins. Also, Takagaki et al. (1980) have reported the detection of antibodies to a continuous region of hen egg white lysozyme (residues 28 to 54) which account for about 0.1 to 1% of the total antilysozyme antibodies. In very recent studies (Kazim and Atassi, 1982) employing the overlapping synthetic peptide approach, peptides that were seemingly inactive in binding of ^{125}I-labeled anti-Hb antibodies in a titration assay were further studied by attempting to isolate from adsorbents of these peptides antibodies specific for Hb. By using large excesses of unlabeled immune IgG, traces of Hb-specific antibodies could be isolated from that of peptide 1-15. Although the amounts of these antibodies were not quantified, they accounted for less than 1 to 2% of the specific antibodies, since no antibody binding could be detected in the titration assay with the immunoadsorbent of this peptide. This indicates the presence of a trace antibody response directed to this region of the α chain, and suggests that other apparently inactive regions of the α chain may also be recognized by amounts of antibodies below the detection level of these methods. Similarly, Dean and Schechter (1979) have reported the isolation of a subpopulation of anti-Hb antibodies directed against the region α129-141, but which represents less than 1% of the total anti-Hb antibodies. Also, recent studies with synthetic reference nonantigenic regions of serum albumin revealed the presence of trace antibody binding accounting for 0.5 to 1.1% of the antiprotein antibodies in the antisera (Sakata and Atassi, 1980a; Atassi, 1982). Thus the presence of trace antibody responses to regions interspersed among immunodominant sites would appear to be a general phenomenon for protein antigens.

Although the regulation of the relative immunodominance of antigenic sites in proteins is poorly understood, the methodology for producing monoclonal antibodies introduced by Köhler and Milstein (1975) should permit amplification in vitro of what would ordinarily constitute trace antibody responses in vivo. The specificities and regulation of such responses could then be studied more readily. This prediction (Atassi, 1980; Kazim

and Atassi, 1982) has recently been confirmed by the description of two monoclonal anti-Mb antibodies whose specificities were directed to regions that are either outside the five antigenic sites of Mb or displaced from the site (H. E. Schmitz, H. Atassi, and M. Z. Atassi, unpublished data, 1983). One of these monoclonals exhibited specificity to the synthetic peptide 121-127 and the other, although essentially reacting with site 2, exhibited a specificity displacement to the left by two residues so that it required residues 54 and 55 as part of the site. Interestingly, such antibodies were below the detection level in the antisera of the parent mice (whose spleens were used for fusion) by immunoadsorbent titrations of ^{125}I-labeled antibodies with peptide adsorbents.

VI. FACTORS THAT DETERMINE AND REGULATE THE ANTIGENICITY OF THE SITES

A. Antigenic Sites Are Inherent in the Three-Dimensional Structure

Until recently, it has been widely believed that the antigenic sites on a protein antigen are located at regions that differ in sequence from the corresponding homologous protein in the immunized host, and therefore are not peculiar to the protein but are defined by the host species in which the antigen is injected. Our determination of the first total protein antigenic structures have shown that these speculations are incorrect.

Having determined the precise antigenic structures of Mb and lysozyme and most of the antigenic structures of Hb and albumin, we have focused our attention on investigating some of the factors responsible for the antigenicity of the sites. Initially, the antigenic structure of Mb was determined with antisera raised in rabbits and goats (Atassi, 1975). Recently, these studies were extended to antisera that were elicited in chicken, cat, pig, and mouse (Twining et al., 1980). It was found that the same five antigenic sites on sperm whale Mb are recognized by rabbit and goat antisera and chicken, cat, pig, and mouse antisera (see Tables 1 to 3). Antisera against hen lysozyme elicited in rabbits, goats, cats, and mice recognize the same three antigenic sites on lysozyme (Atassi, 1978; Yoshida and Atassi, 1983). This suggested that the antigenicity of the sites is inherent in their three-dimensional location (Kazim and Atassi, 1978) and is independent of any sequence differences between the injected protein antigen and the homologous protein of the immunized host. These observations received additional confirmation by the finding that the six major antigenic sites of serum albumin are recognized by both rabbit and mouse antisera against albumin (Sakata and Atassi, 1980a). Also, rabbit, goat, and mouse antibodies against Hb recognized similar antigenic sites (Kazim and Atassi, 1982). Very recently, three antigenic sites on influenza virus hemagglutinin were localized and synthesized (Fig. 9). They were found to be similarly recognized by rabbit, goat, and mouse antivirus antibodies and significantly by antibodies in the sera of humans after natural infection with influenza virus. Sequence analysis of myoglobins by Romero-Herrera et al. (1978) in relation to the antigenic structure of sperm whale Mb determined in our laboratory clearly showed that the sequence variability within the antigenic sites was not different from the variability outside the antigenic sites. Other studies have strongly confirmed the conclusion that antigenic sites are structurally inherent. These studies are summarized below, together with their implications to immune recognition and in particular to autoimmune responses.

The realization that protein antigenic sites are structurally inherent (for a recent review, see Atassi and Kazim, 1978) made it possible to predict and verify by synthesis

Residue Location:	A_{13}	A_{14}	A_{15}	A_{16}	AB_1	B_1	B_2	B_3	B_4	B_5
Site 1 of Mb:	15 Ala	Lys	Val	Glu	Ala	Asp	Val	22 Ala		
Hb α (15-23):	15 Gly	Lys	Val	Gly	Ala	His	Ala	Gly	23 Glu	
Hb β (16-23):	16 Gly	Lys	Val	Asn	...	...	Val	Asp	Glu	23 Val

Figure 10 Diagram showing the sequence and structural location of antigenic site 1 of sperm whale Mb and the corresponding extrapolated regions of the α and β chains of Hb and whose antigenic activity was confirmed by synthesis. (From Kazim and Atassi, 1977a).

the location of two antigenic sites (Fig. 10) of human Hb (one each on the α and β chains) by extrapolation of the three-dimensional location of an antigenic site of sperm whale Mb (Kazim and Atassi, 1977a). The recent localization and synthesis of the remaining antigenic regions on the α chain of human Hb revealed that four out of its five sites coincide with the regions derived from a structural extrapolation of the antigenic sites of Mb to the equivalent regions in the α chain (Kazim and Atassi, 1982). The concept of structurally inherent antigenic sites was further supported by the finding that even the antigenic sites of soybean leghemoglobin, a very distant member of this family of heme proteins, fall within the expected structural locations obtained by extrapolation of the Mb antigenic sites (Hurrell et al., 1978). Moreover, recent findings on bovine and human serum albumins show that their antigenic sites are located at equivalent structural locations (Fig. 8).

The aforementioned findings strongly confirm the proposal that the conformational uniqueness of certain parts of a protein molecule plays a most critical role in the antigenic expression of the protein (Kazim and Atassi, 1978; Atassi and Kazim, 1978).

B. Implication of Structurally Inherent Antigenic Sites in Autoimmune Recognition

Comparison of the primary structures of sperm whale and rabbit myoglobins, which have identical chain lengths, showed that only 5 out of 22 replacements fall within the boundaries of the antigenic sites (Fig. 11). Two antigenic regions, corresponding to sites 2 (residues 56 to 62) and 3 (residues 94 to 99) in sperm whale Mb were identical in the two proteins. Two other sites (sites 1 and 4) had conservative substitutions. Thus in site 1, rabbit Mb has the substitution Ala-15 → Gly and Val-21 → Leu, and in site 4 the replacement Arg-118 → Lys occurs. In the region corresponding to site 5, rabbit Mb has the replacements Lys-145 → Gln and Tyr-151 → Phe. At any rate, at least two antigenic sites in sperm whale Mb were identical in sequence to the corresponding regions in rabbit Mb. Clearly, rabbits reacted by making antibodies to parts of sperm whale Mb that were identical in sequence to rabbit's own Mb. In other words, sequence differences between the protein antigen and its counterpart in the immunized host are not a prerequisite for

SITE 1 of	15							22
Sperm Whale Mb	Ala	Lys	Val	Glu	Ala	Asp	Val	Ala
Rabbit Mb	Gly	Lys	Val	Glu	Ala	Asp	Leu	Ala
SITE 2 of	56						62	
Sperm Whale Mb	Lys	Ala	Ser	Glu	Asp	Leu	Lys	
Rabbit Mb	Lys	Ala	Ser	Glu	Asp	Leu	Lys	
SITE 3 of	94					99		
Sperm Whale Mb	Ala	Thr	Lys	His	Lys	Ile		
Rabbit Mb	Ala	Thr	Lys	His	Lys	Ile		
SITE 4 of	113						119	
Sperm Whale Mb	His	Val	Leu	His	Ser	Arg	His	
Rabbit Mb	His	Val	Leu	His	Ser	Lys	His	
SITE 5 of	145						151	
Sperm Whale Mb	Lys	Tyr	Lys	Glu	Leu	Gly	Tyr	
Rabbit Mb	Gln	Tyr	Lys	Glu	Leu	Gly	Phe	

Figure 11 Diagram showing the primary structures of the five antigenic sites of sperm whale Mb and the corresponding regions of rabbit Mb. The sequences shown occupy identical positions in the respective protein chains. Identical positions having different amino acids in the two chains are indicated by blocks. (From Kazim and Atassi, 1977b.)

antigenicity of the sites (Kazim and Atassi, 1977b). Indeed, we have found that rabbit antisera to sperm whale Mb show considerable immunochemical cross-reactions with rabbit Mb. For more detail, the reference Kazim and Atassi (1977b) should be consulted. Recently, these observations were extended to show that this is a general phenomenon quite independent of the immunized species (Twining et al., 1980). Thus, goat antisera to sperm whale Mb showed autoreactivity with goat Mb, and similarly, chicken antisera reacted with chicken Mb, pig antisera with pig Mb, and mouse antisera with mous Mb (Table 5). More recently, it was shown that rabbit or mouse antisera against bovine or human serum albumin cross-reacted with the animal's own albumin (e.g., see Table 6) (Sakata and Atassi, 1981; Atassi et al., 1982a).

The finding that the antigenicity of the Mb sites is not dependent on sequence differences between the injected Mb and the Mb of the immunized host led to the reasoning that immunization of rabbits with rabbit Mb should generate autoantibodies against this protein (Kazim and Atassi, 1978). Such autoantibodies to rabbit Mb were in fact readily obtained (Fig. 12). Furthermore, an autoimmune T-lymphocyte proliferative response to mouse Mb was also reported and was found to be under genetic control (Yokota et al., 1980). More recently, we have shown that rabbits immunized with their own serum albumin (Sakata and Atassi, 1981) and mice immunized with mouse albumin (Atassi et al., 1982b) each produced autoantibodies against their respective self-serum albumin (Table 6). Complete Freund's adjuvant (CFA) has been shown to have an effect in the stimulation of this autoimmune response (Atassi et al., 1982b). The possibility that, in

Table 5 Autoreactivity of Sperm Whale Mb Antisera with the Animal's Own Myoglobin (Percent Antibodies Bound Relative to Binding by Sperm Whale Mb as 100%[a]

A.	Goat antisera:	G3	G4		
	Goat Mb	39.0	28.8		
B.	Rabbit antisera:	77	80	M8	M9
	Rabbit Mb	40.3	28.8	37.7	41.9
C.	Chicken antisera:	CK1	CK2		
	Chicken Mb	20.4	25.8		
D.	Mouse antisera:	MS2	MS3	MS5	
	Mouse Mb	33.5	21.8	30.4	

[a]Plateau binding values of ^{125}I-labeled antibodies by immunoadsorbent titration studies.
Source: Twining et al. (1980).

the process of isolation or in emulsification with the adjuvant's oil, serum albumin may undergo some trace denaturation and no longer resembles self-albumin has been ruled out in recent studies (Atassi et al., 1982b). The results (Table 6) have shown that autoreactive and autoantibody responses are generated when antigen and CFA are in emulsion and less so when injected into separate sites but on the same side of the animal. The absence of CFA, antigen emulsion in incomplete Freund's adjuvant, or when CFA and antigen are injected into separate sites but on different sides of the animal failed to activate the autoimmune response. These observations indicated that the protein antigen and CFA must act on the same lymph nodes in order to activate autoimmune clones. The findings are highly significant in view of the fact that serum albumin is not an intracellular protein but a normal and most abundant constituent of serum and other body fluids.

The foregoing observations establish that autoimmune recognition does not require the stimulating antigen to be foreign or altered. Thus the capacity for autoimmune recognition is a general phenomenon and strongly confirms that antigenicity of protein antigenic sites is independent of any sequence differences between the protein antigen and its counterpart in the host and that locations of autoimmune antigenic sites are inherent in the three-dimensional structure of the protein (Kazim and Atassi, 1977b; Atassi and Kazim, 1978; Atassi, 1981). Clearly, autoimmune clones are not deleted, even to serum albumin, and several pathological conditions are known to be due to autoimmune recognition of some self proteins, including serum albumin (for a review, see Sakata and Atassi, 1981). Autoimmune clones are part of the overall regulatory circuits of the immune system and can be triggered in many known pathological cases by a mechanism that is little understood but may be related to breaking of T-cell tolerance (Atassi, 1980). In the experimental situation, complete Freund's adjuvant may assist in overcoming tolerance to self proteins (Allison and Denman, 1976; Atassi et al., 1982b) through bypassing T-cell suppression.

Table 6 Adjuvant Effects on Autoreactivity with Mouse Serum Albumin (MSA) of Mouse Antibodies Against Bovine Serum Albumin (BSA) and on Mouse Autoantibody Response to MSA[a,b]

	Antibodies bound (cpm)	
Antigen and mode of injection	Autoreaction with MSA	Reaction with BSA
A. Immunization with BSA		
BSA emulsion in CFA, right thigh	12,950	34,210
BSA in PBS, CFA at separate sites, right thigh	2,130	11,470
BSA in PBS right thigh, CFA left thigh	70	11,040
BSA in PBS	120	6,350
CFA only	140	100
B. Immunization with MSA		
MSA emulsion in CFA right thigh	2,935	n.d
MSA in PBS, CFA at separate sites, right thigh	630	n.d.
MSA in PBS right thigh, CFA left thigh	120	
MSA emulsion in IFA, right thigh	30	
MSA in PBS	0	
None, CFA only	130	

[a]Binding studies were done with the IgG fractions, which were diluted 1/100 (relative to the original antisera). Bound antibodies were determined by reacting with ^{125}I-labeled protein A (10^5 cpm). The results have been corrected for binding of BSA and MSA with preimmune IgG from the same animals (0.8 to 1.1% of total label added) and for binding of these antibodies with unrelated control proteins (myoglobin, lysozyme) which amounted to 0.5 to 0.6% of total label added. Values represent the average of triplicate analyses which varied ±1.3% or less.
[b]MSA, mouse serum albumin; BSA, bovine serum albumin; CFA, complete Freund's adjuvant; IFA, incomplete Freund's adjuvant; PBS, phosphate-buffered saline; n.d., no data.
Source: Atassi et al. (1982b).

C. Effects of Amino Acid Substitutions on Antigenic Sites

During the course of our studies on the antigenic structure of Mb and lysozyme, the binding of a protein antigenic site with its specific antibodies was repeatedly shown to be highly dependent on the chemical characteristics of the residues constituting the site and on the conformational integrity of the site. Thus alterations in the chemical nature of a side chain within a site (e.g., reversal or removal of a charge, creation of a new charge, elimination of a hydrogen bond, etc.), brought about either through chemical modification

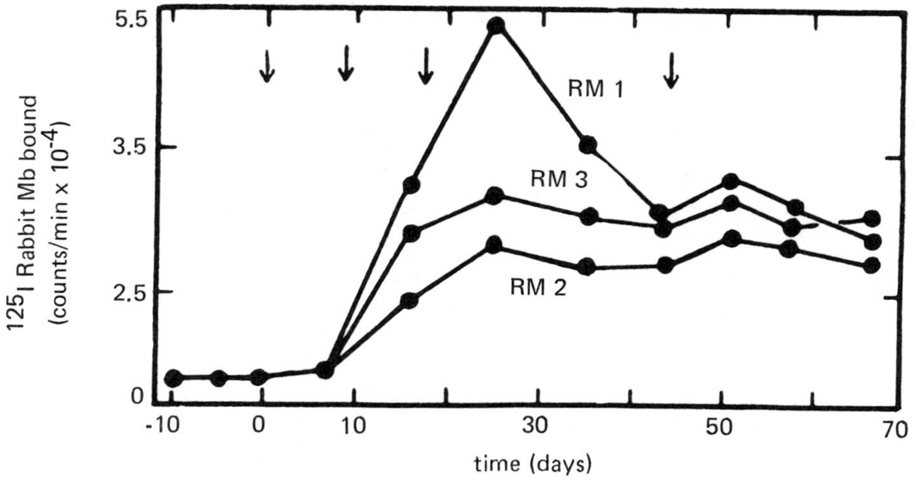

Figure 12 Screening of [^{125}I] rabbit Mb binding by antisera obtained from serial bleedings of three rabbits (RM1, RM2, and RM3) immunized with rabbit Mb. Arrows indicate times of immunization. (From Kazim and Atassi, 1978.)

(e.g., Atassi and Thomas, 1969; Atassi, 1967a, 1968; Atassi et al., 1973b, 1975) or by its evolutionary replacement in a homologous protein, would cause a reduction or even complete elimination of the reactivity of the site (for a review, see Atassi, 1975, 1977a,b, 1978). The presence and extent of reduction will depend on the nature of the chemical change to a site residue, on the location within the site of the residue being substituted or modified, and on its contribution to the overall binding energy of the site.

Apart from substitutions within the sites, whose effect is perhaps readily perceived, substitutions involving residues outside the antigenic sites would also be expected to exert a considerable effect. It is relevant to note that the antibody respone to native protein antigens is directed against their native three-dimensional structure (Atassi, 1967b, 1975, 1978; Atassi and Thomas, 1969). A substitution, even when distant from a site, may cause a conformational readjustment and it is well documented that alterations in the conformation of an antigenic site will influence its reactivity (Atassi, 1967b; Andres and Atassi, 1970; Atassi, 1970; Atassi et al., 1970; Atassi and Skalski, 1969). An example of transmitted conformational effects may be taken from our early studies on Mb and Hb, which showed that conformational changes intentionally imposed on these two proteins by chemical alterations that were clearly outside the antigenic sites (in the heme group) lead in each protein to a reduction in its antigenic reactivity (Atassi, 1967b; Atassi and Skalski, 1969; Andres and Atassi, 1970).

The chemical factors that contribute to and maintain the reactivity of an antigenic site are not, however, limited to those which are localized within the site (Atassi, 1968; Atassi and Thomas, 1969). It must be remembered that protein antigenic sites exist as integral portions of the entire native protein molecule and are, therefore, also subject to chemical influences exerted from regions outside the antigenic sites which are transmitted to the site through mutual interactions. Indeed, antigenic sites recently were described as being in a certain electronic and steric equilibrium with the remainder of the protein molecule (Atassi and Kazim, 1980). Also, it was recently demonstrated (Lee and Atassi,

1977a) for lysozyme that a chemical change outside an antigenic site but within interaction distance from an essential site residue (without detectable accompanying conformational changes) creates an electrostatic inductive effect which exerts a detrimental effect on the reactivity of that site.

These findings prompted an analysis of the residues that comprise the environment of the antigenic sites of Mb and lysozyme, which, when altered, may be expected to influence their reactivity. Recently a description was presented (Atassi and Kazim, 1980) of the residues that constitute the environment of the antigenic sites of lysozyme, together with a somewhat detailed discussion of the general nature of protein binding sites and the forces that affect them. This was followed by identification of the nearest-neighbor residues making up the immediate environment for each residue in the five antigenic sites of Mb (Kazim and Atassi, 1980c). Because of the expected rapid decay of the field of influence with increasing distance between a site residue and a nearest neighbor, it may be considered that the effects will become negligible in most cases when the distance is greater than 7 Å. Therefore, the analysis has been limited to those residues that fall within a 7-Å radius of the antigenic site residues. For detailed tables, see Atassi and Kazim (1980) and Kazim and Atassi (1980c).

Knowledge of the precise locations of the antigenic sites of Mb lysozyme and of the environmental residues around each site residue has made it possible to evaluate the effects of substitutions in the antigenic sites, in the residues close to those sites, and in locations elsewhere in the molecule (Atassi and Kazim, 1980; Kazim and Atassi, 1980c). Recently, the immunochemical cross-reaction of 15 myoglobins at the antibody level was studied in several outbred species of animals (Twining et al., 1980) and in selected inbred mouse strains, both at the antibody level (Twining et al., 1981a) and by the T-cell proliferative response (Atassi et al., 1981a), and the findings showed that the immunochemical cross-reactions can be reasonably rationalized on the basis of amino acid substitutions occurring in the residues of the antigenic sites and in the environmental (nearest-neighbor) residues surrounding the sites (Twining et al., 1980). Furthermore, conformational readjustments can frequently alter the binding energy of an antigenic site (Twining et al., 1980).

From the foregoing it is clear that the structural and immunochemical effects of mutations are quite complex. A penetrating analysis of the immunochemical data from a given set of mutants can be achieved only with knowledge of the antigenic structure as well as the three-dimensional structure of one of these homologous proteins, which can then be used as a reference antigen. It is well to suggest here that caution should be exercised in the interpretation of the immunochemical cross-reactions of protein mutants, an approach that is often applied by some workers to infer the unknown locations of the antigenic sites on these mutants. Other independent approaches (see Sec. II) should be applied in the determination of protein antigenic structures. In general, the results of this approach have been quite disappointing when applied to Hb, cytochrome *c*, and Mb. For example, by analysis of the immunochemistry of cytochrome *c* mutants, certain regions had been proposed to be antigenic sites on these proteins. Recently, these proposals were shown to be incorrect by synthesis of the indicated peptide regions and examination of their antibody binding activities (Atassi, 1981). Also, deductions for locations of antigenic sites on Hb based on analysis of Hb mutants were not substantiated by the comprehensive synthetic strategy already described (Kazim and Atassi, 1982).

D. Heterogeneity of the Response

The response to each antigenic site is heterogeneous, in that a variety of antibodies are produced to each antigenic site. This is not unusual or unexpected, since it is well knwon that even small haptenic determinant groups (e.g., dinitrophenyl) generate a heterogeneous antibody response. What is the basis of this heterogeneity?

1. It is well established that a variety of antibody molecules having similar specificities can be generated by appropriate conservative substitutions in the hypervariable regions. For example, the following simple situation (see Atassi, 1980), in which only three amino acids in the combining site undergo substitutions, can be visualized to generate a large number of antibodies of similar specificity:

Glu	Lys	Val
Asp	Lys	Val
Glu	Arg	Val
Asp	Arg	Val
Asp	Lys	Leu
Glu	Lys	Leu
	etc.	

But even though these antibodies possess similar binding specificity, they will exhibit different binding energies with the antigenic site. Thus one source of heterogeneity is the presence of a variety of antibody molecules exhibiting essentially the same specificity but having different affinities (Atassi, 1980).

2. In the process of the delineation of the antigenic sites of Mb and lysozyme to their precise boundaries, it was found that quite frequently (but not always) a portion of an antigenic site can bind some (but not all) of the antibodies to the whole site (Atassi, 1972; Koketsu and Atassi, 1973, 1974a,b; Atassi and Saplin, 1968). The extent of the binding ability of a portion of an antigenic site was explained by proposing (Atassi, 1972) that "in antibody response to an antigenic site antibodies are made against various parts of the site. Maximal effective response, which is the resultant average, will be directed against the center of the site." However, it should be pointed out that reactivity of a portion of a site can also be explained by the presence of antibodies having identical specificities but differing in affinities (as in factor 1 above). Thus, whereas low-affinity antibodies are unable to bind with a portion of an antigenic site because of insufficient binding energy, higher-affinity antibodies may well be able to do so (Atassi, 1980).

It is possible to distinguish between the two alternatives by analysis of the appropriate binding and inhibition curves. Factors 1 and 2 both contribute to the heterogeneity of the antibody response (Atassi, 1980). This can be demonstrated by the production of monoclonal antibodies to protein antigens. The ability to prepare a large number of monoclonal antibodies to a protein does not mean that these antibodies possess a vast spectrum of specificities. As mentioned under (1), a number of antibodies can be generated with similar binding specificity but having different affinities to a single antigenic site. Also, as mentioned under (2), a number of monoclonal antibodies should be expected that are made against various parts of an antigenic site (i.e., slight displacement to one side or the other of the site recognized by the various monoclonal antibodies). For example, a monoclonal antibody was recently isolated that recognized site 2 of Mb with a

displacement of two residues to the left (i.e., residues 54 and 55 were required as an essential part of site 2 for this antibody).

E. Genetic Control of the Immune Response

Immune responses to various antigens are controlled by genes (Ir genes) coded within the H-2 (major histocompatibility) complex. Most studies of Ir-gene control of immune responses have employed synthetic polymers of a few amino acids. Only recently, these studies were extended to include natural protein antigens. This information has added a new dimension to the understanding of the mechanism of T-cell recognition and antibody production (for a detailed treatment, see Chapter 25).

Studies on the genetic control of immune response to Mb and lysozyme and their respective antigenic sites have been conducted in our laboratory in collaboration with Chella David. The study of Mb and lysozyme offers a major advantage over the study of other proteins because their entire antigenic structures have been determined. We have also studied the genetic control of the immune response to human Hb and bovine serum albumin. These studies have employed congenic strains of mice expressing the independent haplotypes as well as selected recombinant strains and F_1 crosses.

Studies with Mb revealed that the T-lymphocyte proliferative response to intact Mb was under H-2-linked Ir-gene control (Okuda et al., 1978). At least two genes were identified as mapping in the I-A and I-C subregions. In the case of lysozyme it was found (Okuda et al., 1979a) that the antibody and T-lymphocyte proliferative responses are also genetically controlled in mice. Both responses to lysozyme are controlled by two H-2I region loci, one being in the I-A and the other may be in the I-C subregions. More recently, (Krco et al., 1981a–c) studies were undertaken with human Hb to probe genetic control of the immune response to an oligomeric protein and the role of the indiviudal subunits in the expression and regulation of the response. Such investigations have not been done on an oligomeric protein composed of nonidentical subunits. The findings showed that the immune response to Hb is also under genetic control and is determined by two H-2 loci, one being in the K-A interval and the second in the D end. From in vitro challenge with each of the α and β subunits of Hb, it was found that the responses to the two subunits are under separate genetic control. The response to the α chain is controlled by a locus in the K end, while control of the response to the β chain is associated with the H-2D end (Table 7). It was also shown very recently that the subunits exert interchain regulatory influences on one another. The final response to Hb is the net product of these regulatory effects. In the case of bovine serum albumin, its immune response is controlled by genes in the I-A subregion of H-2, with some slight non-H-2 influences (Atassi et al., 1982a).

Studies were also carried out on the genetic control of the immune response in mice to the synthetic antigenic sites of sperm whale Mb. Myoglobin-primed T cells gave in vitro proliferative responses to the synthetic antigenic sites, and the responses were under genetic control (Okuda et al., 1979b). Furthermore, synthetic nonantigenic parts also consistently did not stimulate a T-cell proliferative response in any of the numerous strains examined. From these studies it was concluded (Okuda et al., 1979b) that individual antigenic sites in a molecule are controlled by unique Ir genes (i.e., the response to each antigenic site is under separate genetic control; see Fig. 13).

Significantly, the Mb antigenic sites that are recognized by mouse B cells are also recognized by mouse T cells (Okuda et al., 1979b). No meaningful genetic control of

Table 7 Summary of Ir-Gene Mapping of Immune Responsiveness to Mb, Lysozyme, Hemoglobin, and Its Subunits and Serum Albumin

Protein	MHC subregions regulating responsiveness		
Mb (sperm whale)[a]	A	C	
Mb (mouse, autoimmune)[b]	A		D
Lysozyme[c]	A	C	
Hemoglobin (adult human)[d]	K-A		D
Hemoglobin α chain[d]	K-A		
Hemoglobin β chain[d]			D
Serum albumin (bovine)[e]	A		

[a]Okuda et al., 1978, 1979b.
[b]Yokota et al., 1980.
[c]Okuda et al., 1979a.
[d]Krco et al., 1981a–c.
[e]Atassi et al., 1982a.

antibody affinity was observed (Young et al., 1981b). Autoimmune antibody and T-lymphocyte proliferative responses were readily generated by immunizing an animal with self-Mb. With mouse Mb, the autoimmune T-lymphocyte proliferative response was under genetic control and mapped in the I-A and the H-2D end of the H-2 gene complex (Yokota et al., 1980). The overall response to Mb is regulated by intersite influences which can be of either a cooperative (help) nature or a suppressive nature (Atassi et al., 1981a). Help between sites 1 and 2 and between one or both of these and site 3 was observed in several mouse strains. A suppressive influence by site 5 on the overall proliferative response to Mb was observed in two strains (D2.GD and A.TFR4) (Atassi et al., 1981a). Finally genetic control of the responses to individual antigenic sites on a protein is determined not only by the genetic constitution of the host but also by the chemical properties of the individual sites. The H-2 subregions mapping the responses to given antigenic sites can also recognize other sites, which were previously unrecognizable in a homologous protein, if the chemical properties of these sites are suitably altered (Young et al., 1981a). It was recently found that the genetic control of the immune response to

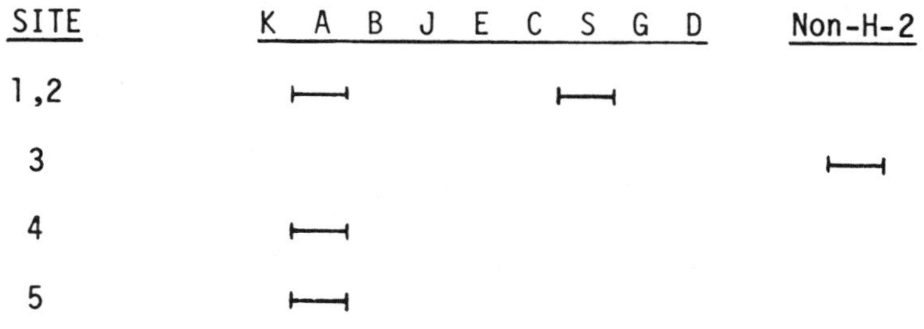

Figure 13 Ir-gene mapping for individual antigenic sites of sperm whale myoglobin. (From Okuda et al., 1979b,)

Mb and its synthetic antigenic sites is dependent on the immunizing dose of antigen (Young and Atassi, 1982). For example, the mouse strains C57Bl/6J, AKR, and SWR/J are low responders to Mb following immunization with 50 μg, responding only to site 4. After immunization with 200 μg of Mb, C57Bl/6J mice are high responders to Mb and respond to antigenic sites 1, 3, 4, and 5; AKR mice are high responders to Mb and respond to antigenic sites 1 and 4; SWR/J mice are high responders to Mb and respond to all five antigenic sites. Also, these studies have made it possible for the first time to separate the response to site 1 from the response to site 2 and thus have conclusively established that sites 1 and 2 are controlled by separate Ir genes (Young and Atassi, 1982).

It was recommended that the determination of the complete antigenic structure of a protein be done using outbred animals (Atassi, 1980). If antisera are elicited in a given congenic strain, only some of the potential antigenic sites will be detected, while the other sites, the response to which is genetically excluded, will go undetected.

F. Cellular Recognition and Regulatory Events in the Immune Response

From studies on synthetic amino acid polymers it was believed that Ir-gene control at the T-cell level was carrier dependent. Findings with the synthetic antigenic sites of Mb indicate that this need not be the only mechanism and is not applicable to natural antigens, where the recognition can operate at the antigenic site level (Okuda et al., 1979b). Even though studies with amino acid polymers provided valuable information on the immune-response-gene phenomenon, there are shortcomings in such studies. Synthetic amino acid polymers have been prepared by base-catalyzed random polymerization of N-carboxyamino acid anhydrides. This reaction would be expected to afford different covalent structures from preparation to preparation. In fact, even a single preparation does not represent a discrete molecular species. This serious chemical heterogeneity would have an effect on the antigenic recognition of these compounds and the mechanism by which they are recognized.

The immune recognition of an antigen is widely believed to depend on phagocytosis of the antigen by macrophages which liberate antigen fragments that are thought to bind to membranes of the immune cells (Unanue and Cerrottini, 1968). More recently, Benacerraf (1978) proposed that the 1a molecule on the surface of macrophages and B cells interacts specifically with unique sequences (in the fragments liberated by macrophages) and thus determines Ir-gene function. This hypothesis, which may be satisfactory to explain many of the phenomena of the immune responses to amino acid polymers or haptens, is totally incapable of explaining the response to native protein antigens. The following shortcomings have been outlined (Atassi, 1980). The immune response to native protein antigens is clearly directed against the native three-dimensional structure (Atassi, 1967b, 1975, 1978). The antigenic sites are discrete and have well-defined boundaries. No hidden parts of the molecule generate an antibody response (Atassi, 1975, 1978). If proteolysis of a protein antigen is an essential step in its immune recognition, it would be expected that the antibody response would be distributed over the whole molecule and the antigenic sites would be more diffuse (Atassi, 1980). In studies on Mb and lysozyme it was found that the general background response (i.e., to regions between the antigenic sites) accounts for 0.1 to 1.4% of the entire antibody population (Atassi, 1975, 1978). The negligible magnitude of this diffuse response indicates that it is not significant in the immune recognition of these proteins. The antigenic sites of Mb and the

sites that have so far been identified and synthesized on the α and β chains of Hb represent continuous exposed portions of the respective polypeptide chain. Because of the accessibility of the Mb antigenic sites and the fact that they contain many potential targets for proteolytic attack, it is virtually impossible to derive these antigenic sites intact by proteolytic fragmentation (Atassi and Saplin, 1968; Atassi, 1975). If proteolysis by macrophage and recognition of the resultant fragments is a critical step in immune recognition, this would unavoidably result in the scission of the antigenic sites, the precise boundaries of which have been unequivocally established by exhaustive chemical analysis followed by synthesis. The hypothesis faces an even greater problem when the antigenic sites of lysozyme are considered. These discontinuous antigenic sites (see Figs. 4 to 6), describing an area on the surface topography of the protein and comprising spatially adjacent residues which are mostly not in direct peptide bond linkage, cannot be generated by fragmentation of the protein by macrophage. Each site can be mimicked only by surface-simulation synthesis in a peptide that does not exist in the protein and which is subject to directional and spacing (conformational) restrictions, and therefore its duplication by the macrophage is not possible. Any satisfactory hypothesis of cellular recognition must reckon with these facts (Atassi, 1980). Thus the determination of the precise antigenic structures of Mb and lysozyme, the six antigenic sites of serum albumin, six of the antigenic sites on the α and β chains of human Hb, and three of the antigenic sites on influenza virus hemagglutinin leads to the unavoidable conclusion that the native protein is recognized as a whole and essentially in its native conformation (Atassi, 1980).

These findings permitted the unambiguous conclusion (Atassi, 1980) that proteolytic fragmentation is not a step in the immune recognition that leads to antibody synthesis in the early stages of the antibody response (about 30 days) to each of the protein antigens discussed here. Furthermore, in the case of Mb, lysozyme, and serum albumin, antibodies have been examined that were obtained from 7 days to a year after the initial immunization, and it was found that the antigenic sites that are recognized on the respective protein are unaltered throughout this period. It may therefore be concluded (Atassi, 1980) that the internalization and proteolytic fragmentation of the antigen by the macrophage is a process that is most likely coincident with, but is unrelated to and functionally distinct from, the well-established collaborative role of the macrophage in immune recognition. Obviously, the internalization and fragmentation of a protein antigen by macrophage is dependent upon, and must follow, the initial recognition by the macrophage of the protein antigen in its native form.

These studies also permitted the proposal (Atassi, 1980; Atassi et al., 1981b) that an antigenic site in a given antigen exerts intramolecular regulatory effects on the response to other antigenic sites. The response to the whole antigen is the net product of these mutual intramolecular regulatory effects. This internal self-regulation of the response to a multivalent antigen was clearly seen in the regulatory effects that Mb antigenic sites exert upon one another (Atassi et al., 1981b) and that a subunit of Hb exerted upon the immune response to the other subunit (Krco et al., 1981a–c). This would mean that the response of a clone of cells that are committed to one antigenic site triggers off chemical signals that are transmitted to and modulate the response of other clones of cells that will be responding to other antigenic sites and the latter clones of cells reciprocating in a like manner (Atassi, 1980). Such interactions would require as yet unrecognized mechanisms of communication.

VII. CONCLUSIONS

Studies in the author's laboratory have resulted in the precise determination of the first complete protein antigenic structures. Thus we have carried out the delineation and synthesis of the antigenic sites of sperm whale Mb, hen egg white lysozyme, the six major antigenic sites of serum albumin, and most of the antigenic sites of Hb. These investigations gave the first unique insight into the molecular features responsible for the immune recognition of protein antigens. Thus protein antigenic sites were invariably found to be surface features, limited in number, small in size, and having discrete boundaries. Accordingly, recognition of, and immune responses to, a protein antigen do not follow its internalization and fragmentation by macrophage (Atassi, 1980). Indeed, we believe that internalization must be preceded and is dependent on recognition of the native protein antigen by macrophage (Atassi, 1980). Fragmentation of antigen by macrophage, which is known to take place, may be more concerned with the clearance of the antigen than with its immune recognition. It is, of course, well established that macrophage plays a collaborative role in immune recognition, and our findings enabled us to conclude that this role is distinct from its function in clearance (Atassi, 1980). The exact nature of the role of macrophage in antigen presentation, especially with protein antigens, requires further investigation.

Antigenic sites are mostly independent of the immunized species, in that essentially the same antigenic sites (with minor displacements) are recognized on a given protein regardless of the host species in which the antisera are raised (Atassi, 1975; Twining et al., 1980, 1981a; Sakata and Atassi, 1980a; Yoshida and Atassi, 1983) and the time antisera are obtained after the initial immunization (Twining et al., 1981b; Sakata and Atassi, 1980a; Yoshida and Atassi, 1983). Accordingly, it is concluded that the immune response is not based on, or directed to, parts of the immunized antigen which differ in sequence from the homologous protein in the immunized animal (Atassi and Kazim, 1978; Atassi, 1980). Rather, the antigenicity of the sites is largely inherent in their three-dimensional locations (Kazim and Atassi, 1977a,b, 1978, 1982). This revealed the role of structurally inherent antigenic sites in the autoimmune response (Kazim and Atassi, 1977b; Yokota et al., 1980; Sakata and Atassi, 1981).

Determination of these antigenic structures has shown not only that the antibody (B-cell) and T-cell immune responses to the protein antigen are under H-2-linked Ir-gene control, but has indicated that each antigenic site is under separate genetic control (Okuda et al., 1979b; Twining et al., 1981a).

The chemical and immunochemical strategy that we have developed with model proteins should be useful in developing immunological therapeutic approaches to antigens of clinical important (e.g., allergens, viruses, toxins, tissue antigens, etc). The systematic localization and synthesis of the antigenic sites on such proteins should afford synthetic peptides that can be employed in the preparation of antibodies for clinical applications.

Another immunological approach (Atassi, 1980) which could be achieved in the more distant future would rely on the identification of the antibody combining site of selected monoclonal antibodies to the antigen. The antibody combining site can then be chemically synthesized (Twining and Atassi, 1978; Kazim and Atassi, 1980a) by the concept of surface-stimulation synthesis (Atassi et al., 1976b). The synthetic peptides, suitably designed, can in principle be used to neutralize permanently the invasive antigen (e.g., a virus). The chemical strategies worked out in this laboratory with model systems have removed any intellectual barriers that may face these approaches.

Finally, the precise determination and synthesis of protein antigenic sites has charted a strategy not only for determining such sites for other proteins but also for

delineating and synthesizing other types of protein binding sites (Atassi, 1978, 1980). Furthermore, the concept of surface-stimulation synthesis which we introduced and developed during our determination of the antigenic structure of lysozyme has provided a new dimension of unlimited versatility in the synthetic mimicking of any type of protein binding site (for a review, see Atassi, 1980). We have shown that the strategy and concepts developed during our determination of the antigenic structures of Mb and lysozyme have enabled us synthetically to mimic binding sites involved in other protein functions (e.g., combining sites of antibodies; see Twining and Atassi, 1978; Kazim and Atassi, 1980a).

ACKNOWLEDGMENTS

The work that resulted in the findings outlined in this chapter was supported by Grants AM-18920 and AI-18657 from the National Institutes of Health and AHA 81-1012 from the American Heart Association.

REFERENCES

Allison, A. C., and Denman, A. M. (1976). Br. Med. Bull. *32*, 124.
Amzel, L. M., Poljak, R. M., Saul, F., Varga, J. M., and Richard, F. F. (1974). Proc. Natl. Acad. Sci. USA *71*, 1427.
Andres, S. F., and Atassi, M. Z. (1970). Biochemistry *9*, 2268.
Atassi, M. Z. (1967a). Biochem. J. *102*, 478.
Atassi, M. Z. (1976b). Biochem. J. *103*, 29.
Atassi, M. Z. (1968). Biochemistry *7*, 3078.
Atassi, M. Z. (1970). Biochim. Biophys. Acta *221*, 612.
Atassi, M. Z. (1972). In *Specific Receptors of Antibodies, Antigens and Cells*, 3rd Int. Convoc. Immunol. S. Karger, Basel, pp. 118–135.
Atassi, M. Z. (1975). Immunochemistry *12*, 423.
Atassi, M. Z. (1977a). In *Immunochemistry of Proteins*, Vol. 1 (M. Z. Atassi, Ed.). Plenum, New York, pp. 1–161.
Atassi, M. Z. (1977b). In *Immunochemistry of Proteins*, Vol. 2 (M. Z. Atassi, Ed.). Plenum, New York, pp. 77–176.
Atassi, M. Z. (1978). Immunochemistry *15*, 909.
Atassi, M. Z. (1979). Crit. Rev. Biochem. *6*, 371.
Atassi, M. Z. (1980). Mol. Cell. Biochem. *32*, 21.
Atassi, M. Z. (1981). Mol. Immunol. *18*, 1021.
Atassi, M. Z. (1982). Biochim. Biophys. Acta *704*, 552.
Atassi, M. Z., and Habeeb, A. F. S. A. (1977). In *Immunochemistry of Proteins*, Vol. 2 (M. Z. Atassi, Ed.). Plenum, New York, pp. 177–264.
Atassi, M. Z., and Kazim, A. L. (1978). In *Immunobiology of Proteins and Peptides*, Vol. 1 (M. Z. Atassi and A. B. Stavitsky, Eds.). Plenum, New York, pp. 19–40.
Atassi, M. Z., and Kazim, A. L. (1980). Biochem. J. *187*, 163.
Atassi, M. Z., and Lee, C.-L. (1978a). Biochem. J. *171*, 419.
Atassi, M. Z., and Lee, C.-L. (1978b). Biochem J. *171*, 429.
Atassi, M. Z., and Pai, R. C. (1975). Immunochemistry *12*, 735.
Atassi, M. Z., and Saplin, B. J. (1968). Biochemistry *7*, 688.
Atassi, M. Z., and Skalski, D. J. (1969). Immunochemistry *6*, 25.
Atassi, M. Z., and Smith, J. A. (1978). Immunochemistry *15*, 609.
Atassi, M. Z., and Thomas, A. V. (1969). Biochemistry *8*, 3385.
Atassi, M. Z., and Webster, R. G. (1983). Proc. Natl. Acad. Sci. USA, *80*, 840.

Atassi, M. Z., Habeeb, A. F. S. A., and Rydstedt, L. (1970). Biochim. Biophys. Acta *200*, 184.

Atassi, M. Z., Habeeb, A. F. S. A., and Ando, K. (1973a). Biochim. Biophys. Acta *303*, 203.

Atassi, M. Z., Perlstein, M. T., and Staub, D. J. (1973b). Biochim. Biophys. Acta *328*, 278.

Atassi, M. Z., Litowich, M. T., and Andres, S. F. (1975). Immunochemistry *12*, 727.

Atassi, M. Z., Koketsu, J., and Habeeb, A. F. S. A. (1976a). Biochim. Biophys. Acta *420*, 358.

Atassi, M. Z., Lee, C.-L., and Pai, R. C. (1976b). Biochim. Biphys. Acta *427*, 745.

Atassi, M. Z., Sakata, S., and Kazim, A. L. (1979). Biochem. J. *179*, 327.

Atassi, M. Z., Twining, S. S., Lehmann, H., and David, C. S. (1981a). Immunol. Commun. *10*, 359.

Atassi, M. Z., Yokota, Y., Twining, S. S., Lehmann, H., and David, C. S. (1981b). Mol. Immunol. *18*, 945.

Atassi, M. Z., Long, P. M., Beisel, K., Sakata, S., Peters, T., Jr., and David, C. S. (1982a). Mol. Immunol. *19*, 313.

Atassi, M. Z., Sakata, S., and Sakata, S. (1982b). Mol. Immunol. *19*, 1509.

Benecerraf, B. (1978). J. Immunol. *120*, 1809.

Benjamini, E. (1977). In *Immunochemistry of Proteins*, Vol. 2 (M. Z. Atassi, Ed.). Plenum, New York, pp. 265–310.

Benjamini, E., Leung, C. Y., and Rennick, D. M. (1978). In *Immunobiology of Proteins and Peptides*, Vol. 1 (M. Z. Atassi and A. B. Stavitsky, Eds.). Plenum, New York, pp. 165–179.

Bernard, C. C. A. (1977). Clin. Exp. Immunol. *29*, 100.

Chou, P. Y., and Fasman, G. D. (1974). Biochemistry *13*, 222.

Dean, J., and Schechter, A. N. (1979). J. Biol. Chem. *254*, 9185.

Eylar, E. H. (1978). In *Immunobiology of Protein and Peptides*, Vol. 1 (M. Z. Atassi, and A. B. Stavitsky, Eds.). Plenum, New York, pp. 259–281.

Fermi, G. (1975). J. Mol. Biol. *97*, 237.

Hopp, T. P., and Woods, K. R. (1981). Proc. Natl. Acad. Sci. USA *78*, 3824.

Hurrell, J. G. R., Smith, J. A., and Leach, S. J. (1978). Immunochemistry *15*, 297.

Kazim, A. L., and Atassi, M. Z. (1977a). Biochem. J. *167*, 275.

Kazim, A. L., and Atassi, M. Z. (1977b). Biochim. Biophys. Acta *494*, 277.

Kazim, A. L., and Atassi, M. Z. (1978). Immunochemistry *15*, 67.

Kazim, A. L., and Atassi, M. Z. (1980a). Biochem. J. *187*, 661.

Kazim, A. L., and Atassi, M. Z. (1980b). Biochem. J. *191*, 261.

Kazim, A. L., and Atassi, M. Z. (1980c). Biochem. J. *191*, 673.

Kazim, A. L., and Atassi, M. Z. (1982). Biochem. J. *203*, 201.

Keck, K. (1975). Nature (Lond.) *254*, 78.

Köhler, G., and Milstein, C. (1975). Nature (Lond.) *256*, 495.

Koketsu, J., and Atassi, M. Z. (1973). Biochim. Biophys. Acta *328*, 289.

Koketsu, J., and Atassi, M. Z. (1974a). Biochim. Biophys. Acta *342*, 21.

Koketsu, J., and Atassi, M. Z. (1974b). Immunochemistry *11*, 1.

Krco, C. J., Kazim, A. L., Atassi, M. Z., and David, C. S. (1981a). J. Immunogenet. *8*, 315.

Krco, C. J., Kazim, A. L., Atassi, M. Z., and David, C. S. (1981b). J. Immunogenet. *8*, 395.

Krco, C. J., Kazim, A. L., Atassi, M. Z., Melvold, R., and David, C. S. (1981c). J. Immunogenet. *8*, 471.

Lee, C. L., and Atassi, M. Z. (1975). Biochim. Biophys. Acta *405*, 464.

Lee, C. L., and Atassi, M. Z. (1976). Biochem. J. *159*, 89.

Lee, C. L., and Atassi, M. Z. (1977a). Biochem. J. *167*, 571.
Lee, C. L., and Atassi, M. Z. (1977b). Biochim. Biophys. Acta *495*, 354.
Okuda, K., Christadoss, P., Twining, S. S., Atassi, M. Z., and David, C. S. (1978). J. Immunol. *121*, 866.
Okuda, K., Sakata, S., Atassi, M. Z., and David, C. S. (1979a). J. Immunogenet. *6*, 447.
Okuda, K., Twining, S. S., David, C. S., and Atassi, M. Z. (1979b). J. Immunol. *123*, 182.
Padlan, E. A., Davies, D. R., Rudikoff, S., and Porter, M. (1976). Immunochemistry *13*, 945.
Perutz, M. F., Muirhead, H., Cox, J. M., and Goaman, L. C. G. (1968). Nature (Lond.) *219*, 131.
Romero-Herrera, A. E., Lehmann, H., Jossey, K. A., and Friday, A. E. (1978). Philos. Trans. R. Soc. Lond. B *283*, 61.
Sakata, S., and Atassi, M. Z. (1980a). Biochim. Biophys. Acta *625*, 159.
Sakata, S., and Atassi, M. Z. (1980b). Mol. Immunol. *17*, 139.
Sakata, S., and Atassi, M. Z. (1981). Mol. Immunol. *18*, 961.
Takagaki, Y., Hirayama, A., Fujio, H., and Amano, T. (1980). Biochemistry *19*, 2498.
Tan-Wilson, A. L., Reichlin, M., and Noble, R. W. (1976). Immunochemistry *13*, 491.
Twining, S. S., and Atassi, M. Z. (1978). J. Biol. Chem. *253*, 5259.
Twining, S. S., and Atassi, M. Z. (1979). J. Immunol. Methods *30*, 139.
Twining, S. S., Lehmann, H., and Atassi, M. Z. (1980). Biochem. J. *191*, 681.
Twining, S. S., David, C. S., and Atassi, M. Z. (1981a). Mol. Immunol. *18*, 447.
Twining, S. S., Lehmann, H., and Atassi, M. Z. (1981b). Mol. Immunol. *18*, 473.
Unanue, E. R., and Cerrottini, J. C. (1968). J. Exp. Med. *127*, 915.
Yokota, S., David, C. S., and Atassi, M. Z. (1980). Mol. Immunol. *17*, 10179.
Yoshida, T., and Atassi, M. Z. (1983). Immunol. Commun, in press.
Young, C. R., and Atassi, M. Z. (1982). J. Immunogenet. *9*, 343.
Young, C. R., O'Connor, G. P., and Atassi, M. Z. (1981a). Immunol. Commun. *10*, 483.
Young, C. R., O'Connor, G. P., and Atassi, M. Z. (1981b). J. Immunogenet. *8*, 387.

3

Immunochemistry of Polysaccharide and Blood Group Antigens

Roger K. Cunningham School of Medicine, State University of New York at Buffalo, Buffalo, New York

I. INTRODUCTION

Carbohydrates are an essential class of compounds of most living things. They serve as a major energy source, as a storage form of energy, and as major structural components of cells and tissues. Among the carbohydrates, the more complex forms of these important compounds can be shown to be antigens. As a rule, small carbohydrates such as monosaccharides are not in themselves antigenic, but when large branched or linear structures are formed, these molecules (oligosaccharides and polysaccharides) are often strongly immunogenic. Nevertheless, polysaccharide antigens are unique in a certain sense. When highly purified they cannot evoke antibody formation in all species (Heidelberger, 1960). Rabbits and guinea pigs, for example, do not respond to purified polysaccharides, although the same animals respond perfectly well to protein antigens (Kabat, 1968). Mice and humans, on the other hand, do respond to purified polysaccharide materials. In spite of their unresponsiveness to purified polysaccharides, rabbits and guinea pigs respond well to polysaccharide antigens that are associated with particles or cells. Thus although these species will not respond to purified pneumococcal capsular carbohydrate, they will respond vigorously to the capsular polysaccharide if they are immunized with intact pneumococci.

There is an additional point to be made concerning the processing of polysaccharide antigens by the immune system. Some polysaccharides constitute a class of antigens known as T independent; that is, these molecules interact directly with B cells without a need for T cells (Zaleski et al., 1983).

Carbohydrates were originally so named because the compounds of this group could be considered as hydrates of carbon with the general structure of two hydrogen to one oxygen to one carbon atom. This general name fails, however, since carbohydrates can contain nitrogen and sulfur, as well as other atoms.

Carbohydrates are classified in three main groups: monosaccharides, oligosaccharides and polysaccharides. Monosaccharides are simple sugars that yield carbon, hydrogen, and oxygen upon hydrolysis. Oligosaccharides and polysaccharides are compound molecules that are assembled by the joining of monosaccharides into larger assemblies. Oligosaccharides are assemblies that, when hydrolyzed under mild conditions, yield two to six monosaccharide subunits. Hence a disaccharide is an oligosaccharide that yields two molecules of monosaccharide; a trisaccharide is an oligosaccharide that is composed of three monosaccharide subunits; and so on.

Polysaccharides comprise a group of compounds that are composed of six or more monosaccharide subunits. These molecules are usually of very high molecular weight and are often associated with other classes of compounds. Oligosaccharides and polysaccharide antigens are often found in heterogenous molecules such as glycoproteins, glycolipids, or lipopolysaccharides. In such molecules, the oligosaccharide or polysaccharide moieties are oriented such that the reducing end of the saccharide chain is bound to the peptide or lipid molecule with which the chain is associated. The nonreducing end of each chain is most distal and will form the antigenic determinant to which antibody will be formed. It is typical that when antibodies are raised to such complex molecules, only the terminal portion of the polysaccharide or oligosaccharide is recognized. Hence the general trend is for the last few monosaccharides that are distal in the chain to form the antigenic determinant that reacts with antibody (Kabat, 1968). Antibodies to midchain portions of polysaccharides are unusual. The linear polysaccharides tend to have few determinants, whereas highly branched structures tend to have many. This is simply a reflection of the fact that highly branched structures have more terminal structures than do linear chains.

Oligosaccharides and polysaccharides are important antigenic determinants in heterophile reactions. The classic heterophile reaction was first noted by the Swedish pathologist Forssman in 1911 (Galton and Goldsmith, 1961). Forssman injected rabbits with suspensions of guinea pig tissue and obtained an antiserum that lysed sheep red cell suspensions in the presence of complement, although surprisingly, it did not lyse guinea pig erythrocytes. This seeming paradox is now known to be due to the fact that guinea pig tissues (but not their erythrocytes) contain a polysaccharide structure (the F antigen) that resembles a similar polysaccharide on the surface of sheep red blood cells.

An important diagnostic test for infectious mononucleosis takes advantage of the fact that a similar heterophile relationship exists between a polysaccharide structure on the surface of the virus of infectious mononucleosis and a separate polysaccharide (not the F antigen) determinant that, by chance, is also present on sheep erythrocytes. The point here is that "accidental" cross-reactions between oligosaccharide and polysaccharide antigens is relatively more common that is found with protein antigens. There are far fewer naturally occurring monosaccharides than naturally occurring amino acids, and thus monosaccharides can be assembled into far fewer unique structures than can the amino acids. A chance occurrence of similar structures is statistically to be expected more often for oligosaccharide and polysaccharide antigens than for protein antigens.

II. BACTERIAL ANTIGENS

Since the earliest days of immunology there has been an awareness that polysaccharide antigens play an important role in combating bacterial and viral infections. Many of the important diagnostic tests used clinically take advantage of the fact that polysaccharide substances abound on the cell surfaces of both gram-negative and gram-positive bacteria

```
α-glu(1-6)glu(1-6)glu(1-6)glu(1-6)glu(1-6)glu(1-6)glu(1-6)glu(1-6)glu(1-6)...
                                           /(1-3)
       α-glu(1-6)...glu(1-6)glu ...(1-6)glu
                               n
```

Figure 1 Structure of dextran.

as well as fungi. It should be emphasized, however, that although a brisk response to bacterial polysaccharides may occur during infection, the resultant antibody may or may not be important in recovery from the infection.

Antigenic bacterial polysaccharides have been used by Kabat and his associates to probe the size of the antibody combining site (Kabat, 1968). These studies employed dextran and levan, which are extracellular products of certain microorganisms. The advantage of these substances is that they are extremely simple, largely linear structures composed entirely of the same monosaccharide subunits. Dextran is a polymer composed of glucose only; the linkage between the subunits is largely α-1,6, which forms linear structures of varying molecular weights. Branching occurs when α-1,3 linkages occur in the chains (Fig. 1). Levan, composed entirely of fructose, is also a largely linear polymer composed of fructose molecules linked α-2,6.

In their native state dextrans have molecular weights in the millions, but in contrast to proteins, these structures have random molecular weights that vary. Mild hydrolysis, however, can be used to produce dextran products of more uniform molecular weight (Kabat and Mayer, 1961). The chief advantage of these materials in immunochemical studies is the fact that the antigenic determinants that react with antidextran are always glucose.

Dextran and antidextran antibodies were used by the Kabat group to determine the size of the antibody combining site in the following way: Because increasing sizes of glucose oligosaccharides could be produced, these were used to inhibit the reaction of dextran with antidextran. Thus molecules composed of increasing numbers of glucose monosaccharide units, ranging from a disaccharide through a heptasaccharide, were mixed with an amount of dextran and antidextran that gave optimum precipitation. The results (Fig. 2) showed that inhibition of such precipitation increase as the size of the oligosaccharide grew from the disaccharide to the hexasaccharide, but that no increase in inhibition occurred when the inhibiting oligosaccharide was longer than six glucose units long. The failure of longer chains to inhibit the precipitation more than the hexasaccharide was taken as evidence that the antibody combining site was six glucose molecules long.

Such experiments are important for an understanding of polysaccharide antigens since they illustrate a fundamental property of polysaccharide antigens and antibodies: The interaction of antibody with antigen occurs over a short region of the chain and as a rule, the determinants that are important in these antigens occur only at the tip of a chain or the tip of a branch to a main chain. In fact, very small differences in the arrangement of monosaccharides at the tip of a polysaccharide chain may change the serologic specificity of the determinant entirely, as will be discussed more fully later when blood group antigens are considered.

The latter point, nevertheless, can also be shown by the dextran–antidextran system. In this case, introducing a change in linkage from the normal α-1,6 (e.g., between the third and fourth glucose units from the end) causes a significant loss in inhibitory

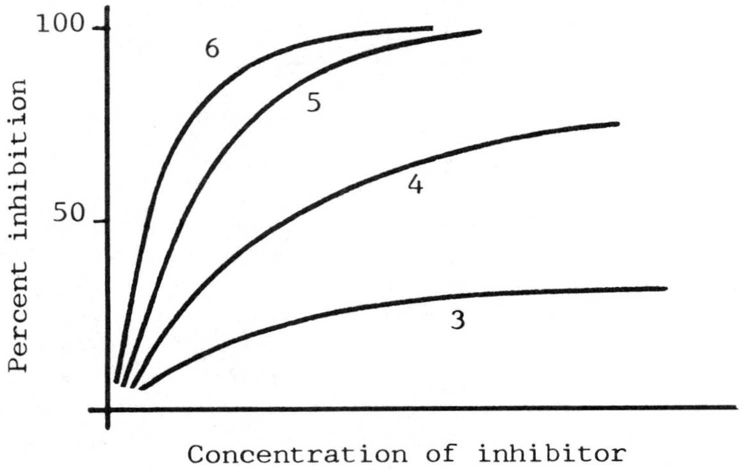

Figure 2 Inhibition of antidextran precipitation of native dextran by oligosaccharides of increasing size: 3, triose; 4, tetraose; 5, pentaose; 6, hexaose.

capacity. The introduction of a single glucose branch in the same area causes a similar loss of inhibitory capacity. It is clear, then, that a seemingly trivial change in the conformation of an antigenic determinant of a polysaccharide can have a profound effect on its capacity to react with antibody.

Similar results have been obtained by a slightly different set of experiments (Kabat, 1968). Rabbits were injected with isomaltose oligosaccharides that had been coupled to bovine serum albumin as a carrier molecule. The two-glucose unit, isomaltose, and the three-glucose unit, isomaltotriose, were derivatized and coupled to the protein carrier as shown in Fig. 3. The resulting conjugate of oligosaccharide yielded an antigenic material that possessed only one terminal glucose residue in the case of isomaltose and two in the case of isomaltotriose. Animals immunized with the conjugated isomaltose responded with antibody that was specific for glucose but which did not react with dextran. In contrast, rabbits injected with the conjugate made with isomaltotriose produced antibody that reacted with dextran. Hence a single glucose residue was insufficient to evoke an antibody that would react with dextran, but antibody made in response to a disaccharide of glucose had a large enough combining site to form a union with dextran. In other words, in the former case the population of antibodies produced to the single glucose determinant did not posses an extent of antibody combining site required to precipitate with dextran. Antibody to the disaccharide structure, however, did possess an extensive

α-glucose-O-CH_2-CHOH-CHOH-CHOH-CHOH-CO-NH-Bovine Serum Albumin

α-isomaltose-O-CH_2-CHOH-CHOH-CHOH-CHOH-CO-NH-Bovine Serum Albumin

Figure 3 Antigens produced by the conjugation of isomaltose and isomaltotriose to bovine serum albumin.

enough interactive interface to form the stronger bond required for precipitation of dextran.

During the course of an infection most mammals will develop a wide variety of antibodies directed against the antigens present on the surface of microorganisms. Although antibodies to proteins are also produced, often the predominating production of antibodies will be found to be directed against polysaccharide or oligosaccharide antigens of the pathogen. An especially interesting, important, and well-studied example can be found in the case of the lipopolysaccharide (LPS) of members of the genus *Salmonella* and related gram-negative genera (Luderitz et al., 1966). Antibodies to *Salmonella* are used in the classification schemes of the various species. Well over 1000 serotypes are known.

Two principal antigenic structures are used to diagnose infection and to classify the organisms. One class of determinants is termed the *flagellar* or *H antigens*. Since these structures are protein, they will not be considered further. The second set of antigens are the *somatic* or *O antigens*. These antigens are highly complex structures that contain an immunodominant polysaccharide structure bound to a complex carrier of mixed chemical composition. LPS, or endotoxin, has potent biological effects that are due to a lipid portion of the endotoxin molecule, called lipid A.

A representative molecule of LPS is shown in Fig. 4. Although this particular diagram depicts the structure of *Salmonella typhimurium*, except for the structure of the specific O antigen, the composition of the remainder is virtually constant for all LPS of gram-negative bacilli. As shown, the basic core of LPS consists of heptose phosphate, 2-keto-3-deoxyoctonate (KDO), D-glucosamine, D-galactose, and D-glucose. The heptose phosphate and KDO constituents form a linkage structure that binds the LPS to the bacterial cell wall. In addition to the core structure, different strains will possess additional monosaccharides attached to the core. Nine different monosaccharides have been identified as forming these additional side chains to the common core. These nine monosaccharides are arranged in groups of four or five that constitute the O antigen of LPS. Furthermore, the O-antigen group may be repeated to form a chain of repeating units that can extend to 30 or more units long.

The O antigens are the major somatic antigens of these Gram-negative bacilli and depending on the particular monosaccharides and their arrangement, each O antigen will possess one or more antigenic determinants. The antigenic determinants are widely shared among these bacteria and serve as a means of classifying them. For example, *Salmonella* of Group D all possess determinant 9, in addition to whatever other determinants may be present. It is the possession of antigen 9 that serves to (arbitrarily) place the organism in a group called D. Similarly, all the *Salmonella* that possess determinant 4 are in group B, and so on.

In this connection, it is interesting that the intense studies on LPS led to the discovery of a new class of sugars, the 3,6-dideoxyhexoses. These are listed in Table 1. When present in a given O antigen these sugars are the immunodominant structures of the particular determinant that is present. These sugars are exceptionally easy to cleave away by mild acid treatment and hence are difficult to study. This has a practical disadvantage in that virtually any treatment of the intact O antigen will cleave away these dideoxyhexoses, making study of their position and linkage difficult. This point is mentioned because a principal method used in determining the structure of determinants is inhibition of the precipitation between O antigen and antibody to a particular determinant by

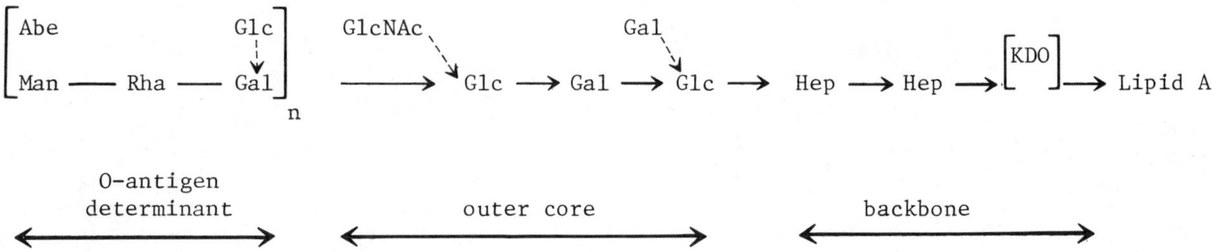

Figure 4 Structure of the lipopolysaccharide (LPS) of *Salmonella typhimurium*. The lipid A end of the molecule is attached to the bacterial cell wall.

Table 1 3,6-Dideoxyhexoses Found in Gram-Negative Bacteria

Tyvelose	3,6-Dideoxy-D-mannose
Abequose	3,6-Dideoxy-D-galactose
Colitose	3,6-Dideoxy-L-galactose
Paratose	3,6-Dideoxy-D-glucose

hydrolysis products of the native polysaccharide. This approach, which is used generally for carbohydrate antigens of all types, employs gentle hydrolysis in acid or alkaline conditions to produce a mixture of cleavage products. By appropriate techniques the mixture of hydrolysis products is separated. The isolated fractions are then examined individiually for the capacity to inhibit the reaction between the native polysaccharide and antibody to it. Fractions that possess the capacity to inhibit this reaction are then examined closely to determine their detailed structure, which is then presumed to be identical with the determinant on the native molecule. Unfortunately, in the case of the 3,6-dideoxyhexoses, the hydrolysis conditions required to separate mono-, di-, and trisaccharide units are sufficient to destroy these sugars completely.

Nevertheless, in spite of the difficulty in studying the 3,6-dideoxyhexose determinants, they are known to be the immunodominant sugar in those O antigens where they occur. For example, as stated earlier, group D organisms all possess antigen 9. Antigen 9 owes its specificity to the presence of tyvelose in the O antigen. The group B antigen, 4, reflects the presence of abequose. Some representative species and the specific sugar in their O antigens are listed in Table 2.

It might be well to digress a bit to make a point about antibodies that arise in response to polysaccharide antigens. As stated earlier, the species of animal responding is often important when purified polysaccharides are injected. Thus the usual practice with antigens that are associated with cells is to inject the animal with whole cells rather than with purified carbohydrates. This means, however, that the animal responding will respond in a wide fashion and produce antibodies of many unwanted specificities. In such cases it is important to absorb the resulting antiserum with appropriate materials to produce a functionally specific antiserum. For example, to produce a specific anti-9

Table 2 Structures of the Repeating O-Antigen Side Chains of Some *Salmonella* Species

Species	Group	Determinants	Composition
S. typhi	D	9, 12	Tyvelose, mannose, rhamnose, galactose, O-acetylglucose
S. typhimurium	B	1, 4, 5, 12	Abequose, mannose, rhamnose, glucose, galactose
S. paratyphi A	A	1, 2, 12	Paratose, mannose, rhamnose, glucose, galactose

serum, one could raise an antibody to *S. typhi* or *S. enteritidis* killed whole cells (Uchida et al., 1963). Both these organisms possess the group D antigen 9 but also possess another antigen determinant, 12. To render such an antiserum specific for antigen 9 it would have to be absorbed: for example, *S. paratyphi A*, which has antigen 12 but does not possess antigen 9. Absorption is nothing more than mixing cells containing the antigen that corresponds to the unwanted antibody with the antiserum and removing the cells after a suitable time of incubation. The "absorbed" serum should then be tested for loss of activity against the cells that were used for absorption.

Although the 3,6-dideoxyhexoses are usually the immunodominant residue when they are present in an O antigen, the final expression of a determinant depends to large extent on the other monosaccharides that accompany these sugars in a given O antigen. To illustrate, abequose is the immunodominant sugar in both determinants 4 and 8, but these are serologically specific; no cross-reactions occur between these determinants. In each case, the specificity of the determinants of these antigens is due to differences in the position of abequose and the other monosaccharides. This illustrates the fact that a determinant is actually composed of a region of a given sequence of sugars; hence the final specificity includes the profile of an immunodominant sugar and neighboring monosaccharides. This is illustrated in Fig. 5, which shows in the case of the O antigen of *S. anatum* that the presence of determinant 3 reflects the combined profile of rhamnose and mannose, whereas determinant 10 results from the same mannose that is recognized in the context of its neighbor on the opposte sixe, acetylgalactose. The final effect of this is that inhibition of the reaction of a polysaccharide with antibody to it is seldom best achieved with the immunodominant monosaccharide alone. The strongest inhibition is obtained when the immunodominant sugar is part of a larger fragment that contains the othe sugar, or sugars, in appropriate configuration.

Another important point that is true for all antigens should be emphasized for polysaccharides; not all animals respond equally well to a given antigen. Often it is necessary to immunize several animals to be able to find a potent response to the antigen of interest or an antiserum that has a sufficient subpopulation of antibodies that have the desired narrow specificity.

Determinant 3

GalOAc-Man-Rhm-(GalOAc-Man-Rhm)$_n$-R

Determinant 10

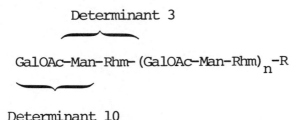

Figure 5 Effect of a "nearest neighbor" on the antigenic determinants of the O antigen of *Salmonella anatum*. See the text for details. GalOAc, O-acetylgalactose; Man, mannose, Rhm, rhamnose.

III. BLOOD GROUP ANTIGENS

An important group of antigens resides on the human erythrocyte. Many of these antigens are polysaccharide in composition, although not all are. Probably the best characterized and most studied of these are the human A, B, and H antigens.

The ABH antigens are under the control of genes that code for the production of protein products which actually synthesize the antigen. Before discussing the actual antigens and their mode of synthesis, however, it might be well to review some basic serologic facts about the ABO blood group system.

The ABO blood group system is an alloantigenic system in humans; that is, virtually all individuals fall into one of four blood groups: O, A, B, or AB (Race and Sanger, 1975). The antigens A and B are detected by two antibodies, anti-A and anti-B, which identify their specific antigens. Hence by means of these two reagents, the human population can be grouped as shown in Table 3. An interesting feature shown in this table is that each individual possesses a naturally occurring antibody for the A or B receptor that he or she does not possess. Thus group O persons, who have neither A nor B antigens, regularly have anti-A and -B antibodies in their serum; persons of group A always possess anti-B, and so on. It should be emphasized that the regular occurrence of antibodies is unique to the ABO blood group system. For antigens of most other systems, immunization is required before antibodies can be produced. Two genes produce the ABO antigens: *A* and *B*. The gene *O* is an amorphic gene that codes for no known product.

After the discovery of the ABO system by Karl Landsteiner in 1900, others soon discovered that the ABO antigens occur in the human body in two forms: a lipid-associated form in which the antigenic oligosaccharide is bound to a portion of the erythrocyte membrane (or other tissue), and a secreted water-soluble form.

Water-soluble A and B antigens have been the forms most intensely studied since these forms are most compatible with serologic analysis. Hence, until recent years, the lipid-associated A and B antigens have been largely left to one side. Attention is now being given to these structures, but the work has had to await developments in lipid biochemistry.

Thus most of our understanding of the A and B antigens and the molecules associated with their structure has been derived from studies on the water-soluble form of the antigens obtained from saliva, gastric, gastric mucosa, and a particularly rich source, the fluid from pseudomucinous ovarian cysts. Unfortunately, the use of water-soluble substances has been accompanied by some very important difficulties. One of the first confusions to arise was an antibody originally called anti-O that seemed to detect a product

Table 3 Human ABO Groups as Determined by Anti-A and Anti-B Sera

Group	Red cell antigens	Antibodies of serum
O	None	Anti-A and anti-B
A	A	Anti-B
B	B	Anti-A
AB	A and B	None

produced by the *O* allele (Kabat, 1956). The putative anti-O antibodies were sometimes present in the serum of some A individuals and could also be found in certain animal sera. The reactions of anti-O were not as clear cut as those of anti-A and anti-B, however. Rather, these reagents reacted more strongly with group O cells than with either A or B. Paradoxically, however, the reaction of anti-O could be inhibited by the saliva of group AB individuals, those who could not possess an *O* gene or produce its product. The fact finally emerged that virtually all individuals possess a gene *H* that codes for an enzyme which produces an antigen, called H, upon which the A and B are imposed. The allele to *H, h,* is amorphic and codes for no known product. It is important to realize that the gene that codes for the H antigen is not part of the ABO system. In fact, the H locus is not associated with the ABO locus. Thus, normally all persons have H antigen present on their erythrocytes, a portion of which becomes converted to A or B antigen in persons with those genotypes. Group O persons, however, since they possess an amorphic gene, *O*, convert none of their H substance. Thus the first evidence arose that independent loci could produce products that interacted to form the final molecule of antigen. However, in the case of secreted blood group substances, further complications were also encountered. The first of these was the discovery of the secretor gene.

Most individuals (about 80%) secrete ABH blood group substances according to their blood group. The remaining persons, however, do not. The ability to secrete blood group substances is controlled by allelic genes called *Se* and *se*. Individuals who are homozygous for *se* do not express ABH antigens in their saliva, tears, and so on, although these antigens are expressed fully in their tissues and on their erythrocytes. It is important to understand that the secretor gene controls only ABH expression and then only in secreted form. Nonsecretors produce a water-soluble polysaccharide just as do secretors, but in nonsecretors the ABH determinants are not synthesized.

An additional complication to the understanding of the synthesis and structure of soluble blood group A, B, and H substances was an apparent interaction of the *ABH* genes, the secretor genes, and an additional blood group system called *Lewis* (see Table 4). The Lewis blood group system originally seemed to be a straightforward Mendelian case of inherited antigens governed by two alleles, Le^a and Le^b. These antigens are identified by anti-Le^a and anti-Le^b sera found in the serum of humans of the appropriate genotype. Anti-Lewis antibodies sometimes occur naturally and can arise following transfusion. However,

Table 4 Effect of the *ABO, H/h, Le,* and *Se* Genes on the Expression of Blood Group Activity on Erythrocytes and in Secretions

	Erythrocyte antigens					Substance in secretion				
Genes present	A	B	H	Le^a	Le^b	A	B	H	Le^a	Le^b
ABO, H, Se, Le	+	+	+	–	+	+	+	+	+	+
ABO, H, se/se, Le	+	+	+	+	–	–	–	–	+	–
ABO, H, Se, le/le	+	+	+	–	–	+	+	+	–	–
ABO, H, se/se, le/le	+	+	+	–	–	–	–	–	–	–
ABO, h/h, Se (or *se/se*), *Le**	–	–	–	+	–	–	–	–	+	–
ABO, h/h, Se (or *se/se*), *le/le**	–	–	–	–	–	–	–	–	–	–

*Bombay phenotype.

some peculiarities were noted. While persons of the presumed genotypes Le^a/Le^a and Le^b/Le^b were identified, adults with the phenotype Le^a/Le^b were not found. Furthermore, an association of secretor and Lewis was apparent from the fact that persons of the *se/se* genotype (i.e., nonsecretors) were all seemingly Le^a/Le^a since their red cells reacted with anti-Le^{a} serum but not with anti-Le^{b}. It is now known that there is, in fact, no Mendelian set of genes that control the expression of Lewis antigens in the expected way. Rather, the genes of the Lewis system are *Le* and *le*. These genes interact with the genes of the AB, Hh, and secretor system to produce the Le^{a} and Le^{b} antigens (Ginsburg et al., 1971; Issitt, 1970; Morgan and Watkins, 1969). This interaction is explained more fully below.

Lewis antigens are present in secretions as well as on the erythrocytes. However, the erythrocyte Lewis antigens are not actually part of the red cell. The Lewis antigens are present on glycoproteins of the serum of each individual. The same glycoprotein also bears the ABH antigens when they are present. Hence, although erythrocytes can be grouped with anti-Lewis reagents, it is important to realize that these antigens are actually glycoproteins and therefore are water-soluble materials. In fact, incubation of cells that lack Lewis antigens in a serum that contains them can convert the cells passively from Lewis negative to Lewis positive.

The present view of the structure of synthesis of the ABH and Lewis blood group antigens postulates that a precusor of the oligosaccharide chain is acted on by the products of four independent sets of genes: *ABO*, Lewis, secretor, and *H* (Fudenberg et al., 1978; Issitt, 1970; Race and Sanger, 1975; Watkins, 1966; Zmijewski, 1968).

In the synthesis of the ABH and Lewis antigens, four principal monosaccharides are involved in making up the chain: N-acetyl-D-galactosamine, D-galactose, L-fucose, and N-acetyl-D-glucosamine. These sugars are added by enzymes coded for by alleles of the genes mentioned above.

Nothing is known of the genes that control the synthesis of the "precursor" substance upon which all the chains are developed. There are two types of precursor chains, termed type 1 and type 2 chains. The structure of type 1 and 2 chains is shown in Fig. 6. Both chains are attached either to a protein backbone in the case of secreted substances or to lipid when glycolipid material is being produced. The only known difference in the chains is that the terminal galactose is linked to the penultimate N-acetylglucosamine in an α-1,3 linkage in type 1 chains, but in an β-1,4 linkage in type 2 chains. This is an important difference since it emphasizes a point about polysaccharide antigens in general. The β-1,4 linkage of galactose in type 2 chains imparts a serologic specificity that is not present when the galactose is linked β-1,3: namely, type 2 precursor chains react with antipneumococcus XIV sera whereas type 1 chains do not. The seemingly minute difference in bonding can, in fact, completely change the capacity of the two terminal sugars to react with a given antibody.

The next step in synthesis involves the addition of fucose to the precursor substance by a fucosyl transferase coded for by the *H* gene. The H fucosyl transferase adds a

```
- - -GalNAc-(β1,3)-Gal-(β1,3)-GluNac-(β1,3)-Gal        type 1

- - -GalNAc-(β1,3)-Gal-(β1,3)-GluNAc-(β1,4)-Gal        type 2
```

Figure 6 Structure of type 1 and 2 precursor chains of the ABH blood group substances.

```
                                         (β1,3)
- - -GalNAc-(β1,3)-Gal-(β1,3)-GluNAc-( or )-Gal                type 1 &
                                         (β1,4)     (α1,2)          type 2
                                                        Fuc

                                                                (H activity)
```

Figure 7 Structures produced by the action of the H-specific fucosyl transferase on precursor chains.

fucose to the terminal galactose of precursor chain 1 and 2 in an α-1,2 linkage. This is shown in Fig. 7. This addition imparts H activity to both chain types. In the case of *h/h* individuals, however, no fucosyl transferase is produced and hence no *H* antigen can be synthesized. Individuals of the *h/h* genotype have a rare phenotype called *Bombay*, in which their erythrocytes lack H and also the appropriate A or B antigens expected from their pedigree. The serum of Bombay individuals contains anti-A, anti-B, and anti-H. Furthermore, their serum does not contain the expected α-1,2-L-fucosyl transferase that would permit synthesis of H.

The secretor gene exerts its influence at the level of the *H* gene but only in certain tissues. The product if the secretor gene apparently is required for expression of the *H* gene. No product has been identified but in those tissues that produce soluble blood group substances when the genotype of an individual is *se/se*, their H-specific fucosyl transferase is not produced there. The enzyme that acts on lipid-bound oligosaccharide, however, is fully active in such persons and their erythrocytes will possess H antigen. The secretor gene thus acts as a regulator gene.

Following the production of H substance, the *A, B* and Lewis (*Le*) genes can all act. It is probable that the enzymes coded for by all these genes operate independently, and with varying efficiency from individual to individual. Thus different amounts of H substance will be converted to A or B and the red blood cells or secretions of different persons possess A and H or B and H. It is rare that all the H active ends of the chains will be converted to A or B. For the vast majority of individuals, the *A* or *B* gene converts only a portion of the available H antigen.

The *Le* gene codes for an α-L-fucosyl transferase that adds a fucose residue to the penultimate N-acetylglucosamine unit in the blood group active chain, as shown in Fig. 8. The addition adds a single fucose molecule to the recipient sugar by an α-1,4 linkage. Hence the specificity of the Lewis-synthesizing enyzme is for α-1,4 addition, in contrast to the H-synthesizing fucosyl transferase, which directs an α-1,2 addition. The addition of fucose in α-1,4 linkage to the N-acetyl-D-galactosamine converts the specificity of type 1 chains to give an Le^a-active molecule. The Lewis specific enzyme, it should be noted , cannot act on the type 2 precursor chains because the 4 position of the N-acetylglucosamine residue is occupied by its linkage to the terminal galactose unit at the end of the chain.

Le^b activity arises when both the H-active fucosyl transferase and the Le-active fucosyl transferase act in concert. In this case, fucosyl residues are added to the N-acetylglucosamine in an α-1,4 linkage and a second fucose is added to the terminal galactose in α-1,2 linkage. The combined presence of the two fucosyl residues substituted on the

<u>type 1 chain</u>

```
— -GalNAc-(β1,3)-Gal-(β1,3)-GluNAc-(β1,3)-Gal
                              |
                           (α1,4)              (Le^a activity)
                              |
                             Fuc
```

<u>type 2 chain</u>

```
- - -GalNAc-(β1,3)-Gal-(β1,3)-GluNAc-(β1,4)-Gal   (Unchanged)
```

Figure 8 Structure of the antigens produced by the action of the Le-specific fucosyl transferase on type 1 and type 2 chains.

adjacent sugars of the main chain conveys a new specificity, Le^b. Both Le^a and Le^b structures possess fucose in α-1,4 linkage to N-acetylglucosamine (of type 1 chains), but the action of the H-active fucosyl transferase completely converts the serologic specificity of what would be an Le^a determinant to Le^b with no residual cross-reactivity. These findings explain why nonsecretors are always Le^a. Since the secretor gene is a gene that controls the action of the *H* gene in secreting tissues, when an individual's genotype is *se/se*, no H-active fucosyl transferase is produced and only the Le-active fucosyl transferase is expressed. Hence only Le^a structures can be produced. However, even if only a single *Se* gene is present in an individual, the α-1,2 fucosyl transferase that is specific for galactose is also produced. The activity of the latter enzyme produces the serologic activity of secretors, Le^b. When no *Le* gene is present, only the H-specific transferase codes for an enzyme, and thus the chain lacks both Le^a and Le^b activity.

The *A* gene codes for an N-acetylgalactosaminyl transferase that adds one molecule of N-acetyl-D-galactosamine in α-1,3 linkage to the terminal galactose of both type 1 and type 2 precursor substances provided that the H-synthesizing fucosyl transferase has already acted. This reaction is depicted in Fig. 9.

The *B* gene codes for a galactosaminyl transferase that adds a galactose residue to the galactose that is terminal in both type 1 and type 2 precursor chains. Again, the enzyme cannot act unless a fucosyl residue has been placed in α-1,2 linkage to the same galactose by the H-synthesizing enzyme. This reaction is shown in Fig. 10.

In group AB individuals, both the A- and B-synthesizing enyzmes are present and because both compete for the same substrate (i.e., H substance), very little H activity remains.

It should be noted that the sole difference in A and B activity resides in the group that is attached to the second carbon of galactose. In A substance the second carbon of galactose is substituted for by an N-acetyl group; in B substance the same position is occupied by an hydroxyl group. These are extremely small structural differences that have enormous serologic impact.

<u>type 1 chains</u>

```
- - -GalNAc-(β1,3)-Gal-(β1,3)-GluNac-(β1,3)-Gal-(α1,3)-GalNAc
                                             \
                                            (α1,2)
                                               \
                                               Fuc      (A active)
```

<u>type 2 chains</u>

```
- - -GalNAc-(β1,3)-Gal-(β1,3)-GluNAc-(β1,4)-Gal-(α1,3)-GalNAc
                                             \
                                            (α1,2)
                                               \
                                               Fuc      (A active)
```

Figure 9 Structures produced by the action of the A-specific N-acetylgalactosaminyl transferase on H substance.

It should also be emphasized that the A and B blood group antigens are actually a combined function of the A and B determinants in conjunction with the H determinant. The reaction of anti-A and anti-B with A or B erythrocytes cannot be inhibited with even extremely high concentrations of N-acetyl-D-galactosamine and galactose, respectively; efficient inhibition depends on the adjacent fucose residue. In other words, efficient inhibition of A or B antigenic activity with their respective antibody can be done only with a trisaccharide that contains fucose and galactose in addition to the A or B determinant sugar. The monosaccharides are not sufficient by themselves. Thus the A and B

<u>type 1 chains</u>

```
- - -GalNAc-(β1,3)-Gal-(β1,3)-GluNAc-(β1,3)-Gal-(α1,3)-Gal
                                             \
                                            (α1,2)
                                               \
                                               Fuc      (B active)
```

<u>type 2 chains</u>

```
- - -GalNAc-(β1,3)-Gal-(β1,3)-GluNAc-(β1,3)-Gal-(α1,4)-Gal
                                             \
                                            (α1,2)
                                               \
                                               Fuc        (B active)
```

Figure 10 Structures produced by the action of the B-specific galactosaminyl transferase on H substance.

antigens are composite antigens in that both A and H determinants are required for their overall activity.

For the enzymatic transfers of the monosaccharides involved in forming the blood group substance chains, the monosaccharides must be in "activated" form; that is, each sugar to be transferred by the enzyme must be associated with a pyrimidine nucleoside in order that interaction with the glycosyl transferase can occur. Thus for every fucosyl transferase the fucose that is to be added to the oligosaccharide chain must be present as a guanosine diphosphate ester of fucose (GCP-L-fucose). Similarly, galactosyl additions require uridine diphospho-D-galactose; and N-acetyl-D-galactosamine additions require uridine diphospho-D-N-acetylgalactosamine.

The glycolipid form of the AB and H antigens have recently been examined (Laine et al., 1974) and found to be structurally the same as that described for secreted ABH substances. The principal difference is that instead of being associated with a peptide backbone, these substances are ultimately attached to ceramide, which is a lipid constituent of the erythrocyte membrane. In the case of the erythrocytic ABH antigens, of course, the *Le/le* and *Se/se* genes play no role.

The P blood group system consists of two common phenotypes, P_1 and P_2. There also exist three other extremely rare phenotypes: P_1^k, P_2^k, and p (Fudenberg, 1978; Race and Sanger, 1975). It is though that the various phenotypes are produced by three antigens, P_1, P, and P^k. The vast majority of individuals possess the antigen P and 75% also possess the antigen P_1 in addition. Persons who possess both P_1 and P are of the P_1 phenotype. Those who only have the P antigen are of the P_2 phenotype. The P^k antigen is expressed only by P^k/P^k homozygous individuals. Erythrocytes from individuals of the rare P^k phenotype do not possess the P antigen, although P_1 may be expressed. Red blood cells from p individuals lack all P antigens.

Since there is no water-soluble form of any P system antigen in humans, the earliest attempts to unravel these structures came from work on hydatid cyst fluid, which contains a material that has P activity. These studies revealed that the antigenic structure of the P antigens was an oligosaccharide that possessed a terminal α-galactose unit which was responsible for producing the immunodominant determinant.

With advances in techniques for working with glycolipid materials from erythrocytes, attempts were made to examine these substances for the presence of P activity. Of the major lipid classes, the glycosphingolipids were found to contain structures with P activity, and in all cases the antigenicity resided in the oligosaccharide portion of the molecule. Furthermore, all of the oligosaccharide chains were found to be linked to the ceramide portion of sphingosine. This is in good agreement with the oligosaccharides known to be associated with glycolipid A, B, and H substances, which are also bound to ceramide.

A partial list of the glycosphingolipids that can be isolated from erythrocytes is shown in Table 5. The simplest structure is lactosyl ceramide, which consists of a terminal nonreducing galactose in β-1,4 linkage to a glucose residue, which, in turn, is linked to ceramide. This simple structure possesses no known natural blood group activity. This is the structure common to all human erythrocytes and its presence is universal in the species. Thus antibodies to this structure are not to be expected following transfusion of the erythrocytes of one individual into another.

The addition of a single additional galactosyl residue linked α-1,4 to the terminal galactose of lactosyl ceramide does, however, produce the blood group antigen P^k. The

Table 5 Structures of Glycosphingolipids Isolated from Human Erythrocytes[a]

Lactosyl ceramide (no activity)	Gal(β,1-4)Glc-ceramide
Trihexosyl ceramide (p^k antigen)	Gal(α,1-4)Gal(β,1-4)Glc-ceramide
Globoside (P antigen)	GalNAc(β,1-3)Gal(α,1-4)Gal(β,1-4)Glc-ceramide
Paragloboside (no activity)	Gal(β,1-4)GlcNAc(β,1-3)Gal(β,1-4)Glc-ceramide
P_1 antigen	Gal(α,1-4)Gal(β,1-4)GlcNAc(β,1-3)Gal(β,1-4)Glc-ceramide

[a]Gal, D-galactose; GLC, D-glucose; GalNAc, N-acetyl-D-galactosamine; GlcNAc, N-acetyl-D-glucosamine; GlcNAc, N-acetyl-D-glucosamine.
Source: Marcus et al. (1977).

resulting structure, trihexosyl ceramide, strongly inhibits the reaction between anti-P^k and erythrocytes that possess the P^k antigen. Anti-P sera, however, are not inhibited by trihexosyl ceramide.

This last point is worth considering since the P antigen differs structurally from P^k only by the fact that one additional monosaccharide, N-acetyl-D-galactosamine, is linked β-1,3 to what otherwise would be a P antigen, forming a structure known as globoside. It is noteworthy that anti-P antibodies occur naturally (i.e., without known antigenic stimulus) in the serum of P^k persons. Here the addition of the single additional N-acetyl-D-galactosaminyl group to the P structure produces an antigen that can be recognized immunologically by P^k persons, but the anti-P produced in these cases does not react with the P^k antigen of the individual who forms the antibody. Thus anti-P antibodies are directed to the distal few monosaccharides of the oligosaccharide, but the reaction of these antibodies is overwhelmingly dependent on the immunodominant N-acetyl-D-galactosaminyl group.

The final member of the P series of antigens, P_1, is strikingly different in structure from the antigens discussed previously. This finding was startling when first made because it seemed to fly in the face of previous genetic studies that postulated that the P antigens were under simple single-locus control. The P locus was originally considered to be occupied by three alleles, P_1, P_2, and p. The p allele was through to be amorphic or silent and hence produced no product by addition reactions to what would be considered P precursor substance. The P_1 allele was thought to produce two antigens, or determinants P_1 and P; the P_2 allele was thought to produce the determinant P. The hypothesis of simple genetic control was difficult to reconcile with the chemical evidence that the P_1 antigen was structurally dissimilar to either P or P^k, and is a derivative of the paragloboside chain.

Examination of p and P^k cells for the presence of the different glycosphingolipids reveals that these cells, in addition to lacking the "normal" P antigen, also lack all but trace amounts of globoside, which is usually the most abundant class of glycosphingolipid of the human erythrocyte. Furthermore, p cells lack trihexosyl ceramide but contain higher than normal concentrations of the trihexosyl ceramide precursor, lactosyl ceramide. It should be mentioned that normal erythrocytes all contain trihexosyl ceramide and thus should exhibit P^k antigenic activity, but in fact, they do not. The reasons for the lack of P^k expression are not understood but probably are due to difficulties with the immunologic assays used to detect such antigens. In P^k cells, however, the quantity of trihexosyl ceramide is presumably high enough to permit detection of this structure by means of appropriate antibody.

Two possible explanations for these chemical results have been put forward. One postulates that because the terminal disaccharides of P_1 and P^k are identical, a single galactosyl transferase adds the final galactose to both chains. This would explain in a simple manner why no antigens are expressed in the p phenotype. If the single galactosyl transferase was missing in these individuals, neither P_1 nor P^k antigens could be synthesized, and the P_2 gene could not be expressed even though an enzyme would be coded for. This would happen because the substrate for this enzyme (trihexosyl ceramide) would not be available. Unfortunately, this simple explanation is compromised by the fact that P_2 persons do not synthesize P_1 antigens. Since these individuals possess paragloboside, which is the precursor to the P_1 antigen, and also presumably have the galactosyl transferase that acts on it, they should express a P_1 antigen.

An alternative explanation is that two different galactosyl transferases act independently to form the P_1 and P^k antigens. P_2 persons would then be missing the enzyme that acts on paragloboside chains and would not produce P_1 antigen. Persons of the p phenotype would be missing both the paragloboside-active transferase and the enzyme that requires lactosyl ceramide as a substrate. However, this idea has been weakened by evidence derived from cell fusion studies showing that the P_1 and P^k genes are on different chromosomes. If this is true, the p phenotype would require that two independent and rare mutations had occurred. Although this is possible, such an occurrence on two separate chromosomes in the same individual is too improbable to be reasonable. The statistically more acceptable idea of closely linked genes being affected by a single mutational event is thus put into question if the two genes are on separate chromosomes. Thus the structural evidence for the P series of antigens awaits further study to determine the exact genetic basis for the serologic differences found in humans that possess the different alloantigens produced.

REFERENCES

Fudenberg, H. H., Pink, J. R. L., Wang, A., and Douglas, S. D. (1978). *Basic Immunogenetics*, 2nd ed. Oxford University Press, New York.

Galton, D. A. G., and Goldsmith, K. L. G. (1961). *Haematology and Blood Groups*. University of Chicago Press, Chicago.

Ginsburg, V., Kobata, A., Hickey, C., and Sawika, T. (1971). In *Glycoproteins of Blood Cells and Plasma* (G. A. Jamieson and T. J. Greenwalt, Eds.). Lippincott, Philadelphia, p. 114.

Heidelberger, M. (1960). In *Progress in the Chemistry of Organic Natural Products*, Vol. 18 (L. Zechmeister, Ed.). Academic Press, New York, p. 503.

Issitt, P. D. (1970). *Applied Blood Group Serology*. Becton, Dickinson, Oxnard, Calif.

Kabat, E. A. (1956). *Blood Group Substances*. Academic Press, New York.

Kabat, E. A. (1968). *Structural Concepts in Immunology and Immunochemistry*. Holt, Rinehart and Winston, New York.

Kabat, E. A., and Mayer, M. (1961). *Experimental Immunochemistry*, 2nd ed. Thomas, Springfield, Ill.

Laine, R. A., Stellner, K., and Hakamori, S. I. (1974). Methods Memb. Biol. *2*, 205.

Luderitz, O., Staub, A. M., and Westphal, O. (1966). Bacteriol. Rev. *30*, 192.

Marcus, D. M., Naiki, M., Kundu, S. K., and Schwarting G. A. (1977). In *Human Blood Groups* (J. F. Mohn, R. W. Plunkett, R. K. Cunningham, and R. M. Lambert, Eds.). S. Karger, Basel, p. 206.

Morgan, W. T. J., and Watkins, W. M. (1969). Br. Med. Bull. *25*, 30.

Race, R. R., and Sanger, R. (1975). *Blood Groups in Man*, 6th ed. Davis, Philadelphia.

Uchida, T., Robbins, P. W., and Luria, S. E. (1963). Biochemistry *2*, 663.

Watkins, W. M. (1966). Science *152*, 172.

Zaleski, M. B., Dubiski, J., Niles, E. G., and Cunningham, R. K. (1983). *Immunogenetics*, Pitman Publishing, Marshfield, Massachusetts.

Zmijewski, C. M. (1968). *Immunohematology*. Appleton-Century-Crofts, New York.

4

Immunochemistry of Tissue-Specific and Tumor Antigens

Felix Milgrom, C. John Abeyounis, and Roger B. Eaton School of Medicine, State University of New York at Buffalo, Buffalo, New York

I. SPECIES SPECIFICITY

A *species-specific antigen* may be defined as an antigen that is present in all members of a given species and which differs from an analogous antigen of other species. The first antigens that were extensively studied for species specificity were serum proteins. Nuttall (1904) performed pioneering experiments on this topic. He produced 30 antisera in rabbits by immunization with whole sera originating from various species. These antisera were then tested against sera from more than 500 species by means of simple test tube precipitation, and the strength of reaction was evaluated. The results obtained by Nuttall may be summarized as follows: (1) the strongest reactions were obtained with the sera of animals belonging to the same species as the donor of the serum used for immunization; (2) strong cross-reactions were obtained with sera from animals of species closely related to the donor of the immunizing serum, and (3) weak or no cross-reactions were noted with sera of animals from species distant from the donor. In this way, cross-reactions were noted within one class of vertebrates in that, for example, antiserum to porcine serum precipitated sera of primates, rodents, and even whales. There were, however, practically no cross-reactions between sera originating from various vertebrate classes; for example, antiserum to human serum did not precipitate sera from birds, reptiles, amphibians, or fish, and antisera to chicken serum did not combine with sera of mammals.

The properties of species-specific antibodies depend to a great extent on the species of animal producing the immune serum. Rabbit, the species that is usually used for immunization, distinguishes quite well serum of rat from serum of mouse, but it hardly distinguishes chicken serum from goose serum. Landsteiner called this phenomenon *faulty perspective*, by which he meant that the immune response of a rabbit injected with

serum of the phylogenetically distant chicken is overwhelmed by cross-reacting antigens common to many birds and, therefore, it cannot distinguish relatively subtle differences between chicken and goose (see Landsteiner, 1946). The discriminating sharpness of an immune serum may be increased by absorption. For example, rabbit antiserum to sheep serum hardly discriminates between sheep and goat; absorption of such an antiserum by goat serum may leave only the sheep-reactive antibodies. Another convenient "trick" used for sharpening the discriminatory properties of species-specific antibodies is called *cross-immunization*. This is based on raising the antiserum in an animal closely related to the donor of the serum used for immunization. For example, chimpanzee injected with human serum may produce antibodies precipitating human serum but not sera of other primates.

Interpsecies cross-reactions between sera originating from various animal species may account for unexpected hypersensitivity reactions observed in testing allergic patients. Strong hypersensitivity reactions may be elicited by sera of some exotic animals with which the patient never had contact. Sensitization engendered by contact with domestic animals can be responsible for cross-reaction with sera of, for example, elephant, opposum, or dolphin.

Most of the traditional investigations on species specificity were conducted using whole serum as an antigen. When it was clearly realized that the serum is composed of some 30 distinguishable proteins, studies on species specificity of these individual proteins were conducted. It was shown that the same general principles that were noted for the whole serum hold also for individual serum proteins such as albumin, gamma globulin, or transferrin; the intensity of cross-reaction depends on biological proximity. In one experiment, rabbit antiserum to bovine serum albumin cross-reacted with serum albumin of man, horse, pig, sheep, dog, cat, guinea pig, hamster, mouse, and rat, but not with albumin of chicken, parakeet, or alligator. Application of double-diffusion gel precipitation for these studies permitted simple analytical evaluation of cross-reactions (see Chap. 17). Rabbit antiserum to bovine serum albumin would give a reaction of partial identity when tested simultaneously against bovine and sheep serum albumin, but since ox and sheep are closely related species, only a very short spur would be observed, which extends from the line formed by bovine serum albumin over the line formed by sheep serum albumin (Fig. 1A). The same antiserum would produce only a weak precipitation line with human serum albumin, and the spur produced by the bovine serum albumin line would be rather thick and long (Fig. 1B). On the other hand, rabbit antiserum to human serum albumin would give a strong precipitation line with human serum albumin but a rather weak line

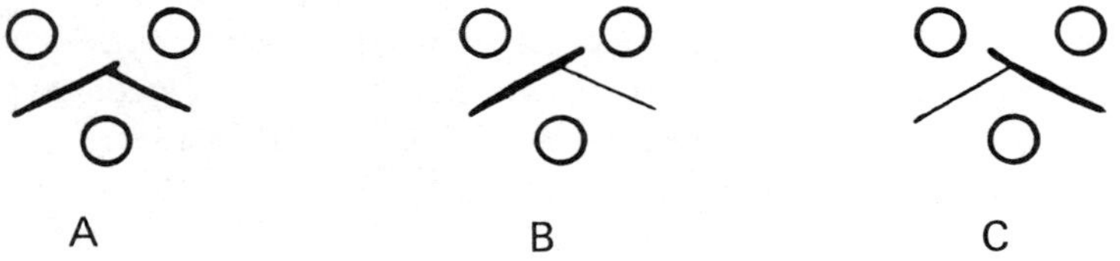

Figure 1 (A) Upper left well: bovine serum albumin; upper right well: sheep serum albumin; lower well: rabbit antiserum to bovine serum albumin. (B) Upper left well: bovine serum albumin; upper right well: human serum albumin; lower well: rabbit antiserum to bovine serum albumin. (C) Upper left well: bovine serum albumin; upper right well: human serum albumin; lower well: rabbit antiserum to human serum albumin.

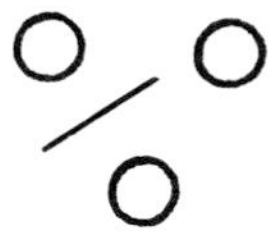

Figure 2 Upper left well: bovine serum albumin; upper right well: bovine gamma globulin; lower well: rabbit antiserum to bovine serum albumin.

with bovine serum albumin, and the spur would extend from the human serum albumin line over the bovine serum albumin line (Fig. 1C). Significantly, antiserum to either bovine or human serum albumin fails to precipitate chicken serum albumin.

It should be mentioned that even though cross-reactions can occur with analogous proteins of various species (e.g., human, bovine, and porcine serum albumin), cross-reactions do not occur with different proteins within the same species (e.g., bovine serum albumin and bovine gamma globulin) (Fig. 2).

In splitting serum proteins by enzymatic digestion, one may obtain fragments that retain species specificity; for example, the Fc fragment of gamma globulin shows clear-cut species specificity very similar to that observed in studies on the whole molecule of gamma globulin.

In addition to serum proteins, many studies were devoted to species specificity of other proteins, such as casein, thyroglobulin, hemoglobin, and cytochromes. For several proteins, amino acid sequencing was conducted that brought some insight into the structures that might be responsible for species specificity. Nisonoff and his associates (1970) conducted illuminating studies on the structure of cytochrome *c*. Among other findings, they noted that the presence of hydrophobic isoleucine in position 58 of human and kangaroo cytochrome *c* correlated with the occurrence of immunogenic determinant responsible for cross-reactions. Similar data are available for other proteins (e.g., gamma globulin). Insulin, which is a small molecule (molecular weight 6000) and rather a weak antigen, also has clear-cut species specificity which depends on a recognized chemical structure.

The surface of cells is also impregnated by species specificity. In the very classic studies of Bordet, the species-specific nature of antigens on the erythrocyte surface was recognized. A great many studies on species specificity of cell-surface antigens were conducted by means of mixed agglutination with cell cultures. In this procedure, monolayer cell cultures were incubated with the proper antisera of rabbit origin and the binding of antibodies was detected by means of an indicator system composed of sheep erythrocytes sensitized by rabbit anti-sheep erythrocyte serum and agglutinated by antiserum to rabbit globulin of sheep or goat origin. Adherence of the indicator erythrocytes to the monolayer was interpreted as a positive reaction. The antisera used in these studies were raised in rabbits by immunization with cell or tissue suspensions containing intact cell-surface antigens. It was noted that the predominant antigens of the cell surface are species-specific antigens. As seen in Table 1, the strongest reactions were noted when an antiserum was tested against cell cultures originating from the same species as the donors of cells or tissues used for immunization (Milgrom et al., 1964a). We may note that this pattern of reaction presented in the table is very similar to that observed with soluble antigens in the classic studies of Nuttall.

There is rather little information available about the cell-surface antigens carrying species specificity. They cannot be brought into solution by simple disruption of cells or

Table 1 Mixed Agglutination with Cell Cultures of Bovine, Hamster, and Human Origin and Antisera to Bovine, Hamster, and Human Tissue

	Titer ($-\log_{10}$) of mixed agglutination with antisera against suspension of tissues from:		
Culture of cells from:	Ox	Hamster	Man
Ox	5	2	3
Hamster	2	5	3
Man	2	3	6

tissues in saline suspension and, therefore, antisera to cell and tissue extracts give very weak or no reactions in mixed agglutination studies with cell cultures. Most of these antigens are destroyed by heating at 80°C, which may well indicate that they are of protein nature. It appears very likely that the results observed in studies such as those presented in the table were due to the reactions of many antigen-antibody systems similar to the results of precipitation studies by Nuttall conducted with a whole serum as antigen.

Separation of individual antigens from the cell surface by chemical means was not attempted. However, immunogenetic studies shed some light on this problem. Studies on man-mouse hybrid cells obtained by cell fusion demonstrated that such cells contain species-specific antigens of both parental species, man and mouse. With deletion of human chromosomes, clones were obtained which, by karyotypic studies, showed only murine but not human chromosomes, and such clones no longer showed any evidence for the presence of human species antigens. On the other hand, a clone that retained only one human chromosome (number 21) showed the presence of species antigens; however, the reactions given by this clone were much weaker than those given by other hybrid cells, let alone human cell cultures. This clone gave positive reactions only with some but not all antisera to human cells. From this study it could be concluded that genes on chromosome 21 code a very limited number of molecules responsible for human species specificity, and it could be inferred that many human chromosomes carry genes coding molecules endowed with species specificity (Chan et al., 1979).

Studies on species specificity found practical application in forensic medicine. Identification of the species origin of tissue debris or blood stains is employed in criminal cases. Study of food products is conducted to detect adulteration. Performing these studies is by no means simple and requires a great deal of expertise and experience.

The properties of species specificity discussed above pertain primarily to proteins in which the antigenic structure reflects the process of evolution in that the analogous proteins (e.g., serum albumins) have similar antigenic structure when they originate from closely related species, but they differ considerably in their antigenic structure when they originate from remote species. In other words, interspecies cross-reactions of protein antigens may be predicted on the basis of their biological proximity.

In sharp contrast to these expected cross-reactions, there exist also unexpected cross-reactions which by-and-large pertain to nonprotein antigens. The term *heterophile antigens* was coined to denote antigens that are similar in their structure, even though this similarity could not be predicted on the basis of their biological proximity (see the review by Kano and Milgrom, 1977). The most important discovery on heterophile antigens was made by Forssman in 1911 when he noted that sera of rabbits immunized with guinea pig

tissues produce agglutination and lysis of sheep erythrocytes. Obviously, this phenomenon could be later explained by the demonstration that guinea pig tissues and sheep erythrocytes share a common antigen. This antigen, frequently referred to as the *Forssman antigen*, is a glycosphingolipid, a ceramide linked with a pentasaccharide (glucose, two molecules of galactose and two molecules of N-acetylgalactosamine) determining its serological specificity. The animal kingdom may be divided into Forssman-positive and Forssman-negative species. Significantly, this division does not reflect biological proximity, as exemplified by the fact that mouse is a Forssman-positive and rat a Forssman-negative animal, or that sheep and goat are Forssman positive whereas deer and ox are Forssman negative. Humans are a Forssman-negative species even though human blood group A substance is related to the Forssman antigen. Obviously, only Forssman-negative animals are capable of producing Forssman antibodies. Of these, rabbit is used most frequently as the convenient animal for raising Forssman antibodies. In some Forssman-negative species, primarily in rabbit, naturally occurring Forssman antibodies may be detected.

Another medically important heterophile antigen-antibody system was described in the 1920s by Hanganutziu and by Deicher independently. They noted that injection of human patients with sera of foreign species origin, primarily with horse sera used for treatment of diphtheria and prevention of tetanus, resulted in the formation of antibodies that acted on erythrocytes of many species. The term *serum sickness antigens and antibodies* was used to name this heterophile system. This was a misnomer since the appearance of these antibodies did not depend on the development of symptoms of serum sickness. More recently, the term *Hanganutziu-Deicher antigen-antibody system* was proposed. Hanganutziu-Deicher antigen(s) is (are) different from Forssman, as epitomized by the fact that this antigen appears also in Forssman-negative animals (e.g., in ox).

If one accepts an oversimplified definition for Hanganutziu-Deicher antibodies as antibodies combining with bovine and sheep erythrocytes as well as with guinea pig kidney, one has to acknowledge the existence of a rather complex and diverse antigen system with which these antibodies combine. Antibodies elicited by injection of humans with foreign species sera such as those of horse or goat combine with an antigen that can be identified as a ganglioside, N-glycolylneuraminic acid, a compound that is present in tissues of several animal species but not in human tissue. There are, however, antibodies present in many pathologic and some human sera which have similar serologic properties but react with a completely different antigen, which appears to be a high-molecular-weight glycoprotein. Antibodies to this antigen are formed due to some hidden immunizing stimuli without any contact of the patient with a foreign species serum.

Another heterophile system was discovered by Paul and Bunnell in 1932. They noted that sera of patients suffering from infectious mononucleosis combine with sheep erythrocytes. It was shown subsequently that these antibodies react also with bovine erythrocytes, which distinguishes them from Forssman antibodies, but do not react with guinea pig tissues, which distinguishes them from both Forssman and Hanganutziu-Deicher antibodies. The Paul-Bunnell antigenic system is composed of at least two separate antigens, referred to as B for its presence in bovine but not sheep erythrocytes, and as BS for its presence in both bovine and sheep erythrocytes. Chemically, the BS antigen is probably a glycoprotein. The chemical nature of the B antigen has not been established.

Paul-Bunnell antibodies can be demonstrated in close to 90% of patients suffering from infectious mononucleosis and are rarely noted in other pathological or in normal

sera. They are almost exclusively of IgM nature and are unrelated to the antibodies against Epstein-Barr virus, which is broadly accepted as the causative agent of infectious mononucleosis.

The mechanism of formation of these antibodies has never been explained satisfactorily. The most plausible explanation for formation of these antibodies proposes that the pathological process results in alteration of physiological constituents of the patient's cells such that the cells acquire the structure of Paul-Bunnell antigen. Since this antigen is foreign to the body of the patient and is released in considerable amounts, the response to this antigen occurs in the form of Paul-Bunnell antibodies. Support for this hypothesis stems from the demonstration of Paul-Bunnell antigen in tissues of a patient who died of septicemia during an early stage of infectious mononucleosis.

Interestingly, Paul-Bunnell antigen may be demonstrated in the spleens of patients suffering from lymphoma and leukemia and may be shown in the circulation of patients suffering from various diseases, including malignancies, syphilis, and lepromatous leprosy. Apparently in these conditions, Paul-Bunnell antigen is released in small amounts and/or nonimmunogenic form, and therefore it fails to elicit formation of antibodies but presumably induces a state of unresponsiveness.

II. TISSUE SPECIFICITY

Antigens characteristic for given cells, tissues, or organs are usually referred to as *tissue-specific antigens*, and many serologists frequently do not bother to distinguish between tissue-specific antigens *sensu stricto* and those characteristic for cells or organs.

The first observation on tissue specificity was made by Uhlenhuth in 1903. He immunized rabbits with homogenates of bovine lens and noted that the resulting antisera would combine with homogenates of lenses not only of bovine origin but also originating from other species. As a matter of fact, these antisera reacted with lens preparation of all species tested, even such remote species as fishes and amphibians. Significantly, these sera also combined with lens preparations originating from rabbits, the antibody-producing species. On the other hand, these sera did not react with preparations of tissues or organs other than lens, even if they were of bovine origin.

On the basis of these experimental results, some investigators believed that a tissue-specific antigen is devoid of species specificity. This assumption proved to be incorrect when studies on tissue specificity progressed. As discussed below, thyroglobulin, which is an outstanding example of a tissue-specific antigen, also has very distinct species specificity. As a matter of fact, species specificity of thyroglobulin may well be compared to that of serum proteins since thyroglobulins originating from related species show strong cross-reactions whereas those from distant species show weak or no cross-reactions. A very similar pattern of reactions was observed in studies on most other tissue-specific antigens.

One could argue that serum proteins are also tissue-specific antigens and consider albumin an antigen characteristic for liver, and gamma globulin an antigen characteristic for plasma cells. In a similar way, one could argue that hormones are tissue-specific antigens, and in these terms, insulin could be considered a pancreas-specific antigen and chorionic gonadotropin a placenta-specific antigen. One has to admit that there is a great deal of justification for this type of argument. On the other hand, since serum proteins and hormones are shed off to the circulation and reach all tissues with the blood, they have not been traditionally discussed as tissue-specific antigens, in contradistinction to

such antigens as those of lens or brain, which remain "sequestered" in their tissues of origin. As a matter of fact, the ubiquitous presence of serum proteins in tissues has interfered a great deal with studies on tissue specificity, in that all tissue extracts are contaminated by serum proteins, and therefore antibodies to these antigens were present in all antisera studied. Only modern serological techniques, including absorption experiments, permitted distinction of tissue specificity by such reagents. It is not a coincidence that the first tissue-specific antigen was described in lens, which is free of contamination by serum.

It is hoped that the reader can appreciate that tissue-specificity studies are intrinsically interesting and not just academic. Often, the first indication of molecules of unique structure have been through the use of antibodies as tools. The definition of the tissue distribution of these molecules immediately generates considerable information about them. Serological techniques can further define the molecule by examining the species distribution of the antigen. Immunohistochemical techniques are particularly helpful in discovering the cellular or even subcellular location of the molecule harboring the antigenic site. Immunochemical studies are often helpful and indeed are sometimes the only means for isolation of the molecule of interest. Even the function of the molecule can be studied directly if antibody binding interferes with (as in the case of enzyme inhibition) or potentiates (as in the case of receptor activation) some normal function of the tissue. Important information regarding some poorly understood human diseases has come from the study of tissue-specific antigens. Finally, on a more general level, the study of well-defined tissue-specific antigens has been useful in examination of the basis of specificity itself and in understanding the basis for certain observed immunoregulatory phenomena.

We have chosen three organs as examples to familiarize the reader with the principles involved in tissue specificity and to demonstrate some of the rewards such studies have yielded. Emphasis has been given to those antigens which are relatively well understood biochemically.

A. Thyroid

The thyroid was one of the first organs extensively studied for tissue-specific antigens and it serves as a good model for classic tissue specificity studies. Furthermore, the molecule responsible for most tissue-specific antigenicity has been identified and has been relatively well characterized. Finally, one of the first experimentally induced autoimmune diseases to be studied involved thyroid-specific antigen(s).

In the pioneering studies of Hektoen et al. in 1927, rabbits were injected with partially purified thyroglobulin (Tg) prepared from seven different mammals, as well as the chicken. The sera were tested by the tube precipitation test against antigen prepared from 33 species of mammals and the chicken. Sera raised against chicken Tg reacted only with chicken Tg; sera raised against mammalian Tg reacted with Tg prepared from each of the 33 mammals, but not with chicken Tg. Sera raised to Tg from other mammals reacted with Tg of various mammalian species, but never with chicken Tg. Negative reactions with homologous* blood proteins indicated tissue specificity.

*The adjectives *homologous* and *heterologous* are used in this text in their original meaning; the former denotes "originating from the same species as the antigen used for immunization," and the latter denotes "originating from a species other than that which provided the antigen used for immunization."

Later studies (reviewed by Shulman, 1974) confirmed the tissue specificity by testing extracts of kidney, liver, spleen, heart, and adrenal. Tissue specificity was demonstrated in a number of ways. (1) Antisera to thyroid extracts reacted in complement fixation tests (CFT) with much higher dilutions of thyroid extract than of other tissue extracts. (2) Absorption of the antisera with homologous serum eliminated the low reactivity with other tissues, but did not affect reactivity with thyroid. On the other hand, absorption with homologous thyroid eliminated reactivity with thyroid extracts prepared from all species. (3) Antisera raised to kidney extracts reacted very weakly with thyroid extract.

The use of the semiquantitative CFT demonstrated that reactions with homologous thyroid extracts were always stronger than reactions with extracts prepared from thyroid of other species. Double-diffusion gel precipitation distinguished two reaction lines with homologous extracts. Only one of these lines merged into a reaction of identity with the single precipitation line formed with heterologous extracts. These tests demonstrated that there were species-restricted as well as cross-reacting antigens in the homologous extracts. This was supported by serum absorption studies, which showed that heterologous Tg absorbed most reactivities with heterologous extracts, but did not absorb all reactivity with homologous extracts.

Purified preparations of Tg have shown that the component responsible for the tissue-specific reactions described above is Tg itself. Tg is a glycoprotein, with a carbohydrate content of 8 to 10%. Purified Tg is heterogeneous due to variations in carbodrate content and in degree of iodination. Tg has a molecular weight of about 690,000 and has the shape of a flexible helix with two turns. There are about 100 disulfide bonds per molecule, and Tg is believed to be composed of two to four subunits.

Tg functions as a storage molecule for the hormonally active thyroxine and triiodothyronine. Tg is synthesized and iodinated in the follicular cells of the thyroid, transported to the follicular lumen in exocytotic vesicles, and released into the lumen for storage. When needed, Tg reenters the follicular cells, where enzymes free the active hormones. The hormones are then released into the circulation.

Tg was one of the first antigens used to induce autoantibodies experimentally. Heterologous sera define a large number of antigenic determinants, but only two to six of these are autoantigenic (Shulman and Witebsky, 1960). This was determined immunochemically by the use of quantitative precipitin curves. By determining the relative contributions of antibody and antigen in precipitates formed near antibody excess, and knowing the molecular weights of each component, the molar ratio of antibody to antigen was calculated. The antibody/antigen ratio determined by using heterologous sera had been determined to be 40:1 to 60:1, but the ratio determined by the use of human autoantibodies was 2:1 to 6:1.

The existence of antigenic determinants reactive only with heteroantibodies was supported by double-diffusion studies using various peptide fragments of Tg as antigen. Although some fragments reacted with both autologous and heterologous sera, fragments derived from trypsin digests reacted only with heteroantibodies.

An organ-specific autoimmune disease of the thyroid, experimental autoimmune thyroiditis (EAT), has been induced in a number of species, usually by the injection of homologous Tg in Freund's complete adjuvant (FCA) (reviewed by Weigle, 1980). The lesion typically begins with infiltration by mononuclear cells and some neutrophils. The course may progress to necrosis of the follicular cells. The precise contributions of the humoral and cellular responses remains controversial, but present evidence indicates that humoral immunity plays a predominant role.

EAT is considered to be a good model for the human disease, Hashimoto's thyroiditis. In patients with this disease, the autoantibodies react only with human and monkey Tg, indicating that the autoantigenic determinants are also the species-restricted ones. The reasons for this are not clear.

In addition to Tg, four other antigens have been described which are involved in autoimmune responses to the thyroid (Pinchera et al., 1980; Shulman, 1974). Antibodies to the *microsomal antigen* were found in virtually all cases of Hashimoto's thyroiditis, as well as most cases of idiopathic myxedema and of Graves' disease. In fact, these antibodies were frequently found in the absence of antibodies to Tg. The antigen was localized by immunofluorescence (IF) to the apical portion of the thyroid cells. It is probably an integral lipoprotein of the membrane of the exocytotic vesicles that transport newly synthesized Tg. The antigen is best solubilized by Triton X-100.

A second antigen is the target of the *thyroid-stimulating antibodies*. This antigen is either the actual receptor for thyroid-stimulating hormone (TSH) or is closely associated with this receptor, because the antibody directed to this antigen stimulated the thyroid by activating adenylate cyclase, and inhibited the binding of TSH. Over 90% of patients with Graves' disease developed demonstrable antibody, which is probably the cause of the hyperthyroidism in this disease. Infants born to mothers with high antibody titers always had neonatal hyperthyroidism of short duration. The antibody was also frequent in patients with Hashimoto's thyroiditis and idiopathic myxedema.

There is a group of ill-defined *thyroid cell-surface antigens* distinct from the microsomal antigen. There is also an antigen other than Tg in the follicular colloid. This antigen, called the *second colloid antigen* or CA-2, was defined by immunofluorescent patterns which differed from the pattern obtained using anti-Tg antibodies. Anti-CA-2 antibodies have been described in patients with thyroid diseases and in dogs with spontaneous thyroiditis.

B. Testis

Immunological study of the reproductive system has been and continues to be an area of intensive research, owing largely to the interests in unexplained human infertility and in the development of contraceptive procedures. The present discussion will restrict itself to reviewing studies on testicular antigens. For information on antigens of the male accessory glands or the female reproductive system, the reader is directed to the references cited below.

Perhaps more than any other tissue, the testis contains antigens to which the host's immune system normally has had little exposure. Many differentiation antigens of the testis do not even exist at the time of maturation of the immune system. Furthermore, the most differentiated elements of the testis are separated from the immune system by one of the body's most effective blood-tissue barriers. This barrier, formed by the Sertoli cells of the seminiferous tissues, lies between the spermatocytes and the more mature spermatids.

Normally, no immune response is produced to endogenous testicular antigens. As might be expected, however, little tolerance to these antigens exists, and antigenic stimulation readily induces an immune response. Because of the inaccessibility of the target antigens in the testis to the effector cells, testicular integrity can coexist with a sensitized immune system. However, the products of the testis, the spermatozoa, may lose this

protection after leaving the seminiferous tubules. Finally, if the immune response is strong enough, and/or there is a weakness in the blood-testis barrier, autoimmune disease of the testis itself may result.

The lack of tolerance to testicular antigens was demonstrated by the existence of *natural antibodies* in immature children, a group who should not have been exposed to testicular antigens. IF studies demonstrated at least eight distinct antibody specificities in these sera, directed primarily to intracellular antigens. Only one of these, AC-1, was not strictly tissue specific, as it was absorbed by adrenal homogenates. It was felt that these antigens, although not present in other mammalian tissues, were shared by microorganisms that induced the responses.

A small proportion of adult males produces antibodies to cell-surface antigens of sperm. Such autoantibodies were detected by agglutination or immobilization tests with spermatozoa, and were almost invariably directed to tissue-specific antigens. The incidence of these antibodies was always higher in males with unexplained infertility than in normal males. Serum titers of antibodies to sperm cell surface correlated with infertility, and serum titers were directly proportional to seminal plasma titers in infertile males. Infertility in these cases was probably due to the reaction of specific antibody in the seminal plasma with mature sperm. Similar antibodies were shown in vitro to interfere with many aspects of fertility, including penetration of spermatozoa into cervical mucus and, in guinea pigs, interference with the acrosomal reaction, attachment to and penetration of the zona pellucida, and attachment to zona-free ova.

Ready induction of autoantibody response was demonstrated by the high (50%) incidence of antisperm antibodies in vasectomized men. The fate of sperm in vasectomized men is uncertain. Presumably, the sperm degenerate and soluble antigens manage to leak out of the testis in some individuals. Phagocytosis by epithelial cells or migrating cells may be responsible for transport of immunogenic material to the immune system.

It is important to point out that whereas in infertile males titers of antisperm antibodies in serum were directly proportional to those in seminal plasma in vasectomized males seminal plasma antibodies were rare. The presence of seminal plasma antibodies in a vasectomized male may interfere with fertility after an otherwise successful re-anastomosis.

Although vasectomy of experimental animals has led to circulating immune complexes, in situ formation of immune complexes, testicular inflammation, and tissue destruction (orchitis), none of these complications have been convincingly demonstrated following vasectomy in humans. Intensive investigations along these lines are continuing.

The injection of testicular material in FCA is followed by the development of experimental allergic orchitis (EAO) in guinea pig, rat, and monkey. An excellent description of the induction, histopathology, and proposed mechanisms of disease has been presented elsewhere (Tung, 1981). EAO eventually involves degeneration of germinal epithelium, aspermatogenesis, and interstitial fibrosis. The induction of testicular lesions by the injection of normal animals indicated that sequestration of testicular antigens is not complete and is ineffective in the face of a strong immune response.

A large number of distinct sperm-specific antigens have been reported. The antigens that are most likely to be involved in infertility are the plasma membrane antigens. These antigens are difficult to isolate in immunogenic form and are relatively poorly characterized.

The T antigen is an aspermatogenic membrane component which induces EAO. T is insoluble in water, but may be solubilized in 8 M urea. It is thermolabile and sensitive to

5% trichloroacetic acid (TCA) as well as periodate. It is a lipoprotein or glycolipoprotein and probably requires some lipid for its antigenicity.

RSA-1 is an integral sialoglycoprotein of the plasma membrane of rabbit spermatocytes and of the membrane of more differentiated cells of the spermatozoan linage. It is most concentrated in the postacrosomal and midpiece regions. RSA-1 has a molecular weight of 13,000. Sialic acid is necessary for its antigenicity.

There are four sperm-specific antigenic enzymes, the functions of which have been at least partially studied. Lactate dehydrogenase-X (LDH-X) is found in primary spermatocytes and more differentiated derivatives. In spermatozoa, it is found in the midpiece. It is autoimmunogenic, and immunization of either male or female animals results in decreased fertility. LDH-X differs from LDH in other tissues in the composition of its four subunits. It has a molecular weight of about 140,000. Similar to LDH, LDH-X functions as an enzyme in the conversion of lactate to pyruvate. No testis-specific function is known.

Acrosin is found in the acrosome, probably on the inner acrosomal membrane. The function of this hydrolytic enzyme is uncertain, but it may be involved in the penetration of the zona pellucida. By means of double diffusion gel precipitation, acrosin was found in testis and ovary but not in any other tissue. Previous reports claimed reduced fertility in animals injected with acrosin, however, recent studies with purified preparations failed to confirm this observation.

Hyaluronidase is present in the sperm acrosome. By double-diffusion gel precipitation it was found only in sperm. It was also species restricted, with the exception of a cross-reaction between closely related species such as bull and ram (Morton, 1977). Enzymatic activity of hyaluronidase was blocked by antiserum. The function of hyaluronidase is uncertain, but it may be involved in the penetration of the cumulus cells, the zona pellucida, or the cervical mucus.

There appears to be a sperm-specific DNA polymerase. Antibodies in the sera of some infertile males inhibit the activity only of DNA polymerase isolated from sperm.

P is a sperm-specific acrosomal antigen. It is aspermatogenic and induces EAO. P is water soluble, thermostable, insoluble in 5% TCA, and is resistant to protease, lipase, ribonuclease, and periodate. It is a protein with a low content of carbohydrates.

S is another sperm-specific acrosomal antigen. It is aspermatogenic and can induce EAO. S is soluble in water and 5% TCA, thermostable, and resistant to lipase and ribonuclease. It is sensitive to protease and periodate. S is a glycoprotein rich in carbohydrates.

A is a thermolabile sperm-specific antigen. It is soluble in water and in 5% TCA.

GP1, GP2, and GP4 are acrosomal antigens isolated from the guinea pig. GP1 and GP4 are cross-reacting antigens which induce EAO. GP2 induces antibodies but not EAO. All are glycoproteins.

Protamine is a sperm-specific nuclear antigen. Antibodies to human protamine do not cross-react with histones, but do cross-react with salmon sperm protamine. Antiprotamine antibodies have not been found in normal males, but have been found in vasectomized males and males with unexplained infertility. Protamines are strongly basic, with very high arginine and cystein content but with no lysine. There are two forms of human protamine. Protamine 1 is composed of 47 amino acids and has a molecular weight 6300. Protamine 2 is composed of 51 amino acids and has a molecular weight of 7700. Protamines may function to stabilize the unusually compact sperm nucleus by the formation of disulfide bonds.

The mouse has been shown to possess a sperm-specific cytochrome *c*. Of the 104 amino acids, 13 differ from somatic cytochrome *c*. It is bound to the inner mitochondrial membrane of primary spermatocytes and more differentiated derivatives.

C. Brain

Tissue specificity of the brain has been studied for over half a century. Included in this section are some antigens that are shared with the peripheral nervous system (PNS), as well as those restricted to the central nervous system (CNS). Brain-specific proteins, glycoproteins, and glycolipids have been described. The antigen responsible for the induction of experimental allergic encephalomyelitis (EAE) is perhaps the best studied of all disease-associated autoantigens. Finally, several human neurological diseases probably involve brain-specific antigens.

Generally, the brain-specific antigens demonstrate wide interspecies cross-reactivity. One may suspect that such cross-reactions represent conservation of structures necessary for normal neural function. Unfortunately, the function of most brain-specific antigens remains unknown.

Two brain-specific antigens were described in 1928 by Witebsky and Steinfeld, using heterologous rabbit antisera to aqueous brain suspensions. One of these antigens was a thermolabile protein, shown by CFT to be both species- and tissue-specific. The other antigen was a thermostable lipid present in the brain of a variety of species, including rabbit. This ethanol-soluble antigen was later shown to be a glycosphingolipid, galactocerebroside (reviewed by Shulman, 1974). The ethanol solubility was due to the lipid moiety, but antigenicity depended on the carbohydrate. Even glucocerebroside did not react with antibodies to the galactocerebroside. Galactocerebroside is found in CNS and PNS myelin. IF studies have demonstrated galactocerebroside on cultured oligodendrocytes and short-term cultures of Schwann cells, making this antigen useful as a cell marker in cultures of nerve tissues.

The gangliosides constitute another group of brain-specific glycosphingolipid antigens. There are about 10 types of gangliosides in normal human brain, differing from each other in the sialic acid content. One of the normally minor gangliosides comprises 90% of the gangliosides in Tay-Sachs disease. Double-diffusion gel precipitation tests have been used to distinguish normal gangliosides from those extracted from tissues of patients with Tay-Sachs disease.

A saline-soluble brain antigen which is resistant to boiling and precipitable in 70% ethanol (BE) was described by Milgrom et al. (1964b). Similar tissue-specific BE antigens have been described in eight other tissues. Brain BE was found in highest concentration in the brain gray matter, with low levels found in the spinal cord. Sciatic nerve was negative. The antigen was not species specific. Brain BE was destroyed by chymotrypsin but not by periodate, although it has been shown to be a glycoprotein. Perhaps the molecule's tissue-specific determinants reside in the protein portion and the carbohydrate contributes to its marked thermostability.

S-100 is an acidic protein, so named because of its solubility in 100% saturated ammonium sulfate. Antibodies to S-100 cross-react with this protein isolated from mammals, fish, birds, and reptiles. S-100 is found throughout the CNS and PNS, with highest levels in the cerebellum and white matter. IF studies have indicated that S-100 is located in glial cells, synaptosomes, and neuronal nuclei. S-100 is a protein composed of three nonidentical subunits, each with a molecular weight of about 7000. S-100 is

thermolabile unless it is stabilized with certain ions, notably calcium. Suggested functions include calcium binding and genomic regulation.

A brain-specific antigen with known function is antigen 14-3-2. This antigen is present in the cytoplasm of some, but not all, neurons of the CNS and PNS. Antibodies to 14-3-2 cross-react with preparations from all mammals and birds tested. It is composed of two 39,000-molecular weight subunits, in the arrangement $\gamma\gamma$ or $\gamma\alpha$. The neural specificity resides in the γ subunit. Antigen 14-3-2 possesses enolase activity, converting 2-phosphoglycerate to phosphoenol pyruvate. Enolase in nonneural tissues is composed of two α subunits. No nerve-specific function is known for the γ-containing variety.

Other nerve-specific antigens include glial fibrillary acid protein, α-2 glycoprotein, GP350 sialoglycoprotein, synaptin (glycoprotein), D1 (protein), D2, D3, P-400, NS-1 (membrane protein), ran-1 (cell-surface protein), and central nervous system basic protein (CNSBP). Most of these will not be discussed further here, and interested readers are directed to the review by Bock (1978). One could not, however, leave the subject of brain-specific antigens without a discussion of the fascinating basic protein antigen, CNSBP.

CNSBP, which is responsible for the induction of EAE, is present in the myelin of the CNS and PNS (where it is generally referred to as P1). A multitude of reports has been published on the induction and immunoregulation of EAE (see reviews by Hashim, 1978, and Paterson et al., 1981). Typically, EAE is induced by a single injection of brain (or spinal cord) homogenates or purified CNSBP with FCA. The disease generally follows an acute, fatal course involving cellular infiltration and demyelination of the CNS; the PNS is generally spared. EAE closely resembles acute disseminated encephalomyelitis. A chronic, relapsing form occurs in strain 13 guinea pigs and has been likened to multiple sclerosis. Cell-mediated immunity is the first to be detected in EAE, and continues to dominate over the humoral response. The pathology is probably a result of this cell-mediated response.

Immunochemical studies have been very important in understanding the induction and immunoregulation of EAE. The amino acid sequences of CNSBP prepared from several species are known. Considerable conservation has been maintained and over 90% of the 170 amino acids of bovine and human CNSBP are identical. CNSBP has a molecular weight of 18,200 and an isoelectric point of 10.5. It contains high amounts of arsinine, lysine, and histidine, one tryptophan per molecule, and no sulfhydryl groups.

An important feature of this molecule is its open configuration in solution and, very likely, in vivo (Hashim, 1978). The primary structure is of overriding importance in the antigenicity of CNSBP. In fact, complete encephalitogenicity is maintained by certain crucial peptide fragments of CNSBP. The particular fragment(s) involved depend(s) on the species being immunized. For guinea pigs, a fragment consisting of only nine amino acid residues (numbers 114 to 122) retains full encephalitogenicity. The amino acid sequence of this fragment is H-Phe-Ser-Trp-Gly-Ala-Glu-Gly-Gln-(Lys or Ars)-OH. Except for the indicated substitution of arsinine for lysine, each amino acid substitution studied has disturbed the encephalitogenicity of the fragment. Similar minimum essential amino acid sequences have been determined for the Lewis rat, rabbits, and monkeys. None of these sequences overlapped, but all contained lysine and/or arsinine, which are basic amino acids with long side chains.

The immunoregulation of the induction and maintenance of the disease process in EAE is an area of intensive investigation. Injection of high doses of CNSBP without FCA

renders the animal resistant to subsequent challenges with CNSBP in FCA. In fact, CNSBP injection with FCA can even suppress ongoing EAE. Certain nonencephalitogenic agents can also specifically protect against EAE. Among these agents are nonencephalitogenic fragments of CNSBP, altered CNSBP or its encephalitogenic fragments, and certain synthetic random copolymers of basic amino acids. Since encephalitogenicity does not depend on secondary or tertiary structure, small peptides have been studied extensively to determine the biochemical basis of this immunoregulation.

It has been found that EAE does not occur without a delayed-type hypersensitivity (DTH) to the particular fragments which are encephalitogenic for the given species. However, DTH to nonencephalitogenic regions of CNSBP can be produced without the induction of disease. One explanation for this has been that some regions of CNSBP which are exposed in the immunizing and DTH testing antigens are either inaccessible or are present in a structurally different form in intact myelin.

Interestingly, some of these fragments, which were not in themselves encephalitogenic but which were reactive in DTH tests, retained the ability to react in DTH tests in animals immunized with the minimum encephalitogenic amino acid sequences. Apparently, although incapable of inducing a response to encephalitogenic determinants (if they could, they would presumably be encephalitogenic themselves), these fragments could cross-react with the reactive sites of the effector cells. These peptides were very effective in protecting against or suppressing EAE. If large doses of these peptides were administered at early stages of the disease, complete reversal of symptoms and permanent protection from future encephalitogenic injections resulted. It has been suggested that these fragments bind directly to the effector cells, causing their irreversible inactivation (Paterson et al., 1981).

Other interesting studies have involved the use of cop-1. This is a synthetic random copolymer of L-Ala, L-Glu, L-Lys, and L-Try in a residue molar ratio of 6.0:1.9:4.7:1.0, and an average molecular weight of 23,000. Cop-1 is not a general immunosuppressant, is nonencephalitogenic, and does not react with antibodies to CNSBP. However, it does cross-react with DTH determinants, shown by its activity in DTH tests performed on animals immunized with encephalitogenic preparations. Cop-1 is active in suppressing ongoing acute and chronic relapsing forms of EAE. Interest has been generated in the use of such nonencephalitogenic preparations in the suppression of human demyelinating diseases.

III. TUMOR SPECIFICITY

Tumor antigens are characteristic for tumor tissue and thus may be considered tissue specific. Tumor antigens result from alterations that frequently occur in the malignant transformation of normal tissue. The alteration that occurs may be quantitative in that a particular normal antigen may decrease or increase in concentration. Such normal antigens that have increased concentration in tumors are termed *tumor-associated antigens* (TAA). Antigenic alteration also may be qualitative in that a new antigen, foreign to the host, may appear. These are termed *tumor-specific antigens* (TSA) and may be present as new cell-surface structures or as new intracellular structures in the cytoplasm or nucleus.

Most information concerning tumor antigens has been obtained in studies on experimental animals using tumors induced by chemical agents, physical agents, or

viruses. These agents include polycyclic hydrocarbon, x-irradiation, DNA viruses, and RNA viruses. This discussion will focus primarily on murine tumors, since much of our present knowledge has been obtained in that area.

TSA were first clearly demonstrated in mice that had been immunized with a methylcholanthrene-induced sarcoma taken from syngeneic mice. The immunized mice rejected subsequent inoculation of the same tumor but would accept grafts of normal skin from the tumor donors (Prehn and Main, 1957). TSA detectable by rejection of transplants are cell-surface antigens and are known as tumor-specific transplantation antigens (TSTA).

Characteristically, TSTA induced by chemical or physical agents show individual specificity. For example, two distinct tumors, A and B, induced by the same chemical carcinogen possess TSTA that are unique for each tumor, even though the tumors are induced in the same animal. Syngeneic animals immunized with A are resistant to A, but not to B, and animals immunized with B are resistant to B, but not to A.

Chemically induced tumors may also have cell-surface TSA that are detectable by serologic techniques such as cytotoxicity or immunofluorescence. These antigens are termed *tumor-specific cell-surface antigens* (TSCSA). This term is used here to denote those cell-surface antigens that are detected primarily by serologic procedures and whose role in tumor resistance is not known. Evidence suggests that TSCSA and TSTA on chemically induced tumors are not identical determinants. Finally, chemically induced tumors may bear antigens on the cell surface that are present on normal embryonic tissue but not in normal adult tissue. Embryonic antigens are detectable by both transplantation tests and by serologic procedures. These antigens appear to be distinct from TSTA and TSCSA. Little information is available on the physicochemical nature of these cell-surface antigens. Similar to histocompatibility antigens (see Chapter 27), the tumor antigens described above are glycoproteins. They may be solubilized by treatment of tumor cells with proteolytic enzymes or 3 M KCl. The active fractions exhibit considerable variation in molecular size and surface charge.

Malignant cells induced by viruses exhibit several types of tumor antigens. These antigens are usually virus specific, not individual specific; they are shared by tumors induced by the same virus, even though the tumors are histologically different and even though the tumors are induced in animals of different species.

DNA viruses that have oncogenic properties are the papovaviruses, which include the papilloma, polyoma, and vesiculating viruses; the adenoviruses; and the herpesviruses. DNA viruses often are not oncogenic in their natural hosts, but may transform cells and produce tumors in other species.

Similar to chemically induced tumors, tumors induced by DNA viruses bear TSTA, TSCSA, and embryonic antigens. The relationship of these antigens to each other is not clear, but it appears that they are distinct determinants. TSTA, but not TSCSA or embryonic antigens, induce tumor resistance. Little information is available concerning the biochemical properties of these determinants; however, they are probably glycoproteins.

In addition to cell-surface TSA, tumors induced by DNA viruses possess TSA located in the nucleus. These TSA, termed T antigen and U antigen, are virus-specified proteins but are not part of the virion. T antigen has a molecular weight of 70,000 and is thought to be involved in viral DNA synthesis. Neither T nor U antigen plays a role in tumor rejection.

The oncogenic RNA viruses (oncornaviruses), unlike the DNA viruses, produce tumors in their natural hosts. Oncornaviruses produce leukemia or sarcoma in a variety of species, including birds (Rous sarcoma virus, avian leukosis virus), rodents (murine leukemia viruses, mouse mammary tumor virus), cats (feline leukemia virus), and even in higher animals such as monkeys and apes. Many oncornaviruses are endogenous. They are part of the host's genome and are transmitted by germ cells. Expression of the viruses can be induced by chemical or physical agents or may occur spontaneously. Much of the knowledge concerning oncornaviruses has been obtained in studies on the murine leukemia viruses (MuLV; Gross, Graffi, Friend, Maloney, and Rauscher viruses).

Tumor cells infected with MuLV exhibit a variety of cell-surface and internal antigens. Foremost among these are the virus envelope antigens (VEA). These antigens are specified by the viral genome and are produced by the host cell as part of its cell membrane. VEA are detectable by serological as well as transplantation tests, and they elicit tumor resistance. It should be stressed that VEA are viral antigens produced by the host cell. The major component of VEA, termed gp70, is a glycoprotein of molecular weight 70,000, consisting of 32% carbohydrate and 68% protein. Several specificities have been associated with gp70: a specificity shared by all MuLV, specificities that are shared by some MuLV, and specificities that are characteristic for individual MuLV.

Internal viral components can also be detected serologically in cells infected with MuLV. These polypeptides, termed p10, p15, and p30 (corresponding to molecular weights of 10,000, 15,000, and 30,000), are major core proteins of the virion. Group-specific and type-specific viral determinants have been described on these polypeptides.

Antigens that are virus-related but are not part of the virion have also been described on the surface of infected cells. The best known example of this type of TSA are the Gross cell-surface antigens (GCSA), found on the cell membrane of all cells infected by the Gross virus. Little biochemical information is available concerning these determinants; however, it has been shown that papain digests of plasma membranes of infected cells contained two major components with molecular weights of 60,000 and 45,000. Antigenic activity was destroyed by heat, repeated freezing and thawing, and even storage at 4°C.

Both VEA and the virus-related cell-surface antigens are detectable by serologic techniques, such as cytotoxicity and immunofluorescence, and by rejection of transplanted tumor cells. Several of the antigens associated with RNA tumor viruses, including GCSA and VEA, appear in soluble form in the circulation and as antigen–antibody complexes in the kidney.

A group of TSA of particular interest are those that may also occur as normal alloantigens. The best known example is the thymus-leukemia (TL) antigen system, described by Old and his co-workers in 1963 (for a review, see Old and Stockert, 1977). TL antigens are expressed on the normal thymocytes of some murine strains (TL positive), but not others (TL negative). Significantly, leukemia cells of TL-negative mice may express TL antigens.

The phenomenon of antigenic modulation was first described in this antigen system. TL-positive leukemia cells become TL-negative in the presence of anti-TL antibodies. Growth of the modulated leukemia cells in an environment lacking anti-TL antibodies leads to reversion of the leukemia cells to the TL-positive state. Subsequently, antigenic modulation has been observed in other antigens. This phenomenon has been suggested as a mechanism by which tumor cells may escape immunological surveillance.

The structure of the TL antigens appears to be similar to that of H-2 antigens (see Chapter 27). It consists of a glycoprotein heavy chain of molecular weight 45,000 and a noncovalently linked protein light chain of molecular weight 12,000. The light chain is probably the murine β_2-microglobulin. TL antigen is bound to the plasma membrane by the C-terminal portion of the heavy chain. It is not clear whether TL antigen exists in the native state as the dimer described above or as a tetramer composed of two heavy and two light chains.

Other examples of normal alloantigens appearing as TSA have been noted. These include the G_{IX} antigen, which is a tissue-restricted antigen present on thymocytes and spermatocytes of some (G_{IX}+) but not of other strains of mice (G_{IX}−). Similar to TL antigens, G_{IX} may be present on leukemia cells of both G_{IX}+ and G_{IX}− mice. It has been shown that the G_{IX} antigen is similar to or identical with gp70 of MuLV (see above). This has provided evidence to support the idea that viral genes can be incorporated into the mammalian genome.

Of particular interest are the recent observations that tumor cells may bear H-2-like specificities that are absent from normal cells of the tumor-bearing animal. For example, it has been shown that immunization of mice with normal tissue from an H-2-incompatible strain rendered the mice resistant to a syngeneic tumor. These studies have led to the proposal that somatic cells of mice bear structural genes for all H-2 specificities, and that expression of a particular specificity is encoded by regulatory genes. A similar concept has been suggested for TL antigens.

TSA in human malignancies have been especially difficult to demonstrate. This is probably due, in large measure, to the genetic heterogeneity of man. Several TAA have been described in recent years. Some have proven to be useful in the diagnosis and management of malignant diseases. These include enzymes such as alkaline phosphatase, and hormones such as human chorionic gonadotropin, which can be an ectopic secretion of certain tumors and is useful in the diagnosis of those malignancies.

Much interest has been focused recently on a group of TAA that are classed as oncofetal antigens. These include pancreatic oncofetal antigen, which appears to be associated with pancreatic malignancies; α-fetoprotein, which is associated with primary hepatoma; and carcinoembryonic antigen (CEA), which is associated principally with gastrointestinal malignancies. This discussion will be limited to the last two antigens mentioned.

Alphafetoprotein (AFP) was first described in the sera of newborn mice and subsequently was demonstrated in other animals, including man. AFP is a serum protein that is found in high concentration in fetal blood (2 to 3 mg/ml in human blood in week 13 of gestation) and in very low concentrations in normal adult blood ($\leqslant$5 ng/ml in human blood).

The relationship of AFP to malignancy was first shown in mice by Abelev (1963), who described AFP in the sera of adult mice bearing primary hepatomas. Later, Tatarinov (1964) demonstrated AFP in the sera of adult human beings with primary hepatoma (for a review, see Sell, 1980). Subsequently, it has been shown that elevated levels of AFP are seen in a significant number of patients with primary hepatoma (70 to 90%) and germ cell tumors of gonadal, retroperitoneal, or mediastinal origin (50 to 70%). However, it should be noted that high levels of AFP are also seen in a variety of nonneoplastic liver diseases.

In addition to diagnosis of certain malignant diseases, determination of AFP levels is most valuable in the diagnosis of developmental abnormalities, such as congenital nephrosis and open spina bifida.

AFP is similar to albumin in physicochemical properties. Indeed, the amino acid composition and sequence of the two molecules are so similar that AFP has been proposed as a precursor molecule of albumin. AFP is a glycoprotein of molecular weight 70,000, with an isoelectric point of 4.7, and the electrophoretic mobility of an alpha globulin. Unlike albumin, AFP contains up to 4% carbohydrate, which is reflected in its electrophoretic heterogeneity. Rat AFP forms two bands on electrophoresis, and two subfractions can be prepared on the basis of binding to concanavalin A. Mouse AFP can be separated into five components by SDS polyacrylamide gel (see Chap. 13) electrophoresis, due to variations in the amount of sialic acid residues in the molecule. Little is known concerning the biological significance of these variants.

CEA was described originally by Gold and Freedman (1964) as an antigen present only in adenocarcinomas of the human intestinal tract and embryonic tissue (for a review, see Sell, 1980). Later studies showed that CEA is found in nonintestinal tumors and in very low concentrations in nonmalignant pathologic tissue, and even in normal adult tissue. Originally, CEA was detected in gel precipitation tests by properly absorbed antisera. Presently, it is studied most often by radioimmunoassay and enzyme immunoassays. By these techniques the antigen is detectable in nanogram quantities in patients' sera.

Significantly elevated levels of CEA are seen in patients with a variety of malignancies as well as in patients with noncancerous diseases, such as chronic lung disease, pancreatitis, and cirrhosis of the liver. Levels above normal are also seen in a significant number of healthy individuals, most of whom are heavy smokers.

Tests for CEA are useful in the prognosis of cancer and in monitoring the effectiveness of cancer therapy. They should not be used for screening or for confirming the diagnosis of malignancy.

CEA is a glycoprotein found in the glycocalyx of the cell surface. It is extractable from tissue by perchloric acid or even physiologic saline. CEA has a molecular weight of approximately 200,000 daltons and has the electrophoretic mobility of a beta globulin. It has a sedimentation constant of 6.2 to 6.8 and an isoelectric point of 3 to 4. CEA contains 50 to 60% carbohydrate by weight, and it exhibits considerable heterogeneity with respect to charge and molecular size. It also gives a broad profile on SDS polyacrylamide gel electrophoresis.

The use of different isolation procedures has resulted in preparations of CEA that differ mainly in carbohydrate content. Interestingly, these CEA variants are remarkably constant in amino acid composition, which suggests that CEA consists of a uniform polypeptide chain. Evidence to support this comes from studies showing the constancy of NM_2-terminal sequence for CEA isolated by a variety of procedures.

Several CEA-related antigens have been extracted from normal and malignant tissue. Some of these give lines of identity with CEA in gel precipitation and others give lines of partial identity. Extensive studies on these antigens show that they differ slightly from CEA in molecular weight, amino acid and carbohydrate composition, degree of glycosylation, and tissue distribution.

It is clear that CEA and related antigens are a complex and heterogeneous group of molecules. Identification of a definite tumor-specific marker is yet to be realized.

Convincing evidence for cell-surface TSA on human tumors has been difficult to obtain. Best results have been studied on malignant melanoma, osteogenic sarcoma, bladder carcinoma, and Burkitt's lymphoma. More recently, TAA have been demonstrated in leukemia cells from patients with acute lymphatic leukemia. There is little information on the biochemical nature of these determinants.

REFERENCES

Abelev, G. I., Perova, S. D., Khramokova, N. I., and Irin, I. S. (1963). *Transplant. 1*, 1974.

Abeyounis, C. J., and Milgrom, F. (1983). In *Immunology of Transplantation, Cancer and Pregnancy* (P. K. Ray, Ed.). Pergamon Press, Elmsford, N.Y.

Bock, E. (1978). J. Neurochem. *30*, 7.

Chan, M. M., Kano, K., Dorman, B., Ruddle, F. H., and Milgrom, F. (1979). Immunogenetics *8*, 265.

Hashim, G. A. (1978). Immunol. Rev. *39*, 60.

Hektoen, L., Fox, H., and Shulhof, K. (1927). J. Infect. Dis. *40*, 641.

Kano, K., and Milgrom, F. (1977). Curr. Top. Microbiol. Immunol. *77*, 43.

Landsteiner, K. (1946). *The Specificity of Serological Reactions*. Harvard University Press, Cambridge, Mass.

Milgrom, F., Kano, K., Barron, A. L., and Witebsky, E. (1964a). J. Immunol. *92*, 8.

Milgrom, F., Tuggac, M., Campbell, W. A., and Witebsky, E. (1964b). J. Immunol. *92*, 82.

Morton, D. B. (1977). In *Clinical and Experimental Immunoreproduction*, Vol. IV: *Immunobiology of Gametes* (M. Edidin, and M. H. Johnson, Eds.). Cambridge University Press, New York, pp. 115–155.

Nisonoff, A., Reichlin, M., and Margoliash, E. (1970). J. Biol. Chem. *245*, 940.

Nuttall, G. H. F. (1904). *Blood Immunity and Blood Relationship*. Cambridge University Press, Cambridge.

Old, L. J., and Stockert, E. (1977). Annu. Rev. Genet. *17*, 127.

O'Rand, M. G. (1980). In *Immunological Aspects of Infertility and Fertility Regulation* (D. S. Dhinda, and G. F. B. Schumaker, Eds.). Elsevier/North-Holland, New York, pp. 33–91.

Paterson, P. Y., Day, E. D., and Whitacre, C. C. (1981). Immunol. Rev. *55*, 89.

Pinchera, A., Fenzi, G. F., Bartalena, L., Chioriato, L., Morcocci, C., and Baschieri, L. (1980). In *Proceedings of the Serono Symposia*, Vol. 33: *Autoimmune Aspects of Endocrine Disorders* (A. Pinchera, D. Doniach, G. F. Fenzi, and L. Bachiere, Eds.). Academic Press, New York, pp. 57–72.

Prehn, R. T., and Main, J. M. (1957). J. Natl. Cancer Inst. *18*, 769.

Sell, S. (1980). *Cancer Markers*. Humana Press, Clifton, N.J.

Shulman, S. (1974). *Tissue Specificity and Autoimmunity*. Springer-Verlag, New York.

Shulman, S., and Witebsky, E. (1960). J. Immunol. *85*, 559.

Tung, K. S. K. (1980). In *Immunological Aspects of Infertility and Fertility Regulation* (D. S. Dhinda, and G. F. B. Schumaker, Eds.). Elsevier/North-Holland, New York, pp. 33–91.

Tung, K. S. K., Teuscher, C., and Ming, A. L. (1981). Immunol. Rev. *55*, 217.

Uhlenhuth, P. T. (1903). In *Festschrift zum Sechzigsten Geburstag von Robert Koch*. Fischer, Jena, pp. 49–74.

Weigle, W. O. (1980). Adv. Immunol. *30*, 159.

Witebsky, E., and Steinfeld, J. (1928). Z. Immunitaetsforsch. *58*, 271.

5

Immunochemistry of Bacterial Antigens

Milton R. J. Salton New York University School of Medicine, New York, New York

I. INTRODUCTION

Interest in the chemical nature of bacterial antigens stems from the early recognition of immunity to certain bacterial pathogens, the detection of antibodies in convalescing patients' sera, and the development of vaccines and in vitro reactions of antibodies with bacterial components for diagnostic and serologic classification systems. Much of our early knowledge of the chemistry of the bacterial cell came from the immunochemical studies of surface antigens, and the classic work of Heidelberger and his colleagues on the pneumococcal capsular polysaccharides established the enormous potential of immunochemical analysis, which is now being realized in almost every conceivable area of biological research.

Apart from the capsular polysaccharides, the flagella (H) and O-somatic antigens, there had been little knowledge of the anatomical origins of many bacterial antigens, and this aspect of the immunochemistry of the bacterial cell had to await the development of the electron microscope and suitable procedures for the subcellular fractionation of microorganisms. With the early focus on the principal surface antigens (capsular, flagella, and somatic) the extent of the antigenic complexity of the bacterial cell was not fully appreciated. The ability to separate bacterial cells into their discrete anatomical entities (e.g., cell walls, membranes, flagella, pili) and establish their structural homogeneity in the electron microscope led to their chemical characterization and rapidly changed the impression of the antigenic simplicity of bacteria. Indeed, it has only been relatively recently that the application of the high-resolution, two-dimensional, or crossed immunoelectrophoresis (CIE) method of antigen analysis (see Chap. 17) has given us a much clearer picture of the antigenic complexity of the major cellular compartments of bacteria, and aspects of this will be discussed in later sections.

The bacterial cell is a complex structure possessing a variety of discrete anatomical entities, including surface appendages such as flagella and fimbriae (pili), capsules, cell walls, outer membranes of Gram-negative bacteria, plasma or cytoplasmic membranes, and in some species spores and specialized intracellular membranous structures such as chromatophores and cytomembranes of nitrifying bacteria (Salton, 1978). These major structures of the bacterial cell may be composed of one or more molecular species, the primary building blocks of which may have the potential to act as determinant groups in the antibody responses elicited by natural infection or by immunization with whole bacteria or subcellular fractions. As immunogens, therefore, the bacterial cell contains an enormous variety of antigens, not only those of the peripheral structures but also the periplasmic proteins of Gram-negative organisms and the array of enzymes, ribosomes, and DNA of the cytosol or cytoplasmic compartment of the cell.

The expression, as antigens, of the large number of potentially immunogenic components in the bacterial cell depends on a multiplicity of factors. The state of morphological integrity of the cell obviously is of prime importance in determining whether the initial responses will be primarily to the major surface components (e.g., capsules, O antigens, wall peptidoglycan, and teichoic acids). Moreover, the physical state of the components is of considerable importance, as for example the difference in the strong immunogenicity of the pneumococcal polysaccharides and lipopolysaccharides (LPSs) in situ on the bacterial cell surfaces versus the poor antibody responses seen with the purified, soluble substances in some animal species. Similar differences have also been seen with the membrane-bound succinylated lipomannan versus the soluble form of this amphiphile (Owen and Salton, 1977). Thus marked differences in immunogenicity of individual components can be expected for the particle-bound state compared to the free, soluble form of a given antigen, and such factors have to be taken into account when the array of antigens in different cellular compartments is being studied.

In the immunochemical analysis of the bacterial cell it is important to realize that it cannot be assumed that antibody responses will be limited to the surface components when whole organisms are used as immunogen. Autolysis occurring in vivo during the course of immunization can result in the "expression" of antigens of the cytoplasmic compartment of the cell, as demonstrated with the gonococci by Smyth and Salton, 1977. Thus immunization of rabbits with gonococcal cells of colony types T3 and T4, which are more prone than other types to lysis, gave rise to antibodies to many more cytoplasmic antigens than was observed on immunization with the more autolysis resistant T1 and T2 colony types (Smyth and Salton, 1977). It would appear likely, therefore, that in the course of bacterial infections, antibodies may also be produced to cytoplasmic as well as surface components of the cell and that detection of these antibody responses may have diagnostic significance and utility.

A vast amount of knowledge and a voluminous literature have accumulated on the immunochemistry of bacterial antigens (Kwapinski, 1969), and any review of this topic will of necessity be highly selective. For a variety of reasons, in many of the earlier studies on the characterization of bacterial antigens, neither the cellular location nor the homogeneity of the preparations have been adequately established. As emphasized in this introduction, the bacterial cell possesses a sizable array of potentially immunogenic molecular species, including proteins, nucleic acids, polysaccharides, glycolipids, lipids, and the unique prokaryotic macromolecules, peptidoglycans, lipopolysaccharides, lipoproteins, teichoic acids, and the membrane amphiphiles (e.g., lipoteichoic acids and lipomannans). This contribution will accordingly be limited to selected examples, emphasizing the chemical nature and cellular locations of various bacterial antigens.

II. CHEMICAL ANATOMY OF THE BACTERIAL CELL

We now have an extensive knowledge of the chemical composition and structure of the major anatomical components of many Gram-positive and Gram-negative bacteria. Tables 1 and 2 summarize, respectively, the major classes of chemical components found in the various structures and compartments of the bacterial cell. Of all the surface components, lipopolysaccharides and the capsular polysaccharides, especially those of the pneumococci, are the most extensively characterized by immunochemical analyses (Heidelberger, 1973). Although flagella antigens have long been used in serologic studies of members of the Enterobacteriaceae in particular, the flagellins have not been subjected to the same systematic chemical and immunochemical characterization as that seen for the pneumococcal polysaccharides. Similarly, pili can be separated into distinct serological groups (e.g., *Escherichia coli* pili, Brinton, 1965) but the chemical basis of the determinant groups or regions of the pilin protein molecules is largely unresolved at present. Other surface components which have been investigated both chemically and serologically in more recent years are the major outer membrane proteins of Gram-negative bacteria. Indeed, Johnston et al. (1976) have proposed a serological classification of *Neisseria gonorrhoeae* based on the immunological reactivity of the various major outer membrane proteins of this organism. Thus, as far as we know, most of the surface appendages and peripheral components of the bacterial cell are good immunogens and the chemical and immunological differences in these surface proteins, polysaccharides, and lipopolysaccharides can provide the basis for serological identification and classification.

As shown in Tables 1 and 2, the unique prokaryotic components are generally confined to the cell walls and membranes. Peptidoglycan is present in the walls of all Gram-positive and Gram-negative organisms with two notable exceptions, *Halobacterium* species and the mycoplasma group (Schleifer and Kandler, 1972; Kandler and Schleifer, 1980). More recently, a novel type of peptidoglycan which contains talosaminouronic acid instead of the usual amino sugar, muramic acid, was found in the cell walls of *Methanobacterium* (König and Kandler, 1979). No muramic acid has been found in *Halobacterium* species, but *H. salinarium* does contain a 194,000-molecular weight envelope glycoprotein which may perform some of the structural functions of wall peptidoglycan (Mescher and Strominger, 1976). This is one of the few reports of an authentic glycoprotein in a bacterium, and although there have been numerous suggestions of the occurrence of glycoproteins, these have usually lacked the definitive evidence such as that presented for the halobacterial component. The apparent infrequent occurrence of glycoproteins in bacteria further distinguishes them from eukaryotic microorganisms possessing carbohydrate structures in their glycoproteins similar to those found in mammalian cells (Kornfeld and Kornfeld, 1980). It is of interest to note that the structure of the *H. salinarium* glycoprotein proposed by Mescher and Strominger (1976) involved the typical linkage of the carbohydrate residue through N-acetylglucosamine and asparagine to the protein. However, the carbohydrate composed of mannose, galactose, glucose, and an unidentified N-acetylamino sugar differed from that usually seen in eukaryotic glycoproteins.

Other unique bacterial wall polymers include the peptidoglycan-associated teichoic and teichuronic acids of Gram-positive organisms, both of which possess their own distinctive immunological specificities. In addition to the teichoic acids, the immunologically important group-specific carbohydrates are attached to the peptidoglycan of the streptococcal cell wall. The only other surface structure possessing components with unusual chemical constituents and features are the LPSs, characteristic molecular complexes of

Table 1 Major Classes of Chemical Components of Surface Structures and Compartments of Gram-Positive Bacteria

Structures or compartments	Distribution	Chemical components
Surface appendages		
Flagella	*Bacillus, Clostridium* sp., and in a few cocci	Proteins (flagellins)
Fimbriae (pili)	Rare in gram-positives; Group A streptocci *Corynebacterium renale*	Type-specific M proteins Protein
Superficial layers		
Regular subunit (RS) layers	*B. polymyxa, Sporosarcina, Clostridia*	Protein subunits
Surface proteins	*Staphylococcus aureus*	Protein A
Capsules and slime	Most genera *Bacillus* spp.	Polysaccharides γ-D-Glutamyl polypeptide and polysaccharides
Cell walls	All groups and species	Peptidoglycan
	Many species have accessory polymers linked to peptidoglycan	Polysaccharides, teichoic acids, teichuronic acids, peptidoglycolipids
Cell membranes	All groups and species	Proteins, lipids, (phospholipids, glycolipids) Lipoteichoic acids or lipomannans
Cytoplasm	All groups and species	Proteins, ribosomes, nucleic acids

Table 2 Major Classes of Chemical Components of Surface Structures and Compartments of Gram-Negative Bacteria

Structures or compartments	Distribution	Chemical components
Surface appendages		
Flagella	Widely found in Enterobacteriacea and in many other groups	Proteins (flagellins)
Fimbriae (pili)	Widely distributed throughout many different genera of gram-negatives	Proteins (pilin)
Superficial layers		
Regular subunit (RS) layers	Spirilla and various groups of rods and cocci	Protein subunits; others undefined
Capsules and slime, Vi antigens	Encapsulated strains found in most groups	Polysaccharides
Cell envelopes		
Outer membranes	All groups and species	Lipopolysaccharides, major proteins (porins), proteins, and lipids
	Many species	Lipoproteins, free and peptidoglycan-associated
	Enteric bacteria	Common polysaccharide-amphiphile (ECA); LPS-linked and free
Rigid layer	All species; possible exceptions stable L forms	Peptidoglycan, lipoprotein covalently linked in some species
Cell membranes (inner membrane)	All groups and species Enterics	Proteins, lipids (phospholipids) ECA?
Cytoplasm	All groups and species	Proteins, ribosomes, nucleic acids

the outer membranes of Gram-negative bacteria (with the exception of one Gram-positive organism, *Listeria monocytogenes*, discussed further below) and the lipoteichoic acids and succinylated lipomannans of the membranes of Gram-positive organisms. Another unusual bacterial structure is the endospore, but whether any of the coat-specific antigens have unique chemical structures is not known at present. Thus the cell walls, outer membranes of Gram-negative bacteria, and plasma membranes of Gram-positive species appear to be the principal reservoirs of the chemically unique, prokaryotic macromolecular components.

A. Capsules

By light microscopy capsules appear to form a thick, viscous layer surrounding the bacterial cell, but in the electron microscope they are seen as dense coats frequently resolved into a fibrillar matrix of densely packed filaments. Most of the evidence on the physical structure of bacterial capsules and surface slime, including critical point drying and ruthenium-red staining, indicates that the capsule is made up of a network of fibers which are probably unlinked except perhaps at the peripheral zone, and such linkages could account for the well-defined edge of the capsule (Roth, 1977). The mechanism whereby the capsule is held to the cell wall surface is not known at present. The interesting recent discovery that phosphatidyl diacyl glycerol is covalently attached to the group C meningococcal capsular polysaccharide would provide hydrophobic ends to the molecules for anchoring in the outer membrane of the Gram-negative meningococcus (Fraser et al., 1982). If a similar mechanism existed for encapsulated Gram-positive organisms, the hydrophobic ends would probably have to be anchored in the plasma membrane and the fibrils extruded through the cell wall. Such a mechanism of retaining macromolecular components on the cell surface would be analogous to the lipid A of LPSs, providing the anchor for the insertion of these molecules into the outer leaflet of the outer membrane of Gram-negative bacteria. It will therefore be of great interest to see if diglycerides are found in other surface capsular substances, as has already been discovered with the membrane amphiphiles, lipoteichoic acid and lipomannan (Shockman and Wicken, 1981).

The vast majority of the capsular substances are polysaccharides composed of two or more monosaccharides and can therefore be classed as heteropolysaccharides. Most of the pneumococcal polysaccharides fall into this category and some 80 or more type-specific substances have been identified. The truly unique capsular components are the γ-D-glutamyl polypeptides of *Bacillus anthracis* and related species, and such homopolymers are quite rare. In some *Bacillus* species the capsule is formed of both the γ-glutamyl polypeptide and a separate capsular polysaccharide, and antibodies specific for each capsular substance were used by Tomcsik (1956) to demonstrate their localization within the capsule structure on the cell surface in an elegant phase-contrast microscopy study.

Both Gram-positive and Gram-negative species produce a wide variety of capsular polysaccharides. Selected examples of these are presented in Tables 3 and 4, together with examples of the more unusual polymers, including the colominic acid of *E. coli*, a homopolymer of neuraminic acid, the polysaccharides of meningococci and group B streptococci (type III) also containing neuraminic acid, together with other monosaccharides and the type B *Hemophilus influenzae* capsular substance composed of equimolar proportions of ribose and ribitol linked by phosphodiester bonds (Crisel et al., 1975). Not surprisingly, antibodies to this uncommon *H. influenzae* capsular antigen cross-react with ribitol teichoic acids of bacterial cell walls. A number of other bacteria produce copious

Table 3 Composition of a Selected Variety of Capsular and Cell Surface Exopolysaccharides of Gram-Positive Bacteria

Organism	Components
Pneumococcus	
Type II	L-Rhamnose, D-glucuronic acid, D-glucose
Type III	D-Glucuronic acid, D-glucose
Type IV	D-Galactose, D-glucose, L-rhamnose, ribitol, phosphate
Type XIV	N-Acetyl-D-glucosamine, D-galactose, D-glucose
Group A *Streptococcus*	Hyaluronic acid
Group B *Streptococcus*, type III	D-Galactose, D-glucose, N-acetyl-D-glucosamine, sialic acid
Streptococcus mutans	Dextrans
Leuconostoc mesenteroides	Dextrans, Levans
Bacillus subtilis	Levan
Staphylococcus aureus	
Type 1	N-Acetylfucosamine, N-acetylgalactosaminuronic acid
Type 2	N-Acetyl-L-alanylglucosaminuronic acid, N-acetylglucosaminuronic acid
Type 4	N-Acetylmannosaminuronic acid, N-acetyl-fucosamine

Source: Data from Heidelberger (1973), Karakawa and Vann (1982), and Sutherland (1977).

quantities of fibrillar exocellular polysaccharides, notable among which are the oral streptococci producing dextrans, *Leuconostoc mesenteroides* and its dextrans and levans, and *Azotobacter vinelandii* and pseudomonads forming alginates, copolymers of mannuronic acid and guluronic acid.

Much work has been performed on the immunodeterminants and cross-reactivities of the pneumococcal capsular polysaccharide antigens and has been extensively reviewed elsewhere (Heidelberger, 1973). Widespread cross-reactions are exhibited by a great variety of exocellular bacterial and other polysaccharides, originating from similar or identical sugar constituents, glycosidic linkages, and other common determinant substituents, such as O-acetyl and pyruvyl groups. Indeed, the high specificity and stereospecificity of antibodies have provided immunochemical means of accurately predicting and determining chemical structures of polysaccharide antigens. In addition to the role of individual sugars as immunodeterminants, recent studies with the type III polysaccharide of group B streptococci have established that the sialic acid residues are critical in the conformation of the polysaccharide and that the tertiary structure confers "native" type III polysaccharide specificity (Kasper et al., 1982). The desialated core type III antigen, composed of galactose, glucose, and N-acetylglucosamine, and the pneumococcal

Table 4 Composition of a Selected Variety of Capsular and Cell Surface Exopolysaccharides of Gram-Negative Bacteria

Organism	Components
E. coli Bos-12	Sialic acid (colominic acid)
E. coli K antigens	
K26	Glucuronic acid, galactose, rhamnose
K30	Glucuronic acid, galactose, mannose
K8	Glucuronic acid, galactosamine, glucosamine, galactose
K87	Glucuronic acid, fucosamine, glucosamine, galactose, glucose
Klebsiella	
Type 1	Fucose, glucose, glucuronic acid, pyruvate
Type 7	Mannose, glucose, galactose, glucuronic acid, pyruvate
Type 52	Rhamnose, galactose, glucuronic acid
Hemophilus influenzae, type B	Ribose, ribitol, phosphate
Neisseria meningitidis	
Type A	N-Acetyl-D-mannosamine phosphate
Type C	Sialic acid
Azotobacter vinelandii	D-Mannuronic acid, L-guluronic acid (alginate)

Source: Data from Heidelberger (1973), Fraser et al. (1982), Ørskov et al. (1977), Sutherland (1977).

XIV polysaccharide are structurally and immunologically identical but show partial identity with the native type III antigen. Thus the sialic acid of this capsular polysaccharide contributes to the immunodeterminant of the native antigen, but this sugar in itself was not immunodominant. It is suggested that hydrogen bonding between the carboxyl group on sialic acid and the hydroxyl group at C-3 of the N-acetylglucosamine residues contributes to the formation of the critical conformation of the polysaccharide molecule (Kasper et al., 1982). These important studies serve to emphasize the role of conformation in determining "native" specificities with polysaccharide antigens, and there is little doubt that further examples of this will be found.

Capsules of *Staphylococcus aureus* have been much less extensively investigated either structurally or chemically. However, Karakawa and Vann (1982) have recently proposed a preliminary classification scheme based on eight antigenically distinct capsular polysaccharides. Whether all of these staphylococcal surface polysaccharides fulfill all of the criteria of capsular structures is unclear at the present time. Indeed, it will be recalled from the interesting work of Wu and Park (1971) that no capsule was detectable by the standard India ink method on a mutant strain of *S. aureus* H which was inagglutinable by antiteichoic acid antibodies due to the production of a new surface polysaccharide antigen composed of N-acetyl-D-fucosamine and N-acetyl-D-mannosaminuronic acid. It was concluded that this antigenic polysaccharide on the outermost surface of this mutant was in the form of a microcapsule which covered the underlying wild-type ribitol teichoic acid shown to be present in the wall.

In the provisional serological grouping of the staphylococcal polysaccharides proposed by Karakawa and Vann (1982), type 1 polysaccharide was composed of N-acetylfucosamine and N-acetylgalactosaminuronic acid, type 2 was a polymer of N-acetyl-L-alanylglucosaminuronic acid and N-acetylglucosaminuronic acid, and type 4 contained N-acetylmannosaminuronic acid and N-acetylfucosamine. The latter corresponds to the mutant strain polysaccharide discovered by Wu and Park (1971). Structures for the other serological types do not appear to be available at the present time.

The discovery by Fraser et al. (1982) of the covalently linked phosphatidyl diacyl glycerol in the group C meningococcal polysaccharide is of considerable interest in that the lipid material imparts micellar behavior to the polysaccharide, but what role it may play in immunogenicity is not clear at present. It will be of interest to discover whether the lipid ends endow the polysaccharide with any adjuvant properties.

Both pneumococcal (as many as 13 types) and meningococcal (groups A and C) capsular antigens have been used in the development of suitable vaccines for protection against infection with these organisms (Austrian, 1977; Gold et al., 1977). The pneumococcal polysaccharides were tested over a wide dosage range of 5 to 1000 μg in adults, and for each type selected for the vaccine, 50 μg of the polysaccharide proved to be a suitable dose (Austrian, 1977). The meningococcal polysaccharides of groups A and C were evaluated over a concentration range of 5 to 200 μg, and results of studies indicated that the group A vaccine can be recommended for use in adults and children over 6 and group C only in adults (Gold et al., 1977). Minimum antibody concentrations to the meningococcal polysaccharides required for protection are not known, nor is the mechanism of natural immunization against both group A and C polysaccharides. The possession of identical or closely related capsular polysaccharides by some *E. coli* and *Bacillus* species may have some significance for the natural immunity.

In contrast to the successes with other bacterial capsular polysaccharides vaccines, Anderson et al. (1977) have concluded that purified "polyribophosphate" of the type B *H. influenzae* capsule will not provoke humoral immunity in young infants. Studies of the immunogenicity of the *H. influenzae* type B polyribophosphate complex in weanling rabbits (Anderson and Smith, 1977) again emphasize the importance of the state of the antigen (cell-bound versus soluble extract) in determining the magnitude of antibody response.

Purification and chemical characterization of capsular polysaccharides can present problems, especially in gram-negative species, where the structural lability of the LPS complexes of the outer membrane may contribute significant contaminants in the preparations. Such contamination with LPS was manifest in the type B *H. influenzae* polyribosophosphate purified by Anderson and Smith (1977). The presence of highly biologically reactive components such as LPSs, with their pyrogenic, adjuvant, and endotoxic properties, and reactivity in the *Limulus* lysate test complicate the evaluation of such intrinsic properties of the capsular polysaccharides of Gram-negative organisms. The problem appears to be less acute for the purification of capsular substances from Gram-positive organisms. However, some of the extraction procedures may also release less firmly bound membrane amphiphiles and the possibility remains that they may be unsuspected contaminants. Difficulties have also been encountered in the characterization of the capsule or slime material produced by certain strains of *S. aureus*, and Seltmann and Beer (1976) have concluded that the isolated capsular material is rather heterogeneous, consisting of a variety of chemically more or less similar molecules and is not simply an acidic polysaccharide.

Although most of the recent studies on the purification of capsular and other cell surface antigens extracted from bacteria utilize standard immunological methods (e.g., double-diffusion or immunoelectrophoresis), very few investigators have examined the homogeneity of their preparations by the highly sensitive systems for resolving complex mixtures of antigens such as the crossed immunoelectrophoresis (CIE) method (e.g., Axelsen et al., 1973). It would seem that this method of immunochemical analysis and lectin affino electrophoresis (Owen and Salton, 1977) has been greatly underutilized in the purification and characterization of acidic polysaccharides and other surface components of bacteria. One of the few instances where this procedure has been used with a bacterial capsular polysaccharide is with the type B antigen of *H. influenzae* by Buckmire (1976).

In summary, the bacterial capsule constitutes the major surface external layer of encapsulated organisms, it is the dominant surface immunogen of the cell, and it masks many of the underlying antigens, including group-specific carbohydrates, M proteins of streptococci, teichoic and teichuronic acids of Gram-positive organisms, and the LPS and other outer membrane antigens of Gram-negative species. The capsules can render the cells inagglutinable with antibodies to the antigens of the wall-associated accessory polymers or the O-somatic polysaccharide antigens and other wall and envelope antigens. Chemically, the capsular substances range from polysaccharide homopolymers to heteropolymers and the unique homopolymer of the γ-D-glutamyl polypeptide of the *Bacillus* species capsule. Many of the capsular and surface slime components of both Gram-positive and Gram-negative bacteria are acidic polysaccharides and have been implicated in resistance to phagocytosis and pathogenicity, and adherence to mammalian cells, and they possess adjuvant properties and antitumor effects by immunomodulation, and in some instances they induce formation of relatively homogeneous antibody populations. Important advances have been made in studies of the biosynthesis of capsular polysaccharides and the γ-D-glutamyl polypeptide, and the key role of lipid intermediates and the bacterial membrane in catalyzing the steps in the assembly of these polymers has been reviewed in detail by Sutherland (1977) and Troy (1979). The relationships of phosphatidyl diglycerides attached to the meningococcal group C polysaccharide (Fraser et al., 1982) and membrane biosynthesis have yet to be resolved. It will be recalled that the C_{55}-isoprenoid is a common lipid carrier for the biosynthesis of a number of external structures of the bacterial cell, including wall peptidoglycan, capsular polysaccharides, and membrane lipomannan (Sutherland, 1977). For the function of capsules in bacterial pathogenicity, and the reason why antibodies are of prime importance in the protection against bacterial infections, see Chap. 31.

B. Cell Walls and Cell Envelopes

The cell walls of Gram-positive organisms and the envelopes of Gram-negative species can be isolated as homogeneous morphological entities of the bacterial cell. They are, however, chemically heterogeneous, and even in the Gram-positive group there may be one or more distinctly different molecular species (e.g., teichoic acids, polysaccharides, teichuronic acids) covalently linked to the basic structural component, the peptidoglycan. The least heterogeneous wall structure is that of *Micrococcus lysodeikticus*, peptidoglycan accounting for 90 to 95% of the wall and the remaining 5 to 10% being accounted for by the teichuronic acid, a polymer of glucose and N-acetylmannosaminuronic acid (Rogers et al., 1980). The peptidoglycan content of the walls can thus range from almost pure peptidoglycan as in *M. lysodeikticus* to 50% or less as in *S. aureus* and *Bacillus*

subtilis, with the ribitol teichoic acids making up the rest of the structure. Other antigenically important peptidoglycan associated polymers include the group carbohydrates of the streptococci and peptidoglycolipids of mycobacteria and related species. The envelopes of Gram-negative bacteria have long been known for their chemical complexity, containing in addition to peptidoglycan (10 to 20%) a variety of outer membrane proteins, the Braun (1975) lipoprotein covalently linked to the peptidoglycan (as in *E. coli*), LPSs, and phospholipids of the outer membrane structure. Thus both Gram-positive and Gram-negative walls and envelopes possess a variety of molecular species with distinctive antigenic potentials.

Peptidoglycan

With few exceptions, peptidoglycan is the universal structural polymer of the bacterial cell wall of Gram-positive organisms and the envelopes of Gram-negative species. This unique prokaryotic wall component is responsible for the rigidity and morphological integrity of the cell, and its enzymatic degradation by exogenous enzymes such as lysozyme or by the cells' own autolytic enzymes can, under the appropriate conditions, result in osmotically fragile protoplasts, spheroplasts, autoplasts, or complete bacteriolysis. The essential details of peptidoglycans of different taxonomic groups of Gram-positive bacteria and the rather uniform type of peptidoglycan of Gram-negative species have been studied extensively by Schleifer and Kandler (1972) and reviewed in detail by these authors. The glycan strand is composed of alternating groups of N-acetylglucosamine and the unique prokaryotic amino sugar N-acetylmuramic acid, and in the majority of peptidoglycans a tetrapeptide (L-Ala-D-iso-Glu-L-Lys-D-Ala or L-Ala-D-iso-Glu-meso-Dap-D-Ala) is linked through an amide bond to the muramic acid residues. The linkage between adjacent peptides can be direct, head to tail as in *M. lysodeikticus*, or through cross-bridges, and the extent of cross-linking may be high (>90%) as in *S. aureus* peptidoglycan or relatively low (40%) as in *E. coli*. The interpeptide bridges are often characteristic for a certain species, such as the glycine bridges in staphylococci. Peptidoglycans from different taxonomic groups of Gram-positive bacteria show many variations with respect to bridge amino acids, the diamino acids, and occasionally replacement of L-alanine by L-serine or glycine as the amino acid linked to the carboxyl group of N-acetylmuramic acid (Schleifer and Kandler, 1972). This is in marked contrast with the peptidoglycans of diverse Gram-negative bacteria uniformly L-Ala-D-iso-Glu-meso-Dap-D-Ala, and the cross-linking is direct, (i.e., ϵ-NH_2 of the Dap linked to the carboxyl group of the D-Ala of the tetrapeptide, as in *E. coli* peptidoglycan).

Until recently, muramic acid had invariably been present in all bacterial peptidoglycans as the universal (or nearly so) prokaryotic amino sugar marker. A new type of peptidoglycan structure, given the name *pseudomurein*, was found in *Methanobacterium* by König and Kandler (1979). Instead of muramic acid, this peptidoglycan contained N-acetyltalosaminuronic acid, $\beta(1\rightarrow3)$-linked to N-acetylglucosamine, and a peptide with the sequence glutamic acid-alanine-lysine-glutamic acid, linked through the carboxyl group of the talosaminuronic acid. It will be of great interest to see if this type of peptidoglycan is characteristic for the walls of Archaebacteria.

Investigations from a number of laboratories have established that wall peptidoglycans are immunogenic and that antibodies can be directed against both the glycan backbone and the peptide subunits (e.g., Rolicka and Park, 1969; Schleifer and Krause, 1971; Seidl and Schleifer, 1977). Specificity of antibodies for the polysaccharide backbone was evident from the best and most specific inhibition observed with the purified

M. lysodeikticus wall tetrasaccharide (Rolicka and Park, 1969). Immunoprecipitation was less effectively inhibited by β-methyl-N-acetylglucosamine and even less by N-acetylmuramic acid. Streptococcus group A-variant antiserum also contained antibodies to glycan as well as pentapeptide (an additional D-alanine is often found on the C terminus of the un-cross-linked peptides of streptococcal peptidoglycan). With various synthetic peptides it was shown that the pentapeptide L-Ala-D-Glu-L-Lys-D-Ala-D-Ala was the antigenic determinant of the peptide moiety of the peptidoglycan, that the D-alanyl-D-alanine terminus was the immunodominant structure of the antigen, and that the contribution of the other amino acids (L-Lys,D-Glu,L-Ala) to the antigenic site was of less importance. Inhibition studies with pentapeptide and N-acetylglucosamine established that *Staphylococcus epidermidis* peptidoglycan and walls of *Lactobacillus acidophilus* reacted with antibodies to pentapeptide. Peptidoglycans from *M. lysodeikticus* and *Corynebacterium poinsettiae* reacted primarily with antibodies to the glycan and they both lack the terminal D-alanyl-D-alanine residues in their peptide moieties (Schleifer and Krause, 1971).

It is evident from these studies and from the known variations in the peptide moieties of the peptidoglycans (Schleifer and Kandler, 1972) that there must be abundant opportunities for other immunodeterminants in the peptidoglycan structures. Indeed, Seidl and Schleifer (1977) have extended earlier observations of antibodies to the pentaglycine interpeptide bridge of *S. aureus* wall peptidoglycan to the synthesis of an albumin-glycylpolypeptide conjugate producing antibodies cross-reacting with the peptidoglycan. Antibodies to the synthetic immunogen reacted strongly with staphylococcal peptidoglycans with pentaglycine bridges and with those in which glycine residues were partially replaced by L-serine. The high specificity of the antibodies for the glycine-bridge structures was demonstrable by the complete absence of precipitin reactions, with peptidoglycans lacking this type of interpeptide bridge. Thus synthetic immunogens of this type and one for the D-alanyl-D-alanine determinant have been invaluable in identifying the immunodominant peptide structures of wall peptidoglycans, and antibodies to the albumin-polyglycine and conjugate have been used in a highly sensitive latex agglutination test for quantitation of peptidoglycan and serological separation of staphylococci and micrococci (Seidl and Schleifer, 1978).

Antibodies to the glycan backbone of peptidoglycans cross-react with the identical structures widely distributed throughout different taxonomic groups. As with other polysaccharide immunogens, the antibodies to glycan have limited heterogeneity (Schleifer and Krause, 1971). It will be of interest to see what cross-reactivities occur with the pseudomureins of the *Methanobacterium*, although earlier inhibition studies suggest that N-acetylglucosamine may be the more important immunodeterminant of the glycan structure.

Finally, one other aspect of peptidoglycan immunology worthy of mention was the discovery that the adjuvant properties of the mycobacterial cells used in Freund's complete adjuvant was eventually traced to the N-acetylmuramyl-tripeptide portion of the wall peptidoglycan (Parant, 1979). The adjuvant and other biological properties could be duplicated with the synthetic N-acetylmuramyl dipeptide (L-Ala-D-Glu-NH_2), designated as MDP. This small molecule apparently does not share the variety of heterogeneous antigenic determinants in the whole mycobacterial cell. Guinea pigs receiving MDP in Freund's incomplete adjuvant failed to become sensitized to tuberculin, water-soluble adjuvant, cell wall oligomer, or even MDP. It can, however, act as a hapten, and MDP is immunogenic when conjugated to carrier protein. Hyperimmune rabbit antipeptidoglycan

sera did not bind [^{14}C]MDP, but the derivative MDP-L-Lys-D-Ala was capable of binding (Parant, 1979). This would be in accord with the strong contribution of the D-alanyl-D-alanine as an immunodeterminant of the peptide moiety of certain peptidoglycans.

Peptidoglycan-Associated Polysaccharides and Teichoic Acids

Other peptidoglycan-associated polymers of antigenic significance in the walls of certain Gram-positive organisms are the group-specific polysaccharides of *Streptococcus* species and the ribitol and glycerol teichoic acids found principally in species of the *Bacillus, Lactobacillus, Staphylococcus*, and *Streptococcus* genera. Much has been written on the chemical structures and immunochemistry of these components and for detailed reviews the reader is referred to Heidelberger (1973), Duckworth (1977), and Knox and Wicken (1973). The immunologic properties of group A streptococcal polysaccharide and its variant have been clearly established and summarized by McCarty (1971). On the other hand, the nature of the "C substance" of the pneumococcus is problematic and, as pointed out by Heidelberger (1973), if it is a single substance, it appears to be both a polysaccharide and a teichoic acid. One of the problems with the accessory wall polymers covalently linked to peptidoglycan is that the extraction procedures probably do not give a clean cission of a single specific bond, thereby avoiding partial degradation of the wall structures. Loss of substituents linked by labile bonds (e.g., ester-linked D-alanine of teichoic acids) and hydrolysis of bonds during extraction have made it difficult to assess chain lengths and amounts of substituents on the "native" peptidoglycan-associated components.

Immunological studies of ribitol teichoic acid of *S. aureus* have established that both α- and β-linked N-acetylglucosamine act as determinants (Nathenson et al., 1966). Other sugar residues (glucose-N-acetylgalactosamine) have been found on *Staphylococcus lactis* teichoic acid, and these would be expected to be strong determinants (Duckworth, 1977). Apart from the staphylococcal teichoic acids, there are many instances of terminal determinants (e.g., in lactobacilli and streptococci), and the specificity also depends on the carbohydrate residues attached to the ribitol- or glycerol-phosphate backbone (Knox and Wicken, 1973).

Cross-reactivity between the *H. influenzae* type B ribose-ribitol phosphodiester-linked polymer suggested the presence of antibodies to the polyolphosphate backbone (Crisel et al., 1975). Indeed, earlier serological studies with group A streptococci by McCarty (1959) established the presence of antibodies to unsubstituted glycerol teichoic acid, and the reactions were inhibitable with synthetic polyglycerophosphate, with average chain lengths of six units. Cross-reactivities were seen with extracts of a number of other Gram-positive organisms, clearly establishing a heterophile type of antigen with production of antibodies specific for the glycerophosphate backbone (McCarty, 1959; Knox and Wicken, 1973). Moreover, the precipitation of glycerol teichoic acid was inhibitable by the glycerol phosphate component of cardiolipin and could account for false-positive reactions for syphilis (Knox and Wicken, 1973). Although there have been suggestions that D-alanine substituents ester-linked to the teichoic acids can serve as immunodeterminants, the evidence is not compelling, and nonspecific inhibition of antibody precipitation has been seen with D,L-alanine methyl ester hydrochloride (Knox and Wicken, 1973). Antistaphylococcal teichoic acid antibodies have been detected in sera from patients with virulent *S. aureus* infections, but it is not known what specific determinants are being recognized.

Lipopolysaccharides and Lipoprotein

As pointed out earlier, the envelope of Gram-negative is structurally and chemically complex, and the Braun (1975) lipoprotein covalently linked to the peptidoglycan probably provides an anchoring structure for the outer membrane assembly. Although the outer membrane contains a variety of major and minor proteins, it is undoubtedly the lipopolysaccharide organized in its outer leaflet which is the immunodominant antigenic component.

LPSs (endotoxins) have long been recognized as the chemical structures carrying the sugar-specific determinants of the O-somatic antigens of gram-negative bacteria. Their structures have been investigated extensively in the enteric bacteria and in diverse groups of gram-negative organisms, and mutations affecting the O-specific polysaccharide side chains have been studied at the genetic, chemical, biochemical, and immunochemical levels. Many comprehensive reviews and monographs have been written on this topic (e.g., Lüderitz, 1981; Jann and Jann, 1977; Wilkinson, 1977). LPS appears to be universally present in Gram-negative bacteria and has also been found in blue-green algae (Wilkinson, 1977). Until recently, no lipopolysaccharides had been found in Gram-positive bacteria, but the studies of Wexler and Oppenheim (1980) established the presence of a lipopolysaccharide-like component in *L. monocytogenes* which gives a Gram-positive reaction and possesses the typical Gram-positive surface profile, as seen in electron micrographs of thin sections. The presence of LPS in this organism thus represents the single current exception to the general absence of this component in Gram-positive organisms. In common with the LPS of many Gram-negative species, the purified listerial LPS contained KDO, lipid A, β-OH fatty acids, gave a positive limulus lysate test, and was endotoxic, pyrogenic, and mitogenic (Wexler and Oppenheim, 1980). Antibodies formed in response to immunization with intact cells of *L. monocytogenes* react with purified LPS. At present, the localization of the listerial LPS in the wall or membrane has not been established. It will be recalled that in Gram negative bacteria the LPS occurs in the outer leaflet of the bilayer of the outer membrane (OM) structure of the envelope.

As with the pneumococcal polysaccharides, the immunological properties of the lipopolysaccharides have been studied and documented extensively (Lüderitz, 1981; Jann and Jann, 1977), and no attempt will be made to review further this important field of bacterial surface antigens already well covered by the major contributors to LPS research. The extensive chemical and serological studies of O-antigenic lipopolysaccharides has provided comprehensive schemes for classification of chemotypes and serological specificites of *Salmonella* species and many other groups of enteric bacteria. Many of the factors contributing to the antigenic specificities of the polysaccharides, including LPS, have been investigated and much is known about determinants, the immunodominant sugars determining O-antigen serological specificity. Important factors in immunodominance include not only the nature of the sugar group or unit but also the anomeric configuration of the linkages: the position of linkage, terminal, or intrachain locations in the polysaccharides. Although the specificities may center around single sugar residues, they can extend along the polysaccharide chains for different lengths. In addition to the immunodominant sugars, other noncarbohydrate substituents on the bacterial polysaccharides can function as immunodeterminants. For example, in *S. typhimurium* O-acetyl groups on abequose form part of the antigenic determinants. Other substituent groups that may function as determinants include acyl, formyl, succinyl, phosphate, glycerol, acetal, and ketal groups (Jann and Jann, 1977). One of the consequences of the relative

simplicity of many polysaccharides with identical oligosaccharide regions, linkages, and substituents is that cross-reactivities are common for both lipopolysaccharides and capsular polysaccharides, and the quantitative aspects of this have provided new immunochemical and structural information about polysaccharide antigens (Heidelberger, 1973; Jann and Jann, 1977).

The bacterial wall and membrane have been a reservoir of unusual chemical entities, and the discovery of an outer membrane lipoprotein by Braun and his colleagues (Braun, 1975) added yet another unique component to the cell envelope of Gram-negative bacteria. This major outer membrane component was linked to the *E. coli* peptidoglycan component through the C-terminal carboxyl group of L-alanine to the ϵ-NH_2 group of α,ϵ-diaminopimelic acid. The cysteinyl residue at the N terminus of the polypeptide chain possessed two unusual features, a fatty acid amide linked to the N-terminal group and a thioether linkage through the SH group to a diglyceride (Braun, 1975). Molecular models of the known amino acid sequence of the lipoprotein have suggested the manner in which it may be assembled in the membrane to present the hydrophobic amino acid residues for interaction with lipids and a more hydrophilic pore, with peripheral clusters of fatty acids (Inouye, 1979). This lipoprotein was detectable as a surface antigen in rough strains of *E. coli* (Braun, 1975) but appears to be covered by O-polysaccharide chains in smooth strains. It was also identifiable as a major envelope antigen in the outer membrane fractions isolated from *E. coli* K12 (Smyth et al., 1978). Similar covalently linked lipoproteins have been detected in other gram-negative species (Inouye, 1979), but at present this type of lipoprotein attached by covalent bonds to the peptidoglycan structure has not been found in Gram-positive bacteria. Not all of the lipoprotein in the *E. coli* envelope is linked to the peptidoglycan since it has also been found in a free form, not requiring proteolytic cleavage from the peptidoglycan for its release (Inouye, 1979).

One other important "surface" antigen that should be mentioned is the enterobacterial common antigen (ECA). Although it is quite distinct from LPS and lipoprotein and has indeed been known for some time ("Kunin antigen" discovered in 1962), it received surprisingly little attention until comparatively recently (Makela and Mayer, 1976). The precise localization of the ECA is uncertain but it is believed that some of it resides on the surface of members of the Enterobacteriaceae. Efforts to define its compartmentalization were inconclusive in that in some cases it was clearly in the envelope fraction, but in many it was also found in cytoplasmic fractions (Makela and Mayer, 1976). It is believed to exist as a free form linked to LPS, and chemical characterization of purified ECA indicates that it is a heteropolymer composed of D-glucosamine and D-mannosaminuronic acid partly esterified with palmitic acid. The ECA material from *Salmonella montevideo* has recently been shown to be a linear polymer of 1,4-linked N-acetyl-D-glucosamine and N-acetyl-D-mannosaminuronic acid esterified with both palmitic and acetic acids (Kiss et al., 1978). From its chemical composition and solubility properties, ECA thus appears to be an amphiphilic molecule. The free form, of low molecular weight (2700), is haptenic, but the LPS-linked form is immunogenic. Although a considerable amount of information is now available on the serology, distribution, and immunogenicity of ECA, it is evident that much work remains to be done on its purification, chemistry, and identification of immunodeterminants and cross-reactivities (Makela and Mayer, 1976). ECA is clearly a different entity from the "CA" detected in many groups of Gram-negative bacteria (Høiby, 1975) and recently purified in aggregated form by Sompolinsky et al. (1980), which appears to be a protein common antigen.

C. Membranes of Gram-Positive Bacteria

The biochemical and immunochemical characterization of the plasma membranes of many Gram-positive bacteria has been greatly facilitated by cell-wall-degrading enzymes such as lysozyme, fungal muramidase (*Chalaropsis* enzyme), lysostaphin, mutanase, and a variety of other bacteriolytic and autolytic enzymes (Salton, 1976). The selective dissolution of the rigid cell walls has enabled their isolation under gentle conditions, thereby minimizing degradation and perturbation of the membranes with loss of "native" components. The membranes of Gram-positive bacteria are essentially lipid-protein (approximately 1:3 ratios) structures and they perform a multiplicity of biochemical and transport functions (Salton, 1978). From the variety of biochemical functions it could be anticipated that many distinct protein antigens would be present in the membranes, and this indeed has been demonstrable by the high-resolving power of the crossed immunoelectrophoresis techniques (Owen and Salton, 1977; Salton, 1978). In addition to the lipid and protein constituents, the plasma membranes of many (if not all) Gram-positive organisms possess a membrane amphiphile of the lipoteichoic acid or lipomann type (Shockman and Wicken, 1981), both of which are immunogenic. It is not known at the present time whether Gram-negative organisms have analogous structures in their plasma or inner membranes although, lipoprotein and LPSs of the outer membrane are classifiable as amphiphilic molecules (Shockman and Wicken, 1981). One other possible candidate for a Gram-negative amphiphile in the inner plasma membrane is the common enterobacterial antigen (ECA), composed of N-acetylglucosamine and N-acetylmannosaminuronic acid (Wicken and Knox, 1980), although the structure of the lipid and the specific localization of ECA in the membranes of the envelope have not been established.

Lipoteichoic Acids and Lipomannans

Recognition that certain bacteria may possess both ribitol and glycerol teichoic acids (Baddiley, 1972; Duckworth, 1977) eventually led to the conclusion that the glycerol type was not in the cell wall but existed in membrane-bound form as a lipoteichoic acid (see Knox and Wicken, 1973). The taxonomic distribution of lipoteichoic acids parallels that of the wall teichoic acids, being prevalent in lactobacilli, staphylococci, strepococci, and *Bacillus* species. The pigmented micrococci, devoid of wall teichoic acids, possess an analogous type of amphiphile, the succinylated lipomannans (Duckworth, 1977), but its possible distribution in other taxonomic groups does not appear to have been explored. *Actinomyces viscosus* apparently possesses yet another type of polysaccharide amphiphile, the structure of which has not been determined unequivocally (Wicken and Knox, 1980).

All of the amphiphiles investigated so far are immunogenic (Wicken and Knox, 1980), but the form of the immunogen (whole cells, isolated membranes, purified amphiphile) required for antibody formation does not appear to have been studied systematically. Precipitating antibodies have been used successfully for quantitation of both lipoteichoic acids and lipomannan (Shockman and Wicken, 1981). Much remains to be establshed on the immunodeterminants of the amphiphiles, such as the roles of substituents, including succinyl residues of the lipomannan and amino acids (D-alanine) of the lipoteichoic acids. Immunochemical discrimination between wall glycerol teichoic acids and lipoteichoic acids coexisting in the same organism presents difficulties (Shockman and Wicken, 1981).

Both the lipoteichoic acid of lactobacilli and the lipomannan of *M. lysodeikticus* are detectable on the cell surface of intact bacteria. In the latter organism where intact, stable protoplasts can be prepared, the lipomannan is exposed as a major antigen on the protoplast membrane surface (Owen and Salton, 1977). Mesosomes also appear to be enriched in the amphiphiles in both *S. aureus* and *M. lysodeikticus*.

Protein Antigens

One of the basic problems that had to be overcome in determining the antigenic complexity of the bacterial membrane was the development of suitable solubilizing procedures which would permit both the retention of biological activities and maximum resolution of the complex mixtures for interaction with antimembrane antibodies. This was finally achieved by using the high-resolution techniques of crossed immunoelectrophoresis (Axelsen et al., 1973) with Triton X-100-solubilized membranes. This method permits the resolution of complex mixtures of antigens as a pattern of immunoprecipitate peaks because of intrinsic differences in the electrophoretic mobilities of the components (anodal migration in the first dimension) and differences in the individual ratios of antigen to specific antibodies in the second direction of electrophoresis into essentially immobile antibodies (pH 8.6) of the reference antimembrane antibodies. A typical pattern of immunoprecipitates for TX-100-solubilized membranes of *M. lysodeikticus* is illustrated in Fig. 1. A total of 28 discrete immunoprecipitates have been detected in this Gram-positive plasma membrane and five of the immunoprecipitates have been identified as enzymes (succinate, malate, and two NADH dehydrogenases, and ATPase) by zymogram

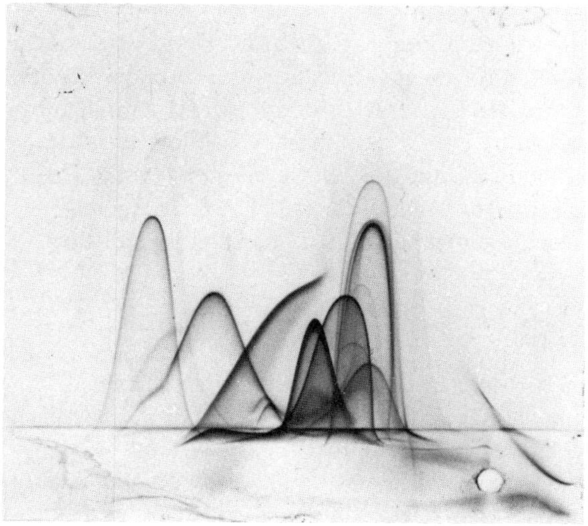

Figure 1 Illustration of the resolution of *M. lysodeikticus* membrane antigens by crossed immunoelectrophoresis of Triton X-100-solubilized membranes (anodal electrophoresis to the left) into the antimembrane antibodies in the reference gel (anodal electrophoresis to the top). Procedures and antigen peak identifications were reported by Owen and Salton, 1977. Immunoprecipitates were stained with Coomassie blue. (From Salton, 1980.)

staining, in addition to the succinylated lipomannan (Owen and Salton, 1977). By adapting the basic CIE procedure to labeling and lectin-affinity techniques, further identifications of putative glycoproteins and nonheme Fe proteins, for example, can be made (Owen, 1981).

The energy-transducing F_1-ATPase of the membranes of *M. lysodeikticus* is a major immunogen, and its asymmetric distribution (on the inner, protoplasmic side of the membrane) and that of the dehydrogenases was established by absorption of membrane antibodies with intact protoplasts (Owen and Salton, 1977). Thus the asymmetric architecture of bacterial membranes and cell surfaces can be studied successfully by combining absorption and CIE procedures and permits the identification of antigens expressed on cell surfaces and/or the outer face of the plasma membrane. Similar studies have been extended to other Gram-positive and Gram-negative organisms (Owen, 1981), and aspects of the latter will be discussed further in the next section.

In addition to identification of major membrane antigens such as F_1-ATPase and dehydrogenases, antibodies to the F_1-ATPase have been used to resolve the problem of the suggested glycoprotein nature of this inner-face membrane component. The carbohydrate present in F_1-ATPase purified to protein homogeneity was shown to be due to co-separating contaminating lipomannan not covalently linked to the ATPase protein (see Salton and Lim in Shockman and Wicken, 1981).

Despite the high sensitivity and resolution of the CIE methodology and its great adaptability to affinity and labeling techniques, several problems remain for the complete identification of membrane antigens. At the moment the procedure is still largely qualitative, since the immunoelectrophoretic analysis is limited by the extent of solubility of the membranes in nondenaturing detergents and because of the unknown degree of recovery of the membrane components as immunoprecipitates. There is little doubt that some of the components are excluded from the agarose gel and that recovering these fractions with sodium dodecyl sulfate (SDS) may reduce or eliminate the reactivity of some of the antigens with antibodies. Another limitation for the identification of membrane enzymes, short of purification, is the restriction imposed by the available zymogram staining procedures based on identification of enzymes by insoluble formazans, colored substrates and products, and detergent and/or antibody inactivation. However, despite all of these limitations, application of these techniques of immunoelectrophoresis has greatly expanded our ability to explore the structure and function of membranes and their antigens.

D. Membranes of Gram-Negative Bacteria

Envelopes of Gram-negative bacteria as isolated by cell disruption procedures usually consist of both the inner (plasma or cytoplasmic) and outer membranes together with the peptidoglycan layer. Some of the mechanical disruption procedures may result in fragmentation and loss of the more fragile plasma membrane structures. The development of suitable methods for the separation of inner and outer membranes, or at least enrichment in either structure, has greatly advanced studies of their biochemistry and immunochemistry (Osborn and Munson, 1974).

Apart from the work done on LPS, lipoprotein, and several outer membrane proteins, very little has been done on individual membrane antigens, although the recent work of Owen (1981) is rapidly expanding our knowledge of several of the *E. coli* enzymes. Of the Gram-negative bacteria, the principal studies relate to the resolution of

the antigens of *E. coli* envelopes, membrane vesicles, and inner and outer membranes; the envelopes of *N. gonorrhoeae*; and the envelope and chromatophores of *Rhodopseudomonas spheroides* (for details and original references, see the reviews by Owen, 1981, 1983). In all of these investigations the basic methodology (CIE) has been similar to that described in the preceding section: display of antigens as immunoprecipitates by CIE, identification of specific enzymes as antigens by zymogram staining, identification of iron-containing proteins by radioactive labeling with ^{59}Fe, and recognition of photosynthetic pigments from chromaphores (Owen, 1981, 1983).

The most intensively studied organism is *E. coli*, starting with the earlier investigations of Smyth et al. (1978) with the resolution of the patterns of antigens for the envelopes and outer membranes of *E. coli* K12, and greatly expanded for the *E. coli* ML308-225 membrane vesicles by Owen and Kaback (1979) and Owen (1981, 1983). The distinctive patterns for the *E. coli* K12 outer and envelope antigens are illustrated in Fig. 2 from the work of Smyth et al. (1978). In the study by Owen and Kaback (1979), a total of 52 antigens were resolved in the membrane vesicles and of these 13 were identifiable by zymogram staining either as enzymes or as other antigens (LPS, lipoprotein). None of the immunoprecipitates stained for periplasmic phosphatases and phosphodiesterase, and there appeared to be minimal contamination of the vesicles with cytoplasmic components, although polynucleotide phosphorylase partitioned between vesicle membranes and cytoplasmic fractions. From absorption studies it has been concluded that the membrane vesicles have the right-side-out orientation and that minimal inversion has occurred (see Owen, 1981, 1982, for discussions). Autoradiography combined with CIE of ^{59}Fe-labeled membranes gave seven discrete antigens, three of which were identified as NADH dehydrogenase, NADPH dehydrogenase, and glutamate dehydrogenase, and two nonheme iron antigens. Of interest was the detection of Braun's lipoprotein with firmly bound iron (Owen, 1981, 1983).

It is evident from these studies that much can be learned about the complexity, compartmentalization, and topography of the membrane antigens by resolution of the antigens using the CIE method, and that the ability to generate antibodies to individual enzyme antigens by excision of immunoprecipitates will greatly facilitate their identification and isolation (Owen, 1981, 1983).

E. Cytoplasm

Perhaps the greatest source of antigens in the bacterial cell is the cytoplasm, with its great variety of enzymes, ribosomal proteins, and nucleic acids. Under normal conditions of immunization with whole bacteria, many of the immunogenic proteins in the cytoplasmic compartment of the cell may not be expressed. Extensive studies by Axelsen et al. (1973) and Høiby (1975) using cell extracts of a wide range of bacterial species have demonstrated the complexity and cross-reactivities of the variety of antigens, which probably include cytosol and surface components of the bacterial cell. Moreover, these studies have established that a number of antigens are common to a wide variety of bacterial species and that one antigen in particular is found in both Gram-positive and Gram-negative species (Høiby, 1975). This common antigen, as isolated from extracts of whole cells, has a high molecular weight (400,000 to 600,000), but in SDS-PAGE its molecular weight is of the order of 62,000 to 65,000 (Sompolinsky et al., 1980). The functions of this protein a.e unknown at the present time and it has been identified only as a major, fast-moving, negatively charged component in bacterial extracts (Høiby, 1975; Smyth and

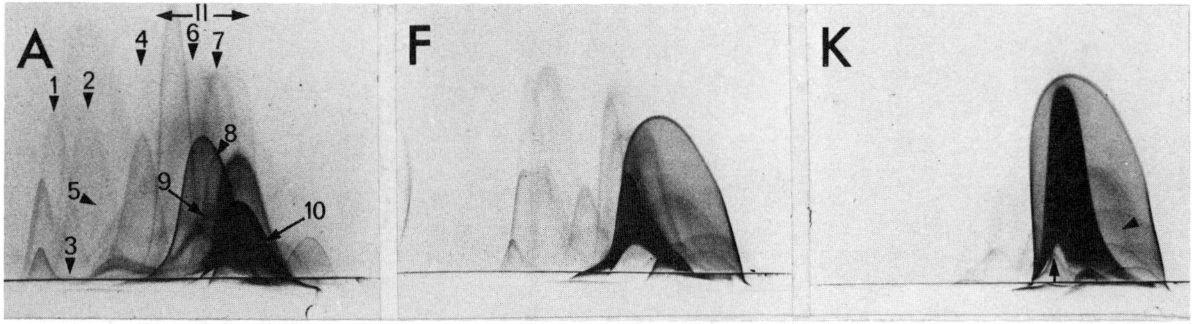

Figure 2 Resolution of the antigens in the isolated inner membranes (A), total envelopes (F), and isolated outer membranes (K) of *E. coli* K12 adapted from the study by Smyth et al. (1978, top panel row of Fig. 2, p. 312). All fractions in this study were solubilized in Triton X-100 and electrophoresed into anti-whole envelope antiserum as described by Smyth et al. (1978). Antigen and enzyme assignments of peaks are as reported by Smyth et al. (1978). Note the large dome-shaped immunoprecipitate in the outer membrane fraction (K) identified as Braun's lipoprotein. (Reproduced with permission of American Society for Microbiology.)

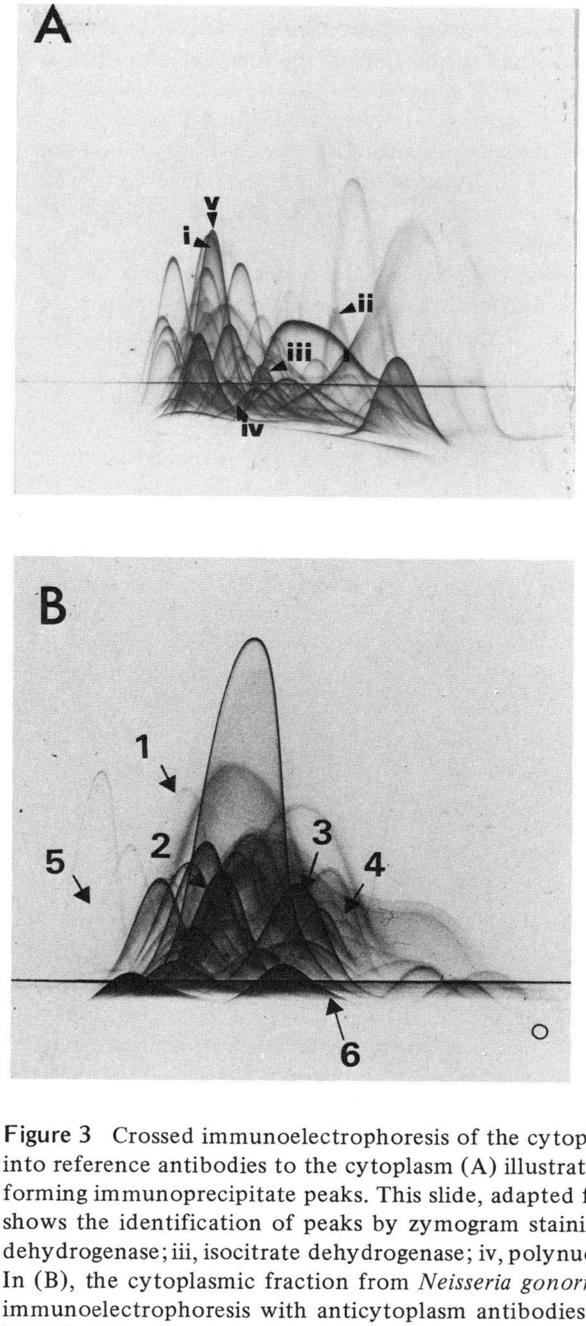

Figure 3 Crossed immunoelectrophoresis of the cytoplasmic fraction of *M. lysodeikticus* into reference antibodies to the cytoplasm (A) illustrates the complex array of antigens forming immunoprecipitate peaks. This slide, adapted from Owen and Salton (1977), also shows the identification of peaks by zymogram staining as follows: i and ii, NADH dehydrogenase; iii, isocitrate dehydrogenase; iv, polynucleotide phosphorylase; v, catalase. In (B), the cytoplasmic fraction from *Neisseria gonorrhoeae* is resolved by crossed immunoelectrophoresis with anticytoplasm antibodies as shown by Smyth and Salton (1977). Note that the pattern of immunoprecipitates characteristic for the gonococcus differs from that seen for *M. lysodeikticus* (A). By zymogram staining immunoprecipitates 1 to 4 stained for NADH dehydrogenase, 5 stained for glucose-6-phosphate dehydrogenase, and 6 was identified as glutamate dehydrogenase. [(A) from Owen and Salton (1977); reproduced with permission from the American Society of Microbiology. (B) from Smyth and Salton (1977); reproduced with permission of John Wiley & Sons, Inc., New York.]

Salton, 1977) and has been detected in membrane and envelope fractions from a variety of bacteria (Smyth et al., 1976, 1978). The identification of the function of such a highly conserved protein, whether it is a cytoplasmic or membrane-associated component, will be of great interest.

The antigenic complexity of the cytoplasmic fractions of both Gram-positive and Gram-negative bacteria has been clearly established by CIE for *M. lysodeikticus* (Owen and Salton, 1977) and *N. gonorrhoeae* (Smyth et al., 1976). The resolution of 90 or more discrete antigens is seen in the CIE patterns presented in Fig. 3, and several of these antigens have been identified as enzymes. Thus catalase and polynucleotide phosphorylase, readily detectable by sensitive zymogram staining techniques, were shown to be major immunogens of the cytoplasmic compartment of *M. lysodeikticus* (Owen and Salton, 1977).

Although the CIE technique has shown excellent resolution of the variety of antigens in both membrane and cytoplasmic compartments, it is evident that the total number of individual proteins detected by this method falls short of expectations from the great variety of biochemical activities and from single- or two-dimensional electrophoresis of denatured polypeptides. It would appear at the present time difficult to avoid the trade-off between maximal identification of nonfunctional denatured polypeptides and minimal detection by resolution of antigens retaining their ability to react with antibodies and in some instances still identifiable as enzymes. One feature reinforcing confidence in the CIE-zymogram analysis of the antigens of either cytoplasmic or membrane compartments of the bacterial cell is that none of the antigens have shown multiple enzyme staining, nor has there been evidence of identity or cross-reactivity between antigens showing the same enzymatic reactions. However, the significance and functions of multiple, discrete antigens possessing the same biochemical activities (e.g., multiple antigenic forms of NADH dehydrogenase) remain to be elucidated.

III. CONCLUSIONS

The bacterial cell possesses a complex array of antigens or potentially immunogenic components both on its surface and within the cell. The expression and complexity of the antigens depend on the state or form of the cell or its subcellular components presented as immunogens. It is meaningful to consider the antigens of both Gram-positive and Gram-negative bacteria in terms of their anatomical entities and cellular compartments. Both groups can possess in certain species the surface appendages, flagella, and pili, although the latter occur most widely in Gram-negative species and are comparatively rare in Gram-positive bacteria. The principal "compartments" delineated in the bacterial cell are the capsular/slime surface structures external to the cell walls in both Gram-positive and Gram-negative groups, the rigid cell wall compartment and underlying plasma or cytoplasmic membrane of Gram-positive organisms, and the envelopes of Gram-negative bacteria, consisting of the outer membrane, peptidoglycan layer, and inner (cytoplasmic) membrane. In both instances the cytoplasmic membrane constitutes the boundary of the cytoplasmic compartment, with its considerable array of potentially immunogenic proteins, ribosomal proteins, and nucleic acids. Gram-negative bacteria possess one additional compartment, the periplasmic space, with its binding proteins and hydrolytic enzymes.

In encapsulated bacteria, the capsular polysaccharides constitute the major surface antigens, and their immunochemical properties have been investigated extensively in pneumococci and other streptococcus species and staphylococci, as well as the enteric K

antigens and capsules of many other taxonomic groups of Gram-negative species. Both homo- and heteropolysaccharides have been investigated extensively and immunodominant sugar specificities identified. The γ-D-glutamyl polypeptide of *Bacillus* species constitutes the unique capsular antigen of the surface of these organisms.

The Gram-positive cell wall is generally structurally simple but may be antigenically heterogeneous, with specific determinants in the peptidoglycan and, where they occur, accessory polymers such as polysaccharides and/or teichoic or teichuronic acids having their own specific sets of determinants. Membranes of Gram-positive bacteria possess as major immunogens the amphiphiles lipoteichoic acid or lipomannan, in addition to a complex variety of protein antigens, some of which can be identified as enzymes. Little is known at present about the immunodeterminant groups or conformational domains of the membrane proteins.

The outer membrane of the Gram-negative envelope carries the major LPS antigens of these organisms, together with other immunogenic major outer membrane proteins and Braun's lipoprotein. With the sole exception of the Gram-positive organism *L. monocytogenes*, LPS appears to be exclusive for Gram-negative organisms. The specific location of the ECA amphiphile in the cell envelope of enteric species has yet to be defined. As with the cytoplasmic membranes of Gram-positive bacteria, the inner membrane of Gram-negative species can be resolved by CIE into a complex array of antigens, a number of which can be enzymatically identified by zymogram staining procedures.

The cytoplasmic compartments of both groups have been shown to be highly complex mixtures of antigens when this fraction is used to generate antibodies and examined by the sensitive CIE method.

Much progress has been made in the characterization, localization, and quantitation of bacterial antigens. Antibodies to many of these antigens will undoubtedly be put to good effect in localizing such interesting components as the ECA and common antigen of other species, as well as using the antibodies as probes for the functions of cell surface and membrane proteins, exploring the determinant domains of membrane enzymes, and discovering more about the molecular asymmetry of the important surface structures of bacteria.

ACKNOWLEDGMENTS

The author's work on bacterial membranes is supported by a grant from the National Science Foundation (PCM-8105126). I wish to thank Josephine Markiewicz for her expert typing and great help throughout the preparation of the manuscript.

REFERENCES

Anderson, P., and Smith, D. H. (1977). J. Infect. Dis. *136*, Suppl., S-63.

Anderson, P., Smith, D. H., Ingram, D. L., Wilkins, J., Wehrle, P. F., and Howie, V. M. (1977). J. Infect. Dis. *136*, Suppl., S-57.

Austrian, R. (1977). J. Infect. Dis. *136*, Suppl., S-57.

Axelsen, N. H., Kroll, J., and Weeke, B., Eds. (1973). Scand. J. Immunol. *2*, Suppl. 1, 15.

Baddiley, J. (1972). Essays Biochem. *8*, 35.

Braun, V. (1975). Biochim. Biophys. Acta *415*, 335.

Brinton, C. C., Jr. (1965). Trans. N.Y. Acad. Sci. *27*, 1003.

Buckmire, F. L. A. (1976). Infect. Immun. *13*, 1733.

Crisel, R. M., Baker, R. S., and Dorman, D. E. (1975). J. Biol. Chem. *250*, 4926.

Duckworth, M. (1977). In *Surface Carbohydrates of the Prokaryotic Cell* (I. W. Sutherland, Ed.). Academic Press, New York, pp. 177–208.

Fraser, B. A., Gotschlich, E. C., Nishimura, O., and Liu, T.-Y. (1982). Semin. Infect. Dis. *4*, 242.

Gold, R., Lepow, M. L., Goldschneider, I., and Gotschlich, E. C. (1977). J. Infect. Dis. *136*, Suppl., S-31.

Heidelberger, M. (1973). In *Research in Immunochemistry and Immunobiology*, Vol. 3 (J. B. G. Kwapinski, Ed.). University Park Press, Baltimore, pp. 1–40.

Høiby, N. (1975). Scand. J. Immunol. *4*, Suppl. 2, 187.

Inouye, M. (1979). In *Bacterial Outer Membranes Biogenesis and Functions* (M. Inouye, Ed.). Wiley, New York, pp. 1–12.

Jann, K., and Jann, B. (1977). In *Surface Carbohydrates of the Prokaryotic Cell* (I. W. Sutherland, Ed.). Academic Press, New York, pp. 247–287.

Johnston, K. H., Holmes, K. K., and Gotschlich, E. C. (1976). J. Exp. Med. *143*, 741.

Kandler, O., and Schleifer, K. H. (1980). Prog. Bot. *42*, 234.

Karakawa, W. W., and Vann, W. F. (1982). *Semin. Infect. Dis. 4*, 285.

Kasper, D. L., Baker, C. J., Edwards, M. S., Nicholson-Weller, A., and Jennings, H. J. (1982). *Semin. Infect. Dis. 4*, 275.

Kiss, P., Rinno, J., Schmidt, G., and Mayer, H. (1978). Eur. J. Biochem. *88*, 2311.

Knox, K. W., and Wicken, A. J. (1973). Bacteriol. Rev. *37*, 215.

König, H., and Kandler, O. (1979). Arch. Microbiol. *123*, 295.

Kornfeld, R., and Kornfeld, S. (1980). In *The Biochemistry of Glycoproteins and Proteoglycans* (W. J. Lennarz, Ed.). Plenum, New York, pp. 1–34.

Kwapinski, J. B. G. (1969). In *Analytical Serology of Microorganisms*, Vol. 2 (J. B. G. Kwapinski, Ed.). Wiley-Interscience, New York, pp. 485–500.

Lüderitz, O. (1981). In *Chemistry and Biological Activities of Bacterial Surface Amphiphiles* (G. D. Shockman and A. J. Wicken, Eds.). Academic Press, New York, p. 41.

McCarty, M. (1959). J. Exp. Med. *109*, 361.

McCarty, M. (1971). Harvey Lect. *65*, 73.

Makela, P. H., and Mayer, H. (1976). Bacteriol. Rev. *40*, 591.

Mescher, M. F., and Strominger, J. L. (1976). Proc. Natl. Acad. Sci. USA *73*, 2687.

Nathenson, S. G., Ishimoto, N., Anderson, J. S., and Strominger, J. L. (1966). J. Biol. Chem. *241*, 651.

Ørskov, I., Ørskov, F., Jann, B., and Jann, K. (1977). Bacteriol. Rev. *41*, 667.

Osborn, M. J., and Munson, R. (1974). In Methods Enzymol. *31*(A), 642.

Owen, P. (1981). In *Organization of Prokaryotic Cell Membranes*, Vol. 1 (B. K. Ghosh, Ed.). CRC Press, Boca Raton, Fla., pp. 73–164.

Owen, P. (1983). In *Electroimmunochemical Analysis of Membrane Proteins* (O. J. Bjerrum, Ed.). Elsevier Biomedical Press, Amsterdam, pp. 55–76, 347–373.

Owen, P., and Kaback, H. R. (1979). Biochemistry *18*, 1413.

Owen, P., and Salton, M. R. J. (1977). J. Bacteriol. *132*, 974.

Parant, M. (1979). Springer Semin. Immunopathol. *2*, 101.

Rogers, H. J., Perkins, H. R., and Ward, J. B. (1980). *Microbial Cell Walls and Membranes.* Chapman & Hall, New York, pp. 1–564.

Rolicka, M., and Park, J. T. (1969). J. Immunol. *103*, 196.

Roth, I. L. (1977). In *Surface Carbohydrates of the Prokaryotic Cell* (I. W. Sutherland, Ed.). Academic Press, New York, pp. 5–26.

Salton, M. R. J. (1976). *Methods Membr. Biol. 6*, 101.

Salton, M. R. J. (1978). In *Relations Between Structure and Function in the Prokaryotic Cell* (Soc. Gen. Microbiol. Symp. 28). Cambridge University Press, Cambridge, pp. 201–223.

Salton, M. R. J. (1980). *Subcell. Biochem. 7*, p. 309.

Schleifer, K. H., and Kandler, O. (1972). Bacteriol. Rev. *36*, 407.

Schleifer, K. H., and Krause, R. M. (1971). Eur. J. Biochem. *19*, 471.

Seidl, P. H., and Schleifer, K. H. (1977). Eur. J. Biochem. *74*, 353.

Seidl, P. H., and Schleifer, K. H. (1978). Arch. Microbiol. *118*, 185.

Seltmann, G., and Beer, W. (1976). In *Staphylococci and Staphylococcal Diseases* (J. Jeljaszewicz, Ed.). G. Fischer, Stuttgart, pp. 377–381.

Shockman, G. D., and Wicken, A. J., Eds. (1981). *Chemistry and Biological Activities of Bacterial Surface Amphiphiles*. Academic Press, New York, pp. 1–390.

Smyth, C. J., and Salton, M. R. J. (1977). In *The Gonococcus* (R. B. Roberts, Ed.). Wiley, New York, pp. 303–331.

Smyth, C. J., Friedman-Kien, A. E., and Salton, M. R. J. (1976). Infect. Immun. *13*, 1273.

Smyth, C. J., Siegel, J., Salton, M. R. J., and Owen, P. (1978). J. Bacteriol. *133*, 306.

Sompolinsky, D., Hertz, J. B., Høiby, N., Jensen, K., Mansa, B., and Samra, Z. (1980). Acta Pathol. Microbiol. Scand. Sect. B *88*, 143.

Sutherland, I. W. (1977). In *Surface Carbohydrates of the Prokaryotic Cell* (I. W. Sutherland, Ed.). Academic Press, New York, pp. 97–175.

Tomcsik, J. (1956). Soc. Gen. Microbiol. Symp. No. 6, pp. 41–67.

Troy, F. A. (1979). Annu. Rev. Microbiol. *33*, 519.

Wexler, H., and Oppenheim, J. D. (1980). In *Bacterial Endotoxins and Host Response* (M. K. Agarwal, Ed.). Elsevier/North-Holland Biomedical Press, Amsterdam, pp. 27–50.

Wicken, A. J., and Knox, K. W. (1980). Biochim. Biophys. Acta *604*, 1.

Wilkinson, S. G. (1977). In *Surface Carbohydrates of the Prokaryotic Cell* (I. W. Sutherland, Ed.). Academic Press, New York, pp. 97–175.

Wu, T. C. M., and Park, J. T. (1971). J. Bacteriol. *108*, 874.

6

Immunochemistry of Viral Antigens

W. G. Laver John Curtin School of Medical Research, Australian National University, Canberra City, Australia

I. INTRODUCTION

At the present time, vaccination is the only effective way to control virus diseases in humans and animals. Viruses, generally, are excellent antigens and infection or vaccination with a particular virus induces humoral and cellular immune reactions which will usually confer protection against that virus for many years.

The procedure is simple and effective and the methods used to destroy virus infectivity (during the production of inactivated vaccines) do not affect the antigenic determinants on the virus. As a consequence, there has been little incentive to investigate the molecular structure of the antigens of most of the viruses that infect humans or animals.

The exception to this is influenza. Unlike other viruses, the influenza virus has the capacity to vary its antigens so remarkably that the specific immunity established in response to infection by a particular strain may give little or no protection against viruses that subsequently arise. Because of this variation, influenza continues to be a major epidemic disease of humans.

Two distinct kinds of antigenic variation have been demonstrated in influenza viruses, antigenic drift and major antigenic shifts. The first consists of relatively minor changes that occur gradually within a family of strains, all of which are clearly related to each other with respect to both internal and surface antigens. Among influenza A strains infecting humans, each successive variant replaces preexisting ones. This may be due to a selective advantage possessed by antigenic variants in overcoming immunological host resistance. Antigenic drift occurs in type B influenza viruses as well as in type A.

The second kind of antigenic variation, which has been described only for influenza type A, involves much more sudden and dramatic antigenic changes. These are referred to as major antigenic shifts. These major shifts, which occur at intervals of 10 to 15 years,

are marked by the appearance of antigenically "new" viruses to which the population has no immunity, and it is these viruses that cause the major pandemics of influenza. Because of this extraordinary capacity of influenza virus to change its antigenic structure and because the virus has the capacity to cause devastating pandemics of influenza which are impossible to control, the molecular structure of the surface antigens of influenza virus has been under intensive investigation during the past few years (for a review, see Webster et al., 1982b) and more is known about the structure of the two surface antigens of influenza than those of any other virus. However, the immunochemistry of plant viruses also has been studied extensively, see, for example, Van Regenmortel (1982), and Benjamini (1977).

This chapter will be concerned mainly with the immunochemistry of the antigens of influenza type A virus.

II. PROPERTIES OF INFLUENZA VIRUS PARTICLES

Particles of influenza virus exist in a great variety of shapes and sizes and, in the electron microscope, are seen to be covered with a layer of surface projections, or "spikes" (Fig. 1). These spikes are the two surface antigens of the virus, the hemagglutinin (HA) and the neuraminidase (NA), which induce the formation of neutralizing antibodies.

Antibodies to the HA neutralize the infectivity of the virus much more efficiently than do antibodies to the NA and for this reason the immunochemistry of the HA has been investigated more extensively. The HA is a triangular, rod-shaped molecule, while the NA has the shape of a mushroom with a square, boxlike head and a long, thin stalk. Both antigens are attached to the lipid of the virus envelope (derived from the plasma membrane of the host cell) by short sequences of hydrophobic amino acids which occur at the base of the HA spike and near the end of the NA stalk (Fig. 1). The HA and NA are glycosylated and some of the carbohydrate side chains possess antigenic activity characteristic of the host cell in which the virus grew.

Within the lipid envelope lies the matrix protein, which is the major protein component of the virus and which is believed to be structural in function. The genome of the virus consists of eight single-stranded RNA molecules of negative sense (i.e., the virion RNA is complementary to the messenger). Each of the RNA segments codes for one (in some cases two) virus protein, but little is known about the replication complex or how the eight RNA segments are packaged in the virus. The RNA is associated with a nucleocapsid protein and three large proteins, P1, P2, and P3, responsible for RNA replication and transcription. At least three virus-coded nonstructural proteins are found in infected cells, but their functions are unknown.

A. The Hemagglutinin

The HA accounts for about 25% of the virion protein. It is responsible for facilitating entry of the virus into cells* and for the stimulation of antibodies that neutralize the virus. The HA monomer is coded by RNA segment 4 and is synthesized as a single polypeptide chain which undergoes post-translational cleavage at three places (Fig. 2). An N-terminal signal sequence is removed and, depending on the host cell and virus strain, the molecule is cleaved into HA1 and HA2, with the removal of one or more intervening residues. HA1 and HA2 (of molecular weights 47,574 and 28,768, respectively, in the A/Memphis/102/72 strain) are joined by a disulfide bond and each HA spike is composed of three of these

*Due to their small radius of curvature the HA "spikes" can easily overcome the electrostatic repulsion fields surrounding negatively charged cells (see Chap. 31).

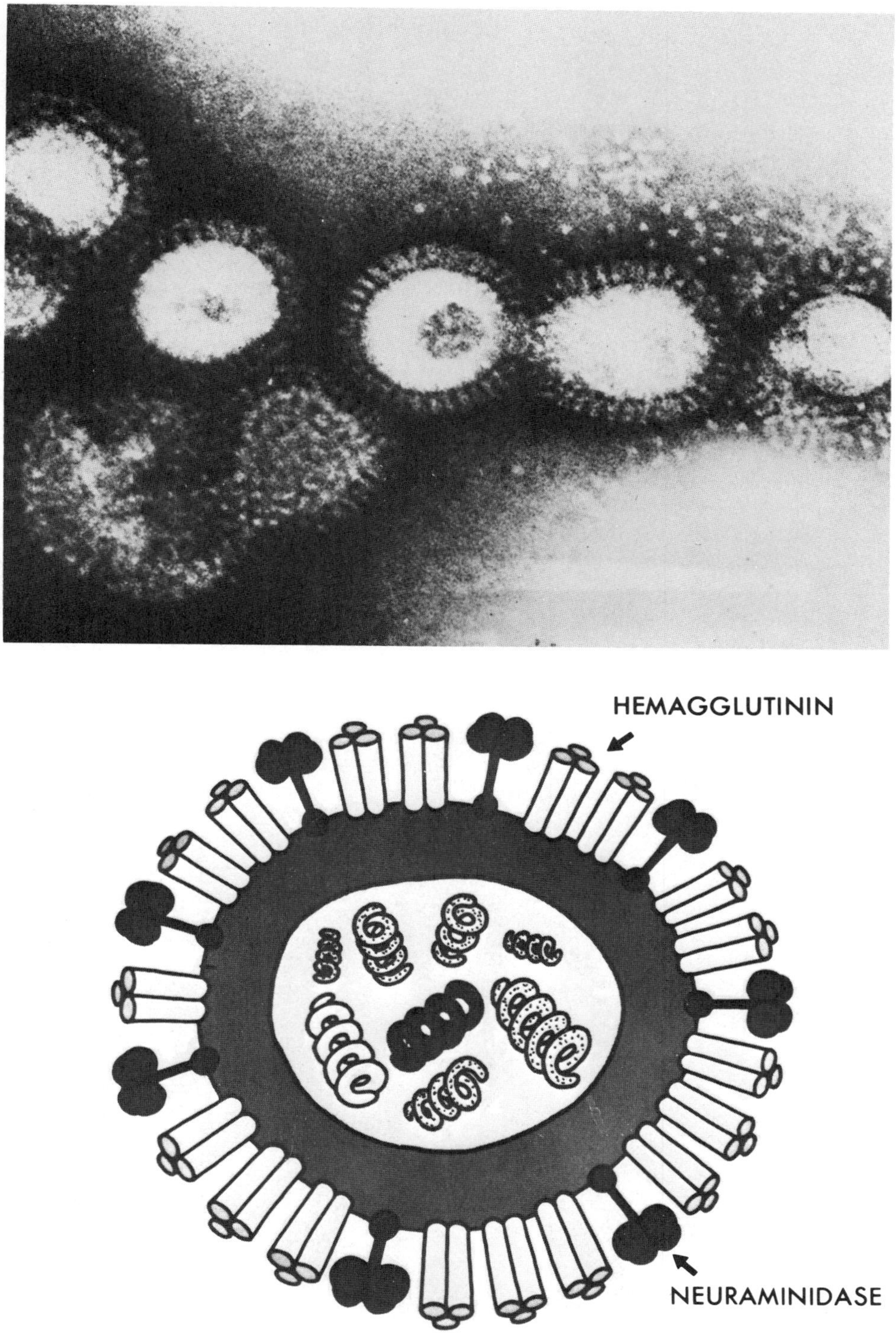

Figure 1 Electron micrograph (top) and diagram showing the surface structure of particles of influenza type A virus. The hemagglutinin spikes are approximately 16 nm in length. The coiled structures inside the virus represent the eight RNA segments of the virus, each of which codes for one (in some cases two, using overlapping reading frames) of the virus proteins.

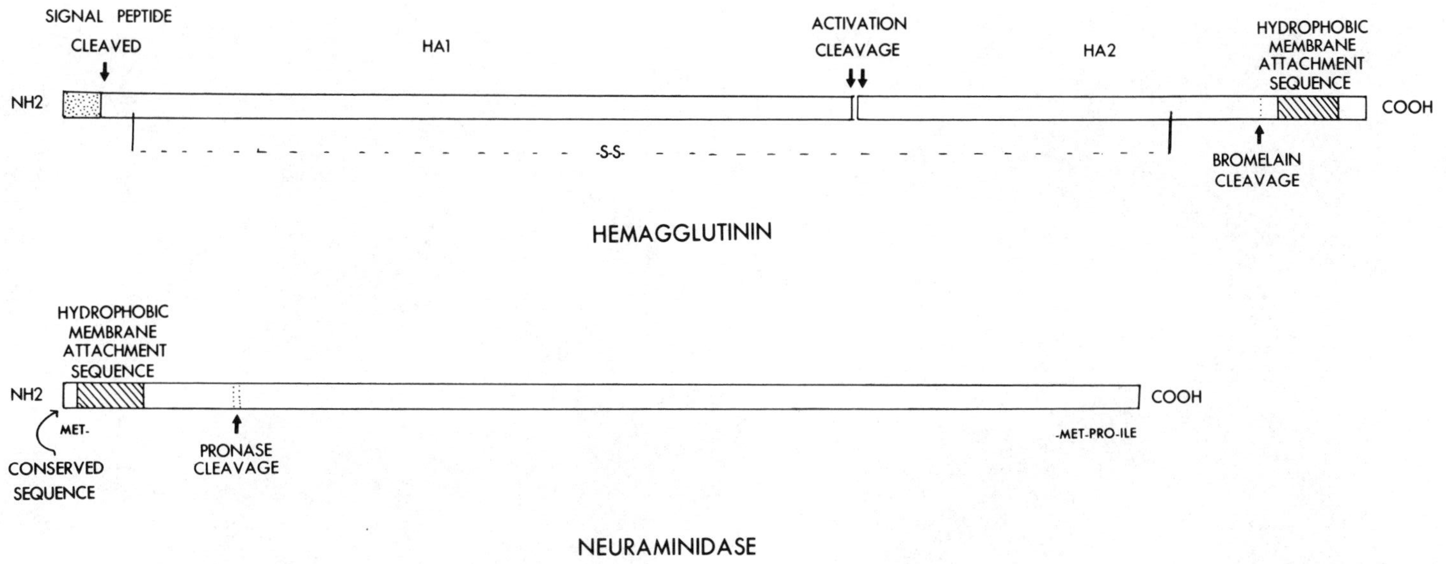

Figure 2 Diagrammatic representation of certain features of the hemagglutinin and neuraminidase polypeptides. *Hemagglutinin*: The HA is synthesized as a single polypeptide. Following its synthesis an N-terminal signal peptide is cleaved off and the molecule is cleaved further into HA1 and HA2 with the removal of one or more intervening amino acids. The latter cleavage is necessary for the virus to be infectious. HA1 and HA2 remain linked by a single disulfide bond and each HA spike contains three of these dimers. A sequence of hydrophobic amino acids near the C terminus of HA2 serves to anchor the HA in the lipid of the virus membrane. Treatment with bromelain removes this hydrophobic region without damaging the rest of the molecule, which, in some cases, can be crystallized. *Neuraminidase*: The neuraminidase is oriented in the virus membrane in the opposite way to the hemagglutinin. No post-translational cleavage of the NA polypeptide occurs, no signal peptide is split off, and even the initiating methionine is retained. No processing at the C terminus takes place; the C-terminal sequence –Met–Pro–Ile predicted from the gene sequence is found in intact NA molecules isolated from virus and in the pronase-released NA heads. A sequence of six polar amino acids at the N terminus of the NA polypeptide, which are totally conserved in at least eight different NA subtypes, is followed by a sequence of hydrophobic amino acids which probably represent the transmembrane region of the NA stalk. This sequence is not conserved at all between subtypes (apart from conservation of hydrophobicity). Pronase cleaves the polypeptide in the positions shown, removing the stalk and releasing the enzymically and antigenically active head of the NA, which, in some cases, can be crystallized.

HA1 + HA2 dimers. A sequence of hydrophobic amino acids near the C terminus of HA2 serves to anchor the HA in the virus membrane.

The HA "spikes" can be isolated in intact form after disruption of the virus particles of the virus particles with detergents such as sodium dodecyl sulfate (SDS). These spikes exist as individual molecules in the presence of the detergent, but when this is removed the HA molecules aggregate by their hydrophobic regions to form rosettes (Fig. 3).

The distal portion of the HA spike may also be isolated after digestion of the virus particles with bromelain. This removes the hydrophobic C-terminal region of HA2 (Fig. 2) which anchors the HA in the viral membrane, and the rest of the HA spike, carrying the receptor binding site and all of the antigenic determinants, is released into the medium.

B. Antigenic Structure of the HA

Early experiments, using polyclonal antisera, on the number and location of antigenic sites on influenza virus HA showed that two distinct groups of sites, strain-specific and cross-reacting, could be distinguished (Laver et al., 1974; Virelizier et al., 1974) and electron micrographs showed that these were located on the side of the HA spike, just below its tip (Fig. 3) (Wrigley et al., 1977).

Further dissection of the antigenic sites on the HA using polyclonal antisera was not possible and, since antigenically active fragments of the HA could not easily be isolated (this will be discussed later) no further information about the antigenic structure of the HA was obtained until monoclonal antibodies to the HA became available.

C. Antigenic Variation in the HA

Major Antigenic Shifts

A major antigenic shift in influenza type A viruses infecting humans occurs when a virus suddenly appears in the human population having a hemagglutinin (and sometimes also a neuraminidase) antigen which is totally urelated, serologically, to the viruses currently circulating. This massive change in antigenic properties involves complete replacement of the RNA segment coding for the HA.

Twelve distinct hemagglutinin subtypes are currently recognized by the World Health Organization (WHO Memorandum, 1980). These subtypes are all quite unrelated, serologically, but some show cross-reactions in cell-mediated immune responses.

Major antigenic shifts occurred in 1957 (H1N1 → H2N2), in 1968 (H2N2 → H3N2), and in 1977 when a 1950 strain of the H1N1 subtype reappeared. (Following the emergence of the H2N2 subtype, H1N1 viruses disappeared from the human population.)

All of the HA subtypes exist in viruses infecting birds and lower mammals, and the way in which the "new" viruses suddenly appear (or reappear) in humans is not known.

Antigenic Drift

Drift is a slow change in the antigenic properties of the HA which takes place over a long period of time. Antigenic drift has been well documented in the H3N2 subtype (Hong Kong influenza) and serologic cross-reactions between members of the Hong Kong series of viruses are shown in Table 1. All are clearly related, but the early and late viruses are vastly different in their antigenic properties.

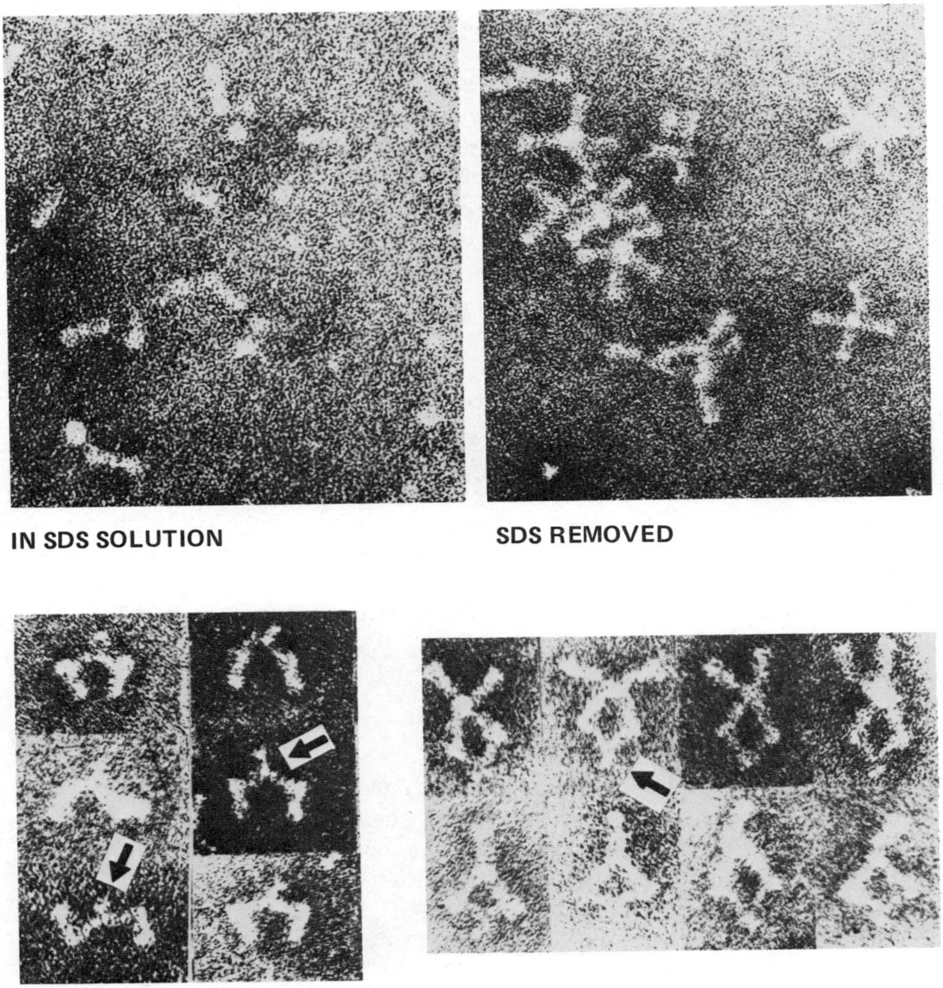

HEMAGGLUTININ MOLECULES

IN SDS SOLUTION

SDS REMOVED

BROMELAIN RELEASED HEMAGGLUTININ PLUS 1GG ANTIBODY

SDS-RELEASED HEMAGGLUTININ PLUS 1GG ANTIBODY

Figure 3 Electron micrographs showing isolated hemagglutinin molecules. The top pictures show HA isolated after disruption of virus with sodium dodecyl sulfate (SDS). The HA exists as individual molecules in the presence of the detergent (left) which aggregate by their hydrophobic ends (these serve to attach the HA to the lipid of the virus envelope) when the detergent is removed (right). The lower pictures show the attachment of IGG antibody molecules to the HA. On the left is bromelain-released HA which has lost its hydrophobic membrane-attachment region and exists as individual molecules in the absence of detergent. On the right is aggregated, detergent-released intact HA. Anti-HA IGG antibody molecules (indicated by arrows) can be seen attached to the HA molecules. The antibody seems to bind to the HA just below the tip of the HA spike. (Courtesy of Robin Valentine and Nick Wrigley.)

Table 1 Antigenic Drift of H3N2 Viruses Between 1968 and 1977 as Demonstrated by Their Cross-Reactions in Hemagglutination-Inhibition Tests

Virus	Ferret Antiserum							
	A/HK/1/68	A/E/42/72	A/PC/1/73	A/Sc/840/74	A/Vic/1/75	A/Tok/1/75	A/E/864/75	A/Tex/1/77
A/Hong Kong/1/68	2560	2560	320	80	20	<20	20	20
A/England/42/72	40	1280	320	160	40	<20	40	20
A/Port Chalmers/1/73	20	640	640	160	40	<20	40	20
A/Scotland/840/74	<20	40	40	640	<20	<20	<20	<20
A/Victoria/1/75	<20	<20	<20	20	1280	80	160	20
A/Tokyo/1/75	<20	<20	<20	<20	320	1280	40	40
A/England/864/75	<20	80	80	20	320	<20	5120	640
A/Texas/1/77	<20	40	80	20	160	<20	5120	1280

Source: Pereira (1979).

III. AMINO ACID SEQUENCE CHANGES IN THE HA POLYPEPTIDES OF NATURAL ANTIGENIC VARIANTS (FIELD STRAINS) OF INFLUENZA VIRUSES

The HA from a number of variants of the A/Hong Kong/68 (H3N2) strain of influenza virus have been sequenced. This virus appeared in humans in 1968 and since then has undergone considerable antigenic drift, so that viruses isolated in 1979 were only distantly related, serologically, with the original 1968 strain (Table 1).

Most of the amino acid sequence changes that occurred between 1968 and 1979 were in the HA1 portion of the molecule, only three changes being found in HA2 (for a review, see Ward, 1981). 33/328 residues in HA1 changed and these changes were distributed fairly evenly along the HA1 polypeptide (Fig. 4), but is is not possible to tell which of the sequence changes were responsible for changes in the antigenic properties of the HA and which were incidental, unselected changes unrelated to drift.

IV. USE OF MONOCLONAL ANTIBODIES TO INVESTIGATE THE ANTIGENIC STRUCTURE OF INFLUENZA VIRUS HA

A large number of monoclonal hybridoma antibodies have been generated which recognize different antigenic sites on the surface of the influenza virus HA molecule. When cells are infected with virus in the presence of one of these monoclonal antibodies, all of the virus particles are very effectively neutralized with the exception of variants present in the inoculum, which have changes in the antigenic site on the HA specifically recognized by the particular monoclonal antibody used. These variants, which occur with a frequency of about 1 in 10^5 virus particles do not bind the monoclonal antibody used and grow freely in its presence. A large number of antigenic variants of the PR8/34 (H1N1) and Hong Kong (H3N2) strains of influenza type A virus were isolated in this way with a number of different monoclonal antibodies (Gerhard and Webster, 1978; Webster et al., 1980).

Table 2 illustrates the selection of variants using three quite distinct monoclonal antibodies. Each of the variants failed to bind the particular monoclonal antibody used for its selection, but variants selected with one monoclonal antibody were inhibited in HI tests to the same titer as wild-type virus when tested with the other two monoclonal antibodies. This suggested that the antigenic sites recognized by these three monoclonal antibodies were different and that the antibodies bound to independent regions on the hemagglutinin molecule (Table 2).

Further analysis of the variants in antibody binding assays with a panel of monoclonal antibodies showed that they could be divided into four distinct groups, suggesting that at least four nonoverlapping antigenic regions exist on the HA molecule. However, different variants selected with a particular monoclonal antibody could, in some cases, be distinguished from each other with other monoclonal antibodies (Table 3). Thus the three variants (V1, V2, and V3) selected with H14/A20 (which fell into one of the nonoverlapping regions on the HA) could be discriminated with the monoclonal antibodies Mem 27/2, Mem 212/1, and Mem 123/4 (Table 3), suggesting that within the nonoverlapping antigenic regions individual sites *do* overlap and that some amino acids may be part of two (or more) sites.

The term *antigenic site* is used in the sense in which it is used by Atassi and Smith (1978). An antigenic site comprises those amino acids which are involved in reaction with antibody. These amino acids may, of course, be widely separated in the primary structure of the polypeptide chain, but are brought into close association by folding of the protein.

	10	20	30	40	50	60
NT68	QDLPGNDNNT	ATLCLGHHAV	PNGTLVKTIT	DDQIEVTNAT	ELVQSSSTGK	ICNNPHRILD
X31	QDLPGNDNST	ATLCLGHHAV	PNGTLVKTIT	DDQIEVTNAT	ELVQSSSTGK	ICNNPHRILD
ENG69	QDLPGNDNST	ATLCLGHHAV	PNGTLVKTIT	NDQIEVTNAT	ELVQSSSTGK	ICNNPHRILD
QU70	QDLPGNDNST	ATLCLGHHAV	PNGTLVKTIT	NDQIEVTNAT	ELVQSSSTGK	ICNNPHRILD
MEM72	QDFPGNDNST	ATLCLGHHAV	PNGTLVKTIT	NDQIEVTNAT	ELVQSSSTGK	ICNNPHRILD
VIC375	QDLPGNDNST	ATLCLGHHAV	PNGTLVKTIT	NDQIEVTNAT	ELVQSSSTGK	ICNNPHRILD
TEX77	QNLPGNDNST	ATLCLGHHAV	PNGTLVKTIT	NDQIEVTNAT	ELVQSSSTGR	ICDSPHRILD
BK79	QNLPGNDNST	ATLCLGHHAV	PNGTLVKTIT	NDQIEVTNAT	ELVQSSSTGR	ICDSPHRILD

	70	80	90	100	110	120
NT68	GIDCTLIDAL	LGDPHCDVFQ	NETWDLFVER	SKAFSNCYPY	DVPDYASLRS	LVASSGTLEF
X31	GIDCTLIDAL	LGDPHCDVFQ	NETWDLFVER	SKAFSNCYPY	DVPDYASLRS	LVASSGTLEF
ENG69	GINCTLIDAL	LGDPHCDVFQ	DETWDLFVER	SKAFSNCYPY	DVPDYASLRS	LVASSGTLEF
QU70	GIDCTLIDAL	LGDPHCDGFQ	NETWDLFVER	SKAFSNCYPY	DVPDYASLRS	LVASSGTLEF
MEM72	GINCTLIDAL	LGDPHCDGFQ	NETWDLFVER	SKAFSNCYPY	DVPDYASLRS	LVASSGTLEF
VIC375	GINCTLIDAL	LGDPHCDGFQ	NEKWDLFVER	SKAFSNCYPY	DVPDYASLRS	LVASSGTLEF
TEX77	GKNCTLIDAL	LGDPHCDGFQ	NEKWDLFVER	SKAFSNCYPY	DVPDYASLRS	LVASSGTLEF
BK79	GKNCTLIDAL	LGDPHCDGFQ	NEKWDLFVER	SKAFSNCYPY	DVPDYASLRS	LVASSGTLEF

	130	140	150	160	170	180
NT68	ITEGFTWTGV	TQNGGSNACK	RGPGSGFFSR	LNWLTKSGST	YPVLNVTMPN	NDNFDKLYIW
X31	ITEGFTWTGV	TQNGGSNACK	RGPGSGFFSR	LNWLTKSGST	YPVLNVTMPN	NDNFDKLYIW
ENG69	ITEGFTWTGV	TQNGGSNACK	RGPDSGFFSR	LNWLTKSGST	YPVLNVTMPN	NDNFDKLYIW
QU70	ITEGFTWTEV	TQNGGSNACK	RGPGSGFFSR	LNWLTKSGST	YPVLNVTMPN	NDNFDKLYIW
MEM72	INEGFTWTGV	TQNGGSNACK	RGPDSGFFSR	LNWLYKSGST	YPVLNVTMPN	NDNFDKLYIW
VIC375	INEGFNWTGV	TQNGGSSACK	RGPDSGFFSR	LNWLYKSGST	YPVQNVTMPN	NDNSDKLYIW
TEX77	INEGFNWTGV	TQNGGSYACK	RGPDNSFFSR	LNWLYKSEST	YPVLNVTMPN	NGNFDKLYIW
BK79	INEGFNWTGV	TQSGGSYACK	RGSDNSFFSR	LNWLYESESK	YPVLNVTMPN	NGNFDKLYIW

	190	200	210	220	230	240
NT68	GVHHPSTNQE	QTSLYVQASG	RVTVSTRRSQ	QTIIPNIGSR	PNVRGLSSRI	SIYWTIVKPG
X31	GIHHPSTNQ	QTSLYVQASG	RVTVSTRRSQ	QTIIPNIGSR	PWVRGLSSRI	SIYWTIVKPG
ENG69	GVHHPSTNQE	QTSLYVQASG	RVTVSTRRSQ	QTIIPNIGSR	PWVRGLSSRI	SIYWTIVKPG
QU70	GVHHPSTNQE	QTSLYVQASG	RVTVSTRRSQ	QTIIPNIGSR	PWVRGQSSRI	SIYWTIVKPG
MEM72	GVHHPSTDQE	QTSLYVQASG	RVTVSTKRSQ	QTIIPNIGSR	PWVRGQSSRI	SIYWTIVKPG
VIC375	GVHHPSTDKE	QTNLYVQASG	KVTVSTKRSQ	QTIIPNVGSR	PWVRGLSSRI	SIYWTIVKPG
TEX77	GVHHPSTDKE	QTNLYVQASG	RVTVSTKRSQ	QTIIPNVGSR	PWVRGLSSGI	SIYWTIVKPG
BK79	GVHHPSTDKE	QTNLYVRASG	RVTVSTKRSQ	QTIIPNIGSR	PWVRGLSSRI	SIYWTIVKPG

	250	260	270	280	290	300
NT68	DVLVINSNGN	LIAPRGYFKM	RTGKSSIMRS	DAPIDTCISE	CITPNGSIPN	DKPFQNVNKI
X31	DVLVINSNGN	LIAPRGYFKM	RTGKSSIMRS	DAPIDTCISE	CITPNGSIPN	DKPFQNVNKI
ENG69	DVLVINSNGN	LIAPRGYFKM	RTGKSSIMRS	DAPIDTCISE	CITPNGSIPN	DKPFQNVNKI
QU70	DVLVINSNGN	LIAPRGYFKM	RTGKSSIMRS	DAPIDTCISE	CITPNGSIPN	DKPFQNVNKI
MEM72	DILVINSNGN	LIAPRGYFKM	RTGKSSIMRS	DAPIGTCISE	CITPNGSIPN	DKPFQNVNKI
VIC375	DILVINSNGN	LIAPRGYFKM	RTGKSSIMRS	DAPIGTCSSE	CITPNGSIPN	DKPFQNVNKI
TEX77	DILLINSNGN	LIAPRGYFKI	RTGKSSIMRS	DAPIGTCSSE	CITPNGSIPN	DKPFQNVNKI
BK79	DILLINSNGN	LIAPRGYFKI	RTGKSSIMRS	DAPIGTCSSE	CITPNGSIPN	DKPFQNVNKI

	310	320	
NT68	TYGACPKYVK	QNTLKLATGM	RNVPEKQT
X31	TYGACPKYVK	QNTLKLATGM	RNVPEKQT
ENG69	TYGACPKYVK	QNTLKLATGM	RNVPEKQT
QU70	TYGACPKYVK	QNTLKLATGM	RNVPEKQT
MEM72	TYGACPKYVK	QNTLKLATGM	RNVPEKQT
VIC375	TYGACPKYVK	QNTLKLATGM	RNVPEKQT
TEX77	TYGACPKYVK	QNTLKLATGM	RNVPEKQT
BK79	TYGACPKYVK	QNTLKLATGM	RNVPEKQT

Figure 4 Amino acid sequences using the single-letter amino acid code of the HA1 polypeptide from eight variants of Hong Kong (H3N2) influenza virus isolated between 1968 and 1979. Residues that have changed from NT68 are boxed. The location of most of these changes on the three-dimensional structure of the HA are shown in Fig. 6. Arrows indicate the single sequence changes found in variants selected with monoclonal antibodies. These are (53) Asn → Lys, (133) Asn → Lys, (143) Pro → Ser, Thr, Leu, His, (144) Gly → Asp, (145) Ser → Lys, (205) Ser → Tyr. The change (226) Leu → Gln appears to arise with remarkable ease. This change has been found in different clones of the same field strain and in monoclonal variants which already show another change and may be unrelated to changes in antigenicity.

Table 2 Cross-Reactions of the Antigenic Variants with the Monoclonal Antibodies Used for Their Selection

Monoclonal antibody to A/Mem/1/71 (H)	Wild-type	HI titer with the following antigenic variants of A/Mem/1/71 virus[a]									
		Variants selected with H14/A21			Variants selected with H14/A20			Variants selected with H14/A2			
		V1	V2	V3	V1	V2	V3	V1	V2	V3	V4
H14/A21	5.0	<[b]	<	<	5.0	5.0	5.0	5.0	5.0	5.0	5.0
H14/A20	4.8	4.8	4.8	4.8	<	<	<	4.8	4.8	4.8	4.8
H14/A2	4.1	4.1	4.1	4.1	4.1	4.1	4.1	2.4	2.4	2.4	2.4

[a]HI titers expressed as $\log_{10}$.
[b]Less than 1.0.

Table 3 Cross-Reactions of the Antigenic Variants with a Panel of Monoclonal Antibodies to A/Mem/1/71 Hemagglutinin

		HI titers of the following variants[a]					
		Selected with H14/A20[b]			Selected with H14/A21		
preparation	Wild-type	V1	V2	V3	V1	V2	V3
Mem 93/1	4.1	<[c]	<	<	+[d]	+	+
Mem 27/2	5.0	5.0	<	<	+	+	+
Mem 212/1	5.1	5.1	2.6	<	+	+	+
Mem 123/4	4.6	4.6	2.5	3.6	+	+	+
H14/B18	4.0	+	+	+	2.5	<	2.5
Mem 200/2	5.0	+	+	+	4.4	3.2	4.4
HK 30/1	4.8	+	+	+	4.4	4.4	4.4
12 Other monoclonal antibody preparations	+	+	+	+	+	+	+

[a]HI titers expressed as $\log_{10}$.
[b]H14/A20 V1, V2, and V3 showed sequence changes of Asn(133) → Lys, Pro(143) → Ser, and Pro(143) → Leu, repsectively.
[c]Less than 1.7.
[d]Titers for wild-type and variants were identical.

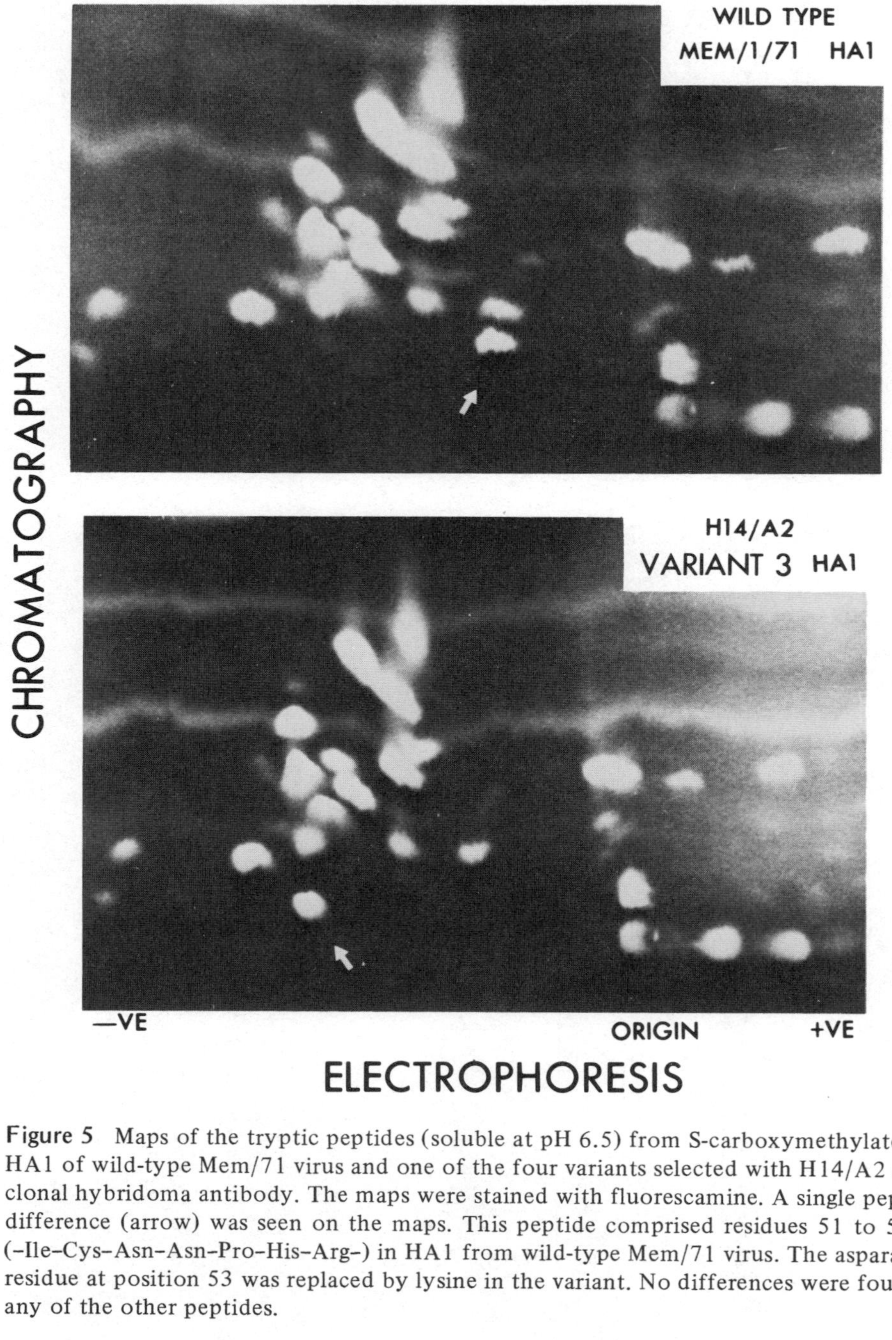

Figure 5 Maps of the tryptic peptides (soluble at pH 6.5) from S-carboxymethylated HA1 of wild-type Mem/71 virus and one of the four variants selected with H14/A2 monoclonal hybridoma antibody. The maps were stained with fluorescamine. A single peptide difference (arrow) was seen on the maps. This peptide comprised residues 51 to 57 (-Ile-Cys-Asn-Asn-Pro-His-Arg-) in HA1 from wild-type Mem/71 virus. The asparagine residue at position 53 was replaced by lysine in the variant. No differences were found in any of the other peptides.

V. SEQUENCE CHANGES IN THE HA OF VARIANTS SELECTED WITH MONOCLONAL ANTIBODIES

Amino acid changes in the hemagglutinin of variants selected with monoclonal antibodies were determined in the following way (Laver et al., 1981). HA molecules were isolated from detergent-disrupted virus particles and the two polypeptides, HA1 and HA2, were separated. These were S-carboxymethylated, digested with trypsin, and the soluble tryptic peptides were separated on filter paper sheets by two-dimensional electrophoresis and chromatography. The peptides were located by staining with fluorescamine, eluted from the paper, hydrolyzed with 6 N HCl, and analyzed for amino acid composition. The complete amino acid sequence of HA1 and HA2 from the wild-type virus was known from other RNA and protein sequence data and the sequence changes in HA1 and HA2 of the variants could be deduced by comparing the amino acid compositions of the peptides from the variants, with the compositions of the homologous peptides from the wild-type virus. This procedure in most cases gave unambiguous results, but if ambiguities existed, the relevant peptides were either sequenced directly, or the portion of the gene corresponding to that peptide was sequenced.

An example of the results obtained using this method is shown in Fig. 5. One of the variant peptides had increased electrophoretic mobility and its analysis showed that this

Table 4 Sequence Changes Found in the HA1 Polypeptide from Antigenic Variants of A/Memphis/1/71 Virus Selected with the Monoclonal Antibodies Listed

Monoclonal antibody	Variant	Sequence change[a]
H14/A2	V1	Asparagine (53) → lysine
	V2	Asparagine (53) → lysine
	V3	Asparagine (53) → lysine
	V4	Asparagine (53) → lysine
H14/A21	V1	Serine (205) → Tyrosine
H14/A20	V1	Asparagine (133) → lysine
	V2	Proline (143) → serine
	V3	Proline (143) → leucine
	V4	Serine (145) → lysine
Mem/212/1	V1	Proline (143) → serine
	V2	Proline (143) → threonine
	V3	Proline (143) → leucine
	V7	Proline (143) → histidine
Mem 27/2	V5	Proline (143) → serine
	V9	Proline (143) → threonine
Mem 123/4	V1	Proline (143) → histidine
	V3	Proline (143) → histidine
	V10	Glycine (144) → aspartic acid

[a]Numbers in parentheses give the position of the amino acid in HA1.

was due to the replacement of an asparagine residue by lysine. In the wild-type HA this peptide occupied residues 51 to 57 in HA1 and had the sequence -Ile-Cys-Asn-Asn-Pro-His-Arg-. However, it was not possible to tell from the composition which of the two asparagine residues had changed to lysine. The region of the gene coding for this part of the HA was therefore sequenced and the results showed that it was the asparagine at position 53 that changed to lysine. All of the other soluble peptides from the variant HA had the same composition as the wild type.

Two insoluble tryptic peptides, comprising residues 110 to 140 and 230 to 255 in the HA1 molecule, were not examined and it is not known if additional changes occurred in these regions. No changes were found in the HA2 region of the HA molecule.

In 10 other variants of Hong Kong (H3N2) influenza virus selected with monoclonal antibodies, the proline residue at position 143 in HA1 changed to serine, threonine, leucine, or histidine. In other variants, asparagine 133 changed to lysine, glycine 144 to aspartic acid, and serine 145 to lysine. All these changes are possible by single base changes in the RNA except the last, which requires a double base change. The changes in these variants are listed in Table 4. Some variants selected with the same monoclonal antibody showed different sequence changes and some variants selected with different monoclonal antibodies showed the same sequence change. The reason for this is not known.

VI. LOCATION OF THE SEQUENCE CHANGES ON THE THREE-DIMENSIONAL STRUCTURE OF THE HA

When the three-dimensional structure of the HA was solved (Wilson et al., 1981; Wiley et al., 1981), it became clear that many of the amino acids that changed during antigenic drift (both in nature and in the laboratory variants) and which were scattered along the HA1 polypeptide came together when the polypeptide folded into its three-dimensional structure. The variable residues formed four distinct clusters on the surface of the HA (Fig. 6), which are assumed to be the four nonoverlapping antigenic regions previously defined using monoclonal antibodies. Thus site C contains the variable residues at positions 53 and 54 and 275 and 278, which are widely separated in the polypeptide chain but come together after it is folded. However, there is no direct evidence at the present time that the amino acids in these regions actually make contact with complementary binding sites on antibody molecules.

VII. MECHANISM OF DRIFT

It is clear that the variants obtained in the selection experiments using monoclonal antibodies have lost an antigenic determinant and that this loss is due to a single change in the amino acid sequence of HA1. What happens when the old site is lost? Is a new one created? What happens when variants are selected, using antibody against the new site? Does the amino acid that changed in the first selection undergo a further change in the second selection?

Examination of the sequence changes that occurred in the field strains between 1968 and 1979 (Fig. 4) shows that once an amino acid changed it did not change again in subsequent variants, suggesting that antigenic drift does not occur by way of a series of sequential changes at the same position in the HA. The only exceptions to this were a number of instances where an altered amino acid apparently changed back to the original one at that position. Because the field strains examined do not represent a direct

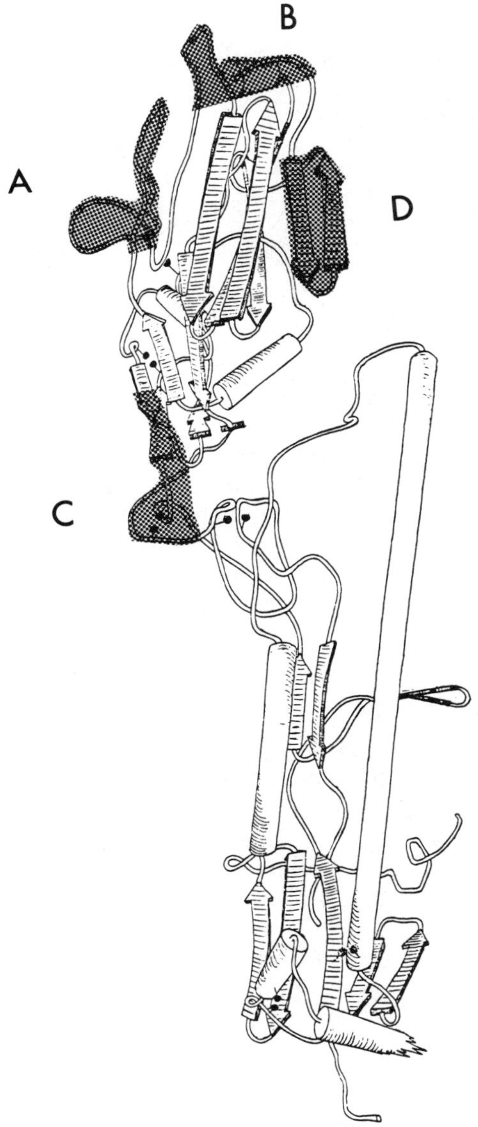

Figure 6 Drawing (by Hidde Pleogh) of the Hong Kong monomer showing folding of the HA1 and HA2 polypeptides (Wilson et al., 1981; Wiley et al., 1981). The four shaded areas show where four independent antigenic areas may be located. Note that in the HA spike, which is composed of three of the monomers, site (D) is buried and may not be involved in antibody binding.

genealogical lineage, however, we do not know if these changes were true reversions or whether in the particular virus examined that amino acid never changed in the first place.

In order to obtain antibody directed specifically against the new antigenic determinant on the variant HA, rabbits were hyperimmunized with the isolated variant HA and

	141 ... 150
FIRST SELECTION WITH MONOCLONE MEM 123/4	-ARG-GLY-PRO-GLY-SER-GLY-PHE-PHE-SER-ARG-
	↓
	-ARG-GLY-HIS-GLY-SER-GLY-PHE-PHE-SER-ARG
	↓
SECOND SELECTION WITH ABSORBED ANTISERUM TO THE FIRST VARIANT	-ARG-GLY-HIS-ASP-SER-GLY-PHE-PHE-SER-ARG-
	(3/3 VARIANTS SHOWED SAME CHANGE)

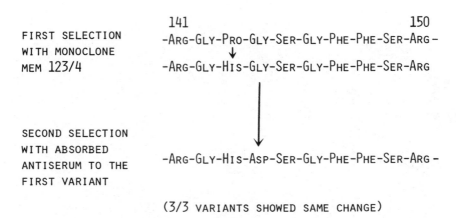

Figure 7 Sequence changes during the sequential selection of antigenic variants of Mem/71 virus. In the first variant, selected in the presence of mouse monoclonal hybridoma antibody to Mem/71 HA, the proline at position 143 in HA1 changed to histidine. In the second selection, the initial variant was grown in the presence of antibody raised against the new antigenic site on this variant created by the sequence change of Pro (143) to His. This antibody was prepared by absorbing with wild-type Mem/71 virus hyperimmune rabbit antiserum raised against the variant HA.

the sera were absorbed with wild-type virus until all of the antibody that bound to wild-type HA was removed. There remained a small amount of unabsorbed antibody that bound to the variant but not to the wild type and was presumably directed specifically against the new antigenic site on the variant HA. In this respect it behaved like "monoclonal" antibody.

Antibody was prepared in this way against the new antigenic site on the variant selected with MEM 123/4 monoclonal antibody, which had a sequence change at position 143 in HA1 of proline to histidine. This antibody was then used to select second-generation variants of the initial variant. However, the amino acid that changed in the second selection was *not* at the same position as the one that changed in the initial selection (Fig. 7). In three out of three second-generation variants, the glycine at position 144 changed to aspartic acid, while the histidine at position 143 was retained in each.

In a similar experiment using a variant selected with Mem 212/1 monoclonal antibody, which had a sequence change at position 143 of proline to threonine, the glycine at position 144 was retained but the threonine at position 143 reverted to proline and the variant regained the antigenic properties of the wild type. This experiment is important because it supports the notion that only a single sequence change in HA1 is needed to abolish totally the binding of a monoclonal antibody to the HA.

VIII. ANTIGENIC SITES ON INFLUENZA VIRUS NEURAMINIDASE

The neuraminidase (NA) is the other surface antigen on particles of influenza virus (Fig. 1). It is a mushroom-shaped molecule with a boxlike head and a long, thin stalk. The NA can be isolated from virus after disruption of the virus particles with detergents such as SDS. Subsequent removal of the detergent allows the NA molecules to aggregate by the hydrophobic regions near the end of the stalk to form rosettes (Fig. 8). The NA can also

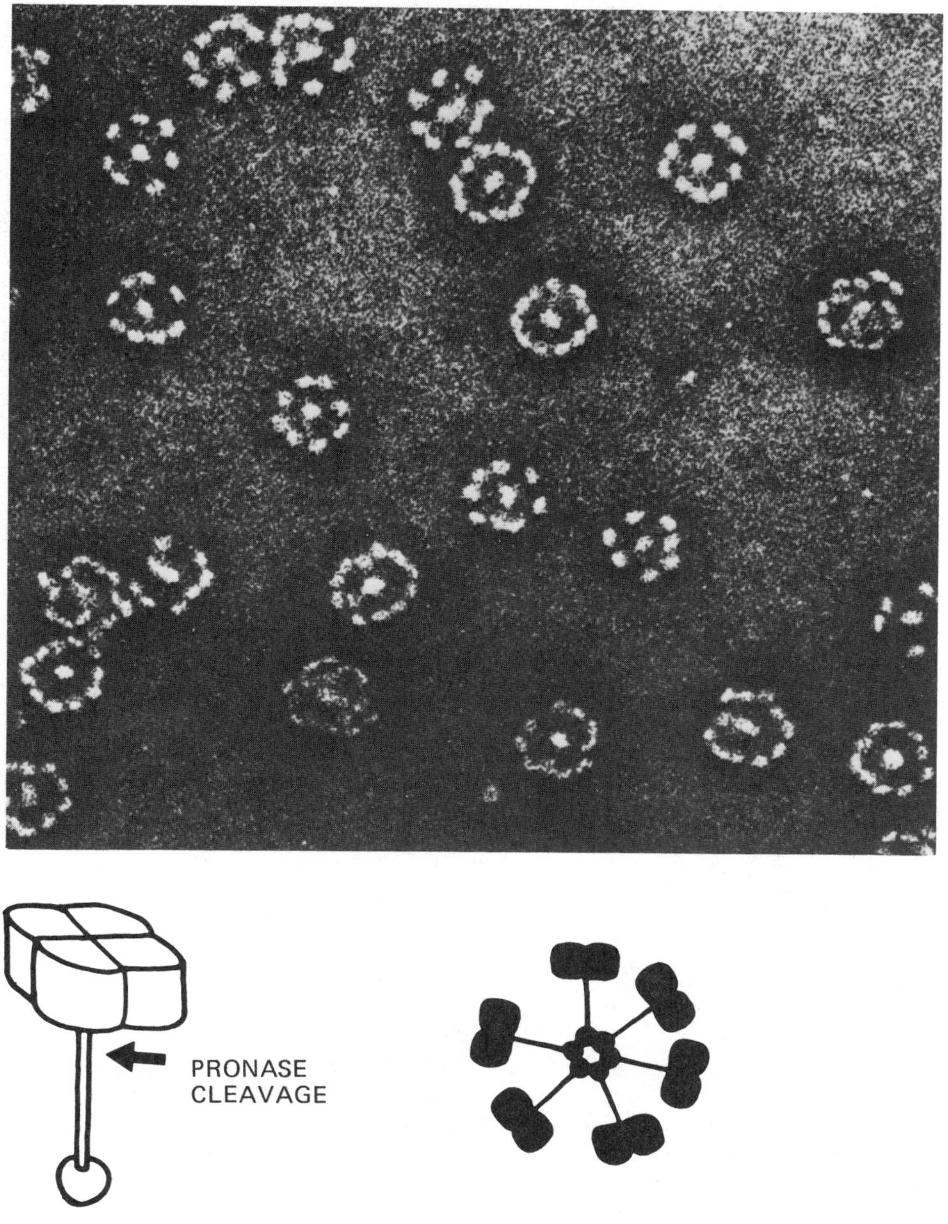

A B

Figure 8 Electron micrograph and diagram showing detergent-released neuraminidase molecules from which the detergent has been removed. The NA molecules have aggregated by the hydrophobic region near the end of the stalk, which served to attach the NA to the lipid of the virus envelope. Treatment of virus particles with pronase releases the head of the NA, which carries the enzymic and antigenic activities of the molecule and, in some cases, can be crystallized. (Electron micrograph courtesy of Nick Wrigley.) (A) Single neuraminidase molecule; (B) Rosette formed by six neuraminidase molecules.

	1	2	3	4	5	6	7	8	9	10	11	12	13	14	15	16	17	18	19	20	21	22	23
					5					10					15					20			
UDORN/72	MET	ASN	PRO	ASN	GLN	LYS	ILE	ILE	THR	ILE	GLY	SER	VAL	SER	LEU	THR	ILE	ALA	THR	ILE	CYS	PHE	LEU
					55					60					65					70			
RI/5+/57																							
TOKYO/3/67																							
AICHI/2/68																							
ENG/42/72																							
UDORN/72	MET	PRO	CYS	GLU	PRO	ILE	ILE	ILE	GLU	ARG	ASN	ILE	THR	GLU	ILE	VAL	TYR	LEU	THR	ASN	THR	THR	ILE
VIC/3/75																							
					105					110					115					120			
RI/5+/57	SER	LYS	ASP	ASN	SER	ILE	ARG	LEU	SER	ALA	GLY	GLY	ASP	ILE	TRP	VAL	THR	ARG	GLU	PRO	TYR	VAL	SER
TOKYO/3/67	SER	LYS	ASP	ASN	SER	ILE	ARG	LEU	SER	ALA	GLY	GLY	ASP	ILE	TRP	VAL	THR	ARG	GLU	PRO	TYR	VAL	SER
AICHI/2/68	SER	LYS	ASP	ASN	SER	ILE	ARG	LEU	SER	ALA	GLY	GLY	ASP	ILE	TRP	VAL	THR	ARG					
ENG/42/72	SER	LYS	ASP	ASN	SER	ILE	ARG	LEU	SER	ALA	GLY	GLY	ASP	ILE	TRP	VAL	THR	ARG	GLU	PRO	TYR	VAL	SER
UDORN/72	SER	LYS	ASP	ASN	SER	ILE	ARG	LEU	SER	ALA	GLY	GLY	ASP	ILE	TRP	VAL	THR	ARG	GLU	PRO	TYR	VAL	SER
VIC/3/75	SER	LYS	ASP	ASN	SER	ILE	ARG												GLU	PRO	TYR	VAL	SER
					155					160					165					170			
RI/5+/57			ILE	PRO	HIS	ARG	THR	LEU	LEU	MET	ASN	GLU	LEU	GLY	VAL	PRO	PHE	HIS	LEU	GLY	THR	LYS	GLN
TOKYO/3/67	ASP	ARG	ILE	PRO	HIS	ARG	THR	LEU	LEU	MET	ASN	GLU	LEU	GLY	VAL	PRO	PHE	HIS	LEU	GLY	THR	ARG	
AICHI/2/68	ASP	ARG	ILE	PRO	HIS	ARG	THR	LEU	LEU	MET	ASN	GLU	LEU	GLY	VAL	PRO	PHE	HIS	LEU	GLY	THR	ARG	
ENG/42/72			ILE	PRO	HIS	ARG	THR	LEU	LEU	MET	ASN	GLU	LEU	GLY	VAL	PRO	PHE	HIS	LEU	GLY	THR	ARG	GLN
UDORN/72	ASP	ARG	THR	PRO	HIS	ARG	THR	LEU	LEU	MET	ASN	GLU	LEU	GLY	VAL	PRO	PHE	HIS	LEU	GLY	THR	ARG	
VIC/3/75	ASP	ARG	THR	PRO	HIS	ARG	THR	LEU	LEU	MET	SER	GLU	LEU	GLY	VAL	PRO	PHE	HIS	LEU	GLY	THR	ARG	
																					HIS ↑		
					205					210					215					220			
R1/5+/57											LEU	VAL	ASP	SER	ILE	GLY	SER	TRP	SER	GLN	ASN	ILE	LEU
TOKYO/3/67					INSOLUBLE PEPTIDE →						LEU	VAL	ASP	SER	ILE	GLY	SER	TRP	SER	GLN	ASN	ILE	LEU
AICHI/2/68											LEU	VAL	ASP	SER	ILE	GLY	SER	TRP	SER	GLN	ASN	ILE	LEU
ENG/42/72											LEU	VAL	ASP	SER	ILE	GLY	SER	TRP	SER	GLN	ASN	ILE	LEU
UDORN/72	ALA	THR	ALA	SER	PHE	ILE	TYR	ASP	GLY	ARG	LEU	VAL	ASP	SER	ILE	GLY	SER	TRP	SER	GLN	ASN	ILE	LEU
VIC/3/75											LEU	VAL	ASP	SER	ILE	GLY	SER	TRP	SER	GLN	ASN	ILE	LEU
					255					260					265					270			
RI/5+/57	ASP	THR	ARG	ILE	LEU	PHE	ILE	LYS	GLU	GLY	LYS	ILE	ILE	HIS	ILE	SER	PRO	LEU	SER	GLY	SER	ALA	GLN
TOKYO/3/67	ASP	THR	ARG	ILE	LEU	PHE	ILE	GLU	GLU	GLY	LYS	ILE	VAL	HIS	ILE	(SER)	PRO	LEU	(SER)	GLY	(SER)	ALA	GLN
AICHI/2/68	ASP	THR	ARG	ILE	LEU	PHE	ILE	GLU	GLU	GLY	LYS	ILE	VAL	HIS	ILE	SER	PRO	LEU	SER	GLY	SER	ALA	GLN
ENG/42/72	ASP	THR	ARG	ILE	LEU	PHE	ILE	GLU	GLU	GLY	LYS	ILE	VAL	HIS	ILE	SER	PRO	LEU	SER	GLY	SER	ALA	GLN
UDORN/72	ASP	THR	LYS	ILE	LEU	PHE	ILE	GLU	GLU	GLY	LYS	ILE	VAL	HIS	ILE	SER	PRO	LEU	SER	GLY	SER	ALA	GLN
VIC/3/75	ASP	THR	LYS	ILE	LEU	PHE	ILE	GLU	GLU	GLY	LYS	ILE	VAL	HIS	ILE	SER	PRO	LEU	SER	GLY	SER	ALA	GLN
					305					310					315					320			
RI/5+/57	PRO	(VAL	VAL)	ASX	ILE	ASX	(MET)	GLX	ASX	TYR	SER	ILE	ASX	SER	SER	TYR	(VAL)	CYS	SER	GLY	LEU	(VAL)	GLY
TOKYO/3/67	PRO	(VAL	VAL)	ASX	ILE	ASX	(MET)	GLX	ASX	TYR	SER	ILE	ASX	SER	SER	TYR	(VAL)	CYS	SER	GLY	LSU	(VAL)	GLY
AICHI/2/68	PRO	(VAL	VAL)	ASX	ILE	ASX	(MET)	GLX	ASX	TYR	SER	ILE	ASX	SER	SER	TYR	(VAL)	CYS	SER	GLY	LEU	(VAL)	GLY
ENG/42/72	PRO	VAL	VAL	ASP	ILE	ASN	VAL	LYS	ASP	TYR	SER	ILE	ASP	SER	SER	TYR	VAL	CYS	SER	GLY	LEU	VAL	GLY
UDORN/72	PRO	VAL	VAL	ASP	ILE	ASN	VAL	LYS	ASP	TYR	SER	ILE	ASP	SER	SER	TYR	VAL	CYS	SER	GLY	LEU	VAL	GLY
VIC/3/75	PRO	VAL	VAL	ASP	ILE	ASN	VAL	LYS	ASP	TYR	SER	ILE	ASP	SER	SER	TYR	VAL	CYS	SER	GLY	LEU	VAL	GLY
																		GLU ↑					
					355					360					365					370			
RI/5+/57	GLY	TRP	ALA	PHE	ASP	ASP	GLY	ASN	ASP	LEU	TRP	MET	GLY	ARG	THR	ILE	ASN	LYS				SER	GLY
TOKYO/3/67	GLY	TRP	ALA	PHE	ASP	ASP	GLY	ASN	ASP	LEU	TRP	MET	GLY	ARG	THR	ILE	SER	LYS	ASP	LEU	ARG	SER	GLY
AICHI/2/68	GLY	TRP	ALA	PHE	ASP	ASP	GLY	ASN	ASP	VAL	TRP	MET	GLY	ARG	THR	ILE	SER	LYS	ASP	LEU	ARG	SER	GLY
ENG/42/72	GLY	TRP	ALA	PHE	ASP	ASP	GLY	ASN	ASP	VAL	TRP	MET	GLY	ARG	THR	ILE	SER	LYS	ASP	SER	ARG	SER	GLY
UDORN/72	GLY	TRP	ALA	PHE	ASP	ASP	GLY	ASN	ASP	VAL	TRP	MET	GLY	ARG	THR	ILE	SER	GLU	ASP	SER	ARG	SER	GLY
VIC/3/75															THR	ILE	SER	GLU	ASP	SER	ARG	SER	GLY
					405					410					415					420			
RI/5+/57	ASP	ASN	ARG													SER	CYS	ILE	ASN	ARG	CYS	PHE	TYR
TOKYO/3/67	ASP	ASN	ARG	SER	GLY	TYR	SER	GLY	ILE	PHE	SER	VAL	GLU	GLY	LYS	SER	CYS	ILE	ASN	ARG	CYS	PHE	TYR
AICHI/2/68	ASP	ASN	ARG	SER	GLY	TYR	SER	GLY	ILE	PHE	SER	VAL	GLU	GLY	LYS	SER	CYS	ILE	ASN	ARG	CYS	PHE	TYR
ENG/42/72	ASP	ASN	ARG													SER	CYS	ILE	ASN	ARG		PHE	TYR
UDORN/72	ASP	ASN	ARG	SER	GLY	TYR	SER	GLY	ILE	PHE	SER	VAL	GLU	GLY	LYS	SER	CYS	ILE	ASN	ARG	CYS	PHE	TYR
VIC/3/75	ASP	ASN	ARG	SER	GLY	TYR	SER	GLY	ILE	PHE	SER	VAL	GLU	GLY	LYS	SER	CYS	ILE	ASN	ARG		PHE	TYR
					455					460					465				469				
RI/5+/57																							
TOKYO/3/67	GLY	THR	TYR	GLY	THR	GLY	SER	TRP	PRO	ASP	GLY	ALA	ASP	ILE	ASN	LEU	MET	PRO	ILE				
AICHI/2/68																							
ENG/42/72																							
UDORN/72	GLY	THR	TYR	GLY	THR	GLY	SER	TRP	PRO	ASP	GLY	ALA	ASP	ILE	ASN	LEU	MET	PRO	ILE				
VIC/3/75																							

Figure 9 Comparative partial amino acid sequences of the NA from six strains of N2 virus isolated between 1957 and 1975. The sequences were deduced from the amino acid compositions of the tryptic peptides from pronase-released NA. Boxes indicate unambiguous differences in the sequences of the different NAs. It is not known which of the five valine residues between positions 301 and 328 of A/Udorn/72 was a methionine in RI/5+/57, Tokyo/67, and Aichi/68. The proline in RI/5+/57 is placed at position 431 since in this strain a tryptic peptide containing residues 429 to 435 was isolated. Sequence changes of variants selected with monoclonal antibodies are shown by arrows above the sequences; they are: (221) Asn → His, (368) Lys → Glu, and (344) Arg → Ile (four variants selected with different monoclonal antibodies showed this change).

	25					30					35					40					45					50
MET	GLN	ILE	ALA	ILE	GLN	VAL	THR	THR	VAL	THR	LEU	HIS	PHE	LYS	GLN	TYR	GLU	CYS	ASP	SER	PRO	ALA	ASN	ASN	GLN	VAL
	75					80					85					90					95					100
↓	PRONASE	↓	ILE	CYS	PRO	GLU	VAL	VAL	GLU	TYR	ARG	ASN	TRP	SER	LYS	PRO	GLN	CYS	GLX	ILE	THR	GLY	PHE	ALA	PRO	PHE
GLU	LYS	GLU	ILE	CYS	PRO	LYS	VAL	VAL	GLU	TYR	ARG	ASN	TRP	SER	LYS	PRO	GLN	CYS	GLX	ILE	THR	GLY	PHE	ALA	PRO	PHE
GLU	LYS	GLU	ILE	(CYS)	PRO	LYS	(VAL)	VAL	GLU	TYR	ARG	ASN	TRP	SER	LYS	PRO	GLN	CYS	GLX	ILE	THR	GLY	PHE	ALA	PRO	PHE
							LEU	VAL	GLU	TYR	ARG	LYS	TRP	SER	LYS	PRO	GLN	CYS	LYS	ILE	THR	GLY	PHE	ALA	PRO	PHE
GLU	LYS	GLU	ILE	CYS	PRO	LYS	LEU	VAL	GLU	TYR	ARG	ASN	TRP	SER	LYS	PRO	GLN	CYS	LYS	ILE	THR	GLY	PHE	ALA	PRO	PHE
GLU	LYS	GLU	ILE	(CYS)	PRO	LYS	LEU	VAL	GLU	TYR	ARG	ASN	TRP	SER	LYS	PRO	GLN	CYS	LYS	ILE	THR	GLY	PHE	ALA	PRO	PHE
	125					130					135					140					145					150
CYS	ASP	PRO	GLY	LYS	(CYS)	TYR	GLN	PHE	ALA	LEU	GLY	GLN	GLY	THR	THR	LEU	ASP	ASN	LYS							
CYS	ASP	PRO	VAL	LYS	CYS	TYR	GLN	PHE	ALA	LEU	GLY	GLN	GLY	THR	THR	LEU	ASP	ASN	LYS	HIS	SER	ASN	ASP	THR		HIS
					CYS	TYR	GLN	PHE	ALA	LEU	GLY	GLN	GLY	THR	THR	LEU	ASP	ASN	LYS	HIS	SER	ASN	ASP	THR	ILE	HIS
CYS	ASP	PRO	GLY	LYS																						
CYS	ASP	PRO	GLY	LYS	CYS	TYR	GLN	PHE	ALA	LEU	GLY	GLN	GLY	THR	THR	LEU	ASP	ASN	LYS	HIS	SER	ASN	ASP	THR	ILE	HIS
CYS	ASP	PRO	GLY	LYS	CYS	TYR	GLN	PHE	ALA	LEU	GLY	GLN	GLY	THR	THR	LEU	ASP	ASN	LYS	HIS	SER	ASN	ASP	THR	ILE	HIS
	175					180					185					190					195					200
VAL	CYS	VAL	GLY	TRP	SER	SER	SER	SER	CYS	HIS	ASP	ALA	LYS													
																				← INSOLUBLE PEPTIDE						
VAL	CYS	ILE	GLY	TRP	SER	SER	SER	SER	CYS	HIS	ASP	GLY	LYS	ALA	TRP	LEU	HIS	VAL	CYS	VAL	THR	GLY	TYR	ASP	LYS	ASN
	225					230					235					240					245					250
ARG	THR	GLN	GLU	SER	GLU	CYS	VAL	CYS	ILE	ASN	GLY	THR	CYS	THR	VAL	VAL	MET	THR	ASP	GLY	SER	ALA	SER	GLY	ARG	ALA
ARG	THR	GLN	GLU	SER	GLU	CYS	VAL	CYS	ILE	ASN	GLY	THR	CYS	THR	VAL	VAL	MET	THR	ASP	GLY	SER	ALA	SER	GLY	ARG	ALA
ARG	THR	GLN	GLU	SER	GLU	CYS	VAL	CYS	ILE	ASN	GLY	THR	CYS	THR	VAL	VAL	MET	THR	ASP	GLY	SER	ALA	SER	GLY	ARG	ALA
ARG	THR	GLN	GLU	SER	GLU	CYS	VAL	CYS	ILE	ASN	GLY	THR	CYS	THR	VAL	VAL	MET	THR	ASP	GLY	SER	ALA	SER	GLY	ARG	ALA
ARG	THR	GLN	GLU	SER	GLU	CYS	VAL	CYS	ILE	ASN	GLY	THR	CYS	THR	VAL	VAL	MET	THR	ASP	GLY	SER	ALA	SER	GLY	ARG	ALA
ARG	THR	GLN	GLU	SER	GLU	CYS	VAL	CYS	ILE	ASN	GLY	THR	CYS	THR	VAL	VAL	MET	THR	ASP	GLY	SER	ALA	SER	GLY	ARG	ALA
	275					280					285					290					295					300
HIS	VAL	GLU	GLU	CYS	SER	CYS	TYR	PRO	ARG	TYR	PRO	ASP	VAL	ARG	CYS	ILE	CYS	ARG	ASP	ASN	TRP	LYS	GLY	SER	ASX	ARG
HIS	VAL	GLU	GLU	CYS	(SER)	CYS	TYR	PRO	ARG	TYR	PRO	GLY	VAL	ARG	CYS	ILE	CYS	ARG	ASP	ASN	TRP	LYS	GLY	SER	ASX	ARG
HIS	VAL	GLU	GLU	CYS	SER	CYS	TYR	PRO	ARG	TYR	PRO	GLY	VAL	ARG	CYS	ILE	CYS	ARG	ASP	ASN	TRP	LYS	GLY	SER	ASX	ARG
HIS	VAL	GLU	GLU	CYS	SER	CYS	TYR	PRO	ARG	TYR	PRO	GLY	VAL	ARG									GLY	SER	ASN	ARG
HIS	VAL	GLU	GLU	CYS	SER	CYS	TYR	PRO	ARG	TYR	PRO	GLY	VAL	ARG	CYS	ILE	CYS	ARG	ASP	ASN	TRP	LYS	GLY	SER	ASN	ARG
HIS	VAL	GLU	GLU	CYS	SER	CYS	TYR	PRO	ARG	TYR	PRO	GLY	VAL	ARG	CYS	ILE	CYS	ARG	ASP	ASN	TRP	LYS	GLY	SER	ASN	ARG
																				ILE ↑						
	325					330					335					340					345					350
ASX	THR	PRO	ARG																		GLY	ASN	PRO	GLY	VAL	LYS
ASX	THR	PRO	ARG	ASP	ASP	ASP	ARG								ASN	PRO	ASN	ASN	GLU	ARG	GLY	(THR	GLN)	GLY	VAL	LYS
ASX	THR	PRO	ARG	ASP	ASP	ASP	ARG														GLY	ASN	GLN	GLY	VAL	LYS
ASP	THR	PRO	ARG												ASN	PRO	ASN	ASN	GLU	ARG	GLY	ASN	HIS	GLY	VAL	LYS
ASP	THR	PRO	ARG	ASN	ASN	ASP	ARG	SER	SER	ASN	SER	TYR	CYS	ARG	ASN	PRO	ASN	ASN	GLU	LYS	GLY	ASN	HIS	GLY	VAL	LYS
ASP	THR	PRO	ARG	ASN	ASN	ASP	ARG	SER	SER	HIS	SER	TYR	CYS	ARG	ASN	PRO	ASN	ASN	GLU	LYS	GLY	ASN	HIS	GLY	VAL	LYS
	375					380					385					390					395					400
TYR	GLU	THR	PHE	LYS	VAL	ILE	GLY	GLY	TRP	SER	THR	PRO	ASN	SER	LYS	SER	GLN	VAL	ASN	ARG	GLN	VAL	ILE		ASP	SER
TYR	GLU	THR	PHE	LYS	VAL	ILE	GLY	GLY	TRP	SER	THR	PRO	ASN	SER	LYS	SER	GLN	ILE	ASN	ARG	GLN	VAL	ILE	(VAL)	ASP	SER
TYR	GLU	THR	PHE	LYS	VAL	ILE	GLY	GLY	TRP	SER	THR	PRO	ASN	SER	LYS	SER	GLN	ILE	ASN	ARG	GLN	VAL	ILE	VAL	ASP	SER
TYR	GLU	THR	PHE	LYS	VAL	ILE	GLY	GLY	TRP	SER	THR	PRO	ASN	SER	LYS	SER	GLN	ILE	ASN	ARG	GLN	VAL	ILE	VAL	ASP	SER
TYR	GLU	THR	PHE	LYS	VAL	ILE	GLY	GLY	TRP	SER	THR	PRO	ASN	SER	LYS	LEU	GLN	ILE	ASN	ARG	GLN	VAL	ILE	VAL	ASP	SER
TYR	GLU	THR	PHE	LYS	VAL	ILE	GLY	GLY	TRP	SER	THR	PRO	ASN	SER	LYS	LEU	GLN	ILE	ASN	ARG	GLN	VAL	ILE	VAL	ASP	SER
	425					430					435					440					445					450
VAL	GLU	LEU	ILE	ARG	GLY	ARG	(PRO	GLN	GLU)	THR	ARG															
VAL	GLU	LEU	ILE	ARG	GLY	ARG	(LYS	GLN	GLU)	THR	ARG	VAL	TRP	TRP	THR	SER	ASN	SER	ILE	VAL	VAL	PHE	CYS	GLY	THR	SER
VAL	GLU	LEU	ILE	ARG	GLY	ARG	(LYS	GLN	GLU)	THR	ARG															
VAL	GLU	LEU	ILE	ARG	GLY	ARG	GLU	GLN	GLU	THR	ARG															
VAL	GLU	LEU	ILE	ARG	GLY	ARG	GLU	GLN	GLU	THR	ARG	VAL	TRP	TRP	THR	SER	ASN	SER	ILE	VAL	VAL	PHE	CYS	GLY	THR	SER
VAL	GLU	LEU	ILE	ARG	GLY	ARG	GLU	GLN	GLU	THR	ARG															

Figure 9 (Continued)

be released from the virus by digesting the virus particles with proteolytic enzymes. These enzymes cleave off the head of the NA (Fig. 8), which carries the antigenic sites and enzyme active site of the NA and, in some cases, can be crystallized (Fig. 10). The NA polypeptide is coded for by RNA segment 6 in the virus and, unlike the HA, is not cleaved following its synthesis and no signal peptide is split off (even the initiating methionine is retained).

The NA is anchored in the lipid of the virus envelope by a hydrophobic sequence near its N terminus (unlike the HA, which is attached by its C terminus), and treatment with pronase cleaves off this hydrophobic region (Fig. 2) together with the rest of the stalk (Blok et al., 1982).

Like the hemagglutinin, the neuraminidase of type A influenza virus undergoes major antigenic shifts and antigenic drift. In 1957, the H1N1 viruses infecting humans

were replaced by viruses of another subtype, Asian (H2N2) influenza virus. Between 1957 and 1968 the N2 neuraminidase underwent antigenic drift. In 1968, a recombinant event apparently occurred and a virus appeared in humans which retained the neuraminidase (and other genes) from an H2N2 virus but which had acquired its hemagglutinin from some other virus related to A/duck/Ukraine/63 and A/equine/Miami/1/63 (H3N8) viruses. This virus, the Hong Kong (H3N2) strain, has been the cause of influenza epidemics in humans since 1968 and the N2 NA has undergone further antigenic drift during this period.

Pronase-released neuraminidase heads from six strains of influenza typ A virus (of the H2N2 and H3N2 subtypes) isolated between 1957 and 1975 were examined for changes in amino acid sequence. In 469 residues 19 changes that occurred during this period were located (Fig. 9). Two regions of the neuraminidase molecule were not examined: residues 188 to 210, which were insoluble, and residues 1 to 73 (or 76), which formed part of the stalk of the neuraminidase and were removed during pronase digestion.

Neuraminidase heads from seven variants of A/Tokyo/3/67 virus, selected with different monoclonal antibodies to the neuraminidase, were also examined for sequence changes. In four of these, a single sequence change at position 344 of arginine to isoleucine was found, in another variant an asparagine residue at position 221 changed to histidine, and in another the lysine at position 368 changed to glutamic acid. This last change also occurred in the field strains isolated in 1972 and 1975. In the seventh monoclonal variant the change could not be found and may be in the insoluble region.

Some of these sequence changes were clustered into two highly variable regions in the neuraminidase comprising residues 344 to 347 and 367 to 370, suggesting that these regions may be involved in antigenic sites on the neuraminidase molecule. An antigenic map of N2 neuraminidase (Webster et al., 1982a) suggested three or four nonoverlapping antigenic areas and the variants with changes at residues 344 and 368 were grouped in one of these antigenic regions.

IX. CELL-MEDIATED IMMUNE RESPONSES

This chapter has been concerned so far with the nature of the antigenic sites on the hemagglutinin and neuraminidase of influenza viruses which induce the formation of serum antibodies that recognize and bind specifically to those sites.

Influenza virus infections also generate cytotoxic thymus-derived lymphocytes (T cells) which kill virus-infected histocompatible target cells. After infection with influenza A viruses of serologically different subtypes, cross-reactive (between subtypes) as well as subtype-specific cytotoxic T cells can be distinguished. Evidence is accumulating that some of the cross-reactive cytotoxic T cells recognize sites on the hemagglutinin which are common to all influenza A viruses but which are not recognized by circulating antibodies and which do not exist on influenza B virus HA (Koszinowski et al., 1980; Wabuke-Bunoti et al., 1981). Helper T (T_h) cells with specificity for the HA are also induced following influenza virus infection. It has been shown that such T_h cell will recognize antigenic determinants present on the HA of different subtypes (Anders et al., 1981). but whether these are the same cross-reactive sites as those recognized by cytotoxic T cells is not known.

X. DISCUSSION

Very little is known about the antigenic sites on virus proteins, but the antigens of influenza viruses are under intensive investigation and the precise structure of antigenic sites on these will soon be known.

One difficulty encountered with the hemagglutinin of influenza virus has been the inability to isolate small peptides which retain antigenic activity characteristic of the intact HA. In the case of some other viruses, such as tobacco mosaic virus (Anderer, 1963; Fearney et al., 1971), MS-2 coliphage (Langbeheim et al., 1976), and hepatitis B virus (Dreesman et al., 1982), small immunologically active synthetic peptides, analogous to amino acid sequences in the virus proteins, could be obtained. Very recently, small peptides corresponding to sequences in one of the proteins of foot-and-mouth disease virus (VPI) have been synthesized, and some of these have been shown to induce protective antibodies in animals (Bittle et al., 1982). However, in the case of the influenza virus hemagglutinin (no information is available for the neuraminidase) small fragments of the molecule which are capable of inducing antibody that will neutralize virus infectivity have not been obtained (Jackson et al., 1980).

On the other hand, the availability of a large number of monoclonal antibodies to individual antigenic sites on the HA and NA of influenza viruses and the ability to select and analyze variants with changes in these sites may provide information that will lead to an understanding of the structure of the sites.

It is more likely, however, that x-ray crystallography data will provide a solution. The three-dimensional structure of Hong Kong (H3) hemagglutinin has been determined by Wiley and his colleagues (Wilson et al., 1981; Wiley et al., 1981) and the variable areas on the molecule have been defined. Similar structural determinations on the HA of variants selected with monoclonal antibodies, in which single sequence changes are known, should show how these changes alter the shape of the site so that it is totally unable to bind the antibody used to select the variant.

Similar studies are being done on the neuraminidase. NA heads from influenza viruses of the N2 subtype isolated in 1957 and 1967 have been crystallized (Fig. 10) and determination of the three-dimensional structure is in progress (Colman and Laver, 1980). One of the variants (selected with S10/1 monoclonal antibody to the NA) showed a change at position 368 of lysine to glutaminc acid (Fig. 9). NA heads from this variant form crystals isomorphous with those of the wild-type NA and x-ray diffraction data have been collected and are being analyzed. An Fab fragment from this monoclonal antibody (S10/1) has also been crystallized (Fig. 10). This separate crystallization of antigen and Fab opens the way to map, for the first time, the complementary surfaces of an antigen-antibody complex (Colman et al., 1981).

XI. CONCLUSION

The hemagglutinin spikes of type A influenza viruses possess four distinct nonoverlapping antigenic regions which are located at the distal end of the molecule. Within each of these regions there appear to be many individual overlapping antigenic sites. Monoclonal antibodies can be generated that bind with high affinity to these individual sites, and variant viruses have been selected, in the presence of different monoclonal antibodies, which do

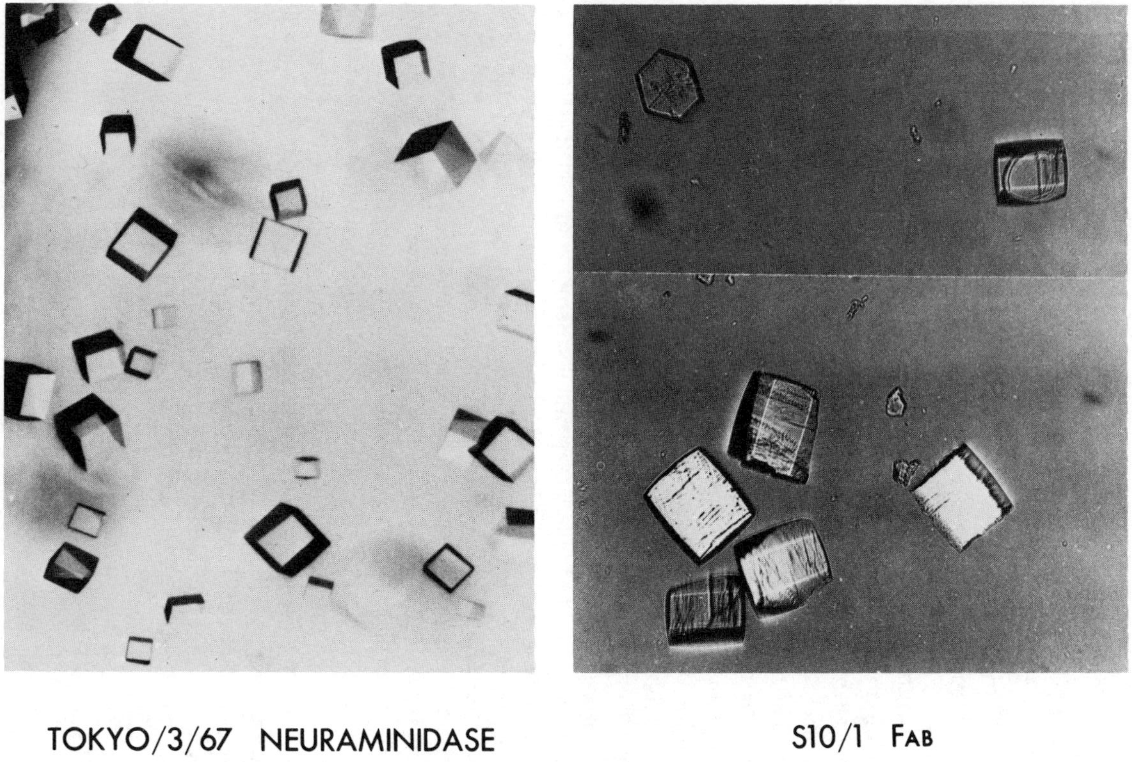

Figure 10 Crystals, suitable for x-ray diffraction analysis, of neuraminidase heads from Tokyo/67 virus and of the Fab fragment of a monoclonal antibody (S10/1) that binds to the NA. Variant NA molecules from a virus selected in the presence of S10/1 monoclonal antibody, with a sequence change at position 368 of lysine to glutamic acid, have also been crystallized. The variant NA does not bind the S10/1 antibody and x-ray diffraction analysis will show how the sequence change affects the shape of the antigenic site.

not bind the antibody used in the selection. These variant viruses have been found to have single changes in the amino acid sequence of the large hemagglutinin polypeptide, HA1. Similar findings have been made for the neuraminidase. X-ray crystallography data now being collected will show how these sequence changes affect the shape of the sites.

Antigenic drift in the HA and NA of type A influenza virus seems to be due to the accumulation of a number of sequence changes in these molecules over a period of time. It does not seem to occur by a series of sequential changes at the same position in the HA and NA polypeptides. Evidence suggests that T lymphocytes recognize different regions on the HA than do B lymphocytes.

REFERENCES

Anderer, F. A. (1963). Biochim. Biophys. Acta *71*, 246.

Anders, E. M., Katz, J. M., Jackson, D. C., and White, D. O. (1981). J. Immunol. *127*, 669.

Atassi, M. Z., and Smith, T. A. (1978). Immunochemistry *15*, 609.

Benjamini, E. (1977). In *Immunochemistry of Proteins*, Vol. 2 (M. Z. Atassi, Ed.), pp. 265–310.

Bittle, J. C., Houghten, R. A., Alexander, H., Shinnick, T. M., Sutcliffe, J. G., Lerner, R. A., Rowlands, D. J., and Brown, F. (1982). Nature (Lond.) *298*, 30.

Blok, J., Air, G. M., Laver, W. G., Ward, C. W., Lilley, G. G., Woods, E. F., and Roxburgh, C. M. (1982). Virology *119*, 109.

Colman, P. M., and Laver, W. G. (1981). In *Proceedings of the 7th Aharan-Katsik-Takzhalsky Conference on Structural Aspects of Recognition and Assembly in Biological Macromolecules* (M. Balaban, J. L. Sussman, W. Traub, and A. Yonath, Eds.), Balaban 155, Rehovot and Philadelphia, pp. 869–872.

Colman, P. M., Gough, K. H., Lilley, G. G., Blagrove, R. J., Webster, R. G., and Laver, W. G. (1981). J. Mol. Biol. *152*, 609.

Dreesman, G. R., Sanchez, Y., Ionescu-Matiu, I., Sparrow, J. T., Six, H. R., Peterson, D. L., Hollinger, F. B., and Melnick, J. L. (1982). Nature (Lond.) *295*, 158.

Fearney, F. J., Leung, C. Y., Young, J. D., and Benjamini, E. (1971). Biochim. Biophys. Acta *243*, 509.

Gerhard, W., and Webster, R. G. (1978). J. Exp. Med. *148*, 383.

Jackson, D. C., Brown, L. E., Russell, R. J., White, D. O., Dopheide, T. A., and Ward, C. W. (1980). In *Structure and Variation in Influenza Virus* (W. G. Laver and G. M. Air, Eds.). Elsevier, New York, pp. 309–320.

Koszinowski, U. H., Allen, H., Gething, M.-J., Waterfield, M. D., and Klenk, H.-D. (1980). J. Exp. Med. *151*, 945.

Langbeheim, H., Arnon, R., and Sela, M. (1976). Proc. Natl. Acad. Sci USA *73*, 4636.

Laver, W. G., Downie, J. C., and Webster, R. G. (1974). Virology *59*, 230.

Laver, W. G., Air, G. M., and Webster, R. G. (1981). J. Mol. Biol. *145*, 339.

Periera, M. S. (1979). Br. Med. Bull. *35*, 9.

Van Regenmortel, M. H. V. (1982). *Serology and Immunochemistry of Plant Viruses*, Academic Press, New York.

Virelizier, J. C., Allison, A. C., and Schild, G. C. (1974). J. Exp. Med. *140*, 1571.

Wabuki-Bunoti, M. A. N., Fan, O. P., and Braciale, T. J. (1981). J. Immunol. *127*, 1122.

Ward, C. W. (1981). Curr. Top. Microbiol. Immunol. *94–95*, 1.

Webster, R. G., Laver, W. G., Air, G. M., Ward, C. W., van Wyke, K. L., and Gerhard, W. (1980). Ann. N.Y. Acad. Sci. *354*, 142.

Webster, R. G., Hinshaw, V. S., and Laver, W. G. (1982a). Virology *117*, 93.

Webster, R. G., Laver, W. G., Air, G. M., and Schild, G. C. (1982b). Nature (Lond.) *296*, 115.

Wiley, D. C., Wilson, I. A., and Skehel, J. J. (1981). Nature (Lond.) *289*, 373.

Wilson, I. A., Skehel, J. J., and Wiley, D. C. (1981). Nature (Lond.) *289*, 366.

World Health Organization Memorandum (1980). Bull. WHO *58*, 585.

Wrigley, N. G., Laver, W. G., and Downie, J. C. (1977). J. Mol. Biol. *109*, 405.

7

Structure and Function of Immunoglobulins

Nickolas J. Calvanico The Medical College of Wisconsin and Wood Veterans Administration Medical Center, Milwaukee, Wisconsin

I. INTRODUCTION

Immunoglobulins are multichain glycoproteins. While all immunoglobulins bear the same basic structure, there is a heterogeneity of class, subclass, phenotype, and individuality superimposed on their apparent homogeneity. These variations account for the specificity and biological function of immunoglobulins, two properties that distinguish them.

The purpose of this chapter is to review the structure and function of immunoglobulins. Space limitations preclude this from being an exhaustive account. It is intended to be an update and accordingly assumes some previous knowledge of basic concepts of immunoglobulin structure. The chapter focuses on human immunoglobulins since most of our structural information has been obtained from them, but immunoglobulins of other species are referenced when appropriate. During the course of this chapter reference will be made to several reviews which will aid the reader in acquiring more in-depth knowledge on the various subjects covered. It should also be noted that overlap may occur between this chapter and several others in this volume. This is impossible to avoid, especially at a time when many of the structural, functional, and cellular facets of immunology are beginning to merge into a unified picture.

II. THE BASIC STRUCTURE

A. General Properties

Immunoglobulins are actually a group of proteins with similar structures, but they in fact exhibit many levels of sequence variability, ranging from a few amino acids to extensive differences. This heterogeneity is reflected in the broad electrophoretic distribution shown by immunoglobulin. Figure 1 depicts the fact that serum immunoglobulins migrate

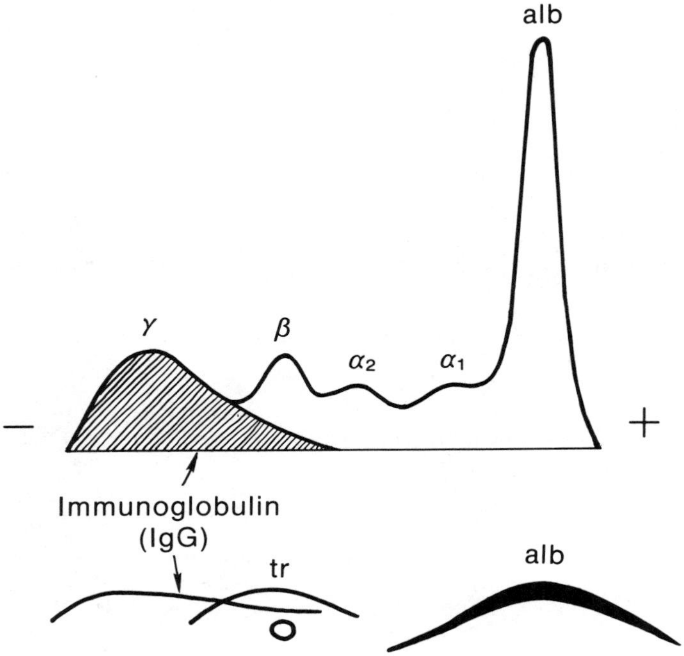

Figure 1 Electrophoresis of serum immunoglobulins. (Top) Schematic representation of a densitometric tracing of serum protein separated by paper electrophoresis. Immunoglobulin distribution is depicted by the shaded area. (Bottom) Represents immunoelectrophoretic precipitin arcs developed with antisera to the three major protein components of serum: albumin (alb), transferrin (tr), and immunoglobulin (IgG). Electrophoretic heterogeneity of IgG is reflected by the long precipitin arc.

electrophoretically from the gamma region to the alpha region. Essentially all protein in the gamma region is immunoglobulin and this region contains approximately 90% of the circulating immunoglobulin, hence the name *gamma globulin*. The hterogeneity of immunoglobulins accounted for the relatively slow progress made in early studies on immunoglobulin structure. This changed when it was realized that homogeneous immunoglobulins were produced in large amounts by malignant clones of plasma cells in patients with multiple myeloma. Similar monoclonal proteins can also be induced artificially in certain strains of mice by intraperitoneal injection of mineral oil. Myeloma proteins are generally considered a normal product of an abnormal cell (Kunkel, 1965). Many myeloma proteins have been shown to bind various antigens, which has greatly facilitated studies on the antigen combining site.

B. Subunit Structure

The basic architecture of all immunoglobulins is the tetrapolypeptide structure diagrammed in Fig. 2. Because of their molecular weights, two of the polypeptides are designated heavy (H) chains and the other two are designated light (L) chains. It should be noted that there are two identical pairs of chains; that is, the molecule is composed of two identical halves, each half consisting of an H-L pair. The chains are stabilized by both covalent (disulfide) and noncovalent forces.

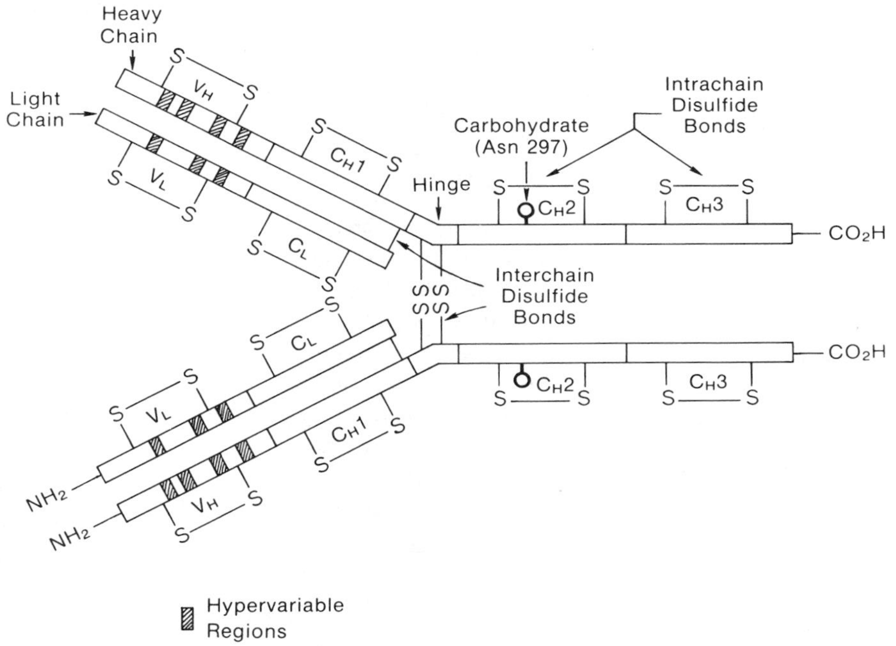

Figure 2 Stick model representation of human IgGl. Variable and constant domains are denoted by V and C. Disulfide bonds, hypervariable regions, and the hinge region are also designated. The single carbohydrate oligosaccharide is attached to Asn 297 in the C_H2 domain. The amino terminus is designated by NH_2- and the carboxyl terminus by -CO_2H.

The H chain defines the class and subclass of the molecule. This is based on antigenic as well as structural properties. There are five classes of immunoglobulins: IgG (γG), IgA (γA), IgM (γM), IgD (γD), and IgE (γE). The H chains for each class are designated by the appropriate Greek letter: γ, α, μ, δ, and ϵ, respectively. The molecular weights and carbohydrate contents of these H chains vary from one class to another. This will be covered in more detail in later sections of this chapter. Additional antigenic determinants and structural differences define subclasses. Thus in humans there are four IgG subclasses (IgGl, IgG2, IgG3, and IgG4), two IgA subclasses (IgA1 and IgA2), and two IgM subclasses (IgM1 and IgM2). It should be emphasized that all molecules of a given subclass bear both class *and* subclass antigenic determinants. For example, an IgG1 and an IgG3 molecule both bear IgG (class) determinants in addition to their IgG1 and IgG3 (subclass) determinants. Heavy chains may also bear determinants resulting from polymorphism in the genes (alleles) coding for their structure; these are referred to as *allotypic determinants*. Many allotypes are known for IgG and they are distributed among all four subclasses. They are referred to as Gm(1), Gm(2), and so on. Two allotypes are known for the IgA2 subclass [A2m(1) and A2m(2)]. Allotypes are discussed in greater detail in Chap. 11.

There are two types of L chains, kappa (κ) and lambda (λ). Four subtypes are known for human λ chains, while only one type of human κ chain is known. Three allotypes are known for κ chains [Km(1), Km(2), and Km(3)], but no allotypic variants of λ chains have been found. Both κ and λ chains have molecular weights of 22,000 to

23,000; neither contains carbohydrate. The empirical formula for immunoglobulins is therefore $H_2\kappa_2$ or $H_2\lambda_2$, where H can be $\gamma1, \gamma2, \gamma3, \gamma4, \alpha1, \alpha2, \mu1, \mu2, \delta$, or ϵ. Variations on the basic structure exist in all classes of immunoglobulins. These will be discussed in later sections.

C. Domain Structure

The primary structures are known for at least one human myeloma H chain of each class and for both types of *Bence Jones* (BJ) *proteins*. Bence Jones proteins are monoclonal L chains frequently present in large concentrations in the serum and urine of patients with multiple myeloma. The BJ proteins are identical to the L chains of the myeloma proteins produced by the patients and represent an overproduction (compared to H chains) by the malignant clones. They are named for Sir Henry Bence Jones (1814-1873), a physician who first described these proteins in the urine of a myeloma patient in 1847. This event is generally considered the beginning of immunochemistry. An interesting account of the history of this finding has been given by Putnam (1978).

In addition to the complete sequences, partial sequences are available for many other myeloma and BJ proteins, both of human and murine origin (Dayhoff, 1976). This wealth of information is due primarily to the development of the automatic protein sequencer. Examination of these sequences leads to several important conclusions concerning the structure of immunoglobulins:

1. Both H and L chains are composed of a series of similarly constructed segments, referred to as *domains*. Each of these domains is composed of 110 to 120 amino acid residues containing one intrachain disulfide bond formed by two cysteine residues approximately 60 residues apart within the domain. For IgG there are four domains. The structural similarity of these domains has led to the concept that all H and L chains evolved from a common precursor similar to a domain.

2. A comparison of sequences from proteins of one class and subclass shows that whereas the N-terminal domain is variable from molecule to molecule, the other domains are identical except for inherited differences (allotypic determinants) of one or two amino acid residues. The *variable domain* is designed as V_H and the *invariant* or *constant domains* are designated as C_H1, C_H2, and C_H3 (proceeding from N to C terminus of the H chain). L chains contain two domains, one variable (V_L) and one constant (C_L). As expected, the V domains contain the antigen binding site and both H and L chains are involved. Since there are two pairs of H-L chains, there are two identical antigen binding sites per molecule. The finding of two types of domains (V and C) led to the concept that at least two structural genes code for each immunoglobulin chain, one V gene and one C gene. It is important to note that there are three families of V genes. One family is linked with C_H genes of all classes, one with C_κ and the third with C_λ (Fig. 3). There is no linkage between these three families. The genetics of immunoglobulin synthesis is discussed in greater detail in Chap. 14.

3. There is a region much smaller than a domain, situated between the C_H1 and C_H2 domains, which contains the half-cystines forming *inter* H and H-L disulfide bonds. This region is known as the *hinge region* because it acts as a fulcrum upon which the two antigen binding arms of the molecule rotate. As will be discussed later, the hinge region is unique for each class and subclass of immunoglobulin.

4. A comparison of sequences shows considerable homology (a position-by-position identity of amino acid residues) between domains of different polypeptide chains as well

VC Product	V-Gene Family Subgroup	C-Region Gene
H chain	V_HI	γ_1, γ_2, γ_3, γ_4
	V_HII	α1, α2
	V_HIII	μ1, μ2
	V_HIV	δ
		ε
λ chain	V_λI	C_λ1, C_λ2, C_λ3, C_λ4
	V_λII	
	V_λIII	
	V_λIV	
	V_λV	
	V_λVI	
κ chain	V_κI	
	V_κII	C_κ
	V_κIII	
	V_κIV	

Figure 3 Linkage of V-gene families and C-region genes. Each V-region subgroup is composed of many genes, whereas each C-region gene is unique (excluding allotypic variants). The three families of V-region genes are not linked to each other, but are linked to their respective C-region genes. The four Cλ genes are defined by the antigenic determinants Oz, Kern, Mcg, and Mz.

as on the same polypeptide chain, supporting the common gene precursor concept. The lowest homology exists between V and C domains (ca. 15%), whereas the degree of homology between any two C domains is at least 30%. This suggests early divergence of genes controlling V and C domains. There is little sequence homology between the hinge region of a given subclass and any other hinge region or domain, suggesting the independent evolution of this segment. A few examples of the degree of homology between C domans will serve to illustrate their similarity. The C_H domains of the four IgG subclasses have greater than 90% homology, while the $C\gamma$ is 33% homologous to the $C\mu$. This contrasts with the 69% homology between the $C\gamma$ regions of human and rabbit IgG, indicating that the divergence of subclasses is relatively recent and that of IgM and IgG relatively early. A similar relationship is found between the two types of L chains. Human C_κ and human C_λ are 39% homologous, while human C_κ and murine C_κ are 60% homologous. Human C_κ and murine C_λ show 29% homology. Thus it would appear that the divergence of κ and λ genes also occurred early. It should be noted that these homologies are maximized by the insertion of gaps into the sequences to obtain the best alignment. This is necessary because of the variability in size of the various chains.

5. A comparison of V-domain sequences reveal two very important points. The first is that the sequences can be grouped according to similarities of sequence. These are referred to as ***subgroups*** and are based on the length of the V domain and the presence of certain amino acid residues at particular positions in the sequence. Thus for humans there are three V_{κ} subgroups ($V_{\kappa I}$, $V_{\kappa II}$, $V_{\kappa III}$), five or six V_{λ} subgroups ($V_{\lambda I}$, $V_{\lambda II}$, etc.), and four V_H subgroups (V_{HI}, V_{HII}, etc.). A detailed discussion of the characteristics of each subgroup is given by Nisonoff et al. (1975). For our purposes it is sufficient to recognize that within a subgroup there is a limited number of amino acid interchanges and that these are usually the result of single base changes in the codon. This is not true when comparing changes between subgroups. This subgrouping is generally thought to indicate that at lease one gene codes for each subgroup and that the generation of additional V genes results from mutational and recombinational events. The exact contribution of germ line genes and mutational events to the final array of V genes in a subgroup is a matter of considerable controversy and is discussed further in Chap. 14. It should be recognized, however, that the criteria for a subgroup are arbitrary and with additional sequence information they may change.

The second important point revealed from a comparison of V-domain sequences is that certain regions have a much greater variability than that displayed by the remainder of the domain. Variability is measured by the number of different amino acids at a given position divided by the frequency of the most common amino acid at that position. A plot of variability versus sequence position gives a bar graph as shown in Figs. 4 and 5. The regions of exceptional variability are referred to as *hypervariable* or *complementarity*

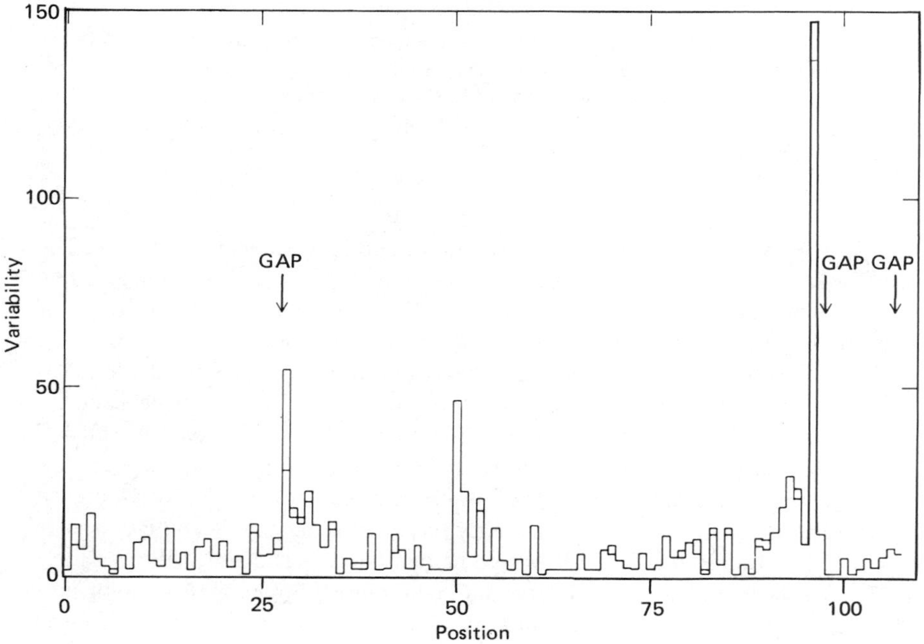

Figure 4 Variability (see text for definition) of amino acids at each position of human light chains. (From Wu and Kabat, 1970.)

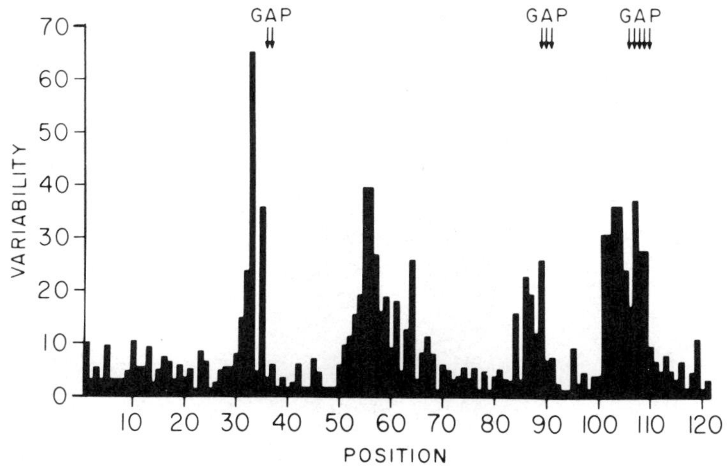

Figure 5 Variability of amino acids at each position of human H chains. (From Capra and Kehoe, 1975).

regions. For L chains there are three hypervariable regions and for H chains there are four hypervariable regions. It has been postulated that these regions are the basis of the antigen binding diversity of immunoglobulins (Wu and Kabat, 1970). X-ray crystallographic studies have indeed shown that five of these hypervariable regions form a pocket in which the antigen is bound. Even with the hypervariable regions, homology between V regions within a subgroup is between 70 and 90%, while intersubgroup comparisons give 50 to 60% homology. A V sequence of one family (e.g., V_{κ}) is 40 to 50% homologous with the V sequence of another family (e.g., V_H).

A third set of antigenic determinants is associated with the antigen binding site of antibodies. Thus each antibody has an individual antigenic specificity referred to as an *idiotype*. It should be noted that in contrast to isotypic and allotypic determinants, which are usually (but not always) present on the constant domains, idiotypic determinants reside in the variable domains. Idiotypes are important in regulation of the immune response and are discussed in more detail in Chap. 12.

D. Subunits and Fragments

A major impetus to modern immunology was initiated with the studies of G. Edelman and R. R. Porter, which elucidated the basic structure of IgG using methods of protein chemistry. These studies laid the groundwork for the subsequent rapid advancement of this field and earned the Nobel prize for both of these investigators.

As stated earlier, immunoglobulins are composed of H and L polypeptide chains held together by disulfide bonds and noncovalent forces. In order to separate the chains, the interchain disulfide bonds must be broken and prevented from reforming. The molecule must then be subjected to denaturing conditions capable of breaking down the noncovalent interactions. Several methods have been used to obtain these results, but the most frequently used procedure is that of reduction with mercaptans such as β-mercaptoethanol and alkylation of the resulting sulfhydryl groups with iodoacetamide.

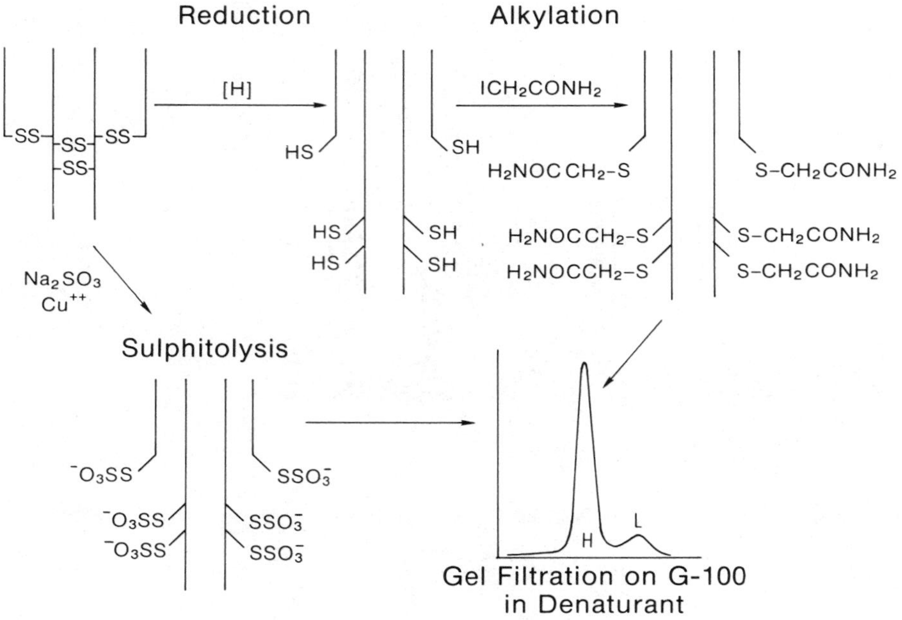

Figure 6 Isolation of H and L chains of immunoglobulins. Cleavage of interchain disulfide bonds can be accomplished by reduction either by mercaptans (followed by alkylation with iodoacetamide) or sulfitolysis. After cleavage, the chains are separated by gel filtration chromatography in denaturant.

Under these conditions only interchain disulfides are reduced, since intrachain disulfides are generally more stable and require denaturants in addition to reducing agents. After removal of the excess reducing and alkylating agents, the modified protein is then subjected to gel filtration in denaturing solvent to separate H and L chains. Several denaturants have been used, including low-pH (3.0 to 4.0), detergents, urea, and guanidine hydrochloride (Fig. 6). Following removal of denaturants or return to neutral pH, H chains become insoluble, whereas L chains usually remain stable in solution.

The limited cleavage of polypeptide bonds with proteolytic enzymes for the preparation of fragments for further study is an important technique in protein chemistry. Papain attacks rabbit IgG in the hinge region and produces two types of fragments. One fragment is capable of binding antigen and is referred to as the *Fab fragment* (antigen binding fragment). The other fragment does not bind antigen, but possesses the sites necessary for many secondary effects of antibodies (discussed later). This fragment is also crystallizable at low ionic strength and is referred to as the *Fc fragment* (crystallizable fragment). The Fab contains the N-terminal half of the H chain (Fd fragment), that is, the V_H and C_H1 domains, and the complete L chain, linked by a disulfide bond. The Fc is a dimer of the two C terminal halves of H chains, held together by disulfide bonds and noncovalent forces. Both fragments are stable; however, the Fc is subject to further proteolysis in the C_H2 domain. The product is referred to as the *Fc′ fragment*, which is actually a dimer of the C_H3 domains stabilized by noncovalent forces. It should be noted that two Fab and one Fc fragment is obtained from the complete papain cleavage of one IgG molecule (Fig. 7).

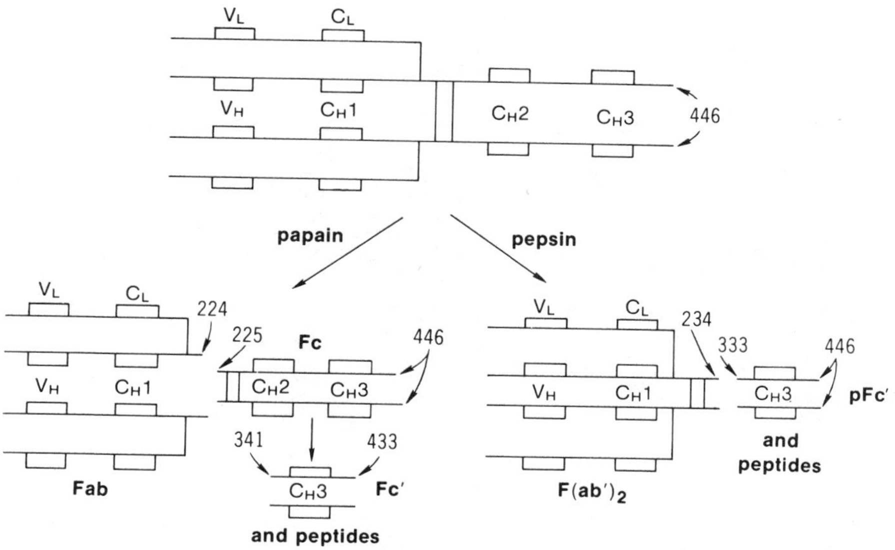

Figure 7 Proteolytic cleavage of IgG by papain and pepsin. Papain yields two Fab and fragments and one Fc fragment. Further cleavage of Fc to Fc′ is also indicated. Pepsin yields primarily the $F(ab')_2$ fragment with small amounts of pFc′. Note the difference in size between the Fc′ and the pFc′.

Pepsin also attacks IgG in the hinge region but, in addition, degrades most of the Fc fragment as well. Under careful regulation, a fragment related to the Fc can be recovered before proteolysis of the Fc fragment is complete. The more extensive digestion of IgG by pepsin is related to the fact that this enzyme is used at a more acidic pH than papain (pH 4 versus pH 7.6) which causes partial denaturation of the substrate. The major product of pepsin digestion is the divalent antigen binding fragment, which is composed of two Fab-like fragments attached by the hinge disulfide. This fragment, termed the *$F(ab')_2$*, results from cleavage in the hinge region on the C-terminal side of the inter-H chain disulfide, whereas Fab fragments result from cleavage on the N-terminal side. It should be noted that because of its bivalency, $F(ab')_2$ fragments are still able to precipitate antigen in contrast to Fab fragments, which only bind antigens. Papain and pepsin cleavage of IgG is diagrammed schematically in Fig. 7.

Space does not permit a full discussion of the cleavage of immunoglobulins by other enzymes. A more comprehensive account may be found elsewhere (Calvanico and Tomasi, 1979). In addition, several references to other specialized procedures will be described in later sections.

E. Three-Dimensional Structure

The three-dimensional model of IgG depicts the molecule as having three globular regions, two Fab and one Fc, linked by the more extended hinge region upon which they independently rotate. This model has been refined by x-ray diffraction studies on a number of crystallized human myeloma proteins, including BJ dimers and whole IgG, Fab, and Fc fragments (Poljak, 1975; Padlan, 1977; Amzel and Poljak, 1979). All have been analyzed at least at the 3.5-Å resolution level and most below 3.0 Å. Several important findings

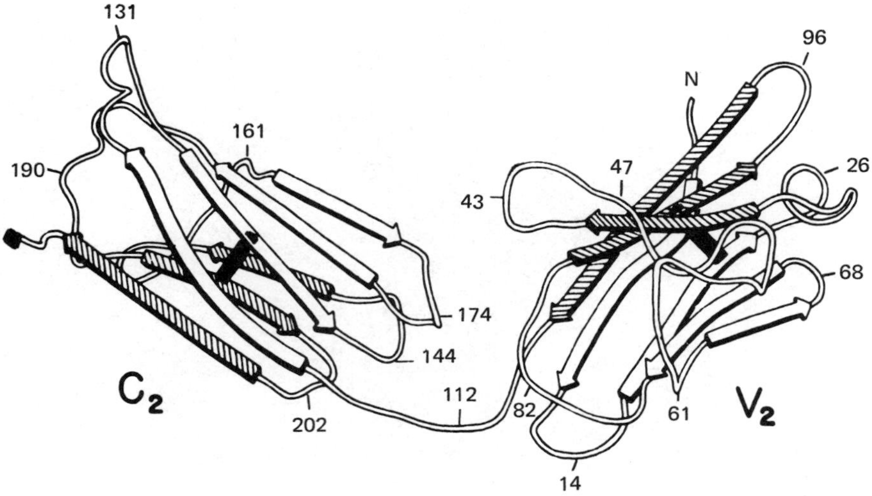

Figure 8 Schematic representation of a V region and a C region of a BJ protein showing the immunoglobulin fold with the four-chain layer (white arrows) and three-chain layer (striated arrows). The numbers designate approximate amino acid residues and the black bars the intrachain disulfide bonds. Note the positions of the hypervariable regions on the loops of the V region. [Reproduced with permission from M. Schiffer et al. (1973), Biochemistry *12*, 4620; copyright 1973 American Chemical Society.]

have emerged from these studies, the most important of which is a confirmation of the domain concept. Thus Fab fragments can be divided into V and C segments with approximate dimensions of 40 × 50 × 40 Å.

In our discussion we shall refer to the homology units as domains (i.e., V_L, V_H, C_H1, C_H2, etc.) and the pairs of domains as globular units (i.e., V_L/V_H, C_H1/C_L, etc.). All domains show the same basic structure of a stacked pair of antiparallel β pleated sheets referred to as the *immunoglobulin fold*. A schematic representation of a BJ protein is shown in Fig. 8. The two layers differ in the number of chain segments (arrows in Fig. 8) contributing to the layer; thus one can speak of four and three chain layers or faces. An interconnecting disulfide bond and tightly packed side chains from hydrophobic amino acid residues lie between the two layers. A series of loops connect the chain segments in the layers. There are three loops on both the N- and C-terminal sides of the layers. In V domains, an extra loop occurs on both sides. The hypervariable region of V domains are contained in the loops on the N-terminal side and are spatially in close proximity.

As stated previously, immunoglobulins are stabilized by disulfide bonds and noncovalent interactions. These interactions occur between paired domains (V_H/V_L, C_H1/C_L, C_H2/C_H2, and C_H3/C_H3) on different chains (i.e., trans interaction) and involve several contact points. In contrast, cis interactions (between globular units) involve a small number of contact points and are weak.

The trans interactions are of two types. In V domains, contact is between three chain layers while in the C domains it is between the four chain layers. In V domains, the interactions are quite extensive ($K_a \geqq 10^8 M^{-1}$) and usually involve less polar residues.

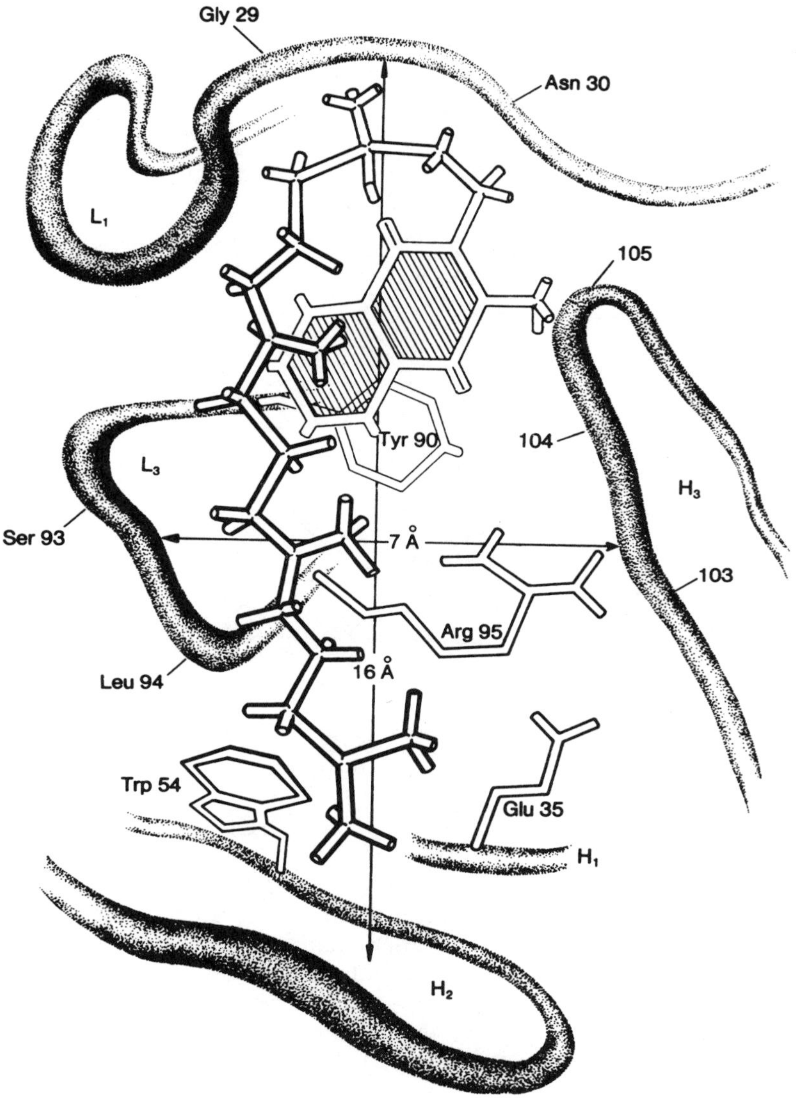

Figure 9 Schematic drawing of K_1OH bound to the binding site of the IgG myeloma protein New. Five contact regions are shown: two light chain hypervariable regions (L_1 and L_3) and three heavy chain hypervariable regions (H_1, H_2, and H_3). The view is one of looking into a shallow groove with L_1, L_3, H_2, and H_3 forming the rim and part of L_3 (Arg 95 and Tyr 90) at the bottom. (From Amzel et al., 1974.)

The C_H2 domains are a special case due to the fact that a carbohydrate chain is attached to the H chain in this domain. The carbohydrate apparently covers part of the contact face, resulting in weak C_H2 interactions.

Cis interactions are between C-terminal loops of one domain and N-terminal loops of the next. They are not extensive and depend on the flexibility of the extended chain

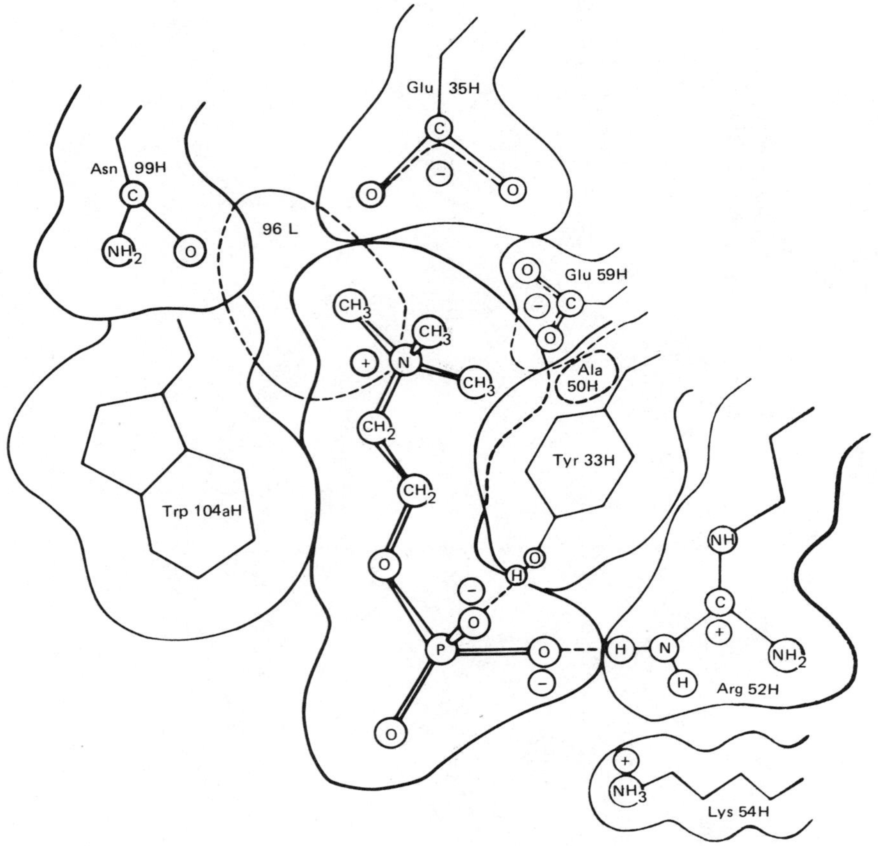

Figure 10 Schematic representation of the binding of phosphorylcholine by the murine myeloma protein M603. A deep cavity (in contrast to the shallow groove of New) is formed by residues of the H_1, H_2, H_3, L_1, and L_3 hypervariable regions. The choline group is oriented toward the interior of the cavity with the phosphate group at the exterior, where it interacts primarily with H_1 and H_2 residues. Note that L_2 is shielded from interacting with phosphorylcholine by L_1 and H_3. [Reprinted with permission from E. A. Padlan et al. (1976), Immunochemistry *13*; copyright 1976 Pergamon Press, Ltd.]

segment (switch region) connecting them. This flexibility between globular units (intrasegmental flexibility) was revealed by x-ray diffraction studies but had been suspected on the basis of proteolytic cleavage products. The lack of extensive cis interactions does not support any models for conformational changes in the Fc being transmitted as a result of antigen binding. This will be discussed in a later section.

Several crystallized IgG fragments and BJ dimers have been shown to bind haptens. One such fragment, Fab New, binds the γ-hydroxyl derivative of vitamin K_1 (K_1OH). The hypervariable regions of the H and L chains form a shallow groove. Figure 9 depicts the binding of K_1OH by Fab New. The contacts are made by the first and third L-chain hypervariable regions (L_1 and L_3) and three H-chain hypervariable regions (H_1, H_2, and H_3).

Stabilization of the V_H/V_L unit results from contacts between the two domains. Thus Fab New contains a core of amino acid residues in the contacting region of the three

chain layers which are invariant (or minimally variant) in V_H and V_L sequences. A similar core of contact residues is present in C_H1 and C_L. The invariance of these residues may explain the ability of H and L chains to recombine in different pairings, a mechanism by which antibody diversity can be expanded. Other interactions between hypervariable residues may explain the observed preferred associations of certain H and L chains.

The three-dimensional structure of a mouse Fab has also been reported. Fab-M603 is from a mouse myeloma IgA κ protein and closely resembles Fab New. This protein binds phosphorylcholine in a wedgelike cavity with the dimensions 12 Å (depth) × 15 Å (width) × 20 Å (length) lined by the L1, L3, H1, H2, and H3 hypervariable regions similar to Fab New. The orientation of phosphorylcholine in the cavity is shown in Fig. 10. The negatively charged phosphate is in close association with the postively charged guanidinium group of Arg 52 (H chain) and the amino group of Lys 54 (H chain). The positively charged trimethylammonium group interacts with the negatively charged carboxyls of Glu 35 and 59 (H chain). Interestingly, the sequences of seven other phosphorylcholine binding mouse myeloma proteins have been examined and shown to preserve these amino acids at the same positions.

By comparison to the Fab fragments, the x-ray diffraction patterns of Fc fragments and whole IgG molecules have been difficult to resolve. This is due to the fact that the hinge area appears to be less ordered and has little electron density. The region around the C_H2-C_H3 contact area is more ordered and recent studies have refined the Fc model

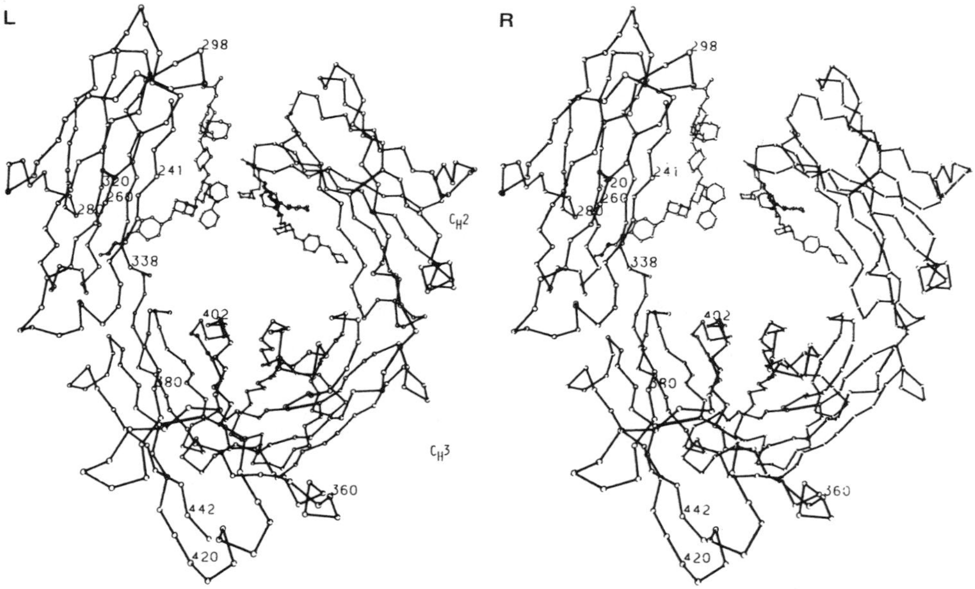

Figure 11 Stereo drawing of the α-carbon backbone of an Fc fragment of human IgG. The carbohydrate moiety is shown attached to position 297 (Asn) in the C_H2 domain. Other residue positions are indicated by the numbers and the disulfide bonds are indicated by heavier lines. Note the differences in proximity of the two halves of the C_H2 and C_H3 domains. [Reprinted with permission from J. Diesenhofer (1981), Biochemistry *20*, 2361; copyright 1981 American Chemical Society.]

at 2.9 Å resolution. As previously indicated, the C_H2 domains are not in contact with each other due to the presence of a carbohydrate moiety attached to Asn 297 (at a loop between the two β pleated sheet layers). The carbohydrate chains afford some stabilization of the C_H2 domains through their contact, but only about 25% as much as the contact between C_H3 domains. Additional stabilization of the Fc results from the extensive contact of the C_H3 domains, which is similar to the C_H1/C_L unit of the Fab New fragment. Regions of the C_H2 unit closest to the C_H3 unit show less disorder than the more remote regions. There also appears to be considerable contact between the C_H2 and C_H3 units. The residues most important in these contacts appear to be conserved in other classes of immunoglobulins and in other species as well. Figure 11 shows a stereo drawing of the alpha carbons ($C\alpha$) of the Fc model.

The Fc part of the molecule does not appear to greatly affect the conformation of the Fab. This has been ascertained from crystallographic studies on the whole IgG molecule (IgG mol) and the Fab made from it. Thus the Fab assumes a conformation close to that found in the whole molecule. These studies confirm others demonstrating the relative independence of the Fab and Fc regions. Sufficient studies are not yet available to construct a generally acceptable three-dimensional model of IgG. Those that are available are in disagreement as to the angle between the two Fab arms (i.e., Y- or T-shaped) and whether or not the Fab interacts with the Fc (i.e., C_H1-C_H2 contact). This would be important in explaining the mechanism by which antigen binding in the Fab is translated into changes in the Fc, allowing the molecule to carry out secondary functions (to be discussed later). In addition, many of the crystallized proteins used for these studies are "abnormal" in that they either have deletions in the hinge region or are cryoglobulins.

III. VARIATIONS OF THE BASIC STRUCTURE

A. The Light Chains

Structure of L Chains

Light chains are common to all classes and subclasses of immunoglobulins. In normal human serum immunoglobulins, the ratio of κ to λ varies for each class (see Tables 2 and 3). The recognition that BJ proteins were free L chains was a milestone in understanding the structure of immunoglobulins. As discussed previously, allotypes are known for human κ chains but not λ. These variations result from amino acid substitutions in the C domains of the respective chains. From sequence analysis of human BJ protein, it is known that positions 153 and 191 of the C_κ domain determine the Inv allotypic determinants. Thus position 153 can be valine or alanine, and position 191 can be leucine or valine (discussed in Chap. 11).

The subtypes of human λ chains are also determined by substitutions in the constant region. Four different subtypes are known for λ chains. They are defined by four isotypic determinants and designated by the BJ protein in which the variants were first recognized. Table 1 lists these four determinants and the corresponding amino acid substitutions. Oz(+) and Kern(+) proteins are detectable with antisera. One-half of normal serum λ chains are Oz(+) and approximately 25% are Kern(+).

Sequence comparisons of the V_L domains show that both κ and λ chains can be grouped by homologies of the first 23 amino acids. For κ chains, there are four homology groups ($V_{\kappa I}$, $V_{\kappa II}$, etc.) and for λ chains there are four or five such groups ($V_{\lambda I}$, $V_{\lambda II}$, etc.). The incidence of the V_κ subgroups among normal human immunoglobulins is 60,

Table 1 Subtypes of λ Chains

Marker	Position	+/–
Oz	193	Lys/Arg
Kern	156	Gly/Ser
Mz	147	Val/Ala
	174	Asn/Lys
Mcg	116	Asp/Ala
	118	Thr/Ser
	167	Lys/Thr

10, 28, and 2%, respectively, for $V_{\kappa I}$ to $V_{\kappa IV}$ (Nisonoff et al., 1975). As discussed previously, it is commonly believed that there is at least one germ line gene for each subgroup. At least three antigenic sites have been recognized on the V_L domain of κ chains, and antisera prepared against these sites have been used to identify subgroups. In contrast to κ chains only limited data have been obtained for V_κ domains, and only two antigenic sites have been uncovered, which do not define any one subgroup. One-half of the λ chains of normal immunoglobulins do not possess a V_λ antigenic site. None of the known V_L antigenic sites are allelic. Only rabbits and mice have allotypic determinants in their V_L domains.

Most of the carbohydrate associated with immunoglobulins is found on the H chains. Occasionally (5 to 10%), however, carbohydrate has been found on BJ proteins or isolated L chains from myeloma proteins of both murine and human origin. Usually, the site of attachment of carbohydrate on L chains is the V_L domain, and the peptide sequence is Asn-x-Ser/Thr. The significance of attached carbohydrate is unknown.

The region between the V_L and C_L domains (the *switch region*) is very susceptible to proteolysis, resulting in free V_L and C_L fragments. While V_L and C_L are usually more stable to further degradation, prolonged exposure of C_L to proteolytic enzymes at 37°C results in extensive digestion, whereas the V_L is not affected. Interestingly, at 55°C the opposite is true: V_L is degraded and C_L is not. This is not the only example of stability of constant domains at elevated temperatures. As will be discussed later, the Fc of both IgM and IgA show similar properties.

The distribution of κ and λ chains in all classes, subclasses, and species differs. Thus, while IgG shows the overall κ/λ ratio of approximately 2.0, the subclass ratios vary from 1 for IgG2 to 8 for IgG4. On the other hand, the ratio for IgD is approximately 0.3. The incidence of κ and λ myelomas for each class or subclass reflects the ratio found in normal immunoglobulins. This indicates that conversion of plasma cells into neoplastic cells is a random process. For species other than humans, the relative amount of each type of chain varies widely. Thus mice have approximately 95% κ chains and horses are essentially 100% λ.

Light chains are bound to H chains by a disulfide bond. The involved L-chain cysteine is terminal on κ chains and penultimate on λ chains. In addition to the κ and λ antigenic determinants, it has been found that hidden antigenic determinants are revealed on L chains after separation from H chains. These determinants have been used to quantitate the amount of free L chains or BJ proteins. The hidden determinant is strongly

antigenic and antisera made against isolated L chains or BJ proteins frequently react primarily with these determinants. Anti-κ or anti-λ specific antisera must be made against intact κ or λ myeloma proteins or their Fab fragments in order to be strong enough to be useful. This is because of a conformational factor associated with these determinants which depends on the H-L interaction.

The most salient characteristic of BJ proteins and the property that originally caused their discovery is their solubility behavior at elevated temperatures. Under defined conditions of concentration, pH, and ionic strength, BJ proteins characteristically precipitate at 50°C and resolublize at 100°C. This is unusual in that most, but not all serum proteins show different thermal behavior; that is, they precipitate at elevated temperatures and do not resolubilize on further elevation of temperature. The fact that not all BJ proteins show this behavior is understandable because it reflects the solubility properties of the V_L domain.

Bence Jones protein may exist as monomers or dimers (or mixtures of both). Occasionally, higher polymers are also formed. κ chains are usually monomeric but are also found as dimers, via either disulfide linkage or noncovalent interactions. λ chains usually exist as disulfide-linked dimers.

Bence Jones proteins bear structural similarities to a component of amyloid. "Primary" amyloid has been shown to be composed of homogeneous BJ protein or the V_L domain of BJ protein. There is a high correlation of the occurrence of amyloid with the occurrence of BJ proteinuria of the $V_{\lambda I}$ subgroup. However, other factors are also involved in the pathogenesis of amyloidosis. Light chain structure has been reviewed by Solomon (1976).

Function of L Chains

The primary function of L chains is to bind antigen. This is accomplished through the V_L domain. Three hypervariable regions are associated with the V domain: 24 to 34, 50 to 56, and 89 to 97. Three-dimensional structure studies show that these sites are intimately involved in the antigen binding cavity, as discussed previously. Earlier studies did not recognize the antigen binding properties of L chains and relegated the function of H-chain stabilization to this component since it was found that, unlike L chains, isolated H chains aggregate and precipitate at neutral pH.

No definitive function has been attributed to the C_L domain. The H-L disulfide bond occurs in this domain, but this does not appear to be essential to the integrity of the molecule since other classes or subclasses exist without this linkage (e.g., IgA2 in humans). The noncovalent interactions between L and H chains are distributed throughout both L-chain domains. Perhaps the C_L function has been lost during evolution. It is possible that the "first antibody" was composed of L-chain dimers in which one chain behaved as the N-terminal half (Fd) of a heavy chain to give a pseudo-Fab fragment. This has been shown to occur in crystallized L-chain dimers where one monomer assumes the conformation of the Fd. The C_L domain may play a role in these conformational changes.

B. The IgG Subclasses

Structure of the IgG Subclasses

The major subclass of human IgG is the IgG1 subclass. This accounts for over half of the circulating immunoglobulin and it is the prototype for the structure of all immunoglobulins.

Table 2 Properties of Human IgG Subclasses

	IgG1	IgG2	IgG3	IgG4
H-chain molecular weight				
Total (× 10^3)	52–54	52–54	60	52–54
Polypeptide (× 10^3)				
Carbohydrate				
Percent (w/w)	2–3	2–3	2–3	2–3
Oligosaccharide units	1	1	1	1
Sedimentation rate, $s_{20,w}$	6.6	6.6	6.6	6.6
Normal serum concentration (mg/ml)	6.63 ± 1.7	3.22 ± 1.08	0.58 ± 0.3	0.46
Percent of total IgG concentration	60.9	29.6	5.3	4.2
κ/λ ratio	2.4	1.1	1.4	8.0
Half-life (days)	21	20	7	21
Inter-H-chain disulfides	2	4	15	2

Source: Data from Morell et al. (1976).

Most of the structural properties attributed to the IgG molecule are derived from studies on this subclass since it comprises approximately 60% of the IgG class. The other three subclasses of IgG have a similar structure but differ in some respects. Table 2 compares the properties of the four IgG subclasses.

The distribution of the subclasses in normal human serum varies widely. The values shown in Table 2 are mean values (±1 SD) and percent of total IgG for young male adults (Morrell et al., 1976). This distribution changes with age, sex, and health status. Figure 12 shows the IgG distribution through the first 2 years of life.

A major difference in the subclasses is the structure of the hinge regions. The number of inter-H-chain disulfide bonds differ for the subclasses. Thus whereas IgG1 and IgG4 have two such bonds, IgG2 has four and IgG4 may have as many as 15. The hinge region of the IgG3 subclass deviates furthest from the IgG1 hinge. Its structure is a quadruplication of a 15-residue segment extending the hinge region to 62 amino acid residues. The H-L disulfide bonds of all subclasses except IgG1 originate at position 131 (Eu numbering), which is not in the hinge region. The H-L disulfide of IgG1 originates within the hinge region at position 220.

The identification of subclasses can be accomplished by detection of antigenic markers or by chemical means. Most subclass and allotype markers are on the C_H2 and C_H3 domains (although all subclass and allotype determinants are not confined to a single subclass) and either of these can be used to define a subclass. Subclass-specific antisera are made with isolated myeloma proteins, but it is very difficult to obtain specific antisera of sufficient strength to be useful, and they are generally unavailable commercially. The alternative is to purify the immunoglobulin and, after mild reduction and alkylation with [^{14}C]iodoacetate to radiolabel the interchain disulfides (conversion of cystine to carboxymethylcysteine), carry out extensive proteolysis. The peptides are then separated by high-voltage electrophoresis and subjected to radioautography. Owing to the different arrangements (and adjacent sequences) of interchain disulfides in the subclasses, each will give a characteristic pattern of radioactive cysteine-containing peptides from which class, subclass, and light chain type can be determined. The various patterns are diagrammed in Fig. 13.

The subclasses also differ with respect to susceptibility to proteolytic enzymes. This property has also been used to help identify subclasses of myeloma proteins. It

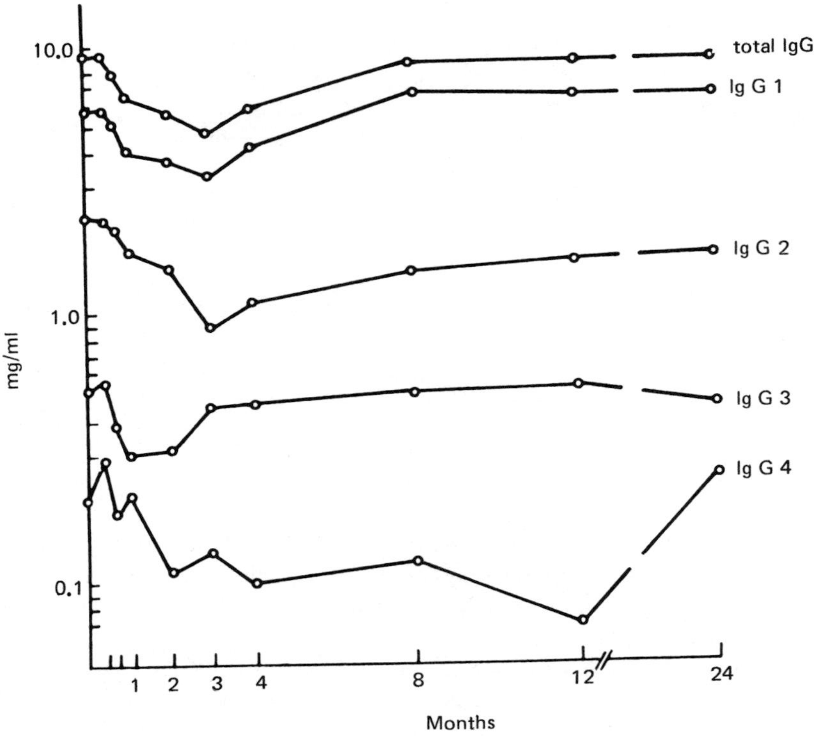

Figure 12 Variations in concentrations of the four IgG subclasses during the first 2 years of life. Note the decrease during the first 3 months of age, during the initial phase of autonomous synthesis and catabolism of maternal supply. By 24 months, levels are close to adult values for IgG1 and IgG3, and about half adult values for IgG2 and IgG4. The study represents the mean values of 95 children. (From Morell et al., 1976.)

should, however, be emphasized that a great degree of variability in susceptibility exists within a subclass and it is difficult to make generalities. Nevertheless, it is generally held that the order of increasing resistance to proteolysis with papain at neutral pH is IgG3 < IgG1 < IgG4 < IgG2. On the other hand, IgG1 and IgG2 are more resistant to digestion by pepsin at pH 4.5, while IgG3 and IgG4 are more susceptible. A variety of proteolytic enzymes have been used to cleave IgG and to compare the susceptibility of the subclasses. These include plasmin, trypsin, subtilisin, chymotrypsin, and elastase. In general, IgG1 and IgG3 have been found susceptible and IgG2 and IgG4 resistant at neutral pH. The products of the digestion are similar to the products described for papain and pepsin: Fab, $F(ab')_2$, Fc, and Fc′. These fragments can be prepared from all subclasses but the conditions required vary. The fragments can be characterized by electrophoretic mobility (Fig. 14). In addition to these fragments, other fragments consisting of one or more domains have been prepared by specialized procedures. With the exception of the C_H2 domain, essentially all enzymatic cleavages have been found to occur at interdomain sites (i.e., in the hinge or switch regions), with little or no degradation within the domains. This supports the crystallographic data, which portray the molecule as being composed of

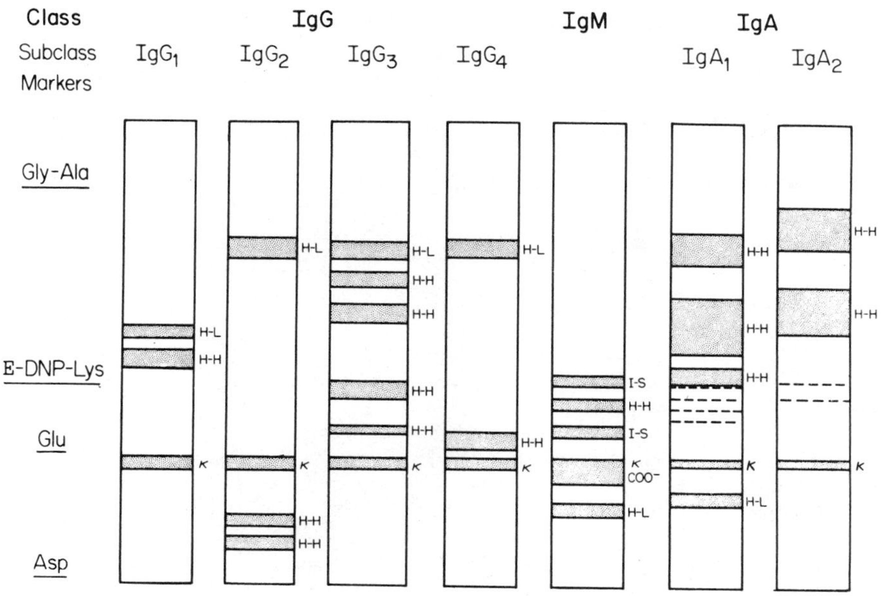

Figure 13 Radioautograph patterns of peptic-tryptic peptides of ^{14}C-labeled carboxymethylcysteine derivatives of immunoglobulins of the IgG subclasses, IgA subclasses, and IgM. The cysteine-containing peptides have been identified as being from inter-H chain (H-H), inter-H-L (H-L), intersubunit (I-S), carboxyl terminal (COO^-), or K chain (K). Separation of peptides was accomplished by electrophoresis at pH 3.5. (From Frangione and Franklin, 1972.)

a series of tightly packed globular domains linked by relatively extended portions of the chain which are more susceptible to enzymatic attack. The preparation of these fragments has allowed detailed analysis of the location of various antigenic determinants on the IgG molecule as well as information on the sites responsible for the many effector functions of antibodies (Calvanico and Tomasi, 1979; Stanworth and Turner, 1978).

The carbohydrate content of the IgG subclasses is similar. Although variations occur, they appear to be random; that is, they occur within the same subclass and are not characteristic for a particular subclass. Essentially all carbohydrate is located on asparagine 297 in the C_H2 region, as discussed previously. The monosaccharides found are mannose, galactose, fucose, N-acetylglucosamine, and sialic acid as the terminal sugar. The ratio of these sugars in the four subclasses is similar. When additional carbohydrate is found on myeloma proteins of the various subclasses, it is usually on the Fab part of the molecule associated with either the L chain (as discussed previously) or the Fd region (i.e., the V_H-C_H1 fragment), or both. Microheterogeneity can also be found in the molecules of an individual myeloma. In general, the carbohydrate content of myeloma proteins reflects the range found with normal IgG. No function has been attributed to the carbohydrate moiety.

Several species have subclasses of IgG. Thus the mouse and rat have four subclasses similar to humans, while the guinea pig and possibly the rabbit have two. The horse appears to have five subclasses of IgG. Space does not permit a complete description of

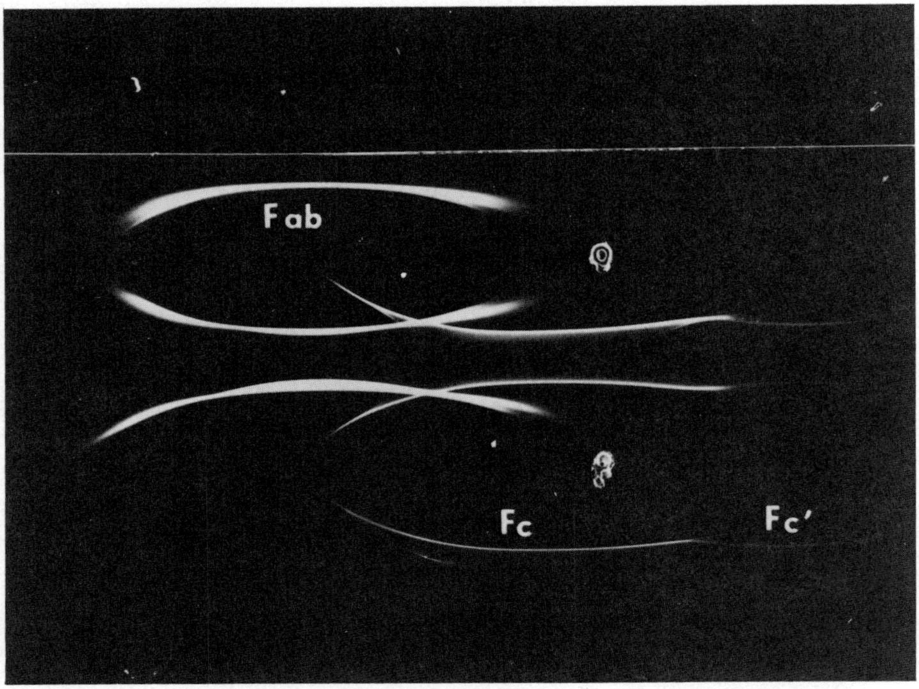

Figure 14 Mobility of Fab, Fc, and Fc′ on immunoelectrophoresis. Wells contain human IgG digested with papain and troughs contain anti-Fab (top), anti-native IgG (middle), and anti-Fc (bottom). Note reactions of nonidentity between Fab and Fc, and partial identity between Fc and Fc′. Undigested IgG would occupy a position intermediate between Fab and Fc. Anode is to the right. (From Calvanico and Tomasi, 1979.)

These immunoglobulins in other species, but in general the variations are similar to those found in human IgG subclasses. Cross-reactivity between IgG of different species varies widely. As expected, the monkey is 100% cross-reactive with humans, while the rat is 15% cross-reactive with humans. A more extensive discussion of structures of immunoglobulins of mammalian species other than humans can be found in Nisonoff et al. (1975). Phylogeny of immunoglobulins is discussed in Chap. 10.

Functions of IgG

Structural changes due to ligand binding. The primary function of all immunoglobulins is their ability to recognize and bind foreign components. As a result of this binding, there is a consequential alteration in the molecule which allows secondary functions to be expressed. These secondary functions vary with class and subclass of immunoglobulins, but usually involve interactions of the immunoglobulin with cells or other proteins. All of these interactions are mediated by structures located on the Fc part of the molecule. The antibody molecule can thus be thought of as a bifunctional protein, binding antigen on one side (i.e., in the Fab part of the molecule) and performing another biologic function at a different site removed from the antigen binding site. The mechanism by which antigen binding in the Fab activates accessory systems is unknown.

Several hypotheses have been advanced (Metzger, 1978). The most appealing model is one in which binding of antigen induces conformational changes in the Fc part of the molecule (i.e., allosteric changes), thus exposing new sites. Metzger (1979) has argued persuasively that all studies to date, whether direct (e.g., x-ray diffraction) or indirect (e.g., optical rotatory dispersion), have failed to demonstrate conclusively changes in the Fc as a result of antigen binding. Most of these studies, however, were carried out with monovalent haptens, which are generally not efficient in initiating secondary functions. With polyvalent antigens of sufficient size, however, small changes have been observed by very sensitive techniques. It should also be emphasized that the extent of change that is required for the initiation of secondary effects has not been determined.

Since it has long been known that antigen-antibody complexes consisting of more than one antibody molecule are frequently required for maximal effector function activity, an alternative hypothesis to allosteric changes has been that aggregation of antibody mediates the necessary transition. This is supported by the fact that heat aggregates or chemically cross-linked Fc regions are often as effective as antigen-antibody complexes in secondary functions. The effect of aggregation on the structure of the Fc region is unknown, but Metzger (1979) argues that aggregation per se is the important event and not a specific conformational change resulting from it. The principal argument against this is the structure of the IgM molecule (see Sec. III.C), which exists in the unliganded state as a polymer and not able to fix complement until it binds antigen. Metzger, however, argues that intra- and intermolecular aggregation cannot be considered equivalent.

One additional fact is well documented with regard to secondary functions. Reduction of the inter-H-chain disulfide bonds results in loss of complement fixing ability. Dependency of complement fixation on intact hinge disulfides, however, is not exhibited by the Fc fragment. In addition, studies with myeloma proteins lacking a hinge region have shown that they act much the same as reduced proteins (i.e., they do not express secondary functions). These results have been interpreted as indicating that a close association between the Fab and Fc regions interferes with effector functions and that the structural rigidity and the spacer effect of an intact hinge region in normal IgG prevents this Fab-Fc interaction. This would also explain the inhibitory effect of reduction of the hinge disulfide. It has been shown that increased segmental flexibility results from loss of the hinge disulfide bond, which would also lead to interaction between the Fab and Fc. Reduction of the isolated Fc would not be expected to have any effect if this is true, and this is what is observed.

In summary, therefore, although no clear-cut explanation of the mechanisms of initiation of secondary functions by ligand binding is acceptable, the observed experimental evidence indicates that both aggregation and conformational changes may be involved.

Complement fixation. The activation of complement (C′) is one of the principal secondary functions of liganded antibody and the prime model for the interaction of antibody with other proteins. Complement activation by IgG can occur through the classical (C1, C4, C2, etc.) or alternative (C3, C5, C6, etc.) pathways. Complement activation leads to a multitude of biologic effects, including the lysis of cells (bacteria, etc.), release of inflammatory mediators, and the recruitment of other cells. The structure and function of complement proteins are discussed in detail in Chap. 23.

Complement activation is not effected equally by the four IgG subclasses. IgG1, IgG2, and IgG3 are all able to fix C′ via the classical pathway while IgG4 is not. IgG1 and IgG3 are more effective than IgG2. It appears that C′ fixation by IgG requires at least two

molecules of antibody cross-linked by a multivalent antigen. Aggregation of IgG may also be accomplished in vitro by heating or by chemical means and shown to be effective in C′ fixation. Although most studies show that the antigens most effective in inducing C′ fixation are multivalent, a few isolated reports indicate that univalent antigen can also initiate C′ fixation. This would suggest that aggregation is not necessary for C′ fixation since univalent antigens would not be able to cross-link antibodies. Some of the conclusions drawn from this work, however, are open to question (Metzger, 1978).

The site of C1 fixation has been localized to the C_H2 domain of IgG1. Modification of two tryptophan residues in the Fc fragment results is complete loss of C′ fixing ability, but less than 20% loss of C1q binding. These results emphasize the importance of differentiating between the *binding* of C1q and the *activation* of C1. Another approach that has been used to determine the position of the C′-fixing site is the use of synthetic polypeptides for activation studies. Such studies have shown that polypeptides in which tryptophan and tyrosine are coupled, as in positions 277 and 278 of human IgG1 (Eu numbering), are capable of consuming CH_{50} units significantly. The peptide sequence Lys-Phe-Asn-Trp-Tyr-Val-Asp-Gly corresponding to positions 274 to 281 of human IgG1 appears to be as effective as heat-aggregated human gammaglobulin in C1-consuming activity. Position 274 (Lys) appears to be critical to this activity. In IgG2, a weak C′-fixing subclass, this position is occupied by G1x rather than Lys. In IgG4, a nonfixing subclass, the same substitution is found at position 274, but in addition, at position 268, His is replaced by Gln. It has been postulated that these substitutions result in inefficient exposure of the C′-fixing site. Other studies, which show that the isolated Fc of IgG4 and IgA, two non-C′-fixing immunoglobulins are capable of interacting with Clq, suggest that the C′-fixing site is present on these molecules but some structural hindrance caused by the Fab region in the native molecule prevents this interaction.

In contrast to the classical pathway, it appears that all four IgG subclasses are able to activate the alternative pathway (i.e., C3 conversion). Aggregation of proteins appears to be an absolute requirement. This is the only known secondary function of IgG that appears to be associated with the Fab region instead of the Fc.

Cytophilic functions. IgG is capable of binding to a variety of cells via specific membrane receptors. Regardless of cell type, interaction and binding of IgG occur through structures present on the Fc part of the molecule. The biological results of this binding depend on the cell type involved. Thus macrophages, lymphocytes, neutrophils, and basophils (mast cells) all bind IgG via Fc_γ receptors.

1. *Macrophages and monocytes.* Antibody bound to these cells acts as an opsonin for phagocytosis of foreign antigens or cells (e.g., bacteria). The mediation of antibody in phagocytosis significantly enhances the process. It is important to note that cytophilic antibody is bound to cells before it binds antigen. Aggregation, however, whether through antigen or chemicophysical methods, increases the binding of IgG. Immune complex binding by macrophage activates them to become cytotoxic and liberate lysosomal enzymes. In the mouse, macrophages have been shown to have at least two Fc_γ receptors, one trypsin-sensitive receptor specific for IgG2a monomers and one trypsin-resistant receptor specific for aggregated IgG1 and IgG2b. It has been postulated that the former is associated with phagocytosis and the latter with antibody-dependent cellular cytotoxicity (ADCC). Guinea pig macrophages also appear to have two Fc_γ receptors, but multiple receptors have not been demonstrated on human macrophages. It has been demonstrated, however, that the four subclasses of IgG vary in cytophilic properties. IgG1 and IgG3 have the highest affinity, while IgG4 is less cytophilic and IgG2 is

noncytophilic unless it is aggregated. The cytophilic site on human IgG appears to be on the C_H3 domain. IgG1 and IgG3 also are the most hydrophobic of the IgGs (see Chap. 31).

2. *Neutrophils.* These cells are also phagocytic, and cytophilic antibody enhances phagocytosis (opsonization) similar to macrophages. The IgG subclass cell binding pattern described for macrophages also applies to neutrophils. There is little information available concerning the properties and specificities of the neutrophil $Fc\gamma$ receptor. Interaction of the $Fc\gamma$ receptor with antibody is postulated to initiate internalization of the bound antigen and also to be necessary for fusion of phagosome and lysosome.

3. *Lymphocytes.* $Fc\gamma$ receptors for IgG can be found on both B and T cells. The role of IgG in the functions of these cells is not well understood. Thus, while IgG can serve as a recognition unit for killer (K) cells for ADCC, the role of IgG bound to B cells is not readily apparent since membrane-bound immunoglobulins of the M and D class (discussed later) act as recognition units for antigens on these cells. Certain T cells also possess $Fc\gamma$ receptors. In humans, $Fc\gamma$-receptor-bearing T cells appear to suppress immunoglobulin production, as assayed by pokeweed mitogen stimulation of B cells. The binding of antigen-antibody complexes by T cell $Fc\gamma$ receptors activates these suppressor cells. A recent issue of *Immunologic Reviews* (Moller, 1981) has focused on the role of Fc-receptor-bearing T cells.

Lymphocytes, similar to monocytes and neutrophils, also vary in their affinity for the IgG subclasses. The pattern of binding resembles that obtained with macrophages and neutrophils.

4. *Basophils and mast cells.* These cells, which release mediators of immediate hypersensitivity, have receptors for two classes of immunoglobulins. They are referred to as *cytotropic antibodies*, and they can sensitize other species (heterocytotropic) or the same species (homocytotropic). Hetercytotropic antibodies are primarily IgG, while homocytotropic antibodies are primarily IgE (Sec. III.D). Some homocytotropic antibodies have been reported to be IgG and are termed *short-term sensitizing* (STS) *antibodies* because they sensitize for only 2 to 4 hr. The IgG4 subclass has frequently been associated with this activity, but the evidence is inconclusive and very controversial. The biological function of heterocytotropic antibody is completely unknown.

5. *Other cells.* Several nonimmunocompetent cell types have been shown to possess $Fc\gamma$ receptors. These receptors function primarily for the transport of IgG. Thus maternofetal transfer of antibodies occurs in utero in humans via $Fc\gamma$ receptors on the placental syncitiotrophoblast. In rodents and ruminants, however, passage of IgG to the fetus occurs by feeding and subsequent transfer of antibody across the intestinal epithelium also via $Fc\gamma$ receptors.

The catabolic rate of immunoglobulins is regulated by structures on the Fc region. The site of catabolism has not been clearly located, but the kidney is clearly implicated. The mechanism by which the Fc regulates the catabolic rate is also unclear, but it is thought that an $Fc\gamma$ receptor mediates protection of the molecule.

Platelets also have $Fc\gamma$ receptors. Binding of antigen-antibody complexes or aggregated immunoglobulins causes the relase of vasoactive amines from these cells.

C. IgM and IgA: The Polymeric Immunoglobulins

IgM and IgA occur as polymers of the basic four polypeptide chain units. The polymeric forms, however, differ in the two classes. IgM forms a closed pentameric ring structure and IgA forms an open-ended polymer with varying numbers of units. In addition, both

of these classes of immunoglobulins have extra polypeptide chains not found associated with the other classes of immunoglobulins. Although there are structural similarities between these classes, there are major differences in their functions.

IgM

Structure. The closed-ring penatmuric structure of IgM is depicted in Fig. 15. Certain features of its structure are noteworthy. First, there is a third type of polypeptide chain incorporated into the structure, referred to as the *joining* or *J chain*. It appears to be involved in the polymerization of 7S IgM subunits (IgMs) into the 19S pentamer (molecular weight 900,000) (Brandtzaeg, 1976). Recently, it has been found that the polymerization requires the activity of a membrane-bound polymerizing metalloenzyme different from the disulfide interchange enzyme originally thought to be involved in this

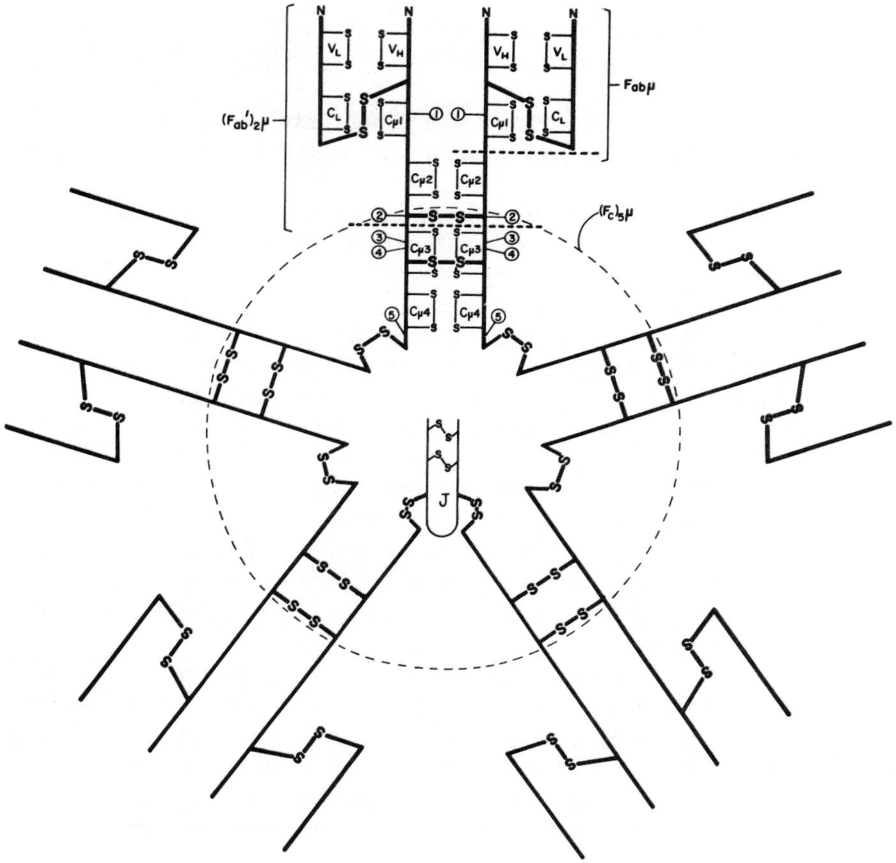

Figure 15 Schematic representation of the pentameric ring structure of IgM. Linkage between IgMs and J chain involve the penultimate cysteine on the μ chain. Positions of the five oligosaccharides (circled numbers) and the five μ-chain domains are shown. Cleavage by trypsin at elevated temperatures yields Fabμ and $(Fc\mu)_5$ (encircled) as indicated, whereas at 37°C only the $F(ab')_2$ is obtained. (From Cathou, 1978).

conversion (Roth and Koshland, 1981). The enzyme is found in significant quantities only in cells differentiated to secrete immunoglobulins. J chain apparently joins only two IgMs; the remaining IgMs are joined to this dimer. Linkage of IgMs, either to each other or to the J chain, appears to occur via the penultimate cysteine of μ chain. The sequence of human J chain has been determined. It consists of 129 amino acids with one carbohydrate chain and has a molecular weight of 16,422 (7.6% carbohydrate). Eight cysteines are found in this polypeptide and two (positions 13 and 15) have been implicated in the binding to the μ chain (Mole et al., 1977).

Another feature of the structure of IgM is the presence of a fifth domain on the μ chain. There are four C domains instead of three, increasing the molecular weight of the μ chain to approximately 70,000 with five carbohydrate units. The hinge region is between $C\mu2$ and $C\mu3$ rather than $C\mu1$ and $C\mu2$, and two inter-H disulfide bonds are present (Fig. 16). The Fc region of IgM is composed of the $C\mu3$ and $C\mu4$ domains.

Proteolytic cleavage of IgM at 37°C usually results in destruction of the $Fc\mu$ fragment and recovery of the $F(ab')_2$ or Fab regions (Fig. 15). Cleavage at 60°C with trypin (or at 37°C in the presence of urea), however, causes destruction of the $C\mu2$ domain and liberation of Fab and $(Fc\mu)_5$ fragments in excellent yields. This technique has facilitated studies on the sequence and function of IgM.

IgM is considered to have two subclasses, but they have not been characterized. An allotypic determinant (Mm1) has been identified on 5 of 23 Waldenström monoclonal IgM proteins.

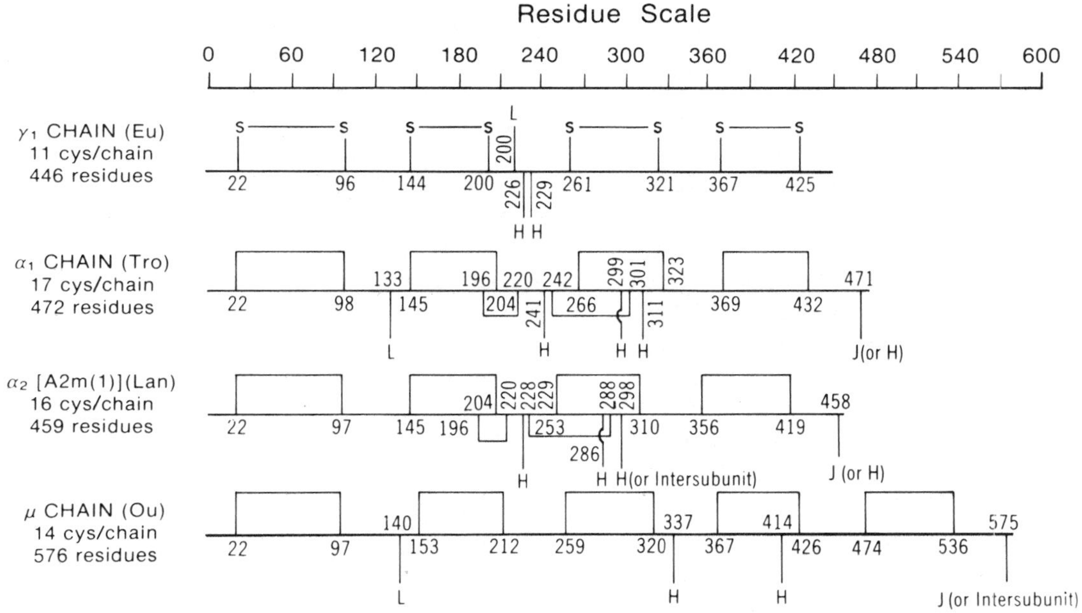

Figure 16 Schematic representation of H chains of IgM(Ou), IgA1 (Tro), and IgA2 (But) compared with IgG1 (Eu). The distribution and linkage of intra- and interchain disulfide bonds are indicated by position number. Rectangles above chains represent domain disulfide loops.

Function of IgM. IgM is the primary antibody response to immunization. Like IgG, IgM is able to fix complement, but more efficiently. In addition, a single molecule of IgM is sufficient to initiate complement fixation, whereas two of IgG are needed. The IgMs are about as efficient as IgG in C1 fixation, but 15-fold less efficient than native IgM on a weight basis. Natural polymers of IgM with sedimentation rates greater than 35S are 100 times more efficient than IgMs in C1 fixation. Heat aggregation of IgM destroys its ability to fix C1, whereas chemical cross-linking does not. As in the case of IgG, IgM C′ fixation is greatly enhanced by antigen binding. Although it has generally been accepted that the occupation of a single antigen binding site of IgM is sufficient to cause *C1 binding*, it should be emphasized that this is *not* equivalent to *C′activation*. Activation of the lytic sequence appears to require occupation of more than one site. The site C1 activation on IgM has been localized to the C-terminal (Cμ4) domain.

As can be seen in Fig. 15, IgM is dekavalent. However, owing to the closeness of its antibody-active sites, interaction with large antigens is strongly subject to steric hindrance, which gives rise to the illusion of pentavalency. Thus, IgM could only be proven to be dekavalent with antigens of Mw 1500 (Edberg et al., 1972).

Few cells have Fcμ receptors. The cytophilic properties of IgM may be somewhat enhanced by cross-linking with anti-IgM $F(ab')_2$ fragments. A population of T lymphocytes, generally considered to act as helper cells, exhibit Fcμ binding. By the IgM-coated ox erythrocyte rosetting assay, approximately 50% of human peripheral blood T cells and a small percentage of B cells have Fcμ receptors. The cytophilic site of IgM appears to be on the same domain as the C′-fixation site (i.e., the Cμ4 region).

In spite of the relative lack of Fcμ receptors on B lymphocytes, it has been found that most B cells have surface IgM. These surface molecules are membrane bound and can only be solubilized in solvents containing surfactants (i.e., they are integral proteins). They function as antigen receptors on B cells and provide one part of the signal required to transform a lymphocyte into an antibody-secreting plasma cell (the other part of the signal is supplied by the T cell), the product of which has the same specificity as the surface IgM. Since IgM is also present in significant quantities in serum, a question arose concerning the difference between membrane-bound and secreted IgM. One difference is that, while secreted IgM is pentameric, membrane-bound IgM is the monomeric IgMs. However, the major difference, accounting for the solubility differences is the amino acid sequence of their C-terminal segments. The secreted form of μ chains (μ_s) has a 20-amino acid tailpiece (tp) attached to the C-terminal domain (Cμ4). The membrane-bound μ chain (μ_s) has a different tp composed of 41 amino acids with a very hydrophobic core. Each of these tps has been shown to be coded for by separate gene segments (exons). The exon coding for the secreted tailpiece is continguous with the Cμ4 exon, whereas the membrane tp exon is considerably removed.

IgA

Structure. As stated, the structure of IgA consists of open-ended polymers (Heremans, 1974). In most species, the prevalent form in the serum consists of a dimer of the 7S IgAs linked by a J chain to the penultimate cysteine similar to IgM. It should also be mentioned that conflicting data exist concerning the amount of J per IgA dimer and IgM pentamer. Some studies indicate that there are two J chains per IgA dimer or trimer and three to four J chains per IgM pentamer (Brantzaeg, 1976). To our knowledge, the amount of J chain present in higher polymers of IgA has not been studied. In contrast to other animals, the prevalent form of IgA in humans is the monomeric IgAs. A distinguishing characteristic of serum IgA is the size of heterogeneity of the polymers present. This

Table 3 Properties of Human IgM and IgA

	IgM	IgA1	IgA2 m(1)	IgA2 m(2)
H-chain molecular weight				
Total (× 10^3)	65–70	56–58	52–54	52–54
Polypeptide (× 10^3)	57	50	50	50
Carbohydrate				
Percent (w/w)	7–12	5–10	5–10	5–10
Oligosaccharide units	5	7	4	5
Sedimentation rate, $s_{20,w}$				
Serum	19–32	6.2–17	6.2–17	6.2–17
Secretions	19	10–11	10–11	10–11
Normal serum concentration	0.5–1.5	0.5–2.0	0–0.2	0.0.02
κ/λ	3.2	1.4	1.6	1.6
Half-life (days)	5.1	5.9	4.5	4.5
Formula				
Serum	$[(IgM)_5\text{-}J]_n$	$(IgA)_n\text{-}J_m$	Same	Same
Secretions	$(IgM)_5\text{-}J(SC)^a$	$(IgA)_2\text{-}J\text{-}SC$	Same	Same
H-L disulfide bonds	+	+	–	+

[a]Noncovalently bound SC.

heterogeneity is one of the factors making it difficult to obtain a preparation truly representative of serum IgA; another is the propensity of this molecule to form complexes with other proteins. Both of these properties are due to the apparent ease with which IgA forms intermolecular disulfide bonds.

Two subclasses of human IgA exist, IgA1 and IgA2. There are two known IgA2 allotypes: A2m(1) and A2m(2). The A2m(1) molecules are distinguished by their lack of H-L disulfide bonds. This deficiency is compensated for by very strong noncovalent interactions which maintain the H-L integrity. The two L chains are disulfide-bonded to each other. A similar arrangement of α and L chains occurs in IgA myeloma proteins of BALB/c mice. Approximately 10 to 20% of normal serum IgA belongs to the IgA2 subclass. The sequences of both $\alpha 1$ and $\alpha 2$ chains have been determined. Figure 16 depicts the disulfide bond distribution of both $\alpha 1$ and $\alpha 2$ chains and compares them with that of μ chains. Both $\alpha 1$ and $\alpha 2$ contain four domains, but $\alpha 2$ chains are slightly shorter due to a deletion of 13 amino acids in the hinge region. There is also one less cysteine in A2m(1) chains since the H-L disulfide is absent. Properties of both subclasses are shown in Table 3.

The most important characteristic of the IgA class of immunoglobulins is its presence in exocrine secretions, usually as the major antibody class. A fourth type of immunoglobulin polypeptide chain, *secretory component* (SC), is associated with IgA in these fluids. This IgA_2-J-SC complex is referred to as *secretory IgA* (SIgA), differentiating it from the serum variety of IgA, which is devoid of SC.

In the limited space of this chapter it is impossible to discuss the many important aspects of the secretory system in general and IgA specifically. For many reasons, IgA is the least understood class of immunoglobulins, yet its importance as a defense system at the mucosal surfaces is well recognized. Several excellent reviews have been written about this system (Tomasi, 1976; Bienenstock and Befus, 1980; Lamm, 1976), and the reader is referred to them for further reading. It is only possible to cover some of the major points here.

The SC is a glycoprotein containing approximately 20% carbohydrate. In humans, SC has a molecular weight of about 70,000, whereas in rabbits it appears to be smaller (ca. 60,000). Rabbit and human SC also differ in their amino-terminal sequences and about one-half of their tryptic peptides. Free SC has been shown to bind polymeric IgA and IgM noncovalently. The J chain appears to be involved in this binding by its influence on the conformation of the IgA dimer, but is not itself bound to SC. In the case of IgA, binding subsequently becomes covalent via two disulfide linkages. In humans, all SIgA has covalently linked SC, whereas in rabbits, which also have two subclasses of IgA (f and g), only the f subclass has covalently linked SC. It has been determined that SC is disulfide-bonded to only one of the two monomer IgAs of human SIgA. Linkage of SC to the α chain is known to be in the Fc region, but the exact location is unknown, although the Cα3 domain does not appear to be involved since removal of this segment with CNBr does not affect the binding of SC. The binding mechanism is also unknown but probably involves an interchange of two intrachain disulfide bonds, one from the SC. The former would seem to implicate either the Cys 242–301 loop or the Cys 266–323 loop in the α1 chain. It is tempting to speculate that Cys 242–301 is involved since it is an "extra loop," while the latter corresponds to the C_H2 domain loop (Fig. 16). The configuration of SIgA may consist of either of two possible arrangements of the IgA dimer: (1) a stacked Y arrangement in which the two IgAs are stacked one upon the other as if to superimpose them; or (2) an end-to-end Y arrangement in which the two IgAs are joined by their Fcα regions. The predominant form appears to be the end-to-end double Y, with a small percentage in the stacked Y configuration. It is believed that the two configurations are interconvertible. The prevalence of the double-Y configuration is supported by both electron microscopy and hydrodynamic data.

Human IgA is resistant to papain digestion, unlike mouse or rabbit IgA. Mild reduction of human IgA renders it susceptible to digestion, but the Fcα is completely degraded to peptides and only $F(ab')_2$ or Fab fragments are recovered. By contrast, mouse and rabbit IgA both yield Fcα fragments with papain digestion. The resistance of IgA to proteolysis appears to be related to the presence of covalently bound SC. Thus human SIgA is more resistant than serum IgA (no SC), and rabbit IgA with covalently bound SC (f subclass) is resistant to proteolysis while the g subclass, which has noncovalently bound SC, is not. The discovery of a bacterial protease capable of splitting human IgA (serum or secretory) into Fab and Fcα fragments has greatly facilitated structure and function studies on IgA (Kornfeld and Plaut, 1981). The enzyme IgA protease cleaves a single peptide in the hinge region of IgA1 molecules only. The bond Pro-Thr (positions 227 and 228) is not present in the IgA2 hinge and hence the enzyme is not effective against this subclass. It is a remarkably unique enzyme in that no other Pro-Thr bond examined (IgG, BJ proteins, etc.) is susceptible, suggesting a recognition of the enzyme for some structure on the IgA molecule.

The SC is synthesized by epithelial cells present in the mucosal tissue or secretory glands. Its major function is to mediate the transport of IgA and IgM across the epithelial cells into the lumen of the gland or to the mucosal surface. This is apparently accomplished through the ability of SC to act as a receptor and to complex with J chain containing dimeric IgA and pentameric IgM (another property shared by these two classes of immunoglobulins) on the basolateral surface of secretory epithelial cells. The SC-IgA or SC-IgM complex is then transported to the apical surface in a vesicle after endocytosis and externalized in the lumen. During the translocation the SC-IgA complex becomes covalently linked by a disulfide interchange reaction. This is not true for SC-IgM, where

the linkage remains noncovalent. This is the only known case of a receptor becoming permanently incorporated into the structure of its ligand. It is not known if SC is an integral membrane component or is itself bound to a membrane receptor.

Another route of transport of IgA is from the blood into the gut lumen via the hepatobiliary path. IgA reaches the blood by lymphatic drainage of the lamina propria of several glandular or mucosal tissues in the body where IgA is synthesized. The major sites are the gut and upper respiratory tract. IgA is synthesized and secreted by local plasma cells as $(IgA)_2$-J dimers and enters the circulation via the thoracic duct. The liver parenchyma has been shown to have SC on the surface, similar to epithelial cells, which acts as a receptor for the $(IgA)_2$-J complex (or IgM-J) in portal blood. The IgA_2-J-SC complex is then transported by a vesicular mechanism through hepatocytes (similar to epithelial cells) into the bile and finally secreted into the intestine. It is not known if other tissue can selectively remove IgA from the blood in a similar fashion. Figure 17 depicts these two routes of entry of IgA to the mucosal surface.

Function. The paradox of IgA is that while it is known to be important in local immunity, its physiologic function is unclear. There are isolated reports on its ability to fix C′ by the classical pathway, but in general, IgA is not recognized for this function.

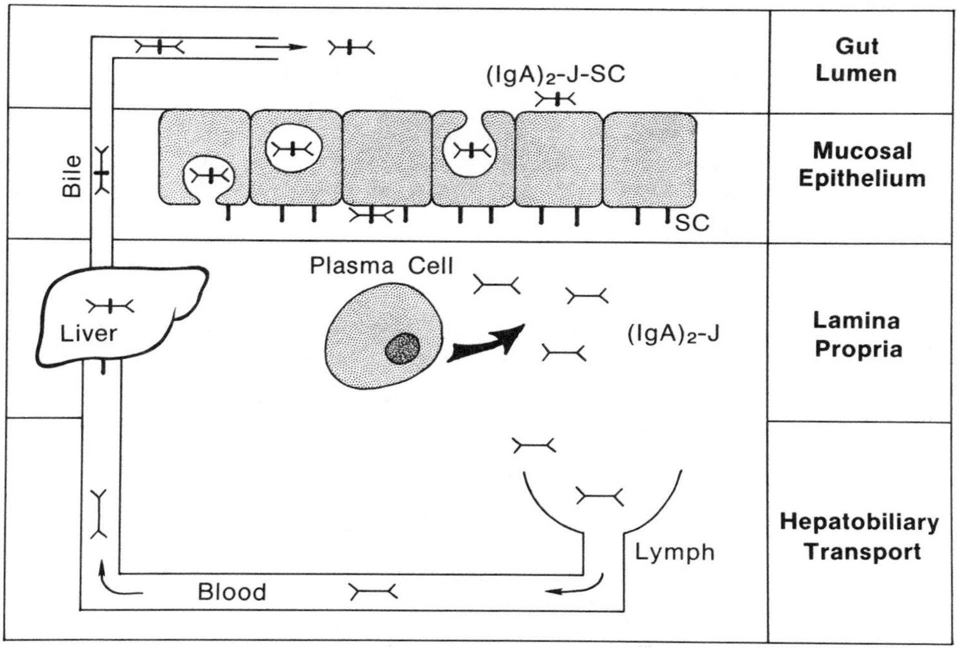

Figure 17 Routes of migration of IgA from intestinal plasma cells to gut lumen. Migration through epithelial cells occurs via a vesicular mechanism after joining with secretory component (SC) on the basolateral side of cell. The hepatobiliary route occurs via lymph node drainage of the gut. The IgA then enters the blood and is removed by SC on the surface of hepatocytes. The $(IgA)_2$-J-SC complex then passes into the gut via the bile.

IgA is known to fix C′ by the alternative (C3) path if aggregated, but the physiological significance of this is unknown. Similarly, IgA is not generally considered to be an opsonin (see Chap. 31) and is not cytophilic for monocytes. There is evidence, however, that IgA may be cytophilic for neutrophils, especially IgA1.

The major role of IgA appears to be the prevention of attachment of microorganisms to the epithelium, thereby preventing colonization and penetration (see Chap. 31). This is probably related to the efficiency of IgA in the agglutination of bacteria and neutralization of their enterotoxins. IgA is also efficient in the neutralization of viruses. Thus breast-fed babies have been shown to have fewer diarrheal infections than bottle-fed infants. This is attributed to the defense of the gut in infants by SIgA antibody against *Escherichia coli, Salmonella,* and *Shigella*. Per oral vaccination against poliovirus is also attributed to locally produced SIgA.

Not only microbes but food proteins as well are prevented from gaining entry into the bloodstream. This antigen exclusion is also considered an important mechanism in the prevention of allergic disease. Thus IgA-deficient individuals are known to have increased frequency of circulating immune complexes with bovine milk or meat components, and a higher incidence of atopic allergy.

In the blood, IgA complexes have been shown to be cleared by the hepatobiliary route. Thus despite the fact that IgA is not very active in phagocytic processes, it is still able to mediate the efficient removal of antigens from the circulation.

D. IgE and IgD: The Trace Immunoglobulins

Both of these classes of immunoglobulins are present in the serum at concentrations at least 100-fold less than those of any other class (Table 4). IgE is present in the lowest concentration of any class or subclass. The low serum concentration of IgE and IgD, however, is due to the fact that these two classes are associated primarily with leukocyte membranes and perform their major functions in this capacity. Aside from these similarities, IgE and IgD have little else in common. Their structures and mode of attachment to cell surfaces are quite different, as well as their biologic effects.

IgE

Structure. No new features need to be introduced regarding the structure of IgE. It consists of the basic four polypeptide chains, but the ϵ chain is similar to μ in that it has five domains, one V and four C. This is depicted in Fig. 17. Unlike IgM, however, IgE

Table 4 Properties of Human IgE and IgD

	IgE	IgD
H-chain molecular weight		
Total ($\times 10^3$)	70–72	65
Polypeptide ($\times 10^3$)	59–61	66
Carbohydrate		
Percent (w/w)	12	12
Oligosaccharide units	6	7–8
Sedimentation rate, $S_{20,w}$	7.9	6.6
Normal serum concentration (mg/ml)	3×10^{-4}	3×10^{-2}
Half-life (days)	2.4	2.8

does not have a propensity to form polymers and does not contain J chain. It should be noted that both μ and ϵ chains have the extra domain in place of the hinge region (C_H2). Major H-chain structural differences between the individual classes and subclasses appear to be confined to this area.

Papain cleaves the ϵ chain between the C_H1 and C_H2 domains, resulting in two Fab fragments and an $Fc\epsilon$ consisting of three domains, C_H2, C_H3, and C_H4. Pepsin cleaves between the C_H2 and C_H3 domains, liberating an $F(ab')_2$ fragment composed of the C_H2 in addition to the V and C_H1 domains. The C_H3 and C_H4 are hydrolyzed to dialyzable peptides and thus lost. It should be noted that the C_H2 domain is present on the Fc and $F(ab')_2$ fragments and not destroyed by either papain or pepsin. Thus the C_H2 domain is recoverable by successive papain and pepsin cleavage, which facilitates studies on its structure and function. It is referred to as the *Fc''_ϵ fragment*.

The ϵ chain contains 15 cysteines, as depicted in Fig. 17. Three of these are involved in interchain bonds, two of which are H-H and the other H-L. The remaining 12 cysteines form intrachain disulfide loops, two of which are in the C_H1 region. Reduction of IgE under relatively mild conditions results in the scission of five of these disulfides. The most labile is the inter-H-L bond and the most resistant is the inter-H bond on the NH_2- terminal side of the C_H2 region (Cys 241). The other inter-H bond (Cys 328) is less resistant to reduction. The two C_H1 intrachain loops are the remaining two disulfides, which are reduced under mild conditions. This is unusual since intrachain disulfides are usually more resistant than interchain disulfides to reduction, and denaturants are generally required for their reduction. The intrachain disfulfides of C_H1 regions of other H chains are not sensitive to reduction in the absence of denaturants. As will be shown later, disulfide bonds are important in the biologic properties of IgE. An excellent source of further reading on the structure and function of IgE is *Immunological Reviews* (Moller, 1978). The sequence of IgE is given in one of the papers of that review (Dorrington and Bennich, 1978).

Function. The binding of IgE to mast cells or basophils potentiates the immunopathological effects of immediate hypersensitivity. This is triggered by the reaction of the cell bound IgE with antigen, causing the release of allergic mediators (vasoactive amines) by mast cell degranulation. Binding of IgE occurs through a cell surface receptor on mast cells, which has been characterized (from rat basophilic leukemia cells) as being a glycoprotein with 30% (w/w) carbohydrate and having a molecular weight of approximately 50,000. It is a single polypeptide chain which probably extends through the membrane and into the cytoplasm of the cell. It binds one molecule of IgE very tightly ($K_a \geqq 10^{10}$ M^{-1}) through the Fc portion of the molecule, apparently through the C_H3 and C_H4 domains. Aggregation of the receptor via antigen cross-linking of the bound IgE is required to initiate the events leading to degranulation. The binding of IgE to receptor is lost if IgE is heated (56°C for 30 min) or mildly reduced. The inter-H-disulfide of Cys 328 appears to be very important to the cytotropic properties of IgE.

The role of IgE in the immunologic defense system has not been clear and millions of atopic people would probably agree that they cannot envision any beneficial function from it. Recent work, however, has indicated that these allergic reactions may be the undesirable side effects of an important function. Evidence has begun to accumulate which indicates that IgE plays a modulating role in cellular immunity, either indirectly through the release of vasoactive amines which can act on H-2 receptors (see Chap. 28), or directly through Fc receptors on other cells.

The role of IgE in the defense against parasitic infections is of particular importance. It has been shown that helminthic infestation results in the production of relatively high IgE levels in both humans and experimental animals, primarily the rat. Protection against *Schistosoma mansoni* has been found to correlate with IgE antibodies which mediate a cellular attack mechanism by macrophages and eosinophils, both of which have been shown to possess Fc receptors for IgE. Interaction of IgE-schistosome antigen complexes with normal rat macrophages causes the release of lysosomal enzymes. This cytotoxic mechanism has been demonstrated in human schistosome infections and other nonparasitic disease. Similarly, eosinophils are also able to carry on IgE-dependent cytotoxic attack on schistosomes. This has been shown to be enhanced by a factor secreted by mast cells to which IgE is bound. The factor has been termed *eosinophil chemotactic factor of anaphylaxis* (ECF-A) and it appears to increase both the expression of the IgE receptor and the cytotoxic activity of eosinophils. The activation of mast cells and eosinophils by IgE is another example of the direct and indirect manifestation of activity of this class of antibody. A similar system exists in rats for IgG2a-mediated eosinophil cytotoxicity, but the eosinophil receptors for this class and IgE are distinct, although present on the same population of cells. This is also true for Fc_α and Fc_ϵ receptors on macrophages. The mast cell Fc_ϵ receptor differs from the macrophage and eosinophil receptors in that the latter bind IgE much less avidly. It should also be noted that the mast cell Fc_ϵ receptor is "armed" with monomeric IgE which is cross-linked by antigen to initiate the cell signal. This contrasts the multivalent IgE complex binding of the macrophage Fc_ϵ receptor required to activate these cells. Lymphocytes have also been shown to possess Fc_ϵ receptors. They appear to be primarily a small population of B cells, although some T cells may also bear Fc_ϵ receptors. The immunological role of these cells is not yet clear but they are thought to be involved in the regulation of IgE synthesis.

IgD

Structure. The complete amino acid sequence of IgD has recently become available (Putnam et al., 1982). The structure and distribution of disulfides in the δ chain are represented schematically in Fig. 18. Similar to IgE, the structure of this class of immunoglobulins does not introduce any new features, but there are certain distinctive characteristics. The most prominent of these is an exceptionally long hinge region. This, coupled with the high carbohydrate content (Table 4), had suggested in earlier studies that the δ chain was composed of four C-region domains, similar to the μ and ε chains. However, sequence studies showed that only three were present. A long hinge region is also found in the IgG3 subclass. In IgD the hinge contains the cysteine (Cys 290) forming the single inter-H disulfide bond uniting the two δ chains, another distinguishing feature of IgD. The 64-residue hinge appears to be composed of two distinctive halves. The N-terminal half is characterized by a high content of galactosamine containing oligosaccharide units (four or five). The α_1 chain also contains galactosamine glycans in its hinge. These two regions (the α_1 and δ hinges) are the only known immunoglobulin structures containing galactosamine glycans. The other half of the δ-chain hinge (the C-terminal side) is highly charged and extremely susceptible to proteolysis, another property highly characteristic of IgD. In fact, the liberation of Fab and Fc fragments from IgD is complete after incubation with papain, in the absence of sulfhydryl activation, within 10 min at 37°C. One of the factors delaying sequence studies on IgD has been the inability to obtain sufficient intact protein. The overall sequence homology of the Fc

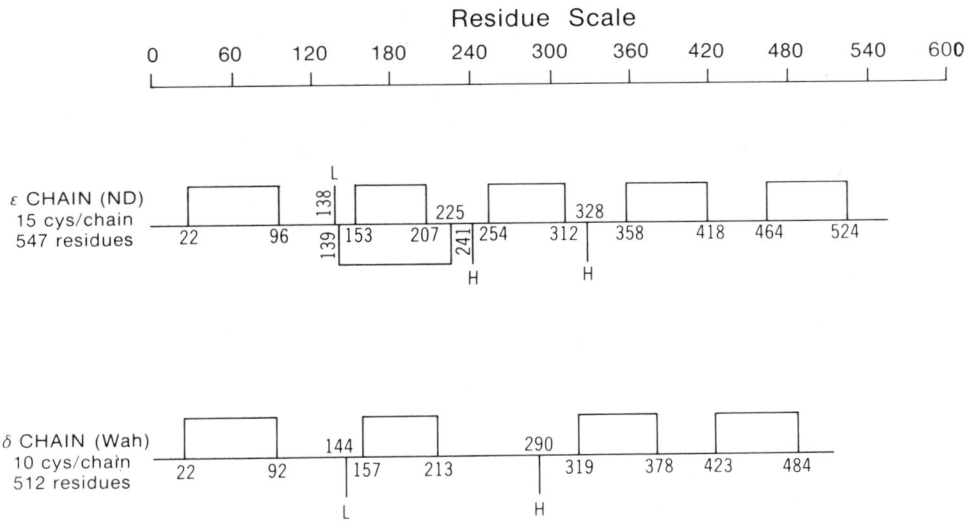

Figure 18 Schematic representation of H chains of IgE (ND) and IgD (Wah). The distribution and linkage of intra- and interchain disulfide bonds are indicated by position number. Rectangles above the chain represent domain disulfide loops. [The author wishes to thank Dr. F. Putnam for generously making the IgD (Wah) sequence available before publication.]

region to the other immunoglobulin classes is about 25%, which is lower then the homology found between the other classes. This has been interpreted as indicating that the δ chain arose early in evolution.

Function. Together with IgM, IgD acts as a lymphocyte receptor on the surface of mature B cells. The majority of splenic B cells show both IgD and IgM on their surface; however, the role of IgD in the B cell response and maturation is unclear.

Like IgM and in contrast to IgE, IgD is a membrane-bound immunoglobulin. This cytotropic property is believed to be associated with the last ($C_\epsilon 3$) domain, and specifically with the tp, similar to IgM. It must be emphasized that both secreted and membrane-bound forms of IgD and IgM have tp, each coded by a different exon. Sequences obtained from plasmacytoma proteins are assumed to reflect the structure of tp associated with the *secreted* forms.

Sequence studies on mouse and human plasmacytoma IgD have indicated that major differences exist between these two species. The tp of human IgD is composed of seven amino acids, whereas the tp of murine IgD has 26 residues. In addition, there is a deletion of 135 residues in murine IgD comprising the $C_\epsilon 2$ domain and the C-terminal half of the hinge region. This deletion includes the inter-H disulfide, which indicates that half molecules of IgD may exist. The greatest homology between murine and human IgD occurs in the $C_\epsilon 3$ domain (ca. 53%), which is similar to that of two heavy chains of the same class from different species.

At present it is not known if the tp associated with the murine plasmacytoma IgD reflects the membrane or secreted form. It corresponds to the product of the genomic locus associated with the membrane tp of IgM (noncontiguous) and is less polar than the

contiguous exon, but its association with the secreted form and its borderline hydrophobicity raises some doubt as to its function. In addition, it has been demonstrated that surface IgD from the same murine plasmacytoma has a higher molecular weight than the secreted form. This has led to the speculation that an as yet unidentified exon codes for the membrane tp of IgD.

REFERENCES

Amzel, L. M., and Poljak, R. J. (1979). Annu. Rev. Biochem. *48*, 961.

Amzel, L. M., Poljak, R. J., Saul, F., Varga, J. M., and Richards, F. F. (1974). Proc. Natl. Acad. Sci. USA *71*, 1427.

Bienenstock, J., and Befus, A. D. (1980). Immunology *41*, 249.

Brandtzaeg, P. (1976). Ric. Clin. Lab. *6*(Suppl. 3), 15.

Calvanico, N. J., and Tomasi, T. B. (1979). *Immunochemistry of Proteins*. Plenum, New York, pp. 1-85.

Capra, J. D., and Kehoe, J. M. (1975). Adv. Immunol. *20*, 1.

Cathou, R. E. (1978). Compr. Immunol. *5*, 37.

Dayhoff, M. O. (1976). *Atlas of Protein Sequence and Structure*. National Biomedical Research Foundation, Washington, D.C.

Diesenhofer, J. (1981). Biochemistry *20*, 2361.

Dorrington, K. J., and Bennich, H. H. (1978). Immunol. Rev. *41*, 1.

Edberg, S. E., Bronson, P. M., and van Oss, C. J. (1972). Immunochemistry. *9*, 273.

Frangione, B., and Franklin, E. C. (1972). FEBS Lett. *20*, 321.

Heremans, J. F. (1974). *The Antigens*. Academic Press, New York, pp. 365–522.

Kornfeld, S. J., and Plaut, A. G. (1981). Rev. Infect. Dis. *3*, 521.

Kunkel, H. G. (1965). Harvey Lect. *59*, 219.

Lamm, M. E. (1976). Adv. Immunol. *22*, 223.

Metzger, H. (1978). Contemp. Top. Mol. Immunol. *7*, 119.

Metzger, H. (1979). Annu. Rev. Pharmacol. Toxicol. *19*, 427.

Mole, J. E., Bhown, A. S., and Bennett, J. C. (1977). Biochemistry *16*, 3507.

Moller, G., Ed. (1978). Immunological Reviews, Vol. 41. (Munksgaard, Copenhagen.)

Moller, G., Ed. (1981). Immunological Reviews, Vol. 56. (Munksgaard, Copenhagen.)

Morrell, A., Skvaril, F., and Barandun, S. (1976). *Clinical Immunobiology*. Academic Press, New York, pp. 37–56.

Nisonoff, A., Hopper, J. E., and Spring, S. B. (1975). *The Antibody Molecule*. Academic Press, New York.

Padlan, E. A. (1977). Q. Rev. Biophys. *10*, 35.

Padlan, E. A., Davies, D. R., Rudikoff, S., and Potter, M. (1976). Immunol. Chem. *13*, 945.

Poljak, R. J. (1975). Adv. Immunol. *21*, 1.

Putnam, F. W. (1978). *Clinical Immunochemistry*. American Association for Clinical Chemistry, Washington, D.C., pp. 45–74.

Putnam, F. W., Takahasi, N., Tetaert, D., Lin, L.-C., and Debrurre, B. (1982). Ann. N.Y. Acad. Sci. *399*, 41.

Roth, R. A., and Koshland, M. E. (1981). J. Biol. Chem. *256*, 4633.

Scheffer, M., Girling, R. L., Ely, K. R., and Edmundson, A. B. (1973). Biochemistry *12*, 4620.

Solomon, A. (1976). N. Engl. J. Med. *294*, 17 and 19.

Stanworth, D. R., and Turner, M. W. (1978). *Handbook of Experimental Immunology*. Blackwell, Oxford, Chap. 6.

Tomasi, T. B. (1976). *The Immune System of Secretions*. Prentice-Hall, Englewood Cliffs, N.J.

Wu, T. T., and Kabat, E. A. (1970). J. Exp. Med. *132*, 211.

8

Monoclonal Antibodies/Myelomas and Hybridomas

Richard B. Bankert, Yi-Her Jou, and George L. Mayers Roswell Park Memorial Institute, Buffalo, New York

I. INTRODUCTION

One of the hallmarks of the humoral immune response is its exquisite specificity. Antibodies that are the effector molecules of the humoral immune response selectivity recognize a vast spectrum of antigenic determinants. Recognition of this myriad of determinants is accomplished by a very diverse repertoire of antibody molecules, each with a specificity for one or perhaps a few antigenic determinants. The heterogeneous antibody pool is a reflection of millions of individual clones of antibody-forming cells; each cell within a clone is secreting structurally identical antibody molecules (i.e., homogeneous antibodies).

Most foreign substances or antigens exhibit multiple antigenic determinants. Thus it is not surprising that the typical immune response is extremely heterogeneous since each different antigenic determinant may stimulate the proliferation and secretion of multiple antibody-forming cell clones. In addition to specificity, the antibodies may differ with respect to several serologically defined markers, such as isotype, allotype, and idiotype. Such variability from one clone to another results in a constantly shifting mix of antibodies in the serum following immunization. This heterogeneity and variability of antibodies exists between animals and is also apparent within a single animal, as the selection of antibodies from the available library may change daily through the duration of the response to a given antigen.

Antibody heterogeneity poses a most interesting and fundamentally important biological enigma: How is it possible to generate such a large diversity within a finite gene pool? Relevant to this issue are the following additional questions: (1) How large is the available or potential gene pool? (2) How is it regulated? (3) What is the genetic and molecular basis of antibody specificity? Resolution of all these questions has been

impeded largely by the very same thing that stimulated them—the antibody heterogeneity itself.

To resolve this problem, several independent approaches have been employed which result in the production of large quantities of homogeneous antibodies. With each of these approaches, one is able to cope with antibody heterogeneity by reducing it to its component parts. By defining the antibody heterogeneity one is in a position to begin to resolve the still pressing questions regarding the expression and regulation of clones within the antibody-forming cell repertoire. Perhaps of equal importance, the unlimited quantities of certain homogeneous antibodies have been shown to represent a valuable clinical resource for both diagnosis and treatment of infectious diseases and cancer. Also, the cells and cell lines producing the homogeneous antibodies are being exploited to study the molecular basis of antibody synthesis and secretion, cell differentiation, and the regulation of cell proliferation.

The three major sources of homogeneous antibodies and antibody-producing cell lines are: antibody-forming cell tumors or myelomas, a few immune responses (primarily to polysaccharide antigens) that produce large quantities of homogeneous antibodies, and antibody-forming hybrid cell lines or hybridomas which result from the fusion of antibody-forming spleen cells with a neoplastic cell (usually a myeloma). The production of monoclonal antibodies via these approaches and the utilization of these antibodies (and cell lines) to advance our understanding of the immune response are discussed in this chapter.

II. CRITERIA FOR IDENTIFYING ANTIBODIES AS HOMOGENEOUS

As the subject of this chapter is homogeneous antibodies (their production and use), it is essential to establish at the outset a precise definition of homogeneous antibodies and an explanation of how one best establishes the homogeneity of an antibody preparation regardless of its source. The term *homogeneous antibody* is applied to antibody which is the biosynthetic product from a single clone of normal or neoplastic antibody-producing cells or cell hybrids (hybridomas). As homogeneous antibody represents the product of only one clone, the term *monoclonal antibody* is also used to describe such a set of antibodies. *Monoclonal* and *homogeneous* are often used synonomously, here and elsewhere. It is recognized that a single cell usually produces one structurally identical antibody (an exception to this one cell–one antibody rule is when a cell switches from producing IgM to producing IgG antibodies). Other exceptions that apply to hybridomas are considered later. It should follow from the one cell–one antibody rule that homogeneous antibodies are therefore identical in their primary and three-dimensional structures. This is essentially correct, but one must qualify this assumption since with very rigorous tests of homogeneity (e.g., isoelectric focusing), microheterogeneity is frequently observed even in the immunoglobulin produced by a clone of neoplastic antibody-forming cells (myelomas).

Homogeneous antibodies can be recognized and defined most commonly by one of three experimental methods. The first method is based on the electrophoretic separation of antibody molecules and the second method is based on a thermodynamic measurement of antigen binding by the antibody (i.e., affinity). The third and most definitive approach is to determine the amino acid sequences of the light and heavy chains of the antibody molecule.

A. Electrophoresis and Isoelectric Focusing

Electrophoresis (movement of macromolecules through an electric field) and immunoelectrophoresis have been used for a long time to establish an estimate of homogeneity of an immunoglobulin preparation (see Chap. 13). The typical electrophoresis pattern of serum immunoglobulin is very broad, without any distinct areas of protein concentrations or peaks. However, the electrophoretic pattern of the serum of a patient with myeloma often results in one very narrow electrophoretic peak. This itself is almost diagnostic of the disease, which is the result of the transformation of a single antibody-forming cell, (plasma cell) or one of its precursors, to a malignant state.

The resolving power of the conventional electrophoresis or immunoelectrophoresis is limited, however, and the development of isoelectric focusing (IEF) represents a major improvement in the field of electrophoresis analysis of antibodies for homogeneity (see Chap. 13). IEF separates the antibodies according to their isoelectric points by electrophoresing them in stable pH gradients. Since IEF is currently one of the most practical and reliable assays to establish antibody homogeneity (or restricted heterogeneity) it is important to understand the principle of this technique. A stable pH gradient increasing progressively from anode (+) to cathode (–) is accomplished by electrolysis of carrier ampholytes in a suitable anticonvective media such as polyacrylamide gel, agarose gel, or Sephadex beads. When antibodies are introduced into the gels, these amphoteric molecules migrate according to their surface charge in an electric field placed across the gel. If the antibody's initial charge is positive, the molecules will migrate to the cathode into regions of higher pH. As the antibody molecules migrate they gradually lose positive charges via deprotonation of carboxyl or amino groups. When the molecules reach a position in the gel where their net electrical charge is zero (i.e., the antibody's isoelectric point or pI), the molecules stop migrating. After the antibodies have been focused (reached their isoelectric point) they are fixed in the gel by either acid precipitation or by immobilization of the focused proteins with glutaraldehyde. The position of the focused antibodies in the gels may be established by several protein staining procedures. The most commonly used stain is Coomassie Brilliant Blue, but a more sensitive staining protocol has recently been reported using silver nitrate to stain the separated proteins. In lieu of using stains one can also reveal the positions of antibodies in the gel using radiolabeled antigens. Antibodies bind the radiolabeled antigens and the excess unbound antigens are washed out of the gel. The gel is then dried and subjected to autoradiography to reveal the bands of localized radioactivity in the gels, which are coincident with the focused antibodies.

Representative isoelectric focusing patterns of antibodies are depicted graphically in Fig. 1. On the right-hand side of this figure is shown the typical multiple-banding pattern of heterogeneous antibodies isolated from immune serum. The isoelectric focusing pattern on the left is representative of monoclonal antibodies. The fact that there are three bands instead of one (which would be expected for a truely homogeneous protein) illustrates the microheterogeneity one can expect in antibodies isolated from a single clone of cells. As many as four or five bands have been observed with antibodies isolated from putative homogeneous antibodies. This microheterogeneity is generally attributed to postsynthetic hydrolysis of amide groups by serum enzymes, yielding negatively charged carboxylate groups, or to slight variations in the content of charged carbohydrates (such as sialic acid).

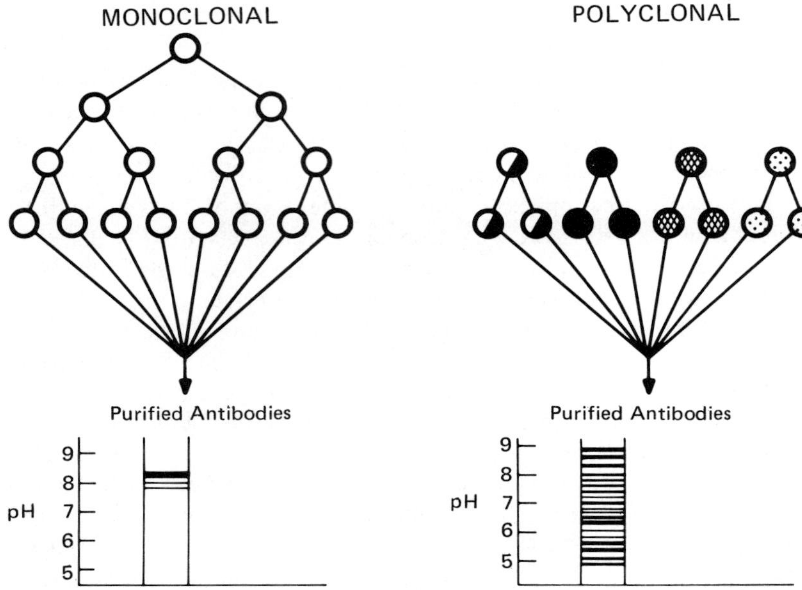

Figure 1 Schematic representation comparing the isoelectric focusing patterns of antibodies produced by a typical heterogeneous population of antibody-forming cells (i.e., polyclonal) to the isoelectric focusing patterns of antibodies derived from a single clone of antibody-producing cells (monoclonal) such as a myeloma or hybridoma. An explanation of the principles of isoelectric focusing is given in the text.

B. Antibody Affinity

Another important criterion for establishing homogeneity is a linear binding relationship for the antibody combining site with a monovalent antigen, which can be written

$$\mathrm{Ab} + \mathrm{Ag} = \mathrm{AgAb}$$

From the law of mass action the equilibrium constant for this reaction is

$$K_A = \frac{[\mathrm{AgAb}]}{[\mathrm{Ab}]\ [\mathrm{Ag}]}$$

where K is the equilibrium or association constant and [Ag], [Ab], and [AgAb] represent the concentrations of antigen, antibody, and antigen-antibody complex at equilibrium, respectively (see Chap. 16). Since the [Ab] at equilibrium is equal to the concentration of antibody initially present, [Ab°], minus the concentration of antibody bound in antigen-antibody complex, [AgAb], we can substitute this in the equation above to arrive at

$$K = \frac{[\mathrm{AgAb}]}{([\mathrm{Ab}^\circ] - [\mathrm{AgAb}])[\mathrm{Ag}]}$$

Rearrangement of the equation above leads to

$$\frac{[\mathrm{AgAb}]}{[\mathrm{Ab}^\circ]} = \frac{K\ [\mathrm{Ag}]}{1 + K\ [\mathrm{Ag}]}$$

The ratio [AgAb] : [Ab°] is the number of moles of antigen bound per mole of antibody and is often designated as r; thus

$$r = \frac{K[Ag]}{1 + K[Ag]}$$

As stated at the beginning, the equation defines the binding of a single binding site with an univalent antigen. Since antibody molecules usually have two combining sites with the same affinity, we can multiply by n, the number of combining sites with the same affinity per antibody molecule, to give

$$r = \frac{nK[Ag]}{1 + K[Ag]}$$

When we rearrange terms to put the equation in a form suitable for graphical solution, referred to as a *Scatchard plot* (see Chap. 16), the equation becomes

$$\frac{r}{[Ag]} = -Kr + nK$$

where r/[Ag] is plotted on the ordinate and r is plotted on the abscissa. If the preparation of antibody is homogeneous, a Scatchard plot of the data should describe a straight line. The slope of the line is the negative value of the association constant K. As the antigen concentration approaches infinity, the antibody combining sites become fully saturated; therefore, the X intercept, n, equals the number of combining sites on the antibody molecule (see Chap. 16, Fig. 1A). Figure 2 presents graphical results for four monoclonal antihapten antibodies produced by hybridomas.

C. Amino Acid Sequencing Analysis

The absolute proof that a protein is homogeneous is a demonstration of a unique amino acid sequence. This would exclude any heterogeneity that might be introduced postsynthetically by the addition of carbohydrate side chains. Until recently, sequencing was a slow, tedious process. With the development of the amino acid sequencer, the determination of the primary structure of a large protein was greatly facilitated, but it could still take a year or two and require hundreds of milligrams of pure protein. Recent developments, primarily by L. Hood at California Institute of Technology, have resulted in much faster sequencing times (i.e., months) and dramatically reduced the amounts of protein required, to 0.1 mg or less. Many technical difficulties still exist for amino acid sequencing, such as preparing and isolating appropriate large fragments and removing blocked amino terminal residues. For isolated immunoglobulins, there may be some microheterogeneity in the sequence, since, although the antibody may have been biosynthetically homogeneous, transaminases in the serum may convert some asparagine and glutamine residues to aspartic acid and glutamic acid residues. Most recently, molecular biologists have developed methods for isolating and sequencing the appropriate RNA and DNA, and from the genetic code deduced the sequence of the antibody. This was especially useful for the hydrophobic carboxy-terminal portion of the membrane-associated immunoglobulin on B lymphocytes. The primary structure of this molecule could not be determined directly by amino acid sequencing, but its sequence was predicted by gene sequence analysis.

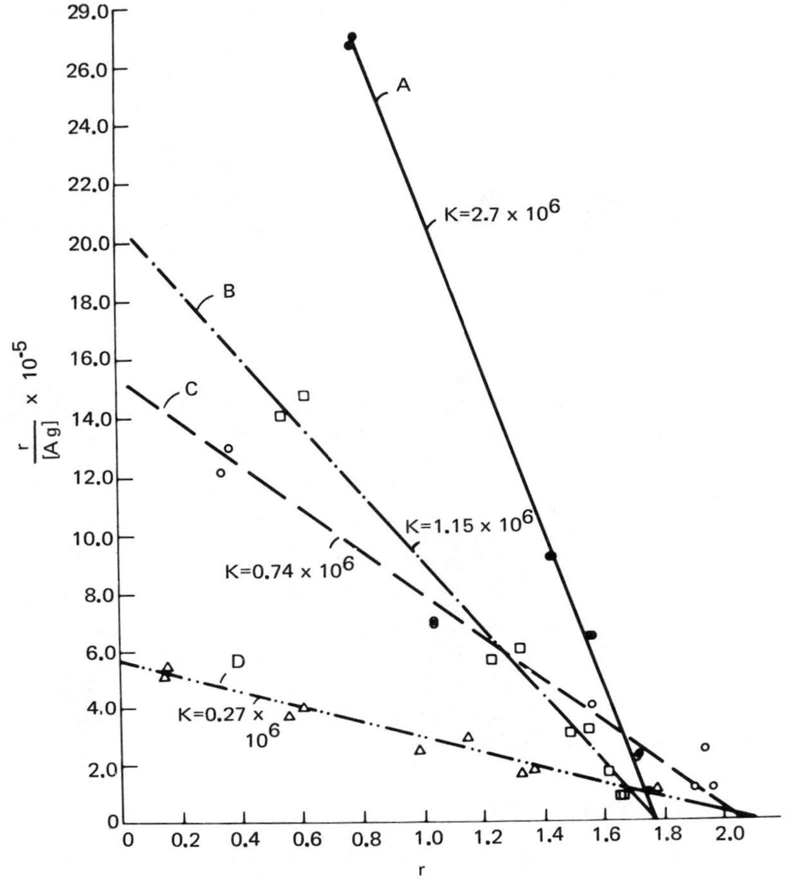

Figure 2 Scatchard plots of the binding data for four different monoclonal antibodies (A, B, C, D) prepared against the hapten 4-azophthalate. The binding of the monoclonal antiphthalate antibodies to [^{125}I] 4-iodophthalate was determined using the method of equilibrium dialysis. The preparations had approximately two binding sites per 160,000 molecular weight. The binding constants K depicted for each in the figure vary from 2.7×10^5 to 2.7×10^6 liters/mole.

III. ANTIBODY-PRODUCING TUMORS AS A SOURCE OF HOMOGENEOUS ANTIBODIES

Tumors of antibody-producing lymphocytes or plasma cells have been recognized in a variety of different vertebrate species, including humans. These tumors represent transformations of the immunocompetent B cells at many different stages of differentiation. The antibodies produced by these cells may be found (1) in (but not secreted from) the cytoplasm of the cell, (2) on the cell surface, or (3) actively secreted by the tumor cells. The first two types of tumors are termed *B-cell lymphomas* and the latter tumors represent transformation of plasma cells and are referred to as *plasmacytomas* or *myelomas*. The plasma cells represent the terminal differentiation state of the B-lymphocyte lineage

and they and their neoplastic counterpart possess a well-developed protein synthesis and secretion apparatus by which they are able to produce and secrete large quantities of homogeneous antibody.

A. Induction of Myelomas with Mineral Oil

While the myelomas arise spontaneously in many species, including cats, dogs, mice, and humans, a remarkable and extremely significant finding was made demonstrating that myelomas could be induced in two inbred strains of mice, BALB/c and NZB, by simply injecting mineral oil into the peritoneal cavity (Potter, 1972). These induced myelomas of mice, which can be continuously transplanted in vivo or cultivated in vitro, have provided the major quantitative source of homogeneous antibody. The wealth of information that has been and is currently being derived from the study of these myeloma cell lines and the homogeneous antibodies produced by them (myeloma proteins) has led to fundamental discoveries and new insights into several areas of immunology and molecular biology.

While immunoglobulins of all known light and heavy chain classes have been identified among myelomas, the mineral-oil-induced myelomas (particularly in the BALB/c strain) produce primarily the IgA class of immunoglobulin (Table 1). The proportion of myelomas making IgA is not an accurate reflection of the relative concentrations of IgA among the other immunoglobulin classes in the serum. For example, there is less than 15% IgA in the serum of normal BALB/c mice and yet about half of all the myelomas secrete IgA. This suggests that the transformation event does not affect the entire plasma cell population of the mouse equally. This could be due either to (1) an inherent susceptibility of cells differentiated to make IgA immunoglobulin or (2) a predilection of IgA-producing cells to migrate into the peritoneal cavity, where the transformation event appears to occur (Potter, 1972). It should be emphasized that while the induction of the myelomas is a very simple and reproducible process, the actual pathogenesis of this transformation is probably very complex and it is at this time a very poorly understood series of events.

Table 1 Frequency of Immunoglobulin Classes Produced by Mineral-Oil-Induced Murine Myelomas

	Percentage:	
Heavy chain isotype	In normal serum[a]	Of myelomas[b]
IgM (μ)	15	1.1
IgA (α)	13	45.5
IgG_1 (γ_1)	44	6.4
IgG_{2a}, IgG_{2b} (γ_{2a}, γ_{2b})	26	12.6
IgG_3 (γ_3)	2	0.7
Multiple classes[c]	—	6.6
Light chain only	—	2.2
None detected	—	24.9

[a]These are percentages of isotypes in conventional BALB/c mouse serum. The total quantity is 5.4 mg/ml. The values are taken from Barth et al. (1965). The κ/λ ratio is 97:3.
[b]Data obtained from Morse et al. (1978).
[c]Most myelomas secreting more than one class are mixtures of tumors.

Table 2 Immunoglobulins Produced by the Most Commonly Used BALB/c Murine Myelomas

	Light chain isotype		
Heavy chain isotype	κ	λ_1	λ_2
IgM (μ)	McPC 774, TEPC 183, ABPC 22[a]	MOPC 104E[a]	
IgA (α) polymer	MOPC 460D[a], MOPC 467[a] TEPC 15, MOPC 511[a], MOPC 167[a], HOPC 8[a], MOPC 320, McPC 603[a], AdjPC 6A, XRPC 24[a]	J558[a]	MOPC 315[a]
IgA (α) two-chain half-polymer	MOPC 47A, AdjPC 6C, MOPC 88		
IgG_1 (γ_1)	MOPC 21, MOPC 31C		
IgG_{2a} (γ_{2a})	MOPC 173, LPC 1, RPC 5, UPC 10[a]	HOPC 1	
IgG_{2b} (γ_{2b})	MOPC 195S, MOPC 141, MPC 11, UPC 120TA		
IgG_3 (γ_3)	J 606[a], FLOPC 21		
IgD (δ)	TEPC 1017, TEPC 1033		
Light chain only	MOPC 41, MOPC 46B, MOPC 47B, MOPC 321, MOPC 63	RPC 20, S176	

Source: This list of mineral-oil-induced tumors was compiled from Potter (1972).
[a]Indicates myelomas that secrete immunoglobulin with a known antigen (or hapten) specificity.

B. Nomenclature

Because of the widespread use of the murine myelomas and the myeloma proteins in immunology, one should be familiar with the names identifying the more commonly used tumor cell lines. Table 2 lists some of the most frequently utilized mineral-oil-induced myelomas, and identifies their heavy and light chain isotypes. Unfortunately, a uniform system of naming the tumors has never been established. Perhaps the most familiar common laboratory name which appears in the literature is MOPC and then a number. The MO stands for mineral oil and the PC for plasma cell. The number and letter at the end (e.g., MOPC-104E) indicate the transplant number and subline. Other prefixes of tumors listed in Table 2 stand for the following: Adj, adjuvant; TE, tetramethylpentadecane; and HO, 7-n-hexyloctadecane, which are the various agents used to induce the tumors. Sometimes the prefix stands for the originator of the tumor (i.e., M for Merwin; Mc for McIntire, and R for Robertson).

As indicated in Tables 1 and 2, there are some myelomas which secrete only light chains. This type of tumor occurs in both mouse and humans. In humans with multiple myeloma, patients may excrete light chains into their urine. The human light chains, termed *Bence Jones protein*, have an unusual thermosolubility property which may be used tentatively to identify them. The classical Bence Jones reaction is as follows. When urine is gradually heated to 45 to 70°C the Bence Jones protein precipitates and then redissolves as the urine is heated to near boiling; as the specimen cools, the Bence Jones protein precipitates again in the indicated temperature range (45 to 70°C), and redissolves as the temperature of the urine falls to 30 to 35°C. Some light-chain-producing myelomas produce large quantities of Bence Jones protein in the urine (20 to 50 mg/ml in mice), which represent a useful source of immunoglobulin light chains for chemical analysis. For a more complete discussion of the pathological and pathophysiological changes associated with myelomas, the reader is encouraged to consult the review by M. Potter (1972).

C. Applications of Myelomas and Myeloma Proteins

The remainder of this section on myelomas will consider the use of the homogeneous proteins and cells to study the structure, antigen binding activity and synthesis of immunoglobulin, and the structure, function, and regulation of immunoglobulin genes.

The scheme of the basic four-chain immunoglobulin molecule was in part determined by the study of immunoglobulins from myelomas. The immunoglobulin structure covered in the preceding chapter should be reviewed in order to understand better the following remarks about antigen binding myeloma proteins and immunoglobulin genes.

Antigen Binding Myelomas

Since 1967 a major effort was made to identify antigen binding myeloma proteins. By a random and tedious process of screening, approximately 5% of the murine myeloma proteins has been shown to bind one or more antigens from a library composed of (1) artificial ligands such as dinitro- and trinitrophenyl, (2) natural antigens isolated from the environment of the mouse including various bacterial carbohydrates, and (3) an assortment of antigens available to investigators from colleagues or commercial sources. Antigen binding myelomas are identified in Tables 2 and 3, and the antigen and/or hapten specificities are listed in Table 3. The fact that many myeloma proteins from different tumors react with the same ligand (e.g., nitrophenyl or phosphorylcholine) is a very useful and fundamentally important finding. For one thing they provided a focus of study to compare the myeloma proteins to antibodies induced by immunization of normal mice. On the basis of these studies (which showed striking similarities, if not identities between the proteins derived from the neoplastic and normal antibody-forming cells) the myeloma proteins were established as bona fide biologically significant antibody molecules and useful probes to study further the structure and functions of these molecules.

Probes to Study Immune Regulation

Biologically, many of the antigen binding myeloma proteins have been used as prototypes for conventional immune responses. The similarities of the myeloma to the corresponding antibody were first recognized by idiotype analysis and isoelectric focusing properties. This information has been applied to study mechanisms of immune response regulation

Table 3 Murine Myelomas that Secrete Immunoglobulin with a Known Antigen or Hapten Specificity

Antigen specificity	Hapten specificity	Myelomas[a]
DNP- and TNP-substituted proteins	Nitrophenyl DNP, TNP	MOPC 46, MOPC 315, MOPC 460, XRPC 25, MOPC 172
Lactobacillus 4 *Pneumococcus* C PS[b] *Trichoderma ascaris* extract	Phosphorylcholine	TEPC 15, HOPC 8, McPC 603, MOPC 167, MOPC 511, S107, S63
Dextran fraction S from *Leuconostoc mesenteroides* NRRL		
B-1355	α, 1-3-linked glucose	MOPC 104E, J558, UPC 102, TEPC 1035
B-512	α, 1-6-linked glucose	QUPC 52, W 3129, PC 3658, PC 3936
Wheat, hardwood, bedding extract, arabinogalactan, and ghatti gum	β, 1-6-linked galactose	J539, SAPC 10, TEPC 191B, XRPC 24, XRPC 44
	N-Acetyl-D-glucosamine	S117
Proteus mirabillis *Salmonella tel aviv* LPC[c]	α-Methyl-D-glucosamine	MOPC 384
Salmonella common flagellar antigen *Pasteurella pneumotropica*	–	MOPC 467
	Dansyl	ABPC 22
Levan, Inulin	–	J606
Hemophilus influenzae B PS	–	SAPC 15

[a]List of antigen-binding myelomas derived in part from Potter (1972, 1977).
[b]PS, polysaccharide.
[c]LPC, lipopolysaccharide.

(see Chap. 25). In another area, the idiotype probes have been applied to compare immune responses in different inbred strains of mice which have revealed polymorphisms in the antibody populations. The genes controlling some of these variable region polymorphisms are linked to the immunoglobulin heavy chain allotype IgC_H complex locus. For a review of these data, see Weigert and Potter (1977).

Chemically Defining the Antigen Combining Site

The primary amino acid sequence analysis of the V_L and V_H domains of the antigen binding myeloma proteins in the BALB/c mouse have helped to define the nature of the antigen combining site. This information has been derived from the groups of myelomas that are listed in Table 3. It is beyond the scope of this chapter to review the contributions of these myelomas individually. An excellent comprehensive review of this literature is available (see Potter, 1977).

An extensive literature exists on the two nitrophenyl binding myelomas, MOPC-315 and MOPC-460, and the phosphorylcholine binding myeloma, McPC-603. This literature forms the basis for the following discussion about the primary and tertiary structure of the binding site. A close look at the V regions of the immunoglobulin light and heavy chains indicates that most of the variability is restricted to three hypervariable regions on each chain. From x-ray crystallography and M-603 Fab′ crystals, a three-dimensional structure was established (Segal et al., 1974). Analysis of the crystal lattice has established that the hapten is bound to amino acids which are restricted to three of the hypervariable regions (or contact residues) in both the light and heavy chains of the immunoglobulin molecule (Padlan et al., 1976; Davies et al., 1975). These regions are termed *complementarily-determining regions* (CDRs). The CDRs for mouse V_L are amino acids 24 to 34, 50 to 56, and 89 to 97 and for mouse V_H are positions 31 to 35, 50 to 65, and 95 to 102. The non-CDR amino acids (framework residues) provide the structure that keeps the binding site intact. A three-dimensional homology of the framework of V domains between distantly related species (mouse and humans) has been revealed by x-ray crystallography and has permitted crystallographers to build hypothetical models of V domains from amino acid sequencing data (for a review see Potter, 1977).

Study of Genes, Cell Function, and Differentiation

The amino acid sequence analysis of light and heavy chain V regions provided the framework for building a classification of immunoglobulins into V_L and V_H isotypes. A number was assigned to each variable region isotype (e.g., VK1 and VK2 for kappa light chain variable region molecules), which ultimately helped the molecular biologists to enumerate and define the V-region structural genes. The initiation of this classification presaged by only a few years the very exciting discoveries on the immunoglobulin gene organization. The key element in all of the initial studies on the gene structure was again the myelomas, which provided the essential materials for generating molecular probes for the specific V- and C-gene sequences. The very large body of amino acid sequence data of myeloma proteins complemented and confirmed the information that was generated regarding the structure of the immunoglobulin gene.

In addition to providing the homogeneous products (immunoglobulins and nucleic acids) the myelomas have been recognized and used as a homogeneous source of cells to study a variety of cell functions. For example, a general model for the synthesis and secretion of immunoglobulin chains has emerged through the analysis of myeloma cell lines adapted to grow in vitro. The immunoglobulins, like other secreted protein molecules, are synthesized on membrane-bound polyribosomes released into the cisternae of the endoplasmic reticulum and exported via the Golgi apparatus through the plasma membrane, possibly by reverse pinocytosis (see Kuehl, 1977, for details). Membrane events associated with activation or tolerance have been studied with the dextran binding

myeloma MOPC-104E and the nitrophenyl binding tumor MOPC-315 (Bankert and Abbas, 1980). Further studies with MOPC-315 indicate that myelomas are capable of responding to the same kinds of immunoregulatory signals that govern the expansion and differentiation of conventional B-cell clones (Lynch et al., 1980). These studies clearly establish myelomas as a useful potential resource for further studies of cell function, regulation, and differentiation.

IV. HOMOGENEOUS ANTIBODIES FOLLOWING IMMUNIZATION

In general, the antibody response to a particular antigen is very heterogeneous; however, occasionally a highly restricted response has been observed. Many times this has involved serendipity, but in some instances strategies were planned to obtain homogeneous antibody preparations. Many of these efforts were aimed at preparing homogeneous antibodies to study protein structure and to characterize the antibody combining site. Studies using myeloma proteins and more recently hybridoma products have fulfilled those purposes. But there is still a requirement for conventional homogeneous antibody responses for in vivo studies such as regulation of the immune response by auto-anti-idiotype antibodies or regulation with idiotype-specific T cells.

A restricted antibody repertoire following immunization was first observed for human antipolysaccharide antibodies to dextran, levan, teichoic acid, and blood group substances. The light chains isolated from these antibodies showed electrophoretic restriction. Extensive studies carried out between 1965 and 1975 on the antistreptococcal antibody response in rabbits and in mice showed that some of the antibody responses to these antigens were relatively restricted. In addition, large amounts of homogeneous antibodies to pneumococcal polysaccharides or *Micrococcus*-specific carbohydrate antigen could also be prepared in rabbits. Many instances of restricted responses to small haptens have also been observed.

A. Homogeneous Response to Bacterial Polysaccharides

It was recognized early that the antigenic determinants for many carbohydrate antigens included the structure and linkage of the terminal nonreducing sugars. There are many examples of terminal antigenic determinants: streptococcal carbohydrate A and C, human blood group substances, the phosphorylcholine determinant on *Streptococcus pneumoniae* strain R36A, and dextran B1355 isolated from *Leuconostoc mesenteroides*. However, some immune responses are initiated by internal antigenic determinants on linear polysaccharides such as the determinant for type III and VII pneumococcal polysaccharides and linear polymers of dextran such as the naturally occurring α(1-6)-linked B512.

Streptococcal Polysaccharides

The antibody response to vaccines of whole heat-killed and papain-treated streptococci has been studied extensively (see Krause, 1970). Papain treatment of the dead bacteria is essential for removal of protein antigens and exposure of the group-specific carbohydrate determinants. Whole bacteria are necessary to achieve a good antibody response because purified polysaccharide isolated from the bacteria is a very poor immunogen. After an intensive course of immunization, consisting of three intravenous injections per week for 4 weeks and after a 6-month rest another series of injections for 4 weeks, rabbits developed

high levels of antibodies with restricted heterogeneity (sometimes as high as 50 mg/ml of antiserum) to the group-specific carbohydrate (group A, group A variant, or group C expressed on the corresponding streptococcal strain). One criterion that was used to establish antibody homogeneity was a demonstration of individual antigenic specificity for the group C carbohydrate. Additional evidence for homogeneity included restricted microzone electrophoresis patterns, common or cross-reactive idiotypic determinants, a monodisperse distribution of light chains in disk electrophoresis, and the expression of only one of two possible light or heavy chain alleles (allelic exclusion) in the monoclonal anticarbohydrate antibody. Isoelectric focusing characteristics confirmed the monoclonal nature of the antibody, and the amino acid sequence for the N-terminal residue 20 to 27 was consistent with a single light chain sequence. The antibody response to *Streptococcus* A in BALB/c mice is also very restricted. As in rabbits, these responses also appear to be under genetic restriction.

Pneumococcal Polysaccharides

A second polysaccharide that has been studied is the pneumococcal polysaccharide present in the capsular portion of the bacteria. Rabbits immunized with formalized pneumococci can produce copious amounts of homogeneous antibody or highly restricted heterogeneity to these capsular polysaccharides. This also involves an intensive immunization protocol in which only 6 to 8% of rabbits responded with a predominant antibody species. The antipolysaccharide antibodies could be isolated on appropriate immunoadsorbents, and in some cases selective elution could yield a homogeneous antibody preparation from an antiserum demonstrating restricted heterogeneity. The binding properties for some of these homogeneous preparations with labeled octasaccharides have been examined in equilibrium dialysis. The binding curves described a straight line, as expected for a homogeneous antibody preparation. The binding constant for some of the first antibodies was about 2.5×10^5 liters/mole, which is in the range commonly seen for antibodies to carbohydrates.

It is interesting to note that there was a temporal change of the antibody species present in the serum of individual rabbits. When one pneumococcal specific antibody species would disappear, new ones were detected in electrophoresis. Binding studies showed that antibodies occurring later in the response had higher binding constants, on the order of 10^7 liters/mole, which is consistent with maturation of the immune responses first described for antidinitrophenyl (DNP) antibodies. In some cases, a particular antibody species persisted for many months.

Other Polysaccharides

Two antipolysaccharide responses have received extensive study in the past 10 years. The response of BALB/c mice to dextran B1355 appears to be highly restricted to the same limited repertoire in every mouse. Antibodies to dextran contain a λ_1 light chain (an isotype limited to less than 3% of the mouse immunoglobulin) and this light chain appeared to be a unique sequence for all antidextran antibodies. The isoelectric focusing pattern for the majority of primary response antidextran antibodies resembled that seen for the dextran binding myeloma protein M104E. While idiotype analysis suggested that there was more than one antidextran antibody, they still appeared to be very similar in structure according to amino acid sequencing studies and isoelectric focusing analysis.

Hybridomas have added insight into this microheterogeneity. Fifteen monoclonal antidextran antibodies have been prepared without two having the identical sequence. Many had almost identical heavy chains varying in only two positions, amino acid residues 100 and 101 of the heavy chain. This led to the proposal of a new gene region, named the *D region*, for *diversity*.

Another widely studied restricted response is elicited by pneumococci strain R36A. This response is not to the polysaccharide per se but to phosphorylcholine (PC), which is attached to the polysaccharide surface. BALB/c mice immunized with this strain of bacteria make predominantly an antibody population that shares its idiotype with myeloma protein TEPC-15, produced in the same strain of mice. This is referred to in the literature as the *T-15 idiotype*. The anti-PC antibodies are essentially all of the IgM class of antibodies. These antibodies, which were also originally considered to be homogeneous, have been found to expressed some heterogeneity in recent studies with monoclonal antibodies, which suggests a need to utilize more than one criterion for indicating an antibody's homogeneity. If, indeed, there is alteration of only two to four amino acids residues in a total of over 1300 residues per antibody molecule, the difficulty of distinguishing between these various molecular species by physical means alone may be next to impossible.

B. Homogeneous Responses to Haptens

Although homogeneous anticarbohydrate antibodies were observed, antibodies produced in response to proteins and haptenic determinants on proteins were generally quite heterogeneous. A notable exception to this was observed by D. Pressman and colleagues, who obtained homogeneous responses to bovine gamma globulin (BGG)-azobenzoate and BGG-azoarsonate in 10 to 15% of their rabbits. A. Nisonoff had one of these antihapten antibodies crystallize during purification.

A second approach to producing homogeneous antibodies has been to restrict the heterogeneity of the antigenic determinants. The antibody combining site often recognizes both the haptenic determinant and the surrounding topography on the protein carrier. With antigens such as multihaptenated BGG and KLH, the haptens are substituents on many amino acid residues, each in a different environment on the protein. Thus the immune system is exposed to a large number of potentially different immunogens. Since it is believed that the restricted heterogeneity of antibodies observed in the carbohydrate response is the result of the repetitive presentation of a restricted number of identical antigenic determinants, several groups have looked at restricting the heterogeneity of hapten determinants by preparing suitably substituted hapten-protein conjugates. For example, a single cysteine in the active site of papain has been modified with α-(N-iodoacetyl)ϵ-N-2,4-dinitrophenyllysine (DNP). The immunogen mono-DNP-papain stimulates a restricted response to DNP, as do mono-DNP-ribonuclease and mono-DNP-insulin. Some have used the small cyclic decapeptide gramicidin S for stimulating homogeneous responses to DNP and adenine monophosphate. In all of these situations (unlike the voluminous homogeneous anticarbohydrate antibody responses) the antihapten antibody levels have been extremely low, usually less than 50 μg/ml of serum. This may be the result of using weakly immunogenic carriers. Recently, we have been examining monosubstituted neurotoxins for the production of high-titer antihapten antisera since the neurotoxin is a potent immunogen, but results so far have been disappointing. There is still a need for methods of initiating homogeneous antihapten antibody responses for use

in physiological studies in vivo where hybridoma technology would not be applicable. The availability of such antigen-antibody systems would be especially useful in studying idiotypic regulation of the immune system. For a more comprehensive coverage of the methods used to induce immune responses exhibiting restricted heterogeneity, (see Krause (1970, Brown and Jaton (1974), and Nisonoff et al. (1975).

V. HYBRIDOMAS AS A SOURCE OF HOMOGENEOUS ANTIBODIES

In 1975, Köhler and Milstein reported that it was possible to produce a cell line secreting antibodies which had a specificity for a preselected antigen. Their method was based on the fusion between murine myeloma cells and spleen cells derived from a mouse that had been immunized with sheep red blood cells (SRBC). They observed that a portion of the resulting cell hybrids grew continuously and secreted anti-SRBC antibodies. This basic methodology has subsequently been employed to prepare hybrid cell lines (termed *hybridomas*) secreting antibodies against a large variety of ligands. The hybridomas can be cloned and continuously cultivated either in tissue culture or by serially passaging the cells (which grow as tumors) in mice.

The implications of Köhler and Milstein's findings are immense, since in principle an unlimited supply of monoclonal antibody can be prepared against virtually any ligand that is capable of eliciting an antibody response in the mouse. Hybridoma-produced antibodies of diagnostic and therapeutic value as well as a whole host of invaluable research reagents with specificity for biologically active peptides, many different proteins, carbohydrates, and haptens have already been produced and many more hybridomas are currently under development. A partial list of the monoclonal antibody specificities is presented in Table 4.

A. Methodology

Theoretically, the hybridoma methodology is quite simple. However, in practice there are a number of technical problems or potential pitfalls that can best be understood by brief consideration of the protocol used to produce hybridomas. A detailed description of the hybridoma technology is available elsewhere (Galfre and Milstein, 1981).

Choice of Antigen-Primed Cells

The standard protocol for generating hybridomas is presented in Fig. 3. The choice and preparation of the two types of cells that are used as the fusion parents (myeloma cells and antigen-primed cells) are crucial to the success in generating antigen-specific hybrids. The most convenient source of large numbers of antigen-primed B cells is spleens from immunized mice. One fundamental advantage of the hybridoma technique is that the ligand used to immunize the mouse does not have to be purified. The purity of the antigen becomes important only if the impure material elicits a weaker immune response or the method of assay does not distinguish between antibodies to the specific reference antigen and antibodies to the impurities. The optimal immunization protocol must be established for each immunogen. A general rule to follow is to select animals that produce the greatest response to the immunization according to an assay of their serum, and to immunize the high-responder mice 1 month later. These animals are sacrificed 3 to 4 days after the last immunization and the spleen cell suspension is prepared for fusion.

Table 4 Partial List of Specificities of Monoclonal Antibodies Produced by Hybridoma Technology

Biologically active proteins/peptides		
Interferon	Renin	Nonhistone proteins
Interleukin 2	Alkaline phosphatase	Ricin
Osteoclast-activating factor	Human growth hormone	Lysozyme
Clotting factors	Thyroid-stimulating hormone[a]	Creatinine kinase[a]
Human chorionic gonadotropin[a]	Human serum albumin[a]	Hemoglobulin S, A
Prostatic acid phosphatase[a]	Estrophilin	
Viruses, bacteria, and fungi		
Group B *Streptococcus* types II, III	Polyoma	Herpes simplex I
Mycobacterium tuberculosis	Respiratory syncytial	Measles
Influenza A, B	SV40	Moloney leukemia
Para influenza I 350/220	Rabies	Epstein-Barr gp
Murine leukemia gp70[a]/gp15[a]	Vesicular stomatitis	*Streptococcus*
Lymphocytic choriomeningitis	Hepatitis B[a]	*Cryptococus neoformans*
Parasites		
Schistosoma japonica	*Toxoplasma gondii*	*Plasmodium yoelii*
Trypanosomes	*Plasmodium falciparum*	
Haptens, proteins, and carbohydrates		
Phthalate/isophthalate	Ovalbumin	Pyridine
Nitroiodophenylacetate	Collagen	Dinitrophenyl[a]
Trinitrophenyl	Fibronectin[a]	Phosphorylcholine
Dextran B-1355	Pneumococcal polysaccharide	Phenylarsonate

Receptors		
Acetylcholine receptor	E-rosette receptor[a]	β-Adrenergic receptor
Human estrogen receptor	Concanavalin A receptor	Thyrotropin receptor
Human-tumor-associated antigens		
Melanoma	Neuroblastoma[a]	Colorectal carcinoma
Alpha feto-protein[a]	Carcinoembryonic antigen[a]	Mammary carcinoma
Lung (adenocarcinoma and undifferentiated)		
Common acute lymphoblastic leukemia[a]		
Immunoglobulin markers		
Isotype[a]	Allotype[a]	Idiotype[a]
Cell surface antigens		
HLA[a]	β_2 Microglobulin	Glycophorin A
DR[a]	B-cell subsets[a]	Human group A
H-2[a]		
IA[a]	T-cell subsets[a]	Macrophages[a]
I-J[a]	Glial cells	Granulocytes
Forssman[a]	NK cells	Sheep erythrocytes[a]
Miscellaneous		
RNA	Digitoxin	DNA

[a]Indicates monoclonal antibodies that are currently available commercially.

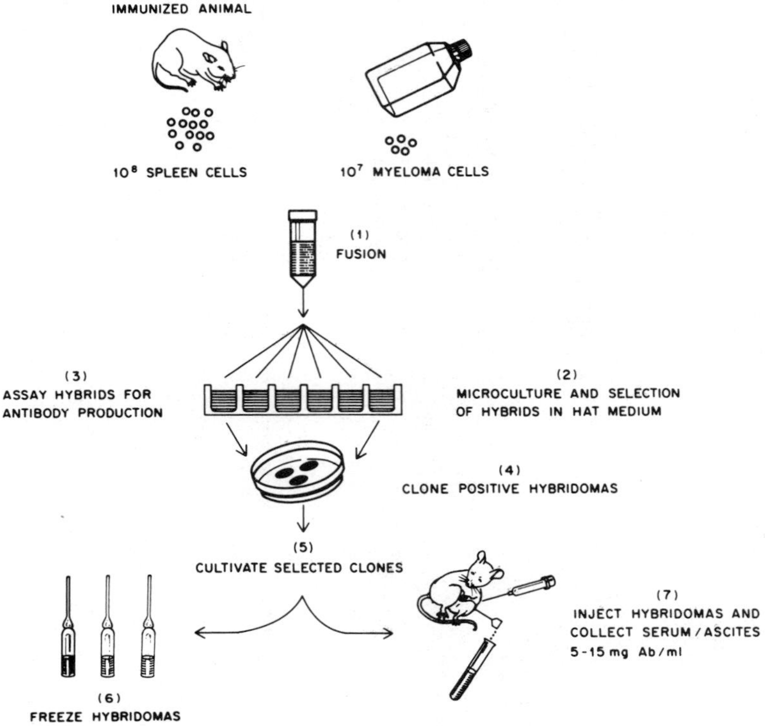

Figure 3 Outline of the protocol used to produce antibody-forming cell hybrids (hybridomas). In the first step antigen-primed lymphocytes (usually murine spleen cells) are mixed with HAT-sensitive murine myeloma cells in a ratio of 10:1. Polyethylene glycol is added to the mixed cell suspension to promote cell fusion. After fusion the cells are distributed into microculture wells and cultivated in HAT selection medium. In about 2 weeks the microcultures are screened for the production of antibodies that are specific for the immunizing ligand. Antibody-producing hybridomas are cloned either in agarose or by limiting dilutions in microculture. Selected clones are cultivated either in vitro or by serial transfer in histocompatible mice. Clones should also be preserved by freezing the cells and storing aliquots in liquid nitrogen. This is particularly important since clones may stop secreting antibody after cultivation of the cells either in vitro or in vivo.

Drug-Resistant Myeloma Cell Lines

Principle of selection in hypoxanthine aminopterine thymidine medium. In order to select for the growth of hybrid cells from the mixture of cells containing myeloma cells, spleen cells, and myeloma-spleen cell hybrids, myeloma cell lines with drug-resistant genetic markers have been established. Either 8-azaguanine/6-thioguanine-resistant [loss of hypoxanthine guanine phosphoribosyl transferase (HGPRT)] or 5-bromodeoxyuridine-resistant (loss of thymidine kinase activity) cell lines are used for this selection. The technique is based on the following principle. Variant mouse myeloma cells lacking the HGPRT enzyme cannot utilize exogenous hypoxanthine to synthesize purines via an alternate biosynthetic pathway. Therefore, these cells die when cultivated in the presence of aminopterin, which blocks the endogenous synthesis of both purines and pyrimidines.

Table 5 Some HAT-Sensitive Myeloma Cell Lines Used for Making Hybridomas

Species	Cell line	Common name	Immunoglobulin chain excreted by the myeloma cell line
Mouse[a]	P3-X63-Ag8	P3	κ, γ_1
	P3/NSI/1-Ag4-1	NS-1	κ[b]
	MPC11-45.6.TG1.7	MPC11	κ, γ_{2b}
	Sp2/0 Ag14	Sp2/0	None
	S194/5.XX0.BU.1	S194	None
	P3-X63-Ag8.653	653	None
	FO	FO	None
Rat[c]	210.RCY3.Ag.1.2.3	R210 or Y3	κ
	YB 2/3.0 Ag 20		None
Human[d]	SK0-007		λ, ϵ
	GM1500 6TGA1 2		κ, γ_2

[a]Derived from mineral-oil-induced myeloma of BALB/c mouse strain.
[b]Kappa chain is intracellular in the NS-1 cell line, but it is secreted in hybridomas.
[c]Derived from Lou rat strain myeloma.
[d]Derived from patients with multiple myeloma.

The spleen cells are not killed directly by the aminopterin, but these cells do not grow in tissue culture and they either die or they are rapidly overgrown by the hybrids, which double about every 16 to 24 hr. The myeloma-spleen cell hybrid contains the spleen cell's HGPRT enzyme and the myeloma's property of "immortality." These cells, and these cells only, survive in the presence of aminopterin by utilizing the exogenous hypoxanthine and thymidine, which is provided along with aminopterin in the selective hypoxanthine aminopterine thymidine (HAT) medium.

Myeloma cell lines available. A list of HAT-sensitive myeloma cell lines which have been utilized to make hybridomas is given in Table 5. Note that some of the myeloma cell lines secrete an immunoglobulin heavy and/or light chain of their own. The myeloma-spleen cell hybrids resulting from the use of these immunoglobulin-producing cell lines may (and usually do) secrete two immunoglobulin light and two immunoglobulin heavy chains. This problem is further compounded by the fact that hybrid molecules are produced, that is, individual immunoglobulin molecules containing polypeptide chains from both the myeloma and the spleen cell. The hybrid immunoglobulin molecules are heterogeneous with respect to their biological and physicochemical properties and with respect to their antigen binding activity. This problem is now resolved by using one of the somatic variant myeloma lines that produce neither light nor heavy chain molecules (i.e., Sp2/0, S194, 653, FO, or YB2/3.0). These nonproducing myelomas are now routinely used to generate hybridomas that synthesize only the antibody produced by the spleen cell parent. One should be aware, however, that many of the older hybridomas and some of the commercially prepared hybridoma-produced monoclonal antibodies available currently may have myeloma light and heavy chains together with a variety of hybrid half-molecules in addition to the desired antibody. One should also recognize and monitor periodically for the possibility that putative nonproducing myelomas have reverted back to the wild-type immunoglobulin-producing cell lines.

It is also apparent from Table 5 that HAT-sensitive myelomas are now available from three different species. Whenever possible the myeloma should be of the same species as the donor of the antigen-primed lymphocytes. While successful hybrids between mouse and rat have been produced, most other interspecies fusions result in very unstable hybrids. With mouse-human and mouse-rabbit fusions, chromosomes are progressively lost. Although this is a major limitation in the production of antibody producing cell lines, the observation of chromosomal loss has been exploited by molecular biologists to segregate chromosomes and thereby assign genes to a particular chromosome. This type of approach led to the assignment of heavy and light chain immunoglobulin genes in humans to the appropriate chromosome number. The chromosomal constitutions of intrastrain hybrids (i.e., mouse × mouse, rat × rat, and human × human) are much more stable.

Cell Fusion

After selecting and preparing the most appropriate source of antigen-primed lymphocytes and HAT-sensitivie myeloma cell line, the two cell suspensions are mixed together in a tube with an agent such as Sendai virus or preferably polyethylene glycol (PEG), which promotes cell fusion (step 1 of the fusion protocol, Fig. 3). After fusion, which takes only a matter of minutes, the cell suspension is placed into microcultures using a limiting dilution such that only about 10% of the wells in the microculture plate ultimately contain viable growing hybrids. An alternative is to plate the cell suspension in soft agar such that the hybrids grow as clones.

Screening for Antibody Production

In about 10 days to 2 weeks after cell fusion (or as soon as active cell growth is observed in the wells) the hybrids must be assayed for antibody production (step 3 in Fig. 3). The screening should be done as quickly as possible to prevent the overgrowth of the antibody-producing hybrids by irrelevant hybrids and to circumvent the need for the continued maintenance of the unwanted hybrids.

The screening of the hybrids for the production of antibodies of the desired specificity is the most tedious and time-consuming aspect of the entire hybridoma production scheme. The screening assay must be sensitive enough to detect as little as 100 ng of antibody/per milliliter, but it should also be able to screen large numbers of hybrid culture supernatants very rapidly and require as little antigen as possible. The most commonly used screening procedures detect antibody in the supernatants of the cultures via indirect binding assays. The supernatants to be tested are incubated with antigen to allow the putative antibody time to bind. The antigen-antibody complex is separated from the unreacted portion of the supernatants, and (in the mouse system) antibody binding to the antigen is detected with either ^{125}I-labeled (for a radioimmunoassay) or enzyme-linked (for an enzyme immunoassay) antibody to mouse immunoglobulin (see Chaps. 19 and 20).

Another approach to screening hybrid supernatants is the hemolytic assay, in which antibodies bind to antigen that is attached to erythrocytes. Hemolysis of the antibody target red cell complexes may occur with the addition of guinea pig serum as a source of complement. However, most immunoglobulin classes do not cause direct hemolysis; therefore, the addition of a second facilitating antibody (anti-mouse immunoglobulin) prior to the addition of complement is required to produce hemolysis (see Chap. 23).

Several special adaptations of these assays have substantially improved the efficiency and speed of the screening protocols. For example, a replica-plating hemolytic assay makes it possible to assay several thousand cultures in one afternoon (Bankert et al., 1980). The autoradiography of plates as an alternative to γ-counting after binding ^{125}I-labeled second antibody is another example of how screening techniques have been modified to increase the speed and reduce the tedium of assaying large numbers of samples.

Cloning, Freezing, and Propagating

The hybrids that are identified as antibody producers should again be cloned and retested for the production of antibody. At this point cells from each clone should be cryopreserved in liquid nitrogen. The clones are also cultivated either in vitro or in vivo. One must ensure that the hybridoma and the host animal are histocompatible prior to attempting to transplant the cell line in vivo. For example, if the hybridoma is a product of fusion between C3H mouse spleen cells and BALB/c myeloma cells, then the hybridoma must be passaged in C3H $\times$ BALB/c F_1 hybrid mice. Tumors are normally injected intraperitoneally and as much as 15 mg of monoclonal antibody can be isolated from 1 ml of sera or ascites from tumor-bearing mice. Substantially less antibody (ca. 10 to 50 μg/ml) can be isolated from tissue culture supernatants of actively growing hybridomas.

B. Clinical Applications

Perhaps the most significant benefit of monoclonal antibodies to medicine is that they represent inexhaustible supplies of clinical laboratory reagents which can be used worldwide for everything from diagnosis of various infectious diseases to tissue typing for transplantation centers. A very real potential also exists for the use of monoclonal antibodies therapeutically. It has already been established experimentally, for example, that the passive administration of monoclonal antibodies to rabies protects mice from a lethal challenge with rabies virus. Antibodies that bind preferentially to tumors (see Table 4) could be utilized selectively to target chemotherapeutic agents to tumors either by directly coupling the drugs to the antibody or by incorporating the drugs into liposomes and coupling the antibodies to the surface of the liposomes. It has also been suggested that monoclonal antibodies (which recognize various functionally different subsets of lymphocytes) could be utilized selectively to enrich or deplete a particular subset of lymphocytes in order to manipulate their response in patients. Theoretically, an antibody specific for T cells mediating a graft versus host response could be used selectively to deplete this T-cell subset from a mixed population of cells from the bone marrow. These cells could then be transplanted [without a risk of inducing graft versus host (GVH) reactions] to reconstitute patients who have received high doses of irradiation as part of a cancer treatment modality.

C. Research Applications

One of the earliest and most productive uses of hybridoma-produced monoclonal antibodies in research has been in the isolation and characterization of membrane macromolecules, including histocompatibility antigens, differentiation antigens, receptors, and putative tumor-specific antigens (Table 4). With the availability of large quantities of

antibodies, it is possible to prepare affinity columns composed of insoluble granulated matrices or beads to which are attached the monoclonal antibodies. Preparation of solubilized membranes containing hundreds of macromolecules are passed through the affinity column, and only the macromolecule with the antigenic determinant that is complementary to the antibody sticks to the column. This macromolecule is then eluted from the affinity column by altering the pH or salt concentration and/or the surface tension of the eluant (see also Chap. 13).

The hybridoma technology is now being applied extensively to define and enumerate the actual or potentially available repertoire of B cells specific for a given ligand. The immortalization and expansion of individual antibody-forming cell clonotypes have made it possible to study their induction and regulation at the cellular level and the mechanisms of generating diversity at the DNA level.

The list of uses of monoclonal antibodies to isolate and study biologically relevant substances (such as nonhistone chromosomal proteins, various receptors, neurotransmittors, cytoplasmic organelles, and lymphokines) is increasing rapidly. In addition to using the homogeneous antibodies produced by the hybridomas, the antibody-secreting cells have been recognized and used as a resource to study cell function: protein synthesis and secretion, transmembrane signaling, and differentiation. Moreover, the study of cell functions and biologically active synthetic products of T cells and macrophages have also been made possible by the discovery that these cells can be fused to neoplastic cell lines of the same lineage, resulting in functionally stable T-T cell or macrophage-macrophage hybrid cell lines.

D. Comparison of Monoclonal and Conventional Antibodies

By comparing the properties of monoclonal antibodies to conventional antibodies it becomes easier to recognize and understand the advantages and the limitations of each. As indicated in the introduction to this chapter, the conventional immune response to most antigens is extremely heterogeneous and the composition in the mix of antibodies often changes daily in any given animal. Therefore, it is impossible to establish any uniformity in the composition of a conventional antiserum which is composed of antibody molecules that differ in physicochemical characteristics as well as in their biological activities.

Standardizing Reagents

The most readily apparent advantage of the hybridoma technology is that it can provide an unlimited quantity of homogeneous antibody with a defined specificity. This makes it possible to prepare and to distribute antibody reagents worldwide, and the results of an immunoassay performed with these reagents in Philadelphia can be expected to be comparable to those produced in Tokyo.

Cross-Reactivity

At times the monospecificity exhibited by the hybridoma-produced antibody can be a disadvantage. For example, a monoclonal antibody is unable to distinguish between a group of different molecules that bear common or cross-reactive determinants. The heterogeneous antiserum therefore expresses a greater ability to distinguish between molecules with cross-reactive determinants, since the antibodies that bind the cross-reactive determinants represent only a portion of the whole antibody population in the antiserum. If necessary, the cross-reacting antibodies can be removed by adsorption with

the molecules possessing the cross-reactive determinants. This obviously would not be possible with the monoclonal antibody. If the cross-reactivity of the monoclonal antibody is a problem, one must attempt to make another monoclonal antibody which does not cross-react with the common determinants in question.

From the foregoing discussion it should be recognized and remembered that although the monoclonal antibody is monospecific, it clearly can and often does exhibit cross-reactivity. A most obvious and yet sometimes overlooked corollary to this fact is that the binding of a monoclonal antibody to a soluble ligand (or to a ligand present in a cell membrane) does not necessarily indicate identity or even similarity with the structure of the original immunogen. In a radioimmunoassay, for example, when an unknown molecule can compete effectively with the original immunogen for binding to a monoclonal antibody, this may very well reflect only the presence of a single common or cross-reactive determinant, not a complete structural identity between the two macromolecules. In contrast, if an unknown macromolecule can completely inhibit the binding of the immunogen to the heterogeneous antiserum, the chances are much greater that complete structural identity/similarity exists between the unknown material and the immunogen.

Precipitin Reactions

Another limitation of monoclonal antibodies is that they often fail to precipitate antigens. This is understandable since with a restricted specificity and with only two combining sites per molecule the antibodies can combine with no more than two antigenic molecules. Since precipitation requires extensive cross-linking to form a lattice, most monoclonals will not precipitate the antigen unless the reference antigenic determinant is repeated several times on the antigen. It is possible to overcome this problem by using a mixture of monoclonal antibodies or to enhance cross-linking with a second antibody (anti-immunoglobulin).

Antibody Affinities

The homogeneity of the monoclonal antibody may present other practical problems which would be encountered by conventional antiserum. If the antibody is to be used for the preparation of an affinity column to isolate macromolecules, the success of this approach will depend in part on an appropriate antibody affinity. If the affinity is too low, the macromolecules will not bind to the antibody-coated beads. On the other hand, if the antibody affinity is too high, it may become impossible to elute the molecules from the beads without denaturing them. With regard to monoclonal antibodies and affinity there is evidence to suggest that the hybridomas represent a stochastic sample of the different affinities from within the available repertoire (Bankert et al., 1981). That is, monoclonal antibodies are neither inherently low- or high-affinity antibodies, and one should be able to select a monoclonal antibody that has a suitable affinity.

Special Physicochemical and Biological Properties

The physicochemical homogeneity of monoclonal antibodies may result in molecules that are unusually susceptible to denaturation by freezing or iodination. The monoclonal antibodies often precipitate at or near their isoelectric point and they may elute from a diethylaminoethyl (DEAE)-cellulose column at an atypical pH. These problems are unique to homogeneous antibodies and they must be anticipated in working with them.

Since monoclonal antibodies, in contrast to conventional antisera, are all of the same isotype they may not be able to carry out a desired biological activity such as complement fixation. For example, an antibody unable to fix complement would be of little value where one is attempting selectively to kill a particular subset of cells. Therefore, if a cytotoxic antibody is required, it may be necessary to generate several monoclonal antibodies in order to select one that will fix complement and thereby lyse (kill) the desired cell type.

The limitations regarding the use of monoclonal antibodies that are presented here are intended to serve as caveats for the investigator or clinician who encounters the need to work with monoclonal antibodies and the hybridoma technology. These comments are not intended to diminish the tremendous positive impact that the hybridoma technology has had and will continue to have on basic scientific and clinical problems (see Galfre and Milstein, 1981; Raschke, 1980).

E. Alternatives to Cell Fusion for the Production of Monoclonal Antibodies

Other techniques have been employed in the production of monoclonal antibodies. However, they all suffer from disadvantages which do not apply to the hybridoma technology and will therefore only be considered briefly.

One approach is based on the finding that the Epstein-Barr virus (EBV) transforms human B lymphocytes in vitro (reviewed in Raschke, 1980). By using peripheral blood lymphocytes from donors with high natural antibody levels to a hapten, EBV transformation of the hapten-specific lymphocytes was accomplished. This resulted in a cell line secreting antihapten antibody. The successful transformation of antibody-secreting cells with this technique is very low, however, and requires an enrichment of the antigen-specific cells. Another limitation is that it can be applied only to primate lymphocytes.

Monoclonal antibodies have also been induced and studied by transferring a limiting amount of lymphocytes to irradiated recipients followed by the clonal expansion of the antibody-forming cells either in the intact animal or in vitro (Segal and Klinman, 1978). The in vitro expansion of antibody-producing clones has been used very effectively to monitor and quantitate the expression of individual clonotypes. However, the major limitation of this technique is the very low yield of antibody that is produced. The most popular protocol of this general approach is referred to as the *splenic focus assay*. Limiting dilutions of spleen cells are transferred to syngeneic irradiated recipients, and the next day the recipients' spleen is diced and the splenic fragments are pulsed with antigen and placed in microculture wells. The cultures may secrete antibody for several weeks and sometimes for as long as 3 months. However, the total amount of antibody produced is still only nanogram to microgram quantities. According to antigen binding studies, isoelectric focusing analysis, and idiotype analysis, the antibody produced in the fragment cultures is monoclonal, provided that a low enough donor cell number was used at the outset.

VI. CONCLUSION

The discovery and exploitation of the various sources of homogeneous antibodies have made it possible to study the chemistry of the antibody molecules and the biology of the antibody-producing plasma cell. Heretofore, these investigations had been frustrated by the tremendous heterogeneity of antibodies, which was a reflection of an equally heterogeneous clonal repertoire of antibody-forming cells. This problem was overcome when it

was found that large quantities of homogeneous immunoglobulins could be recovered from neoplastic plasma cells (myelomas) and that homogeneous or nearly homogeneous antibody responses could be elicited by special antigens which have identical and repeating simple structural units (such as bacterial polysaccharides). Most of our early data on the structure and function of antibodies were derived from the study of these two sources of homogeneous antibody protein (see Chap. 7). The recognition that myelomas could be induced in mice by the injection of mineral oil and the discovery that a small portion of these mineral-oil-induced myelomas produced antibodies with a known specificity facilitated and extended the studies of antibodies. The studies of homogeneous antigen-binding myeloma proteins have established many new insights into the three-dimensional structure of the antigen combining site by x-ray crystallography. Comparing the structure of V domains and the combined use of nuclear magnetic resonance and electron spin resonance spectroscopy may make it possible to define the precise chemical structure, size, and shape of antigen combining sites of immunoglobulins in solution that cannot be resolved by x-ray studies.

In addition to the immunochemical data derived from the study of antigen binding myeloma proteins, many fundamental biological aspects of antibodies have been probed. In particular, the myelomas have been an invaluable resource for the study of the genetic basis of antibody diversity, the regulation of the expression of that repertoire, and the relationship between antigen binding, idiotypy, and amino acid sequence (see Chap. 14).

Although one cannot underestimate the contribution that the myelomas have made to our understanding of the biology and chemistry of antibodies, continued research in this area was limited by the relatively small number of antigen binding myelomas that were available for study. It took several years and a major effort to generate and identify the 30 or 40 myelomas that were producing antibodies with specificity for the various polysaccharide and hapten ligands.

Using the hybridoma technology, a single laboratory can, in a few months, obtain an equal number of antibody-producing cell lines (hybrids or hybridomas), and contrary to the mineral-oil-induced myelomas, the specificity of the hybridomas can be selected by the investigator. The outcome of the hybridoma technology is that one can generate unlimited quantities of homogeneous monoclonal antibodies, and the ability to produce a hybridoma with any given specificity is restricted only by the limits of the immunological repertoire of the mouse or rat. That is, if a mouse or rat can produce a significant antibody response to a selected antigen, one should be able to produce a hybridoma that secretes antibodies specific for that antigen—thanks to Georges Köhler and Cesar Milstein. Monoclonal antibodies have now been used in virtually every area of biology, biochemistry, and medicine to identify, quantify, classify, or purify elements of physiologically significant biological systems.

The opportunities for future research in this area appear to be unlimited. The immediate opportunities include (1) improvements in the technology and resolution of the remaining technical problems (such as stabilizing antibody production, obtaining hybridomas producing antibodies against poorly immunogenic antigens, and automating the screening technology) and (2) the continued production of monoclonal antibodies (particularly of human origin) against a wide variety of antigens of general biological and medical importance.

The ability to immortalize antibody-producing cells by cell fusion is now being applied to other immunocompetent cells. For example, stable T-cell hybrids have been established and shown to maintain cell function (Kappler et al., 1981; Irigoyen et al.,

1981). Given the appropriate HAT-sensitive neoplastic cell line, one should be able to establish a complete library of antigen-specific B- and T-cell lines at various stages of differentiation. Many efforts along these lines are currently under way. The ultimate goals of many of these studies are to establish the biochemical mechanisms by which B and T cells are activated to differentiate and divide and to explore the genetic and molecular events that are involved in the expression and regulation of effector molecules produced by the various B- and T-cell subsets.

REFERENCES

Bankert, R. B., and Abbas, A. K. (1980). In *Progress in Myeloma* (M. Potter, Ed.). Elsevier/North-Holland, Amsterdam, pp. 109–128.

Bankert, R. B., DesSoye, D., and Powers, L. (1980). J. Immunol. Methods *35*, 23.

Bankert, R. B., Mazzaferro, D., and Mayers, G. L. (1981). Hybridoma *1*, 47.

Barth, W. F., McLaughlin, C. L., and Fahey, J. (1965). J. Immunol. *95*, 781.

Braun, D. G., and Jaton, J.-C. (1974). Curr. Top. Microbiol. Immunol. *66*, 29.

Davies, D.R., Padlan, E. A., and Segal, D.M. (1975). Annu. Rev. Biochem. *44*, 639.

Galfre, G., and Milstein, C. (1981). Methods Enzymol. *73*, 1.

Irigoyen, O., Rozzolo, P. V., Thomas, Y., Rogozinski, L., and Chess, L. (1981). J. Exp. Med. *154*, 1827.

Kappler, J. W., Skidmore, B., White, J., and Marrack, P. (1981). J. Exp. Med. *153*, 1198.

Köhler, G., and Milstein, C. (1975). Nature (Lond.) *256*, 495.

Krause, R. M. (1970). Adv. Immunol. *12*, 1.

Kuehl, W. M. (1977). Curr. Top. Microbiol. Immunol. *76*, 1.

Lynch, R. G., Rohrer, J. W., Gerbel, H. M., and Odermatt, B. (1980). In *Progress in Myeloma* (M. Potter, Ed.). Elsevier/North-Holland, Amsterdam, pp. 129–150.

Morse, H. C., Ribbet, R., Asofsky, R., and Weigert, M. (1978). J. Immunol. *121*, 1969.

Nisonoff, A., Hopper, J. E., and Spring, S. B. (1975). In *The Antibody Molecule* (F. J. Dixon and H. G. Kunkel, Eds.). Academic Press, New York, pp. 407–440.

Padlan, E. A., Davies, D. R., Rudikoff, S., and Potter, M. (1976). Immunochemistry *13*, 945.

Potter, M. (1972). Physiol. Rev. *52*, 631.

Potter, M. (1977). Adv. Immunol. *25*, 141.

Raschke, W. C. (1980). Biochim. Biophys. Acta *605*, 113.

Segal, N. H., and Klinman, N. R. (1978). Adv. Immunol. *26*, 255.

Segal, D. M., Padlan, E. A., Cohen, G. H., Rudikoff, S., Potter, M., and Davies, D. R. (1974). Proc. Natl. Acad. Sci. USA *71*, 4298.

Weigert, M., and Potter, M. (1977). Immunogenetics *4*, 401.

9

β_2 Microglobulins

Erik P. Lillehoj* and M. D. Poulik William Beaumont Hospital, Royal Oak, Michigan

I. INTRODUCTION

During the last 20 years, plasma membrane components have been shown to mediate important events during eucaryotic cell–cell interactions by way of recognition processes at the cell surface. Recent technological advances in the methodologies of isolation and characterization of cell surface proteins has contributed enormously to our understanding of these recognition phenomena. Many such surface receptor molecules have been purified and characterized for both experimental and diagnostic purposes: for example, surface immunoglobulins, Fc receptors, virus receptors, and hormone receptors. Major histocompatibility complex (MHC) gene products are also included in this class of molecules and represent some of the best characterized cell surface components (for a review, see Klein, 1979). Although MHC-coded proteins have not been regarded as receptor molecules per se, recent studies suggest that both class I (classical transplantation or serologically defined antigens: HLA-A, -B, -C in humans and H-2K, D, L in mouse) and class II (HLA-D in humans and H-2I in mouse) glycoproteins may serve a receptor function during the immune response (Zinkernagel and Doherty, 1975; see also Chaps. 26 and 27).

Class I MHC gene products consist of two polypeptides in noncovalent association (Klein, 1979): a transmembranal glycoprotein of molecular weight 44,000 to 46,000 expressing alloantigenic determinants of the MHC (heavy chain) and a peripheral membrane protein of molecular weight 12,000, β_2 microglobulin (β_2m). Despite the fact that both polypeptides are expressed primarily on the lymphocyte plasma membrane, they are also identified as soluble components in serum and other biological fluids. Indeed, human

For a brief discussion of the connection between Chaps. 9, 24, 25, 26, 27, and 28, see the appendix, "Histocompatibility Antigens, Lymphocyte Interactions, and the Immune Response."

*Current affiliation: Laboratory of Immunogenetics, National Institutes of Health, Bethesda, Maryland.

β_2m (hβ_2m) was first identified and isolated by Berggård and Bearn (1968) from urine of patients with renal dysfunction. Since this original investigation, urine has consistently served as the primary source for purification of β_2m from humans and other animal species, including dog, rat, mouse, rabbit, guinea pig, chicken, and bovine (Poulik, 1976). Human β_2m has also been identified in colostrum, milk, saliva, seminal fluid, amniotic fluid, synovial fluid, cerebrospinal fluid, pleural fluid, and ascites (Poulik et al., 1979a). However, these sources are not as rich in β_2m as is urine from patients with renal tubular disease. Urine specimens from cases of Wilson's disease, cadmium poisoning, renal tubular acidosis, and refractory anemia were starting materials for the initial studies of Berggård and Bearn (1968) and Smithies and Poulik (1972). Urinary sources additionally include patients with leukemia, Balkan nephropathy, and renal allografts (Lillehoj and Poulik, 1979). The latter appears best for the isolation of hβ_2m.

II. BIOCHEMICAL PROPERTIES OF β_2M

β_2 Microglobulin is considered by many to be the most thoroughly investigated of all eucaryotic plasma membrane proteins. Biochemical, immunological, physiological, and cytological parameters of this protein have been extensively described and many reviews have attempted to coalesce this information (Lillehoj and Poulik, 1979; Poulik, 1976; Poulik et al., 1979a). Berggård and Bearn (1968) have provided the most complete physicochemical characterization of hβ_2m. From amino acid composition data these authors calculated a molecular weight of 11,800 compared to the observed molecular weight of 11,600 obtained by gel filtration chromatography. Molecular weights and isoelectric points of $\beta_2$ms from a variety of animal species are compared in Table 1. As can be seen, all proteins are similar in size, with an average molecular weight of 11,678. In contrast, the electrical charge characteristics of $\beta_2$ms are more variable, as reflected by respective isoelectric point values. On paper or agar gel electrophoresis, hβ_2m migrates as a slow beta globulin, while rat, rabbit, and mouse $\beta_2$ms demonstrate even slower mobilities. Conversely, the electrophoretic mobility of dog β_2m is much faster than hβ_2m. A certain degree of charge hetergeneity between $\beta_2$ms may result from varying susceptibility to protein deamidation and/or proteolysis particularly with urinary $\beta_2$ms. However, the major course of charge differences presumably reflects variability at the primary structural level (see below). In contrast to interspecies charge heterogeneity,

Table 1 Molecular Weight and Isoelectric Point Comparisons of β_2m from Various Animal Species

Animal species	Molecular weight	Isoelectric point
Human	11,800[a]	5.7
Rabbit	11,666[a]	6.0
Dog	11,745[b]	n.d.[c]
Mouse	11,668[a]	7.5
Rat	11,750	7.4
Cow	12,000	7.1
Guinea pig	11,398[a]	6.7
Chicken	11,400	5.0

[a]Calculated from amino acid sequence.
[b]Calculated from amino acid composition.
[c]n.d., not determined.

detection of genetic polymorphism of $\beta_2$ms from some species by electrophoretic methods has been unsuccessful until recently. Nevertheless, electrophoretic evidence has been presented indicating that human, guinea pig, rat, and mouse β_2m displays charge heterogeneity, but these charge variations have been shown to represent true genetic polymorphism rather than post-translational modifications (Goding and Walker, 1980). Complete amino acid sequence analysis of mouse β_2m indicates allelic variations of residue 85 (Gates et al., 1981). Genetic polymorphism of mouse β_2m appears to be linked to the H-3 minor histocompatibility locus on chromosome 2 (Goding, 1981; see Chap. 27).

Complete amino acid sequence data are available for human (Cunningham et al., 1973), rabbit, guinea pig, and mouse (Gates et al., 1981) β_2m. Both human and mouse β_2m have been shown to be synthesized in cell lysates as precursors containing an amino-terminal signal sequence of approximately 19 residues. Partial amino-terminal sequences have been published for dog, rat, and cow β_2m (Poulik et al., 1979b). Comparison of these sequences reveals substantial interspecies homology. Most of the observed differences can be accounted for by single DNA base pair changes. For example, comparison of complete sequence data of rabbit (Gates et al., 1981) and human (Cunningham et al., 1973) β_2m reveals that most of the amino acid variations throughout the entire molecule can be attributed to single base pair alterations. Interestingly, most of those requiring multiple base changes are found in the C-terminal third of the protein. The primary structural similarity among various $\beta_2$ms suggest that this protein has been well conserved through evolution, although nonmammalian $\beta_2$ms have not been studied in this respect. In contrast, amino acid composition data are available for both chicken and turkey β_2m (Table 2). The avian homologues of β_2m show similarities that distinguish them from

Table 2 Comparison of Amino Acid Composition of β_2m from Various Animal Species

	Human[a]	Mouse[b]	Cow[c]	Chicken[d]	Turkey[e]
Asx	12	10	11	14	9
Thr	5	8	2	6	5
Ser	10	6	8	7	13
Glx	11	11	12	14	13
Pro	5	9	9	8	4
Gly	3	2	3	7	13
Ala	2	3	1	7	7
Val	7	5	5	8	4
1/2 Cys	2	2	2	3	2
Met	1	5	0	2	0
Ile	5	7	6	3	3
Leu	7	3	8	7	6
Tyr	6	5	6	3	1
Phe	5	4	4	4	3
Lys	7	9	9	3	6
His	4	5	4	1	1
Arg	5	3	5	4	3
Trp	2	2	2[f]	2	2

[a]Calculated from amino acid sequence from Cunningham et al. (1973).
[b]Calculated from amino acid sequence from Gates et al. (1981).
[c]From Winkler and Sanders (1977).
[d]From Groves and Greenberg (1977).
[e]Lillihoj and Poulik (1982).
[f]Assumed value.

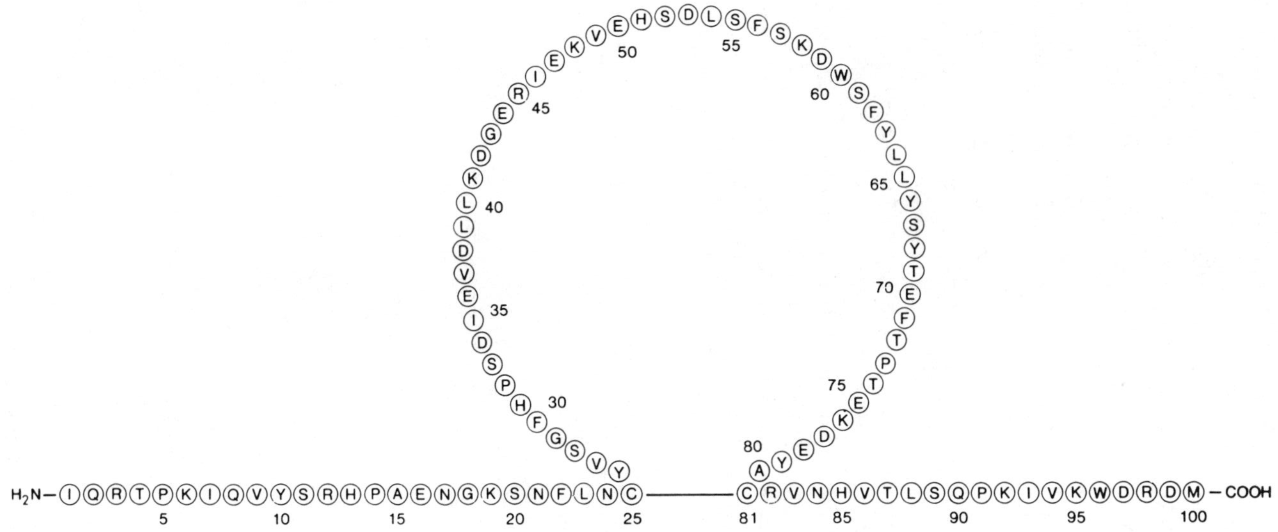

Figure 1 Amino acid sequence of human β_2 microglobulin. Single-letter denomination of amino acids is used. A, alanine; R, arginine; N, asparagine; D, aspartic acid; C, cysteine; Q, glutamine; E, glutamic acid; G, glycine; H, histidine; I, isoleucine; L, leucine; K, lysine; M, methionine; F, phenylalanine; P, proline; S, serine; T, threonine; W, tryptophan; Y, tyrosine; V, valine. (From Cunningham et al., 1973).

mammalian $\beta_2$ms such as relatively increased amounts of alanine and decreased levels of isoleucine, tyrosine, lysine, and histidine. Avian $\beta_2$ms also possess important differences in residue composition. Chicken β_2m possesses three cysteine residues compared to two for turkey and mammalian $\beta_2$ms. All $\beta_2$ms probably possess a single disulfide loop similar to the immunoglobulin domain and this has been confirmed in human (Cunningham et al., 1973), rabbit, mouse (Gates et al., 1981), guinea pig, and rat β_2m. The presence of a free sulfhydryl group in chicken β_2m suggests the possibility of disulfide linkage to the histocompatibility heavy chain which would further substantiate the immunoglobulin-like structure of class I MHC antigens that has been proposed to exist at both the primary (Martinko et al., 1980; Orr et al., 1979) and tertiary structural levels. Furthermore, turkey β_2m appears more similar to some mammalian $\beta_2$ms on the basis of the number of certain residues. Both cow and turkey β_2m lack methionine. Human and turkey β_2m possess similar amounts of proline and serine. It is presently unknown whether this represents an adventitious association, although the effect of proline on the secondary and tertiary structure of proteins make this seem unlikely.

The first partial N-terminal amino acid sequences of hβ_2m, provided by Smithies and Poulik (1972), demonstrated substantial homology with the residue sequence of heavy chain constant domains of myeloma immunoglobulin Eu (IgGl (Fig. 1). Although the greatest homology exists with the third heavy chain constant region, considerable sequence similarity is found with all constant domains. With recent elucidation of amino acid sequence of several HLA (Orr et al., 1979) and H-2 (Martinko et al., 1980) antigens, it has also become apparent that β_2m shares sequence homology with these glycoproteins. However, unlike immunoglobulins, only the α_3 domain of the histocompatibility molecule displays this similarity. Therefore, the original suggestion of Smithies and Poulik (1972) that both β_2m and immunoglobulins have a common ancestral gene may now be extended to include major transplantation antigens.

III. IMMUNOLOGICAL PROPERTIES OF β_2M

Although the amino acid sequence homology between β_2m and immunoglobulins is well recognized, controversy exists concerning immunological cross-reaction between them. Evidence has been published that some anti-immunoglobulin antisera react with β_2m. However, this has not been confirmed. On the basis of the immunoglobulin-fold domain structure of β_2m, immunoglobulins, and histocompatibility antigens, it was proposed that "active sites" are located at spatially homologous positions of these proteins. Tertiary structural similarities may indeed be sufficient for certain antisera to recognize similar conformations. Despite the phylogenetic conservation of β_2m primary structure, evidence indicating immunological reaction across specific lines is also controversial. Poulik and Bloom (1973), using rabbit anti-hβ_2m serum, demonstrated that human and chimpanzee β_2m possess common antigenic determinants detectable by immunofluorescence, and cross-reaction was shown between hβ_2m and chimpanzee and gorilla β_2m by a sensitive radioimmunoassay (RIA). With the same RIA, however, rabbit anti-hβ_2m did not recognize gibbon, monkey, mouse, or rat β_2m (M. Poulik and M. Weiss, unpublished data, 1975). Gordon and Kindt (1976) have presented evidence that goat anti-hβ_2m antiserum recognizes both human and rabbit β_2m. Finally, results suggesting cross-reaction between rat and mouse β_2m have been provided. As with all immunogens, the choice of animal species, route of immunization, and genetic makeup of the animal determines the specificity of the antiserum.

IV. PHYSIOLOGICAL PROPERTIES OF β_2M

The physiological parameters of hβ_2m have received special attention in recent years as diagnostic and prognostic value for assessment of various disease states. In normal adult human serum the concentration of β_2m varies between 1.0 and 2.5 μg/ml, and normal adult urinary excretory rates are 30 to 140 μg/day (Evrin and Wibell, 1972). Approximately 2% of hβ_2m is associated with soluble HLA antigens, whereas a greater proportion of these high-molecular-weight complexes are found in rat serum. Although neither serum nor urinary concentrations of hβ_2m differ between adult males and females, age does appear to influence the relative level of this protein. Cord and fetal sera analyzed for β_2m by immunodiffusion revealed cord serum levels to be 3.0 μg/ml, while fetal serum levels were 7.8 μg/ml. Fetal serum β_2m content increases during gestation, reaching a maximum value at 20 weeks and returning to normal levels by 36 weeks. Maternal plasma β_2m levels remain relatively constant at 1.5 μg/ml between 13 and 24 weeks of gestation but later increase to 2.8 μg/ml. Corresponding concentrations in amniotic fluid increased from 4.5 μg/ml at 13 weeks to 13.8 μg/ml at 24 weeks of gestation and then significantly decreased during the third trimester. Urine of newborns contained about 500% more β_2m than that in maternal serum. Consequently, β_2m in amniotic fluid may be primarily of fetal origin. Urinary β_2m levels in healthy infants ranging in age from birth to 1 year gradually decrease during the first 4 months of life to a level almost 5% of the initial value. A higher proportion of total low-molecular-weight urinary proteins in newborn infants exists as compared to older children, suggesting that this change reflects an increasing reabsorption of protein by cells of the renal tubule during the first few months of life. In summary, changing serum β_2m levels in the developing fetus remain relatively high at birth and return to adult levels shortly after delivery. It is not unreasonable to propose that the dynamics of fetal serum β_2m concentration reflect fundamental cellular processes during embryogenesis: for example, cell proliferation, differentiation, and attrition.

β_2-Microglobulin levels in the circulation and urine have been extensively analyzed as markers of renal tubular function. Most diseases associated with tubular dysfunction are characterized by low-molecular-weight proteinuria in which β_2m may represent the predominant protein. One such disorder, Balkan nephropathy, occurs at a high frequency in areas of eastern Europe. Urinary content of β_2m has been proposed to be of diagnostic value for early detection of this disease. Evaluation of β_2m and albumin excretion in apparently "healthy" residents from areas of endemic nephropathy has revealed that 19% of them had increased β_2m excretion alone, which is indicative of tubular damage, and 15% of the individuals had elevated β_2m and albumin excretion, suggestive of more severe kidney disease. The etiology of Balkan nephropathy is unclear, but contributing factors may include exposure to renal tubular nephrotoxins, viral infection, and /or low-molecular-weight protein overload in the kidneys. The high incidence of Balkan nephropathy and urinary tract tumors in the same population suggests that β_2 microglobulinemia due to overproduction and membrane release from tumor cells may saturate tubule cell reabsorption and thus cause β_2 microglobulinuria.

Environmental exposure to heavy metals, particularly cadmium (Cd), also induces renal damage and subsequent proteinuria. A direct relationship was demonstrated between β_2m excretion and blood Cd levels in Japanese inhabitants of areas with Cd-contaminated food and water. In another study, a correlation was shown between airborne Cd and β_2 microglobulinuria in industrial employees. These observations have been

substantiated by others and it is now believed that elevated β_2m excretion after Cd exposure is diagnostic of renal tubular disease.

Serum and urinary content of β_2m have additionally received special interest for monitoring renal allografts. Bernier and Post (1973) compared pre- and post-transplantation concentrations of β_2m in patients afflicted with terminal renal failure. Serum levels were greatly increased prior to kidney transplantation but generally decreased after surgery. Fermin and collaborators (1974) have also observed high urine β_2m concentrations in kidney transplant recipients. Continued proteinuria following renal allografting has been proposed to be the consequence of spreading of the original disease to the new kidney or de novo development of new lesions due to tissue incompatibility. β_2 Microglobulinuria in post-transplant patients may also be the result of ischemic damage, although some workers have been unable to correlate the length of ischemia with the degree of proteinuria. Nevertheless, several workers have shown that serum β_2m levels are a more sensitive indicator of transplant rejection than is serum creatinine. Acute and chronic rejection crises could be differentiated on the basis of the protein/creatinine excretion ratio.

Although the use of serum and urine β_2m measurements as an index of renal function holds promise, it is important to caution that this is valid only in cases of kidney dysfunction in the absence of other disease. Circulating and urinary β_2m levels as a measure of kidney function may be misleading in more complex clinical situations, particularly malignant and inflammatory disease. In these cases, elevated β_2m concentrations may occur independent of renal dysfunction, probably as a consequence of increased plasma membrane turnover caused by both cell death and tissue regeneration. Indeed, many workers have documented significant β_2 microglobulinuria or β_2 microglobulinemia during malignancy. Enhanced β_2m excretion in cancer patients has been reported by Hemmingsen and Skaarup (1977). These authors described renal malignancy in association with elevated excretion of both high-molecular-weight (albumin, transferrin, haptoglobin, IgG, IgA, IgM) and low-molecular-weight (immunoglobulin light chains, β_2m) proteins, suggested to be the result of either glomerulonephritis after renal deposition of tumor antigen-antitumor antibody complexes or direct tissue damage from tumor growth in the kidney. Urine levels of these high-molecular-weight proteins were also greater in patients with localized nonrenal epithelial carcinoma. In contrast, cases of disseminated extrarenal epithelial malignancy possessed higher urinary levels of immunoglobulin light chains, β_2m, and lysozyme.

β_2 Microglobulinemia in cancer patients has been extensively investigated in recent years. Many workers have demonstrated enhanced serum β_2m concentrations in patients with lymphoproliferative disorders (Späti et al., 1980). Shuster et al. (1976) observed significantly elevated serum β_2m levels during neoplasia, particularly lymphosarcoma, plasma cell malignancy, and solid tumors developing in the lung. Cooper and associates (1978) have classified non-Hodgkin's lymphomas into two groups on the basis of plasma β_2m and C-reactive protein (CRP) levels. One group had raised β_2m but normal CRP concentrations, while the second showed consistently raised CRP and sometimes elevated β_2m levels. Concurrent measurements of carcinoembryonic antigen (CEA), a well-known tumor marker, and β_2m have been made in several malignant diseases. Other workers reported increased levels of both proteins in females with genital cancer, although no correlation between these proteins was noted in individual patients. CEA and β_2m were shown to be of diagnostic value in the analysis of pleural effusions due to malignant disease. Differentiation of malignant from nonmalignant ascites can be made on the basis

of the ascites/serum ratio of β_2m. Elevated plasma concentrations of β_2m have also been reported in patients with primary bronchial cancer, breast cancer, gastrointestinal malignancy, and bladder cancer.

In spite of these encouraging results suggestive of β_2m as a potential tumor marker, several studies have emphasized the need for controlled evaluation in these circumstances. Adami et al. (1979) were unable to discern a significant different in serum β_2m levels in patients with breast cancer compared to age-matched controls. Furthermore, these authors failed to observe significant alterations in circulating β_2m levels following surgical removal of the tumor. Other workers have been unable to detect strikingly large increases of serum β_2m content in patients afflicted with cancer of the prostate or gastrointestinal tract. A comprehensive statistical analysis of serum β_2m concentrations during lung and gastrointestinal malignancy by Poulik and co-workers (1980) strongly suggests that the potential of β_2m as a tumor marker is severely restricted. These authors emphasized the need for careful control of renal function in cancer patients, particularly when analyzing low-molecular-weight proteins, since many antibiotics and chemotherapeutic drugs are known nephrotoxins.

β_2-Microglobulin levels as an index of various nonmalignant inflammatory diseases have received considerable attention in recent years. Patients with rheumatoid arthritis and Sjögren's syndrome were shown to possess abnormally high levels of β_2m in saliva, synovial fluid, and serum. Moreover, the degree of lymphocyte infiltration into the salivary gland with the amount of salivary β_2m implied local production of this low-molecular-weight globulin. Increased circulating β_2m content has also been reported in cases of rheumatoid arthritis without Sjögren's syndrome (Amor, et al., 1978). However, poor correlation was found between plasma β_2m levels and rheumatoid arthritis. Other inflammatory disorders associated with elevated β_2m concentration include systemic lupus erythematosus, Crohn's disease, anklyosing spondylitis, Reiter's syndrome, hepatitis, and cirrhosis. Elevated concentrations of β_2m in cerebrospinal fluid (CSF) have been noted in a variety of neurological disorders, including meningitis, encephalitis, and central nervous system tumors (Schaub et al., 1978). Schaub and co-workers (1978) were unable to correlate CSF and plasma β_2m content, suggesting independent control of synthesis in these two biological fluids. From the available reports one could suggest that the elevated β_2m levels in various immunopathies is due to production and release from infiltrating inflammatory cells, while high β_2m values during malignancy are the result of production by both infiltrating cells and tumor cells.

V. CYTOLOGICAL PROPERTIES OF β_2M

The presence of β_2m in different biological fluids is presumed to originate from the plasma membrane and many investigators have studied β_2m synthesis and turnover by cells grown in tissue culture. Nilsson and colleagues (1974) have provided evidence that β_2m is synthesized by plasma cells, glioma cells, and cell lines derived from cervix, pharynx, and urinary bladder tumor tissues. Quantitative determinations of cell associated β_2m have been performed in several laboratories. Conway (1981) has recently presented evidence that β_2m production by Raji cells is constant at various cell culture densities, implying regulatory mechanism of β_2m synthesis. Peterson and co-workers (1972) estimated 6×10^7 molecules per cell as the upper limit of β_2m on human peripheral blood lymphocytes. More definitive investigations of cellular β_2m content have been described by Evrin and Pertoft (1973). These authors measured β_2m from a variety of

human cell types after sonication, detergent, or acid cytolysis. Platelets were shown to possess 5.4×10^4 molecules per cell, mononuclear cells 1.8×10^6 molecules per cell, and polymorphonuclear leukocytes 6.9×10^5 molecules per cell.

To avoid complications of variation in cell surface area, some authors have studied the density of membrane β_2m as a more reliable parameter for quantitative analysis. Nilsson et al. (1974) observed that human lymphoblastoid cell lines and peripheral blood lymphocytes expressed equal densities, whereas malignant hematopoietic cell lines generally possessed lower densities of β_2m. Epithelial and mesenchymal cell lines possessed intermediate densities. Furthermore, these investigations found no difference between normal and mitogen-stimulated lymphocytes. Welsh and colleagues (1977) reported that normal human lymphoid and nonlymphoid cells both express tantamount densities of β_2m, but compared to normal cells, malignant lymphocytes possessed diminished densities of both β_2m and histocompatibility antigens. Other studies have shown that most human cell lines express similar levels of surface β_2m, while absolute amounts may vary with cell surface area. Additionally, T and B lymphocytes, whether normal or blastogenic, displayed similar densities of β_2m. The investigations described above concern β_2m on the plasma membrane, but it has been demonstrated that between 12 and 25% of total cellular β_2m is located intracellularly and 85 and 91% of surface β_2m is associated with membrane structures, including transplantation antigens. It is well known that measurement of cell surface components are extremely difficult and burdened with a number of methodological problems. Consequently, all values listed above should be considered as estimates at best.

Human β_2m has been demonstrated on the plasma membrane of all cell types examined with the exception of erythrocytes and the Daudi cell line. Daudi cells express an abnormally small amount of HLA antigens, but it has been shown that hybrid sublines established by fusion of Daudi and Raji cells possess more HLA determinants than the sum of the parental cells. It was proposed that Daudi cells do not have a functional β_2m gene and that the β_2m genetic locus provided by the Raji cells codes for a gene product capable of interacting with both Daudi and Raji HLA antigen heavy chains. Placental trophoblasts have been reported to lack histocompatibility antigens (Faulk and Temple, 1976), which suggests that these cells may also be devoid of surface β_2m. Trowsdale and co-workers (1980), however, have demonstrated β_2m but not HLA antigens in detergent lysates of two human choriocarcinoma cell lines. These authors suggested that the absence of HLA determinants on choriocarcinomas (and trophoblasts) is thus not due to absence of β_2m as with Daudi cells. Other investigators, in contrast, have detected H-2 as well as HLA antigens on trophoblasts.

β_2 Microglobulin was first identified as a plasma membrane protein by employing immunofluorescent techniques. Complement-dependent lymphocytotoxicity mediated by rabbit anti-hβ_2m antiserum was observed by Poulik (1973) and Poulik and Bloom (1973). Inhibition of anti-HLA alloantisera cytotoxic activity occurs after pretreatment of cells with turkey anti-hβ_2m, which is unable to fix mammalian complement or the Fab$'$ fragments of anti-β_2m antibodies. Nakamuro et al. (1973) were the first to demonstrate an association between β_2m and HLA antigens at the molecular level. These workers isolated a low-molecular-weight protein with "HLA common activity" by chromatographic and electrophoretic procedures. Two observations suggested that the HLA common activity was truly β_2m: (1) its amino acid composition was essentially equivalent to purified human urinary β_2m, and (2) both the isoelectric point and molecular weight were identical to an "HLA common portion fragment" isolated by gel filtration of a purified HLA

antigen preparation. These initial publications described a noncovalent association between β_2m and HLA antigens and were subsequently confirmed and extended by a variety of methods (Peterson et al., 1972; Östberg et al., 1974). Both proteins have been co-isolated from plasma and spent cell culture media (Cresswell et al., 1974; Poulik and Bloom, 1973; Tanigaki et al., 1973). Furthermore, both β_2m and HLA antigens have been copurified from viable cells after papain solubilization of membrane proteins (Cresswell et al., 1974), or membrane dissolution with hypertonic salt (Poulik et al., 1974) or detergent (Cresswell and Dawson, 1975; Grey et al., 1973) solution. Identical results have been described in studies of the lymphocyte membrane of mice, guinea pigs, rats, rabbits, and chickens. These studies lead to the inescapable conclusion that β_2m forms an integral subunit of major histocompatibility antigens from many animal species.

Exchange between soluble and cell-bound β_2m has been described which may in part account for earlier observations on the cytophilic activity of this protein. Furthermore, binding of β_2m to histocompatibility antigens is not restricted by species barriers. Heterologous β_2m-transplantation antigen associations have been described both on interspecies hybrid cells and with soluble histocompatibility gene products. Murine plasma proteins lacking alloantigenic activity and noncovalently bound to β_2m have been described. High-molecular-weight β_2m elution peaks from gel filtration colums have similarly been reported in human, rat, and guinea pig serum. As with the cell surface associations, htereologous β_2m binding to serum proteins may occur (Lögdberg et al., 1981).

Other workers have searched for an association between β_2m and membrane components other than H-2 and HLA antigens. Vitetta et al. (1976) were the first to demonstrate such an alliance with murine thymus leukemia (TL) antigens. Additionally, Qa antigens coded by genes linked to the murine MHC are noncovalently associated with β_2m. The histocompatibility Y-chromosome (H-Y) antigen present on cells from the heterogametic sex is also bound to β_2m (Fellous et al., 1978). Evidence to implicate an alliance between β_2m and nonalloantigens on murine lymphocytes has been reported. Some investigators have also obtained evidence for a β_2m-tumor antigen association on the plasma membrane of the malignant cells. Thomson and associates (1979) have undertaken a systematic study of tumor antigens on malignant cells in both experimental and human systems. Using anti-hβ_2m affinity chromatography, these workers succeeded in co-isolating histocompatibility and tumor-specific antigens from malignant melanoma, breast adenocarcinoma, and hepatoma tissues. Other investigators have also described an association of melanoma tumor antigen activity with β_2m. Thus, while some tumor antigens may have β_2m as an integral part of their tertiary structure, this does not appear to be a general feature of cancer cells.

Experimental evidence exists to suggest that β_2m may additionally be associated with plasma membrane components involved in lymphocyte transformation. It is apparent that anti-β_2m serum is mitogenic for human, mouse, and guinea pig lymphocytes, but the mechanism of this effect has yet to be completely elucidated. Anti-β_2m serum has been shown to increase [^{3}H] thymidine uptake by human and mouse B lymphocytes, whereas thymocytes and splenic T lymphocytes are not affected (Rigdén et al., 1976). However, goat anti-guinea pig β_2m serum stimulates guinea pig lymph node T lymphocytes in the presence of macrophages. Other studies have suggested that both T and B lymphocytes are affected. This response does not appear to be mediated by Fc receptors since Fab′ and F(ab′)$_2$ fragments of anti-hβ_2m antiserum are also mitogenic. Bach et al. (1973) observed that low concentrations of anti-β_2m antiserum enhanced phytohemagglutinin (PHA) stimulation of human lymphocytes, whereas high concentrations

were inhibitory. The suppressive effect of anti-hβ_2m antiserum on PHA-induced lymphocyte blastogenesis was questioned by Lindblom et al. (1974). Inhibition of concanalin A and pokeweed mitogen-induced lymphocyte stimulation by anti-hβ_2m antiserum was demonstrated in guinea pigs using antiserum against β_2m from this species (Lögdberg et al., 1979).

β_2 Microglobulin may also serve as a subunit of cell surface components involved in the mixed lymphocyte culture reaction (MLC) and cell-mediated lympholysis (CML). Bach and co-workers (1973) have shown that both responses are abrogated in the presence of anti-β_2m antiserum. More important, Lindblom et al. (1974) was able to block the MLR with Fab′ fragment of anti-β_2m antibody. This antibody was demonstrated to impede the MLR recognition phase by specifically affecting responder but not stimulator cells. These results do not contradict data published by Lightbody et al. (1974) showing that effector T lymphocytes generated in the MLR are insensitive to the effect of anti-β_2m antiserum in their ability to recognize and lyse target cells in CML. Finally, antiserum against both human and guinea pig β_2m inhibits PPD-induced lymphocyte transformation. Anti-guinea pig β_2m antiserum also suppressed ovalbumin (Lödgberg et al., 1979) and lysozyme-mediated transformation. These results have been interpreted to implicate β_2m as an integral subunit of lymphocyte surface receptors and recognition components, but one must also consider the possibility that nonspecific membrane perturbation is sufficient to induce functional alteration independent of physical association with β_2m. Other substances that have been implicated to be structurally united with β_2m include the allogeneic effector factor, chemotactic factor, and colony stimulating factor. Currently, it is not known which of these associations represent specific functional interactions or whether β_2m is a "sticky" polypeptide with nonspecific affinity for a variety of proteins.

VI. FUNCTION OF β_2M

The current popularity in research of β_2m is primarily a consequence of its unknown biological role. The results discussed above strongly suggest that this low-molecular-weight globulin is involved in lymphocyte transformation. Indeed, the amino acid sequence homology shared by β_2m and immunoglobulins leads one to speculate that this polypeptide may have once served as a primitive antigen receptor. Reports of β_2m binding to protozoa and gram-positive bacteria may relate to this possibility. Amino acid sequence similarity between the α_3 domain of transplantation antigens and immunoglobulins further suggests a recognition function of class I MHC gene products. It is presently not known to what extent β_2m contributes to the phenomenon of associative recognition whereby histocompatibility requirements between cytotoxic T lymphocytes and target cells expressing foreign antigens must be made before cytolysis. β_2-Microglobulin may serve as a bridge between target cell histocompatibility and foreign antigens. Alternatively, it may preserve structural features of the transplantation molecule, allowing them to make direct contact with nonself components. Two observations may pertain to the second possibility. Welsh et al. (1977) have suggested that synthesis of β_2m is partly essential for expression of HLA antigens on the basis of hybridization experiments between Raji and Daudi cell lines. Furthermore, Lancet and colleagues (1979) reported that the expression of alloantigenic determinants on HLA glycoproteins is critically dependent on association with β_2m. The function of histocompatibility antigens may be lost after removal of β_2m.

In contrast, the function of β_2m may not necessarily be restricted to an association with histocompatibility antigens since each appears to be independently expressed during embryogenesis. The murine F9 antigen encoded by the *T/t* locus can be detected within a few hours after fertilization and subsequently increases in amount on the embryonic cell surface. This antigen is suggested to be related evolutionarily to H-2K and D antigens since cells that express the F9 molecule generally do not possess H-2 antigens, and vice versa. A similar reciprocal relationship between H-2D and TL antigens has been described. However, hβ_2m can be identified at all stages of embryonic growth, although the relative amount of this protein gradually increases during the first 20 to 24 weeks of development and then decreases.

It is therefore proposed that β_2m may serve a fundamental role required by several types of cell surface antigens. In this respect, Bonner (1979) has recently presented a theory describing histocompatibility antigens as cell surface polymorphisms (CSPs) that direct fetal development. Self-recognition is suggested to be controlled at the molecular level by CSP interactions. The function of β_2m may involve formation of an interdomain pair with CSPs of adjacent cells resulting in cell surface modulations that signal self-recognition and cytodifferentiation. It is interesting to note that many plasma membrane components suggested to be involved in cell–cell interactions (e.g., H-2, TL, and Qa antigens) are glycoproteins of approximately 45,000 molecular weight which share β_2m as a common subunit (Flaherty, 1980). Although the F9 antigen is of similar molecular weight and subunit structure, the small noncovalently bound polypeptide is not β_2m (Dubois et al., 1976). A small component distinct from β_2m but associated with a human cell surface antigen that may be homologous to the murine TL antigen has also been described. It is yet to be established whether these small polypeptides are related functionally to β_2m.

REFERENCES

Adami, H., Hallgren, R., and Lundqvist, G. (1979). Clin. Chim. Acta *93*, 43.

Amor, B., Georgiadis, A., Kahan, A., and Delbarre, F. (1978). Pathol. Biol. Paris *26*, 377.

Bach, M., Huang, S., Hong, R., and Poulik, M. (1973). Science *182*, 1350.

Berggård, I., and Bearn, A. (1968). J. Biol. Chem. *243*, 4095.

Bernier, G., and Post, R. (1973). Transplantation *15*, 176.

Bonner, J. (1979). Birth Defects *15*, 55.

Conway, T. (1981). Fed. Proc. *40*, 1054.

Cooper, E., Bunning, R., Illingworth, S., Späti, B., and Child, J. (1978). Biomedicine *29*, 154.

Cresswell, P., and Dawson, J. (1975). J. Immunol. *114*, 523.

Cresswell, P., Springer, T., Strominger, J., Turner, M., Grey, H., and Kubo, T. (1974). Proc. Natl. Acad. Sci. USA *71*, 2123.

Cunningham, B., Wang, J., Berggård, I., and Peterson, P. (1973). Biochemistry *12*, 4811.

Dubois, P., Fellous, M., Gachelin, G., Jacob, F., Kemler, R., Pressman, D., and Tanigaki, N. (1976). Transplantation *22*, 467.

Evrin, P., and Pertoft, H. (1973). J. Immunol. *111*, 1147.

Evrin, P., and Wibell, L. (1972). Scand. J. Clin. Lab. Invest. *29*, 69.

Faulk, W., and Temple, A. (1976). Nature (Lond.) *262*, 799.

Fellous, M., Günther, E., Kemler, R., Wiels, J., Berger, K., Guenet, J., Jakob, H., and Kacob, F. (1978). J. Exp. Med. *148*, 58.

Fermin, E., Johnson, C., Eckel, R., and Bernier, G. (1974). J. Lab. Clin. Med. *83*, 681.

Flaherty, L. (1980). In *Biological Basis of Immunodeficiency* (E. Gelfand and H. Dosch, Eds.). Raven Press, New York, pp. 99–109.

Gates, F., Coligan, J., and Kindt, T. (1981). Proc. Natl. Acad. Sci. USA *78*, 554.
Goding, J. (1981). J. Immunol. *126*, 1644.
Goding, J., and Walker, I. (1980). Proc. Natl. Acad. Sci. USA *77*, 7395.
Gordon, S., and Kindt, T. (1976). Scand. J. Immunol. *5*, 505.
Grey, H., Kubo, R., Colon, S., Poulik, M., Cresswell, P., Springer, T., Turner, M., and Strominger, J. (1973). J. Exp. Med. *138*, 1608.
Groves, M., and Greenburg, R. (1977). Biochem. Biophys. Res. Commun. *77*, 320.
Hemmingsen, L., and Skaarup, P. (1977). Scand. J. Urol. Nephrol. *11*, 41.
Klein, J. (1979). Science *203*, 516.
Lancet, D., Parham, P., and Strominger, J. (1979). Proc. Natl. Acad. Sci. USA *76*, 3844.
Lightbody, J., Urbani, L., and Poulik, M. (1974). Nature (Lond.) *250*, 227.
Lillehoj, E., and Poulik, M. (1979). Pathobiol. Ann. *9*, 49.
Lillehoj, E., Krutzsch, H., and Poulik, M. (1982). Mol. Immunol. *19*, 817.
Lindblom, J., Östberg, L., and Peterson, P. (1974). Tissue Antigens *4*, 186.
Lögdberg, L., Cigén, R., and Berggård, I. (1979). Scand. J. Immunol. *9*, 263.
Lögdberg, L., Cigén, R., and Björck, L. (1981). Scand. J. Immunol. *13*, 237.
Martinko, J., Uehara, H., Ewenstein, B., Kindt, T., Coligan, J., and Nathenson, S. (1980). Biochemistry *19*, 6188.
Nakamuro, K., Tanigaki, N., and Pressman, D. (1973). Proc. Natl. Acad. Sci. USA *70*, 2863.
Nilsson, K., Evrin, P., and Welsh, K. (1974). Transplant. Rev. *21*, 53.
Orr, H., Lopez de Castro, J., Parham, P., Ploegh, H., and Strominger, J. (1979). Proc. Natl. Acad. Sci. USA *76*, 4395.
Östberg, L., Lindblom, J., and Peterson, P. (1974). Nature (Lond.) *249*, 463.
Peterson, P., Cunningham, B., Berggård, I., and Edelman, G. (1972). Proc. Natl. Acad. Sci. USA *69*, 1697.
Poulik, M. (1973). Immunol. Commun. *2*, 403.
Poulik, M. (1976). In *The Trace Components of Plasma: Isolation and Clinical Significance*, Vol. 5 (G. Jamieson and T. Greenwalt, Eds.). Liss, New York, pp. 155–177.
Poulik, M., and Bloom, A. (1973). J. Immunol. *110*, 1430.
Poulik, M., Ferrone, S., Pelligrino, M., Sevier, D., Oh, S., and Reisfeld, R. (1974). Transplant. Rev. *21*, 106.
Poulik, M., Gold, P., and Shuster, J. (1979a). CRC Crit. Rev. Clin. Lab. Sci. *10*, 225.
Poulik, M., Shinnick, C., and Smithies, O. (1979b). Mol. Immunol. *16*, 731.
Poulik, M., Perry, D., and Sekine, T. (1980). Vox Sang. *38*, 328.
Rigdén, O., Persson, U., and Möller, E. (1976). Immunol. Commun. *5*, 553.
Schaub, C., Bluet-Pajot, M., Szikla, G., Lornet-Videau, C., Mounier, T., and Talairach, J. (1978). Pathol. Biol. Paris *26*, 381.
Shuster, J., Gold, P., and Poulik, M. (1976). Clin. Chim. Acta *67*, 307.
Smithies, O., and Poulik, M. (1972). Science *175*, 187.
Späti, B., Child, J., Kerruish, S., and Cooper, E. (1980). Acta Haematol. *64*, 79.
Tanigaki, N., Nakamuro, K., Appella, E., Poulik, M., and Pressman, D. (1973). Biochem. Biophys. Res. Commun. *55*, 1234.
Thomson, D., Rauch, J., Weatherhead, J., Friedlander, R., O'Connor, R., Grosser, N., Shuster, J., and Gold, P. (1979). Br. J. Cancer *37*, 753.
Trowsdale, J., Travers, P., Bodmer, W., and Patillo, R. (1980). J. Exp. Med. *152*, 11s.
Vitetta, E., Poulik, M., Klein, J., and Uhr, J. (1976). J. Exp. Med. *144*, 179.
Welsh, K., Dorval, G., Nilsson, K., Clements, G., and Wigzell, H. (1977). Scand. J. Immunol. *6*, 265.
Winkler, M., and Sanders, B. (1977). Immunochem. *14*, 615.
Zinkernagel, R., and Doherty, P. (1975). J. Exp. Med. *141*, 1427.

10

Phylogeny of Immunoglobulins

Gary William Litman Sloan-Kettering Institute, Rye, New York

I. INTRODUCTION

The immunoglobulins represent an extensive multigenic family whose history dates to an early period in the evolution of vertebrates (Litman and Kehoe, 1978). In several mammalian species, where the natural occurrence of monoclonal gammopathies and ability to induce large quantities of homogeneous immunoglobulin experimentally have permitted extensive analyses to be carried out at both the protein and DNA levels, it is apparent that the level of structural polymorphism exceeds that of any other protein family studied to date. Although lower vertebrate immunoglobulins have not been characterized as extensively, it is possible to recognize the major patterns and trends in the phylogenetic development of antibody and relate some of the basic observations to probable genetic events which influenced the expansion and diversification of this multigenic family. This chapter does not intend to provide detailed structural descriptions of the numerous immunoglobulin-like humoral recognition mediators which have been characterized in lower vertebrate species, but rather will discuss those observations which have contributed most to our general understanding of the evolutionary development of the immunoglobulins. Many of the statements and illustrations in this chapter represent consensus views and readers are encouraged to consult original sources for more detailed accounts and, in some instances, alternative classifications and interpretations. As may have been predicted, the evolutionary development of the immunoglobulins, as inferred through studies involving the modern representatives of phylogenetically ancient species, did not consistently involve the progressive acquisition of structural and functional complexity but rather reflects many unique departures which provide considerable insight into the genetic control and functions of the antibody gene family.

II. MEDIATORS OF SPECIFIC RECOGNITION

The hemolymph of a large number of invertebrates contains proteins which can be isolated to immunochemical and physicochemical homogeneity and exhibit remarkable specificity for closely related structural determinants located on the surface of a variety of different cell types. When subjected to detailed analysis, however, it is apparent that both the subunit composition and intersubunit interactions lack any resemblance to mammalian immunoglobulin (Marchalonis and Edelman, 1968). In some instances, the proteins have stringent requirements for metal ions or exhibit unique sensitivity to temperature. Although the proteins typically are not inducible, they generally are felt to function in selective recognition of foreign substances and/or facilitation of phagocytosis. In terms of both subunit structure and specific binding function, they resemble plant lectins and lack even remote primary structure homology with immunoglobulins (Kaplan et al., 1977). As yet, no role has been ascribed to these proteins in the highly specific cell-mediated recognition, resembling higher vertebrate major histocompatibility complex (MHC)-associated function, which has been demonstrated in a number of invertebrates (Hildemann et al., 1979).

Functional homologues of these invertebrate recognition macromolecules may be present in more recently evolved species (Harisdangkul et al., 1972). Noninducible proteins exhibiting specificity for certain carbohydrates have been detected in the serum of elasmobranchs and teleosts and an inducible, C-reactive-like protein has been described in a recently evolved teleost. From the standpoints of both structure and function, these proteins resemble the inducible, specific recognition molecules which have been described in the lampreys and hagfishes, the present surviving Agnatha (Litman and Kehoe, 1978).

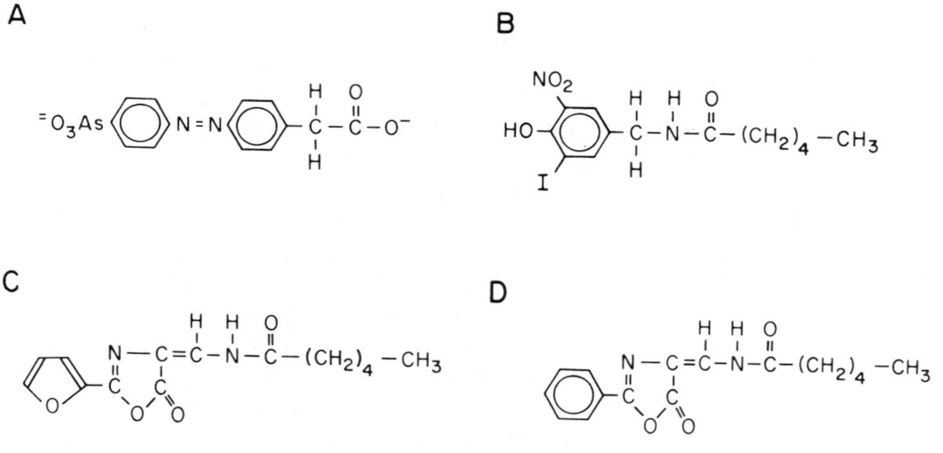

Figure 1 Structures of antigens that are recognized selectively by *Heterodontus*. The haptenated phage neutralization assay, on which these results are based, estimates the ability of antibody (primarily the HMW form) to neutralize the infectivity of bacteriophage that have been conjugated with a given haptenic structure. The primary antigenic structures are: (A) azobenzene arsonate, (B) nitroiodophenol, (C) furyloxazolone, and (D) phenyloxazolone. The caproyl form of the haptens illustrated in (B)–(D) simulates the ϵ-amino group of lysine, which represents the principal binding site for the reactive forms of the hapten used to derivatize the protein or bacterial carriers.

In the latter cases, recognition appears to be mediated by proteins lacking the disulfide bonding patterns and solution stability characteristic of higher vertebrate immunoglobulin. Primary structure data, which could definitively establish the relationships of these proteins to immunoglobulins, is not available (Litman and Marchalonis, 1982). It is, however, tempting to speculate that noninducible humoral recognition arose in invertebrates, and homologues of these proteins may exist in both inducible and noninducible forms in lower as well as higher vertebrate species.

Immunoglobulin-mediated recognition has been found in all contemporary species originating from ancestors of the jawed vertebrates. Specific responses for various erythrocytes, bacteria, viruses, bacterial vaccines, and simple haptens have been described (Litman and Kehoe, 1978; Litman and Marchalonis, 1982). With *Heterodontus francisci,* horned shark, a modern representative of a phylogenetically ancient shark and the earliest member of this group studied to date, hapten-specific responses to the determinants illustrated in Fig. 1 have been demonstrated using an assay that measures the capacity of antibody to neutralize the infectivity of haptenated bacteriophage (Litman et al., 1982b; Litman et al., 1982c). Based on these findings, it is apparent that at least four unique antibody populations exist and it is likely that the response repertoire is considerably more extensive in this species. In their earliest structurally recognizable form, immunoglobulins are associated with specific recognition to what must represent a wide range of antigenic structures; the molecular basis for this apparent specificity and nature of interindividual variation in hapten-specific recognition are discussed below.

III. STRUCTURE OF LOWER VERTEBRATE IMMUNOGLOBULIN (see also Chap. 7)

A. Subunit Organization of the IgM-Like Immunoglobulins

In elasmobranchs, depending on the immunogen employed, antibody activity distributes in high-molecular-weight (HMW) and in some instances low-molecular-weight (LMW) serum fractions. When purified to homogeneity, antibody has been found associated with immunoglobulins that bear a general relationship to the pentameric and monomeric forms of IgM (Litman and Kehoe, 1978; Litman and Marchalonis, 1982). Partial reduction of disulfides converts the HMW → LMW form, which, depending on the disulfide cleavage conditions employed, may continue to exhibit antigen binding activity. More extensive reduction results in a breakdown into heavy and light chains which can be separated, following denaturation, on the basis of molecular weight and are indistinguishable from μ-type heavy and κ or λ light chains. Interchain, heavy ↔ light and heavy ↔ heavy, interactions are stabilized by noncovalent bonds which exhibit considerable interspecies variation in intensity. The heavy chains from the HMW and LMW immunoglobulin classes are indistinguishable by peptide mapping and immunochemical analyses, as are the corresponding light chains. The heavy chain size (70,000 molecular weight), disulfide content, and carbohydrate composition are consistent with a polypeptide belonging to the mammalian μ-heavy-chain class and have led to the designation of these proteins as an IgM type (Litman and Kehoe, 1978; Litman and Marchalonis, 1982). In at least one elasmobranch species, a variant heavy chain exhibiting a molecular weight of about 50,000 has been found to constitute a distinct, additional LMW immunoglobulin class. Limited proteolysis of the IgM-like proteins yields a major Fab (fragment antibody)-like component which contains the entire light chain and a portion of the heavy chain. The

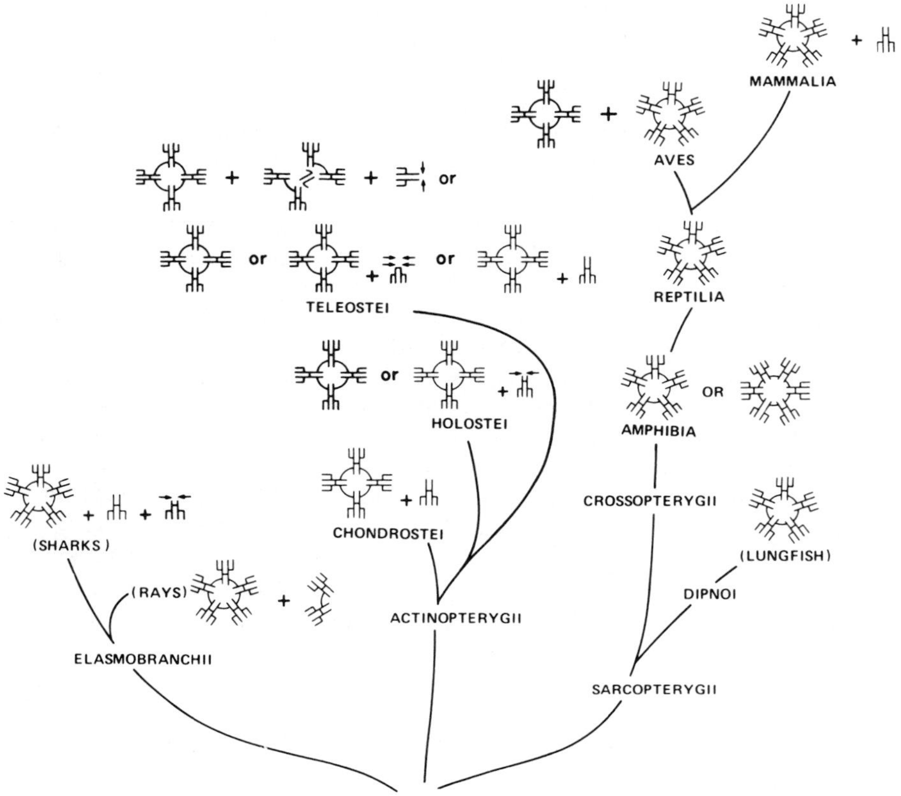

Figure 2 Phylogenetic distribution of the major forms of the IgM-like immunoglobulins. The classes, subclasses, superorders, orders, and common designations are representative of the assignments used in the text. The models illustrated provide the minimal structural formulations consistent with physicochemical analyses. Interchain and intersubunit disulfide bonds are represented by the short or curved lines connecting a basic heavy and light chain structure; their number and placement are not intended to correspond to actual covalent bonding patterns, but rather represent the presence of such linkages. The lengths of heavy chains (e.g., shark HMW) are intended to represent a five-domain structure or corresponding, deleted structures consisting of only four (→ ←) or three domains (⇉ ⇇), respectively. Phylogenetic relationships are defined further in (Litman and Kehoe, 1978).

similarity in sensitivity to proteolytic digestion implies organization into compact globular structures separated by accessible linker regions and, taken together with the presence of a similar β-sheet secondary structure, stresses a basic preservation of immunoglobulin conformation. Several physical studies, however, have suggested that the rotational flexibility of the proteins about the interchain disulfides may be significantly more constrained than in the immunoglobulins of higher vertebrate species.

Molecules resembling the elasmobranch IgM-type proteins, in terms of heavy and light chain mass as well as interchain disulfide bonding, are found in all other vertebrates representing subsequent evolutionary developments (Fig. 2) (Litman and Kehoe, 1978;

Litman and Marchalonis, 1982). With the exception of a trimeric form, all polymer states, monomer → hexamer, have been described and, in some instances, mass differences have been observed in the heavy chains of the LMW immunoglobulin class. Since immunochemical analyses typically have shown these mass differences to represent antigenic deficiencies, the deletions are not considered indicative of distinct classes. In addition, clearance studies have established that the deleted heavy chains are not the products of in vitro proteolysis or in vivo catabolism. It is noteworthy that in certain holosteans and teleosts, only a single tetrameric form is present. In the teleosts, the complexity of polymeric variants correlates roughly with the relative phylogenetic position of the species; that is, only a tetrameric form is detected in the lower teleosts while highly complex distributions are evident in the more recently evolved forms (Lobb and Clem, 1981). Variations in polymer composition are known to influence the functional valence (Karush, 1978) of antibody as well as the physiological distribution of immunoglobulins. In this way, the chain length differences could relate to additional aspects of secondary biological function as well as extravascular distribution. Regulation of polymeric assembly may be mediated by J-chain-like molecules which have been found associated with the HMW immunoglobulins of species as phylogenetically distant as the elasmobranchs (McCumber and Clem, 1976). The deletion of the domain(s) participating in intersubunit disulfide bond formation would prevent the polymeric assembly and could provide one possible functional explanation for the heavy chain mass deletion associated with some of the LMW immunoglobulins.

B. Constant Region Primary Structure

As suggested above, the HMW (and in some instances LMW) immunoglobulins of the lower vertebrates bear a general physicochemical resemblance to mammalian IgM. At the level of the Aves, immunochemical cross-reactivity between HMW immunoglobulin and mammalian IgM has been established and the similar amino acid compositions of mammalian μ-type heavy chains and the heavy chains of lower vertebrate IgM-like immunoglobulins suggest that the phylogenetic emergence of IgM parallels its ontogenetic appearance (Litman and Kehoe, 1978; Litman and Marchalonis, 1982). In order to evaluate this relationship further, we have isolated a CNBr fragment corresponding to the constant region of *Heterodontus* immunoglobulin and determined a significant portion of its primary structure (Litman et al., 1982b). Using a procedure that statistically defines the closest homologous alignment between two sequences and estimates the minimal number of nucleotide base changes required to convert one amino acid sequence into another, we have compared the *Heterodontus* sequence to prototypic mammalian heavy chains (Fig. 3). Over the region compared, about 37% of the positions show absolute identity, stressing the phylogenetic relationship between the protein families. It is interesting to note that the μ chain, which on the basis of the general physicochemical properties indicated above appeared most homologous to the *Heterodontus* heavy chain, requires 46 minimal base changes, whereas only 42 minimal base changes are required to convert this particular portion of the *Heterodontus* sequence into the corresponding sequence of the γ-chain prototype. In relating these observations to the physicochemical classification, it should be noted that the particular region being compared, which corresponds to the first constant region domain of a μ chain, was not found to exhibit the highest degree of interspecies homology when equivalent portions of mammalian IgMs were compared (Kehry et al., 1979). It remains possible that the C-terminal domains of shark and

	1				5					10					15					20					25
Heterodontus	D	Y	I	P	Q	I	T	S	V	S	W	K	K	G	E	E	P	I	T	T	G	L	K	T	Y
Human IgM	D	F	L	P	D	S	I	T	F	S	W	K	Y	L	N	N	S	D	I	S	S	T	R	G	F
Human IgG	D	Y	F	P	E	P	V	T	V	S	W	N	S	G	A	–	L	–	T	S	G	V	H	T	F

	26				30					35					40					45	46
Heterodontus	P	S	V	L	N	K	K	G	T	Y	T	L	I	S	–	V	L	E	X	S	K
Human IgM	P	S	V	L	–	R	G	G	K	Y	A	A	T	S	Q	V	L	L	P	S	K
Human IgG	P	A	V	L	Q	S	S	G	L	Y	S	L	S	S	V	V	T	V	P	S	S

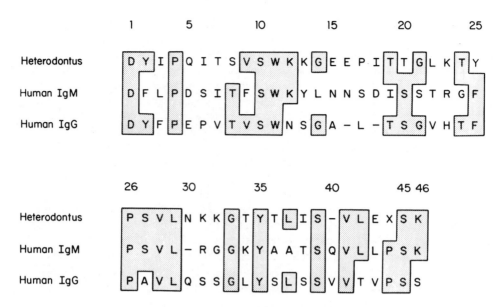

Figure 3 The primary structure of a 46-residue portion of a major fragment derived from the *Heterodontus* HMW immunoglobulin heavy chain by digestion with CNBr. The amino acid sequence was determined by direct N-terminal Edman sequencing of the parent peptide as well as its constituent peptides derived by digestion with chymotrypsin, and *Staphylococcus aureus* V8 protease. Shared residues are highlighted by shading. Note that over the sequence region compared, the prototype sequences share more residues with *Heterodontus* than they share with each other. The introduction of insertions and deletions (relatively infrequent genetic events) used to maximize homology are considered in deriving the alignment scores referred to in the text (Litman et al., 1982b). An unidentified residue at position 44 in the *Heterodontus* sequence is indicated as (X). Alignment of *Heterodontus*, Cμ1, and Cγ1 sequences follows Litman and Marchalonis (1982).

mammalian immunoglobulin heavy chains may be more homologous. The majority of positions in the human μ chain and shark heavy chain which exhibit absolute identity are shared by other immunoglobulin classes. Based on these findings, it is apparent that over an extended period of evolutionary time strong selective pressures have been exerted to preserve certain residues which are critical in determining polypeptide conformations essential for immunoglobulin primary and secondary function. If the general physicochemical properties are used as the basis for classifying the lower vertebrate immunoglobulins as IgM-like, it is interesting to note that the serum concentration of this protein is about 100-fold greater in elasmobranchs versus mammals. Thus, while the IgM class has retained its primary position in ontogenetic development, other more specialized immunoglobulin classes exhibiting additional and perhaps more efficient functions have arisen and may account for the diminishing reliance on immunoglobulins of the IgM family as the exclusive mediators of antibody recognition.

C. Additional Immunoglobulin Classes

Below the phylogenetic level represented by the Dipnoi, all species appear to express only a single immunoglobulin class which exhibits extensive inter- and, in some instances,

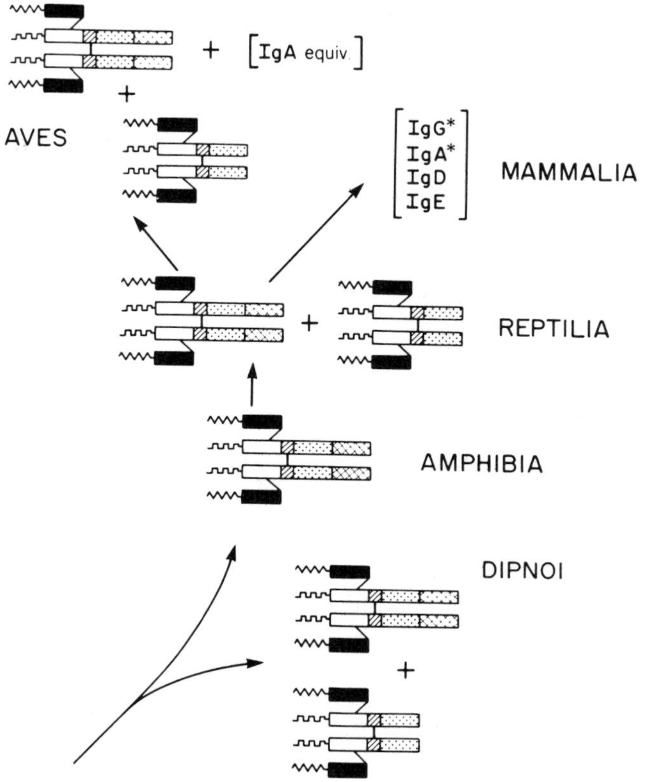

Figure 4 Evolution of additional immunoglobulin classes. Phylogenetic distribution of the additional major forms of immunoglobulin which are distinct from the IgM-like proteins illustrated in Fig. 2. The class or order designations are representative of assignments used in the text. The models illustrated provide the minimal structural formulations consistent with physicochemical analyses. Shading of distinct regions implies the presence of unique structural domains that differ in primary structure; the hinge region containing the interheavy chain disulfide linkage ▨ may exhibit considerable length variation. Equivalent shading of homologous domains in different species does not indicate a primary structure identity but rather proposes a direct, general evolutionary relationship between the structures. Designation of "IgA equivalent" refers to the IgA-like protein and its secretory counterpart which has been described in avians. The presence of additional isotypic heavy chain forms is denoted by (*).

intraspecies variation in both polymer composition and heavy chain length (Litman and Kehoe, 1978; Litman and Marchalonis, 1982). At the level of the Dipnoi, additional, distinct heavy chain classes are encountered and a schematic representation of their distribution is provided in Fig. 4. The assignment of the heavy chains to a unique class is based on the detection of antigenic differences, relative to the IgM-like proteins, which do not relate to simple mass deletion as well as the findings of additional differences in amino acid composition and chromato-electrophoretic distribution of tryptic and/or chymotryptic peptides. Heavy chain mass and interchain noncovalent bonding patterns further support these assignments. In the lungfishes, evidence has been provided for two

distinct immunoglobulin classes, in addition to the typical HMW, IgM-like immunoglobulin discussed above. In terms of serum concentration, it appears as if an immunoglobulin of molecular weight 120,000 (LMW), consisting of two heavy chains of molecular weight 37,000 and typical light chains of molecular weight 22,500, represents the principal immunoglobulin class. The heavy chains most probably contain a variable (V) and two constant regions. In addition, an intermediate-molecular-weight (IMW) immunoglobulin possessing a heavy chain of molecular weight 65,000, consisting of one V and possibly three distinct constant regions, has been described. Both a higher carbohydrate content and an extended hinge region versus mammalian γ_1 heavy chains, which also consist of four domains, could account for the increased size of the IMW heavy chain; as discussed in Chap. 7, differences in hinge region structure may account for significant variation in the length of immunoglobulin heavy chains. An IMW protein also has been detected and characterized extensively in several anuran amphibians. Only a HMW, IgM-like form has been detected in the serum of one urodele amphibian; however, analyses of hapten-specific antibody elicited during hyperimmunization of another urodele species have indicated the presence of both HMW and IMW immunoglobulins. The apparent absence of IMW immunoglobulins in the former species suggests that regulatory events rather than a genetic deficiency may govern expression of IMW heavy chain genes in a unique manner (Warr et al., 1982).

At least some reptiles appear to synthesize all three immunoglobulin classes, and certain avians may express HMW and IMW immunoglobulin, whereas others synthesize HMW, IMW, and LMW immunoglobulin forms. Based on the order of ontogenetic expression of mammalian immunoglobulin heavy chain constant region genes, $\mu \rightarrow \delta$, it has been proposed that the IMW form may represent a phylogenetic precursor of the δ chain; however, analyses of antigenic cross-reactivity have failed to confirm this relationship or to detect shared cross-reactivity with mammalian γ chains. Antigenic cross-reactivity between IMW proteins isolated from amphibians, reptiles, and avians, however, has been demonstrated (Hädge et al., 1980). An IgA-like protein has also been detected in various avian secretions and is discussed below; the unique antigenic character of the protein qualifies it as an additional immunoglobulin class in these species. A homologue of the γ chain is encountered at the phylogenetic level of the early mammals, represented in a contemporary monotreme, although it must be stressed that extensive primary structure studies are required to confirm such assignments (Litman and Marchalonis, 1982).*

In summary, evidence has been presented for divergence of immunoglobulin classes in species belonging to the evolutionary line, which leads eventually to the terrestrial mammals and includes the lungfishes, amphibians, reptiles, and avians. No pattern is obvious in the distribution of the different immunoglobulin forms during this extended evolutionary period; however, it is assumed that the many variations reflected in contemporary species relate to unique primary and secondary biological functions of these proteins. Both IMW and LMW have been shown to be capable of fixing homologous and heterologous complement and binding to cell surfaces (Litman and Kehoe, 1978). The sizes of the proteins, which relate to variation in their Fc (fragment crystalline) regions, suggest that they may exhibit extravascular distribution patterns which differ from those of the IgM-like proteins. The appearance of additional immunoglobulin classes and an immunoglobulin class shift mechanism may relate to an apparently unique property of the immunoglobulin gene family, somatic rearrangement with concomitant somatic variation. This point, which explains certain aspects of antibody diversity in higher vertebrates, is also discussed below.

*Recently, an I_SD-like immunoglobulin has been detected on the surface of avian lymphocytes, see Chen, C-L. H., Lehmeyer, J. E., and Cooper, M. D. (1982). J. Immunol.: 2580.

D. J Chain and Secretory Component

In addition to heavy and light chains, two other polypeptides have been shown to be covalently (and in certain instances noncovalently) associated with the immunoglobulins. A polypeptide resembling the J chain associated with mammalian polymeric immunoglobulin, in terms of both mass and electrophoretic mobility, has been detected in several lower vertebrate polymeric, IgM-like immunoglobulins. A 1:1 covalent association between the HMW immunoglobulin of an elasmobranch and a J-like chain of molecular weight about 12,200 has been described (McCumber and Clem, 1976). Peptide mapping of the isolated J chain and the immunoglobulin heavy and light chains has established that the polypeptide does not represent a breakdown product. The failure to detect J chain in the HMW immunoglobulins of a number of lower vertebrates (Litman and Kehoe, 1978) may be explained on the basis of variation in electrophoretic mobility—the basis for detection of this component—rather than its actual physical absence. Another polypeptide resembling secretory component has been detected in association with the polymeric IgA equivalent identified in the secretions of an avian species (Litman and Kehoe, 1978). Immunoglobulin also has been detected in the secretions of both a reptile and an amphibian, although in both instances it appeared to resemble serum IgM rather than IgA or a unique secretory form of immunoglobulin. In at least one teleost species, however, an additional, nonimmunoglobulin heavy or light chain polypeptide has been found associated with the polymeric immunoglobulin present in the mucosal secretions; its relationship to higher vertebrate secretory component has not been demonstrated (Lobb and Clem, 1981). Functionally, it appears as if J chain is active in some aspect of polymeric organization, whereas the secretory component may stabilize immunoglobulins against proteolytic degradation in secretory compartments. These postulated roles would be consistent with the early evolutionary appearance of polymeric immunoglobulins and the expansion of immunoglobulin function and class divergence which has occurred at several points in vertebrate phylogenetic development.

IV. ANTIBODY DIVERSITY (see also Chaps. 7 and 14)

A. Structural Approaches

As indicated above, inducible, specific humoral immune recognition of several different classes of antigen has been demonstrated in the modern Agnatha (lampreys and hagfishes). Similarly, diversity in the immune response repertoire of a representative elasmobranch is evident from the results of the antigenic (hapten) cross-reactivity studies illustrated in Fig. 1. The complex electrophoretic polymorphisms (spectrotypes) associated with HMW, IMW, and LMW immunoglobulins isolated from a number of different lower vertebrate species are consistent with the observed functional diversity. Fully reduced and carboxyamidomethylated *Heterodontus* immunoglobulin light chains exhibit complex electrophoretic heterogeneity, while the same technique indicates that the corresponding heavy chains are relatively homogeneous (Litman et al., 1982). Similarly, the N (amino)-terminal amino acid sequence of elasmobranch heavy chains, which exhibit a significant homology with mammalian V_H (variable region, heavy chain) sequences, revealed no evidence for substitutions at individual positions through the first 27 residues as has been reported for unblocked V_HIII (variable region heavy chain, subgroup III) sequences of avian and certain mammalian immunoglobulins (Fig. 5). Elasmobranch light chains are less homologous with higher vertebrate prototypes, exhibit residue alternatives at a limited number of

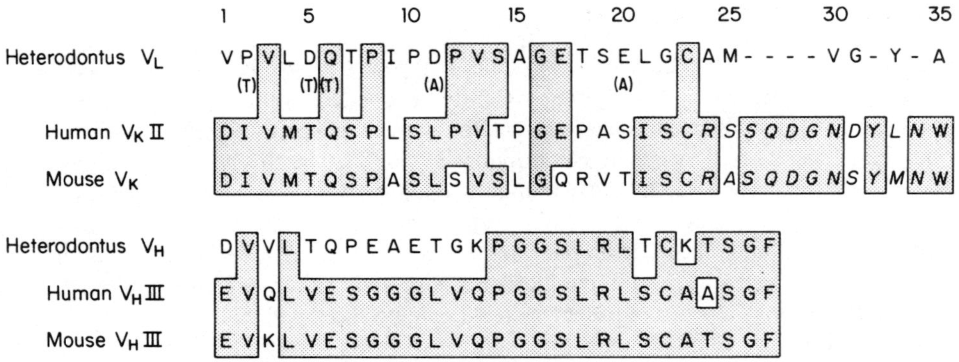

Figure 5 The primary structure of the N-terminal regions of *Heterodontus* heavy and light chains corresponding to the V_H and V_L regions of mammalian immunoglobulins. Human and mouse V_K and V_HIII sequences are included as references. Shading is used to underscore absolute sequence identity. Alternate residues at individual positions are indicated below the major sequence in parentheses. The location of hypervariable regions in the prototypes is indicated by italicized letters. The failure to quantitatively recover residues at certain positions in the *Heterodontus* V_L sequence is indicated by (–) and may represent the occurrence of multiple residue alternates.

N-terminal positions, and may contain a localized region of extensive substitutions equivalent to the hypervariable regions present in murine and human immunoglobulins (Fig. 5). Although extensive, systematic analyses have not been carried out, it is generally accepted that the conformation, depth, and net hydrophobic character of lower vertebrate immunoglobulin antigen combining sites closely resemble their mammalian counterparts (Litman and Kehoe, 1978; Litman and Marchalonis, 1982).

B. Combined Functional-Structural Approaches

A number of techniques, alluded to in other portions of this text, have been used to characterize the specificity and interindividual relatedness of lower vertebrate antibody elicited to specific haptenic determinants. The results of these studies relate directly to basic questions concerning the diversity of the lower vertebrate immune response and will be discussed in relative order of phylogenetic position of the species that have been studied. As evidenced by responses to synthetic haptens and streptococcal carbohydrates, elasmobranchs most probably possess an extensive antibody repertoire. When the fine specificity, defined as the ability to specifically recognize a haptenic determinant (usually the homologous immunogen) versus a series of homologues, of antibody elicited in unrelated *Heterodontus* is compared, however, little interindividual variation is apparent (Mäkelä and Litman, 1980). Furthermore, relative to inbred mammals, little change in affinity and fine specificity is apparent upon prolonged immunization. The results of these studies have been confirmed using large numbers of animals and additional haptens and have been extended to include a more recently evolved marine teleost, *Pseudopleuronectes americanus*. Greater affinity, fine specificity, and interindividual variation of hapten-specific antibody were observed in the more recently evolved species (Litman et al., 1982c). Analyses of the idiotype of hapten-specific antibody elicited in *Cyprinus carpio*, a limited teleost, revealed little

interindividual variation in a large experimental group (Machulla et al., 1980). Individual variation in both the idiotype and spectrotype of hapten or cell surface specific antibody in outbred *Xenopus*, an anuran amphibian, has been described; however, it is apparent that this variation is less than would be encountered in equivalent studies of mammalian antibody (Brandt et al., 1980). Fewer antibody types were encountered in the *Xenopus* response to phosphorylcholine versus dinitrophenol. Isogenic *Xenopus* exhibit extensive sharing of both spectrotypes and idiotypes within a family. The inheritance of idiotypes and spectrotypes through several generations of cloned animals suggests that the response in this species may reflect the predominant expression of germline genes (Dupasquier, 1982). Although the response repertoire of avians to streptococcal antigens is extensive, reduced spectrotypic complexity and affinity maturation in the response to other complex haptens has been observed. Although the results of these studies depend on the hapten selected and detection methods employed, there is good evidence for extensive antibody diversity at early levels of phylogenetic development and the progressive acquisition of a larger number of specific immunoglobulin genes or some mechanism(s) amplifying this diversity in subsequent phylogenetic developments. It is generally conceded that the level of antibody diversity in lower vertebrate species is significantly less than that encountered in the equivalent immune response of mammals (Dupasquier, 1982).

V. GENE STRUCTURE (see also Chap. 14)

To date, the molecular genetic studies of immunoglobulin evolution have largely been confined to inter- and intraspecies analyses of human and murine immunoglobulin constant and variable region genes. One of the principal descriptions, relating directly to the previous discussion of variation in heavy chain structure, involves the coding of individual domains and certain hinge regions by individual genetic regions, *exons*, separated by noncoding intervening sequences, *introns* (Sakano et al., 1979). Regions encoding separate individual classes are separated by 15 to 20 kilobases. The interspersed arrangement supports the probable origin of the constant as well as variable region genes by gene duplication and provides a relatively uncomplicated mechanism for the gain and loss of immunoglobulin domain equivalents in a large number of phylogenetically diverse vertebrates. Class divergence apparently took place within the individual domain exons and may have involved scrambling of individual exons between classes and subclasses as well as by correction of homologous, nonallelic chromosomal segments through a process known as gene conversion. In both humans and mice, the individual classes and subclasses that form the constant region gene family are situated 3′ to the segmented region that forms the variable region. The segmented μ-chain gene, which probably appeared first in phylogenetic development, is located at the most 5′ position in the constant region gene cluster. A specific genetic region, termed the *switch region*, is involved directly with the heavy-chain-class switch mechanism and exhibits significant interspecies homology (Yamawaki-Kataoka et al., 1981). Repeat sequences, located adjacent to switch genes, may be involved in the switch mechanism and have also been conserved phylogenetically (Kirsch et al., 1981). As indicated above, the expansion of immunoglobulin classes, which probably increased the range of secondary biological function, also appears to be related to antibody diversity. Work from several laboratories has established a strong correlation between somatic change in immunoglobulin variable regions and the shift in immunoglobulin class from μ to γ or α, and so on (Crews et al., 1981).

DNA sequence analysis has suggested that smaller sequence units led to the formation of the V region through a duplication process (Ohno and Matsunaga, 1982). Similarities between the V_H and V_L (variable region, light chain) subsequence regions suggest a common origin for these two gene families, as does the identification of similar 3′ noncoding sequences which are involved in V-D (diversity segment)-J (joining segment) and V-J recombination. Insertion of extra nucleotides between D and J and possibly V and D by terminal deoxynucleotidyl transferase as part of the joining process may account for additional diversity. The latter event, together with intragenic (e.g., D-D) recombination and hypermutation in the vicinity of V coding regions, may represent the principal mechanisms for generating somatic change in V-region structure. Enumeration of immunoglobulin variable region genes in mice and humans has been facilitated by cross-hybridization of V_H, V_K (variable region kappa), and J_K (joining region kappa) probes derived from one species with genomic DNA derived from the other species. Depending on the methods used, different estimates for the number of V_H and V_L regions have been obtained, although the simultaneous detection of pseudogenes limits, to a degree, the confidence in these assignments. In the mouse as many as 250 V_H genes may exist and the greater number of hybridizing components in restriction enzyme digests of human versus mouse genomic DNA suggests that the human multigenic family may be more complex. As many as 400 to 2000 murine V_K genes contrast with the 15 to 20 estimated to be present in the human, implying that major shifts in copy number can occur over limited periods of evoluiontary time (Bentley and Rabbitts, 1981). The human J_K gene family has been shown to be more extensive than that found in the mouse, suggesting that a greater capacity to effect combinatorial gene rearrangement may limit the need for a germ line gene commitment as extensive as that suggested by the murine V_K estimates.

We have extended analyses of the V_H gene family to include an early mammalian form, *Tupaia*, a tree shrew, and *Caiman*, a relatively early reptilian (Litman et al., 1982a). Genomic DNAs isolated from these species and BALB/c mouse were digested with restriction endonucleases, the fragments electrophoretically resolved, transferred to nitrocellulose, and hybridized with a probe representing the V_H region of a murine phosphorylcholine binding antibody (Fig. 6). Although the hybridization intensities are not as great as with the homologous DNA, the signals are unambiguous and suggest a significant relationship between the lower and higher vertebrate V_H regions, which in the case of *Caiman* have been characterized completely at the DNA sequence level (Litman et al., 1983). The relatively fewer hybridizing components in *Tupaia* suggest that either the gene family has contracted extensively in this species or that portions of the gene family have diverged appreciably and lack sufficient homology for detection by this method. It is possible that the heterologous murine V_H probe detects only a fraction of the number of components in the multigenic *Caiman* family and that its complexity may exceed that found in mammals. Alternatively, it is possible that relatively few components represent functional genes, and extensive numbers of presumably nonfunctional pseudogenes may be present. While it has been argued that such structures, which are known to be translocated extensively, may play an indirect role in preventing the "accidental" loss of a specific gene, the description of a recent pseudogene duplication suggests that they may function more directly in a fashion as yet unknown. The existence of large numbers of V_H-related sequences in species such as *Caiman* would support the notion that the humoral immune mechanism in lower vertebrates may rely more on a germ line response mode than on somatic rearrangements for generating diversity.

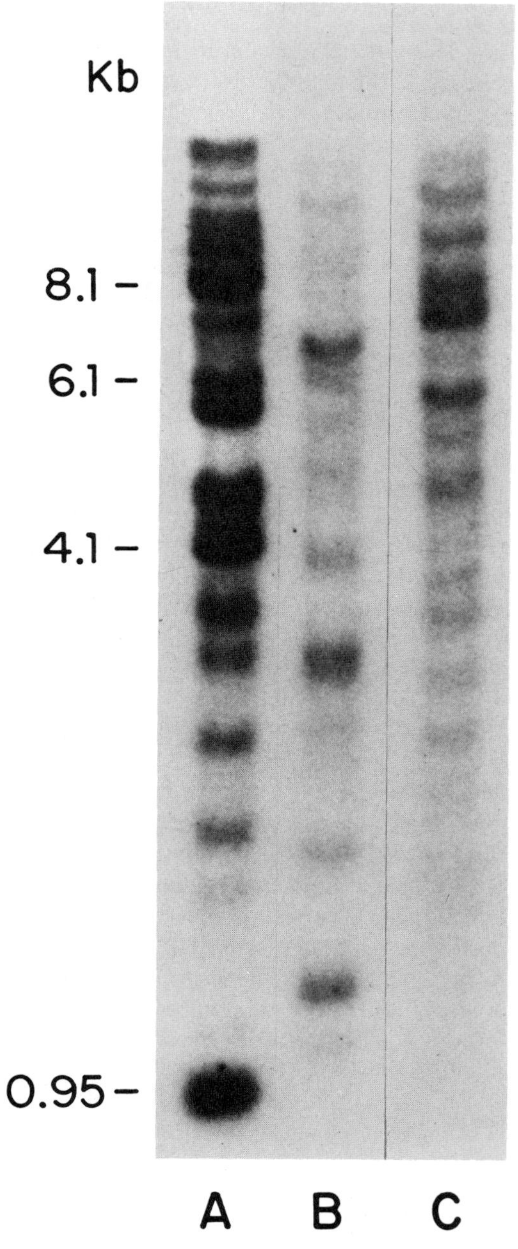

Figure 6 Hybridization pattern of a ^{32}P-labeled DNA probe representing the V_H region of a murine phosphorylcholine binding antibody with 10 μg of restriction enzyme (Sst 1)-digested (A) *Mus* (mouse); (B) *Tupaia* (tree shrew), and (C) *Caiman* (caiman) DNA. The DNA fragments were separated by electrophoresis in 0.8% agarose and transferred to nitrocellulose. Low-stringency hybridization-washing conditions are required to detect the V_H regions in heterologous species (Litman et al., 1982a).

VI. CELL SURFACE IMMUNOGLOBULIN AND β_2 MICROGLOBULIN

Immunoglobulin may be detected readily, using conventional procedures, on the surface of both B and T cells in elasmobranchs and teleosts. The high density presence of immunoglobulin suggests that at least one type of cell surface recognition is mediated by conventional antibody, and it remains likely that one family of cell surface recognition proteins present on higher vertebrate T cells bears a structural relationship to serum immunoglobulin. The limited physicochemical-immunochemical characterizations of the immunoglobulin detected on lower vertebrate B and T cells suggest that these proteins may exhibit some of the unique properties related to detergent binding and solubility which have been described recently for mammalian cell surface immunoglobulin (Warr, 1980). It remains possible that a region(s) of these proteins is encoded by a separate DNA portion and that an equivalent RNA splicing mechanism is responsible for biosynthesis of the membrane versus secreted immunoglobulin form.

As discussed in Chap. 9, β_2 microglobulin (β_2m) closely resembles a free immunoglobulin domain. This protein, of approximately 12,000 molecular weight and containing a single intrachain disulfide loop and exhibiting appreciable amino acid sequence homology to immunoglobulin domains, is associated with the structural domain of transplantation antigens (e.g., H-2, HLA), which is situated closest to the plasma membrane. It is also found associated with membrane of the Tla/Qa family of cell surface antigens which are expressed characteristically during different phases of lymphoid cell differentiation. At the level of the avians, a polypeptide resembling β_2m is found associated with the heavy chain of a histocompatibility antigen which is homologous, at the subunit structure and amino acid sequence levels, to products of the mammalian MHC. Several reports have described the occurrence of a polypeptide exhibiting antigenic cross-reactivity with β_2m in the plasma and on the cells of various vertebrate species, including elasmobranchs. A more recent study has documented the quantitatively variable distribution of a β_2m-like polypeptide in a number of vertebrate and several invertebrate species (Shalev et al., 1981). While it is essential that structural studies be carried out to confirm the assignment of the antigen as a β_2m homologue, the presence at distant levels of phylogenetic development of β_2m, which is felt to exhibit a functional stabilizing effect on transplantation antigens, may provide a link between the allogeneic recognition described in invertebrates and that mediated by the MHC and its homologues in higher vertebrate species. Since specific cell-mediated immune recognition probably arose prior to antibody-mediated recognition, a gene encoding the ancestral form of β_2m may have duplicated, translocated, and diverged into the immunoglobulin and transplantation antigen-like genes. Significant evidence exists for a general resemblance among all three structures in terms of intrachain disulfide bonding and conserved amino acid sequence patterns.

VII. CONCLUSION

Although noninducible and inducible macromolecules which are capable of recognizing certain determinants found on the surfaces of higher vertebrate cells have been described, respectively, in various invertebrates and the Agnatha (hagfishes, lampreys), the relationship of these proteins to immunoglobulin has not been established. Immunoglobulin exhibiting significant homology to mammalian immunoglobulin at the amino acid sequence level can be recognized in the elasmobranchs (e.g., sharks, rays). The organization of the prototypic HMW immunoglobulin resembles mammalian IgM and has led to

classification of these proteins as IgM-like. A variety of polymeric arrangements of the HMW immunoglobulin have been recognized, and differences in heavy chain length are apparent in the LMW immunoglobulins of elasmobranchs and the bony fishes. The pattern of variation in heavy chain length is consistent with the known genetic organization of mammalian immunoglobulins as well as the evolution of this class of protein by gene duplication.

Additional distinct classes of heavy chain can be recognized first in the Dipnoi (lungfishes). As with the IgM-like proteins, essentially little variation in light chain structure is associated with the evolution of the additional immunoglobulin classes; however, variation in heavy chain length has been observed. In contrast to variation in polymer size, which influences principally certain aspects of antigen binding, heavy chain length differences probably affect secondary biological function. An evolutionary precursor of the γ chain, which in terms of concentration is the predominant immunoglobulin class in mammals, has not been recognized, employing conventional immunochemical-physicochemical approaches, below the level of an early egg-laying mammal. It is possible that a polymeric immunoglobulin resembling IgA is present in avians and that functional homologues of the secretory component may be present in at least one higher bony fish. Immunoglobulin found on the surface of lower vertebrate T and B cells may differ structurally from their serum counterparts and represent additional, unique clases.

The ability of lower vertebrates to mount hapten-specific immune responses to a variety of structurally related haptens suggests that an extensive repertoire of antibody may be present at a relatively early level of phylogenetic development. Increases in antibody affinity, fine specificity, and interindividual variation appear to have occurred during phylogenetic development. By comparison to mammals, however, there appears to be reduced interindividual variation in the immune response to certain antigens. The presence of variable regions and hypervariable segments has been documented in immunoglobulins isolated from elasmobranchs and, in general, the antigen combining sites of lower vertebrate hapten-specific antibodies resemble mammalian counterparts. Studies at the nucleic acid level suggest that certain lower vertebrate species possess extensive V_H-related gene families.

At the amino acid level, immunoglobulins, β_2m, and transplantation antigens exhibit a significant level of homology in certain structural regions. These findings suggest a common origin for three different classes of molecule which are involved with selective recognition. The possibility remains that a gene encoding the equivalent of an immunoglobulin domain or subregion of the domain underwent duplication, translocation, and diversification, eventually giving rise to complex gene families which function as the primary mechanisms for selective recognition in adaptive immunity. See also Chaps. 9, 24-28, and the appendix.

REFERENCES

Bentley, D. L., and Rabbitts, T. H. (1981). Cell *24*, 613.

Brandt, D. C., Griessen, M., Dupasquier, L., and Jaton, J. C. (1980). Eur. J. Immunol. *10*, 731.

Crews, S., Griffin, J., Huang, H., Calame, K., and Hood, L. (1981). Cell *25*, 59.

Dupasquier, (1982). Nature (Lond.) *296*, 311.

Hädge, D., Fiebig, H., Puskas, E., and Ambrosius, H. (1980). Dev. Comp. Immunol. *4*, 725.

Harisdangkul, V., Kabat, E. A., McDonough, R. J., and Sigel, M. M. (1972). J. Immunol. *108*, 1259.
Hildemann, W. H., Johnson, I. S., and Jokiel, P. L. (1979). Science *204*, 420.
Kaplan, R., Li, S. S.-L., and Kehoe, J. M. (1977). Biochemistry *16*, 4297.
Karush, F. (1978). In *Comprehensive Immunology*, Vol. 3 (G. W. Litman and R. A. Good, Eds.). Plenum, New York, p. 85.
Kehry, M., Sibley, C., Fuhrman, J., Schilling, J., and Hood, L. E. (1979). Proc. Natl. Acad. Sci. USA *76*, 2932.
Kirsch, I. R., Ravetch, J. V., Kwan, S.-P., Max, E. E., Ney, R. L., and Leder, P. (1981). Nature (Lond.) *293*, 585.
Litman, G. W., and Kehoe, J. M. (1978). In *Comprehensive Immunology*, Vol. 5 (G. W. Litman and R. A. Good, Eds.). Plenum, New York, p. 205.
Litman, G. W., and Marchalonis, J. J. (1982). In *Immune Regulation*: *Evolutionary and Biological Significance* (L. N. Ruben and M. E. Gershwin, Eds.). Marcel Dekker, New York, p. 29.
Litman, G. W., Berger, L., and Jahn, C. L. (1982a). Nucl. Acids Res. *10*, 3371.
Litman, G. W., Erickson, B. W., Lederman, L., and Mäkelä, O. (1982b). Mol. Cell. Biochem. *45*, 49.
Litman, G. W., Stolen, J., Sarvas, M. O. and Mäkelä, O. (1982c). J. Immunol. *9*, 465.
Litman, G. W., Berger, L., Murphy, K., Litman, R., Hinds, K., Jahn, C. L., and Erickson, B. W. (1983). Nature *303*, 349.
Lobb, C. J., and Clem, L. W. (1981). Dev. Comp. Immunol. *5*, 587.
McCumber, L. J., and Clem, L. W. (1976). Immunochemistry *13*, 479.
Machulla, H. K. G., Richter, R. F., and Ambrosius, H. (1980). Immunol. Lett. *1*, 329.
Mäkelä, O., and Litman, G. W. (1980). Nature (Lond.) *287*, 639.
Marchalonis, J., and Edelman, G. M. (1968). J. Mol. Biol. *32*, 453.
Ohno, S., and Matsunaga, T. (1982). Proc. Natl. Acad. Sci. USA *79*, 2338.
Sakano, H., Rogers, J. H., Hüppi, Brack, C., Traunecker, A., Maki, R., Wall, R., and Tonegawa, S. (1979). Nature (Lond.) *277*, 627.
Shalev, A., Greenberg, A. H., Logdberg, L., and Björck, L. (1981). J. Immunol. *127*, 1186.
Warr, G. W. (1980). Contemp. Top. Immunobiol. *9*, 141.
Warr, G. W., Ruben, L. N., and Edwards, B. F. (1982). Immunol. Lett. *4*, 99.
Yamawaki-Kataoka, Y., Miyata, T., and Honjo, T. (1981). Nucl. Acids Res. *9*, 1365.

11

Immunoglobulin Allotypes

Arthur G. Steinberg Case Western Reserve University, Cleveland, Ohio

I. INTRODUCTION

The human allotype systems [Gm, Inv(Km), A_2m] have proven useful for studies in population genetics, immunology, transplantation, and legal medicine (paternity tests and analysis of blood stains). The first human allotype system, the Gm system, was discovered by Grubb (1956); the Inv (renamed Km) system was discovered by Ropartz et al. (1961); and the most recently identified and the least known, the A_2m system, was discovered by two groups independently (Kunkel et al., 1969; Vyas and Fudenberg, 1969). Table 1 contains a list of the presently recognized Gm and Km allotypes and the two systems of nomenclature currently applied to them. The allotypes in all three systems are inherited as codominant alleles. [See Grubb (1970), for a detailed review of the Gm and Inv allotypes found prior to 1970. Those found after 1970 are reviewed later in this chapter.] A panel that met in Rouen suggested that the nomenclature indicate the IgG heavy chain subclass carrying the allotype (see Am. J. Hum. Genet. *29*, 117, 1977). For example, the Gm allotypes 1, 2, 3, and 17 are carried by $\gamma 1$ heavy chains. The panel recommended that these be written Glm(1), Glm(2), and so on, rather than Gm(1), Gm(2), and so on. Similarly, the panel recommended that the allotypes on the $\gamma 3$ chain be written G3m(5) and so on. The panel suggested also that the haplotypes be written as $Gm^{1,17;5,13,14}$ if Gm(23), which is on $\gamma 2$, has not been tested for, or $Gm^{1,17;..;5,13,14}$ if Gm(23) has been tested for but is not present, rather than $Gm^{1,5,13,14,17}$ for both cases as the WHO 1965 working group suggested. The nomenclature of the WHO 1965 working group is used in this chapter.

The Gm allotypes are carried by the $\gamma 1$, $\gamma 2$, and $\gamma 3$ heavy chains for IgG; no allotypes have been recognized on $\gamma 4$ (see Chap. 7 for a description of Ig). The Km(Inv) antigens are carried by the κ chains of Ig, and the A_2m antigens are carried by the α chains of IgA_2.

Table 1 List of Currently Recognized Gm and Km(Inv) Antigens

	Gm nomenclature	
Heavy chain subclass	Numeric	Alphanumeric
$\gamma 1$	1	a
	2	x
	3	f
	17	z
$\gamma 2$	23	n
$\gamma 3$	5	b1
	6	c3
	10	b5
	11	b0
	13	b3
	14	b4
	15	s
	16	t
	21	g
	24	c5
	26	u
	27	v
	28	L1
	Km(Inv) nomenclature	
Light chain subclass	Numeric	Alphanumeric
κ	1	1
	2	2
	3	3

All of the allotypes mentioned above are present in the constant domain(s) of the chains carrying them. The Hv(1) antigen (Wang et al., 1978) may prove to be an exception to the rule that the human allotypes are present in the constant regions of the chains that carry them. It was shown to occur in IgG, IgA, and IgM myeloma proteins (only two IgD and no IgE proteins were tested, hence it may occur in these also) at about equal frequencies (0.16 to 0.29); to be present in isolated heavy (H) chains, but not light (L) chains; to be present in the Fab fragment of IgG and IgM, but not in the Fc fragments of these molecules; and to be inherited as an autosomal dominant. These data, as Wang et al. (1978) state, are consistent with an allotype present in the variable (V) region of H chains chains. A problem with this interpretation resides in the absence of linkage between this marker and the Gm antigens (Pandey et al., 1981).

II. DETECTION

The antigen (Ag)-antibody (Ab) complex in all three systems is soluble; hence an agglutination inhibition method of detection is used. The Gm antigens carried by the $\gamma 1$ and

$\gamma 3$ chains of IgG are detected by coating group O, Rh_0 (D) cells with an IgG anti-D serum carrying the antigens of interest. The antigen on $\gamma 2$ of IgG [Gm(23)], and A2m antigens, and the Hv(1) antigen are detected by binding appropriate myeloma proteins (e.g., those carrying the antigens) to group O cells. The system of detection is the same from this point on in all cases and may be outlined as follows:

		Agglutination
Control 1:	SRBC[a] + antiallotype + saline	+
Control 2:	SRBC + antiallotype + known positive serum	-
Control 3:	SRBC + antiallotype + known negative serum	+
Control 4:	SRBC + saline + serum to be tested	- or +
Test:	SRBC + antiallotype + serum to be tested	+ or -

[a]Sensitized red blood cells (i.e., cells coated with an Ig).

Control 1 is designed to show that the RBC are adequately coated and that the antiallotype antiserum is active. Controls 2 and 3 are used to establish that the test is working. Control 4 indicates whether the serum to be tested has an anti-IgG antibody (agglutination occurs). If an Ab is present, it must be removed (by heating at 63°C for 10 min, which destroys IgM; or by dialyzing against cold distilled water, which precipitates IgM) before the serum can be tested. Controls 1 to 3 need be run only once for a set of tests for a given allotype. Control 4 must be run for each sample to be tested.

The test is negative if the cells are agglutinated (i.e., the serum lacks the allotype detected by the antiallotype antiserum); the test is positive if the cells are not agglutinated (i.e., the serum has the allotype detected by the anti-allotype antiserum).

The concentrations at which the reagents (cells, anti-D, antiallotype) are to be used and the dilution at which sera are to be tested (1/16 is used in my laboratory, because of convenience, to conserve the serum; and because this dilution will rarely yield false negative reactions) are determined empirically.

III. ORIGIN OF THE ANTI-GM AND ANTI-KM ANTIBODIES

The first anti-Gm antibodies (Ab) were obtained from patients with rheumatoid arthritis (RA) (Grubb, 1956). Ropartz et al., 1961 showed that antibodies from normal donors may be used to detect Gm and Inv(Km) antigens. The Ab from RA patients is invariably IgM, whereas those from normal donors, although almost always IgM, may on rare occasions be IgG (see Steinberg, 1967, for a review).

The antibody in healthy donors seems to be induced as an immunological response to Gm- or Inv-incompatible maternal IgG present in the fetus and the newborn, or to multiple transfusions or to deliberate immunization (reviewed in Steinberg, 1967).

As much as 5% of sera from donors who do not have rheumatoid arthritis may have anti-gammaglobulin antibodies. The proportion found depends on the incomplete antisera [almost invariably anti-D (anti-Rh_0)] used to detect the antibody, and hence different studies report different frequencies. The frequency is highest among children 6 months to 5 years of age (ca. 11% in our experience) and appears to remain constant thereafter at about 5% until old age. In one study, virtually all the antibodies (94%) found in the serum of children less than 5 years of age were specific (i.e., could be used to detect a Gm or Inv factor), whereas only 79% of the antibodies from children 5 to 16

years old and only 14% of those from adults were specific (reviewed in Steinberg, 1967). We have no explanation for these observations.

IV. INHERITANCE

A. Gm

The Gm antigens are inherited in sets called haplotypes, because each set has at least one antigen from those on $\gamma 1$ and at least one from those on $\gamma 3$. Since only one antigen [Gm(23), Table 1] is known on $\gamma 2$, a haplotype may or may not have a detectable representative of $\gamma 2$. No allotypes are known for $\gamma 4$. The haplotype arrays differ among the various races. Some examples are shown in Table 2 (the allotypes within a set of parentheses represent a haplotype). [See Steinberg and Cook (1981), for a detailed review and for the data on which Table 2 is based.] Relatively few populations have been tested for Gm(23), but these tests show that among Australian Aborigines, Caucasoids, and Mongoloids, Gm(23), when present, is almost invariably associated with Gm(5). Gm(23) does occur with Gm(1) in the absence of Gm(5), albeit rarely.

Gm(26), Gm(27), and Gm(28) were reported after Grubb's review was published in 1970. They have not been extensively studied. Gm(26) is almost invariably present when Gm(15) is absent, and vice versa, it is almost invariably absent when Gm(15) is present. However, among Jews from the Cochin region in India, 22% of the $Gm^{1,17,21}$ haplotypes did not have Gm(26) (Steinberg et al., 1980). The limited number of tests done with Gm(27) indicate that it is present (with one known exception) when Gm(24) is absent.

Gm(28) in Caucasoids and Mongoloids is, with rare exceptions, present when Gm(21) is present and is absent when Gm(21) is absent. This relationship does not hold among Negroids or Khoisans (Rivat et al., 1978); Gm(21) does not occur among Negroids of unmixed ancestry, nevertheless about 50% are Gm(28) positive. Gm(21) does occur among the Khoisan; however, Gm(28) was present in the absence of Gm(21) in about 38% of 40 samples (Rivat et al., 1978).

Examples of data from a Caucasoid and from a Mongoloid family, to show the transmission of haplotypes, are presented in Table 3. As indicated in the table, all samples were tested for Gm(1,2,3,5,6,10,11,13,14,17,21,24,26). The phenotypes of both parents in the Caucasoid family are the same and indicate that the parents are heterozygous for the haplotypes $Gm^{1,17,21,26}$ and $Gm^{3,5,10,11,13,14,26}$ (see Tables 2 and 3). This is confirmed by the last two sets of offspring. The four Gm(3,5,10,11,13,14,26) offspring are homozygous for the $Gm^{3,5,10,11,13,14,26}$ haplotype and the three Gm(1,17,21,26) offspring are homozygous for the $Gm^{1,17,21,26}$ haplotype.

The phenotype of the father of the Mongoloid family indicates that he is heterozygous for the haplotypes $Gm^{1,2,17,21,26}$ and $Gm^{1,10,11,13,17}$. The mother's phenotype indicates that she is heterozygous for $Gm^{1,17,21,26}$ and $Gm^{1,10,11,13,17}$. The latter haplotype shows Gm(15) and Gm(16) also, when tests are done for these allotypes (see Table 2). The phenotype of the first child confirms the presence of the $Gm^{1,10,11,13,17}$ haplotype in the mother. This child inherited the $Gm^{1,2,17,21,26}$ haplotype from its father [its mother does not have Gm(2)] and the $Gm^{1,10,11,13,17}$ haplotype and establishes that the haplotype does not carry Gm(26). More extensive tests in families and populations establish that Gm(26) is usually absent when Gm(15) is present. The last child confirms that each parent is heterozygous.

The reader will note in Table 2 that only Caucasoids and Ainu have a haplotype that lacks Gm(1); that Gm(3) does not occur among Negroids of unmixed ancestry; that

Table 2 Gm Haplotypes Commonly Present in Each of Several Races[a]

Race	Haplotypes
Caucasoid	(1,17,21,26) (1,2,17,21,26), (3,5,10,11,13,14,26)
Negroid	(1,5,10,11,13,14,17,26), (1,5,10,11,14,17,26), (1,5,6,11,17,24,26), (1,5,6,10,11,14,17,26)
Mongoloid	(1,17,21,26), (1,2,17,21,26), (1,10,11,13,17),[b] (1,3,5,10,11,13,14,26)
Ainu	(1,17,21,26), (1,2,17,21,26), (1,10,11,13,17),[b] (2,17,21,26)
Khoisan	
San (Bushmen)	(1,17,21,26), (1,10,11,13,17),[b] (1,5,10,11,13,14,17,26), (1,5,13,14,17,21,26),[c] (1,5,11,17,26)
Khoikhoi (Hottentots)	(1,2,17,21,26), (1,5,10,11,14,17,26), (1,10,11,13,17),[b] (1,5,10,11,13,14,17,26)
Pigmy	(1,5,6,11,17,24,26), (1,5,10,11,13,14,17,26)
Micronesian	(1,17,21,26), (1,3,5,10,11,13,14,26)
Melanesian	
New Guinea	(1,17,21,26), (1,2,17,21,26), (1,3,5,10,11,13,14,26), (1,5,10,11,13,14,17,26)
Bougainville	(1,17,21,26), (1,2,17,21,26), (1,3,5,10,11,13,14,26)

[a]When tested for Gm(1,2,3,5,6,10,11,13,14,17,21,24,26). The allotypes within a set of parentheses represent a haplotype.
[b]When tests for Gm(15) and Gm(16) are done, the haplotype is $Gm^{1,10,11,13,15,16,17}$ among Mongoloids and Ainu, but it may be also $Gm^{1,10,11,13,15,17}$ among the Khoisan.
[c]Not tested for Gm(10) and Gm(11). Gm(10) is probably present, but Gm(11) may not be.

Table 3 Example of Family Data from a Caucasoid and from a Mongoloid Family to Show the Pattern of Inheritance of Some Haplotypes[a]

	Phenotype	Probable genotype
	Caucasoid	
Father	1,3,5,10,11,13,14,17,21,26	1,17,21,26/3,5,10,11,13,14,26
Mother	1,3,5,10,11,13,14,17,21,26	1,17,21,26/3,5,10,11,13,14,26
Children (no.)		
(2)	1,3,5,10,11,13,14,17,21,26	1,17,21,26/3,5,10,11,23,14,26
(4)	3,5,10,11,13,14,26	3,5,10,11,13,14,26/3,5,10,11,13,14,26
(3)	1,17,21,26	1,17,21,26/1,17,21,26
	Mongoloid	
Father	1,2,10,11,13,14,17,21,26	1,2,17,21,26/1,10,11,13,17
Mother	1,10,11,13,17,21,26	1,17,21,26/1,10,11,13,17
Children (no.)		
(1)	1,2,10,11,13,17,21,26	1,2,17,21,26/1,10,11,13,17
(1)	1,10,11,13,17	1,10,11,13,17/1,10,11,13,17
(1)	1,2,17,21,26	1,17,21,26/1,2,17,21,26

[a]All samples were tested for Gm(1,2,3,5,6,10,11,13,14,17,21,24,26). Only the positive results are indicated.

Gm(6) and Gm(24) are confined to Negroids; that Gm(15) and Gm(16) are confined to Mongoloids, Ainu, and Khoisan (see the second footnote to Table 2); that Gm(3) and Gm(17) do not occur in the same haplotype (they involve a substitution at residue 214 in the $\gamma 1$ chain, see later); and that members of the pairs formed by Gm(6) and Gm(13), Gm(24) and Gm(10), and Gm(15) and Gm(26) are mutually exclusive. No amino acid sequence data, such as for Gm(3) and Gm(17), are available for the latter sets of mutual exclusions.

B. A_2m

The two known allotypes in the A_2m system are inherited as codominant alleles. There has been little testing for the $A_2m(1)$ allotype and even less for the $A_2m(2)$ allotype (reviewed in Steinberg and Cook, 1981). The available data indicate that the frequency of the A_2m^1 allele is $\geqslant 0.95$ among Caucasoids and Australian Aborigines. It is in general less frequent among Mongoloids; however, from published data the range is from as low as 0.125 (among 24 Chinese of unspecified origin) to as high as 0.988 among 40 Amerindians (also of unspecified origin) (see Steinberg and Cook, 1981). Its frequency is

<0.30 among Negroids. The A_2m locus is closely linked to the Gm loci and therefore an A_2m allele forms part of a haplotype containing Gm alleles. The A_2m portion of the haplotype may help to distinguish haplotypes from different races that are identical in their Gm portions. Thus $Gm^{1,5,10,11,13,17}$ is almost always $A_2m(1)$ in Melanesians in Papua, New Guinea, while it is almost always $A_2m(2)$ in Africans.

C. Km (Originally Inv)

The three Km allotypes (1,2,3) are codominant and are inherited via three alleles, Km^1, $Km^{1,2}$, and Km^3. The Km locus is not linked to the Gm locus (reviewed in Steinberg, 1967) and therefore presumably not to the A_2m locus. The Km^3 allele is most frequent in virtually all populations and the Km^1 allele is the least frequent (Steinberg and Cook, 1981). $Km^{1,2}$ ($Inv^{1,2}$) is more frequent among Negroids (30 to 40%) and Mongoloids (20 to 30%) than among Caucasoids (<20% and usually less than 10%). There is some evidence of weak clines of the $Km^{1,2}$ allele on a world scale (Fig. 1), and therefore of the Km^3 (Inv^3) allele on a world scale, but there is a remarkably strong cline among the Melanesians on Bougainville (Fig. 2).

Despite the clines, Km is not as useful as Gm for population studies, because the differences among races are quantitative for Km and not qualitative as for Gm (see Table 2).

D. Unusual Gm Phenotypes Requiring Unusual Haplotypes

Many unusual Gm phenotypes that require unusual haplotypes to explain them have been reported. Those that had been reported prior to 1977 are listed and discussed in the monograph by Steinberg and Cook (1981). Several others have been reported since then. I will discuss only those that provide information about the IgG molecules.

A healthy woman living and working on a large farm, the mother of three children at the time of the report and now the mother of six children, was found to have no $\gamma 1$ or $\gamma 3$ allotypes. This phenotype required the postulation of a Gm blank haplotype (Gm^-). The haplotype was found to be common among members of a large kindred, but this woman was the only one homozygous for it. It was subsequently shown that although the woman had a normal concentration of IgG, the composition of the IgG was unusual. There was a considerable increase in the concentration of G2 and a complete absence of normal $\gamma 1$ or $\gamma 3$ chains. They were replaced by a "hybrid" chain composed of the Fd portion of $\gamma 3$ and the Fc portion of $\gamma 1$ (Steinberg et al., 1970). The switch from $\gamma 3$ to $\gamma 1$ chain occurs in the Fc portion of the $\gamma 3$ chain; hence the hinge region is $\gamma 3$ (see Werner and Steinberg, 1974, for details). The molecule may have arisen as the result of an unequal crossover or of a deletion. In either case, the structure indicates that the sequence of the genes coding for the constant regions of these two heavy chains is $\gamma 3$, $\gamma 1$.

It is interesting to note that the donor is a healthy woman, despite the absence of normal $\gamma 1$ and $\gamma 3$ chains. Healthy individuals with no detectable G3 (hence no $\gamma 3$) (Steinberg et al., 1980 and earlier reports cited in that paper) or no detectable G1 (hence no $\gamma 1$) or no detectable IgG2 (hence no $\gamma 2$ chains) have been reported (van Loghem et al., 1980). It therefore appears that none of these chains is indispensable for health. The conclusion is that various different H chains are sufficiently similar in structure to permit them to substitute for each other in function. This raises the interesting but at present unanswerable question of why and how the four subclasses of IgG are maintained.

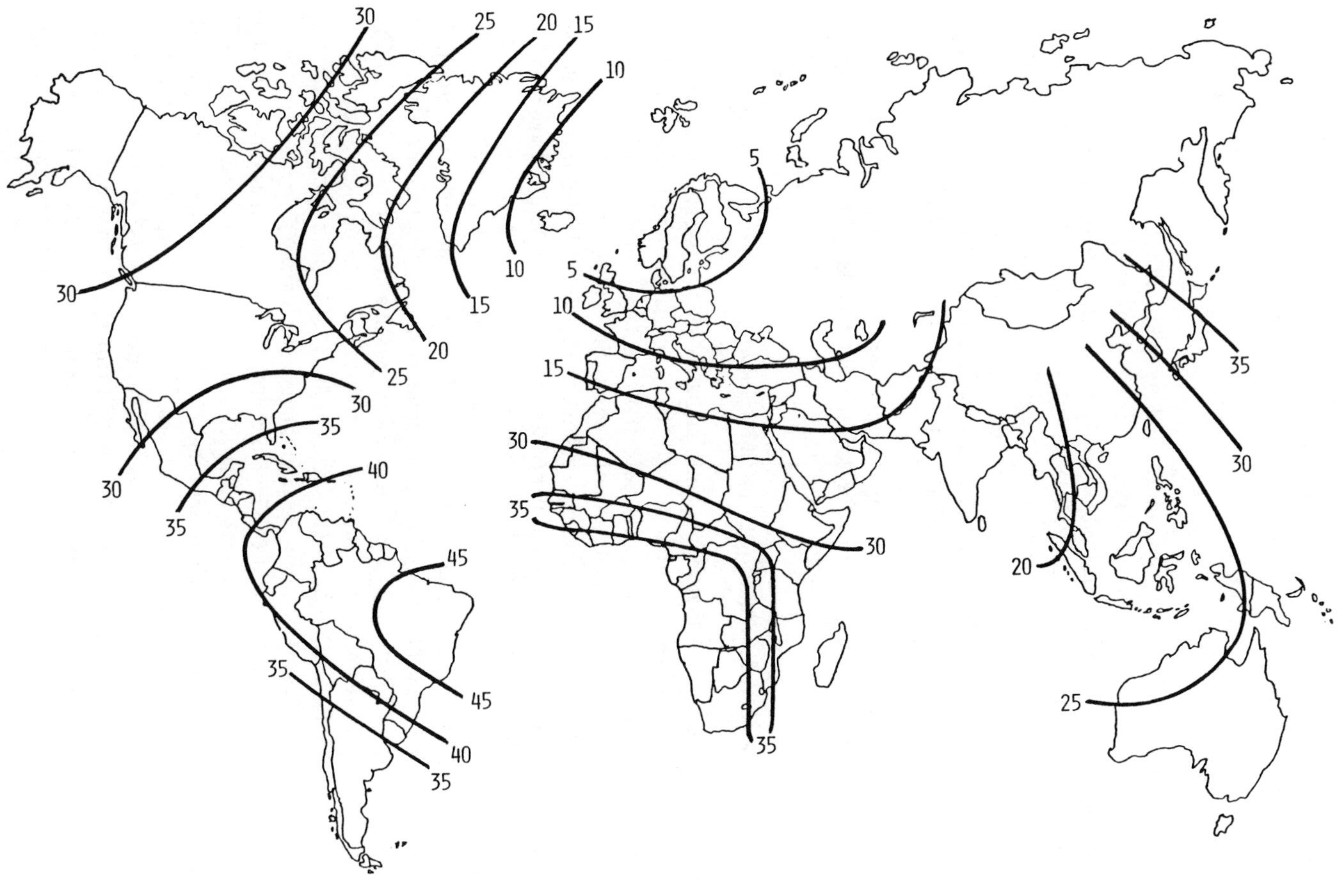

Figure 1 Distribution of $Km^{1,2}$ ($Inv^{1,2}$) in the world. (From Steinberg and Cook, 1981.)

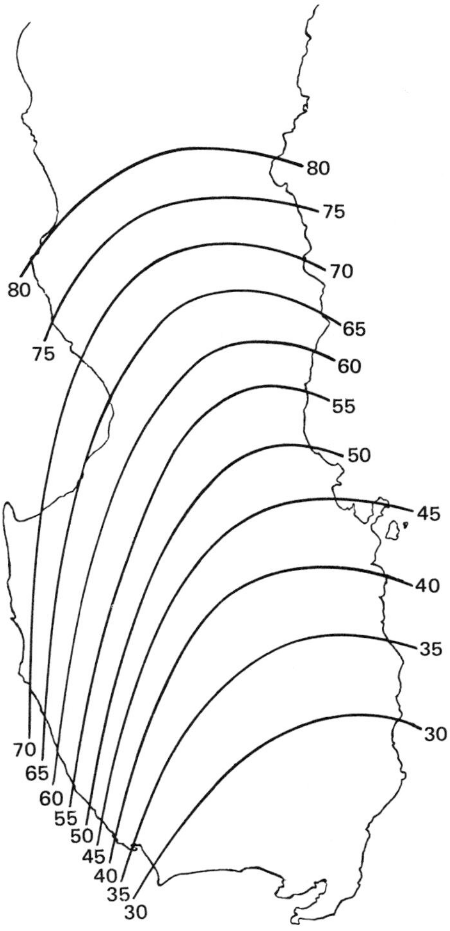

Figure 2 Distribution of $Km^{1,2}$($Inv^{1,2}$) on Bougainville, Solomon Islands. (From Steinberg and Cook, 1981.)

A $Gm^{1,3,17,21}$ haplotype (reviewed in Steinberg and Cook, 1981) has been shown to be most likely the result of a duplication of the gene controlling the constant region of the $\gamma 1$ chain (Natvig et al., 1981). Recall that Gm(3) and Gm(17) involve substitutions of the same amino acid residue. Surprisingly, this duplication has arisen on at least three independent occasions. No homozygote for the duplication has been reported. All of the more than 50 known heterozygotes for the duplication were healthy when observed. Matsumoto et al. (1981) have described a well-documented example of a probable duplication of the constant region of the $\gamma 3$ chain. The duplication probably arose in a parental gamete that gave rise to one of a pair of fraternal male twins. The Gm data are presented in Table 4. No evidence of chimerism was found in the analysis of 10 marker systems on red blood cells, nor by HLA typing or by chromosomal analysis.

Table 4 Gm and Am Phenotypes and Probable Genotypes of Family Mit.[a]

	Phenotype	Probable genotype
Father	1,2,17;..;10,11,13,15,16,21	1,2,17;21/1,17;10,11,13,15,16,21
Mother	1,3,17;23;5,10,11,13,14,15,16	3;23;5,10,11,13,14/1,17;10,11,13,15,16,21
Child 1	1,2,17;..;10,11,13,15,16,21	1,2,17;21/1,17;10,11,13,15,16
Twin 1	1,2,3,17;23;5,10,11,13,14,21	1,2,17;21/1,3;23;5,10,11,13,14
Twin 2	1,2,3,17;23;5,10,11,13,14,15,16,23	1,2,17;21/1,3;23;5,10,11,13,14/17;10,11,13,15,16

[a]The samples were tested for Gm(1,2,3,5,10,11,13,14,15,16,17,21,23).
Source: Matsumoto et al. (1981).

V. LINKAGE RELATIONS

The most striking linkage relation of the Gm antigens is the complete linkage disequilibrium of the allotypes within a haplotype. Consider the Gm(1) and Gm(5) allotypes in the haplotypes among Caucasoids (see Table 2). Gm(1) is on $\gamma 1$ and Gm(5) is on $\gamma 3$. The frequency of Gm(1) is about 0.3, and of Gm(5) about 0.7. They would be expected to occur in a single haplotype with a frequency of 0.21 if they were distributed at random with respect to each other. Nevertheless, there is no haplotype among Caucasoids that contains both Gm(1) and Gm(5). The antigens are not of recent origin, because they occur in all races (Negroids are completely homozygous for them) and antigens similar to, if not identical with, them have been found in baboons and in nonhuman anthropoids. No explanation for this complete linkage disequilibrium is available. As far as I know, the Gm system is the sole example of complete linkage disequilibrium.

Gedde-Dahl et al. (1975) have established that the Gm loci are linked to the Pi (protease inhibitor, also called $\alpha 1$ antitripsin inhibitor) locus. Their more recent data show that the recombination frequency in males is 0.19 with 95% confidence interval 0.14 to 0.27, and in females 0.26 with 95% confidence interval 0.20 to 0.34. The recombination frequencies in families in which the phase (coupling or repulsion) was known for the dihybrid parent were 0.20 for males and 0.28 for females (i.e., close to the values based on all the data). The recombination frequency among males heterozygous for the Z allele of the Pi locus (complete absence of active enzyme) is 0.11 in males (i.e., greatly reduced) and 0.26 among females (i.e., essentially no effect). Since these results were reproduced in two different sets of data, they are probably real and not due to accidents of sampling. No explanation to account for this interesting observation is presently available.

Croce et al. (1979) have presented convincing evidence that the loci determing the γ, μ, and α heavy chains of Ig are on chromosome 14. (Presumably, the δ and ϵ loci are also on chromosome 14, but no tests were done for these heavy chains.) Hence the Gm loci and the Pi locus are on chromosome 14. Unfortunately, none of the other markers known to be on chromosome 14 is polymorphic, so it is unlikely that confirmation of the report by Croce et al. (1979) will be provided in the near future by studies of recombination based on family data.*

VI. RELATION TO Ig STRUCTURE

A. Gm

The Gm antigens are carried by the constant domains (CH) of the heavy chains of IgG (see Table 1) and they have been, to a considerable extent, localized to specific domains (Table 5) (reviewed in Natvig and Kunkel, 1973). The allotypes in the CH_1 and CH_3 domains are detectable in the Fab and pFc′ fragments, respectively (see Chap. 7). The allotypes indicated as being in the CH_2 domain have been placed there because they were not detected in either of the other two domains. While the assignments to CH_2 are

*Early work, based on family studies, showed that the Gm and Km(Inv) loci are at least 30 centimorgans apart and probably not linked. Recent chromosome studies have located the kappa light chain genes (i.e., V and C) on the short arm of chromosome 2, thus confirming the absence of linkage between Gm and Km. The lambda light chain genes have been located on the long arm of chromosome 22.

Table 5 Location of the Gm Allotypes in the CH Domains

Subclass and domain	Allotypes
$\gamma 1$	
CH_1	3,17
CH_3	1,2
$\gamma 2$ CH_2	23
$\gamma 3$	
CH_2	5,15,16,21
CH_3	6,10,11,13,14,24,26,27,28

probably correct, the fact that some allotypes in the CH_3 domain have been shown to depend on the tertiary structure of the molecule indicates that the assignments need further testing. Thus Gm(1) which is known to be due to substitutions at residues 356 and 358 cannot be detected in a CH_3 domain that lacks the 14 C-terminal amino acid residues, 433 through 466 (reviewed in Natvig and Kunkel, 1973).

The amino acid residues associated with each of the four allotypes carried by the $\gamma 1$ chain are known (Table 6), but none of the allotypes on the $\gamma 3$ chains has yet been associated with specific amino acid substitutions. The amino acid sequence of the CH_2 and CH_3 domains of the three $\gamma 3$ proteins and residues 293 to 301 of two others have been reported. Residues that differ among them are listed in Table 7. The sequences or amino acid contents of the C-terminal octadecapeptide from two additional Gm(5) and three additional Gm(21) $\gamma 3$ proteins are known (Werner and Steinberg, 1974). Residues 435 and 436 are Arg-Phe in the $\gamma 3$ Gm(5) proteins and Arg-Tyr in the $\gamma 3$ GM(21) proteins. They are His-Tyr in $\gamma 1$, $\gamma 2$, and $\gamma 4$ proteins.

The heavy chain disease proteins Wis and Zuc are each deficient for a large portion of the V region and for the CH_1 homology region. They have the same Gm phenotype, yet they seem to differ in residue 384 (Table 7), but this remains to be proven. The difference at residue 384 may be due to the abnormality of these proteins or to technical errors in sequencing.

It is of interest to compare protein Gol (also an abnormal protein) with proteins Wis and Zuc. The probable phenotype of the latter two proteins is Gm(5,10,11,13,14,26,27).

Table 6 Amino Acid Residues Associated with the Allotypes on the $\gamma 1$ Chain

	Residues		
Gm Allotype	Number(s)	+	–
1	356 and 358	Asp-Leu	Glu-Met
2	431	Gly	Ala
3	214	Arg	Lys
17	214	Lys	Arg

Table 7 Residues in the CH_2 and CH_3 Domains of γ3 That Differ Among Proteins That Have Been Sequenced

		Residues					
Protein	Gm Phenotype	296	339	384	392	435	436
Wis	(5,10,11,13,14,26)[a]	Phe	Thr	Ser	Asn	Arg	Phe
Zuc	(5,13,14)[b]	Phe	Thr	Asx?[c]	Asn?	Arg	Phe
Gol	(10,11,13,15,16,27)	Tyr	Ala	?[c]	?	His	Tyr
?[c]	(5)[d]	Phe	?	?	?	?	?
?	(21)[d]	Tyr	?	?	?	?	?
Her	(21)[d]	?	Thr	Asx	Asx	Arg	Tyr

[a]Owing to a typographic error in the original publication, Gm(14) was omitted from the phenotype. The protein is also Gm(26) (A.G. Steinberg, unpublished data, 1980). It has not been tested for Gm(27,28).
[b]The Gm data in the table are those published. The protein is Gm(5,10,11,13,14,26) (A.G. Steinberg, unpublished data, 1980).
[c]? = uncertainty if an amino acid is shown, or not reported if none is shown.
[d]This is all the information that was provided. It is highly probable that the phenotype of the CH_3 domain is Gm(26,27).

The phenotype of Gol is Gm(10,11,13,15,16,27). Hence proteins Wis and Zuc differ from Gol in five allotypes [i.e., Gm(5,14,15,16,26]. Gm(5,15,16) have been assigned to the CH_2 domain (Table 5) because they are not detected in the isolated pFc′ fragment which is essentially the CH_3 domain and because they are not in the CH_1 domain. Gm(14) and Gm(26) are in the CH_3 domain. Despite the three differences in allotypes in the CH_2 domain, only two residues (296 and 339) in this domain differ in the published sequences. Two differences may also be noted in the CH_3 domain (residues 435 and 436), possibly corresponding to the two allotype differences indicated for this region. The amino acids at residues 435 and 436 in the γ3 chain are Arg-Phe when the phenotype of the CH_3 domain is Gm(26,27), His-Tyr when the phenotype is Gm(27), and Arg-Tyr when the phenotype is Gm(26,27). The reader will have noted that Gm(27) is present regardless of the amino acids at residues 435 and 436; hence it is probably not associated with changes in this region of the γ3 chain. More differences may be present, because residues 384 and 392 in Gol are not known.

B. Inv (Km)

It was shown in 1966 that Inv(1,2) is associated with a leucyl residue at position 191 of the κ light chains of Ig and that Inv(3) is associated with a valyl residue at position 191. These findings raised the interesting questions of how a single amino acid substitution can lead to the replacement of two antigens [Inv(1,2)] with a third [Inv(3)] (or vice verse), and of what is the amino acid substitution or substitutions associated with the Inv(1) phenotype. These questions went unanswered until 1974 because of the infrequency of the *Inv*[1] allele (less than 0.005 among Caucasoids) and therefore of appropriate myeloma or Bence Jones proteins. When such a protein was discovered it was shown that Inv(1) is associated with Val at position 153 and Leu at 191 (Steinberg et al., 1974). These results and those associated with the Kern and Oz isotypes in the λ chain are presented in Table 8 (see Rivat et al., 1971, for population studies and a review of Kern and Oz).

Table 8 Amino Acid Residues Substitutions Associated with the Inv(Km) Allotypes in the κ Chain and the Kern (k) and $0_z(0)$ Isotypes in the λ Chain

κ Chain					λ Chain			
Allotypes			Residues		Isotypes		Residues	
1	2	3	153	191	K	0	154	191
+	+	–	Ala	Leu	+	–	Gly	Arg
–	–	+	Ala	Val	–	+	Ser	Lys
+	–	–	Val	Leu	–	–	Ser	Arg
	?		Val	Val	?		Gly	Lys

It has been hypothesized that the κ and λ chains evolved from a common ancestral chain (see Putnam et al., 1967, for a review; see also Chap. 10). The parallelism between the location of the changes in the constant regions of the κ and λ chains is consistent with this hypothesis. Crystallographic models of λ light chains show that residues 154 and 191 are less than 10 Å apart (reviewed in Steinberg et al., 1974). It seems likely that the κ-light-chain tertiary structure is similar to that of the λ chain. The data in Table 8 in conjunction with the crystallographic data indicate that the anti-Inv(2) antiserum recognizes a complex involving positions 153 and 191 and that the antigenic site is therefore dependent on tertiary folding.

It seems that the anti-Inv(2) reagent is sterically hindered and cannot recognize Leu at 191 when Ala in position 153 is replaced by the bulkier amino acid Val. The anti-Inv(1) antibody probably does not encompass position 153 and therefore does not distinguish the Val/Ala substitution at this residue. This could explain a number of observations. Inv(2) antigenicity is much more sensitive to manipulation of the IgG molecule than is Inv(1) or Inv(3). Treatment of IgG with succinic anhydride leads to loss of activity of Inv(1) and Inv(2) and also Gm(1), Gm(2), Gm(3), Gm(5), Gm(8), Gm(17), and Gm(21), but not of Gm(11) or Gm(13). Inv(2) activity is lost at a lower concentration of succinic anhydride than that of any of the other antigens listed above and it takes a 12-fold greater concentration of anhydride to abolish Inv(1) antigenicity than to abolish Inv(2) activity. Similarly, it has been shown that digestion of IgG with papain in the presence of cysteine completely destroys Inv(2) but not Inv(1) activity in the Fab fragment. It appears from these experimences that the configuration of the portion of the molecule associated with Inv(2) expression is easily distorted.

Peptic cleavage of κ Bence Jones proteins into the variant region (V_L) and the constant region (C_L) regions of the molecule leads to loss of Inv antigenicity, despite the presence of intact amino acid sequences in the V_c region. Hence the Inv antigens in Bence Jones proteins are dependent on the tertiary structure of the molecule. On the other hand, Inv antigens in IgG molecules depend on the quaternary structure of the molecule—that is, an interaction between the heavy and light chains (see Steinberg and Rostenberg, 1969, for a review). This finding agrees with the crystallographic model presented by Poljak et al., (1974), in which the C_L and CH_1 homology regions are very close, forming the C domain of Fab. It is not known whether Inv antigenicity in intact IgG molecules involves an interaction between the variant and constant regions of the light chains as well as an interaction with the heavy chains. It should be noted that the molarity at which the Inv(1), Inv(2), and Inv(3) antigens may be detected in light chains

isolated from IgG (normal or myeloma) is much less than that of the parent immunoglobulin, while that for Bence Jones proteins is the same as for the intact IgG molecule (Steinberg and Rostenberg, 1969). There is no explanation for the implication that the light chains of Bence Jones origin have a different tertiary structure from that for light chains isolated from IgG molecules by reduction and alkylation. Reduction and alkylation of Bence Jones proteins does not affect the concentration at which the Inv antigens may be detected.

The Inv antigenicity of homologous Bence Jones proteins is much greater than that of intact molecules of the IgG2 and IgG4 subclasses (Steinberg and Rostenberg, 1969). Apparently, folding of the light chain near residue 191 in IgG2 and IgG4 molecules differs from that in IgG1 and IgG3 molecules. Perhaps the region around 191 is less exposed in the presence of $\gamma 2$ and $\gamma 4$ heavy chains.

C. A2m

The A2m antigens A2m(1) and A2m(2) are carried by the constant domain of the $\alpha 2$ heavy chain. Putnam and his colleagues have reported the sequences of the constant region of the H chains of A2m(1) and A2m(2) myeloma proteins and of an $\alpha 1$ chain (see Tsuzukida et al., 1979, and earlier reports cited therein). Tsuzukida et al. compared the sequences of these three chains. The results of the comparisons, based on their Fig. 1, are shown in Table 9. Putnam and his colleagues concluded that residues 428, 458, and 467 are on the surface of the folded molecule and that these account for the nA2m(2) (none 2) isoallotype reported by others. (An *isoallotype* is an antigen that is an isotype on at least one subclass chain and an allotype on at least one other chain.) Tsuzukida et al. (1979) postulate that the substitutions at residues 212 and 221, combined with the presence of GlcN (N-acetylglucosamine) at residue 211 in A2m(2) but not in A2m(1), account for the allotype difference [i.e., A2m(1) versus A2m(2)].

The H chain in A2m(2) proteins is covalently bound to the L chain via Cys at residue 220, but the H chain in A2m(1) proteins is not covalently bound to the L chain even though Cys is present at residue 220 in A2m(1). It is surprising that the A2m(1) α chain is not covalently bound to the L chain because its sequence is identical with the $\alpha 1$ chain from residues 199 to 222, whereas the A2m(2) α chain differs from the $\alpha 1$ chain at residues 212 and 221 (Pro-Ser and Pro-Arg). The molecules have different configurations in this region and some portion of this difference must be due to the presence of a disulfide bond between the H and L chains in A2m(2) and not in A2m(1). This leads me to wonder if the A2m(2) allotype is detectable in the isolated H chain or if it requires the quaternary structure for its expression, as do Gm(3), Gm(17), and the Inv(Km) antigens.

Mårtensson (and subsequently many others) reported that myeloma proteins from heterozygotes showed allotypes from only one chromosome of the pair (reviewed in Grubb, 1970). For example, an IgG myeloma protein from Caucasoids heterozygous for haplotypes $Gm^{1,17,21,26}$ and $Gm^{1,5,10,11,13,14,26}$ may be Gm(1,17), Gm(21,26), Gm(3), Gm(5,10,11,13,14,26), or none of these. Gm(1,17) or Gm(3) is expressed when the H chain is $\gamma 1$; Gm(21,26) or Gm(5,10,11,13,14,26) is expressed when the H chain is $\gamma 3$ (see Table 1); none of these allotypes is expressed when the H chain is $\gamma 2$ or $\gamma 4$. [Note: Gm(23) might be expressed when the H chain is $\gamma 2$.] This knowledge makes it possible to determine the subclass of an IgG myeloma protein by testing for Gm(1,3,5,21). The first two allotypes are on $\gamma 1$, the latter two are on $\gamma 3$. (These allotypes have been selected because antibodies against these are readily available. Various other combinations

Table 9 List of the Amino Acid Residues That Differ Among the α1 Chain and α2 Chain Chains of the A2m(1) and A2m(2) Allotypes[a]

Homology region	Residue no.	α1	α2 A2m(1)	α2 A2m(2)
CH1	133	Cys	Asp	Asp
	136	Gln	Pro	Pro
	137	Pro	Gln	Gln
	143	Ile	Val	Val
	166	Gly	Asn	Asn
	197	Leu	Pro	Pro
	198	Ala	Asp	Asp
	212	Pro	<u>Pro</u>	<u>Ser</u>
	221	Pro	<u>Pro</u>	<u>Ser</u>
CH2	277	Val	Ala	Ala
	319	Lys	Glu	Glu
	327	Thr	His	His
	330	Ser	Leu	Leu
	337	Thr	Asn	Asn
	338	Leu	Ile	Ile
	339	Ser	Thr	Thr
CH3	411	Phe	<u>Phe</u>	<u>Tyr</u>
	428	Asp	<u>Asp</u>	<u>Glu</u>
	458	Val	<u>Val</u>	<u>Ile</u>
	467	Val	<u>Val</u>	Ala

[a]The residues that differ between the allotypes are underlined (see 212, 221, 411, 428, 458, and 467).

of allotypes may be used.) If the protein is positive for Gm(1) or Gm(3), the H chain is γ1; if the protein is positive for Gm(5) or Gm(21), the H chain is γ3; if the protein is negative for all four allotypes, the H chain is γ2 or γ4. The latter may be distinguished by testing for the isoallotype nGm(1) (i.e., none 1). Anti-nGm(1) reacts with G1 proteins that are Gm(-1) and with all G2 and G3 proteins, but not with G4 proteins. Hence a positive test with an IgG protein that is negative for Gm(1,3,5,21) indicates that the H chain of the protein is γ2; a negative test indicates that the H chain of the protein is γ4.

The phenomenon discovered by Mårtensson has been labeled *allelic exclusion*. We now know that a myeloma protein is composed of a single molecular type; the H chain therefore belongs to one subclass. The allotypes detected in a myeloma protein are those on the H chain expressed in the protein. We know also that the H chain synthesized by a plasma cell is determined by the association of the DNA of a V region (and other regions) with the DNA of one of the nine C_H regions (see Chap. 14). I suggested in 1969 (Steinberg et al., 1970) that the phenomenon be called *specific gene activation*. The data at the molecular level show that name is a more accurate description of what happens. However, *allelic exclusion* is used by all investigators and will continue to be used even though the implication it carries concerning the molecular nature of the phenomenon is incorrect. The phenomenon of allelic exclusion is not limited to myeloma cells. Fluorescent antibody staining of normal plasma cells has shown that they also produce only one kind of H chain and one kind of L chain.

VII. RELATION TO DISEASE

The distribution of the allotypes and haplotypes among the human races and populations raises the question of what selective force might be involved. I wrote in 1967: "Since the antigens are associated with antibody-forming molecules, it is tempting to think that molecules with different Gm and Inv types differ in their capacities to form effective antibodies against various pathogens . . ." (Steinberg, 1967). Since then several investigations of the possible association of the allotypes and haplotypes with disease have been undertaken. A review of some of these studies follows. (References will be omitted, because of a restriction on the number of references that may be cited. They will be made available on request.)

A. Response to Flagellin

The response of 113 Caucasoids to flagellin from *Salmonella adelaide* in relation to the Gm and Inv types was determined by investigators who claim to have found an association between the magnitude of response to flagellin and the $Gm^{1,21}$ haplotype, but their analysis leaves room for doubt. I found on reanalysis of their data that contrary to the authors' claim, the frequency of Gm(1,21) among high primary responders is not significantly different from that among low primary responders.

B. Response to O and H Antigens of *Salmonella typhi*

The association between each of four polymorphisms (Hp, Gc, ABO, and Gm) and the ability to respond to vaccination with the O and H antigens of *S. typhi* was investigated. An association between response on the one hand and Gc, Gm, and possibly ABO on the other was indicated. As a check on the observation, the frequencies of the polymorphisms among patients with typhoid fever were examined and compared with the values found among controls. Significant differences were found for all four polymorphisms! Clearly, the investigation needs to be repeated to establish its validity. Association with three out of four tested polymorphisms is a surprising finding.

C. Malignant Melanoma

German investigators studied the relation between melanoma and ABO, Rh, Hp, Gc, Gm, and Inv(Km) polymorphisms. In part, this was a repetition of an earlier study concerning the ABO blood groups. The new data for ABO did not confirm an earlier finding of an association between group O and melanoma, nor did they show any association between Rh, Hp, Gc, or Inv and melanoma. They did show an astonishing excess of samples with the Gm(1,2) phenotype. [Only Gm(1) and Gm(2) were tested.] Among the 164 patients, 55.5% were Gm(1,2) whereas only 21.9% of the 531 controls were Gm(1,2). Furthermore, Gm(1,2) was the most frequent phenotype among the patients whereas it was the least frequent among the controls.

The relation between melanoma and Gm(1,2) was subsequently examined by other investigators. In one study samples from 71 patients with melanoma and from 400 controls were tested for Gm(1) and Gm(2). The investigators found 15.5% of the patients and 16.3% of the controls to be Gm(1,2) (i.e., no difference between the two groups). In another study the investigators did not find an excess of Gm(1,2) among melanoma patients (11% among 185 patients and 11% among 295 controls). They found that the concentration of IgG4 was more frequently abnormal (either higher or lower than the control value) among patients with more advanced disease. I doubt that this has any significance.

Still another group of investigators examined the association between melanoma and the ABO, MNSs, Rh, P, K, Fy, and Jk blood groups, the aP, PGM, EsD, and 6PGD red blood cell enzymes, and the Gm and Inv allotypes. No association with melanoma was found for any of the polymorphisms except for an increased frequency of blood group O and of Gm(5). However, neither of the latter two comparisons was corrected for the fact that a total of 13 polymorphisms had been examined. In addition, the frequency of the Gm^5 haplotype among the patients (0.715) is not significantly different from that among the controls (0.697). I believe that it is reasonable to conclude that no association has been demonstrated.

The most recent investigation of the possible association between melanoma and Gm, of which I am aware, was reported by Pandey et al. (1981). The authors report an increased frequency of Gm(2) (30%) in the phenotypes of patients compared to that in the phenotypes of controls (18%). I have analyzed the samples for goodness of fit to the Hardy-Weinberg (HW) distribution. The fit of the 89 samples from the patients to the HW distribution is satisfactory ($\chi^2_{(2)} = 0.806$; P = 0.668), but that of the 906 controls is not ($\chi^2_{(2)} = 6.649$; P = 0.036). This casts doubt on the validity of the comparison between the data for the samples from the controls and those for the patients.

In summary, it does not seem likely that there is an association between melanoma and any of the polymorphisms investigated thus far.

D. Malaria

Gm allotypes were investigated among five populations in Sardinia. Four of the populations were from coastal regions, the fifth was from a mountainous region. The samples were examined for Gm(1,2,3,5,6,10,11,13,15,16,17,21,23,24). The data for the four lowland populations were homogeneous, but they were significantly different from those for the mountain population. The difference seems to be due to the significantly greater frequency of Gm(23) among the lowland populations (0.655) compared to that among the mountain population (0.518). The authors suggest that this difference may indicate selection for Gm(23) in the lowland area, where malaria had been endemic until relatively recently. Investigators studying populations in Lebanon also found a difference in Gm(23) frequencies between people residing in lowland areas and those residing in highland areas. Once again the difference is ascribed to the possible selective pressure of endemic malaria. The attribution may be correct, but further tests are required.

E. Response to Tetanus Toxoid

It has been reported, in an abstract, that the frequency of $Gm^{3,5}$ is significantly elevated among low responders to immunization with tetanus toxoid as contrasted to the frequency of $Gm^{3,5}$ among high responders. Another group compared the level of antibody in infants before and after immunization with diphtheria-pertussis-tetanus vaccine, *Haemophilus influenzae* type B polyribose phosphate vaccine, and with meningococcal groups A and C polysaccharide vaccines. At most only 20 white infants and 33 black infants were involved for each of these. The data are incompletely presented and not easily examined. No association between Gm and response to tetanus was observed. The authors claim that a significant association exists between Inv(1) and immune response to *H. influenzae* and to meningococcus C polysaccharides in white infants, but not in black infants. It seems unlikely that these data are biologically meaningful.

A third group compared the level of response to tetanus toxoid among 85 Papuans with their Gm phenotypes. Five of the 85 had the phenotype Gm(1,3,5,11,13,17,21) and these had a higher average response than did the 76 with the phenotype Gm(1,17,21) and the four with the phenotype Gm(1,5,11,13,17,21). Here the presence of Gm(3) and Gm(5) is associated with a high level of response, while it was reported in the abstract that $Gm^{3,5}$ in Caucasoids is associated with a low level of response. At best the data are suggestive, but they do not indicate strong associations, if any.

F. Autoimmune Diseases

Gm(2) was reported, in 1980, to be significantly more frequent among patients with Graves' disease, myasthemia gravis, and Hashimoto's disease. An earlier report (1977) by another group found Gm(3,5) to be significantly more frequent among patients with Graves' disease. A report published in 1981 by some of those associated with the 1980 report no longer mentions an association with Gm(2) but now claims, on the basis of family studies, linkage between a disease susceptibility gene and the Gm loci. At best one may say that the data are suggestive and indicate that more work is required.

There have been single reports concerning associations of Gm with neuroblastoma, with polyarthritis caused by Ross River virus infection, and with myasthemia gravis. Suffice it to say that significant associations were reported in each case, that none is strong, and that each requires confirmation.

An association between liver cirrhosis, negative for hepatitis B surface (HBS) antigens, and Gm phenotypes showing heterozygosity has been reported. Once again, without going into details, it is necessary to indicate that there are flaws in the data.

I have written at such length about the allotypes and disease because I wanted to indicate the paucity and the weakness of the data in this area. I am led also to question the value of further efforts seeking such associations. Weak associations are not likely to lead to biological insights or to applications to medicine, and it seems unlikely that strong associations will be found.

G. Allotypes in Other Animals

Ig allotypes have been found in chickens (Foppoli et al., 1979), mice (Herzenberg and Herzenberg, 1978), rats (Gutman and Weisman, 1971), rabbits (Kindt, 1975), cattle (Wegrzyn and Wegrzyn, 1978), and pigs (Rasmusen, 1965). The reader is referred to these references for details, because they are not pertinent to this chapter. It remains only to mention that there are extensive cross-reactions between reagents against the human allotypes and antigens in nonhuman primates and in baboons. (See Steinberg et al., 1977, for data on baboons and references to papers dealing with nonhuman primates.)

VIII. APPLICATIONS

A. Transfusion and Transplantation

Incompatibility for the Gm and Inv allotypes is of no significance in blood transfusion or transplantation. No reactions occur when incompatible serum is transfused, even when the recipient has an antibody against the incompatible allotype. This is not true, however, for the A2m allotypes. The first antibodies were found in patients who had transfusion reactions which were eventually shown to be due to the formation of anti-IgA and

anti-A2m antibodies, the recipients being IgA deficient. This is not a common phenomenon, but neither is it vanishingly rare.

The primary uses of the allotypes in transfusion and in bone marrow transplants are to detect the presence of donor serum in transfusion and to demonstrate that the donor's marrow is functioning in the host in the case of transplantation. As little as 0.03 mg/ml of the IgG1 or IgG3 subclasses containing Gm allotypes may be easily detected. Hence dilution of the donor's serum or low activity of the donor's marrow are not barriers to detecting the presence of the donor's serum or the activity of the donor's marrow.

B. Medicolegal Uses

Paternity Tests

The same principles apply to the use of the allotypes in paternity tests as in the case of any other polymorphism. The accused male is excluded if he does not share with the child allotypes not present in the mother or if the child does not have a haplotype that the male must transmit (i.e., a haplotype for which the male is homozygous). Caution is necessary because unusual haplotypes do occur. The Gm^{-} haplotype and the $Gm^{1,17,21}$ haplotype without Gm(26) have been mentioned. The Gm phenotypes of the immediate family of the individual homozygous for the Gm^{-} haplotype and of a family that may have a $Gm^{1,13,15,16,17,26}$ haplotype are shown in Table 10.

Table 10 Gm Phenotypes and Probable Genotypes for Two Families with Unusual Haplotypes

A. *Gm* haplotype

	Gm phenotype[a]	Probable genotype
Father	1,21	1,21/–
Mother	3,5,13,14	3,5,13,14/–
Children		
3[b]	3,5,13,14	3,5,13,14/–
1	1,21	1,21/–
1	–	–/–

[a]All samples were tested for Gm(1,2,3,5,6,13,14,21).
[b]Number of children with the indicated phenotype.

B. Exceptional Gm(26) transmission

	Gm phenotypes[a]	Probable genotype[b]
Father	1,2,17,21,26	1,2,17,21,26/1,2,17,21,26, or 1,17,21,26, or 1,17,26
Mother	1,13,15,16,17,21,26	1,13,15,16,17/1,17,21,26
Children		
1	1,13,15,16,17,26	?
1	1,2,17,21,26	1,2,17,21,26/1,17,21,26

[a]All samples were tested for Gm(1,2,3,5,6,13,14,15,16,17,21,26).
[b]See the text for further details.

If the parents and one or all three children with the Gm(3,5,13,14) phenotype in the family with the Gm^- haplotype (Table 10A), were examined for paternity exclusion, the male would, in the ordinary course, be excluded. His phenotype indicates that he is homozygous for the $Gm^{1,21}$ haplotype and none of these three children has Gm(1) or Gm(21). Interestingly, the Gm(1,21) child would ordinarily not be accepted as the offspring of the woman who appears to be homozygous for the $Gm^{3,5,13,14}$ haplotype. The Gm(-) child establishes the presence of a Gm^- haplotype.

The data for the family in Table 10B would indicate, on face value, that the father is homozygous for Gm(21) and therefore that the first child is extramarital since it does not have Gm(21). The child's phenotype suggests homozygosity for $Gm^{1,13,15,16,17}$ except for the presence of Gm(26), which is usually (invariably?) absent from a haplotype containing Gm(15). The presence of Gm(26) in the child's phenotype suggests that the father is heterozygous for an unusual haplotype, possibly $Gm^{1,17,26}$.

In summary, exclusions based on this system, or any others for that matter, should be arrived at with great caution.

Identification

IgG is remarkably stable and so are the Gm allotypes. Therefore, the Gm allotypes of IgG in dried blood or any other body fluids containing IgG, such as semen, may be determined after dissolving the IgG in saline. The Gm type of blood stains stored at room temperature for as long as 3 years have been successfully determined.

IX. CONCLUSION

The study of the Gm and Inv allotypes has provided the clearest example, as far as I am aware, of the dependence of an antigen on the quaternary structure of a molecule [e.g., Gm(3)] and that an antigen may be dependent on amino acid substitutions almost 40 residues apart [Inv(Km)(2)]. Studies of the origin of the anti-Gm and anti-Inv antbodies in normal donors showed for the first time that the fetus may make antibodies against antigens from its mother. The unique qualitative difference of haplotype arrays among races has provided a powrful tool for population studies. The tenacious survival of IgG and the Gm allotypes in dried samples has made them useful in forensic medicine.

ADDENDUM

Two interesting associations have been reported since this chapter was written in 1981. The first concerns the association between the Gm phenotype and the ability to form antibodies against osteosarcaoma-associated antigens (OSAA). Nineteen of 22 patients (86%) with the phenotype Gm(1,3,5,13,14,17,21) formed antibodies against OSAA whereas only 1 of 18 patients (5%) with the phenotype Gm(3,5,13,14) did so. (No phenotype was found to be associated with susceptibility to osteosarcoma.) In the light of the review presented in an earlier portion of this chapter, it is clear that these findings require confirmation.

The second concerns the association between the immunoglobulin loci (but not the allotypes) and translocations present in tumor cells. It was reported in the mid-1970s that tumor cells from 30 of 33 patients with Burkitt lymphoma (BL) had a translocation in which a portion of the long arm of chromosome 8 was translocated to the long arm of chromosome 14. The translocation was shown in 1981 to be reciprocal; that is, a portion

of the long arm of chromosome 14 is exchanged with a portion of the long arm of 8. Chromosome 8 is broken at band q24 and 14 is broken at band q23. The point of interest in the present context is that the heavy chain loci are associated with band q23 of chromosome 14. The malignant cells of some patients with BL have translocations involving chromosomes 2 and 8 or 8 and 22. The break in chromosome 8 is always at band q24, the break in chromosome 2 is at band p12, and that in chromosome 22 is at band q11. The κ chain locus is at or near band 2p12 and the λ chain loci are at band 22q11.

Lenoir et al. (1982) have reported that the Ig from cultures of BL cells with a 2:8 translocation has κ chains, while the Ig from cells with an 8:2 translocation has λ chains. The Ig from cells with an 8:14 translocation has κ or λ chains. See Lenoir et al. (1982) for references to the earlier literature.

REFERENCES

Croce, C. M., Shander, M., Martinis, J., Cicurel, L., D'Ancona, G. G., Dolby, T. W., and Koprowski, H. (1979). Proc. Natl. Acad. Sci. USA *76*, 3416.

Foppoli, J. M., Ch'ng, L.-K., Benedict, A. A., Ivanyi, J., Derka, J., and Wakeland, E. K. (1979). Immunogenetics *8*, 384.

Gedde-Dahl, Jr., T., Cook, P. J. L., Fagerhol, M. K., and Pierce, J. A. (1975). Ann. Hum. Genet. *39*, 43.

Grubb, R. (156). Acta Pathol. Microb. Scand. *39*, 195.

Grubb, R. (1970). *The Genetic Markers of Human Immunoglobulins*. Springer-Verlag, New York.

Gutman, G. A., and Weissman, I. L. (1971). J. Immunol. *170*, 1330.

Herzenberg, L. A., and Herzenberg, L. A. (1978). In *Handbook of Experimental Immunology*, (D. M. Weir, Ed.). Blackwell, Oxford, Chap. 12.

Kindt. T. J. (1975). Adv. Immunol. *21*, 35.

Lenoir, G. M., Preud'homme, L. L., Bernheim, A., and Berger, R. (1982). Nature (Lond.) *298*, 474.

Matsumoto, H., Miyazaki, T., Nakao, Y., Tsumi, K., Fujita, H., Ohta, M., Oda, N., and Abe, K. (1981). Jap. J. Hum. Genet. *26*, 217.

Natvig, J. B., and Kunkel, H. G. (1973). Adv. Immunol. *16*, 1.

Natvig, J. B., Michaelsen, T. E., and Kunkel, H. G. (1971). J. Exp. Med. *133*, 1004.

Pandey, J. P., Johnson, A. H., Fudenberg, H. H., Amos, D. B., Gutterman, J. U., and Hersh, E. M. (1981). Hum. Immunol. *2*, 185.

Pandy, J. P., Shannon, B. T., Tsang, K. Y., Fudenberg, H. H., and Camblin, J. G. (1982). J. Exp. Med. *155*, 1228.

Poljak, R. J., Amzel, L. M., Chen, B. L., Phizackerley, R. P., and Saul, F. (1974). Proc. Natl. Acad. Sci. USA *71*, 3440.

Putnam, F. W., Titani, K., Wikler, M., and Shinoda, T. (1967). Cold Spring Harbor Symp. Quant. Biol. *32*, 9.

Rasmusen, B. A. (1965). Science *165*, 1742.

Rivat, L., Rivat, C., Ropartz, C., and Hess, M. (1971). Rev. Eur. Etud. Clin. Biol. *16*, 777.

Rivat, L., Cavelier, B., and Bonneau, J. C. (1978a). Ann. Immunol. (Inst. Pasteur) *129*, 735.

Rivat, L., Rivat, C., Cook, C. E., and Steinberg, A. G. (1978b). Ann. Immunol. Inst. Pasteur *129C*, 33.

Ropartz, C., Lenoir, J., and Rivat, L. (1961). Nature (Lond.) *189*, 586.

Steinberg, A. G. (1967). In *Advances in Immunogenetics* (T. J. Greenwalt, Ed.). Lippincott, Philadelphia, pp. 75–98.

Steinberg, A. G., and Cook, C. E. (1981). *The Distribution of the Human Immunoglobulin Allotypes*. Oxford.
Steinberg, A. G., and Rostenberg, I. (1969). Science *164*, 1072.
Steinberg, A. G., Terry, W. D., and Morrell, A. R. (1970). In *Protides of Biological Fluids*, Vol. 17 (J. Peeters, Ed.). Pergamon Press, Oxford, pp. 111–116.
Steinberg, A. G., Milstein, C. P., McLaughlin, C. L., and Solomon, A. (1974). Immunogenetics *1*, 108.
Steinberg, A. G., Olivier, T. J., and Buettner-Janusch, J. (1977). Am. J. Phys. Anthropol. *47*, 21.
Steinberg, A. G., Levene, C., Yodfat, Y., Fidel, J., Brautbar, D., and Cohen, T. (1980). Am. J. Med. Genet. *5*, 75.
Tsuzukida, Y., Wang, C., and Putnam, F. W. (1979). Proc. Natl. Acad. Sci. USA *76*(3), 1104.
van Loghem, E., Sukernik, R. I., Osipova, L. P., Zegers, B. J. M., Matsumoto, H., de Lange, G., and Lefranc, G. (1980). J. Immunogenet. *7*, 285.
Vyas, G. N., and Fudenberg, H. H. (1969). Proc. Natl. Acad. Sci. USA *64*, 1211.
Wang, A. C., Mathur, S., Pandey, J. (1978). Science *200*, 327.
Wegrzyn, J., and Wegrzyn, Z. (1978). Biochem. Genet. *9*, 59.
Werner, B. G., and Steinberg, A. G. (1974). Immunogenetics *1*, 254.

12

Functional Immune Network

Constantin A. Bona The Mount Sinai Medical Center, New York, New York

I. INTRODUCTION

Between 1960 and 1975, immunology was dominated by Burnet's selection theory (1959). According to this theory, the immune system is composed of several million individual clones. All the clones together account for the entire antigenic repertoire. Antigen selects among the various clones one particular clone which bears the corresponding antigen binding receptor, thereby stimulating its proliferation and maturation into a plasma cell. The clonal theory provided a satisfactory explanation for specificity of the antibody response, generation of diversity, immunological memory, tolerance, and so on. However, this theory formulated before 1960 does not encompass more recent immunological discoveries such as T- and B-cell dichotomy, recognition of self within immune system, regulation of clonal expression, and particularly the idiotypic phenomenon discovered by Kunkel et al. and Oudin et al. in 1963.

Idiotypes (Id) are phenotypic markers which distinguish variable regions of Ig molecule one from another. Idiotypes are defined in immunochemical terms by anti-Id antibodies which can be obtained across xenogeneous, homologous, or autologous barriers. They have structural correlates in the hypervariable region which contribute to the antigen binding specificity of an Ig molecule. Each antibody molecule bears an individual idiotype (IdI) characteristic of an Ig molecule produced by an individual within a strain or among outbred animals. In other circumstances, the same idiotypic specificity can be borne among antibodies of several individuals against various antigens. These idiotypic specificities are called cross-reactive idiotypes (IdX). In an elegant study performed on anti-α1-3 dextran monoclonal antibodies, Clevinger et al. (1980) have shown that IdI are located in the D region, whereas IdX is located in the second hypervariable region. The idiotypic determinants are also expressed on the antigen binding receptor of B and T cells,

which cooperate in an immune response induced by a particular antigen. Thus idiotypes are not only useful markers of V regions but also of T- and B-cell clones.

The discovery of idiotypes provided a theoretical basis for the formulation of Jerne's network theory proposed in 1974. This theory envisaged the immune system as a web of V domains. By contrast to clonal theory, which considered the immune system composed of individual clones waiting to be stimulated by antigen, the network theory considered that lymphocyte clones behave as individuals within a social community and that they speak one to another by using idiotype dictionary. The network theory is based on the dual character of Ig molecule, which binds the antigenic determinants (epitope) by virtue of its combining site (paratope) and which can elicit an immune response by virtue of its idiotope(s).

After birth, in nonimmunized animals, there is an equilibrium between the clones that produce antibodies (Ab_1) and those that produce anti-Id antibodies (Ab_2). This steady state can be upset by antigen. In the earliest phase of the immune response, antigen clears Ab_1, and thus clones producing Ab_2 cease the synthesis of Ab_2 because their natural stimulus represented by idiotypes of Ab_1 is lacking. Then the Ab_1 clones that escape from the control of Ab_2 clones can be stimulated by antigen to proliferate and to maturate in antibody-forming cells. When Ab_1 reaches an immunogenic threshold, it stimulates Ab_2 clones to synthesize anti-Id antibodies. These, in turn, exert a suppressive effect on Ab_1 clones.

Thus the major characteristics of network concept are as follows:

1. The immune system is composed of two types of clones: one devoted to recognition of the epitopes and another devoted to recognition of the idiotopes.
2. The idiotype network is vectorial. Ab_1 stimulates the synthesis of Ab_2, which in turn inhibits the production of Ab_1 and elicits the synthesis of Ab_3, and so on.
3. The basic pattern of idiotype network is the suppression, since only it ensures the steady state, namely, the equilibrium between clones in nonimmunized animals as well as in various phases of the immune response.

II. SOURCES OF NETWORK THEORY

The network concept emerged from several important immunobiological findings in the early 1970s. It is also supported by other observations which were reported later.

A. Autoimmunogenicity of Idiotypes

First evidence of autoimmunogenicity of idiotypic determinants was provided by Rodkey (1974). He isolated anti-trimethylammonium (TMA) antibodies from a rabbit. Then he injected these antibodies back into the same rabbit. The rabbit synthesized antibodies against its own anti-TMA antibodies. The spontaneous auto-anti-Id antibodies during the immune response elicited by immunization with conventional antigens such as phosphocholine, bacterial levan, *Micrococcus lysodeikticus* polysaccharide, trinitrophenyl (TNP) conjugates, and tobacco mosaic virus (TMV) were reported later. These results clearly show that idiotypic determinants of Ig molecules function as immunogens in an autologous system and that for each idiotype of an antibody molecule the host possesses the ability to produce an anti-Id antibody.

Various genetic studies concluded that the V-gene repertoire in mammals is about 10^7 to 3×10^7 (Kunkel, 1970). This estimation suggests that the V-gene repertoire is

sufficiently large to cover both the epitope and idiotype dictionaries. Indeed, later it was shown that across the syngeneic or allogeneic barriers can be produced not only anti-Id antibodies (Ab_2) but also anti(anti-Id) antibodies (Ab_3) and even anti[anti(anti-Id)] antibodies (Ab_4) (Wikler et al., 1979; Bona et al., 1981).

B. Idiotypic Determinants Expressed on the Receptor of B and T Lymphocytes Serve as Sites of Regulation of Lymphocyte Functions

The Ig receptor of B cells bears isotypic, allotypic, and idiotypic specificities of humoral antibodies. The presence of these antigenic determinants on the receptor of B cells is due to V-D-J-C rearrangements of Ig-structural genes.

Idiotypic determinants have also been identified on surface of T cells. The idiotypes associated with the surface of T cells were identified by using several methods as follows:

1. By inhibition of antigen binding ability of T cells by anti-idiotype antibodies
2. By staining with fluoresceinated or enzyme-labeled anti-Id antibodies
3. By binding of anti-Id antibodies to purified antigenic binding receptor isolated from Ig^- cells (data reviewed by Bona and Hiernaux, 1981)

It should be mentioned that polyclonal anti-Id antibodies were used in the vast majority of the studies that identified idiotypic determinants on T-cell receptor. Shared idiotypes between T and B cells led to the conclusion that the same set of V genes encode for the specificity of the antigen binding receptor of T and B clones.

However, recent studies on the rearrangements of V_H, D, and J_H genes in T cells showed only a D-J joining (Kurosawa et al., 1981). These new results suggest that the antigen binding receptor of T cells is encoded for

1. Entirely different set of V genes as those encoded for V region of Ig molecules, or
2. Alternatively, that in T cells the V genes are not rearranged as in B cells and they are located far from D-J genes

There are numerous data which have shown that anti-Id antibodies can suppress or enhance the function of T cells (data reviewed by Bona and Hiernaux, 1981). These results indicate that idiotypic determinants borne by T and B lymphocytes function as sites for regulating lymphocyte function. These two major findings represented by autoimmunogenicity of idiotypes and the presence of idiotypic determinants on the antigen binding receptor of T and B clones provide experimental evidence for idiotype network theory.

III. ANATOMY OF IDIOTYPE NETWORK

The understanding of the immune network is based on the analysis of the structure, immunochemical, and physiological properties of various members of an immune network pathway.

A. Polymorphism of Idiotypes (Ab_1)

Antibodies produced in response to the majority of natural and synthetic antigens are heterogeneous. After the discovery of idiotypy, it was admitted that an idiotype is characteristic for a single combining site and therefore represents the phenotypic marker of one V-region gene.

Table 1 Shared Idiotypes Between *E. coli* 113 LPS Binding and *S. tranaroa* LPS Binding Monoclonal Proteins Detected by the ELISA Technique

Microplates coated with 50 μg of:[a]	OD(405 nm) Alkaline phosphatase-labeled monoclonal antibodies[b]				
	LPS6.1	LPS7	LPS8	MOPC384	MOPC870
Ab_2 A48 93-17	0.03	0.01	0.01	0.02	0.03
MSLPS 60-25	0.11	1.45	0.28	0.10	0.23
MSLPS 61-6	0.25	1.25	0.96	0.68	0.63
MSLPS 2-10	0.04	0.62	0.10	0.05	0.03
Anti-384Id	0.05	0.45	0.11	0.40	0.12

[a] Ab_2 A48 93-17 is a monoclonal antibody specific for A48Id. MSLPS 60-25, 61-6, and 2-10 are monoclonal antibodies obtained by fusion of SP2/0 cells with spleen cells from BALB/c mice immunized with LPS7 monoclonal antibodies. Anti-384Id is a polyclonal syngeneic anti-Id antibody obtained by immunization with MOPC384.

[b] LPS6.1, LPS7, and LPS8 are monoclonal antibodies specific for the glucose and galactose immunodominant sugars of *E. coli* 113. They also bind to *Neisseria lactamica* LPS. MOPC384 and MOPC870 are BALB/c binding myeloma proteins specific for the α-methyl-D-galactoside immunodominant sugar of *S. tranaroa, S. televiv,* and *P. mirabilis* LPS.

However, soon it was shown that the same idiotype can be shared by antibodies belonging to various isotypes, indicating that the same V-region gene can be associated with various C-region genes. Furthermore, later it was shown that the same family of idiotypes identified by a particular polyclonal anti-Id antibody can be shared by various clonotypes. Thus heteroclitic anti-NP antibodies have a common idiotype (NP^b) and IEF analysis of anti-NP response in C57BL/6 mice revealed four clusters of bands, indicating the existence of several clonotypes which share NP idiotypes. These important data obtained by Mäkelë and his colleagues were reviewed (Bona and Cazenave, 1981).

The finding of shared idiotypes among various Ig classes does not contradict the rule that one V-region gene codes for one paratope and one idiotope. However, Oudin and Cazenave (1971) have shown by studying the idiotype of rabbit antiovalbumin response that antibodies specific for different epitopes borne by chicken, duck, and turkey ovalbumin bear a common idiotype. Shared idiotypes among antibodies with various antigenic specificities or unknown specificities were described in numerous antigen systems. Recently, we have demonstrated that monoclonal antibodies specific for glucose and galactose, the immunodominant sugars of *Escherichia coli* 0113LPS share some idiotypes with MOPC384 and MOPC870 myeloma proteins, which bind to α-methyl-D-galactoside the immunodominant sugar of *Salmonella tranaroa, S. televiv*, and *Proteus mirabilis* (Table 1). These findings indicate that idiotypes are not markers of a single paratope or of a single V-region gene. A paratope encoded for by a germ line gene or by a V gene derived from a somatic mutation can share idiotopes with paratopes specific for other epitopes.

B. Antiidiotypic Antibodies (Ab_2)

Antiidiotype antibodies are immunochemical probes by which the idiotypes are defined. However, they represent a heterogeneous family regarding their binding activity, immunochemical, and functional properties.

Types of Ab_2s

The Ab_2 family is composed of four major categories.

Hapten-Inhibitable Anti-Id Antibodies. These anti-Id antibodies are directed against paratope-associated idiotypes. The binding of epitope to paratope induces alterations of the three-dimensional configuration of the polypeptide chain of the V region. These alterations will be mirrored in the immunochemical reactivity of the idiotype antigenicity and will prevent the binding of anti-idiotype. This category of anti-Id antibodies are used as probes to define IdXs or IdIs associated with the combining sites.

Hapten-Noninhibitable Anti-Id Antibodies. These antibodies are specific for the antigenic determinants borne by the framework fragments of V regions. The binding of epitope to the paratope does not induce alterations in the reactivity of idiotypes since the framework fragments are not involved in the combining site. Indeed, in affinity labeling experiments, it was clearly shown that the haptens interact with the hypervariable segments of V region. Although these antibodies are not directed against Id associated with the combining site, they can display the same regulatory functions as hapten-inhibitable and anti-Id antibodies. Indeed, the binding of antibodies to any antigenic determinant of Ig receptor (i.e., isotype, allotype, or idiotype) can inhibit or stimulate the lymphocytic functions (see data reviewed recently in Bona and Cazenave, 1981).

Epibodies. These anti-Id antibodies represent a rare category of antibodies which recognize cross-reactive structures on epitopes and idiotopes. Recently, we have found that A/J antibodies specific for the V_H region of human IgM rheumatoid factor bound to human Fc fragment.

Homobodies. These anti-Id antibodies generally represent a small fraction of a heterogeneous Ab_2 family which displays antigen binding activity. Such antibodies were identified in various experimental systems. Thus it was shown that anti-Id antibodies (Ab_2) against antibodies (Ab_1) specific for insulin, retinol binding receptor, alprenelol, or formylpeptide can inhibit the binding of these substances to their corresponding cell receptors. In addition, it was shown that the binding of the anti-Id antibodies to the cell receptor for insulin or β-adrenergic receptors for alprenelol mimicked insulin or aprenelol-like effects.

In another elegant experiment, Urbain et al. (1981) have shown that a rabbit anti-Id antibody against rabbit anti-TMV was able to elicit in mouse an anti-TMV response in the absence of antigen. The existence of epibody and homobody indicates that some anti-Id antibodies represent the internal image of the antigen within the immune system.

According to the internal image concept, idiotypes (Ab_1) are positive imprints of the antigen and therefore the anti-Id antibodies (Ab_2) are topochemical copies of the antigen. Internal concept was not defined in immunochemical and molecular terms yet. However, this concept predicts that a homobody would block the ability of antigen to bind to the corresponding receptor and that the antigen would display the homobody bound to the cell receptor for antigen. Indeed, in the insulin system mentioned above, it was shown that Ab_2 homobody [i.e., anti(anti-insulin antibody)] successfully competed with insulin for binding to the insulin receptor.

From the internal image concept, several important conclusions can be drawn:

1. This concept implies that the antigen is never really foreign since the idiotypes are their positive imprints or that the idiotypes cross-react with the antigens.

2. That the clones of the immune system can be triggered by self elements without intervention of foreign antigens. This conclusion is supported by recent studies which have shown that the administration of Ab_1 (idiotype) can enhance the corresponding Id^+ fraction of anti-NP response in mouse (Reth et al., 1981). Furthermore, we have shown that administration at birth of minute amounts of A48Id-bearing monoclonal antibodies led to a significant increase of $A48Id^+$ anti-levan clones. This clone is silent in adult BALB/c mice, and $A48Id^+$ anti-bacterial levan antibodies can never be detected following immunization with levan or levan conjugates.

3. Internal image concept implies that an immunoglobulin molecule can bind an antigen as well as an idiotype, suggesting that there are not two + or - entities of clones, one that recognizes the epitopes and another than recognizes the idiotopes. This is particularly important to an understanding of the mechanism of communication between clones since at clonal level the receptor of a clone cannot distinguish between a similar structure shared by an epitope or an idiotope. It appears, therefore, that the immune system is composed of a single world of clones which recognizes the epitopes as well as the idiotopes since the antigen repertoire is mirrored in the idiotype repertoire.

Function of Ab_2s

Anti-Id antibodies possess regulatory properties and they can either suppress or enhance the lymphocytic functions.

Suppressive Effects. The suppression of antibody response by anti-Id antibodies was discovered by Pawlak et al. (1973) in phenylarsonate (Ars) system. A fraction of anti-Ars antibodies bear an IdX recognized by rabbit anti-Id antibodies. The injection of rabbit and anti-IdX Ars antibodies in A/J mice prevented the activation of clones able to synthesis IdX Ars antibodies. Idiotype suppression was obtained in various antigen systems as phosphocholine, TMA, TNP, inulin, and dextran. A long-lasting suppression can also be obtained by injection of anti-Id antibodies at birth (*neonatal suppression*) or by in utero exposure (*maternal idiotype suppression*) (data reviewed in Bona and Cazenave, 1981).

Anti-Id antibodies play a physiological role of regulation of expression of Id^+ clones. In several antigenic systems it was shown that the appearance of auto-anti-Id antibodies during a conventional immune response led to the suppression of Id^+ component of the immune response. The inverse fluctuation between the level of idiotypes and anti-Id antibody clearly suggests that anti-Id antibodies play an important role in the regulation of Id^+ component of the antibody response.

Enhancing Effects. In certain circumstances the anti-Id antibodies stimulate the idiotype component instead of suppressing it. Indeed, in several experimental systems it was shown that they can prime, in lieu of antigen, the precursors of T and B cells.

Thus it was documented that anti-Id antibodies in the absence of antigen can induce proliferation of MLC reactive cells, stimulate T cytolytic cells, or can prime DTH reactive, helper, and suppressor T cells (data reviewed by Eichmann, 1978, and Bona, 1981).

Similarly, the B cells can be primed by anti-Id antibodies. Eichmann and Rajewsky (see Eichmann, 1978) have shown that injection of guinea pig IgG_1 anti-A5A Id antibodies can replace the priming of A/J mice by *Streptococcus* A antigen for a secondary immune response.

Table 2 Effect of Anti-Id Antibody and T Cells in a Bacterial Levan System

A. Effect of administration of 0.01 μg of anti-A48 idiotype antibodies at birth

Pretreatment at birth	Immunization with bacterial levan 1 month later	Anti BL PFC Total	% A48Id
–	20 μg	3600 ± 0.125[a]	6 ± 3
Anti-A48, 0.01 μg	20 μg	3508 ± 0.123	46 ± 14
Anti-384, 0.01 μg	20 μg	3150 ± 0.043	3 ± 1

B. Effect of administration of A/J anti-E109Id antibodies in adults

Pretreatment of BALB/c mice with:[b]	Immunization	Anti-BL PFC/10^7 Total	% E109IdX	Anti-TNP PFC
–	TNP-Ficoll and BL	2569 ± 361	39 ± 8	4194 ± 339
A/J anti-E109 IdX	–	432 ± 147	13 ± 10	7359 ± 2150

C. Effect of T cells from C.B20 mice immunized with E109Id on the anti-inulin PFC response of B cells of BALB/c mice primed with inulin-B abortus

10^6 T cells from C.B20 mice immunized with:[c]	10^6 B cells from BALB/c mice immunized with:	Anti-In PFC/10^6 Total	% E109Id
–	–	34 ± 6	67 ± 9
–	In-BA	32 ± 13	15 ± 7
E109	In-BA	158 ± 50	71 ± 24
X24	In-BA	35 ± 5	35 ± 5

[a]Mean + SEM for $\log_{10}$ PFC/spleen detected for five mice 5 days after immunization with BL.
[b]Adult BALB/c mice were injected three times at 3-day intervals with 0.1 ml of A/J anti-109 antiserum and immunized 7 days later with BL and TNP-Ficoll. The response was tested 5 days later.
[c]EPC109 is an inulin binding myeloma protein, whereas XPC24 binds galactan.

In the levan-inulin system, we have studied the effect of parenteral administration of anti-Id antibodies in adults and neonates. We have found that:

1. Injection of adult BALB/c mice with anti-E109 IdX inulin antibodies causes the suppression of synthesis of E109 IdX inulin antibodies by itself. However, the cells from anti-E109 IdX-pretreated mice incubated in vitro with NWSM mitogen showed an increased level of E109 IdX molecules compared with normal cells.

2. Incubation of T cells from C.B20 mice immunized with E109 IdX with B cells from BALB/c mice immunized with inulin conjugate were able to elicit an anti-inulin PFC response in the absence of antigen. These results indicated that E109 Id-specific T cells interacted with the E109 Id-bearing B cells, and this interaction allowed the maturation of precursors into E109 Id anti-inulin antibody-secreting cells.

3. Administration at birth of minute amounts of anti-A48Id antibodies followed by immunization with levan 1 month later led to the activation of A48Id anti-levan silent clone. In this system we have found that antigen stimulation is required for the maturation of precursors into antibody-forming cells (Table 2).

Our results demonstrate that anti-Id antibodies can prime the precursors of dominant as well as silent idiotype clones. The effect of anti-Id antibodies is specific and anti-Id antibodies do not function as polyclonal activators in vivo. They interact only with precursors that bear the corresponding idiotype. Anti-Id antibodies cannot by themselves cause the maturation of precursors in antibody-forming cells. A second signal provided by antigen, B-cell mitogen, or helper factor is required for maturation into plasma cells.

C. Anti(Anti-Id) Antibodies (Ab_3)

Ab_2 antibodies bear idiotypic determinants which are able to induce anti(anti-Id) antibodies. These molecules act as a buffering system in the idiotypic network pathway. Production of Ab_3 antibodies have been obtained by injection of Ab_2 in normal allotype-matched rabbits and in syngeneic mice (data reviewed by Bona and Hiernaux, 1981). Ab_3s represent a normal component of an idiotypic network pathway. BALB/c mice immunized with levan binding A48 protein produce anti-A48Id antibodies. By 3 weeks after immunization, the titer of Ab_2 had fallen and Ab_3s were detected (Bona et al., 1981). This observation of spontaneously occurring Ab_3 antibodies in response to immunization with Ab_1 is the first demonstration that anti(anti-Id) antibodies represent a physiological component of the idiotype network. The analysis of various experimental data indicates that Ab_3 family is also heterogeneous.

This family of antibodies can be composed of several subsets.

True Ab_3

These anti(anti-Id) antibodies are produced by immunization with anti-Id antibodies, are able to elicit synthesis of anti[anti(anti-Id)] antibodies, and lack the ability to bind to antigen. The idiotypes of Ab_3 should be similar or different from those of Ab_1. The experimental data obtained in rabbits showed that the majority of Ab_4s bind to Ab_1 (Wikler et al., 1979). By contrast, in mice we found that only 60% of Ab_4 displayed Ab_2-like binding activity. Thus some Ab_3 was distinct from Ab_1 (Bona et al., 1981).

Parallel Set (Ab_1^1 in Urbain's Terminology)

Immunization with heterologous, homologous, or syngeneic anti-Id antibodies in certain conditions leads to the occurrence of antibodies which share the idiotypes of Ab_1 but lack the antigen binding activity of Ab_1. Such parallel sets have been described for subsequent immunization of rabbits with anti-Id antibodies against anti-*M. lysodeikticus* polysaccharide antibodies and in mice following immunization with syngeneic anti-Id antibodies against MOPC460 and A48 myeloma proteins, homologous anti-Id antibodies specific for IdX^+ anti-arsonate antibodies, or pig anti-Id antibodies specific for anti-H-2 antibodies. In nonimmunized mice, Ig molecules lacking TNP activity which bear 460Id were also identified (Table 3). Two hypothesis can be entertained to explain the inability of parallel sets to bind the antigens:

1. Parallel sets of antibodies are specific for unrelated antigens or for other antigenic determinants which belong to the same antigenic family for which the Ab_1—the initiator of the network pathway—is specific.
2. Parallel sets have a low affinity for polysaccharide antigens or haptens. If the parallel sets were cross-reactive with Ab_1 but had an energy of binding which was 1 or 2 kcal

Table 3 Antibody with Unknown Binding Activity Which Shares the Idiotypes of Ab_1 and Ab_3 Antibodies (Parallel Sets)

Species	Induction by administration of	Shared idiotypes	References
Mice	BALB/c anti-A48Id antibodies	A48Id	Bona et al. (1981), J. Exp. Med. *153*, 951
Mice	BALB/c anti-460Id monoclonal antibodies	460Id	Bona et al. (1979), J. Exp. Med. *149*, 815
Mice	Rabbit anti-11-4-1 (H-2^k) monoclonal antibodies	11-4-1Id	Bluestone et al. (1981), J. Exp. Med. *154*, 1305
Mice	Rat anti-Ars IdX	Ars IdX	Wisoki et al. (1981), J. Supramolec. Biol. *5*, 40
Rabbit	Rabbit anti-Id antibodies against antipolysaccharide antibodies	Id of anti-polysaccharide	Urbain et al. (1977), Proc. Natl. Acad. Sci. *74*, 5126
Rabbit	Rabbit anti-Id antibodies against anti-RNase antibodies	Id of anti-RNase	Cazenave, (1977), Proc. Natl. Acad. Sci. *74*, 5122
Mice	Occurred spontaneously	IdXA	Bona et al. (1979), J. Immunol. *123*, 1484
Mice	Occurred spontaneously	460Id[a]	Dzierzak et al. (1981), J. Exp. Med. *154*, 1442

[a]These molecules were detected in a higher amount (125 μg/ml) only with peculiar rabbit hapten-inhibitable anti-460Id antibodies. Molecules bearing 460Id without TNP activity cannot be detected in nonimmunized mice with monoclonal F651, FD5-1, and CD5-3 hapten-inhibitable anti-460Id antibodies.

Table 4 Ab_1-Like Properties of Anti[anti(antiallotype b6)] Antibodies

b4/b6 lymphocytes stimulated with 100 μg of Ig fraction of antibodies[a]	[^{3}H] thymidine incorporation	
	cpm	Stimulation index
–	168 ± 17	–
Anti-b4	1677 ± 191	10.0
Anti-b6 (Ab_1)	725 ± 168	4.3
Anti(anti-b6) (Ab_2)	256 ± 86	1.5
Anti[anti(anti-b6)] Ab_3	787 ± 55	4.7
Antialkaline phosphatase	141 ± 17	<1
NWSM[b]	1322 ± 302	7.9

[a]2×10^5 peripheral blood lymphocytes have been incubated with various antisera.
[b]NWSM was used at 50 μg/ml.
Source: Bona and Hiernaux (1981).

lower than an Ab_1 has a low affinity. The existence of parallel sets of antibodies that produced subsequent immunization with or spontaneous occurrence of anti-Id antibodies can explain various phenomena, such as the increase of "nonspecific" resistance during bacteria infections or indirect idiotype suppression, that we described several years ago (data reviewed by Bona, 1981).

Ab_1-Like Ab_3 Antibody

These anti(anti-Id) antibodies which occur subsequent to immunization with anti-Id antibodies share the antigen binding activity and the idiotypes of Ab_1. They are homologous or perhaps even identical to Ab_1. We have the opportunity to identify such antibodies by studying the blastogenic properties of Ab_3 antibodies produced in a b6 allotype system by J. Rolland. In this system rabbit b6-bearing Ig was the antigen, anti-b4 allotype antibodies the Ab_1, anti-Id (anti-b6 allotype) antibodies the Ab_2, and anti[anti-(anti-allotype)] antibodies the Ab_3. We found that Ab_1 and Ab_3 induced proliferation when they were incubated with b4/b6 lymphocytes (Table 4) (Bona and Hiernaux, 1981). These results clearly indicated that Ab_3 possesses the same binding activity and the same stimulatory functions as Ab_1, which in turn were anti-allotype antibodies. Therefore, the administration of Ab_2 in certain circumstances induce the synthesis of an Ab_3 antibody which possesses the immunochemical and functional characteristics of Ab_1.

D. Anti[Anti(Anti-Id)] Antibodies (Ab_4)

Ab_4s have been obtained only into two systems: in rabbits in a system where anti-polysaccharide antibodies were Ab_1s (Wikler et al., 1979) and in mice with the A48 levan binding myeloma protein as Ab_1 (Bona et al., 1981). Study of Ab_4s obtained in these systems indicated that they look like Ab_2 and have the ability to recognize the idiotype of Ab_1. Therefore, these results indicated that the immune network is not open-ended.

This short analysis of immunochemical and functional properties of various members of an idiotype network pathway shows that each member is composed of a heterogeneous family of antibodies.

IV. CELLULAR BASIS OF THE NETWORK

Jerne formulated the network theory in terms of V domains of antibody molecules. On the basis of our current knowledge, it is obvious that the immune network represents communication between B clones as well as between various sets of T regulatory clones. Therefore, a clear-cut symmetry should exist between Ab_1 antibodies and Ab_1 B and T cells, Ab_2 antibodies and Ab_2 B and T cells, and so on.

Studies carried out analyzing antibody responses against phosphocholine, phenylarsonate, hen lysozyme, TNP, TMA, and so on, have contributed to the understanding of cellular basis of the immune network. Based on these studies we can define at least two categories of T cells.

Ab_1-T cells equivalent to Ab_1 antibodies are antigen-specific T cells which bear Id determinants shared with antibodies with the same antigen specificity. This family of T cells is composed of Ab_1-T-helper and Ab_1-T-suppressor cells. Ab_2-T cells equivalent to Ab_2 antibodies are specific for idiotypic determinants of Ab_1 antibodies. This family of T cells is composed of Ab_2-helper (Ab_2 T_H) and Ab_2-suppressor (Ab_2-T_s) cells.

A. T-Helper Cells

Synthesis of antibodies by B cells in response to stimulation with T-dependent antigens or haptens requires T help. Two major categories of T-helper cells cooperate with B cells: antigen-specific T-helper cells (T_H MHC) which trigger the activation of precursors of B cells, and T cells specific for antigenic determinants of Ig receptor of B cells (T_H Ig). This last category of cells induces the differentiation and maturation of activated lymphocytes in plasma cells secreting a particular isotype, allotype, and idiotype.

Ab_1-T_H Cells

These T cells are antigen specific and MHC restricted since they recognize the antigen associated to genetically similar presenting cells. They are of Lyt 1^+ phenotype. Antigen binding receptor of Ab_1-T_H cells share idiotypic determinants with Ig receptor of their corresponding B cells. Since these cells express Id determinants on their receptor, they can be stimulated by anti-Id antibodies in lieu of the antigen. Such cells were described in various antigenic systems (see data reviewed by Eichmann, 1978, and Bona and Hiernaux, 1981). They exert their effect on B cells through soluble factors. These helper factors bind to antigen as well as to anti-idiotype columns.

Ab_2-T_H Cells

These T_H cells are specific for idiotypic determinants of Ig receptor of B cells and they can be stimulated by immunization with antibody bearing a particular idiotype. They are of Lyt 1^+ phenotype. It is still unclear if they are MHC restricted or if they can also be stimulated by antigen. They exert their effect on B cells and the cooperation is V_H restricted. Ab_2-T_H cells artificially induced by immunization with idiotype (A5A Id and E109 Id) can stimulate B cells bearing the corresponding idiotype in the absence of antigen. This stimulation can be explained by the binding of the receptor of these cells to Ig receptor of B cells. However, it is not clear if these cells can directly activate the B-cell precursors in vivo, or if they exert their effect on activated B cells and contribute to the differentiation and maturation into plasma cells. The frequency of these cells increases during the immune response, suggesting that they can play a fine-tuning role in the

expression of clones bearing Id determinants. It is obvious that these cells play a crucial role in anti-Id antibody responses and that their absence in nude mice is responsible for the inability of athymic mice to develop an autoanti-Id antibody response. Ab_2-helper T cells were described in various systems (data reviewed in Bona and Hiernaux, 1981). It would be important to establish if these two subsets of T cells exert a sequential effect on the activation, expansion, and maturation of a particular clone or if they have different targets.

B. T-Suppressor Cells

One of the most remarkable features of the immune response is that subsequent to recognition of antigen, individual lymphocytes proliferate and mature through a differentiation process. It was therefore necessary that mammals develop a regulatory mechanism to prevent the uncontrolled growth of stimulated clones. This control is carried out by T-suppressor cells. The T-suppressor-cell population is divided into two major groups: antigen-specific T-suppressor cells equivalent to Ab_1 (i.e., Ab_1-T_S) and idiotypic-specific T-suppressor cells equivalent to Ab_2 (i.e., Ab_2-T_S).

Ab_1-T_S Cells

This subset of T_S cells are antigen specific and they can be positively selected on antigen-coated plates. The initial event in the activation of these cells is the stimulation of an Lyt 1^+-suppressor inducer cell which is probably a member of a T-helper-cell subset bearing Lyt 1 alloantigens. This Lyt 1^+ cell inducer exerts its effect on an Lyt 123^+ precursor of suppressor. This activated suppressor expresses I-J and Lyt 23 alloantigens, which secrete a factor (SF_1) that can bind to the antigen that initiated the response. In several antigenic systems, it was shown that these T cells express idiotypic determinants and exert their effect on T-helper-cells (data reviewed by Eichmann, 1978).

Ab_2-T_S Cells

This subset of cells possesses a receptor specific for idiotype. As we might expect, the stimulus of these cells is the idiotype. Indeed, it has been shown that these cells can be expended by immunization with Ab_1 in saline or coupled with syngeneic cells. There is recent evidence obtained in DTH reactions that T_SF_1 factor produced by Ab_1-T_S cells can also activate Ab_2-T_S cells. These Ab_2-T_S cells can exert their effect directly on B cells, on T-helper cells, or on other effector cells, such as T cells that mediate DTH, GVH or allograft rejection. These T cells express the Lyt 2 phenotype and share idiotypic determinants of anti-Id antibodies (data reviewed by Bona and Hiernaux, 1981).

The findings described above suggest that there are symmetrical counterparts for Ab_1 and Ab_2 antibodies regarding both T-helper and T-suppressor cells. Ab_1 and Ab_2 regulatory T cells are distinct subsets regarding the specificity of their receptor, the target cells, and their function in the immune response. No available data exist on the property of Ab_3-T cells. These data taken collectively show that the immune network represents a very complex of web domains of T and B clones.

V. IDIOTYPE NETWORK MODELS

Several models aimed at understanding details of the functional immune network have emerged from the original network theory formulated by Jerne in 1974.

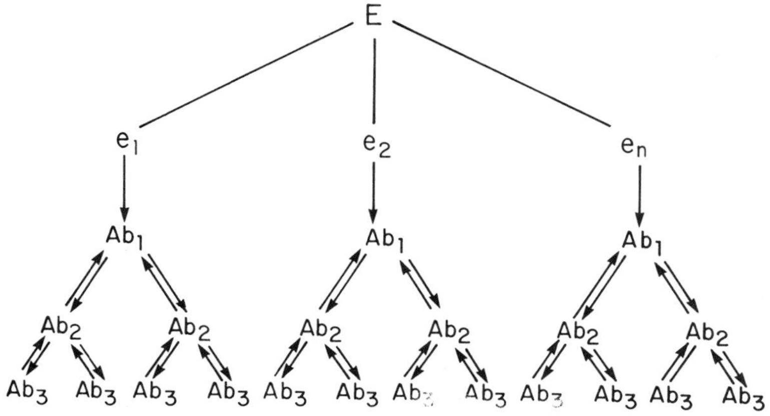

Figure 1 Open-ended idiotype network: E, epitype; e, epitope; Ab_1, antiepitope antibody; Ab_2, anti-idiotype antibody; Ab_3, anti-(anti-idiotype) antibody.

A. Open-Ended Network

An open-ended network model was proposed by Richter (1975). He envisioned the immune system as a functional network of V domains based on stimulatory and inhibitory interactions. According to this model, antigen elicits a cascade of anti-Id complementary reactions. When Ab_1 reaches an immunogenic threshold, it initiates the synthesis of Ab_2, which in turn elicits the synthesis of Ab_3, and so on. This cascade is open ended (Fig. 1). Thus the stimulatory signal provided by antigen is propagated along the idiotype network pathway until suppressive interactions stop it. To prevent aberrant proliferation of clones, Richter proposed that the concentration of Ab_2 required for inhibition of Ab_1 clones was lower than the concentration of Ab_2 required for the stimulation of Ab_3 clones.

This model attempts to explain the three types of immune responses as follows:

1. Low dose tolerance results when antigen stimulates the production of Ab_1, which, in turn, stimulates the synthesis of Ab_2. However, at low concentrations of antigen, Ab_2 never reaches the immunogenic threshold required for induction of Ab_3.
2. Normal immune response results when Ab_1 stimulates production of Ab_2, which induces the synthesis of Ab_3. Ab_3 is present at a concentration that inhibits the synthesis of Ab_2. Inhibition of Ab_2 then allows the synthesis of Ab_1.
3. High dose tolerance results from a configuration of Ab_1 and Ab_3 where Ab_2 and Ab_4 suppress the proliferation of Ab_1 and Ab_3.

The open-ended model predicts that antigen can elicit an unlimited number of anti-Id antibodies (Ab_n). This suggests that the immune system is devoted to recognizing idiotopes, not antigens. Obviously, this can be true only if the idiotopes and antigens are cross-reactive.

Furthermore, recent experimental data have shown that (1) anti-Id antibodies are not detected in high-dose-tolerant animals and (2) that Ab_4 looks like Ab_2, suggesting that the network is not open-ended.

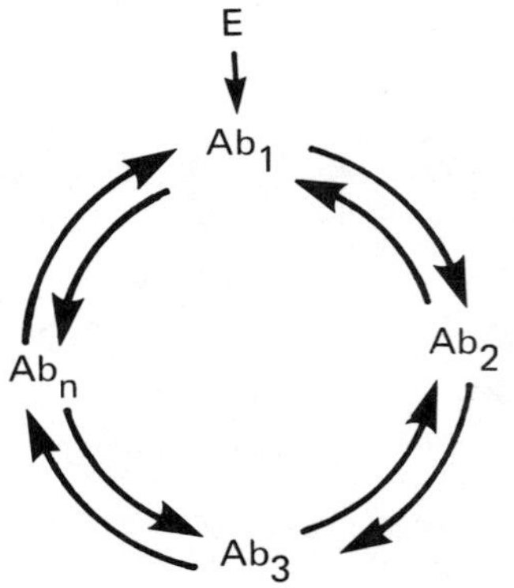

Figure 2 Cyclic idiotypic network. (See Fig. 1 for abbreviations.)

B. Cyclical Network

By contrast to the vectorial idiotype network model proposed by Richter, Hiernaux (1977) considers that the network is composed of short cycles of small interacting cycles. According to this model, antigen induces the production of even and odd cycles of antibodies which interact one with another. Ab_1 recognizes antigen, whereas Ab_3 and Ab_5 lack the ability to recognize antigen. They share idiotypes with Ab_1 and function as nonspecific immunoglobulin. Ab_2 and Ab_4 suppress the production of Ab_1 and Ab_3 and are the internal image of the antigen (Fig. 2). This model considers that antigen drives the system toward a new stable configuration in which odd elements are stimulated and even elements are suppressed.

C. Idiotype Network as a System of Complementary Sets

These models proposed by Hoffman (1975), Urbain et al. (1979), and recently by Roitt et al. (1981) are based on the T cell/B cell dichotomy of the lymphocyte population. The immune network in these models is envisaged as interaction between positive and negative sets.

Hoffman's model is based on several assumptions:

1. An immunoglobulin can function as an anti-epitope as well as an anti-idiotope.
2. The set of V domains that recognizes the antigen is a positive set, whereas the set of V domains that recognizes the idiotopes is a negative set. These sets stimulate each other because their receptors are complementary.
3. Stimulation results from cross-linking of receptors and is carried out by B-cell products that are divalent. T-cell products are monovalent and cannot cross-link receptors themselves. However, they can block the receptors of complementary specificity.

Hoffman's model suggests different interactions between positive and negative sets during various states of the immune system. Thus in nonimmunized animals, the concentration or frequency of positive and negative sets are so low that the reciprocal stimulation is not significant. Immunization will lead to the dominance of positive B and T cells, which will suppress both B and T negative sets. This immune state establishes immune

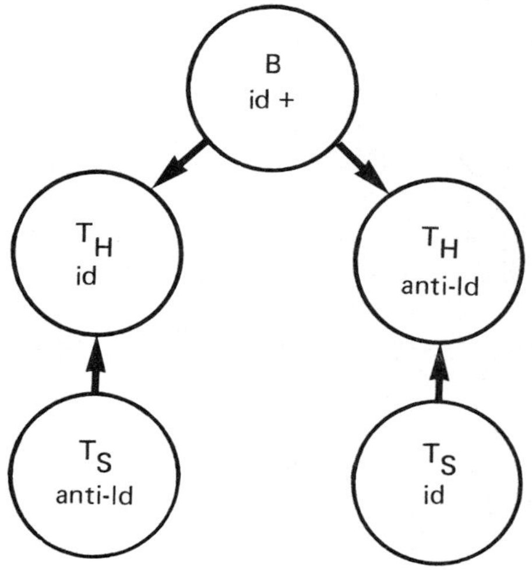

Figure 3 Complementary-sets idiotype network: B, bone marrow-derived lymphocyte; T_H, T-helper cells; T_S, T-suppressor cells.

memory. In the anti-immune state, the B and T negative sets are stimulated, whereas B and T positive sets are suppressed. Finally, a tolerant state is achieved when the concentration of positive set is equal to the sum of the concentration of the complementary negative sets plus the concentration of the self-epitopes.

The major feature of this model is that it points out that the reciprocal interaction between the positive and negative sets (or between Ab_1 and Ab_2 cells) generates stable steady states which characterize various phases of the immune response. The evolution of the population of clones from the virgin state (i.e., nonimmune state) toward an immune state, an anti-immune state, and tolerance depends on the perturbation of the virgin state by antigen, idiotype, or anti-idiotype antibody.

Urbain's (1979) and Roitt's (1981) models are based on multifunctional properties of T-cell populations and on the existence of different subsets of regulatory T cells which exert their effects on the Ig-bearing B cell. These models suggest that the Id^+ B cell is polygomous since it is the subject of the effects of at least four major categories of T cells: T-helper or T-suppressor cells reacting with antigen which share the idiotypes with B cells, and anti-idiotype T-helper or T-suppressor cells (Fig. 3).

These models actually extended the positive and negative sets at the level of T-cell subsets, B cells being a positive set. These models do not consider the obvious regulatory role of anti-Id antibodies.

D. Regulatory Idiotype Network

Recently, we have proposed a limited immune network model, based on the concept of regulatory idiotopes. We consider as regulatory idiotopes the set of antigenic determinants of a variable region which function as autoimmunogens. These idiotypes can be represented or highly represented in the sera of nonimmunized animals because their expression is favored by a genetically determined positive selection or by preexisting idiotype-specific regulatory cells. A functional immune network model should overcome some of the major difficulties of the formal network models.

Vectorial Aspects

Jerne's model envisaged the immune network as a vectorial system in which suppression is the basic pattern. Ab_1 will stimulate the synthesis of Ab_2, which will suppress Ab_1, and so on. The notion that paratope-idiotope interactions are unidirectional and that the idiotypes rather than the antigens stimulate the cells bearing the corresponding antigen binding receptor cannot be reconciled with various experimental findings. Thus the T-cell clones can be activated in vitro solely by antigen associated with macrophages in the absence of antiidiotypes. Furthermore, in several antigenic systems, it was shown that anti-idiotype antibodies can stimulate T cells in lieu of antigen.

Representation of Anti-Idiotypic versus Antiepitopic Repertoire

If we consider that Ab_1 expresses at least two idiotopes, there should be at least two Ab_2s for each idiotope of Ab_1, four Ab_3 for the idiotopes of Ab_2s, and so on. Thus it appears then the immune system is devoted to recognizing idiotopes instead of epitopes.

Idiotope Function

The number of idiotopes that function as autoimmunogens in an autologous system is smaller than those which function as antigens across allogeneic or xenogeneic barriers. Consequently, we proposed a model in which only a few idiotopes function in the autologous system. This model is based on a study of the properties of four members of an immune network pathway which was carried out by Wikler et al. (1979) in rabbit in a homologous system and by ourselves (Bona et al., 1981) in mouse in a syngeneic system.

In the system that we studied, the Ab_1 was ABPC48 (A48-BALB/c myeloma protein, which binds β-2,6-fructosan). The immunization of syngeneic mice with A48 protein induced the synthesis of Ab_1 (i.e., anti-A48Id antibodies). The anti-A48Id antibodies produced in syngeneic strains share a cross-reactive idiotype, clearly indicating that these antibodies have their own idiotypes. Immunization with Ab_2 elicited the production of an Ab_3 that lacked the ability to bind β-2,6-fructosan but which shares A48Id. In addition, we have found that stimulation of Ab_3-producing animals with β-2,6-fructosan led to the activation of $A48Id^+$ anti-β-2,6-fructosan silent clone (i.e., BALB/c mice immunized with β-2,6-fructosan do not produce A48Id-bearing antibodies). Thus the

Table 5 Properties of Various Members of A48 Idiotype Network Pathway[a,b]

	Ligand			
Microplates coated with:	Ab_1	Ab_2	Ab_3	Ab_4
Bacterial levan	++	–	–	–
A48 protein (Ab_1)	n.d.	++	–	+
Anti-A48Id (Ab_2)	++	n.d.	++	–
Anti(anti-A48Id) (Ab_3)	–	++	n.d.	++
Anti[anti(anti-A48Id)] (Ab_4)	+	–	++	n.d.

[a]The degree of binding activity in RIA, HA, and HIA designated by a scale of – to ++.
[b]n.d. not done.

outcome of immunization with Ab_2 is the stimulation of clones expressing A48Id and which are able to make antibodies which may or may not bind to β-2,6-fructosan.

Furthermore, the immunization with Ab_3 produced an Ab_4 of which at least 60% bound to A48Id (Table 5). We explain these results based on the concept of regulatory idiotopes. Only regulatory idiotopes expressed on a family of Ab_1 molecules which are autoimmunodominant antigens can elicit the production of Ab_2. Therefore, the Ab_2 obtained in a syngeneic system are specific for the regulatory idiotopes. The Ab_2s are likely to express conventional idiotopes rather than regulatory idiotopes since they are not selected by cells specific for their idiotopes. The immunization with Ab_2 will lead to the synthesis of Ab_3, which will share the idiotopes of Ab_1 (i.e., regulatory idiotopes) and, therefore, the immunization with Ab_3 will have as outcome the production of an anti-Id antibody, i.e., Ab_4 directed against regulatory idiotopes. Indeed, the experimental data showed that Ab_4 bound to Ab_1 as well as to Ab_3. Thus the Ab_3s should be regarded as a family of antibodies possessing Ab_1 regulatory idiotopes, one of them displaying an Ab_1 binding cross-reactivity, the other lacking antigen binding activity or binding to unrelated antigens.

This functional model considers that only the first degree of idiotype-anti-idiotype interaction is physiologically significant since a single Ab_2 can activate or suppress the clones producing antibodies that express regulatory idiotopes (i.e., Ab_1, Ab_3a, b, c, etc.). We described this system as a ping-pong effect in which the Ab_1 and Ab_3 side has many players and Ab_2 side has only one (Bona et al., 1981) (Fig. 4).

This model predicts that regulatory idiotypes exist because of their obvious potential for mediating regulatory process at the T-cell level. Furthermore, it suggests that the immune system is not highly and randomly interconnected. Rather, it is composed of a miriade of mininetworks characteristic for each antigen.

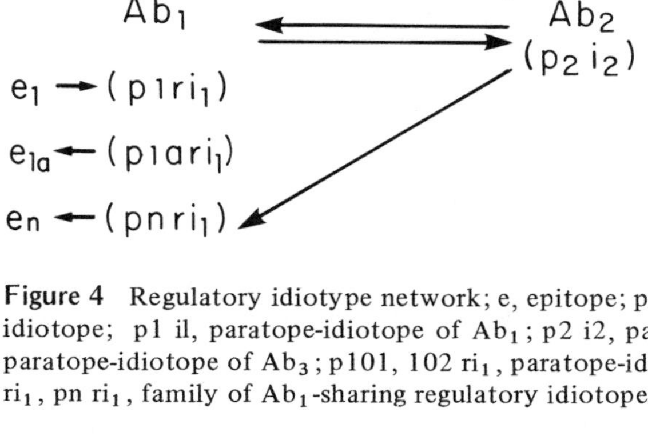

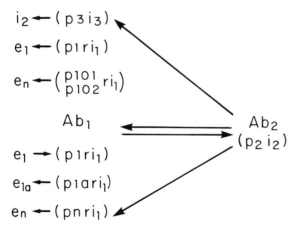

Figure 4 Regulatory idiotype network; e, epitope; p, paratope; i, idiotope; ri, regulatory idiotope; p1 il, paratope-idiotope of Ab_1; p2 i2, paratope-idiotope of Ab_2; p3 i3, paratope-idiotope of Ab_3; p101, 102 ri_1, paratope-idiotope of parallel sets; p_1 ri_1, pi_a, ri_1, pn ri_1, family of Ab_1-sharing regulatory idiotopes.

VI. ANTI-TRINITROPHENYL (TNP) IMMUNE RESPONSE AS AN EXAMPLE OF FUNCTIONAL IMMUNE MININETWORK

As envisioned by IEF and splenic foci assays, anti-TNP antibodies in various species are highly heterogeneous. In BALB/c mice, a fraction of these anti-TNP antibodies share the idiotypes of DNP and TNP binding MOPC460 and MOPC315 myeloma proteins. A vigorous anti-TNP antibody response can be obtained in normal mice by immunization with T-dependent antigens such as TNP-KLH or with T-independent TNP conjugates such as TNP-B-abortus, TNP-LPS, and TNP-NWSM (TI_1 antigens) or TNP-Ficoll and TNP-levan (TI_2 antigens). Several idiotype-determined regulatory mechanisms have been extensively investigated in the TNP system.

A. Spontaneous Occurrence of Auto-Anti-Id Antibodies in Response to Immunization with TNP-Ficoll

Siskind and his colleagues (1979) studied the anti-TNP PFC response in AKR mice immunized with TNP-Ficoll, which is a TI_2 antigen. They observed a significant decline in the number of anti-TNP PFCs between days 4 and 7 after immunization. However, this decline of anti-PFC response was apparent only because the number of PFCs did not decrease when PFC were measured in the presence of 10^{-7} and 10^{-8}M TNP-ϵ-aminocaproic acid. This hapten-augmentable phenomenon was interpreted by the authors as being due to auto-anti-Id antibodies. These antibodies inhibit anti-TNP PFC, but when the hapten is added to agarose, the anti-Id antibodies are displaced and therefore the number of PFCs increases.

The anti-idiotype antibody origin of this hapten-augmentable PFC response is consistent with several experimental findings:

1. Injection of 7-day serum into normal animals accelerates the occurrence of hapten-augmentable PFC.
2. Injection of 7-day serum into 4-day primed normal animals or nude primed mice inhibits the response.
3. The inhibiting activity of these sera was removed after passage over columns coated with anti-mouse Ig antibodies and AKR-anti-TNP antibodies but not on columns coated with rabbit anti-TNP antibodies or AKR anti-dansyl antibodies.
4. The hapten-augmentable PFC response was not observed in nude mice. These results suggest that following immunization with TNP-Ficoll AKR mice developed a T-dependent auto-anti-Id antibody response. These anti-Id antibodies interact with the TNP binding receptor of B cells and block the secretion of anti-TNP antibodies. Since these anti-Id antibodies are directed against the combining site, they can be displaced by hapten, which results in hapten-augmentable anti-TNP-PFC. In DBA/2 mice, Cowdery and Steinberg (1981) were able also to demonstrate the presence of auto-anti-Id antibodies in response to immunization with TNP-Ficoll. Anti-Id antibodies were observed in this strain between 10 and 15 days after immunization. These IgG anti-Id antibodies bound to 5 days IgM anti-TNP antibodies. These results clearly showed that DBA/2 mice by 10 to 15 days developed IgG auto-anti-Id antibodies in response to IgM anti-TNP antibodies produced in the early phase of the primary immune response.

By using a very elegant and complicated experimental protocol, Gorczynsky and his colleagues (1980) provided suggestive evidence on the production of auto-anti-Id

antibodies against IdI of anti-TNP antibodies. The rationale of this experimental protocol was that anti-TNP antibody response is very heterogeneous and then anti-Id antibodies will be produced selectively against IdIs. Anti-TNP antibodies from individual CBA mice were injected in three syngeneic normal mice, and then serum or cells of these recipients were tested regarding their suppressive activity on TNP-PFC response in tissue culture by cells of the same individual donor. The serum obtained in these three recipients inhibited the anti-TNP response of cells from the same donor but not the TNP responding cells from other donors. Since the only known differences between anti-TNP antibodies from various CBA individuals reside in their individual idiotypes, it appears that the suppressive effect observed by the authors can be attributed to anti-Id antibodies. In addition, these authors have shown that the serum from a mouse immunized with TNP-KLH is able to suppress the anti-TNP response of its own cells after 1 month's residence in four irradiated recipients. Interestingly, serum harvested from the recipient mice was able to inhibit the anti-TNP response in vitro of their own cells.

These results suggest that immunization with anti-TNP antibodies in syngeneic system elicited the synthesis of anti-Id antibodies against individual idiotypes of various donors of anti-TNP antibodies. The auto-anti-Id antibodies have the same cross-specificity as the syngeneic anti-Id antibodies, as assessed by the ability of serum of normal recipients to inhibit the TNP response of the donor of serum as well as of the serum of the donor to inhibit the response of its own cells after a 1-month residence in other recipient. The intriguing inhibition of the TNP-responsive cells of the recipient with its own serum observed by the authors can be explained either by the induction of an anti-Id response in the recipients against an IdX shared by anti-TNP antibodies produced by various CBA individuals or, alternatively, by an indirect idiotype suppression type of mechanism that we have described in the TNP-levan system.

These results taken collectively show that anti-TNP antibodies produced in various strains of mice in response to immunization with T-dependent or T-independent TNP conjugates stimulate the spontaneous occurrence of auto-anti-idiotype antibodies specific for IdI and IdX. These antibodies regulate the expression of TNP-reactive clones.

B. Regulation of 460Id Clones

Several myeloma proteins with DNP and TNP binding activity were isolated in mice. MOPC460, a BALB/c IgA_k myeloma protein, was more extensively studied since it was found that the 460Id of this protein is expressed on anti-TNP antibodies.

Study of the regulation of 460Id-bearing clones in BALB/c mice has several advantages:

1. The 460 Id is defined immunochemically and structurally.
2. The 460Id is a cross-reactive idiotype shared on the anti-TNP antibodies of *Ighc*^{a} mice and is silent in other strains of mice.
3. 460Id in *IghC*^{a} strains is a *minor component* of anti-TNP antibody response, and therefore may be subject to regulatory influences.
4. 460Id anti-TNP antibodies can be obtained in response to both T-dependent and T-independent conjugates.

Based on these advantages, in 1978 we started a systematic study of idiotypic regulation of 460Id component of anti-TNP response (data reviewed by Bona and Paul, 1980).

Table 6 Properties of 460Id-specific T-Suppressor Cells

B cells from normal BALB/c mice incubated with:	Number of experiments	Percent of 460Id + anti-TNP response
–	11	48 ± 7
T cells	10	10 ± 3
T cells absorbed on 460 plates	5	56 ± 7
T cells recovered from 460 plates	6	10 ± 4
T cells absorbed on E109 plates	3	14 ± 4
T cells recovered from E109 plates	3	52 ± 7
T cells treated with anti-Thy 1.2 + C	1	27
T cells treated with anti-Lyt 1.2 + C	4	20 ± 5
T cells treated with anti-Lyt 2.2 + C	7	53 ± 8
T cells C.B20	2	24 ± 5
B cells treated with anti Lyt 1.2	3	43 ± 6
B cells treated with anti-Lyt 1.2 + T cells treated with anti-Lyt 1.2	3	25 ± 5
B cells treated with anti-Lyt 1.2 + T cells treated with anti-Lyt 2.2	3	35 ± 7
B cells from C57/J × BALB/c F_1 incubated with:		
–	2	49 ± 7
T cells (F_1)	2	10 ± 3
T cells (F_1 treated with anti-Qa1 + C	2	50 ± 8

In vitro studies of the expression of 460Id on anti-TNP antibodies obtained in response to in vitro stimulation with TNP-NWSM, TNP-LPS, TNP-BA (TI_1), TNP-levan, TNP-Ficoll, and TNP-dextran (TI_2) have shown that only about 20% of total anti-TNP PFC expressed 460Id. This low response corresponds to a low frequency of 460Id-bearing cells in normal BALB/c mice. Indeed, Bernabe et al. (1981) have shown that only 2 to 3×10^3 of Ig-secreting B clones produce 460Id. This frequency is lower compared to that of anti-TNP precursors.

Surprisingly, we have found that the proportion of 460Id-producing cells increased significantly after depletion of T cells and that, conversely, the addition of T cells to B-cell cultures stimulated by NWSM-TNP caused a strong inhibition of 460Id component. This observation suggested that the 460Id component of anti-TNP response was under the regulation of naturally occurring T-suppressor cells. In further experiments, we characterized the properties of these suppressor cells (see Table 6).

In the first set of experiments, we demonstrated that these suppressor cells were specific for 460Id. Indeed, after incubation in petri dishes coated with MOPC460 protein, T-cell-enriched populations lost their suppressor activity. By contrast, the T cells

recovered from these dishes which adhered to 460Id displayed a higher suppressor activity, estimated on a per cell basis, than that of a nylon-enriched T-cell population. The conclusion of these experiments was that naturally occurring T-suppressor cells were specific for 460Id (i.e., Ab_2-T_S set).

In a second set of experiments, we established the phenotype of 460Id Ab_2-T_S. The suppressor activity of these cells was ablated by treatment with anti-Thy 1.2, anti-Lyt 2.2, and anti-Qal antisera + C, indicating that they expressed Thy 1.2, Lyt 2.2, and Qal alloantigens. Recently, Urbain and his colleagues (personal communication) have shown that 460Id Ab_2-T_s cells are sensitive to cyclophosphamide since they obtained a 460Id response in a nonresponsive BALB/c substrain following pretreatment with cyclophosphamide before immunization with TNP-Ficoll.

In the third set of experiments, we have studied the genetic control of the occurrence of 460Id Ab_2-T_S. These cells were identified in BALB/c (H-2^d $IghV^a$, $IghC^a$), BAB.14(H-2^d, $IghV^a$, $IghC^b$), and C58/J(H-2^k, $IghV^a$, $IghC^a$), but not in C.B20(H-2^d, $IghV^b$, $IghC^b$) or CAL.20 (H-2^d, $IghV^e$, $IghC^e$). These results suggested that the occurrence of these cells is not under the control of the MHC gene complex. Rather, their appearance is associated with the 460Id V gene of the *$IghC^a$* gene complex.

This also suggests that the stimulus for the appearance of these suppressor cells can be 460Id V-gene products. However, we were unable to obtain evidence for 460Id molecules in the serum of nonimmunized mice utilizing two monoclonal anti-460Id antibodies [i.e., syngeneic F6(51) and homologous FD5.1 anti-460Id antibodies]. Recently, Dzierzak et al. (1981) identified a parallel 460Id set of Ig molecules by using TNP-inhibitable rabbit anti-460Id antibodies.

These antibodies react with molecules found in the serum nonimmunized mice which lack the ability to bind to TNP. These 460Id molecules devoid of anti-TNP specificity are a good candidate for the natural stimulus of 460Id Ab_2-T_S. These 460Id Ab_2-T_S exert their effect directly on 460Id-reactive B clones. However, we cannot exclude that they can inhibit the function of 460Id-bearing T-helper cells since these cells have not yet been identified.

We demonstrated the direct effect of 460Id Ab_2-T suppressor on B cells in two kinds of experiments:

1. On normal B cells, we found that Lyt 1^--selected 460Id Ab_2-T_S strongly inhibited the ability of Lyt 1^--selected B cells to develop a $460Id^+$ anti-TNP response in subsequent in vitro stimulation with TNP-NWSM. Thus, in the absence of Lyt 1 cells, the 460Id Ab_2 T_S cells still displayed their suppressive activity.
2. On tumor B cells secreting 460Id anti-TNP antibodies, we have found that 460Id Ab_2-T_S cells were able to inhibit the secretion of antibodies without altering the growth of cells.

These findings suggest the existence of a steady-state immune network of the 460Id system in which the expression of precursors of Ab_1 clones characterized by the 460Id V-gene marker is under the control of naturally occurring 460Id-specific T-suppressor cells. The communication between various clones based on 460Id links is very fragile and can be perturbed by other components of the network.

We have investigated the perturbation of the 460Id network by studying the effects of administration at birth or in adults of 460Id protein or anti-460Id antibodies. The results of this investigation are illustrated in Table 7.

Table 7 Effect of Treatment with 460Id and Anti-460Id on 460Id + Anti-TNP Response[a]

Pretreatment	Age of mice: Pretreatment	Age of mice: Immunization	Anti-TNP PFC/10^6: Total	Anti-TNP PFC/10^6: Percent 460Id
–	–	4 weeks	473 ± 36	18 ± 2
M460 in saline (10 μg)	1 day	4 weeks	714 ± 139	24 ± 5
Anti-460Id in saline (0.01 μg)	1 day	4 weeks	458 ± 27	34 ± 4
–	–	12 weeks	305 ± 15	23 ± 2.0
M460 in FCA (75 μg × 8)	4 weeks	12 weeks	245 ± 46	2 ± 0.9
–	–	12 weeks	519 ± 49	32 ± 3
Anti-460Id in FCA (75 μg × 8)	4 weeks	12 weeks	537 ± 24	52 ± 4

[a]Mice were immunized with 20 μg of TNP-Ficoll and their PFC response was tested 5 days later.

1. The effect of treatment at birth with small amounts of 460Id: Administration of 10 μg of MOPC460 proteins at birth followed by immunization with TNP-Ficoll 1 month later significantly increased the 460Id component of anti-TNP antibody response.
2. The effect of hyperimmunization of adult BALB/c mice with 460Id: Hyperimmunization of BALB/c mice with MOPC460Id antibodies led to the production of anti-460Id antibodies. In response to immunization with TNP conjugates these mice developed an anti-TNP response which does not differ in magnitude from that of normal mice, and 460Id$^+$-bearing molecules could not be detected. These results are supported by findings regarding the analysis of the frequency of 460Id-secreting clones, which is below the detectable level (Bernabe et al. 1981).
3. The effect of administration of minute amounts of anti-460Id antibodies at birth: The injection of 0.01 μg of anti-460Id antibodies at birth followed by immunization with TNP-Ficoll 1 month later significantly increased the 460Id component.
4. The effect of hyperimmunization of adult BALB/c mice with anti-460Id antibodies in FCA.

These mice produced a high titer of anti(anti-460Id) antibodies which recognized a cross-reactive idiotype shared by anti-460Id antibodies produced in several strains of mice. These Ab_3-producing mice showed a high increase in the 460Id component of anti-TNP antibodies in response to immunization with TNP-NWSM or TNP-levan. This increase corresponds to a higher frequency of 460Id-secreting cells in these mice. The increased frequency of the precursors is related to the elimination of 460Id Ab_2-T_s by Ab_3 antibodies since the antigen binding receptor of T_s shares the idiotypic specificity of Ab_2 antibodies (Bona and Paul, 1980).

Finally, it was shown that administration of BALB/c anti-460Id in various strains of mice that do not bear the IghC^a allotype can activate the 460Id clones that are silent in these mice, suggesting that 460Id B-cell precursors are found not only in IghC^a strains but also in strains of mice that bear other allotypes (Cazenave et al., 1980). The cellular

basis of perturbation of the 460Id functional steady-state network is not yet completely understood. However, the results presented above illustrate the fragility of 460Id links between clones as well as the flexibility of the immune system, which can respond in various ways to stimuli that constitute the elements of the network.

C. Idiotypically Determined Regulatory Networks of Tumor Cells

Studies of the effects of the immune regulatory mechanism on the growth of MOPC315 myeloma in syngeneic hosts provided another experimental model to use to investigate the functions of various members of a network pathway. MOPC315 is a BALB/c myeloma which secretes a TNP-DNP binding protein. This myeloma line as well as other human or murine B-cell tumors is composed of a heterogeneous population represented by cells in various stages of differentiation, from stem cells to plasmoid cells.

The various stages of differentiation can be investigated using the following methods:

1. The stem cells can be studied using spleen colony assay. MOPC315 cells injected in normal mouse lodge in the spleen, grow, and by 14 days form a colony on the surface of the spleen.
2. The lymphocytes bearing IgA receptor can be visualized by rosette techniques using TNP-SRBC as indicator or by fluorescence with fluoresceinated anti-314Id antibodies.
3. The plasma cells can be examined by indirect PFC assay with TNP-SRBC and anti-murine IgA antibodies.

Lynch and his colleagues (see review by Lynch et al., 1979) have shown that the linear differentiation of stem cells to plasma cells is susceptible to immune regulatory signals. Thus the growth of MOPC315 cells in Millipore chambers grafted in the peritoneal cavity of mice can be enhanced by carrier-specific helper T cells. Indeed, if these chambers are grafted in a host with helper T cells specific for SRBC and the MOPC315 cells and SRBC-TNP, an enhancement of the growth of myeloma cells was observed. Based on this observation, Lynch suggested that the growth of myeloma cells was stimulated by carrier-specific T-helper factors which bound to Ig surface receptors of myeloma cells via TNP-SRBC.

Studies performed by Lynch and his colleagues (1979) and Abbas (1979) have shown that various immune signals can alter the secretion of MOPC314 protein or inhibit the growth of tumor cells. Secretion of IgA-TNP binding M315 protein can be inhibited by in vitro incubation of cells with TNP-protein conjugates or with 315Id-specific suppressor T cells. Both procedures lead to a specific inhibition of the secretion of M315 protein without altering the synthesis of other structural proteins and without inhibiting the growth of cells.

By contrast, the anti-315Id immunity induced in vivo confers an idiotype-specific graft resistance to MOPC315 tumor cells in normal animals. This finding indicates that the idiotype expressed on the receptor of tumor B-lymphocyte functions as a tumor-associated antigen as well as a transplantation antigen. T cells play an important role since the adult thymectomy resulted in a total abrogation of graft resistance to MOPC315. In addition, this graft resistance to MOPC315 tumor cells can be transferred in normal animals by 315 idiotype T cells, which expressed various antigen markers characteristic of suppressor T cells. These results indicate that essentially the same immune network mechanisms regulate the differentiation and expansion of both normal and tumor B clones.

Studies of the perturbation of the steady state of micronetworks which link various clones by idiotypic determinants expressed on their receptors—perturbation which can be caused by antigen, idiotype, and anti-idiotype—can represent an important factor in the elucidation of the pathogenic mechanisms of various infectious and autoimmune diseases and in our understanding of the growth of tumors.

ACKNOWLEDGMENTS

Supported by research grants from the U.S. Public Health Service (AI 18316-01), American Cancer Society (IM-275), and National Science Foundation (PCM 8110578).

I am grateful to Dr. Elizabeth Bikoff for reading this manuscript and to Anita Spingarn for excellent typing assistance.

REFERENCES

Abbas, A. K. (1979). Immunol. Rev. *48*, 245.

Bernabe, R. R., Coutinho, A., Martinez, A., and Cazenave, P. A. (1981). J. Exp. Med. *154*, 552.

Bona, C. (1981). In *Compendium in Immunology*, Vol. 2. (L. Schwartz, Eds.). Van Nostrand Reinhold, New York, pp. 233–237.

Bona, C., and Hiernaux, J. (1981). C. R. C. Crit. Rev. in Immunol. *3*, 33.

Bona, C., and Cazenave, P. A., Eds. (1981). *Lymphocytic Regulation by Antibodies*. Wiley, New York

Bona, C., and Paul, W. E. (1980). In *Regulatory T Lymphocytes* (B. Pernis and H. J. Vogel, Eds.). Academic Press, New York, pp. 292–300.

Bona, C., Heber-Katz, E., and Paul, W. E. (1981). J. Exp. Med. *153*, 951.

Burnet, F. M. (1959). *The Clonal Selection Theory of Acquired Immunity*. Cambridge University Press, New York, pp. 10–120.

Cazenave, P. A., LeGuern, C., Bona, C., and Buttin, G. (1980). In *Membranes, Receptors, and the Immune Response* (E. P. Cohen and H. Köhler, Eds.). Allan R. Liss, New York, pp. 354–381.

Clevinger, B., Schilling, J., Hood, L., and Davie, J. M. (1980). J. Exp. Med. *151*, 1059.

Cowdery, J. S., and Steinberg, A. D. (1981). J. Immunol. *126*, 2136.

Eichmann, K. (1978). Adv. Immunol. *26*, 195.

Gorczynski, R. M., Kennedy, M., Khomasurya, B., MacRae, S., and Cunningham, A. J. (1980). Eur. J. Immunol. *10*, 781.

Hiernaux, J. (1977). Immunochemistry *14*, 733.

Hoffmann, G. W. (1975). Eur. J. Immunol. *5*, 638.

Jerne, N. K. (1974). Ann. Immunol. (Inst. Pasteur) *125C*, 373.

Kunkel, H. G. (1970). Fed. Proc. *29*, 55.

Kunkel, H. G., Mannik, M., and Williams, R. C. (1963). Science *140*, 1218.

Kurosawa, Y., von Boehmer, H., Haas, W., Sakeno, H., Jrauneker, A., and Tonegawa, S. (1981). Nature (Lond.) *290*, 565.

Lynch, R. G., Rohrer, J. W., Odermatt, B., Gebel, H. M., Antry, J. R., and Hoover, J. R. (1979). Immunol. Rev. *48*, 45.

Oudin, J., and Cazenave, P. A. (1971). Proc. Natl. Acad. Sci. USA *68*, 2616.

Oudin, J., and Michel, M. (1963). C. R. Acad. Sci. (Paris) *257*, 805.

Pawlak, L. L., Hart, D. A., and Nisonoff, A. (1973). J. Exp. Med. *134*, 1442.

Reth, M., Kehoe, G., and Rajewsky, K. (1981). Nature (Lond.) *290*, 257.

Richter, P. H. (1975). Eur. J. Immunol. *5*, 350.

Rodkey, S. L. (1974). J. Exp. Med. *139*, 712.

Roitt, I. M., Male, D. K., Guarnotta, G., de Corvalho, L. P., Cooke, A., Hay, F. C., Lydyard, P. M., and Thenavale, Y. (1981). Lancet, 1041.

Schroter, A. I., Goidl, E. A., Thorbeche, J., and Siskind, G. W. (1973). J. Inf. Med. *180*, 138.

Urbain, J. Collignon, C., Franssen, J. D., Mariame, B., Leo, O., Urbain-Vansanten, G., van de Walle, P., Wikler, M., and Wuilmart, C. (1979). Ann. Immunol. (Inst. Pasteur) *130C*, 281.

Urbain, J. Cazenave, P. A., Wikler, M., Franssen, J. D., Mariame, B., and Leo, O. (1981). In *Immunology 1980* (M. Fougereau and J. Dausset, Eds.). Academic Press, New York, pp. 81-93.

Wikler, M., Franssen, J. D., Collingnon, C., Leo, O., Mariame, B., Van de Walle, P., de Groote, D., and Urbain, J. (1979). J. Exp. Med. *150*, 184.

13

Methods of Immunoglobulin Isolation and Characterization

Carel J. van Oss State University of New York at Buffalo, Buffalo, New York

I. INTRODUCTION

A. Scope of This Chapter

Isolation of immunoglobulins is treated here from a *preparative* point of view. From an *analytical* viewpoint, the standard smaller-scale physicochemical methods will be briefly mentioned and referenced in the appropriate places. Analytical immunological and immunochemical methods of immunoglobulin characterization are treated in other chapters. A number of physicochemical methods of immunoglobulin characterization are treated in Sec. X. Unless stated otherwise, all references are to *human* immunoglobulins.

B. Isolation Parameters

With a view to the isolation of immunoglobulins, as well as of all other biopolymers, one must make use of those physicochemical properties or parameters that are peculair to the polymer in question, and that are quantitatively different from the physicochemical properties of the other accompanying (but unwanted) polymers. There are five fundamentally different physicochemical parameters of biopolymers: *solubility, electric charge, surface tension, size and shape,* and *ligand specificity*.

C. Isolation Strategies

Different biopolymers may have one or two of these parameters in common, but they never quantitatively share all five of them. Thus the most successful strategy for isolating any given biopolymer (e.g., an immunoglobulin) from a mixture consists of a *combination* of two or more methods, each of which discriminates for two different physicochemical parameters. IgM in whole serum for example, has its size in common with α_2

macroglobulin and its electric charge with many other betaglobulins. Thus a two-step procedure, in which one step isolates the proteins of 800,000 to 900,000 molecular weight while a second step selects the betaglobulins, should in principle result in virtually pure IgM. In theory this is indeed so, but in practice a few more steps are desirable: for example, an initial precipitating step, to eliminate 90% of the irrelevant proteins, as well as one or two ultrafiltration steps, to counteract the considerable dilution that accompanies practically all separation processes. A drastic improvement in purification can usually be obtained by selecting a *different* isolation parameter as a basis for the next purification step. Repetition of the *same* procedure seldom leads to significantly enhanced purity.

II. SOLUBILITY

A. Precipitation by Protein-Protein Interaction

Antibodies, being proteins and thus amphoteres, have positively as well as negatively charged sites, which can cause them to bind to each other and form large insoluble complexes. Three conditions favor such protein-protein interactions: (1) a pH close to the isoelectric point of the protein (which creates a situation where the positive charges equal the negative charges of the protein molecule, leading to optimal protein-protein interaction); (2) a low ionic strength of the medium (which removes the ions that would otherwise shield the protein's charges, thus giving rise to protein-protein interaction, especially of "euglobulins"); and (3) a low dielectric constant of the medium (which enhances the attractive coulombic interaction energy between the protein dipoles), generally achieved through the addition of water-miscible organic solvents, such as ethanol or ether (Schultze and Heremans, 1966). The major drawback of precipitation through protein-protein interactions lies in the (partly irreversible) protein denaturation and/or aggregation to which they give rise. Thus human IgG, the major component of Cohn's fraction II, produced on a large scale by precipitation from pooled plasma through the admixture of cold ethanol, comprises up to 15% of (10S) dimers as well as a smaller proportion of higher polymers. Precipitation of immunoglobulins as euglobulins (e.g., a sizable fraction of the IgM) through lowered ionic strength also causes a certain degree of irreversible denaturation. Resolubilizing euglobulins, precipitated at low ionic strength by the addition of neutral salts, is termed *salting in*.

Precipitation induced by protein-complex formation through the addition of heavy metal (e.g., Zn^{2+}, Cu^{2+}) or organic ions (e.g., Rivanol) (Schultze and Heremans, 1966) also tends to give rise to some degree of denaturation.

B. Precipitation by Dehydration

Precipitation of proteins by dehydration, when done by the admixture of concentrated salt solutions, is called *salting out*. Salting out is most effective with completely ionized neutral salts that have plurivalent coions (i.e., ions of the same sign of charge as that of the net charge of the proteins to be precipitated).

Sulfates fit these requirements most closely, and mainly owing to its high degree of solubility, $(NH_4)_2SO_4$ is by far the salt most used for salting out. Salting out with $(NH_4)_2SO_4$ is a useful first step in the isolation of immunoglobulins from whole serum, as it can significantly reduce the total load of undesired proteins in one step. The immunoglobulins can all be precipitated in the presence of one-third saturated $(NH_4)_2SO_4$

[saturated $(NH_4)_2SO_4$ is $\approx$ 4 M]. Upon precipitation of this immunoglobulin-containing fraction, one eliminates 75% of the (irrelevant) proteins in the supernatant. The most soluble protein, serum albumin, precipitates completely only in the pressence of 100% saturated $(NH_4)_2SO_4$ (Schultze and Heremans, 1966). Salting out of proteins is an unusually mild precipitation method, causing essentially no protein denaturation.

The principal drawback of the salting-out step is the necessity for removing considerable amounts of salt still present in the redissolved precipitate, requiring relatively long periods of dialysis. Another more serious drawback is the following: When used for the isolation of soluble immune complexes (see Chap. 18) or for the isolation of Ag-Ab complexes for affinity determination (see Chap. 16), the salting-out method may cause the loss of a sizable proportion of some complexes, as these salt concentrations favor the dissociation of certain (e.g., DNA/anti-DNA) Ag-Ab complexes of low to medium affinity (de Groot et al., 1980; Smeenk and Aarden, 1980).

One of the best nonionic protein-precipitating solutes is polyethylene glycol of molecular weight 6000 (PEG-6000), which can be used for the selective precipitation of IgG (Polson et al., 1964; Schultze and Heremans, 1966), although it is somewhat less suitable than $(NH_4)_2SO_4$ for use as a first step in the isolation of immunoglobulins from whole serum. However, PEG-6000 lacks the drawback of $(NH_4)_2SO_4$—that of dissociating certain Ag-Ab complexes. Indeed, precipitation with PEG-6000 is most useful in the isolation of small immune complexes (Zubler and Lambert, 1978); see also Chap. 18). Although PEG-6000, like the other polyethylene glycols and $(NH_4)_2SO_4$, is a strong dehydrating agent, steric exclusion of protein molecules from the interstices between its coils contributes strongly to the protein-precipitating effect (Polson, 1977).

III. ELECTRIC CHARGE

A. Ion Exchange Chromatography

As a γ-grobluin, IgG is among the serum proteins with the lowest electric charge and thus especially easy to isolate by ion exchange chromatography, without actually binding it to the ion exchange material. For that purpose, the positively charged diethylaminoethyl (DEAE) cellulose and DEAE-Sephadex (cross-linked dextran with DEAE ligands) are the materials of choice. At pH 7.5 essentially all IgG, but none of the other serum proteins, is positively charged, as is the anion exchanger. Thus all other serum proteins are bound to the column, and IgG passes through, devoid of other serum proteins [see Table 1; for more details on the methodology, see, e.g., van Oss (1980)]. For other purposes (e.g., for the separation of other immunoglobulin fractions, see below), negatively charged (cation) exchangers may be useful, such as carboxy methyl (CM) cellulose.

Chromatofocusing is a novel variety of ion exchange chromatography, first described by Sluyterman and Wijdenes (1977). It differs from the ion exchange procedure used in separating many proteins from a mixture (employing a pH gradient for elution) primarily in the use of ampholyte buffers (see Sec. III.C) for elution, which gives rise to a certain localized concentrating effect of each fraction, hence the name "focusing" (see also "Isoelectric Focusing," Sec. III.C). There are as yet few published applications to this method of immunoglobulin isolation (Richey and Beadling, 1981), but the method certainly should be considered as a likely improvement over ordinary ion exchange methods.

Although ion exchange is one of the simplest methods for isolating immunoglobulins (especially IgG), it is not the most efficient one, as aspecific adsorption is still

Table 1 Flow Sheet for the Preparation of IgG from Normal Human Serum[a]

SERUM

↓

Precipitate with 1/3 saturated $(NH_4)_2SO_4$ at room temperature

↓

Dialyze precipitate first against H_2O, then against pH 7.5, 0.0175 M phosphate buffer (4°C)

↓

Centrifuge discard residue

↓

Ion exchange chromatography of supernatant, DEAE-cellulose, pH 7.5, 0.0175 M (4°C)

↓

Elution with pH 7.5, 0.0175 M phosphate buffer (4°C)

↓

Concentrate by ultrafiltration[b]

↓

IgG (essentially the only protein not bound)

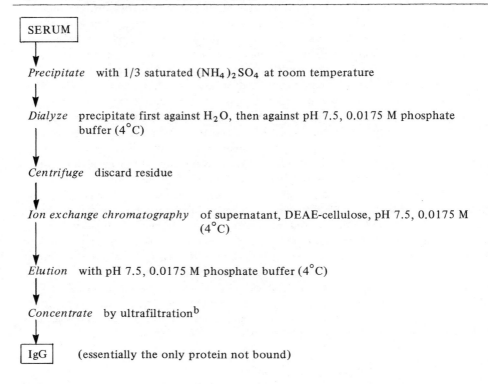

[a]The procedure is applicable to IgG myeloma sera with the proviso that care be taken to ensure that the resin is not overloaded. Alternatively, a procedure employing preparative zone electrophoresis, followed by gel filtration of the γ spike thus obtained, will frequently yield high-purity IgG.
[b]Ultrafilter membranes that may be used are Amicon UM-10, PM-10, or PM-30, or analogous commercially available membranes, or they may be prepared in the laboratory, such as the CA-50 membrane (van Oss and Bronson, 1982).
Source: van Oss, 1981.

an important source of immunoglobulin loss. The yield of IgG-class antibodies obtained by DEAE-cellulose ion exchange from a given antiserum usually is not more than 60 to 80%; somewhat better yields are obtainable when the more hydrophilic, but also more expensive DEAE-Sephadex is used.

B. Electrophoresis

Zone (or block) electrophoresis in slabs of packed particles, such as starch (van Oss, 1980) or polymer particles (e.g., of Pevikon, polystyrene, or glass beads) or of (0.5%) agarose gels (Jaton et al., 1979), is a simple and efficient method for separating serum globulins according to charge, from quantitites of the order of 50 to 500 mg of total protein. The lowest losses due to aspecific adsorption are obtainable with the most hydrophilic carrier materials [i.e., starch blocks (with a 60 to 80% yield) or agarose gels (with a 70 to 80% yield)]. Because of a considerable Jouleian heat development, the

electrophoresis is best done in a cold room, at +4°C, at fairly low field strengths, so that a separation takes 24 hr. At the conclusion of an electrophoretic run, the block is cut up in narrow (5 to 10-mm-wide) slices and each slice is placed into a sintered (coarse) glass funnel, which in its turn stands in a 50-ml polypropylene centrifuge tube, and spun at 1000*g* for 5 min (van Oss, 1980). When 0.5% agarose blocks are used, the individual slices are placed in centrifuge tubes, frozen to -20°C, and kept at that temperature overnight. After thawing, the soluble protein fraction is separated from the agarose pellet by centrifugation (Jaton et al., 1979). In both cases the globulin yield can be much enhanced by additional washing and recentrifugation of the slurry in sintered glass funnels or of the agarose pellets.

Most free liquid electrophoretic methods have no advantage over the solid support methods discussed above for preparative immunoglobulin separations. Free-flow electrophoresis between two closely spaced flat plates, endless-belt electrophoresis, stable-flow electrophoresis, and horizontal rotating-cylinder electrophoresis (van Oss, 1979) all are much more complicated to use than block electrophoresis and are capable of processing only small amounts of protein. The same holds for static liquid column electrophoresis, stabilized by means of a density gradient. However, a number of these free liquid electrophoretic methods are useful for the separation of cells, for which solid supports cannot be used (van Oss, 1979; see also Chap. 32). The only free liquid electrophoretic method that has advantages for the isolation of immunoglobulins is the vertical cylindrical rotating continuous-flow method; (Mattock et al., 1980; see also van Oss, 1979); by this process the γ-globulins of up to 1 liter of serum can be separated per hour.

C. Other Electrokinetic Methods

Isoelectric Focusing

While the electrophoretic transport of charged molecules in an electric field in a homogeneous buffer of a given pH is continuous and has to be stopped if one wishes to avoid losing the molecules through migration into the electrode compartment of the opposite charge, electromigration of a charged amphoteric molecule through a continuous pH gradient automatically ceases when that amphoteric molecule has reached the place in the gradient where the pH is the same as its isoelectric point. Thus isoelectric focusing is the sorting out of amphoteric molecules, according to their different isoelectric points, by electromigration through a pH gradient. Generally speaking, prolonged electromigration of a mixture of amphoteric molecules in a pH gradient tends to enhance the resolution of each of its constituents, hence the designation "isoelectric *focusing*." The development that made the general application of isoelectric focusing a practical reality was the synthesis of many different *carrier ampholytes*, each having several acidic and basic groups with closely spaced pK values per molecule with many different pH values. In protein separation, isoelectric focusing allows an improvement in resolution over electrophoresis of about one order of magnitude. This is largely because, with time, bands separated by isoelectric focusing reach and maintain an optimal sharpness (due to the focusing effect), whereas with electrophoresis separated bands tend to broaden continuously because of diffusion. For the separation of proteins by isoelectric focusing, one must be mindful of the fact that the carrier ampholytes most used are isomers and homologues of aliphatic polyaminopolycarboxylic acids, and are thus difficult to distinguish from proteins by the most commonly used colorimetric or spectropotlometric methods. However, as these carrier ampholytes have an average molecular weight of about 800 whereas that of most

proteins is above 10,000, proteins generally can be easily separated from the ampholyte molecules by means of gel filtration (e.g., with Sephadex G-50 or by ultrafiltration; see Sec. V).

Preparative isoelectric focusing (Righetti, 1979) is practiced in liquid density gradients (stationary as well as continuously flowing), free fluid (Quast, 1979), flat beds, or cylinders stabilized with granulated porous carriers, or with continuous gels (Chrambach and Nguyen, 1979). In the latter case special attention should be paid to the various elution procedures (Chrambach and Nguyen, 1979); see also Sec. X).

Isotachophoresis

Like isoelectric focusing, isotachophoresis is practiced in a buffer system of which the composition is nonconstant with respect to location (see above), but whereas isoelectric focusing is best done in a buffer system consisting of an essentially continuous pH gradient, isotachophoresis needs several buffers with discontinuous conductivities. In isotachophoresis the sample mixture is placed in the sample compartment, the *terminating electrolyte* in the cathode compartment, and the *leading electrolyte* in the anode compartment. The effective mobility of the leading electrolyte is the highest and that of the terminating electrolyte the lowest. The effective mobilities of the components in the sample mixture should be intermediate between those extremes. Because of the possibility of achieving a total separation between different ionic species, once a steady state is reached, isotachophoresis affords an extremely high resolution (see Chrambach and Nguyen, 1979; Bier and Allgyer, 1979; van Oss, 1979). Up to the present time, this method has been used more frequently for analytical purposes (see Sec.X) than for preparative isolations.

Molecular Sieve Electrophoresis

Molecular sieve electrophoresis, electrophoresis in gels with pore sizes of the same order of magnitude as the globulins that have to be separated (see Chrambach and Nguyen, 1979), is used primarily for the separation of proteins according to size. It is discussed in Sec. V.

IV. SURFACE TENSION

A. Surface Tensions and van der Waals Interactions

As indicated in Chap. 6 [eq. (1)], the total van der Waals free energy of interaction between two different particles or molecules 1 and 2, immersed in liquid 3, is

$$\Delta F_{vdw} = \Delta F_{132} = \frac{-A_{132}}{12\pi d_0^2} \tag{1a}$$

in which A_{132} is the effective Hamaker coefficient of the system and d_0 the equilibrium distance between the two parallel particles or molecules. From Hamaker's combining rule (Hamaker, 1937),

$$A_{132} = A_{12} + A_{33} - A_{13} - A_{23} \tag{2}$$

and Berthelot's combining rule,

$$A_{ij} = \sqrt{A_{ii}A_{jj}} \tag{3}$$

where i or j stand for 1, 2, or 3, it can easily be demonstrated (Visser, 1972) that A_{132} becomes negative when

$$A_{11} > A_{33} > A_{22} \tag{4a}$$

and when

$$A_{11} < A_{33} < A_{22} \tag{4b}$$

As, in analogy with Eq. (1),

$$\Delta F_{ii} = \frac{A_{ii}}{12\pi d_0} \tag{1b}$$

and as

$$\Delta F_{ii} = -2\gamma_{iv} \tag{5}$$

where v stands for vapor, it is clear (van Oss et al., 1980a) that A_{132} also must be negative when

$$\gamma_{1v} < \gamma_{3v} < \gamma_{2v} \tag{6a}$$

and when

$$\gamma_{1v} > \gamma_{3v} > \gamma_{2v} \tag{6b}$$

The total free energy of interaction ΔF_{132} can in practice be obtained more easily via

$$\Delta F_{132} = \gamma_{12} - \gamma_{13} - \gamma_{23} \tag{7}$$

Whereas Hamaker coefficients A_{ii} are as yet only rarely easy to determine with any degree of accuracy, liquid-vapor (γ_{3v}) as well as solid-vapor (γ_{1v}, γ_{2v}), surface tensions are readily obtained with appreciable accuracy by a variety of methods (Neumann et al., 1980). The Hamaker coefficient A_{132} of a system will therefore become negative (so that the total van der Waals interaction between components 1 and 2 immersed in liquid 3 becomes repulsive) when the surface tension of the liquid medium 3 acquires a value intermediate between the surface tensions of 1 and 2 (van Oss et al., 1980a). The van der Waals interaction between components 1 and 2 is attractive as long as the surface tension of the liquid medium is either higher or lower than the surface tensions of *both* 1 and 2.

It is thus possible, in a liquid medium, to take the van der Waals interaction between two different components attractive or repulsive at will, by adjusting the surface tension of the liquid. This principle is applied in practice in hydrophobic (and reversed phase) as well as in affinity chromatography (see below).

B. Hydrophobic and Reversed-Phase Chromatography

Hydrophobic interactions are simply the van der Waals attraction between various macromolecules and/or particles with lower surface tensions than water (i.e., they are more or less "hydrophobic") in water, thus under conditions where

$$\gamma_{1v} \text{ (as well as } \gamma_{2v}) < \gamma_{3v} \tag{8}$$

(van Oss et al., 1980b). After the attachment of proteins, which are rather hydrophilic but which do have a slightly lower surface tension than water when not denatured (van Oss et al., 1981a), to fairly hydrophobic carriers in water, the proteins can be *detached*

by lowering the surface tension of the liquid medium to a value in between the surface tensions of the protein and the carrier [Eqs. (6a) or (6b)]—thus turning the van der Waals attraction into a *repulsion*—through the addition of water-miscible surface-tension-lowering solvents, such as ethylene glycol or dimethyl sulfoxide. Upon a gradual lowering of the surface tension of the liquid effected in this manner, human serum proteins are eluted in the order of their decreasing surface tension; for example, first α_2 macroglobulin, then serum albumin, α_2 HS glycoprotein, β_1C globulin (complement component C3), IgG, and finally transferrin (van Oss et al., 1979, 1980a).

Attachment of very hydrophilic proteins to hydrophilic carriers can be enhanced by *raising* the surface tension of the aqueous medium [see Eq. (8), e.g., by increasing the salt content of the water]. *Elution* of such very hydrophilic proteins can then be effected with water of lower ionic strength (van Oss et al., 1979). This approach was especially successful in the isolation of monoclonal as well as of normal IgA by Doellgast and Plaut (1976).

It would seem that IgG3 and IgG1 are more hydrophobic than IgG2 and IgG4 (Weening et al., 1976; Absolom et al, 1982; D. R. Absolom et al., unpublished data, 1982), so that hydrophobic chromatography may be a promising approach for the fractionation of human IgG subclasses.

Reversed-phase chromatography has a mechanism that is essentially identical to that of hydrophic chromatography (van Oss et al., 1980a). This method [under the name *reversed-phase* or *high-performance liquid chromatography* (RP-HPLC)] is used increasingly for protein separation, with preparative columns, in high-performance liquid chromatography devices (Regneir and Gooding, 1980).

V. SIZE AND SHAPE

A. Gel Filtration Chromatography

Gel filtration (or pore-exclusion chromatography) is the method of choice for the preparative isolation of a narrow-molecular-weight range of proteins from a complex biopolymer mixture with a wide range of molecular weights. The method consists of passing a polymer mixture (in solution) through a column packed with spherical gel beads, such that the larger polymers cannot penetrate the pores of the gel beads and emerge first from the columns, in the *void volume*. The smaller polymers become trapped in the pores of the gel beads; the lower their molecular weight, the later they are eluted. Molecular symmetry or asymmetry also play a role in the entrapment of polymers in the pores of the gel beads, so that the method normally should not be used for precise molecular weight determinations.

For immunoglobulin isolation Sephadex (Pharmacia, Piscataway, New Jersey) beads made of cross-linked dextran are used most (van Oss, 1980a). Sephadex G-200 [which *excludes* globular proteins of molecular weight (MV) >200,000] see Tables 2 and 3), or G-150 (excludes MV >150,000) are used most, whereas for the isolation of immunoglobulin fractions, G-100 (excludes MV >100,000) and G-50 (excludes MV >50,000) are indicated. Sephadex G-100, for instance, is useful for the fractionation of fragments with a molecular weight below 100,000. For special purposes (e.g., the isolation of IgD; see Sec. VII) special gel beads made of a three-dimensional polyacrylamide lattice, filled with an interstitial agarose gel, can be used (LKB, Rockville, Maryland; IBF, Clichy, France–Fisher Scientific; e.g., Ultrogel AcA34). Sephadex G-25 is useful for desalting

Table 2 Flow Sheet for the Preparation of IgA from Normal Human Serum[a]

SERUM

↓

Delipidate e.g., by high-speed centrifugation

↓

Gel filtration (4°C), G-200 Sephadex in 2% NaCl, pH 8

↓

Analyze fractions for IgA by immunodiffusion and pool accordingly; discard all others

↓

Concentrate by ultrafiltration[b]

↓

Recycle the preceding three steps two or three times

↓

Starch block electrophoresis[c] (4°C) in 0.05 M, pH 8.6 barbital buffer

↓

Analyze fractions for IgA by immunodiffusion and pool accordingly[d]

↓

Concentrate by ultrafiltration[b]

↓

IgA[e]

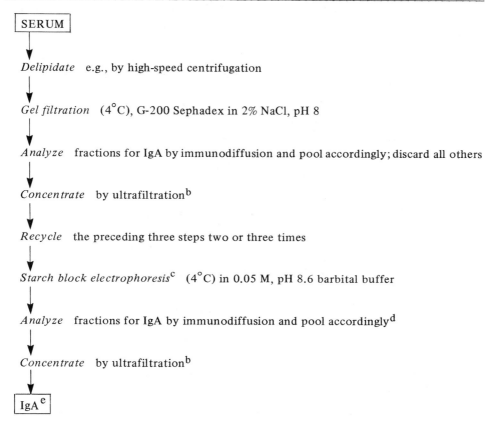

[a]The preparation of monoclonal IgA should proceed differently, involving primarily separation by charge (starch block electrophoresis or DEAE-cellulose chromatography) *followed* by gel filtration (Sephadex, G-200).
[b]See footnote b in Table 1.
[c]van Oss (1980, 1981).
[d]It still may be necessary to remove traces of IgD and IgE by affinity absorption.
[e]sIgA can be isolated in the same manner, but the second and third gel filtration steps should be done with Sepharose 6B.
Source: Adapted from Heremans (1974) and van Oss (1981).

protein solutions. The yield of immunoglobulin recovery, as far as the gel-filtration step is concerned, can, with care, be as high as 90%.

B. Molecular-Sieve Electrophoresis

Molecular-sieve electrophoresis, electrokinetic transport of proteins in gels with pores that are only slightly larger than the proteins (see, e.g., Chrambach, and Nguyen, 1979), aims at the separation of proteins according to their molecular size, but of course simultaneously effects a fractionation according to the electric charge of the proteins (van Oss,

Table 3 Flow Sheet for the Preparation of IgM from Normal Human Serum[a]

SERUM

↓

Precipitate at 1/3 saturated $(NH_4)_2SO_4$ (room temperature); discard supernatant

↓

Wash precipitate three times with 1/3 saturated $(NH_4)_2SO_4$ (room temperature); discard supernatant

↓

Redissolve precipitate in distilled water equal to 1/10 of the starting volume (room temperature)

↓

Dialyze (4°C) against 0.15 m NaCl *until* dialysis bath is free of SO_4, as determined with $BaCl_2$

↓

Gel filtration (4°C), G-200 Sephadex in 0.15 M NaCl

↓

First protein peak collected (appears in void volume) as determined by UV absorption at 280 nm; contains IgM and α_2M, some β and α lipoprotein

↓

Concentrate by ultrafiltration[b] (room temperature)

↓

Dialyze (4°C) against barbital buffer, 0.05 M, pH 8.6

↓

Starch block electrophoresis[c] (4°C) in 0.05 M, pH 8.6 barbital buffer

↓

Elution of peak cathodal to the point of application (room temperature)

↓

Concentrate by ultrafiltration (room temperature);[b] if possible, add 0.1 to 1% serum albumin, for stability.

↓

IgM Product should be evaluated by immunoelectrophoresis as well as by analytical ultracentrifugation. A repeat cycle (from gel filtration) may be necessary. A small amount of polymerized IgM will remain as a contaminant. (If required, the polymer may be removed by gel filtration on a sepharose column.)

[a]The preparation of monoclonal IgM from Waldenström's macroglobulinia patients sera starts directly at *gel filtration*; prior concentration of the globulin fraction by $(NH_4)_2SO_4$ precipitation is not needed.
[b]See footnote b in Table 1.
[c]van Oss (1980).
Source: van Oss (1981).

1979). The only way to eliminate the influence of different electric charges of different proteins, and to use gel electrophoresis exclusively to separate proteins according to size, is to bestow an identical (and rather high) negative electrical charge to all proteins, great and small, by the admixture of sodium dodecyl sulfate (SDS) to the entire system (Stegemann, 1979). For preparative purposes a drawback of SDS-gel electrophoresis is the difficulty of removing the SDS from the proteins after the separation. One way of removing the SDS is by ion exchange chromatography at a pH below that of the iso-electric point of the protein (e.g., with DEAE-cellulose or DEAE-Sephadex), if necessary in the presence of 6 to 8 M urea, although this approach tends to give rise to loss of protein. A better way of SDS removal is to use gel electrophoresis or gel isoelectric focusing (Stegemann, 1979).

C. Ultrafiltration

Ultrafiltration with anisotropic ("skimmed") membranes is the method of choice for the reconcentration of intermediate and final fractions, after various protein-fractionation steps, which always entail considerable dilution (see Tables 1 to 3). Various commercially available protein-stopping membranes with appropriate pore sizes may be used (e.g., from Amicon, Lexington, Massachusetts; Beckman, Anaheim, California; Millipore, Bedford, Massachusetts). One can also rather easily (and much more cheaply) cast one's own (cellulose acetate) membranes (van Oss and Bronson, 1982), which, partly because they do not need to be dried for shipment, have much faster flow rates than those of any of the commercial membranes. For normal immunoglobulin concentration the CA-50 cellulose acetate membrane is the best; membranes with smaller pore sizes may be more indicated for the concentration of smaller fractions, such as L chains: the CA-35 membrane would be suitable for this (van Oss and Bronson, 1982). The preparation of one membrane has been described (CNA-15) which retains *all* IgM, but passes a fair portion (ca. 25%) of the IgG comprised in an immunoglobulin mixture (van Oss and Bronson, 1982).

D. Sedimentation

One other method for the preparative separation of immunoglobulins according to size (i.e., mainly for the separation between IgG and IgM) is (generally sucrose)-density gradient ultracentrifugation. After 16 hr at 40,000 rpm, in tubes with a sucrose (10 to 40%) gradient, in a swinging bucket rotor, IgM is found at the bottom of the tubes and IgG in their middle region. However, the separation is rarely complete and not exquisitely reproducible.

In an analytical ultracentrifuge, with a Tiselius cell (containing a porous partition), a proportion (30 to 50%) of the IgG present in a mixture can be obtained free from IgM. This process can be monitored by visual inspection so that one can determine with precision at which stage all the IgM has passed the porous barrier, leaving part of the IgG (now devoid of IgM) behind that barrier.

VI. LIGAND SPECIFICITY

Affinity chromatography, which is based on ligand specificity, such as exists between antigen between antigen and antibody, is a generalized outgrowth of the older technique of immunoadsorption. For the isolation of immunoglobulins, variants of the old immuno-adsorption technique are still very much applicable.

Table 4 Purification of Antibodies by Affinity Chromatography

Principle: Antigen, covalently linked to a solid matrix, specifically combines with antibody in whole serum or immunoglobulin fractions. Nonantibody proteins are washed away and specific antibody is eluted at low pH (glycine-HCl, pH 2.5) or with hapten.

This method is the one most widely used for specific purification of antibodies. Various solid adsorbents and matrices are commercially available. Examples of solid matrices are polymerized proteins[a,b] useful for batchwise application, or agarose beads (Sepharose), useful for affinity chromatography. Immobilized antibodies can be used similarly to isolate antigens.[b]

Antibody	Immunoadsorbent	Eluant	Yield (%)	References
Anti-ρ-azophenyl-β-lactoside	Sepharose-ρ-aminophenyl-β-lactoside	Lactose; ρ-nitrophenyl-β-lactoside	>90	c
Antihemoglobin A_1 (HbA_1)	Sepharose-HbA_1; α chain, β chain	Acetic acid (1 M)	—	c
Anti-ρ-azobenzene arsonate	Poly rabbit serum albumin (poly RSA)-ρ-azobenzene arsonate	Benzene arsonate	85	a,c
Anti-bovine serum albumin (BSA)	Poly RSA-BSA	Glycine HCl, pH 2.3	90	a,c
		Propionic acid, pH 3.5	>90	d
		50% Dimethyl sulfoxide, pH 9.5	>90	d
Anti-IgG	Poly IgG	Glycine Hcl, pH 2.8	90	b
Antidextran	Sephadex G-10, pH 7.2	60% Dimethyl sulfoxide, pH 7.2	>90	e
Blood group antibodies (anti-A, anti-K, anti-D)	Human red cells	47% Dimethyl sulfoxide, pH 9.5	>90	f

[a]Onoue et al. (1965).
[b]Avrameas and Ternynck (1969).
[c]van Oss (1981).
[d]van Oss (1979, 1980a).
[e]van Oss et al. (1982).
[f]van Oss et al. (1981b).
Source: van Oss (1981).

To begin with, immunoadsorbents are useful for the removal of, for example, unwanted immunoglobulins (see Sec. VII and Table 4), especially in the purification of IgA, IgD, and IgE. Specific immunoadsorbents made with various insolubilized antigens are especially useful for the isolation of immunoglobulins, with one given antibody activity (see Table 4).

The specific attachment step in immunoadsorption is almost invariably quite readily obtained, under physiological conditions, in aqueous media. Exactly because of the (usually) rather pronounced strength of the bond between antigens and antibodies, the subsequent detachment or dissociation (i.e., the elution) is harder to achieve. There are, however, a number of approaches that favor a change of the antigen-antibody interaction into the direction of dissociation;* for example:

1. Increase the temperature
2. Apply excess of low-molecular-weight hapten or hapten-like solutes
3. Denature or deform the antigen and/or antibody molecules
4. Destroy (e.g., enzymatically) the carrier or the spacer
5. Apply considerable dilution
6. Apply an electric field
7. Dissociate by reversing *both* the coulombic and the van der Waals attraction

A particularly interesting ligand for immunoglobulin isolation is the lectin (see Chap. 15) protein A from *Staphylocuccus aureus*, in that columns with that ligand specifically bind to human IgG, except for IgG_3 (van Loghem et al., 1982). Protein A columns are now increasingly used therapeutically, in extracorporeal hemoperfusion, for periodic removal of patients' IgG, in ex vivo treatment of immune complex diseases and also in the experimental treatment of various malignancies (Gribnau et al., 1982). Such columns can also be used in the fractionation of some IgG subclasses (Gribnau et al., 1982). In the same manner, concanavalin A (see Chap. 15) ligands can be used for the removal of IgA and IgM (Gribnau et al., 1982).

VII. ISOLATION OF IMMUNOGLOBULINS

A. Immunoglobulin G

Isolation of human IgG from whole serum is best done by ion exchange chromatography with the anion exchanger DEAE-cellulose (or DEAE-Sephadex). In order to decrease the burden of non-IgG protein and to eliminate any hemoglobulin that might be present in the serum, a preliminary salting-out step with 33% saturated $(NH_4)_2SO_4$ is recommended. IgG is the only serum protein not bound to DEAE at pH 7.5. The pH 7.5 eluate, which is virtually pure IgG, only needs to be concentrated by ultrafiltration (see Table 1; van Oss, 1980, 1981).

Subclasses of human IgG may be separated by affinity chromatography, with specific antisubclass antibodies as ligands (see Sec. VI), or use may be made of the fact that IgG3 and IgG1 are more hydrophobic than IgG2 and IgG4 (using hydrophobic chromatography, see Sec. IV.B) and of the fact that IgG3 does *not* bind to protein A (van Loghem et al., 1982). Protein A affinity chromatography can be used for the isolation of some of the rat and mouse (Ey et al., 1978) IgG subclasses. On an industrial scale IgG is recovered from fraction II of Cohn and Edsall's cold ethanol precipitation method 6 (see Schultze and Heremans, 1966).

*For approaches 1 to 7, see Absolom (1981); for 2, 3, and 7, see Table 4; for 6, see Haff (1981); for 7, see van Oss et al. (1979, 1980a, 1981a, 1982); and for 2 and 3, see Onoue et al. (1965) and Avrameas and Ternynck (1969).

B. Immunoglobulin A

To quote the discoverer of IgA and of many of its functions, Heremans (1974): "The isolation of serum IgA from serum or exudates is besieged by many difficulties." In the case of IgA isolation it is best *not* to start with a precipitation with $(NH_4)_2SO_4$, as this tends to enhance the formation of IgA dimers.

As IgA is exceptionally hydrophilic, it can also be isolated from other serum proteins by (salt-mediated) hydrophobic chromatography (Doellgast and Plaut, 1976; see Sec. IV). Secretory IgA can be isolated in the same manner as IgA (see Table 2), but, in view of the higher molecular weight of sIgA, the second (and third) gel filtration steps should be done with Sepharose 6B. For the isolation of J chain and secretory component, see Sec. VII.G. IgA is found in fractions II and III of Cohn and Edsall's cold ethanol precipitation method 6 (see Schultze and Heremans, 1966).

C. Immunoglobulin M

IgM is rather easy to isolate from serum, by first obtaining the macroglobulin fraction by gel filtration, followed by the removal of α_2 macroglobulin by block electrophoresis (Table 3). Pure IgM has a strong tendency to form complexes (see Sec. II.B); to obviate complex (and even precipitate) formation of purified IgM preparations it is advisable to conserve pure IgM solutions at +4°C, in the presence of, for example, 0.1 to 1% serum albumin. Monomeric IgM (IgMs) can be obtained by mild reduction at neutral pH with 0.015 M 2-mercaptoethylamine (Nisonoff et al., 1975).

D. Immunoglobulin D

IgD can be purified by gel filtration on Ultrogel AcA 34, from which it elutes in a position between IgG and IgM, followed by DEAE ion exchange chromatography, where it elutes at pH 8.0, with a 0.035 M phosphate buffer (Jefferis, 1976). The fractions should be monitored for IgD and ultrafiltration steps used for reconcentration where appropriate (see the third footnote in Table 1) (van Oss and Bronson, 1982).

E. Immunoglobulin E

IgE is difficult to isolate because of its exceedingly low concentration in plasma. Precipitation with 40% saturated $(NH_4)_2SO_4$ (at pH 7 at room temperature) should be the first step, followed by ion exchange chromatography on DEAE-Sephadex A-50, with elution at pH 8 between 0.1 and 1 M (Tris buffer), with a final series of thrice-repeated gel filtrations on Sephadex G-150 (Bennich and Johansson, 1971). All fractions should be monitored for IgE, and ultrafiltration steps should be applied at regular intervals between steps to reconcentrate the active fractions (see the third footnote in Table 1) (van Oss and Bronson, 1982).

F. Monoclonal Immunoglobulins

All the isolation procedures discussed above are also valid for monoclonal immunoglobulins. However, in all cases of monoclonal immunoglobulins, care should be taken that the chromatography columns used are not overloaded. The isolation of monoclonal IgA should be done somewhat differently from the separation of IgA from normal serum (see the second footnote in Table 2). In many cases of monoclonal immunoglobulins polymers

may be encountered, so that gel filtration eluates especially should be monitored with the possibility in mind that active fractions may occur in several different molecular size cutoffs. With monoclonal IgM the preliminary precipitation with $(NH_4)_2SO_4$ is not necessary (see the first footnote in Table 3).

VIII. ISOLATION OF IMMUNOGLOBULIN CHAINS

A. Products of Enzymatic Cleavage

Fab and Fc fragments are obtained by papain digestion of IgG (ca. 1 mg of papain per 100 mg of IgG), for 18 hr at 37°C (in the presence of 0.01 M cysteine, at neutral pH, and with the addition of a drop of toluene to prevent spoilage). $F(ab')_2$ pieces are obtained by pepsin digestion of IgG at pH 4.5 (in 1 M acetate buffer and 1 drop of toluene) for 8 hr at 37°C; the reaction is stopped by increasing the pH to 8.

Undigested molecules should be removed from papain digests by gel filtration on Sephadex G-150. Fab fractions are then obtained by ion exchange on DEAE (cellulose or Sephadex) columns, while Fc fragments are best isolated by ion exchange on carboxymethyl (cellulose or Sephadex) (Deutsch, 1967). Undesired pepsin fragments are best simply removed by dialysis, or the $F(ab')_2$ may be purifed by precipitation with Na_2SO_4 or $(NH_4)_2SO_4$. Most recipes for enzymatic fragments of rabbit IgG also work for human IgG.

It is more difficult to obtain Fc fragments of IgA, but Fab and $F(ab')_2$ can be obtained with papain and pepsin, respectively, in much the same manner as that used for IgG (Nisonoff et al., 1975). With a bacterial endopeptidase (IgA protease), IgA as well as sIgA can be cleaved to form Fc and Fab fragments (Plaut et al., 1974). With IgM it is also rather difficult to obtain Fc fragments, although with trypsin some success has been noted; trypsin also appears to be the enzyme of choice for obtaining Fab_μ (Nisonoff et al., 1975). Tryptic digestion is also the preferred way for obtaining Fab and Fc fragments of IgD (Spiegelberg et al., 1970). IgE yields an Fc fragment upon digestion with papain which does, however, tend to degrade further. Fab is not easily obtained with IgE: short-term digestion is called for and even then the Fab obtained is labile. With pepsin, an $F(ab')_2$ fragment is obtained which also still contains part of the Fc fraction (Nisonoff et al., 1975). With trypsin, a slightly smaller $F(ab')_2$ fragment is obtained (Bennich and Johansson, 1971).

B. H and L Chains and Domains

Dissociation of immunoglobulins into chains is achieved by reduction (with, e.g., 0.1 M 2-mercaptoethanol or dithiothreitol) in the presence of 8 to 10 M urea at pH 8, followed by alkylation with 0.2 M iodoacetamide (Edelman and Marchalonis, 1967; Nisonoff et al., 1975). To prevent reassociation and/or aggregation, the chains should be kept in the presence of 8 to 10 M urea or 0.5 to 1 M propionic acid. Separation of the chains can be done by CM-cellulose ion exchange (in urea), or by gel filtration with Sephadex G-100 in 0.5 M propionic acid (Edelman and Marchalonis, 1967). Treatment of IgG with 0.1 M mercaptoethylamine, followed by addition of p-chloromercuribenzoate and lowering of the pM to 2.5, results in half-molecules of IgG (Edelman and Marchalonis, 1967).

The greater part (98%) of IgA2, of the Am group A2m (1) (see Chap. 11), has no disulfide bond between H and L chains; these H and L chains can thus be separated without any other treatment, by gel filtration on Sephadex G-200, in the presence of 8 M

urea (Grey et al., 1968). Separated H and L chains of antibodies with a given specificity, spontaneously and preferentially recombine, re-forming antibody molecules with most of the original binding activity (Roholt et al., 1965). Some (but not all) of the immunoglobulin *domains* can be obtained by digestion with trypsin or plasmin of H or L chains or fragments at neutral pH, after prior treatment of these chains or fragments at pH 2.5 (Dorrington and Painter, 1974).

C. Secretory Component of IgA

Human secretory component of IgA (SC) can be prepared from human colostrum whey, by precipitation with 70% saturated $(NH_4)_2SO_4$, gel filtration on Sephadex G-200, reprecipitation with 70% saturated $(NH_4)_2SO_4$, and ion exchange chromatography on DEAE-cellulose (Brandtzaeg, 1974).

D. J Chain of IgA or IgM

The isolation of J chain from IgM or from dimeric IgA or sIgA is not easy, because it represents only 1.5% of the total weight of IgM and 4.5 to 4.0% of dimeric IgA and sIgA, respectively. From IgM it can be isolated after the reduction and alkylation needed to form IgMs (see above), after which it can be found in the 15,000 to 25,000 molecular weight fraction by gel filtration on Sephadex G-100, and further separated from possible single L chains by electrophoresis or by ion exchange chromatography; J chain is more strongly negatively charged than are L chains (Nisonoff et al., 1975). J chains can also be obtained from α chains (obtained via reduction and alkylation of dimeric IgA or sIgA that were isolated by gel filtration on Sephadex G-100), followed by further cleavage with 50nM dithiothreitol and alkylation with 120 mM iodoacetic acid and further gel filtration (in 1 M acetic acid) on Sephadex G-100. J chain is quite hydrophilic, asymmetrical (axial ratio ca. 18:1) and has a molecular weight of about 15,000 (Wilde and Koshland, 1973).

IX. DETERMINATION OF MOLAR ANTIBODY CONCENTRATION

A. Introduction

For all quantitative determinations pertaining to antigen-antibody interactions (especially for determinations of thermodynamic parameters, see Chap. 16), it is necessary to know the (molar) antigen and antibody concentrations (usually bound as well as free). In most cases the determination of antigen concentrations is a simple matter, as one can (usually) start with the pure antigenic material and, if necessary, one can have it tagged with, for example, a radioisotope. Determinations of the molar concentration of a specific anti*body*, be it in whole antiserum or in an isolated immunoglobulin fraction, in the presence of a vast excess of irrelevant immunoglobulins and other proteins, is more complicated. Basically, one can determine the specific antibody concentration (1) through interaction with a known amount of antigen, or (2) after specific adsorption to and elution from the insolubilized antigen.

B. Interaction with a Known Amount of Antigen

Although antigens and antibodies can interact nonstoichiometrically, in the most diverse proportions (see Chap. 16 and 17), there are two conditions under which the antigen-antibody ratio is known, or can be simply ascertained. These are (1) at antigen excess,

when all specific antibody-active sites are bound (see, e.g., Chap. 17; Steward and Petty, 1972), or (2) by precipitation done at the optimal antigen/antibody ratio (with neither free antigen nor free antibody in the supernatant), in which the total amount of antibody equals the total amount of precipitate minus the amount of antigen added. When determinations are done at antigen excess, in most cases soluble complexes are obtained (see Chaps. 17 and 18). Here the amount of antigen added must be known, as well as the amount of antigen in the complexes, which can be isolated by, for example, precipitation with $(NH_4)_2SO_4$, with PEG-6000 (see Sec. II), or with anti-immunoglobulin antibodies (Steward and Petty, 1972). Bound and free (radiotagged) antigen can also be determined by gel electrophoresis (Steward and Petty, 1972). As indicated above (Sec. II.B), precipitation of soluble antigen-antibody complexes with $(NH_4)_2SO_4$, while useful in many cases, may cause the *dissociation* of certain types of complexes (de Groot et al., 1980; Smeenk and Aarden, 1980). Finally, via radioimmunoassay (Chap. 19) and enzyme immunoassay (Chap. 20), it also is possible to determine specific antibody concentrations once antigen concentrations are known. To obtain the *molar* antibody concentration, it is of course necessary to know the molecular weight of the antibody in question, and in case of antibody heterogeneity, to effect a further fractionation (see below).

C. Affinity Procedures

The other way of knowing the exact concentration of the specific antibody one is working with is to isolate it by immunoadsorption on (and subsequent elution from) the (insolubilized) antigen in question. In such a case the eluted protein consists solely of specific antibody, and the concentration of that protein is the concentration of the specific antibody in question.

To obtain the *molar* antibody concentration of the preparation, one must of course know of which immunoglobulin(s) it consists. If (dekavalent) IgM as well as other (divalent or tetravalent) immunoglobulin classes are present, an additional separation according to size (e.g., by gel filtration; see Sec. V.A) may be required.

X. CHARACTERIZATION OF IMMUNOGLOBULINS

A. Characterization Methods Treated Elsewhere in This Work

Basically, the characterization of immunoglobulins is a vast subject that deserves a chapter by itself. However, in this work many aspects of immunoglobulins that essentially pertain to their characterization are treated in other chapters. (Chapter 7 on the structure and function of the immunoglobulins, and Chapter 8 on monoclonal immunoglobulins, touch on the subject; so do the chapters on allotypes (Chap. 11) and idiotypes (Chap. 12). The chapters on thermodynamics of antigen-antibody interaction (Chap. 16), agglutination and precipitation (Chap. 17), as well as those on radio, enzyme, and fluorescent immunoassay methods (Chaps. 19 to 21) and immunoelectronmicroscopy (Chap. 22) deal mainly with various aspects of immunoglobulin characterization. A chapter such as the one on the structure and function of complement proteins (Chap. 23) also touches on important aspects of immunoglobulin characterization. Even the chapters on macrophage and PMN receptors (Chap. 29), and on cell adhesion and phagocytic engulfment (Chap. 31), deal with various aspects of immunoglobulin characterization; in the latter chapter methodology as well as results are given of the determination of the surface tensions of the major immunoglobulins (e.g., of IgG1, IgG2, IgG3, IgG4, IgA, and IgM; see also Sec. X.E).

This final section is thus limited to (1) the immunoglobulin characterization methods, of which the principles have already been touched on above, under the treatment of the various isolation methods of immunoglobulins (see Secs. X.B to X.E); and (2) a few other methods, not treated elsewhere, that pertain to the characterization of the surface tension and of the primary, secondary, and tertiary configurations of the immunoglobulin molecules treated in Secs. X.F to X.H).

B. Electrokinetic Methods

Analytical polyacrylamide gel electrophoresis gives the best resolution when done with discontinuous buffers or in "multiphasic buffer systems"; for the most advanced state of the art of this method, see Chrambach (1979) (see Sec. III.B). However, *analytical isoelectric focusing* is fundamentally capable of even higher resolution (with respect to charge). The method has recently been well summarized by Braun et al. (1979), giving examples of multiclonal and oligoclonal antibody responses (see also Righetti, 1979). There is, however, a limitation to the method, in that it is not possible to work with immunoglobulins that tend to precipitate at low ionic strength (i.e., with "euglobulins"; see Sec. II.A).

In such cases it is preferable to use *analytical isotachophoresis*. This technique has been well described by Bier and Allgyer (1979), and its application to immunoglobulin (especially "euglobulin") characterization by Ziegler and Köhler (1979) (see also Sec. III.C).

Electrophoresis in gels with graded porosity permits the distinction of a rather wide range of molecular weights (Gianazza and Righetti, 1979). This technique is especially powerful in the simultaneous discrimination between antigens, antibodies, and soluble immune complexes (see Chap. 18).

Bidimensional combinations (Giannazza and Righetti, 1979; Sidman, 1981) of, for example, isoelectric focusing and SDS-gel electrophoresis (see Sec. V.B), are among the characterization methods that permit the highest resolution at present attainable (see Sec. I.C). Even in monoclonal immunoglobulins there is some microheterogeneity, demonstrable by the "high-resolution bidimensional electrophoresis" method (O'Farrell, 1975), using gel isoelectric focusing in one direction and SDS gel electrophoresis in the other. This permits distinguishing between electric charge and molecular weight heterogeneities. In their study of monoclonal immunoglobulins (see Chap. 8), Latner et al. (1980) could thus show that microheterogeneities reside only in the H chains and not in the L chains. The microheterogeneities of monoclonal γ, α, and μ chains are based mainly on charge and not on size differences, while in monoclonal δ chains heterogeneities exist in both charge and size. The charge heterogeneities of monoclonal α, μ, and δ chains (but not of γ chains) and due partly to sialic acid moieties. For actual molecular weight determinations in the SDS-gel electrophoresis direction, as well as in graded porosity gel electrophoresis, one always must interpolate between two polymers of known molecular weight, and even then errors can arise when the asymmetries of the known and the unknown proteins are dissimilar. For accurate molecular weight determinations, see Sec. X.D).

C. Chromatographic Methods

High-pressure liquid chromatography (HPLC; also high-performance liquid chromatography) is, in a variety of modes, being applied more and more to the characterization of small and large peptides and of proteins (Regnier and Gooding, 1980).

Reversed-phase (or hydrophobic) *liquid chromatography* (RPLC) is probably one of the most useful new techniques, as it separates proteins and peptides according to their relative hydrophobicity, which is not easily duplicated by other analytical methods (see Sec. IV.B).

Gel permeation HPLC is a convenient approach to the estimation of the molecular weight of proteins. To that end one must have a number of polymers at one's disposal, to serve as molecular weight markers, and one must arrange these markers in such a way that the unknown protein(s) emerge from the column in between two or more known ones. This also holds for ordinary *gel filtration* (see Sec. V.A) when used as a method for determining molecular size. In all these cases it must be remembered that a significantly greater (or lesser) deviation from the symmetry of the unknown protein(s) than of the molecular weight markers will lead to erroneous results, even if the interpolation method is adhered to properly. The ultimate criterion of molecular size determination of proteins still resides in direct measurement of molecular weights *and* asymmetries by analytical ultracentrifugation (see Sec. X.D).

Ion exchange HPLC, for the characterization of immunoglobulins and/or immunoglobulin chains, is feasible, but in the face of the existence of electrophoretic, immunoelectrophoretic, gel electrophoretic, and other electrokinetic methods (see Sec. X.B), this approach has little to commend it.

D. Sedimentation

With all the secondary methods for determining molecular weights of immunoglobulins, immunoglobulin chains, and other polypeptides and proteins that are used increasingly, such as gel filtration chromatography (or gel permeation HPLC) or SDS-gel electrophoresis (see Secs. X.B and X.C), one must not lose sight of the fact that in all these cases control biopolymers of known molecular weights must be used for interpolation and estimation of the molecular sizes of the unknown molecules or chains. Such interpolations indeed at best yield only estimations, as it is unlikely that the unknown has the same degree and type of asymmetry as the known molecules. In the final analysis the molecular weights of the known molecules have been determined earlier by analytical ultracentrifugation. Thus there are always reasons why one may want to (or have to) go back to a sedimentation technique for the precise and unequivocal determination of the molecular weight, as well as of the degree of asymmetry, of an immunoglobulin, an immunoglobulin chain, or any other protein or polypeptide. It thus remains useful to recapitulate what can and what cannot be determined by various analytical sedimentation techniques (see, e.g., Bowen, 1970).

Sedimentation rates have the considerable advantage that they can be determined even in heterogeneous mixtures (using, e.g., schlieren optics). The sedimentation constant(s) obtained have the drawback that the *molecular weight* (M) can be obtained from them [via the Svedberg equation: $M = RTs/(1 - \bar{v}\rho)D$] (see Bowen, 1970; Cantor and Schimmel, 1980) only if the diffusion constant (D) is also known. However, nowadays diffusion constants of a given protein can also be determined in heterogeneous mixtures, as long as an antibody to that protein is available; double diffusion at right angles (Allison and Humphrey, 1960) will then yield D (see also Chap. 17). Once both M and D, or both s and D, or both s and M, are known, the friction factor ratio f/f_0, expressing the ratio between the actual diffusion coefficient and the diffusion coefficient a molecule of the same molecular weight would have if it were completely spherical, and therefrom the

asymmetry ratio can be determined (Svedberg and Pederson, 1940; Bowen, 1970; Cantor and Schimmel, 1980). If a protein is available in the purified state, M can also be directly determined (without knowing D) by *sedimentation equilibrium* or *approach-to-equilibrium methods* (Bowen, 1970). Sedimentation *rate* determination is a useful check for molecular weight homogeneity (provided that at least 0.5 ml, containing 2 to 5 mg protein, is available). However, small amounts of contaminants (<2%) cannot be detected by that approach.

E. Surface Tension

From the order in which immunoglobulins and other plasma proteins emerge from a hydrophobic (or reversed-phase) chromatography column (see Sec. IV.B), one can obtain a qualitative estimation of the surface tension of these proteins. For the quantitative determination of the surface tension of proteins (γ_{PrV}), two methods may be used: contact angle measurement or protein adsorption determination.

Contact angle measurement is by far the simplest method for obtaining γ_{PrV} (van Oss et al., 1981a). One must, however, take great care not to denature the protein by exposure of too thin a layer to the air interface (Absolom et al., 1981). To that effect, a thick hydrated layer of concentrated protein must be deposited by ultrafiltration on an anisotropic membrane. The contact angle θ, made by a drop of saline water, deposited on top of the (only slightly dried) hydrated protein layer (attached to the membrane) can then be measured, and from the contact angle θ, γ_{PrV} may be obtained [e.g., by means of tables (Neumann et al., 1980) based on an equation of state].

Protein adsorption from solutions onto different polymer surfaces of various surface tensions while immersed in liquids of the various surface tensions will also yield γ_{PrV} (van Oss et al., 1981a). This is done by finding (by interpolation) the one liquid in which the degree of adsorption of a given protein (in $\mu g/cm^2$) is the same for *all* different polymer surfaces. The surface tension γ_{LV} of that liquid is equal to γ_{PrV}. This method is more laborious than the contact angle method, but while totally different from and unrelated to the contact angle method, it serves the useful purpose of giving the same value of γ_{PrV} for the same protein (within ca. $\pm$ 0.3 erg/cm^2).

Charge-shift electrophoresis is an electrophoretic method, using three types (anionic, nonionic, and cationic) of surfactant as probes for hydrophobic moieties on protein molecules (see, e.g., Sidman, 1981). When the hydrophobic tail of surfactants can bind to a hydrophobic moiety of a protein, anionic surfactants will make the protein more negatively charged, cationic surfactants less negatively charged, and nonionic surfactants should have no influence. The use of all three types of surfactant is advised, to detect whether one of them attaches to the protein via an ionic (instead of a hydrophobic, or van der Waals) interaction, in which case one of the ionic ones would not cause a change in the protein's electrophoretic mobility. If none of the three types of surfactant causes a change in electrophoretic mobility of a protein, that protein is unlikely to have a prominent hydrophobic moiety. The method has yielded interesting results in the study of cell membrane proteins, but does not show great promise in the characterization of the (very hydrophilic) plasma proteins. Still, for further study of the relatively hydrophobic Fc tails of some immunoglobulins (e.g., of IgG1 and IgG3), the method might be useful.

Surface tensions of protein solutions yield no information about the surface tension of the dissolved protein itself, but simply reflect the degree of adsorption of the protein at the air-liquid interface in the measuring device, and the concomitant denaturation of the

protein (Absolom et al., 1981). However, changes in the surface tension of a solution of a synthetic polypeptide (poly-L-lysine), as a function of temperature and pH, correlate well with changes in the secondary conformation of that polymer, as measured by circular dichroism. At the pK value of poly-L-lysine (pH 10.4) a minimum was found in the surface tension of the solution (at all temperatures), implying that at that pH the surface concentration of hydrophobic groups of the polypeptide molecule is at a maximum (Neumann et al., 1973; see also Sec. X.G).

The *surface tensions of proteins* (γ_{PrV}) in the hydrated native state is generally the deciding factor in the degree to which a protein becomes aspecifically adsorbed to various surfaces (see Sec. IV.A). This explains the aspecific opsonization of hydrophobic bacteria by IgG1 and IgG3 (which have rather hydrophobic Fc tails) (see Chap. 31). From the surface tension of a protein, its van der Waals (or Hamaker) coefficient can be calculated, via Eqs. (1a) and (7) (see Sec. IV.A; see also Chap. 16).

F. Primary Configuration

The amino acid sequence of a peptide chain and the way in which two or more peptides chains are covalently bound together form the primary configuration of a polypeptide or protein. Several decades ago the determination of the amino acid sequence of peptides was an exceedingly laborious enterprise. With the advent of automatic protein sequencers (Beckman Instruments, Palo Alto, California) (Putnam, 1973), determination of the primary configuration of even quite large proteins became a matter of routine. For the determination of the primary configuration of immunoglobulins, sequencing is possible only with monoclonal immunoglobulins. A number of amino acid sequences of human (monoclonal) γ, α, μ, δ, ϵ, κ and λ chains, showing the variable as well as the constant parts, compiled by A. L. Grossberg, are given by van Oss (1981) (see also Nisonoff et al., 1975). Automatic protein synthesizers, combined with protein sequencers, form increasingly important tools in the immunochemical study of the complete antigenic makeup of proteins (see Chaps. 1 and 2). Knowledge of the primary configuration of protein antigens as well as of immunoglobulins remains of crucial importance in the understanding of their spatial (secondary and tertiary) configurations, as well as of their immunochemistry (Chap. 2).

G. Secondary Configuration

The secondary configuration of proteins alludes to their first-order spatial folding, for example, in α helices, β-pleated sheets, or random coil conformation (Phillips and North, 1973). The secondary configuration is usually determined by one of two related optical methods: circular dichroism (CD) and optical rotatory dispersion (ORD). CD (usually the method of choice) measures the difference between right and left circularly polarized light, expressed as the difference in optical rotation in degrees (per mole of protein solution the light traverses) versus the wavelength. In ORD the optical rotation of light passing through a column of protein solution is plotted versus the wavelength. ORD spectra essentially depict the first derivative of CD spectra (of a given protein) (Cantor and Schimmel, 1980). CD spectra allow differentiation between α helices, β sheets, and random coils. Surface tension measurements of simple polypeptides (such as poly-L-lysine) also allow the measurement of conversion form α helix to β sheet in good agreement wtih CD determinations (Neumann et al., 1973; see Sec. X.E).

H. Tertiary Configuration

The actual folding of the α helix and/or β sheet of a protein into its final three-dimensional structure results in the tertiary configuration. A variety of methods generally are used in conjunction with each other for the complete unraveling of the tertiary configuration. To begin with, the primary configuration of a protein must be known, and at least a fairly good knowledge of the secondary configuration. Then the method of choice is x-ray diffraction of crystals of the pure protein (Cantor and Schimmel, 1980; Nisonoff et al., 1975). The high resolution of this method is due to the exceedingly small wavelength of x-rays (ca. 1 Å). X-rays are scattered by the electrons within the atoms of the crystal, giving rise to diffraction patterns, due to interference among the scattered radiations from the various atoms. The diffraction patterns are analyzed via Fourier transforms, to yield, after much trial and error, an electron density pattern, corresponding to the pattern corresponding to the atoms in the protein crystal, in one given layer. Many analyses of this type ultimately permit the construction of a three-dimensional representation of the atoms in the protein. For this it is essential to know not only the primary configuration of the protein, but also its real three-dimensional shape, obtained from, for example, hydrodynamic sedimentation data (see Sec. D) or electron microscopy (see Chap. 22; see also Nisonoff et al., 1975).

REFERENCES

Absolom, D. R. (1981). Sep. Purif. Methods *10*, 239.

Absolom, D. R., van Oss, C. J., Zingg, W., and Neumann, A. W. (1981). Biochim. Biophys. Acta *670*, 74.

Absolom, D. R., van Oss, C. J., Zingg, W., and Neumann, A. W. (1982). J. Reticuloendothel. Soc. *31*, 59.

Allison, A. C., and Humphrey, J. H. (1960). Immunology *3*, 95.

Avrameas, S., and Ternynck, J. (1969). Immunochemistry *6*, 53.

Bennich, H., and Johansson, S. G. O. (1971). Adv. Immunol. *13*, 1.

Bier, M., and Allgyer, T. T. (1979). In *Electrokinetic Separation Methods* (P. G. Righetti, C. J. van Oss, and J. W. Vanderhoff, Eds.). Elsevier, Amsterdam, pp. 443–469.

Bowen, T. J. (1970). *An Introduction to Ultracentrifugation*. Wiley-Interscience, New York.

Brandtzaeg, P. (1974). Scand. J. Immunol. *3*, 579.

Braun, D. G., Hild, K., and Ziegler, G. (1979). In *Immunological Methods*, Vol. 1 (I. Lefkovits and B. Pernis, Eds.). Academic Press, New York, pp. 107–121.

Cantor, C. R., and Schimmel, P. R. (1980). *Biophysical Chemistry*, Part II, W. H. Freeman, San Francisco.

Chrambach, A. (1979). In *Electrokinetic Separation Methods* (P. G. Righetti, C. J. van Oss, and J. W. Vanderhoff, Eds.). Elsevier, Amsterdam, pp. 275–292.

Chrambach, A., and Nguyen, N. Y. (1979). In *Electrokinetic Separation Methods* (P. G. Righetti, C. J. van Oss, and J. W. Vanderhoff, Eds.). Elsevier, Amsterdam, pp. 337–368.

de Groot, E. R., Lamers, M. C., Aarden, L. A., Smeenk, R. J. T., and van Oss, C. J. (1980). Immunol. Commun. *9*, 515.

Deutsch, H. F. (1967). Methods of Immunol. Immunochem. *1*, 1.

Doellgast, G. J., and Plaut, A. G. (1976). Immunochemistry *13*, 135.

Dorrington, K. J., and Painter, H. (1974). In *Progress in Immunology II*, Vol. 1 (J. Brent and J. Holborow, Eds.). North-Holland, Amsterdam, pp. 75–84.

Edelman, G. N., and Marchalonis, J. J. (1967). Methods Immunol. Immunochem. *1*, 405.

Ey, P. L., Prowse, S. J., and Jenkin, C. R. (1978). Immunochemistry *15*, 429.

Gianazza, E., and Righetti, P. G. (1979). In *Electrokinetic Separation Methods* (P. G. Righetti, C. J. van Oss, and J. W. Vanderhoff, Eds.). Elsevier, Amsterdam, pp. 293–311.

Grey, H. M., Abel, C. A., Yount, W. J., and Kunkel, H. G. (1968). J. Exp. Med. *128*, 1223.

Gribnau, T. C. J., Visser, J., and Nivard, R. J. F., Eds. (1982). *Affinity Chromatography and Related Techniques*. Elsevier, Amsterdam, pp. 539–555.

Haff, L. A. (1981). Electrophoresis *2*, 287.

Hamaker, H. C. (1937). Physica *4*, 1058.

Heremans, J. F. (1974). In *The Antigens*, Vol. 2 (M. Sela, Ed.). Academic Press, New York, pp. 365–522.

Jaton, J. C., Brandt, D. C., and Vassali, P. (1979). In *Immunological Methods* (I. Lefkovits and B. Pernis, Eds.). Academic Press, New York, pp. 43–67.

Jefferis, R. (1976). J. Immunol. Methods *9*, 231.

Latner, A. L., Marshall, T., and Gambie, M. (1980). Electrophoresis *1*, 82.

Mattock, P., Aitchison, G. F., and Thomson, A. R. (1980). Sep. Purif. Methods *9*, 1.

Neumann, A. W., Moscarello, M. A., and Epand, R. M. (1973). Biopolymers *12*, 1945.

Neumann, A. W., Absolom, D. R., Francis, D. W., and van Oss, C. J. (1980). Sep. Purif. Methods *9*, 69.

Nisonoff, A., Hopper, J. E., and Spring, S. B. (1975). *The Antibody Molecule*. Academic Press, New York.

O'Farrell, P. H. (1975). J. Biol. Chem. *250*, 4007.

Onoue, K., Yagi, Y., and Pressman, D. (1965). Immunochemistry *2*, 181.

Phillips, D. C., and North, A. C. T. (1973). *Protein Structure*. Oxford University Press, London.

Polson, A. (1977). Prep. Biochem. *7*, 192.

Polson, A., Potgieter, G. M., Largier, J. F., Mears, G. E. F., and Joubert, F. J. (1964). Biochim. Biophys. Acta *82*, 463.

Plaut, A. G., Genco, R. J., and Tomasi, T. B. (1974). J. Immunol. *113*, 289.

Putnam, F. W. (1973). *Fractions*, 1.

Quast, R. (1979). In *Electrokinetic Separation Methods* (P. G. Righetti, C. J. van Oss, and J. W. Vanderhoff, Eds.). Elsevier, Amsterdam, pp. 221–227.

Regnier, F. E., and Gooding, K. M. (1980). Anal. Biochem. *103*, 1.

Richey, J., and Beadling, L. (1981). Am. Lab. *13*, 100.

Righetti, P. G. (1979). In *Electrokinetic Separation Methods* (P. G. Righetti, C. J. van Oss, and J. W. Vanderhoff, Eds.). Elsevier, Amsterdam, pp. 389–441.

Roholt, O. A., Radzimski, G., and Pressman, D. (1965). J. Exp. Med. *122*, 785.

Schultze, H. E., and Heremans, J. F. (1966). *Molecular Biology of Human Proteins*, Vol. 1. Elsevier, Amsterdam.

Sidman, C. (1981). In *Immunological Methods*, Vol. 2 (I. Lefkovits and B. Pernis, Eds.). Academic Press, New York, pp. 57–74.

Sluyterman, L. A. A., and Wijdenes, J. (1977). In *Electrofocusing and Isotachophoresis* (B. J. Radola and D. Graesslin, Eds.). De Gruyter, Berlin, 463–466.

Smeenk, R. J. T., and Aarden, L. A. (1980). J. Immunol. Methods *39*, 165.

Spiegelberg, H. L., Prahl, J. W., and Grey, H. M. (1970). Biochemistry *9*, 2115.

Stegemann, H. (1979). In *Electrokinetic Separation Methods* (P. G. Righetti, C. J. van Oss, and J. W. Vanderhoff, Eds.). Elsevier, Amsterdam, pp. 313–336.

Steward, M. W., and Petty, R. E. (1972). Immunology *22*, 747.

Svedberg, T., and Pedersen, K. O. (1940). *The Ultracentrifuge*. Clarendon, Oxford (repr. Johnson, New York, 1959), pp. 38–42.

van Loghem, E., Frangione, B., Recht, B., and Franklin, E. C. (1982). Scand J. Immunol., *15*, 275.

van Oss, C. J. (1979). Sep. Purif. Methods *8*, 119.
van Oss, C. J. (1980). In *Methods in Immunodiagnosis* (N. R. Rose and P. E. Bigazzi, Eds.). Wiley, New York, pp. 229–244.
van Oss, C. J. (1981). In *Blood Banking*, Vol. 2 (T. J. Greenwalt and E. A. Steane, Eds.). CRC Press, Boca Raton, Fla., pp. 181–205.
van Oss, C. J., and Bronson, P. M. (1982). In *Treatise on Analytical Chemistry*, Part I, Vol. 5 (P. J. Elving, E. Grushka, and I. M. Kolthoff, Eds.). Wiley-Interscience, New York, pp. 239–249.
van Oss, C. J., Absolom, D. R., and Neumann, A. W. (1979). Sep. Sci. Technol. *14*, 305.
van Oss, C. J., Absolom, D. R., and Neumann, A. W. (1980a). Colloids Surfaces *1*, 45.
van Oss, C. J., Absolom, D. R., and Neumann, A. W. (1980b). Colloids Polym. Sci. *258*, 424.
van Oss, C. J., Absolom, D. R., Neumann, A. W., and Zingg, W. (1981a). Biochim. Biophys. Acta *670*, 64.
van Oss, C. J., Beckers, D., Engelfriet, C. P., Absolom, D. R., and Neumann, A. W. (1981b). Vox Sang. *40*, 367.
van Oss, C. J., Absolom, D. R., and Neumann, A. W. (1982). In *Affinity Chromatography and Related Techniques* (T. C. J. Gribnau, J. Visser, and R. J. F. Nivard, Eds.). Elsevier, Amsterdam, pp. 29–37.
Visser, J. (1972). Adv. Colloid Interface Sci. *3*, 331.
Weening, R. S., Roos, D., Weemaes, C. M. R., Homan-Müller, J. W. T., and van Schaik, M. L. J. (1976). J. Lab. Clin. Med. *88*, 757.
Wilde, C. E., and Koshland, M. E. (1973). Biochemistry *12*, 3218;
Ziegler, A., and Köhler, G. (1979). In *Immunological Methods*, Vol. 1 (I. Lefkovits and B. Pernis, Eds.). Academic Press, New York, pp. 131–136.
Zubler, R. H., and Lambert, P. H. (1978). Prog. Allergy *24*, 1.

14

Origins of Immunoglobulin Diversity

Philip Early School of Medicine, University of California, San Francisco, California

I. INTRODUCTION

Very rapid advances are occurring in our understanding of the mechanisms used by the cells of the immune system to generate the broad spectrum of functionally distinct immunoglobulin molecules. Much of this progress has been due to the application of techniques from molecular biology. This chapter pertains to the results of these studies as they affect current views of the origins of immunoglobulin variable region diversity, isotype differences, and the membrane-bound and secreted forms of immunoglobulins.

It is assumed that the reader is familiar with the structural and functional properties of immunoglobulins (see Chap. 7). Texts on the molecular biology of immunoglobulin diversity tend to become outdated quickly. The bibliography has therefore been chosen from original papers which made key advances or which contain the most recent work on a particular question. Many of the references are to studies using the BALB/c strain of inbred mouse. Knowledge of human immunoglobulin genes is not yet as complete as it is for the mouse, but the data indicate that the mechanisms for generating immunoglobulin diversity in the two species are substantially the same. Table 1 compares the human and murine immunoglobulin gene families.

II. VARIABLE REGION DIVERSITY

A. Germ Line Genes

An immunoglobulin molecule (Fig. 1) is a heterodimer of two identical light (L) and two identical heavy (H) chains (see Chap. 7). A lymphocyte synthesizing immunoglobulin contains one gene transcribing RNA which encodes the L chain and another gene transcribing RNA encoding the H chain. The transcription of immunoglobulin genes is regulated

Table 1 Human and Murine Immunoglobulin Genes

Humans	Mouse
Three C_λ gene segments known, possibly six more	Four λ loci, each with one V_λ, one J_λ, and one C_λ gene segment
One κ locus, containing multiple V_κ gene segments, four J_κ gene segements, and one C_κ gene segment	One κ locus, containing multtiple V_κ gene segments, four J_κ gene segments, and one C_κ gene segment
One H-chain locus, containing multiple V_H gene segments, multiple D gene segments, six J_H gene segments, and at least nine C_H gene segments: μ, δ, $\gamma 1$, $\gamma 2$, $\gamma 3$, $\gamma 4$, ϵ, $\alpha 1$, and $\alpha 2$	One H-chain locus, containing multiple V_H gene segments, multiple D gene segments, four J_H gene segments, and eight C_H gene segments: μ, δ, $\gamma 3$, $\gamma 1$, $\gamma 2b$, $\gamma 2a$, ϵ, and α

by DNA rearrangement. In this sense, immunoglobulin genes are somatically created: they are only found as complete units in the DNA of lymphoid cells.

The DNA of all cells in vertebrates contains arrays of immunoglobulin gene segments. These gene segments are assembled into immunoglobulin genes in lymphocytes. The assembly process will be considered in detail in the next section, but first we will look at the organization of the immunoglobulin gene segments as they are found in germ line DNA (the DNA an organism inherits from its parents). Just as there are three general types of immunoglobulin polypeptide chains—the heavy chains and the λ and κ light chains—so there are three unlinked families of immunoglobulin genes encoding them. In humans, H-chain genes are on chromosome 14, λ genes on chromosome 22, and κ genes on chromosome 2 (Kirsch et al., 1982; Rowley, 1982). In mice, the corresponding chromosomes are 12, 18, and 6. The organization of the three immunoglobulin families in mice is depicted schematically in the top portion of Fig. 2.

The simplest known immunoglobulin genes are the mouse λ genes. They were also the first immunoglobulin genes to be analyzed using molecular biology techniques (Brack et al., 1978). There are four known loci of gene segments encoding mouse λ chains. The structure of each locus is the same, indicating that they probably diverged from one another fairly recently in evolution. Each λ locus consists of three gene segments: a V_λ-gene segment which encodes most of the variable region (the antigen binding portion) of the λ light chain, including the first two hypervariable regions and all but perhaps two or three amino acid residues of the third hypervariable region; a J_λ gene segment which encodes the remainder of the variable region; and a C_λ gene segment which encodes the constant region of the λ light chain. The J_λ and C_λ gene segments are about 1000 nucleotide pairs apart in germ line DNA, while the distance between the V_λ and J_λ gene segments has not been precisely determined but is apparently much larger.

The human λ locus has not been analyzed in as much detail as that of the mouse, but its organization displays some intriguing differences. At least nine C_λ-like sequences have been identified, six of them linked tandemly on a 50-kilobase (kb) (50,000 nucleotide pairs of DNA) locus (Hieter et al., 1981). It is not known if each of these C_λ gene segments is associated with its own J_λ gene segment, nor is it known if all of the C_λ-like sequences are functional gene segments. Some may be pseudogenes, evolutionarily related to C_λ gene segments, but no longer capable of encoding part of a functional

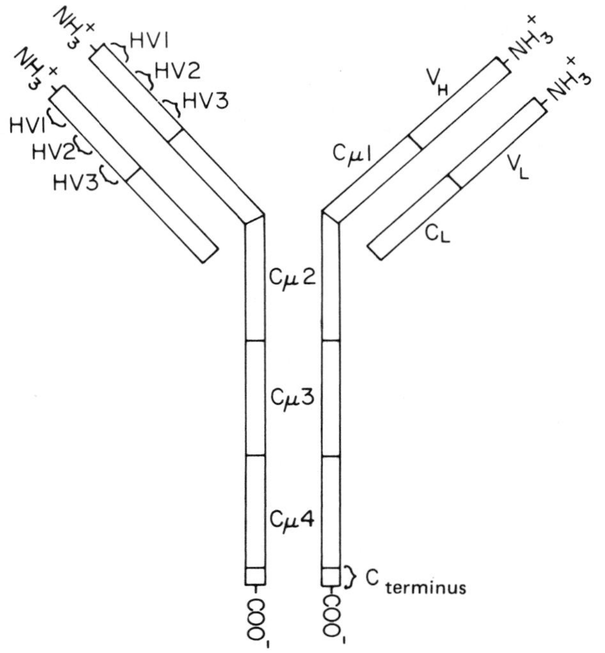

Figure 1 Structure of an immunoglobulin. The molecule shown is the membrane-bound form of murine IgM. It contains two identical light chain polypeptides, each consisting of one variable (V_L) and one constant region (C_L) domain, and two identical heavy chain polypeptides with one variable and four constant region domains ($C\mu$ 1 to 4). The variable regions contain hypervariable regions (HV1 to 3) which are the major sites of antibody–antigen interaction. The V_H and V_L regions are encoded by joined V-D-J and V-J gene segments, respectively. The C_H region is encoded by one of several possible C_H gene segments (in the example shown, $C\mu$), the choice being determined by the results of class switching. The carboxyl terminus of the heavy chain is either hydrophobic (membrane-bound) or hydrophilic (secreted), depending on which exon is included in the messenger RNA as the result of RNA processing. [From P. Early and L. Hood (1981), in *Genetic Engineering,* Vol. 3 (J. K. Setlow and A. Hollaender, Eds.), Plenum, New York, p. 157.]

λ light chain. On the other hand, a higher proportion of L chains are of the λ type in humans than in mouse, so it may be that the greater complexity of the human λ locus provides a greater diversity of λ light chains than in the mouse.

κ gene segments are more diverse than λ gene segments in the mouse. Like λ, the κ locus contains V, J, and C gene segments. The difference lies in the numbers: there are four J_κ gene segments linked to a single C_κ gene segment (Max et al., 1979; Sakano et al., 1979). The number of V_κ gene segments is not accurately known, but estimates run from 50 to a few hundred. The organization of the human κ genes appears to be identical to that of the murine κ genes (Hieter et al., 1980).

The heavy chain genes are the most complex of the three immunoglobulin gene families. The variable regions of heavy chains are encoded by three types of gene

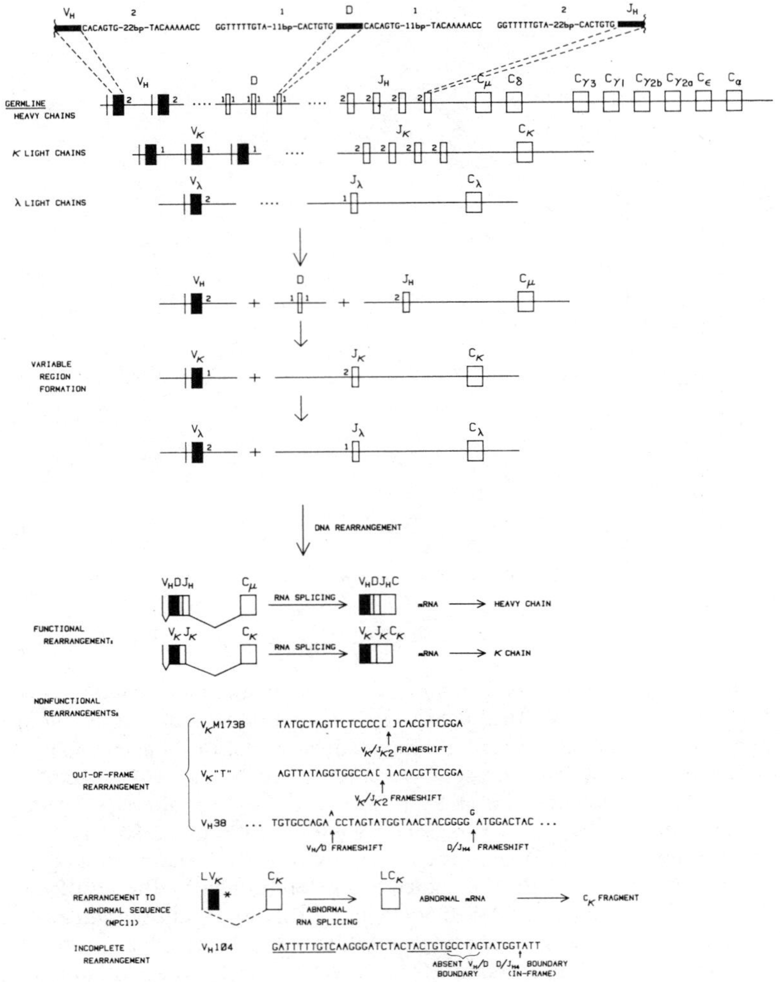

Figure 2 Formation of active immunoglobulin genes by DNA rearrangement. The upper part of the figure depicts the three germ line gene families encoding the immunoglobulins of mice. V, D, and J gene segments encode the variable regions (the antigen binding portions) of immunoglobulins. These gene segments are flanked by conserved nucleotide sequences with long (22-bp) or short (11-bp) spacers. The sequences with long spacers (denoted "2") can join to sequences with short spacers (denoted "1") in the process of creating complete immunoglobulin genes. DNA rearrangements can either be functional, leading to immunoglobulin synthesis, or nonfunctional, in which case the gene may be transcribed, but no immunoglobulin is produced. Nonfunctional rearrangements may disrupt translational reading frames (three examples from mouse), be incomplete (the mouse B-cell V_H104 gene, which contains D and J_H gene segments, but lacks a V_H gene segment), or may occur to an inappropriate sequence (the MPC 11 rearrangement in a mouse plasmacytoma, in which a V_κ gene segment has joined to a sequence lacking the RNA splice signals required for production of normal messenger RNA).

segments: V_H, D, and J_H (Early et al., 1980a; Sakano et al., 1981). The first and second hypervariable regions of the heavy chain are encoded by the V_H gene segment. The third hypervariable region is encoded by the D gene segment and the boundaries of the V_H and J_H gene segments. As depicted in Fig. 2, the H-chain locus contains blocks of V_H and D gene segments. The precise linkages and distances between these gene segments are not currently known, although recombination frequencies suggest that they may occupy several percent of a chromosome. The numbers of V_H and D gene segments are not known in either humans or mouse, but reasonable estimates are 20 to a few hundred V_H gene segments and somewhat fewer D gene segments. The murine H chain locus contains four J_H gene segments linked to the C_H locus, while the human locus has six J_H segments (Ravetch et al., 1981). There are multiple C_H gene segments, each encoding the heavy chain for a different class or isotype of immunoglobulin. The organization of the C_H gene segments is considered in more detail in Sec. III.

B. DNA Rearrangement: V-D-J Joining

The process of DNA rearrangement which creates a functional immunoglobulin gene is called *V-J joining* or *V-D-J joining*, with the latter term referring to heavy chain genes. The immediate result of the DNA rearrangement is the fusion of gene segments encoding the variable region of the immunoglobulin chains, as illustrated in Fig. 2. This process is a crucial one for the generation of immunoglobulin diversity.

Immunoglobulin gene segments from all families (including both human and mouse) have been found to contain highly conserved nucleotide sequences at the sites of DNA rearrangement (Brack et al., 1978; Max et al., 1979; Sakano et al., 1979, 1981; Early et al., 1980a). These sequences are shown at the top of Fig. 2 for the heavy chain gene family (only the strand of DNA with 5′ linkages to the left is represented, although the DNA is actually double stranded). These conserved sequences display an interesting pattern which suggests their involvement in directing DNA rearrangement. Each block of conserved nucleotides (CACAGTG/TACAAAAACC or its inverse complement GGTTTTTGTA/CACTGTG) is interrupted by a block of apparently random "spacer" nucleotides which are either 11 or 22 nucleotides in length. Immunoglobulin gene segments which normally join to one another have spacers of different lengths. The D gene segments of heavy chains have 11 nucleotide spacers on both sides. V_H and J_H gene segments, both of which have 22 nucleotide spacers, join to D gene segments (Early et al., 1980a; Sakano et al., 1981). The details of the mechanism involved in V-J or V-D-J joining are topics of current research (Höchtl et al., 1982).

As a result of this DNA rearrangement, an immunoglobulin gene becomes transcriptionally active. This is somewhat surprising, since the "promoter" for the gene, the region 5′ to the V gene segment where transcription of RNA begins, does not appear to be altered by the rearrangement. Instead, there is evidence that another region associated with the C gene segment in germ line DNA activates efficient transcription from the promoter 5′ to the V gene segment (Mather and Perry, 1981). This activation is possible only after V-J or V-D-J joining. It is intriguing that certain translocations characteristic of Burkitt's and other lymphomas involve the chromosomes carrying the immunoglobulin genes (Kirsch et al., 1982; Rowley, 1982). A hypothesis under current investigation is that these translocations may bring potential cancer-causing genes under the abnormal stimulating influence of elements which usually function by activating the transcription of immunoglobulin genes after DNA rearrangement.

DNA rearrangement apparently occurs by the same mechanism in each immunoglobulin gene family, although it may occur in the heavy chain genes before the light chain genes. Cell lines transformed with the Abelson tumor virus have been isolated from mice. These lines frequently contain rearranged heavy chain genes but not rearranged light chain genes and are considered to be transformed "pre-B" cells, the earliest well-characterized precursor cells in the B-cell lineage. Some of these cell lines are still capable of immunoglobulin gene rearrangement during culture (Alt et al., 1981).

Although a single lymphocyte synthesizes immunoglobulin with only one combination of light and heavy chain variable regions (allelic exclusion), there is less constraint on the number of transcriptionally active immunoglobulin genes. This is due to the presence of rearranged genes which are apparently nonfunctional. As depicted at the bottom of Fig. 2, gene segments sometimes join out of frame, so that a complete polypeptide chain cannot be translated from the nucleotide sequence. In other instances, the rearrangement process may not be complete or may occur to an abnormal sequence. Immunoglobulin genes with these mistakes are often transcribed at levels comparable to properly rearranged genes, although if polypeptides are translated from these messenger RNAs, they are usually degraded rapidly or remain confined to the cytoplasm.

The results of DNA rearrangement are profound for the generation immunoglobulin diversity. As far as we know, V-J and V-D-J joining are random processes, not driven by antigen. The combinatorial association of perhaps 100 V_H gene segments with 20 D gene segments and six J_H gene segments could generate 12,000 different heavy chain variable regions in humans. Even minor differences in variable region sequences can create measurable differences in antigen binding (Cook et al., 1982), so the 12,000 heavy chain variable regions created by the combinations of gene segments would be expected to have a wide range of specificities. Light chain diversity is probably somewhat less, since light chain variable region are combinations of only two gene segments, V and J.

In addition to making use of large numbers of combinations of gene segments, the DNA rearrangement process also creates sequence diversity at the sites where gene segments are joined to one another. Some of this diversity results in nonfunctional genes, as the examples in Fig. 2 show. However, "junctional" diversity also occurs in functional genes. This is illustrated in Fig. 3 for five independent cases of joining mouse D gene

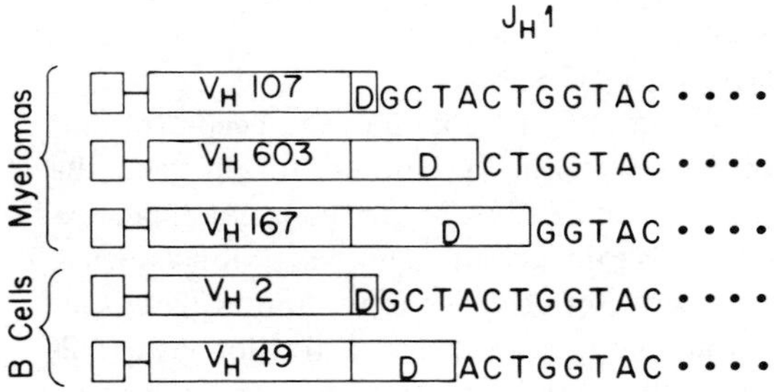

Figure 3 J_H junctions in rearranged genes. These five examples were chosen from genes in mouse myelomas and normal B cells which included the J_H1 gene segment. Four of the five show different sites of D-J joining. [From L. Hood, M. Davis, P. Early, K. Calame, S. Kim, S. Crews, and Huang H. Huang (1981), Cold Spring Harbor Symp. Quant. Biol. *45*, 887.]

segments to the same J_H gene segment, J_H1. There are four different D/J_H1 boundaries in these examples, resulting in different amino acid sequences for the J_H part of the third hypervariable region. Changes in this region have been shown to affect antigen binding (Cook et al., 1982). As well as varying the exact boundary of a gene segment during joining, the DNA rearrangement process can interpolate nucleotides between gene segments (Sakano et al., 1981). The origin of these nucleotides is unclear, but they also affect the amino acid sequence of the third hypervariable region.

Estimating the total amount of immunoglobulin diversity created by the DNA rearrangement process requires consideration of all these factors. A reasonable estimate of the number of junctional variants in combinations of any two gene segments might be about five. Thus the total number of different heavy chains in humans would be about 5 × 12,000, or 300,000. If the number of light chains is about 1/100 of this number (since light chains lack D segments), the number of L × H chain combinations would be about 3000 × 300,000 = 900,000,000. This may be an overestimate of the total number of immunoglobulins, since there may be restrictions on which combinations of light and heavy chain variable regions form functional immunoglobulins. However, the numbers of gene segments and the extent of junctional diversity may well exceed the estimates given above. Clearly, the evolution of an unusual mode of gene expression utilizing DNA rearrangement enables the immunoglobulin gene families to generate a very large spectrum of diverse proteins from a comparatively modest amount of DNA.

C. Somatic Mutation

There is another major source of variable region diversity which is not a consequence of DNA rearrangement. This is termed *somatic mutation*, referring to changes in the variable region gene sequence which occur in somatic cells (lymphocytes) after V-J or V-D-J joining. Figure 4 illustrates the pattern of diversity created by both V-D-J joining and

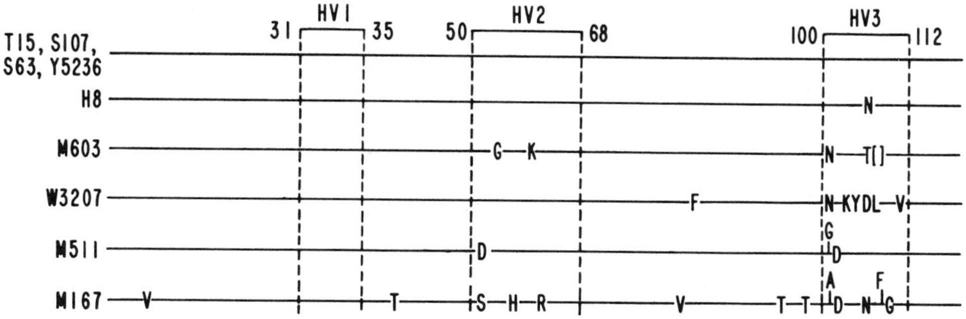

Figure 4 Patterns of diversity in a group of related mouse heavy chain variable regions. These protein sequences are all from mouse myeloma proteins binding the hapten phosphorylcholine. Only differences from the prototype T15 sequence are shown, in the one-letter amino acid code. The chains show a high concentration of diversity in the third hypervariable region, which is the site of V-D and D-J joining. Insertions and deletions of amino acids occur only at the V-D and D-J boundaries. Other changes are scattered throughout the sequences, with some concentration in the second hypervariable region. These changes are due to somatic mutation. (From Early et al., 1980a.)

somatic mutation in a group of related heavy chain variable regions. Junctional and combinatorial diversity is concentrated in the third hypervariable region. The scattered amino acid replacements observed in the remainder of these variable regions are due to somatic mutations, since all of these sequences derive from the same germ line V_H gene segment (Crews et al., 1981). Changes due to somatic mutation can markedly affect antigen binding (Cook et al., 1982).

The mechanism (or mechanisms) of somatic mutation in immunoglobulin genes is not well understood. In part, it may simply be the result of antigen- or idiotype-driven selection of spontaneously occurring mutants. However, special mechanisms may produce high levels of mutation in immunoglobulin genes. A case has been made for a special hypermutational mechanism confined to the locus of the rearranged variable region gene (Kim et al., 1981). Another possibility is gene conversion (recombination) between members of a group of related germ line V gene segments (Clarke et al., 1982).

It is difficult to estimate to what extent somatic mutation contributes to the spectrum of functional diversity in immunoglobulins. A single germ line V gene segment could potentially generate an almost infinite number of variants. The immune system evidently does not rely on this degree of somatic mutation, since most immunoglobulin gene families contain multiple gene segments. It seems likely that the germ line gene segments establish a framework of specificities for commonly encountered antigens, and that somatic mutation is used to "fine-tune" these specificities or, in rare cases, to meet the challenge of completely novel antigens.

III. IMMUNOGLOBULIN ISOTYPE DIVERSITY

A. Germ Line Genes

Immunoglobulins show a second type of diversity which does not affect their specificity for antigen. This is isotype or class diversity, which is the consequence of substituting one heavy chain constant region for another. Different immunoglobulin isotypes have different effector functions, such as complement binding, or are adapted to function in different environments (IgA in the gut, for example). Changes in isotype occur in a regular order during lymphocyte differentiation (Fig. 5). Immunoglobulin is first detectable in the cytoplasm of immature lymphocytes, where it is always of the IgM class. IgM is next displayed on the cell surface, frequently in conjunction with IgD. At a later stage, the resting B cell can be triggered by antigen to undergo a final phase of differentiation. This entails a shift to the synthesis of secreted immunoglobulin, and frequently a switch from the synthesis of IgM and IgD to the synthesis of other classes of immunoglobulins.

The "class switch" from one isotype to another is possible because the heavy chain locus contains multiple C_H gene segments. Class switching does not involve changes in light chains or in the heavy chain variable region, although increased levels of somatic mutation have been observed in the variable regions of immunoglobulins produced by cells which have undergone class switching (Gearhart et al., 1981).

The complete organizational pattern of the C_H gene segments has been determined for the mouse (Shimizu et al., 1982). The linkage of these gene segments is depicted in the upper part of Fig. 2. The actual distances between the gene segments are J_H, 6.5 kilobases (kb)-C_μ-4.5 kb-C_δ-55 kb-C_γ3-34 kb-C_γ1-21 kb-C_γ2b-15 kb-C_γ2a-14 kb-C_ϵ-12 kb-C_α. The total size of the C_H locus is thus about 200 kb, or roughly 0.2% of the chromosome. The human C_H locus contains a minimum of nine functional

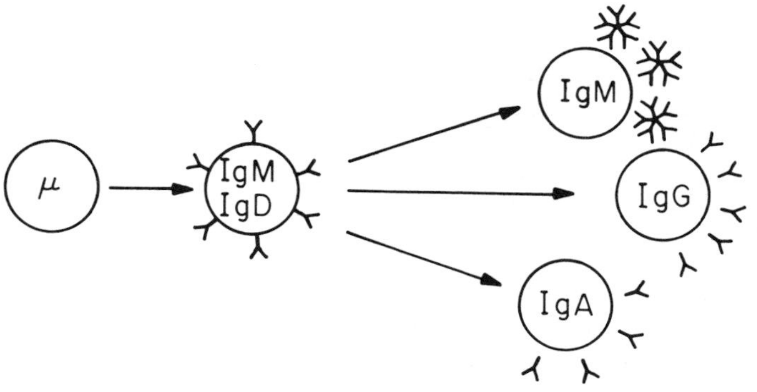

Figure 5 B-cell differentiation. The diagram illustrates the progression of immunoglobulin synthesis in the maturing B cell and its progency. Cytoplasmic μ chains are first detected, followed by membrane-bound IgM and IgD, and then secreted immunoglobulins of a variety of classes. [From P. Early and L. Hood (1981), in *Genetic Engineering*, Vol. 3 (J. K. Setlow and A. Hollaender, Eds.). Plenum, New York, p. 157.

gene segments: μ, δ, $\gamma 1$, $\gamma 2$, $\gamma 3$, $\gamma 4$, ϵ, $\alpha 1$, and $\alpha 2$. The C_μ and C_δ gene segments in humans are closely linked, as they are in mouse (Ravetch et al., 1981). The $\gamma 2$ and $\gamma 4$ gene segments are 19 kb apart (Ellison and Hood, 1982), and the C_ϵ–C_α locus in humans has been duplicated, with one C_ϵ gene segment becoming a pseudogene (Max et al., 1982).

B. DNA Rearrangement: Class Switching

When DNA from mouse plasmacytomas is examined, it is found that cells which have undergone class switching have lost some of their C_H gene segments (Honjo and Kataoka, 1978). Which C_H gene segments are lost can be predicted from the class of immunoglobulin being synthesized by the plasmacytoma. A tumor producing IgA has usually lost all C_H gene segments except C_α, while a tumor producing IgG1 has lost only the C_μ, C_δ, and $C_\gamma 3$ gene segments. These results led Honjo and Kataoka to propose a model for class switching. They suggested that the order of the C_H gene segments is 5′-C_μ-$C_\gamma 3$-$C_\gamma 1$-$C_\gamma 2b$-$C_\gamma 2a$-C_α-3′, which later analysis by gene cloning has shown to be correct (Honjo and Kataoka did not have the reagents needed to determine the location of the C_δ and C_ϵ gene segments). According to their model, class switching occurs by deleting DNA between the heavy chain variable region coding sequence and the new C_H gene segment being expressed (see Fig. 2). In order for C_H gene segments to be lost from the DNA, the deletion process has to affect both chromosome homologues carrying the heavy chain genes, or the plasmacytoma must be aneuploid. Later studies have shown both possibilities to occur.

The first analysis of a cloned gene encoding an α heavy chain supported Honjo and Kataoka's model (Davis et al., 1980a). In this gene, the rearranged V-D-J region is linked to the C_α gene segment (Fig. 6). However, the V-D-J region shows clear evidence for its original association with the C_μ gene segment. A large segment of the intron between J_H and C_α is the same DNA found between J_H and C_μ in the germ line. (Introns are sequences that are removed from RNA transcripts during the maturation of messenger RNAs.) There is an abrupt transition in the middle of this intron to DNA which is found

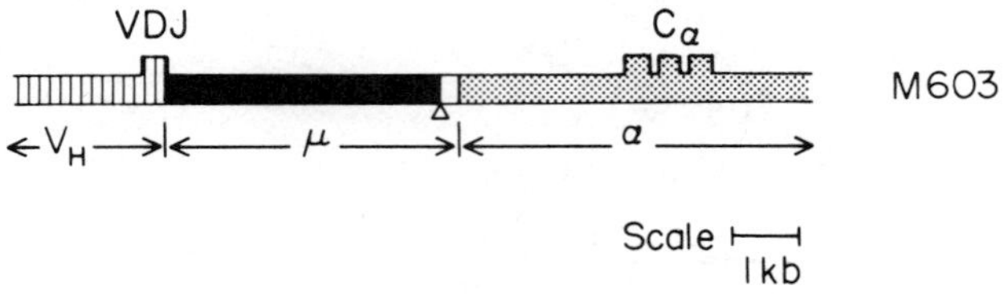

Figure 6 Class switching by DNA rearrangement in an α gene. The exons (raised boxes) encoding the heavy chain variable region and the C_α region are separated by an intron which is derived partially from DNA associated with the C_μ gene segment in the germ line (dark area) and partially derived from DNA associated with C_α (stippled area). The point where these two pieces of DNA join is the "switch point" for the rearrangement. [From P. Early and L. Hood (1981), in *Genetic Engineering*, Vol. 3 (J. K. Setlow and A. Hollaender, Eds.), Plenum, New York, p. 157.

5′ to the C_α gene segment in the germline. Similar evidence for DNA rearrangement in class switching has been found in several other heavy chain genes from mouse plasmacytomas. The introns 5′ to C_H gene segments contain highly repeated DNA sequences which may play a role in the rearrangements associated with class switching (Davis et al., 1980b). These sequences are not related to the conserved sequences associated with V-J and V-D-J joining, suggesting that immunoglobulin genes employ two quite different mechanisms of DNA rearrangement in the generation of variable region diversity and in the generation of isotype or class diversity.

C. RNA Processing

It is not presently clear if all class switching requires DNA rearrangement. In at least one case, a murine cell line has been isolated which can switch from IgM to Ig2b synthesis in culture without losing any copies of the C_μ gene segment (Alt et al., 1982). One possibility is that this cell line synthesizes a very large RNA transcript, beginning with V-D-J and reading through the C_H locus to at least the $C_{\gamma 2b}$ gene segment. This transcript could be processed by RNA splicing to yield a messenger RNA with V-D-J and $C_{\gamma 2b}$ sequences.

Further research will determine whether an RNA processing mechanism is in fact used for class switching, and whether this mechanism is as common as DNA rearrangement. Another intriguing line of investigation is the possibility that the class of immunoglobulin to which a cell switches is under some type of external control (Davis et al., 1980b). Cells synthesizing certain classes of immunoglobulin are not uniformly distributed among the lymphoid organs. Peyer's patches, for example, are enriched in IgA-bearing cells. This may be the result of "homing" processes by these cells, or it may reflect the ability of the local environment to influence the direction of class switching.

IV. MEMBRANE-BOUND AND SECRETED IMMUNOGLOBULINS

Resting B cells display immunoglobulin on their cell surfaces (Fig. 5). Most commonly this membrane-bound immunoglobulin is IgM or IgD, although other classes are present on smaller numbers of lymphocytes. After triggering by antigen, the lymphocyte synthesizes mostly secreted immunoglobulin, which may be IgM, IgD, or one of the other

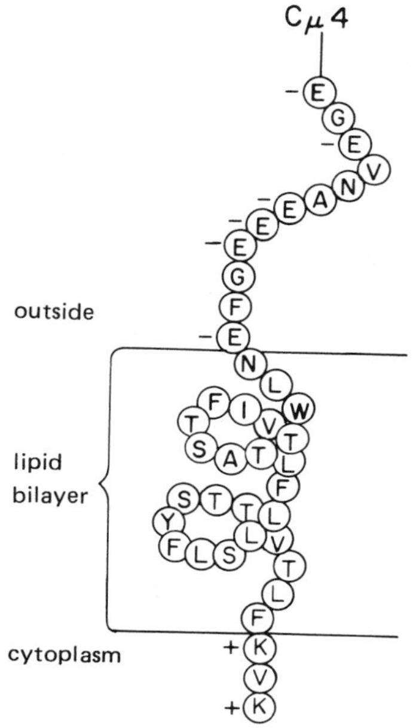

Figure 7 Amino acid sequence of the carboxyl terminus of membrane-bound mouse IgM, as predicted from the sequence of messenger RNA. The one-letter amino acid code is used. Polar amino acid residues are shown with (+) or (−) charges: K is lysine, E is glutamic acid. The model for a transmembrane hydrophobic "anchor" is in accordance with the structures of other known transmembrane proteins. (From Rogers et al., 1980.)

classes. The transition from synthesis of membrane-bound to secreted immunoglobulin is the result of a change in the processing pattern of RNA transcripts from the heavy chain locus. This change alters the carboxyl-terminal "tail" of the heavy chain (Early et al., 1980b; Rogers et al., 1980).

Secreted murine IgM contains a 20-amino acid hydrophilic carboxyl terminus while membrane-bound IgM contains a 41-amino acid hydrophobic carboxyl terminus which serves as an "anchor" to the cell membrane (Fig. 7). Antigen binding to membrane-bound immunoglobulin plays an important role in the transformation from B cell to plasma cell, but it is not known if the membrane-bound immunoglobulin is a conduit for the signal that triggers this event or if other molecules are involved.

The other classes of immunoglobulin have membrane-bound and secreted forms analogous to those of IgM. The most extensively studied example is IgD (Cheng et al., 1982), but this is also true for IgG (Tyler et al., 1982). The hydrophobic "anchors" on all classes of membrane-bound immunoglobulins show extensive homology of their amino acid sequences. These regions may share a common function, perhaps including interaction with other membrane proteins.

The mechanism for the transition from membrane-bound to secreted immunoglobulin synthesis involves RNA processing, not DNA rearrangement. As depicted in Fig. 8, the mouse μ gene contains four C_μ exons, one encoding each domain of the constant region. The $C_\mu 4$ exon also encodes the carboxyl terminus of secreted μ chains. In cells synthesizing secreted IgM, RNA transcripts are processed with a 3′ end following the $C_\mu 4$ exon,

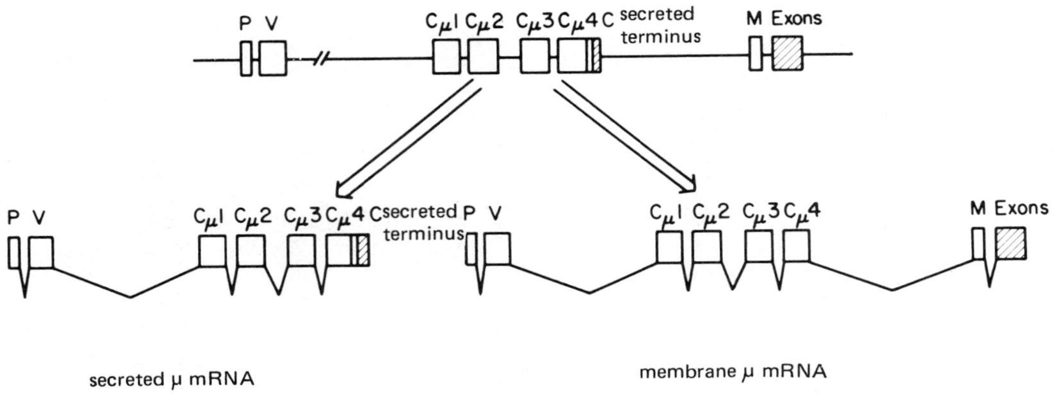

Figure 8 RNA processing pathways leading to the synthesis of alternative forms (membrane-bound or secreted) of murine IgM. The single C_μ gene segment includes exons encoding both a secreted carboxyl terminus and a membrane-bound carboxyl terminus. Depending on the route followed by RNA splicing (bent lines), a μ chain may be sunthesized with either secretory or membrane-binding carboxyl termini. (From Early et al., 1980b).

and RNA splicing produces messenger RNA which encodes secreted μ chains. In cells synthesizing membrane-bound IgM, RNA transcripts are processed with longer 3′ ends, allowing RNA splicing to replace the nucleotides encoding the secreted μ carboxyl terminus with exons encoding the membrane-bound carboxyl terminus. These exons (termed M exons in Fig. 8) are located about 2 kb 3′ to the $C_\mu 4$ exon. The exact step of RNA processing which is regulated in the membrane-to-secreted transition is not yet known: it may be termination of transcription or addition of poly A to the 3′ end of RNA, or it may be the splicing step from $C_\mu 4$ to the M exons.

The pattern of RNA synthesis for δ chains is more complex than for μ chains. There are three possible carboxyl termini for mouse IgD: one for the membrane-bound form, one for the secreted form, and one for a form that has not yet been identified (Cheng et al., 1982). To make matters even more complicated, IgM and IgD with the same variable region are often synthesized simultaneously by resting B cells (Fig. 5). This is apparently possible because the C_μ and C_δ gene segments are only 4.5 kb apart. RNA transcripts from the murine V-D-J-C_μ-C_δ locus can be processed to yield five possible messenger RNAs, two encoding μ chains and three encoding δ chains. The mechanism that permits this degree of control of RNA synthesis remains to be elucidated. This mechanism may be related to one allowing "switching" to other immunoglobulin classes without DNA rearrangement (Alt et al., 1982).

V. CONCLUSIONS

In this chapter the various mechanisms used by the immunoglobulin gene families to generate a very large repertoire of functionally different immunoglobulins are discussed. These mechanisms include DNA rearrangement (V-J and V-D-J joining) and somatic mutation to generate a wide variety of antigen binding variable regions. The isotype or class of an immunoglobulin can be altered by a second type of DNA rearrangement, and

possibly by RNA processing. Finally, immunoglobulins can be synthesized in alternative forms which are either bound to the cell surface, where they serve as receptors for antigen, or are secreted to bind antigen extracellularly. The synthesis of these forms is determined by RNA processing.

REFERENCES

Alt, F., Rosenberg, N., Lewis, S., Thomas, E., and Baltimore, D. (1981). Cell *27*, 381.

Alt, F. W., Rosenberg, N., Casanova, R. J., Thomas, E., and Baltimore, D. (1982). Nature (Lond.) *296*, 325.

Brack, C., Hirama, M., Lenhard-Schuller, R., and Tonegawa, S. (1978). Cell *15*, 1.

Cheng, H.-L., Blattner, F. R., Fitzmaurice, L., Mushinski, J. F., and Tucker, P. W. (1982). Nature (Lond.) *296*, 410.

Clarke, S. H., Claflin, J. L., and Rudikoff, S. (1982). Proc. Natl. Acad. Sci. USA *79*, 3280.

Cook, W. D., Rudikoff, S., Giusti, A. M., and Scharff, M. D. (1982). Proc. Natl. Acad. Sci. USA *79*, 1240.

Crews, S., Griffin, J., Huang, H., Calame, K., and Hood, L. (1981). Cell *25*, 59.

Davis, M. M., Calame, K., Early, P. W., Livant, D. L., Joho, R., Weissman, I. L., and Hood, L. (1980a). Nature (Lond.) *283*, 733.

Davis, M. M., Kim, S. K., and Hood, L. E. (1980b). Science *209*, 1360.

Early, P., Huang, H., Davis, M., Calame, K., and Hood, L. (1980a). Cell *19*, 981.

Early, P., Rogers, J., Davis, M., Calame, K., Bond, M., Wall, R., and Hood, L. (1980b). Cell *20*, 313.

Ellison, J., and Hood, L. (1982). Proc. Natl. Acad. Sci. USA *79*, 1984.

Gearhart, P. J., Johnson, N. D., Douglas, R., and Hood, L. (1981). Nature (Lond.) *291*, 29.

Hieter, P. A., Max, E. E., Seidman, J. G., Maizel, J. V., Jr., and Leder, P. (1980). Cell *22*, 197.

Hieter, P. A., Hollis, G. F., Korsmeyer, S. J., Waldmann, T. A., and Leder, P. (1981). Nature (Lond.) *294*, 536.

Höchtl, J., Müller, C. R., and Zachau, H. G. (1982). Proc. Natl. Acad. Sci. USA *79*, 1383.

Honjo, T., and Kataoka, T. (1978). Proc. Natl. Acad. Sci. USA *75*, 2140.

Kim, S., Davis, M., Sinn, E., Patten, P., and Hood, L. (1981). Cell *27*, 573.

Kirsch, I. R., Morton, C. C., Nakahara, D., and Leder, P. (1982). Science *216*, 301.

Mather, E. L., and Pettery, R. P. (1981). Nucleic Acids Res. *9*, 6855.

Max, E. E., Seidman, J. G., and Leder, P. (1979). Proc. Natl. Acad. Sci. USA *76*, 3450.

Max, E. E., Battey, J., Ney, R., Kirsch, I. R., and Leder, P. (1982). Cell *29*, 691.

Ravetch, J. V., Siebenlist, U., Korsmeyer, S., Waldmann, T., and Leder, P. (1981). Cell *27*, 583.

Rogers, J., Early, P., Carter, C., Calame, K., Bond, M., Hood, L., and Wall, R. (1980). Cell *20*, 303.

Rowley, J. D. (1982). Science *216*, 749.

Sakano, H., Hüppi, K., Heinrich, G., and Tonegawa, S. (1979). Nature (Lond.) *280*, 288.

Sakano, H., Kurosawa, Y., Weigert, M., and Tonegawa, S. (1981). Nature (Lond.) *290*, 562.

Shimizu, A., Takahashi, N., Yaoita, Y., and Honjo, T. (1982). Cell *28*, 499.

Tyler, B. M., Cowman, A. F., Gerondakis, S. D., Adams, J. M., and Bernard, O. (1982). Proc. Natl. Acad. Sci. USA *72*, 2008.

15

Lectins

Miercio E. A. Pereira   Tufts University School of Medicine, Boston, Massachusetts

I. INTRODUCTION

Lectins are divalent or multivalent carbohydrate binding proteins of nonimmune origin that can agglutinate cells or other materials having more than one saccharide of appropriate complementarity. Polyvalent enzymes, such as galactose oxidase, agglutinate cells under appropriate conditions and therefore are considered to be lectins. They are widely distributed among living organisms: in plants, fungi, and bacteria; in invertebrates such as snails, horseshoe crabs, and sponges; in fish; in chicken; and in mammals.

Lectins were discovered in 1888 by Stillmark, who observed that crude extracts of castor beans (*Ricinus communis*) contained a toxic principle named *ricin* that agglutinated human and some animal red blood cells. The hemagglutinating ability was first regarded as a manifestation of toxicity, but immunochemical and chromatographic evidence obtained later indicated that a nontoxic agglutinin could be separated from a nonagglutinating highly toxic protein. Landsteiner and Raubitschek in 1908 demonstrated that some plant lectins are species specific; that is, the relative hemagglutinating activities of various seed extracts are quite different for red cells from several animal species. The recognition that lectins may also interact selectively with human blood group antigens came about 40 years later (Boyd, 1970). In 1945, Boyd reported that extracts of lima beans (*Phaseolus lunatus*) agglutinated red cells of human type A, but not those of type B or O. Renkonen in 1948 made similar observations, reporting several blood-group-specific extracts among 57 species of plants belonging to 28 different genera. But most lectins are not blood group specific, as indicated by a survey of 2663 plant species which revealed that over 700 of them agglutinated A, B, and O erythrocytes equally well, while only 90 were blood group specific.

The first structural studies on blood group substances with the aid of lectins was carried out in the early 1950s by Watkins (1972), who provided the first evidence that an α-L-fucosyl and an α-N-acetyl-D-galactosaminyl residues were the immunodominant groups of the blood group H and A determinants, respectively. Both blood-type-specific and nonspecific lectins are now widely employed for the detection, isolation, and characterization of carbohydrate-containing materials such as polysaccharides, glycoproteins, and glycolipids.

Lectins are invaluable reagents in immunologic research (Lis and Sharon, 1977). They were used as antigens soon after their discovery. In the early 1900s, Ehrlich, using crude extracts of ricin and abrin, the latter a toxic lectin from jequirity beans (*Abrus precatorius*), established fundamental immunological principles of specificity and of the neutralization of a toxin by antitoxin. Mice became immune to a lethal dose of ricin by repeated small subcutaneous injections of the lectin, and their serum acquired the capacity to neutralize the toxicity of ricin. The reaction was specific since the antiserum would not neutralize the toxic effects of abrin, nor would antiabrin neutralize the toxicity of ricin. Furthermore, there was a quantitative relationship between the amount of antiserum and the amount of toxin it could neutralize.

The widespread use of lectins in biological research today is due to their specific sugar-binding properties. In immunology they are studied in connection with their interactions with lymphocytes, which trigger a number of important biologic phenomena such as blast transformation and cell division, activation of T-suppressor cells, and alteration of the movement of receptors on cell membranes producing patching and capping. The stimulation of lymphocytes by lectins results in the release of soluble substances with a variety of biologic actions, including lymphokines such as migration-inhibitory factor, lymphotoxin, and interferon, as well as immunoglobulin and prostaglandins.

II. ISOLATION AND PURIFICATION

For some purposes, lectins may be used as crude seed or tissue extracts. However, it is often desirable to work with purified preparations, since the crude material may contain agglutinins of different sugar specificities, or they may be contaminated with enzymes or with other materials that might interfere with binding between the lectin and its carbohydrate receptor.

Isolation of lectins generally begins with a buffer extraction of finely ground seed meal or animal organ. Preextraction with organic solvents (e.g., methanol or petroleum ether) is commonly employed to remove lipids and other interfering substances. Virtually all lectin purification protocols employ affinity chromatography based on the ability of lectins to bind sugar specifically and reversibly. A carbohydrate ligand with which the lectin interacts is insolubilized, the lectin is adsorbed as the extract passes through the adsorbent, and elution of bound lectin is accomplished generally with a sugar that competes for lectin sites with the specific adsorbent. Commercially available adsorbents such as Sepharose (polymer of D-galactose), Sephadex (polymer of D-glucose), or chitin (polymer of β,1 → 4 linked N-acetyl-D-glucosamine) may be employed. In other cases, mono- or oligosaccharides, or glycoproteins such as hog gastric mucin or desialylated fetuin coupled to Sepharose, are used. Since the adsorbents used for the isolation of lectins may also bind glycosidases, which are frequently present in crude extracts, lectins purified by affinity chromatography may be contaminated with such enzymes.

Table 1 Relative Affinity of Mono- and Oligosaccharides to L-Fuc Binding Lectins

Inhibitor	Nanomoles for 50% inhibition: Ulex I[a]	Nanomoles for 50% inhibition: *Lotus* lectins[b] (unfractionated)	Nanomoles for 50% inhibition: Eel lectin[c]	Relative potency: Ulex lectin I	Relative potency: *Lotus* lectins (unfractionated)	Relative potency: Eel lectin
2′-Fucosyllactose	65	150		77	1.2	
Lactodifucotetraose	180	80		28	2.3	
Methyl α-L-Fuc	800	50	300	6.0	3.6	4.0
Lacto-N-fucopentaose I	2000	580 (0% inhibition)		2.5	–	
L-Fuc	5000	180	1200	1.0	1.0	1.0
Lacto-N-difucohexaose I	380 (16% inhibition)	600 (0% inhibition)		–	–	
Lacto-N-fucopentaose III	1500 (0% inhibition)	450		–	0.4	
Lacto-N-fucopentaose II	6000 (0% inhibition)	700		–	0.3	
3-Fucosyllactose	2000 (8% inhibition)	575		–	0.3	

[a]Pereira et al. (1978), Arch. Biochem. Biophys. *185*, 108.
[b]Pereira and Kabat (1974), Biochemistry *13*, 3184.
[c]Springer et al. (1964), Biochemistry *3*, 1078.

In many cases, the purified lectins do not occur singly, but as a group of closely related proteins, called isolectins, with similar biochemical and biological properties. For example, three isolectins specific for α-L-fucose are found in the seeds of *Lotus tetragonolobus*; they differ in their isoelectric points, association constants for α-L-fucose, and in their relative affinities for α-L-fucosyl oligosaccharides (Table 1) (Pereira and Kabat, 1979).

Groups of lectins differing markedly in biological activity and in molecular properties are also present in several plants (Lis and Sharon, 1981). The roots of the common pokeweed, *Phytolacca americana*, contain five lectins; although each of these binds to desialylated human blood cell glycopeptides to about the same extent, they differ in mitogenic properties, in amino acid and carbohydrate composition, and in molecular size. The lectin from the red kidney bean, *Phaseolus vulgaris* (PHA), is composed of two different subunits combined into five different forms of noncovalently bound tetramers. The PHA isolectins (termed E_4, L_1E_3, L_2E_2, L_3E, and L_4), each of molecular weight 115,000, differ in their biological properties: E_4 is the most potent erythroagglutinin, L_4 is the most potent lymphocyte mitogen, and intermediate forms possess both properties. Similar mixed tetramers of two functionally distinct subunits have been observed in a plant seed lectin from *Bandeirae simplicifolia* and in a slime mold lectin.

III. PHYSICOCHEMICAL PROPERTIES

The biochemical and biophysical parameters of purified lectins vary considerably (Table 2) (Goldstein and Hayes, 1978). Many of these proteins are relatively rich in aspartic acid, serine, and threonine, which may comprise as much as 30% of their amino acid content, and are low in, or completely devoid of, sulfur-containing amino acids. In contrast, lectins such as those from wheat germ, potato, and pokeweed are very rich in cysteine (20, 11.5, and 18% of the total amino acid residues, respectively). The high content of disulfide bonds in wheat germ agglutinin confers this protein with stability to heat, to proteolytic enzymes, and to denaturing agents such as detergents, urea, alkali, and acids.

Many of the lectins that have been characterized are glycoproteins, with sugar contents as high as 50% for the potato glycoprotein, but others, like concanavalin A (ConA), peanut and wheat germ agglutinins, and the three lectins from the sponge *Aaptos papillata* are devoid of covalent by bound sugars.

The molecular weights range from 11,000 for the blood group B-specific lectin from the bacterium *Streptomyces* sp. to 335,000 for the lectin from the horseshoe crab *Limulus polyphemus*. Some lectins show a tendency to aggregate; for example, the molecular weight (MW) of ConA at physiological pH is 102,000, and at pH below 6 it is 51,000. Upon storage at room temperature, soybean agglutinin (MW 120,000) undergoes irreversible self-association to high-molecular-weight aggregates. All lectins consist of a variable number of subunits (Table 2), whose structures may be affected by various chemical modifications; thus ConA is a tetramer at physiological pH, but succinylated ConA is a dimer. Oxidation of tryptophan in wheat germ agglutinin by N-bromosuccinimide causes the formation of subunits that possess sugar-binding activity but cannot reassociate and are devoid of hemagglutinating activity. Most subunits are identical, but lectins made up on nonidentical subunits are known. Thus the garden pea lectin is a tetramer comprised of two subunits of molecular weight 17,000 to 18,000 and 5800.

Table 2 Physicochemical and Biological Properties of D-GalNAc Binding Proteins

		DB[b]		LBL[c]				
	HP[a]	A	B	II	III	BS-IA$_4$[d]	SBA[e]	RHBL[f]
Molecular weight × 10^{-3}	79	113	109	247.1	124.4	114	120	260
Subunits: number	6	4	4	8	4	4	4	6
Molecular weight × 10^{-3}	13	26.5	26	31	31	32	30	40, 48
K × 10^{-3} M^{-1}	5.0	9.0	9.0	0.93	1.0		30	
Number of sugar binding sites	6			4	2		2	
Mitogenic activity	–	–	–	+	-		+	-
Sugar content (%)	7	3.8–4.7	1.3	5	5	6.7	5.7	10
Metalloprotein	–	–	–	Mn^{2+}, Ca^{2+}	Mn^{2+}, Ca^{2+}		–	Ca^{2+}

[a]HP (*Helix pomatia*), Hammarstrom et al. (1975), Scand. J. Immunol. *1*, 295.
[b]DB (*Dolichos fiblorus*), Carter and Etzler (1975), J. Biol. Chem. *250*, 2756.
[c]LBL (lima bean), Galbraith and Goldstein (1972), Biochemistry, *11*, 3976.
[d]BS-IA$_4$ (*Bandeirae simplicifolia*), Murphy and Goldstine (1977), J. Biol. Chem. *252*, 4379.
[e]SBA (soybean agglutinin), Lotan et al. (1974), J. Biol. Chem. *249*, 1219.
[f]RHBL (rabbit liver lectin), Kawasaki and Ashwell (1976), J. Biol. Chem. *251*, 1296.

	72										80										90			
ConA	Ala	Thr	Ser	( )	( )	Val	Ser	Tyr	Asp	Val	Asp	Leu	Asn	Asp	Val	Leu	Pro	Glu	Trp	Val	Arg	Val	Gly	Leu
	α1									α10										α20				
Lens	Val	Thr	Ser	Tyr	Thr	Leu	Asn	Glu	Val	Val	Pro	Leu	Lys	Asp	Val	Val	Pro	Glu	Trp	Val	Arg	Lle	Gly	Phe

							100										110							
ConA	Ser	Ala	Ser	Thr	Gly	Leu	Tyr	Lys	Glu	Thr	Asn	Thr	Lle	Leu	Ser	Trp	Ser	Phe	Thr	Ser	Lys	Leu	Lys	Ser
						α30										α40								
Lens	Ser	Ala	Thr	Thr	Gly	Ala	Glu	Phe	Ala	Ala	Gin	Glu	Val	His	Ser	Trp	Ser	Phe	Asn	Ser	Gin	Leu	Gly	His

			120										130										140	
ConA	Asn	Ser	Thr	His	Gln	Thr	Asp	Ala	Leu	His	Phe	Met	Phe	Asn	Gln	Phe	Ser	Lys	Asp	Gln	Lys	Asp	Leu	Lle
		α50				β1									β10									
Lens	Thr	Ser	Lys	Ser	( )	Thr	Glu	Thr	Thr	Ser	Phe	Ser	Ile	Thr	Lys	Phe	Ser	Pro	Asp	Gln	Gln	Asn	Leu	Lle

									150										160					
ConA	Leu	Gln	Gly	Asp	Ala	Thr	Thr	Gly	Thr	Asp	Gly	Asn	Leu	Glu	Leu	Thr	Arg	Val	Ser	Ser	Asn	Gly	Ser	Pro
	β20										β30											β40		
Lens	Phe	Gln	Gly	Asp	Gly	Tyr	Thr	Gly	Lys	Glu	Gly	( )	Leu	Thr	Leu	Thr	Lys	Val	Ser	Lys	Glu	Gly	Thr	Gly

Figure 1 Amino acid sequences of the lectins ConA and *Lens culinaris*. The sequences in the boxed positions are identical. Deletions between positions 74 and 75 in ConA and between 121 and 122 in the β chain of the *L. culinaris* lectin are introduced to maximize the homology. (Modified from Forier et al., 1978.)

Although there is usually one sugar-binding site per subunit of a lectin, wheat germ has two such sites per subunit, while soybean agglutinin, pea lectin, and lentil lectin have two binding sites per four subunits (Table 2). The subunits of the same lectin usually have the same sugar specificity, and the binding sites are homogeneous and noninteracting. However, as already mentioned, the subunits of PHA and *B. simplicifolia* differ in their sugar specificity.

Despite the differences in carbohydrate specificity exhibited by ConA, the lectins from pea, lentil, soybean, and peanut, and by the subunits of the two phytohemagglutinins from red kidney beans, they nevertheless display some structural similarity in their primary structure (Forier et al., 1978). Extensive homologies exist between the 25 residues of the amino terminal sequence of the β chains of lentil, pea, and of the other lectins. The entire primary structure of the α subunit from lentil lectin, which consists of 52 amino acid residues, is homologous with the region between positions 72 and 121 from ConA. A total of 43 identical positions is found for the 94 residues of lentil α and β chains and ConA (Fig. 1). Both lentil and ConA show a striking preservation of charged residues, as well as of aromatic amino acids (Fig. 1). Thus the extensive homology between different lectins obtained from a single plant family, the leguminosae, strongly suggest a common ancestry for the genes coding for these proteins. Moreover, the homologies suggest that lectins may have an important physiological role in plants.

IV. CARBOHYDRATE SPECIFICITY

Since the uniqueness of lectins relates to their sugar-binding properties, precise knowledge of the nature of their specific combining sites is important to understand their reactivity with cell surface receptors and with soluble glycoproteins and glycolipids (Pereira and Kabat, 1979a). This is usually established by hapten inhibition technique, in which different sugars are tested for their ability to inhibit either hemagglutination or glycoprotein (or polysaccharide) precipitation by the lectin. The lectin combining site is considered to be most complementary to the sugar that inhibits at the lowest concentrations. The value of the inhibition assay in determining the specificity, dimensions, and shape of the lectin combining site is illustrated by the findings with ConA, which show a direct parallel between the inhibitory capacity and the association constants of several glycosides in the range 70 to 6×10^4 M^{-1} (Loontjens et al., 1975). In addition, the combining site size of ConA, as determined by the precipitin inhibition assay, correlates with that computed by nuclear magnetic resonance and by x-ray diffraction. Thus, although the inhibition assay is an indirect measure of the hapten-lectin interaction, since it involves a competition between the low-molecular-weight sugar and the cell surface receptor, polysaccharide, or glycoprotein for the lectin combining site, it gives estimates comparable with those based on direct measurement of the interaction.

Immunochemical studies on the specificity of a large number of purified lectins allow their classification into small number of groups (Table 3) (Goldstein and Hayes, 1978). Lectins with a similar specificity toward monosaccharides may differ in their affinity for particular disaccharides, oligosaccharides, or glycoproteins. For example, the lectin from *L. tetragonolobus* and the lectin I from *Ulex europaeus* are both specifically inhibited by fucosyl glycosides, but they are inhibited to different extents by oligosaccharides from human blood group substances (Table 1). ConA and lentil lectin both interact specifically with mannose and glucose derivatives, but they differ in their binding to glycoproteins isolated from human erythrocyte membranes. ConA binds to only one

Table 3 Specificity of Various Purified Lectins

Lectin	Blood group specificity	Best estimate of size and complementarity
L-Fuc binding lectins		
Lotus tetragonolobus	H	LFucα1→2DGalβ1→4(Fucα1→3(DGlcNAcβ1→
Ulex lectin I	H	LFucα1→2DGalβ1→4DGlcNAcα1→
Anguilla anguilla; eel	H	LFucα1→
D-GalNAc binding lectins		
Helix pomatia	A	DGalNAcα1→
Dolichos biflorus	A	DGalNAcα1→3(LFucα1→2)DGalβ1→
Phaseolus lunatus; lima bean	A	DGalNAcα1→
Bandeiraea simplicifolia BS-IA	A	DGalNAcα1→
Glycine max; soybean	None	DGalNAcα1→3DGalβ1→3DGlcNAc
Rabbit liver	None	DGalNAcα1→
D-Gal binding lectins		
Bandeiraea simplicifolia BS-IB	B	DGalα1→3DGal
Evonymus europaeus	B + H	DGalα1→3(LFucα1→2)DGalβ1→3 or 4DGlcNAc
Ricinus communis I (RCA 1); castor bean	None	DGalβ1→
Ricinus communis II (RCA II)	None	DGalβ1→
Arachis hypogaea; peanut	None	DGalβ1→3DGalNAC
Axinella lectin I	None	DGalβ1→6
Axinella lectin II	None	DGalβ1→6
D-GlcNAc binding lectins		
Triticum vulgaris; wheat germ	None	DGlcNAcβ1→4DGlcNAcβ1→4DGlcNAc
Aaptos lectin I	None	DGlcNAcβ1→4DGlcNAcβ1→4DGlcNAcβ1→ 4DGlcNAc
Aaptos lectin II	None	DGlNAcβ1→4DGlcNAcβ1→4DGlcNAc
Bandeiraea simplicifolia BS-II	None	DGlcNAcα1→
Solanum tuberosum; potato	None	DGlcNAcβ1→4DGlcNAcβ1→4DGlcNAcβ1→ 4DGlcNAc
D-Man binding lectins		
Canavalia ensiformis; ConA	None	DManα1→
Pisum sativum, pea	None	DManα1→
Lens culinaris, lentil	None	DManα1→

Source: Adapted from Pereira and Kabat (1979a).

of the membrane glycoproteins (band 3), while lentil lectin binds to this glycoprotein and to glycophorin.

Most lectins are inhibited by monosaccharides, although some lectins bind only to oligosaccharides. Glycopeptides from erythrocyte membranes, fetuin, and immunoglobulin A, but not monosaccharides, bind to the lectin from the commercial mushroom (*Agaricus bisporus*). The lectin in potato tubers is best inhibited by β,1 → 4-linked pentamer of N-acetyl-D-glucosamine, which is 2.5, 25, and 50 times more active than the corresponding tetramer, trimer, and dimer, respectively; N-acetyl-D-glucosamine by itself does not inhibit at concentrations up to 200 mM. Similarly, the lectin II from *U. europaeus* is inhibited by oligomers of N-acetyl-D-glucosamine but not by the monomer.

Some lectins are specific for only one anomeric configuration, whereas others interact equally well with both anomeric configurations. Thus the lectins from the seeds of *L. tetragonolobus* and *U. europaeus* I are all very specific for the α anomer of L-fucose, whereas the galactose-binding lectins found in hemolymph of the insect *Rhodnius prolixus* and in castor beans are almost completely devoid of anomeric specificity. There are no cases of lectins specific for β-glycosides.

Most lectins interact with more than one monosaccharide. For instance, lectins that exhibit a primary specificity for mannose also bind glucose and, to a lesser extent, N-acetyl-D-glucosamine. Wheat germ agglutinin binds to N-acetylglucosamine, N-acetylgalactosamine, and sialic acid. A substantial number of lectins exhibit preferential affinity for N-acetylgalactosamine but cross-react with galactose (e.g., soybean agglutinin), whereas others react primarily with galactose and, to a lesser extent, with N-acetylgalactosamine (e.g., *R. communis* agglutinin). On the other hand, some lectins interact only with N-acetylgalactosamine (e.g., *Dolichos biflorus*) and galactose (e.g., peanut agglutinin).

Lectins have sites that seem to vary in size from a lower limit of a monosaccharide unit plus a glycosidic linkage to that of at least a tetrasaccharide. Most N-acetylglycosamine-binding proteins are in the latter category.

In general, lectins interact with the terminal nonreducing glycosyl groups of polysaccharides and glycoproteins. ConA is an exception since, in addition to its interaction with terminal α-mannosyl and α-glucosyl residues, it also binds internal 2-α-mannose. Also, some lectins, like *A. bisporus* lectin and the blood group N-specific lectin of *Vicea graminea*, apparently recognize carbohydrate sequences together with the amino acid or peptide to which the latter are linked.

V. BIOLOGICAL ACTIVITIES

Lectins interact with cells (Nicholson, 1974; Lis and Sharon, 1977) triggering a number of important biologic phenomena, which include agglutination, mitogenic stimulation, redistribution of cell surface components, insulin-like activity on fat cells, activation of suppressor T cells, modification of the activity of membrane enzymes, blocking of fertilization of ova by spermatozoa, toxicity in vitro and in vivo, and inhibition of fungal growth. The biological effects are mediated through the carbohydrate combining sites of the lectin, which interact with complementary receptors on the cell surface. The surface structures to which lectins bind are the carbohydrate moieties of glycoproteins or glycolipids which are exposed on the outer cell membrane. The various biological activities, however, are not exhibited by all lectins.

Table 4 Activity of Lectins of Various Specificities for *T. cruzi* Developmental Stages

Lectin	Blood group specificity	Developmental forms of the trypanosome		
		Epimastigote	Trypomastigote	Amastigote
I. DGlcNac binding lectins				
Wheat germ	None	25.0	>1000.0	>800.0
Aaptos papillata lectin II	None	71.0	381.0	>762.0
Bandeirae simplicifolia	None	115.0	>1275.0	>1275.0
II. DGalNAc binding lectins				
Wistaria floribunda	None	37.0×10^{-3}	2263.0	>6035.0
Sophora japonica	A + B	125.0	>1000.0	1000.0
Phaseolus vulgaris	None	>1520.0	300.0	300.0
Glycine max	None	250.0	250.0	250.0
Bauhinea purpurea	None	93.5	375.0	>1500.0
Helix pomatia	A	32.0	>718.0	718.0
III. DGal binding lectins				
Peanut agglutinin	None	>2400.0	>2400.0	300.0
Ricinus communis I	None	162.0	325.0	325.0
Ricinus communis II (ricin)	None	7.8	15.0	3.9
Geodia cydodonium	None	50.0	50.0	50.0
Artocarpus integrifolia (jacaline)	None	82.0	41.0	82.0
Axinella polypoides	None	64.0	64.0	–
IV. DMan binding lectins				
ConA	None	62.0	125.0	62.0
Lens culinaris	None	375.0	128.0	375.0
Pisum sativum	None	237.0	950.0	950.0
V. Sialic acid binding lectin				
Limulus polyphemus	None	187.0	1500.0	>1500.0

Source: Adapted from Pereira et al. (1980).

Agglutination is the most easily detectable manifestations of the interaction of a lectin with cells. It results from multiple cross-bridges between apposing cells afforded by the multivalent lectin. The agglutinating activity of lectins is usually performed with erythrocytes, using the serial dilution technique; the end point is determined either visually or spectrophotometrically. It is very difficult to interpret agglutination because of the multiple factors that influence the phenomenon. For instance, cell surface properties such as number, accessibility, and distribution of receptor sites, mobility of sites (membrane fluidity), electric charge, surface rigidity and surface structures (e.g., microvillae), as well as cytoplasmic components such as microtubule-microfilament system, metabolic state of the cells, and so on, all affect lectin agglutination. As a consequence, lectins may bind to cells without causing agglutination. Almost all cells examined, as well as many subcellular organelles, bind lectins. Binding is best demonstrated with the aid of

Table 5 Mitogenic Lectins

Lectin	References
DGal binding lectins	
Abrus precatorius	Closs et al. (1975), J. Immunol. *115*, 1045.
Axinella polypoides	Bretting et al. (1980), J. Immunol. *120*, 1032.
Agaricus compestria	Presant and Korfeld (1972), J. Biol. Chem. *247*, 6937.
Bauhinia carronii	Curtain and Simmons (1972), Int. Arch. Allergy *42*, 225.
D-GalNA binding lectins	
Lima bean	Ruddon et al. (1974), Proc. Natl. Acad. Sci. USA *71*, 1848.
Maclura pomifera	Horejsi and Kocourek (1978), Biochim. Biophys. Acta *358*, 299.
Phaseolus coccineus	Angelisova and Haskovec (1978), Eur. J. Immunol. *83*, 163.
Phytohemoagglutinin	Nowell (1960), Cancer Res. *20*, 462.
Robinia pseudoaccacia	Schumann et al. (1973). Int. Arch. Allergy Appl. Immunol. *45*, 331.
Sophora japonica	Terao and Osawa (1973), J. Biochem. *74*, 199.
Wistaria floribunda (mitogen)	Barker and Farries (1967), Nature (Lond.) *215*, 659.
Wax bean	Sela et al. (1973), Biochim. Biophys. Acta *310*, 273.
D-Man binding lectin	
ConA	Novogrodsky and Katchalsky (1971), Biochim. Biophys. Acta *228*, 579.
Lentil	Toyoshima et al. (1970), Biochim. Biophys. Acta *221*, 514.
Pea	Trowbridge (1974), Proc. Natl. Acad. Sci. USA *70*, 3650.
Vicia faba	Wang et al. (1974), J. Mol. Biol. *88*, 259.
D-GlcNAc binding lectin	
Pokeweed mitogen	Farnes et al. (1964), Lancet *2*, 1100.
Wheat germ mitogen	Brown et al. (1976), J. Immunol. *117*, 1976.

lectins labeled with radioactive isotopes or with compounds that are visible under a microscope, such as fluorescein, ferritin, peroxidase, and hemocyanin. Nevertheless, agglutination of cells by lectins is a very useful criterion to analyze morphologically similar but functionally distinct cell populations. There are marked differences in agglutinability between normal and malignant cells, between embryonal and adult cells, between mitotic and interphase cells, and between the developmental forms of the protozoan *Trypanosoma cruzi* (Table 4), a unicellular organism that causes Chagas' disease (Pereira et al., 1980).

One of the most interesting effects of the interaction of lectins with cells is mitogenic stimulation (Lis and Sharon, 1977), that is, the conversion of resting lymphocytes into actively growing and dividing blast cells. Mitogenic lectins are polyclonal activators in that they stimulate a large proportion (usually 30 to 60%) of lymphocytes from normal subjects, as well as newborns, independent of immunization. Many lectins of various sugar specificities are mitogenic (Table 5).

Most lectins stimulate only the thymus-dependent class of lymphocytes (T cells) and are inactive or inhibitory for the other set of lymphocytes (B cells). However, the pokeweed mitogen Pa-1 stimulates both T and B cells. Various agents other than lectins also stimulate lymphocytes, including antigens, antibodies to lymphocyte surface components, bacterial endo- and exotoxins, calcium ionophores, and oxidizing agents such as periodate and the enzyme galactose oxidase.

The mechanism of mitogenic stimulation is unknown. It is clear, however, that the initial step is the binding of the lectin to cell surface carbohydrates since the mitogenic activity can be inhibited by sugars for which the lectins are specific or by treating the lymphocytes with appropriate glycosidases to remove the sugar with which the lectin interacts. However, in certain instances, lectins are mitogenic only after glycosidase digestion of the lymphocytes because such treatment exposes sugar structures to which the lectin is specific. For instance, mouse lymphocytes do not respond to stimulation by soybean agglutinin unless they have been treated with neuraminidase, a process that exposes cell surface galactosyl residues with which the lectin reacts and triggers the mitogenic activity. As a consequence, the response to soybean agglutinin is almost completely abolished by the enzymatic removal of the galactosyl residues from the neuraminidase-treated cells. The valency of lectins has been shown to be important for mitogenic activity. Thus, of the two naturally occurring lima bean lectins, the tetravalent one (MW 247,000) is a potent mitogen for human lymphocytes, while the divalent one (MW 124,000) shows only a very weak mitogenic activity. Similarly, soybean agglutinin in its native form, which is divalent, is not mitogenic to a variety of lymphocytes, but it becomes active after polymerization with glutaraldehyde. However, polyvalency may not be essential for mitogenic activity since the mitogenic potency of ConA is not altered by conversion of the tetravalent lectin into a divalent one by succinylation.

VI. APPLICATIONS

Lectins are invaluable reagents in immunological, chemical, and biological research, as well as in clinical medicine. In all cases, the methods rely on the specific sugar-binding properties of the lectins.

A. Isolation of Glycoproteins

Affinity chromatography on lectins is being employed for the detection, isolation, and characterization of both soluble glycoproteins and membrane glycoproteins. Lectins are also useful in detecting glycoproteins on gels. The specific interaction of a lectin with an organic material is taken as evidence that the material contains carbohydrate. For instance, the binding of HLA-A antigens of human lymphocytes to columns of immobilized lentil lectin has shown that these substances are glycoproteins (Snary et al., 1974).

Affinity chromatography on lectins is also useful for the fractionation of glycoproteins which differ only slightly in their carbohydrate composition or in the structure of their oligosaccharide units. For example, the blood group A, B, and H determinants (Watkins, 1972; Lloyd, 1976) have been each found to be of four kinds, two of which are:

A.

$$\begin{array}{l} \qquad\qquad\qquad \text{LFuc1} \\ \qquad\qquad\qquad\quad \downarrow \alpha \\ \qquad\qquad\qquad\quad 2 \\ \text{DGalNAc}\alpha 1 \rightarrow 3\text{DGal}\beta 1 \rightarrow 3\text{DGlcNAc}\beta 1 \rightarrow 3 \end{array}$$

B. DGal substitutes for DGalNAc above

or

A.

$$\begin{array}{c} \text{LFuc1} \\ \downarrow \alpha \\ 2 \\ \text{DGalNAc}\alpha 1\rightarrow 3\text{Gal}\beta 1\rightarrow 4\text{DGlcNAc}\beta 1\rightarrow 3 \end{array}$$

B. DGal substitutes for DGalNAc above

The other two A and B determinants have a second LFuc linked to the DGlcNAc on carbons 4 and 3, respectively. The four H determinants have the same structures, but lack nonreducing DGalNAc or DGal. The A, B, and H specificities thus involve substituents on two types of structures, DGalβ1→3DGlcNAc and DGalβ1→4GlcNAc called type 1 and type 2 determinants, respectively.

The lectins from *L. tetragonolobus* are specific for H oligosaccharides of type 2, while the lectin from *D. biflorus* is selective for A oligosaccharides of both type 1 and type 2 determinants. Therefore, H and A blood group substances are specifically precipitated by *Lotus* and *Dolichos* lectins, respectively. The blood group substances extracted from hog stomachs have A, H, or AH serological activities. The commercial preparation of hog gastric mucin is a mixture obtained by pooling hog stomachs and therefore contains individual molecules having A, H, and both A and H activities. Since the analytical composition of these substances is very similar, chemical fractionation procedures employing phenol extraction and fractional ethanol precipitation yields products possessing all serological activities. However, individual activities can easily be obtained from the commercial hog gastric mucin by affinity chromatography with the *Lotus* and *Dolichos* lectins attached to Sepharose as affinity adsorbents (Pereira and Kabat, 1976). Similarly, molecular variants of other glycoproteins, such as α-fetoprotein, fetuin, ceruloplasmin, and carcinoembryonic antigen, have been resolved by lectin-affinity chromatography, which underlies heterogeneity in their sugar moiety.

B. Cell Separation

The use of lectins to separate intact cells bearing different surface carbohydrates is just beginning to be explored. Most experiments are done with mammalian cells, particularly with lymphocytes, but in principle, any population of cells, including bacteria and protozoa, can be sorted out into subclasses or subpopulations by lectin affinity, provided that there exists differences in cell surface sugars.

Two approaches have been used for separating cells with the aid of lectins. In one approach, cells are agglutinated selectively with lectins of a particular specificity and the agglutinated cells separated from the unagglutinated ones by differential centrifugation in serum; clumps settle to the bottom of the tube and the single, unagglutinated cells remain suspended, permitting separate collection of the two fractions (Reisner et al., 1976). The agglutinated cells are then dissociated in a solution of suitable sugar inhibitor and washed with appropriate buffers. This approach was first applied in 1949 in the separation of leukocytes from erythrocytes with lectins from the red kidney bean, *P. vulgaris*, and more recently the principle has been applied to fractionation of mouse and human thymocytes into medullary (immunologically mature) and cortical (immunologically immature) cells by the use of peanut lectin, which agglutinates only the latter.

In a similar manner, mouse spleen B cells, which bind soybean agglutinin, were separated from T cells, which do not bind the lectin.

In the other approach, cells can be separated by affinity chromatography, in which the lectin is immobilized on a solid support either by nonspecific adsorption or by covalent attachment (Edelman and Rutishauser, 1974). Nylon fibers derivatized with ConA have been used to isolate tumor cells that were shown to be more immunogenic than the parent tumor cell population (Killion and Kollmorgen, 1976). In other studies, cells were fractionated on lectin immobilized on agarose beads. For instance, a column of *Helix pomatia* lectin coupled to Sepharose 4B selectively retained neuraminidase-treated T cells and thus could be used to separate T and B cells (Hellstrom et al., 1976). Human peripheral blood T lymphocytes were separated by chromatography on Sepharose-bound wheat germ agglutinin into two subpopulations that differ in their response in the mixed lymphocyte reaction and to mitogens.

Another approach to separate cells by affinity chromatography utilizes a glycoprotein (hog gastric mucin blood group A and H substance) covalently attached to Sepharose (Pereira and Kabat, 1979b). Lectins of specificities capable of reacting with the glycoprotein are passed through the column as crude extracts; the column is saturated with lectin and is then washed free of unrelated proteins, a first step that essentially leaves purified lectin attached to the column with many lectin sites available. The lectin $\rightleftarrows$ glycoprotein-Sepharose is then used as an affinity probe for isolating and fractionating cells. Thus, in this technique, crude lectin preparations may be used, separate coupling of each lectin to Sepharose is not required, nonspecific adsorption is minimal, and cells bound to the column are readily eluted (together with the lectin) with a specific sugar. In addition, columns of different specificity can be prepared by simply adsorbing a lectin to a single glycoprotein-Sepharose conjugate.

C. Characterization of Cell Surfaces

The majority of exposed plasma membrane proteins are glycoproteins, and there is increasing evidence that glycoproteins as well as glycolipids are of major importance in the maturation and regulation of functions displayed by cells, such as those of the immune system. On the other hand, much of the work concerning structure and function of cell surface carbohydrate is derived from studies with lectins.

Information on the lectin receptor sites on the cell surface and their homogeneity, as well as on the strength of binding of the lectin to the cells, is provided with studies using radioactively labeled lectins. Microscopic methods have also been used to determine lectin receptor sites. In this case, the lectin is bound to molecules such as ferritin and hemocyamin, which can easily be distinguished in the electron microscope. In the peroxidase method, a peroxidase-lectin complex is bound to the cell and quantitated by measuring the activity of the bound enzyme. Insight into the relative disposition of the receptor sites for different lectins may also be obtained from competition experiments.

The number of receptor sites for lectins on lymphocytes and trypanosomes is usually in the range 10^6 to 10^7 per cell, and on erythrocytes it is frequently 10^5 to 10^6.

Sialic acid has been discovered on the surface membrane of *T. cruzi* with the aid of lectins, as the binding of wheat germ agglutinin to the trypanosome was completely abolished by neuraminidase treatment of the cells; N-acetylneuraminic acid and N-glycolylneuraminic acid were identified in the digests by gas-liquid chromatography (Pereira et al., 1980).

Lectins are of great importance in studies of receptor redistribution, originally described with antibodies to lymphocyte surface immunoglobulin, and on receptor rotation (Nicholson, 1974). These studies provide strong evidence for the fluid mosaic model of membrane structure, which suggests that membranes consist of proteins and glycoproteins floating in a lipid matrix.

Strong evidence for the asymmetry of surface membranes is also provided with the aid of lectins. Thus experiments with labeled lectins and intact cells show that lectin receptors are found exclusively on the outer membrane surfaces. Asymmetric distribution of lectin receptors has also been found on intracellular membranes.

D. Structural Studies on Carbohydrates

As already mentioned, the specificity of the lectin combining sites is ascertained by the hapten inhibition technique using model sugar compounds. As the sites are characterized, they, in turn, contribute to the identification of newly isolated oligosaccharides, particularly when structure determination by chemical methods is ambiguous (Pereira and Kabat, 1979a). Using milk and blood group oligosaccharides of known structure as probe, the *Lotus* lectins have been found to be highly specific for type 2 chains containing fucose residues on C-2 of the galactose of DGalβ1→4DGlcNAc, but not for similarly substituted type 1 chains of structure DGalβ1→3DGlcNAc. A second fucose on DGlcNAc of the type 2 chain further increases activity. This striking specificity of the *Lotus* lectins for mono- and difucosyl type 2 determinants has proven specially useful in establishing structures of certain disubstituted blood group oligosaccharides. For example, for the compound

(a) DGal2←1αLFuc
1
β↓
4
DGlcNAc3←1αLFuc
1
β↓
6
DGalβ1→3N-acetylgalactosaminitol

isolated from a human ovarian cyst fraction, methylation analysis can only establish that there is a 1→3 substituted DGlcNAc. Thus there is an alternative structure for oligosaccharide (a):

(b) DGal2←1αLFuc
β↓
3
DGlcNAc4←1αLFuc
1
β↓
6
DGalβ1→3N-acetylgalactosaminitol

The structure of the compound was established unambiguously as being (a) (type 2 chain), because it reacted strongly with the *Lotus* lectin, while structure (b) (type 1 chain) was excluded since it should be noninhibitory.

E. Clinical Uses

The mitogenic properties of lectins have been used in clinical medicine to recognize congenital and acquired immunologic deficiencies, to detect sensitization caused by infectious agents, or in some autoimmune diseases, to monitor the effects of various immunosuppressive and immunotherapeutic manipulations, and in the diagnosis of genetic diseases with chromosomal defects (Oppenheim et al., 1975). In blood banks, lectins are used to detect secretors—humans who secrete blood group substances A, B, or H in saliva or other body fluids—and to type blood. Thus the lectins of *L. tetragonolobus* and *U. europaeus* serve to identify O cells due to the unavailability of natural anti-H(O) antibodies. The lectin of *D. biflorus* agglutinates A_1 but not A_2 erythrocytes and thus is used to identify A_1 cells, and similarly, the *V. graminea* lectin is used to distinguish between M and N erythrocytes, as it agglutinates only the latter.

Another application of lectins is in the differential diagnosis of polyagglutination, a condition in which the red cells are agglutinated by antibodies present in the sera of nearly all human adults (Bird, 1978). In the T type, which accompanies certain bacterial and viral infections that produce neuraminidase, the erythrocytes are specifically agglutinated by peanut lectin. In another variety of polyagglutinable red cells, termed Tn, the cells are agglutinated by *D. biflorus* lectin, but not by peanut. Other types of erythrocyte polyagglutination are also characterized by their differential agglutination with lectins. The importance of this condition relates to its occurrence with hemolytic anemia and also because it interferes with blood typing.

VII. FUNCTION

The wide distribution of lectins in living organisms, being detectable intracellularly as well as on cell surfaces and extracellularly, and their ability to discriminate among different sugars in solution and on cell surfaces, suggest that lectins have a number of biological actions (Barondes, 1981).

The ubiquitous presence of lectins on cell surfaces seems to indicate that they may be involved in intercellular adhesion. For instance, in cellular slime molds, a galactose binding protein appears to control the adhesion of aggregating amoebae, and a cell aggregating factor extracted from certain sponges is specifically inhibited by glucuronic acid. In other cases, lections isolated from tissues seem to mediate intercellular adhesion of different cell types. For example, some lectins, such as those in root hairs, apparently bind complementary saccharides in a fairly discriminating way, which leads to a highly specific symbiosis between nitrogen-fixing bacteria and the roots of leguminous plants. There is evidence that lectins acting in this way participate in both the prevention of plant infection, by binding to saccharides on bacteria or fungi, and in the promotion of bacterial infection of vertebrate cells. As an example of the latter case, *Escherichia coli* contains a lectin that binds D-mannose and that presumably mediates attachment of this microorganism to animal cells, which could be a requirement for infection; similarly, the binding of *Vibrio cholera* to intestinal cell surfaces is inhibited by L-fucose. In higher plants, the attachment of pollen to the surface of the stigma, and in sea urchin, the attachment of sperm to eggs also seem to be mediated by protein-carbohydrate interaction.

Another postulated role for the lectins is in the binding of soluble glycoproteins to the cell surface as a prelude to their pinocytosis (Ashwell and Harford, 1982). Desialylated serum glycoproteins, like orosomucoid, have drastically reduced survival times in the circulation compared to native forms of the same protein. A specific receptor on hepatocytes mediates their clearance by recognition of galactose residues made terminal by removal of sialic acid. The newly exposed galactose residues are implicated as recognition determinants by the findings that treatment of asialo-orosomucoid with galactose oxidase or β-galactosidase, or enzymatic replacement of the missing sialic acid residues prolonged survival in the circulation. This galactose binding protein has been isolated from the liver of several mammals, including humans, and shown to be a lectin because of its hemagglutinating properties and because of its specific interaction with galactose and N-acetylgalactosamine (Table 2). Although it is generally presumed that the hepatic lectin participates in the regulation of mammalian plasma glycoprotein homeostatis, unequivocal evidence is still lacking.

REFERENCES

Ashwell, G., and Harford, J. (1982). Annu. Rev. Biochem. *51*, 531.

Barondes, S. H. (1981). Annu. Rev. Biochem. *50*, 207.

Bird, G. W. G. (1978). Proc. XVII Int. Soc. Hematol., Paris, p. 87.

Boyd, W. C. (1970). Ann. N.Y. Acad. Sci. *169*, 168.

Edelman, G. M., and Rutishauser, U. (1974). In *Methods in Enzymology*, Vol. 34, Part B (W. B. Jakoby and M. Wilchek, Eds.). Academic Press, New York.

Forier, A., De Neve, R., Kanarek, L., and Strosberg, A. D. (1978). Proc. Natl. Acad. Sci. USA *75*, 1136.

Goldstein, I. J., and Hayes, C. E. (1978). In *Adv. Carbohydr. Chem. Biochem.*, Vol. 35. Academic Press, New York.

Hellstrom, U., Dillner, M. L., Hammarstrom, S., and Perlman, P. (1976). J. Exp. Med. *144*, 1381.

Killion, J. J., and Kollmorgen, G. M. (1976). Nature (Lond.) *259*, 674.

Lis, H., and Sharon, N. (1977). In *The Antigens*, Vol. 4 (M. Sela, Ed.). Academic Press, New York.

Lis, H., and Sharon, N. (1981). In *The Biochemistry of Plants*, Vol. 6 (P. K. Stumpf and E. E. Cohn, Eds.). Academic Press, New York.

Lloyd, K. O. (1976). Int. Rev. Sci. (Ser. 2) *7*, 251.

Loontjens, F. G., Van Wauwe, J. P., and De Bruyne, C. K. (1975). Carbohydr. Res. *44*, 150.

Nicholson, G. L. (1974). Int. Rev. Cytol. *39*, 89.

Oppenheim, J. J., Dougherty, S., Chan, S. P., and Baker, J. (1975). In *Laboratory Diagnosis of Immunologic Disorders* (G. N. Vyas, D. P. Stites, and G. Brecher, Eds.). Grune & Stratton, New York.

Pereira, M. E. A., and Kabat, E. A. (1976). J. Exp. Med. *143*, 422.

Pereira, M. E. A., and Kabat, E. A. (1979a). Crit. Rev. Immunol. *1*, 33.

Pereira, M. E. A., and Kabat, E. A. (1979b). J. Cell Biol. *82*, 185.

Pereira, M. E. A., Loures, M. A., Villalta, F., and Andrade, A. F. B. (1980). J. Exp. Med. *152*, 1375.

Reisner, Y., Linker-Israeli, M., and Sharon, N. (1976). Cell. Immunol. *25*, 129.

Snary, D., Goodfellow, P., Hayman, M., Bodmer, W. F., and Crumpton, M. (1974). Nature (Lond.), *247*, 457.

Watkins, W. M. (1972). In *The Glycoproteins*, 2nd ed., Part B (A. Gottschalk, Ed.). Elsevier, Amsterdam.

16

Nature and Thermodynamics of Antigen-Antibody Interactions

Carel J. van Oss School of Medicine and School of Engineering and Applied Sciences, State University of New York at Buffalo, Buffalo, New York

Darryl R. Absolom Research Institute, The Hospital for Sick Children, University of Toronto, Toronto, Ontario, Canada

I. INTRODUCTION

From a chemical as well as from a biochemical point of view, antigen-antibody (Ag-Ab) interactions are somewhat unusual. The bonds involved are of the weak interaction variety (mainly coulombic and van der Waals), and not covalent. While the valencies of Abs are well known (they are either divalent or dekavalent, dependent on the antibody class), the valencies of most Ags are less well established. Also, different valency sites of a given Ag often are completely different from each other as far as the chemical composition of the various antigenic determinants is concerned. Thus plurivalency of such Ags only has meaning vis-à-vis a whole serum, which comprises various families of Abs, each family having a chemically completely different antibody-active site. To make matters worse, even within one family of Abs that are all directed specifically toward one given antigenic determinant, the antibody-active sites, while chemically fairly similar, generally still differ enough to manifest a wide array of different affinities to that determinant. Finally, most Ags can combine with their Abs in virtually any proportion at all, so that the concept of stoichiometry is in most cases irrelevant to Ag-Ab interactions. The role of entropy in Ag-Ab reactions also is far from clear; although the formation of Ag-Ab lattices would seem to have to go together with an increase in order, the formation of Ag-Ab complexes more often than not is accompanied by a measurable increase in entropy of the system. It thus is not surprising that the thermodynamics of Ag-Ab reactions are fraught with difficulties that complicate the discipline of immunochemistry even more than most other fields.

II. NATURE OF AG–AB BONDS

A. Van der Waals Bonds

Between all atoms that are brought closely enough together, there is a mutual attraction, caused by the interaction between the fluctuating dipole occurring in one atom, and a second dipole which the first dipole induces in a second atom. The interatomic and intermolecular forces that originate in this manner are called van der Waals-London, or dispersion forces. Van der Waals-London interactions between atoms, when occurring in large molecules or even particles, are to a significant degree additive, so that their free energy of interaction ΔF (e.g., in the configuration of two semi-infinite flat parallel bodies) may be expressed as

$$\Delta F_{vdw} = \frac{A}{12\pi d_0^2} \tag{1}$$

where A is the Hamaker coefficient (which is linked to physical properties of the interacting materials, including those of the liquid, when the interaction takes place in a liquid medium and d_0 is the equilibrium distance between the two parallel bodies or molecules). For a number of materials the Hamaker coefficients can be measured and/or calculated (Visser, 1972); van der Waals-London interactions can also be estimated via the Lifshitz approach (Visser, 1972; Israelachvili, 1974).

In addition to the universally occurring van der Waals-London interactions, in more polar liquids (such as water) and between more polar molecules, there also are interactions between permanent dipoles (van der Waals-Keesom forces) and interactions between permanent and induced ipoles (van der Waals-Debye forces). All these van der Waals interactions, taken together, henceforth will be called van der Waals forces (or van der Waals interactions). There is a direct way to determine the (total) free energy ΔF_{132} of the van der Waals interaction between two materials 1 and 2 (in a liquid medium 3):

$$\Delta F_{132} = \gamma_{12} - \gamma_{13} - \gamma_{23} \tag{2}$$

where γ stands for the interfacial tensions between the materials indicated by the subscripts and where $\Delta F_{132} = \Delta F_{vdw}$ from Eq. (1). Once the surface tensions ($\gamma_{1V}, \gamma_{2V}, \gamma_{3V}$, where V stands for vapor) of the various materials are known, the necessary elements of Eq. (2) can be estimated. Material 1 then stands for the antigenic determinant, material 2 for the antibody-active site, and 3 for the aqueous medium.

Due to the marked increase in the van der Waals attraction at short distances of interaction d_0 [see Eq. (1)], van der Waals attractions are stronger the better the steric fit between an antigenic determinant and its complementary antibody-active site. The overall distance between the antigenic determinant and the antibody-active site, as between other macromolecules and/or particles, attracted to each other via van der Waals forces, can be as short as about 2 Å (Israelachvili, 1974; Pressman and Grossberg, 1973; van Oss et al., 1980a), and in ideal cases probably about 1.8 Å (van Oss et al., 1980a).

Van der Waals-Keesom (orientation) and van der Waals Debye (induction) forces obey the same rules, but short-range ("hydrophobic") interaction forces must be treated separately. The sum of van der Waals interactions plus short-range interactions then represents the total interfacial (or "hydrophobic") interactions (van Oss et al., 1986a).

B. Electrostatic Bonds

The electrostatic, or Coulombic, interactions between Ag and Ab are caused by attractive forces between one or more ionized sides on the antigenic determinant and oppositely charged ions on the antbody-active site. These typically are the COO^- and the NH_4^+ groups on polar amino acids of the Ag and Ab molecules. In certain hapten-Ab systems the number and position of the ionized sites have been determined (Pressman and Grossberg, 1973). Due to the shielding effect of the diffuse ionic double layers surrounding the charged sites, which effect varies strongly with the ambient ionic strength and with the distance between charged sites the calculation of ionic interaction energies is a complicated operation (Gabler, 1978). However, as a first approximation, the free energy ΔF_{el} of attraction between a COO^- and a NH_4^+ group, at a distance of about 3 Å, in a medium with an ionic strength $\mu = 0.15$, is of the order of about -7 kcal/mol (Gabler, 1978) or about -10 mJ/m^2, which binding energy is of the same order of magnitude as that of a typical van der Waals attraction between Ag and Ab (van Oss and Neumann, 1977; van Oss and Grossberg, 1979; see below).

As might be expected, by analogy with cell–cell interaction and cell-adhesion phenomena, purely (or mainly) electrostatic Ag–Ab interactions can occur not only through negatively charged sites on one determinant attracting positive sites on the other, but also via the binding of negatively charged antigenic determinants to equally negatively charged antibody-active sites, by means of linkage through, for example, Ca^{2+} ions. A case in point, involving synthetic polypeptides (comprising polyglutamic acid moieties), has been described by Liberti (1975). Ag–Ab complexes of that type can be dissociated with the complexing agent ethylenediaminetetraacetic acid (EDTA). No naturally occurring Ags that involve Ca^{2+} bridging have been described as yet, to our knowledge.

As in the case of van der Waals forces, electrostatic attractions are at a maximum at the shortest distances. Thus here also, precision of fit and especially precise juxtaposition of oppositely charged ions on antigenic determinant and antibody-active site, favor strong electrostatic bonding.

C. Other Interactions

Hydrogen Bonding

Hydrogen bonds are dipole interactions between, for example, OH and C=O, NH and C=O, and NH and OH groups. The binding energy is of the same order of magnitude as that of van der Waals and electrostatic interactions. Because these dipole–dipole interactions have a significant energy only when they can react over short distances (of the order of 2 to 3 Å), a rather exquisite precision of "fit" plus many unusually successful chance encounters are required for them to become dominant. However, in Ag-Ab interactions, hydrogen bonding is known to play a direct role in some cases, for example, in the case of certain orthosubstituted benzoate haptens, reacting with anti-para-azobenzoate antibody (Pressman and Grossberg, 1973). Somewhat more indirectly, hydrogen bonds play an important role in hydrophobic interactions.

Hydrophobic Interactions

"Hydrophobic interactions" play a significant role in various Ag-Ab interactions; their effect is short-range and is, in conjunction with van der Waals forces, of an interfacial nature (van Oss et al., 1986a, 1986b).

Covalent Bonds

No cases of covalent bonding are known to occur in Ag–Ab interactions.

D. Combined van der Waals, Hydrophobic, and Electrostatic Interactions

There are Ag-Ab interactions that involve only van der Waals and hydrophobic (or interfacial) interactions, for example, in the cases where the Ag is a neutral polysaccharide such as dextran (van Oss and Neumann, 1977), and in the case of the hapten 3-azopyridine (P3) in the reaction between rabbit anti-P3, and P3 coupled to rabbit serum albumin (Pressman and Goldberg, 1973; van Oss et al., 1979). There also are instances in which virtually only electrostatic bonds play a role, for example, in the case of DNA-anti-DNA, where the Ag is strongly (negatively) charged and thus quite hydrophilic, while the Ab-active site has an equally pronounced positive charge; the interfacial attraction between two strongly hydrophilic components in an aqueous medium is negligible (de Groot et al., 1980).

However, in the cases of many polysaccharide or glycoprotein Ags and of most polypeptide Ags, the Ag-Ab bond is brought about by a *combination* of interfacial and electrostatic interactions. Thus in these cases the "lock and key" mechanism of the Ag–Ab interaction proposed by Ehrlich in the nineteenth century consists of the best steric fit for optimal van der Waals attraction, combined with the most precise juxtaposition of oppositely charged ions for maximum electrostatic attraction. Also, in preponderantly electrostatic bonding, as with bovine serum albumin/anti-bovine serum albumin (van Oss et al., 1982b), as soon as Ag and Ab are brought close together by electrostatic attraction, in most cases an additional van der Waals attraction will subsequently be established between the antigenic determinant and antibody-active site, and possibly also between neighboring moieties of Ag and Ab.

It is important to realize the dual nature of the attractive forces in all these cases, when one attempts to dissociate such Ag–Ab complexes (see below).

E. Nonstoichiometry of the Ag–Ab Reaction

As is especially apparent from the formation of soluble as well as insoluble Ag–Ab complexes (see Chap. 17), Ags and Abs (as well as other complex-forming substances, such as cationic and anionic surfactants) can combine in a wide range of proportions. Ag–Ab reactions thus are essentially nonstoichiometric.

F. Valency

The nonstoichiometry of Ag–Ab reactions creates a special difficulty for the determination of the valency of Ags as well as of Abs. The valency of Ags can be determined only with complexes formed in an excess of Ab, while the valency of Abs can only be determined with complexes formed in an excess of Ag. Still, precipitates obtained at optimal Ag/Ab ratios (see Chap. 17) tend to comprise close to stoichiometric Ag–Ab proportions, with perhaps a slight excess of Ag.

Antibodies of the IgG, IgA, IgD, and IgE classes are divalent; IgM-class antibodies are dekavalent (see Chap. 7). It is especially important to measure the valency of antibodies of the IgM class with Ags (or haptens) with a molecular weight of not much more than 1000; with larger Ags steric hindrance effects cause a reduction in the apparent valency of IgM (Edberg et al., 1972). Secretory IgA is tetravalent.

Most protein Ags are plurivalent only vis-à-vis a complete anti*serum* elicited against them, comprising antibodies against each of the antigenic determinants. Each different valency site of protein Ags generally is an antigenic determinant with a completely different configuration from all the other valency sites. A given monoclonal Ab can react with only one valency site of such a protein Ag. Some repeating types of biopolymer may be plurivalent, with all the antigenic determinants being identical to each other (e.g., DNA), or they may have only two or three different groups of antigenic determinants that are identical to one another within each group (DNA also can be an example of this type of Ag). On the other hand, other repeating biopolymers may be monovalent, as is dextran in the ideal, totally unbranched form; its dominant antigenic determinant is the terminal nonreducing sugar (Kabat, 1976).

The different antigenic determinants of globular proteins tend to be situated near their carboxy terminal and at prominent places on the outer periphery of their tertiary configuration (see Chap. 2). From known valencies of globular proteins and comparable biopolymers (e.g., viruses), it would seem that there (very roughly) is one antigenic determinant for about every 35 to 40 amino acids. As a first approximation one may thus estimate the valency N of a given globular protein to be

$$N \approx \left(\frac{M_w}{5000}\right)^{2/3} \tag{3}$$

in which M_w is the protein's molecular weight.

G. Size of Binding Sites

The size of antigenic determinants is (for polysaccharides) close to that of penta- to hexasaccharides, and for proteins close to that of penta- to hexapeptides, yielding a specific surface area that may vary between 250 and 400 Å^2 (Kabat, 1976) (see also Chaps. 1 to 3), and a possible total surface area of contact, and thus of the interaction (due to the participation of neighboring saccharide moieties or amino acids, mainly in van der Waals interactions), of up to 600 Å^2.

III. THERMODYNAMICS

A. Principles

In a general manner, the Ag-Ab interaction may be summarized as

$$Ag + Ab \rightleftharpoons Ag\text{-}Ab + x \text{ calories} \tag{4}$$

The equilibrium constants of that reaction are

$$K_{ass} = \frac{[Ag\text{-}Ab]}{[Ag] \cdot [Ab]} \tag{5a}$$

and

$$K_{diss} = \frac{[Ag] \cdot [Ab]}{[Ag\text{-}Ab]} \tag{5b}$$

where K_{ass} and K_{diss} stand for the association and dissociation constants, respectively, and the terms in square brackets indicate (molar) concentrations. $K_{ass} = 1/K_{diss}$ only in

ideal cases. In practice one will find in many instances that the energy needed to prevent association (corresponding to K_{ass}) is less than the energy required to dissociate already existing complexes (corresponding to K_{diss}), for example, because of the continuing formation of additional van der Waals bonds, after the initial Ag-Ab complex formation (van Oss and Neumann, 1977; van Oss et al., 1979; 1980a, 1982b; see also Karush, 1976). On account of the (generally) high molecular weights of Ag and Ab, and thus the relatively low molar concentrations of these reagents, molar *concentrations* may be used here, instead of the more accurate (but more difficult to measure) *activities*. Depending on the method of measurement used (see below), one obtains either K_{ass} or K_{diss}. In view of the heterogeneity and multiplicity of antigenic determinants in most Ags, and of the heterogeneity of antibodies in antisera directed against such Ags, Eqs. (4) and (5) are considerable oversimplifications. Thus K_{ass} and K_{diss} are mainly to be considered as practical constants, reflecting the average of all the subreactions involved, in all cases except in the ideal situation of the reaction of a monovalent hapten with a monoclonal antibody.

Nevertheless, once the practical equilibrium constant K has been determined, the total free energy change ΔF of the complete reaction can be derived:

$$\Delta F_{total} = -RT \ln K \tag{6}$$

where R is the gas constant (1.986×10^{-3} kcal, 8.3144×10^{7} ergs, or 8.3144 J, per degree Kelvin per mole), and T is the absolute temperature in degrees Kelvin. It should be emphasized however that Eq. (6) is only valid for standard conditions, i.e., for those cases where unit molar concentrations* of both Ag and Ab were used, and is formally expressed as:

$$\Delta F^{\circ} = -RT \ln K^{\circ} \tag{6a}$$

(where the superscripts ° indicate that standard conditions prevailed). When the binding energy between single Ag and Ab molecules is to be derived, the gas constant R must be divided by Avogadro's number, i.e., by 6.022×10^{23}, to obtain Boltzmann's constant k, so that then:

$$\Delta F^{\circ} = -kT \ln K^{\circ} \tag{6b}$$

(where k is 1.38×10^{-16} ergs per degree, or 1.38×10^{-23} J per degree). It should, however, be stressed that all (association) equilibrium constants K (Eqs. 6, 6a-b, 7) should be expressed in L/mole. By measuring K at two different temperatures, the enthalpy ΔH can be calculated:

$$\frac{d \ln K}{dT} = \frac{\Delta H}{RT^2} \tag{7}$$

*These are the salient "standard conditions" in immunochemistry; unless otherwise stated, ambient pressure, temperature, and pH are held to be kept constant. See also the note on "standard conditions" at the end of Table 1.

which allows the derivation of the entropy ΔS, *via* the van't Hoff equation:

$$\Delta F = \Delta H - T \Delta S \tag{8}$$

Another way of obtaining ΔH is via microcalorimetry (see below), and if K is also determined, Eqs. (6b) and (8) also yield ΔF and ΔS. ΔF as derived from K usually is expressed in kcal per mole, but it can also be expressed in units of ergs/cm^2 (or mJ/m^2), provided that the surface area of the antigenic determinant(s) [and of the antibody-active site(s)] is known. ΔF comprises $\Delta F_{vdw} + F_{el}$. If one measures ΔH by equilibrium determinations at different temperatures, via Eq. (7), it always is advisable to do such determinations at least at *three* different temperatures (van Oss et al., 1982b).

B. Affinity and Avidity

Much confusion reigns concerning the definitions of Ab affinity and Ab avidity. It is best to follow Steward (1974), and to define *affinity* as a thermodynamic expression of the binding energy of an antibody-active site for its homologous antigenic determinant. To quote Steward (1974): "Experimentally this term has its most precise application in monovalent hapten-anti-hapten systems"—and, we may add, especially when the antibody is monoclonal. *Avidity*, although it is based on affinity, in addition involves factors such as Ab valency, Ag valency, Ab-heterogeneity, and differences in antigenic determinants of a given Ag. Thus ΔF_{total} [Eq. (6)] stands for the *affinity* of a monoclonal antihapten Ab toward its homologous monovalent hapten, and in general for the *affinity* of a monoclonal Ab toward a solitary nonrepeating antigenic determinant on a given Ag. However, ΔF_{total} [Eq. (6)] can also express the *avidity* of a family of antibodies (of one immunoglobulin class, e.g., IgG) toward their homologous (plurivalent) Ag. To distinguish between these two thermodynamic expressions it may be advisable to label them ΔF^{aff}_{total} and ΔF^{av}_{total}, respectively; Karush (1976) calls these *intrinsic affinity* and *functional affinity*, respectively.

C. Kinetics

Studies of the kinetics of the reaction

$$Ag + Ab \underset{k_{21}}{\overset{k_{12}}{\rightleftharpoons}} Ag\text{–}Ab \tag{4a}$$

which can be determined with temperature-jump relaxation and stopped-flow techniques (Froese and Sehon, 1974), have shown (in hapten-Ab systems) that the rate constants of association are very high (of the order of $k_{12} \approx 10^6$ to 10^8 M^{-1}/sec) and that in most cases they tend to be of that order of magnitude. Binding constants appear to be determined mainly by the (much more variable and much slower) dissociation rate constants k_{21} (see Table 1; see also Pecht, 1982).

IV. BINDING AND DISSOCIATION

A. Interfacial Interactions

It is not yet generally realized that van der Waals interactions in liquids, between two different materials, can be repulsive as well as attractive. For materials 1 and 2, interacting in liquid medium 3, we can rewrite Eq. (1):

Table 1 Representative Kinetic and Thermodynamic Values for Hapten/Antihapten[a] and Ag-Ab Interactions

Antibody	Hapten	k_{12} (M^{-1}/sec $\times 10^{-8}$)	k_{21} (sec^{-1})	$K^{\circ *}_{ass}$ (M^{-1} $\times 10^{-6}$)	ΔF° (kcal/mole)	ΔH° (kcal/mole)	ΔS° (e.u./mole)	Source
Anti-p-nitrophenyl	DHNDS-NP[b]	1.8[a]	760[a]	0.6[c]	-8.0	–	–	a
Anti-dinitrophenyl	ε-N-DNP-aminocaproate	1.0[a]	1.1[a]	90[d]	-11.2	-9.5	–	a
Antidintrophenyl	ε-N-DNP-L-lysine	–	–	23[d]	-10.3	-19.6	-30.4	a
Anti-dinitrophenyl	2,4-Dinitroaniline	–	–	0.3[d]	-7.3	--8.7	-5.2	a
Anti-p-azobenzene arsonate	Terephtal-anilide-p-p′-diarsonate	–	–	0.3[c]	-7.3	-0.8	+22	a
Anti-D-phenyl-(p-axobenzoyl-amino)acetate	D-Phenyl-[p-(p-dimethyl-aminobenzeneazo)	–	–	0.3[c]	-7.4	-7.3	+0.7	a
Anti-p-azophenyl-β-lactoside	p-(p-Dimethylamino-benzeneazo)phenyl-β-lactoside	–	–	0.16[c]	-7.1	-9.7	-8.8	a
Anti-SU_p[e]	p-(p-aminobenzeneazo)-hippurate	–	–	34[c]	-10.2	-21.6	-38	a
Monoclonal anti-DNP-lysine	DNP-lysine				-8.1	-16.6	-28.3	f

Antibody	Antigen	$K^{\circ *}_{ass}$	ΔF°	ΔH°	ΔS°	Source
Rabbit antidextran (IgM)	Dextran	0.08	-6.6	—	—	g
Rabbit antidextran (IgG)	Dextran	0.1	-6.7	—	—	g
Goat anti-bovine serum albumin (IgG)	Bovine serum albumin	50	-10.3	-6.5	+13	h
Human anti-B (IgM)	Human B erythrocytes	100 to 140	-10.8 to -11.0	—	—	i
Human anti-A sera	Human A erythrocytes	0.2 to 0.6	-6.8 to -7.8	-0.6 to -12.5	-16 to +24	j
Human anti-D (IgG)	Human D (Rh_0) erythrocytes	300 to 1200	-10.6 to -12.3	+9.4 to +11.0	+74	k

[a]From van Oss and Grossberg (1979) and van Oss (1981), where the individual references are given. In Pecht (1982) many more rate constants can be found for various systems.
[b]DHNDS-NP, 4,5-dihydroxy-3-(p-nitrophenylazo)-2,7-naphthalenedisulfonate.
[c]Measured by equilibrium dialysis.
[d]Measured by fluorescence quenching.
[e]SU_p = $-(CH_2)_4NHCOCH\text{-}SCH_2CONH\text{-}C_6H_4\text{-}N{=}N\text{-}C_6H_4\text{-}CONHOH_2COO^-$.
[f]Johnston et al. (1974); via equilibrium dialysis, fluorescence quenching, and flow calorimetry.
[g]Edberg et al. (1972); by $(NH_4)_2SO_4$ precipitation.
[h]van Oss et al. (1982b); by precipitation at optimal ratio—the figures in parentheses are obtained by affinity diffusion, with the same system.
[i]Economidou et al. (1967); by equilibrium measurement.
[j]Steane (1974).
[k]Green (1982); by equilibrium measurements, with protein A as indicator. This is a typical example of the exceptional endothermic behavior of the D/anti-D ("warm" antibody) reaction; this phenomenon is best explained by a phase change accompanying the reaction (e.g., "melting" of a lipid moiety), which fits in well with the apparently irreversible denaturation of the D-antigenic determinant following the D/anti-D reaction (van Oss et al., 1981).

***It should be noted that in most cases K°_{ass} is not a dimensionless number. Its computation from concentrations that are (of necessity, in the case of proteins, such as antibodies) considerably less than one molar, and therefore distinctly non-"standard," may well lead to a significant distortion of its actual value, and thus of the value of the free energy of association derived from it.**

$$\Delta F_{132} = \frac{A_{132}}{12\pi d_0^2} \tag{1a}$$

where 1 and 2 may stand, for example, for the antigenic determinant and the antibody-active site, respectively. A_{132} is composed of

$$A_{132} = A_{12} + A_{33} - A_{13} - A_{23} \tag{9}$$

in which the A_{ij} are the various Hamaker coefficients of van der Waals interaction between Ag and Ab molecules (A_{12}), liquid molecules among each other (A_{33}), Ag and liquid molecules (A_{13}), and Ab and liquid molecules (A_{23}), respectively. It has been shown by Visser (1972) that $(A_{13} + A_{23}) > (A_{12} + A_{33})$, which makes $A_{132} < 0$, when

$$A_{11} > A_{33} > A_{22} \tag{10a}$$

and when

$$A_{11} < A_{33} < A_{22} \tag{10b}$$

It can be shown (van Oss et al., 1980a) that in actual practice this is the case when

$$\gamma_{1v} < \gamma_{3v} < \gamma_{2v} \tag{11a}$$

$$\gamma_{1v} > \gamma_{3v} > \gamma_{2v} \tag{11b}$$

where γ_{iv} is the surface tension of the material in question versus vapor (V). It thus suffices to reduce the surface tension γ_{3v} of the liquid medium to a value intermediate between the surface tensions of the Ag and the Ab (or more precisely between the surface tensions of the antigenic determinant and the antibody-active site), to effect a van der Waals repulsion between Ag and Ab. Indeed, Ag–Ab complexes that are bound solely by van der Waals bonds can be completely dissociated in this manner at neutral pH (van Oss et al., 1979, 1980a). The surface tensions of cells and macromolecules can be determined via the contact angle method (van Oss et al., 1975). Such macroscopic surface tensions will not in all cases reflect the microscopic surface tensions of given antigenic determinants and/or antibody-active sites, although, for example, in the cases of dextran (van Oss and Neumann, 1977), DNA (de Groot et al., 1980), and p-azopyridine (van Oss et al., 1979), the microscopic and macroscopic surface tensions may be measured.

Van der Waals *binding* can take place when the two reacting sites have similar, as well as when they have different surface tensions, as long as the surface tension of the liquid medium is either larger or smaller than that of *both* sites. Van der Waals *repulsion* can only take place when the two reacting sites have *different* surface tensions, and then only when the surface tension of the liquid medium has a value intermediate between those of the surface tensions of the two reacting sites. Zero van der Waals interaction prevails when one of the sites has the same surface tension as that of the liquid medium.

With the exception of Eq. 1a, the hydrogen bond components of interfacial interactions obey the same laws as van der Waals interactions, but must be computed separately (van Oss et al., 1986a, 1986b).

B. Electrostatic Interactions

As Ags frequently and Abs always are amphoteric proteins, the value and the sign of the electrostatic component of the Ag–Ab interaction are strongly pH-dependent. Ags that are negatively charged at neutral pH tend to have homologous Abs that are positively charged, and vice versa. Thus for *attraction* one requires a pH that is intermediate

between the isoelectic points of the antigenic determinant and the antibody-active site. A case in point is that of the BSA-anti-BSA system, in which no precipitate is formed at all at pH 9.5 (van Oss et al., 1982b). However, once a BSA-anti-BSA precipitate is formed at neutral pH, it cannot subsequently be dissociated at pH 9.5, as by then the secondary van der Waals bonds, which are impervious to pH changes, have prevailed (van Oss, et al., 1979, 1982b).

Electrostatic *repulsion* (when the Ag is a protein) can be achieved at low as well as at high pH. When the Ag is strongly negatively charged (e.g., DNA), only high pH causes electrostatic repulsion (de Groot et al., 1980), while low pH can only lead to an approach to zero electrostatic interaction. When the avidity of the electrostatic attraction component between Ag and Ab is low, shielding of the charges with neutral salt ions is also possible, leading to close to zero electrostatic interaction (de Groot et al., 1980).

C. Conditions Favoring Ag–Ab Binding

Interfacial Interactions

As most Ag–Ab interactions are exothermic [see Eq. (4)], low temperatures generally favor Ag–Ab binding.

One well-known exception to this trend is the reaction between anti-D (Rh_0) and D (Rh_0)-positive erythrocytes, which is markedly stronger at higher temperatures [see Sec. V.C and Table 1, and, e.g., Green (1982)].

Although van der Waals (especially van der Waals-London) interactions are not strongly temperature dependent (Mahanty and Ninham, 1976), at higher temperatures van der Waals interactions tend to predominate over electrostatic attractions and compensate for their decline. This often causes the *total* binding energy (ΔF) to remain constant over a rather wide range of temperatures (Mukkur, 1980; van Oss et al., 1982b).

A high surface tension (γ_{3v}) of the liquid medium will lead to a strongly negative value of ΔF [Eq. (2)], thus favoring binding. Water has a high surface tension: $\gamma_{3v} \approx$ 72.5 ergs/cm^2; admixture of neutral salts tends to increase that value even further. Chicken Abs, which more avidly form immune precipitates at ionic strength $\mu = 1.5$ than at the physiological μ 0.15 (Benedict, 1967), probably bind predominantly via interfacial interactions.

Electrostatic Interactions

Neutral pH and a relatively low ionic strength increase Coulombic Ag–Ab binding (see Sec. IV.B). Low temperatures are especially favorable for enhancing the electrostatic component of the Ag–Ab bond; the electrostatic bond furnishes the major contribution to the exothermic aspect of the Ag–Ab interaction [Eq. (4); van Oss et al., 1982b].

D. Conditions Favoring Ag–Ab Dissociation

Interfacial Interactions

Lowering the surface tension(γ_{3v})of the liquid medium usually leads to a positive value of ΔF [see Eqs. (11a) and (11b)] and thus to dissociation (van Oss et al., 1979). With time, Ag-Ab bonds tend to become stronger, probably through an increasingly close fit [and thus a decrease in the overall equilibrium d_0; see Eq. (1)]. It thus requires a lower value of the surface tension(γ_{3v})of the liquid medium to achieve *dissociation* of a given

Ag-Ab complex than is needed just to prevent its *formation* (van Oss et al., 1979). This effect may be considerably enhanced if the progressive exclusion of liquid (with time) between antigenic determinant and antibody-active site causes not only a decrease in d_0, but a transition of ΔF_{vdw} [Eq. (1)] from a trinary system (which includes the interaction of Ag and Ab molecules with those of the liquid) to a binary one, where the antigenic determinant interacts directly with the antibody-active site without further intervention of (now expelled) liquid molecules (van Oss and Neumann, 1977).

Electrostatic Interactions

Increased temperatures favor Ag-Ab dissociation [Eq. (4)]. High as well as low pH favors dissociation of electrostatic Ag-Ab bonds, with high pH (generally of the order of 9.5 to 10) being preferable and less denaturing than low pH (2 to 4). In the cases of weak to medium-strong electrostatic interactions, addition of salts (1 to 5 M) also will favor dissociation, but high-avidity electrostatic bonds dissociate only at extremes of pH (e.g., pH 12; see de Groot et al., 1980). With purely electrostatic bonds *no* higher pH (or higher ionic strength) is required to *dissociate* an Ag-Ab complex than is needed to prevent its *formation* (see above; Smeenk et al., 1983).

Combined Interfacial and Electrostatic Interactions

In practically all Ag-Ab interactions in which the Ag is a protein, as well as in several other types of Ag-Ab interaction, the bond between Ag and Ab is based on a *combination* of interfacial and electrostatic interactions. In these cases neither of the two measures indicated above, applied *alone*, will lead to dissociation. Only by *combining* the two approaches discussed above (i.e., lowering the surface tension of the liquid medium, *and* raising its pH) can such combined bonds be totally dissociated, without denaturation (van Oss et al., 1979). Propionic acid is especially effective, as it lowers the surface tension *and* the pH. Here also increased temperatures will generally favor dissociation.

In the cases where the electrostatic component of the interaction is of medium or low avidity, an increased ionic strength often suffices for dissociation of the electrostatic part of the bond (de Groot et al., 1980); when the salts used for increasing the ionic strength at the same time can significantly lower the surface tension of the liquid medium, *total* dissociation of all but the highest avidity complexes may be achieved at or near neutral pH. Examples of such salts are tetramethyl and tetraethyl ammonium chloride (Chadha and Sulkowski, 1981) and chloroquine diphosphate (Edwards et al., 1982). At least part of the dissociating power of some "chaotropic" salts such as guanidine HCl and KCNS may also be ascribed to these combined effects (see below). Ionic surfactants may serve the same purpose, but these compounds tend to be rather difficult to remove from Ag and/or Ab, after their separation.

Other Approaches

In the cases of complexes formed between high-molecular-weight Ags with Abs, dissociation may be achieved by addition of excess hapten, due to a shift in equilibrium to the left [see Eq. (4)], provided that a hapten is available which corresponds to the antigenic determinant.

One other group of approaches is based on reversible (or even irreversible) denaturation of the antigenic determinant and/or antibody-active site, by means of, for example, very low pH, or hydrogen bond dissociating agents, such as 6 to 8 M urea, or "*chaotropic*"

salts. Such procedures may cause changes in the tertiary configuration of the macromolecules (both Ag and Ab), which tend to decrease the "fit" between Ag and Ab, and to increase the distance d_0 (and also the distance between opposing charges), thus favoring at least partial dissociation; often, however, at the cost of some degree of irreversible denaturation. Particularly in the case of the addition of "chaotropic" salts, other effects also play a role, such as ion shielding (see earlier in this section) and lowering of the surface tension of the medium (see earlier in this section) which is quite pronounced with guanidine HCl, and by no means negligible with KCNS.

V. MEASUREMENT METHODS

A. Theory

Methods for measuring the equilibrium constant of antigen–antibody interactions usually depend on the accurate determination of free and bound forms of the antigen under conditions in which the total antibody concentration is kept constant, or is known if varied. The methods for determining these concentrations are described later in this chapter. Results of the binding experiments are generally analyzed by means of standard Scatchard plots (Steward, 1977; Kabat, 1976) derived from the law of mass action [Eq. (5a)]. The following is an outline of the equations frequently used for thermodynamic calculations using the data from such experiments. The derivation is based on the work of Hardie and van Regenmortel (1975) and gives rise to two forms of the Scatchard plot which may be used to determine antigen valence, antibody valence, or the equilibrium constant. The derivation makes use of the concept of free or bound antigen or antibody binding sites. Let

A = total Ag concentration, mol/liter
s = antigen valence
sA = total number of Ag binding sites, mole sites/liter
B = total Ab concentration, mol/liter
n = antibody valency
nB = total number of Ab binding sites, mole sites/liter
y = bound Ab concentration, mol/liter
x = bound Ag concentration, mol/liter
C = free Ag concentration, mol/liter

At equilibrium the total number of occupied Ab sites (= ny) must *equal* the total number of occupied Ag sites (= sA) (i.e., A = ny/s). Thus

$$sA - sx = sA - ny = \text{free Ag mole sites/liter}$$

$$nB - ny = nB - sx = \text{free Ab mole sites/liter}$$

and it follows from equilibrium considerations that

$$(sA - sx) + (nB - sx) \underset{k_{21}}{\overset{k_{12}}{\rightleftharpoons}} sx \tag{12}$$

From the law of mass action it follows that

$$K_{ass} = \frac{k_{12}}{k_{21}} = \frac{sx}{(sA - sx)\ (nB - sx)} \tag{13}$$

or (by using an equivalent statement) that

$$K_{ass} = \frac{k_{12}}{k_{21}} = \frac{ny}{(sA - ny)\ (nB - ny)} \tag{14}$$

Dividing Eq. (13) by B/B, one obtains

$$K_{ass} = \frac{sx/B}{(sA - sx)\ (n - sx/B)} = \frac{x/B}{(A - x)\ (n - sx/B)}$$

$$= \frac{r}{C(n - sr)} \tag{15}$$

$$\frac{r}{C} = K_{ass}(n - sr) \tag{16}$$

where r = x/B and C = A - x. Dividing Eq. (14) by A/A, one obtains

$$K_{ass} = \frac{ny/A}{(s - ny/A)\ (nB - ny)} = \frac{y/A}{(s - ny/A)\ (B - y)}$$

$$= \frac{f}{(s - nf)\ (d)} \tag{17}$$

$$\frac{f}{d} = K_{ass}(s - nf) \tag{18}$$

where f = y/A and d = B - y.

Equation (16) is commonly known as the *Scatchard form* of the law of mass action. By plotting r/C versus sr for a range of antigen concentrations, a straight line theoretically is established from which both the (1) limiting Ab valency, as well as the (2) equilibrium constant (expressed as the association constant) can be obtained provided that the Ag valency is known. Ab valency is obtained from the intercept on the abscissa (= n) since

$$\frac{r}{C} \rightarrow 0 \text{ (excess [Ag])}$$

$$n - sr \rightarrow 0$$

$$n \rightarrow sr \text{ (where s is the known Ag valency)}$$

At low Ag concentrations, curves of this type often deviate from linearity. The association constant is obtained from the slope of the straight line ($\approx -nK_{ass}$). When the antigen valency = 1, Eq. (16) is reduced to the following form: $r/C = K_{ass}(n - r)$, which is one of the most widely used expressions in immunochemistry, where C represents, in this special case, the free or unbound dialyzable antigen molecules as in hapten-Ab experiments (see Fig. 1A).

Consideration of Eq. (18) reveals that by plotting f/d versus nf for a range of Ab concentrations, it is possible to determine experimentally the number of binding sites of a multivalent Ag. The advantage of this method is that it is not dependent on the valency of Ab. The Ag valency is obtained from the intercept on the abscissa, since

$$\frac{f}{d} \rightarrow 0 \text{ (large [Ab] excess)}$$

$$s - nf \to 0 \quad \text{and} \quad nf \to s$$

The value of s, on the assumption of bivalent Ab binding, will be twice as large as the value based on univalent binding.

Equation 16 may be rewritten as follows:

$$\frac{1}{C} = \frac{nK_{ass}}{r} - sK_{ass}$$

$$\frac{1}{nCK_{ass}} = \frac{1}{r} - \frac{s}{n}$$

$$\frac{1}{r} = \frac{1}{n} \frac{1}{CK_{ass}} + \frac{s}{n} \qquad (19)$$

This is called the ***Langmuir plot*** because the form of the equation resembles the form of a Langmuir adsorption isotherm, thus serving to emphasize the similarity between Ag–Ab binding and adsorption phenomena.

By plotting 1/r versus 1/C for a range of Ag concentration, a straight line relationship is obtained from which the slope = $1/nK_{ass}$ and the intercept = 1/n, thus permitting the experimental determination of Ab valency and the equilibrium constant (see Fig. 1B).

Both the Scatchard and Langmuir equations theoretically, under conditions of Ag excess, should give rise to straight line plots. However, with most Ab systems, even with isolated and purified Abs, there is usually some deviation from linearity. This implies heterogeneity of binding affinities which may be caused by (1) heterogeneity of antibody affinities for a particular antigenic determinant, (2) heterogeneity of antigenic determinants, and (3) interference of antibody association by other factors, such as steric hindrance or negative cooperativity of the antigenic determinants in the binding process. Thus in the case of the Scatchard plot, the graph curves toward the x axis and in the Langmuir plot the graph curves toward the y axis (see Figs. 1A and 1B). For a well-founded warning on the pitfalls of extrapolating precise Ab valencies from Scatchard plots, see Klotz (1982).

In order to assess the degree of heterogeneity of the binding constants it is generally assumed that the variation in the Ag-Ab interaction can be described by some type of distribution function. The indices of heterogeneity indicate a range and distribution of the association constants (K_{ass}), about some *average instrinsic association* constant, K_{ave}, and Eqs. (16) and (18) may be rewritten as follows:

$$\frac{r}{C} = K_{ass}(n - sr)$$

$$\frac{r}{n - sr)} = C \cdot K_{ass} \qquad (20)$$

$$\frac{f}{d} = K_{ass}(s - nf)$$

$$\frac{f}{s - nf} = d \cdot K_{ass} \qquad (21)$$

The distribution functions are generally assumed to follow either a Gaussian error function or more conveniently a Sips distribution function by assuming the product

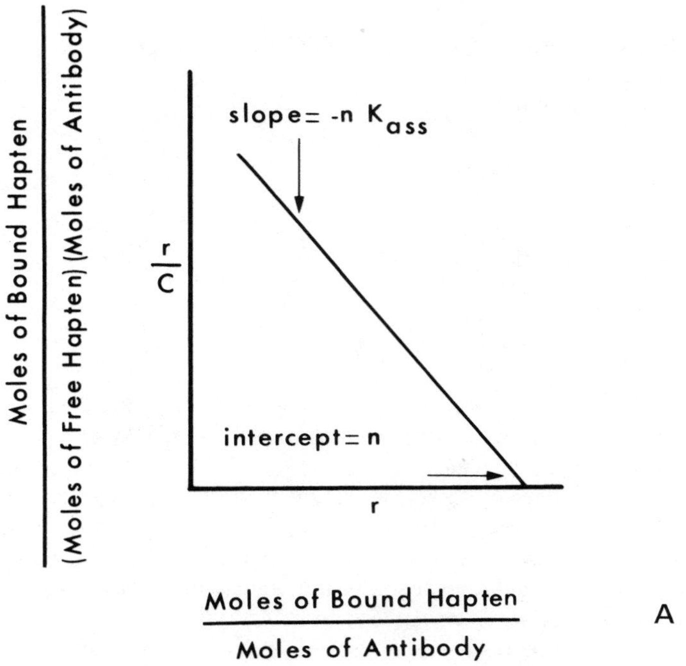
Moles of Bound Hapten
(Moles of Free Hapten)(Moles of Antibody)
r/C
slope= -n K_ass
intercept= n
r
Moles of Bound Hapten
Moles of Antibody
A

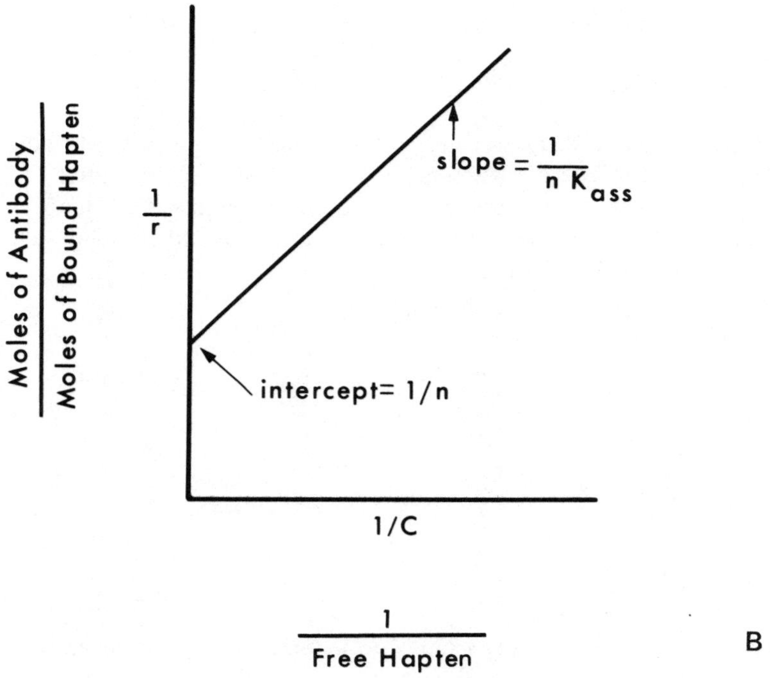
Moles of Antibody
Moles of Bound Hapten
1/r
slope = 1/(n K_ass)
intercept= 1/n
1/C
1
Free Hapten
B

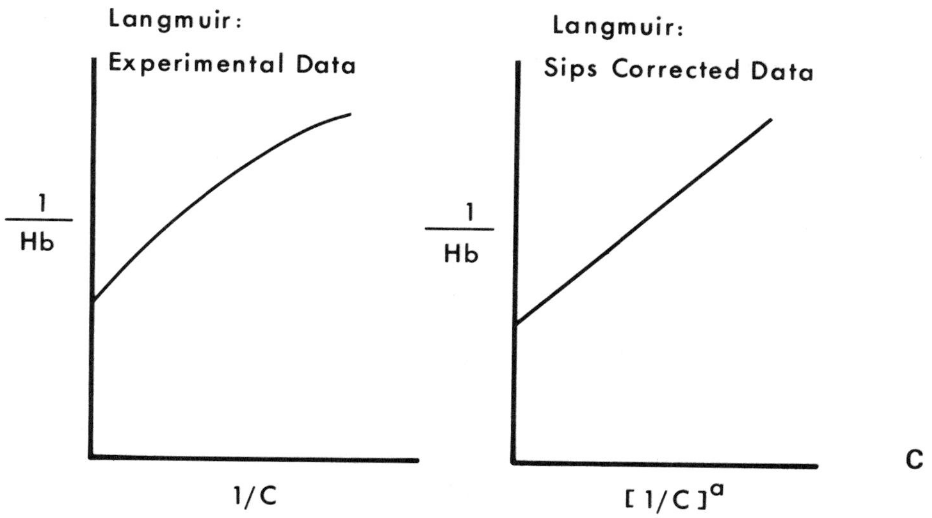

Figure 1 (A) Scatchard plot. (B) Langmuir plot. (C) Sips heterogeneity index; a is the exponent of 1/C required to reduce the curvature of the experimental data, plotted as a Langmuir plot, to a straight line.

$C \cdot K_{ass}$ (or $d \cdot K_{ass}$) as an exponential function $(C \cdot K_{ass})^a$ or $(d \cdot K_{ass})^a$, while $0 < a < 1$, and where a is the Sips heterogeneity index.

Thus from Eq. (20),

$$\frac{r}{n - sr} = (C \cdot K_{ass})^a$$

$$\log \frac{r}{n - sr} = a \log C + a \log K_{ass}$$

and from Eq. (21),

$$\frac{f}{s - nf} = (d \cdot K_{ass})^a$$

$$\log \frac{f}{s - nf} = a \log d + a \log K_{ass}$$

Thus having determined the value of n and s when carrying out these experiments, then by plotting log [r/(n − sr)] versus log C (or alternatively log [f/(s − nf)] versus log d) one obtains the best linear fit and calculates from the slope of this line Sips heterogeneity index, *a* (see, e.g., Steward, 1977). The closer the determined value of *a* is to 1, the greater the extent of homogeneity in terms of association constants. Decreasing values of *a* correspond to increasing degrees of heterogeneity (see Fig. 1C). By combining hapten binding (Scatchard) curves obtained with two or more homogeneous antibodies to the same hapten, Roholt et al. (1972) could make an interesting analysis of the influence of precisely known heterogeneity on the shape of binding curves in general. Figure 2 illustrates the difference between theoretical and experimental Scatchard and Langmuir plots.

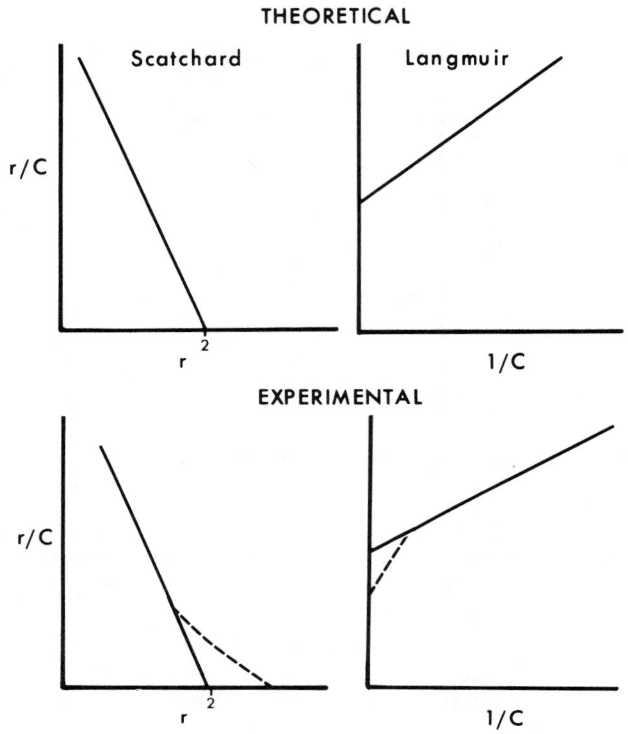

Figure 2 Comparison between the theoretical and experimental Scatchard and Langmuir plots, showing (bottom) the manner in which the actual deviating course of the plots (dashed lines) is disregarded, and the main (solid) lines are simply continued to obtain the intercepts.

B. Experimental Methods for Determining Antibody-Affinity Constants

Introduction

Many techniques are presently available for determining the binding constants of a particular Ag-Ab interaction. The thermodynamic estimations of Ab affinity, however, require the accurate determination of the concentration of the free and bound Ag or Ab under conditions that do not disturb the equilibrium of the Ag-Ab interaction. After the primary interaction has occurred, secondary and/or tertiary manifestations of the Ag-Ab interaction (e.g., agglutination or precipitation) may or may not occur depending on the nature of the antibody involved. In view of the many variables that influence the secondary and tertiary manifestations, studies attempting to delineate nature of the Ag-Ab reaction require that techniques which can accurately quantitate the primary interaction be employed. In general, quantitation of the required free and bound antigen or antibody concentrations is achieved either physically through the separation of the bound and free components by dialysis, selective precipitation, centrifugation, gel filtration, or chemically by utilizing changes in the properties of the antigen or antibody (e.g., fluorescence) which occur as a result of binding.

Equilibrium Dialysis and Ultrafiltration

When the antigen is much smaller than the antibody molecule (particularly when hapten-antibody interactions are studied) equilibrium dialysis or ultrafiltration with the aid of a membrane that is permeable to the antigen (or hapten), and impermeable to both the antibody molecules and the antigen- (or hapten) antibody complexes are the methods most often used. Because of antibody heterogeneity such experiments are generally performed using a range of antigen concentrations and a constant antibody concentration, or vice versa.

At equilibrium the concentration of unbound antigen (or hapten) will be the same on both sides of the partitioning membrane. The amount of hapten on the antibody side of the membrane will represent the total concentration of the free plus bound hapten. By subtraction the amount of bound antigen (hapten) may be calculated and hence the equilibrium constants from the data by the methods described above (Pressman and Grossberg, 1973; Kabat, 1976; Steward, 1977).

Precipitation Methods

When the antigen is soluble under conditions where the antibody molecule and hence antigen-antibody complexes are insoluble [e.g., 50% saturated $(NH_4)_2SO_4$ solution, 8% polyethylene glycol; see Chap. 19], usable data can be obtained. The required concentrations of free and bound components can readily be determined by evaluating the precipitates and supernatants after precipitation. This technique is generally employed when the antigen can be radiolabeled (Edberg et al., 1972). The method is ideal for rapid estimations of affinity, has the advantage of not requiring purified antibody, and gives results that compare well with those obtained using equilibrium dialysis. For purely electrostatic Ag-Ab complexes, such as DNA/anti-DNA, the use of high salt concentrations is not indicated, as the medium- and low-affinity complexes tend to dissociate at high ionic strengths (de Groot et al., 1980; see also Chap. 13, Sec. II.B).

At equivalent Ag/Ab ratios (which can be determined by double diffusion precipitation in a gel; see Chap. 17), equilibrium constants, and thus ΔF°, may be obtained under standard conditions, by precipitate formation in tubes. It then suffices to have either Ag or Ab radioactively or otherwise labeled to derive [Ag-Ab], as well as [Ag] and [Ab], because at equivalence [Ag] = [Ab] (van Oss et al., 1982b).

Analytical Ultracentrifugation

Analytical ultracentrifugation has been used to yield equilibrium constant values for a variety of antigen-antibody systems (Kabat, 1976). The method relies on the fact that antigens, antibodies, and the Ag-Ab complexes have different sedimentation velocities so that, upon separation in a gravitation field, they will separate and the quantity of each can be quantitated optically (e.g., with schlieren optics) as the surface area under each peak.

Equilibrium Gel Filtration or Sieving

This technique relies on the molecular weights of the antigen, antibody, and Ag-Ab being different. Consequently, they will elute separately upon gel filtration, thereby permitting an evaluation of the various concentrations. This technique requires that the eluant contain a constant concentration of one of the reagents so as not to disturb the equilibrium; it is generally performed in antigen excess.

Insoluble Antigen

When one is dealing with insoluble antigens [e.g., blood group antigens on the surfaces of erythrocytes, or antigen covalently coupled to an insoluble matrix (e.g., polystyrene latex)], the amounts of free or bound radio- or fluorescent-labeled antibody can easily be obtained by simple centrifugation, to obtain K values.

Affinity Methods

Affinity methods, such as affinity electrophoresis (Horejsi, 1979; Bøg-Hansen and Takeo, 1980) and affinity diffusion (van Oss et al., 1982a,b), depend on the degree to which, for example, Ag, when migrating into an Ab-containing gel, slows down as a function of Ab concentration. The *dissociation* constant K_d can be obtained according to

$$\frac{D_0}{D_i} = 1 + \frac{C_i}{K_d} \tag{22}$$

where D_0 is the electrophoretic mobility (or diffusion coefficient) of Ag in the gel without Ab and D_i the electrophoretic mobility (or diffusion coefficient) of Ag in the gel at Ab concentration C_i. When D_0/D_i is plotted on the ordinate versus C_i on the abscissa, one finds $-K_d$ at the intercept of the (straight) line of the function with the abscissa, or $1/K_d$ as the slope of the line. The zero Ab concentration diffusion coefficient (D_0) of Ag is found by extrapolation to zero Ab concentration (van Oss et al., 1982a). One can obtain K_a from $1/K_d$ [Eq. (5)] only in ideal (totally homogeneous) systems, with Ags with only one specificity of antigenic determinant, and with monoclonal Abs (in which case one would not expect precipitation to occur). With heterogeneous Ag–Ab systems one finds K_a values (at least with affinity electrophoresis and diffusion methods visualized by an advancing precipitation front) that are several decimal orders of magnitude lower than those found via precipitation in tubes, as one tends mainly to measure the components with the highest K_d value (van Oss et al., 1982b). A further reason for the different values found with affinity and equilibrium systems lies in the fact that affinity methods are kinetic systems, where one does not wait for equilibrium to set in, so that secondary (interfacial) interactions (see Secs. III.A and IV.D) may not have had time enough to occur. Curiously enough, however, at least in the BSA/anti-BSA system, affinity diffusion results also reflect mainly interfacial interactions, as they occur even at pH 9.5, where electrostatic interactions in this system are largely inoperative (van Oss et al., 1982b). These considerations do not necessarily apply to nonprecipitating affinity methods.

Fluorescence

In recent years the spectrofluorophotometer has become increasingly important in biological research. Fluorescence provides a marker by which the behavior of a fluorescent molecule in a complex mixture can be studied. Fluorescence intensity is strongly influenced by local molecular environment. Changes in fluorescence can provide sensitive and precise information about intermolecular complex formation (e.g., antigen–antibody interactions) (Kabat, 1976).

Fluorescence Quenching

Antibody molecules have a natural fluorescence derived mainly from tryptophan residues in the molecule. When some nonfluorescing haptens bind with the antibody, they

cause this fluorescence to decrease. From the relative extents of quenching, the amount of hapten bound by a specifically purified antibody can be calculated and a value for K_{ass} obtained (see e.g., Kabat, 1976). A similar principle can be employed in the case of haptens whose fluorescence is enhanced when they are bound to an antibody. In contrast to the above, however, this method allows impure antibody preparations to be used and values of K_{ass} for weakly bound haptens can also be determined. Kinetic measurements by the temperature-jump method can also be done by spectrofluorometry (Haselkorn et al., 1974).

Fluorescence Polarization

Fluorescent solutions which are excited with vertically polarized or natural light emit partially polarized fluorescence as viewed at right angles to the incident beam. Polarization of fluorescence is due to the fixed relationship between molecular orientation and absorption and emission of fluorescence. If a small fluorescent hapten is bound by an antibody, there will be marked reduction in the rotational freedom of the hapten and a corresponding increase in polarization. Again it is possible to obtain relative values of the extent of fluorescence polarization for 100% unbound antigen and to compare these with the extent of polarization for bound haptens and thereby establish equilibrium constants. The common labeling reagents for protein antigens are fluorescein isothiocyanate, rhodamine, and dimethylaminonapthalenesulfonyl chloride.

Immunoassay Methods

Immunoassay methods such as radioimmunoassay and enzyme immunoassay can also be applied to obtain K_a values (see, e.g., Thorell and Larson, 1978; see also Chaps. 19 and 20).

Hapten Inhibition Methods

Several methods exist for obtaining relative values of K for both strongly and weakly bound haptens, relative to a reference hapten. If the absolute value of K for the reference hapten is known from measurements by one of the methods described above, absolute K values can be calculated from the relative K values. Such methods include hapten inhibition of antigen-antibody precipitation (Kabat, 1976), hapten inhibition of complement fixation, and several other competitive reactions where the hapten competes with the antigen for binding to the antibody.

Calorimetry

Calorimetry or microcalorimetry is the direct way of measuring Ag–Ab binding enthalpies ΔH [in contrast to the two-temperature method, using Eq. (7); see also van Oss et al., (1982b); Kabat (1976)]. By the "batch" microcalorimetric approach one measures, for example, the heat output (in μcal/sec) versus time during the course of an Ag–Ab reaction (Steiner and Kitzinger, 1956). Much more versatile and simpler to operate is the flow calorimeter (Sturtevant, 1969), which also records microcalories per second, versus time, but which is much more suitable for operation with small volumes (10 ml or less), with great ease of mixing and allowing direct determination of ΔH of the reagents, reacting in many different ratios, in a short time (see also Johnston et al., 1974).

Interfacial Free Energies

Interfacial free energies, obtained by measurement (or estimation) of the surface tensions of the antigenic determinant and the antibody-active site of a purely van der Waals Ag–Ab system, can give the upper and the lower limits of the free energy of binding, for example, in the dextran-antidextran system (van Oss and Neumann, 1977). Measurement of the surface tensions of double-stranded DNA and of anti-DNA antibodies permits one to determine that the van der Waals component of the reaction energy of the DNA/anti-DNA is so low as to be practically negligible; that system can thus be regarded as a unique example of a purely electrostatic Ag-Ab system (de Groot et al., 1980). The methods for measurement of all the surface tensions needed to obtain the relevant interfacial tensions leading to the ΔF values of the interacting systems are discussed in Chap. 13 [see, e.g., Eq. (17)].

Other Methods

Other methods that utilize disturbances in quantum levels of atoms to indicate energy changes brought about by antigen–antibody binding include nuclear magnetic resonance and electron spin resonance (Kabat, 1976).

Via dilatometry, the volume changes (ΔV) occurring during Ag–Ab reactions can be measured. These volume changes go through a maximum when optimal Ag-Ab ratios are reached in nonstoichiometric precipitating systems, whereas hapten-antihapten systems reach a plateau value when saturation is reached. There is a certain proportionality between plateau ΔV values and K_a (Ohta et al., 1970).

C. Some Orders of Magnitude of Ag–Ab Interaction Parameters

In Table 1, a number of representative kinetic and thermodynamic values are given for hapten–antihapten (van Oss, 1981) and Ag-Ab interactions: for example, kinetic constants ($k_{1,2}$, $k_{2,1}$), association constants (K_a), free energies of association (ΔF), enthalpy associated with formation (ΔH), and the entropy (ΔS) accompanying the formation of Ag-Ab complexes.

In general, K_a can vary from 10^3 to 10^{10} liters/mol and ΔF from -4 to -13 kcal/mol (van Oss and Grossberg, 1979). As most Ag-Ab reactions are exothermic, ΔH tends to be negative (usually from 0 to -10 kcal/mol). ΔS can vary from -40 to +80 entropy units per degree mole (or cal/degree mole); more often than not, ΔS is positive.

The energies of formation found in Ag-Ab interactions (and in hapten-antihapten interactions) agree well with the order of magnitude of the (relatively weak) physical bonds involved in such reactions. Covalent bonds have higher energies of formation (ΔF of the order of -10 to -1000 kcal/mol). The exothermicity of formation of the Ag-Ab reaction, as manifested by a negative ΔH value, is for the most part due to the electrostatic component of the interaction.

In the exceptional case where ΔH is *positive* (see Table 1), as in the D/anti-D reaction (Green, 1982), one must suspect the occurrence of a phase change (e.g., the "melting" of a lipid moiety), which agrees well with the apparently irreversible denaturation of the D-antigenic site following the D/anti-D reaction (van Oss et al., 1981).

In most typical Ag-Ab interactions, as the temperature increases, ΔH becomes less negative, while $T\ \Delta S$ proportionally increases, leaving ΔF approximately unchanged [see

Eq. (8)] and Sec. IV.C] (Mukkur, 1980; van Oss et al., 1982b). One may call this an enthalpy-entropy compensation (Mukkur, 1980), but it is more descriptive of the actual physicochemical interactions between Ag and Ab to interpret this phenomenon as an electrostatic bond-van der Waals bond compensation (van Oss et al., 1982b), the effect of the electrostatic bond being more enthalpic, and that of the van der Waals bond more entropic.

The fact that the entropy of the Ag–Ab complex formation often has a positive value (notwithstanding an increase in "order" involved in the Ag–Ab combination) is best explained by the expulsion (and thus the randomization) of water molecules that earlier participated in forming (organized) layers of water of hydration attached to the antigenic determinant as well as on the antibody-active site. Water of hydration is released after electrostatic as well as after van der Waals bond formation between Ag and Ab. The increase in positive ΔS (or decrease in negative ΔS) with an increase in temperature accompanies the increase in van der Waals or "hydrophobic" bonding (van Oss et al., 1980b, 1982b).

Finally, if instead of expressing the Ag–Ab binding energy ΔF in the usual kcal/mol [Eq. (6)], one wishes to calculate that energy in cgs units (ergs/cm^2) or in SI units (mJ/m^2), instead of Eq. (6) one must apply

$$\Delta F = -kT \ln K_a \qquad (6e)$$

where k is the Boltzmann constant ($= 1.38 \times 10^{-16}$ ergs/deg); see also Eq. (6d). For instance, for the dextran-antidextran reaction (see Table 1), where $K_a = 10^5\ M^{-1}$, $\Delta F = 4.654 \times 10^{-13}$ erg/molecule. If one assumes a value of about 300 Å^2 for the surface area of interaction (van Oss and Grossberg, 1979), this yields $\Delta F = -15.5$ ergs/cm^2. Also, if one knows ΔF in ergs/cm^2, for example from the fact that the surface tension γ_{LV} must be lowered from 72.5 to 50 ergs/cm^2 to dissociate an Ag–Ab complex, which corresponds [see Eq. (17); Chap. 13] to an energy of dissociation $\Delta F_d = 16.4$ ergs/cm^2 and thus $\Delta F_a = -16.4$ ergs/cm^2 [Eqs. (5a) and (5b)], then one can, in this case assuming a surface area of interaction per molecule of about 250 Å^2 (van Oss et al., 1982b), recalculate ΔF [by first finding $\ln K_a$ via Eq. (6a)], in terms of kcal/mol, using Eq. (6); that is, $\Delta F_{vdw} = -6.2$ kcal/mol. In reality, $|\Delta F_d|$ and $|\Delta F_a|$ are rarely identical; there usually is a hysteresis effect, due to secondary binding, making $|\Delta F_d| > |\Delta F_a|$ (see Secs. III.A and IV.D). The extrusion of water, which causes a decrease in the intermolecular distance between Ag and Ab (see Eq. 1a), *and* a transition from a trinary to a binary van der Waals interaction (Sec. IV.D), further contributes to the increase in $|\Delta F_d|$ (van Oss and Neumann, 1977).

REFERENCES

Benedict, A. A. (1967). Methods Immunol. Immunochem. *1*, 229.

Bøg-Hansen, T. C., and Takeo, K. (1980). Electrophoresis *1*, 67.

Chadha, K. C., and Sulkowski, E. (1981). Prep. Biochem. *11*, 467.

de Groot, E. R., Lamers, M. C., Aarden, L. A., Smeenk, R. J. T., and van Oss, C. J. (1980). Immunol. Commun. *9*, 515.

Economidou, J., Hughes-Jones, N. C., and Gardner, B. (1967). Immunology *13*, 235.

Edberg, S. C., Bronson, P. M., and van Oss, C. J. (1972). Immunochemistry *9*, 273.

Edwards, J. M., Moulds, J. J., and Judd, W. J. (1982). Transfusion *22*, 59.

Froese, A., and Sehon, A. H. (1975). Contemp. Top. Mol. Immunol. *4*, 23.

Gabler, R. (1978). *Electrical Interactions in Molecular Biophysics.* Academic Press, New York, p. 245.

Green, F. A. (1982). Immunol. Commun. *11*, 25.

Hardie, G., and van Regenmortel, M. H. V. (1975). Immunochemistry *12*, 903.
Haselkorn, D., Friedman, S., Givol, D., and Pecht, I. (1974). Biochemistry *13*, 2210.
Horejsi, V. (1979). J. Chromatogr. *178*, 1.
Hughes-Jones, N. C. (1967). Immunology *12*, 565.
Israelachvili, J. N. (1974). Q. Rev. Biophys. *6*, 341.
Johnston, M. F. M., Barisas, B. G., and Sturtevant, J. M. (1974). Biochemistry *13*, 390.
Kabat, E. A. (1976). *Structural Concepts in Immunology and Immunochemistry*. Holt, Rhinehart and Winston, New York.
Karush, F. (1976). Contemp. Top. Mol. Immunol. *5*, 217.
Klotz, I. M. (1982). Science *217*, 1247.
Liberti, P. A. (1975). Immunochemistry *12*, 303.
Mahanty, J., and Ninham, B. W. (1976). *Dispersion Forces*. Academic Press, New York, pp. 82–86.
Mukkur, T. K. S. (1980). Trends Biochem. Sci. *5*(3), 72.
Ohta, Y., Gill, T. J., and Leung, C. S. (1970). Biochemistry *9*, 2708.
Pecht, I. (1982). In *The Antigens* (M. Sela, Ed.), Vol. 6. Academic Press, New York, pp. 1–68.
Pressman, D., and Grossberg, A. L. (1973). *The Structural Basis of Antibody Specificity*. W. A. Benjamin, Reading, Mass.
Roholt, O. A., Grossberg, A. L., Yagi, Y., and Pressman, D. (1972). Immunochemistry *9*, 961.
Smeenk, R. J. T., Aarden, L. A., and van Oss, C. J. (1983). Immunol. Commun *12*, 177
Steane, E. A. (1974). Thermodynamic Studies of Erythrocyte Antigen–Antibody Interaction. Ph.D. dissertation, George Washington University, p. 139.
Steiner, R. F., and Kitzinger, C. (1956). J. Biol. Chem. *222*, 271.
Steward, M. W. (1974). *Immunochemistry*, Wiley, New York, pp. 42, 43.
Steward, M. W. (1977). In *Immunochemistry, An Advanced Textbook* (L. E. Glynn, and M. W. Steward Eds.). Wiley, New York, pp. 233–262.
Sturtevant, J. M. (1969). Fractions, No. 1, 1.
Thorell, J. I., and Larson, S. M. (1978). *Radioimmunoassay and Related Techniques*. Mosby, St. Louis, pp. 14–18.
van Oss, C. J. (1981). In *Blood Banking*, Vol. 2 (T. J. Greenwalt and E. A. Steane, Eds.). CRC Press, Boca Raton, Fla., pp. 181–205.
van Oss, C. J., and Grossberg, A. L. (1979). In *Principles of Immunology* (N. R. Rose, F. Milgrom, and C. J. van Oss, Eds.). Macmillan, New York, pp. 65–79.
van Oss, C. J., and Neumann, A. W. (1977). Immunol. Commun. *6*, 341.
van Oss, C. J., Absolom, D. R., Grossberg, A. L., and Neumann, A. W. (1979). Immunol. Commun. *8*, 11.
van Oss, C. J., Absolom, D. R., and Neumann, A. W. (1980a). Colloids Surfaces *1*, 45.
van Oss, C. J., Beckers, D., Engelfriet. C. P., Absolom, D. R., and Neumann, A. W. (1981). Vox Sang. *40*, 367.
van Oss, C. J., Bronson, P. M., and Absolom, D. R. (1982a). Immunol. Commun. *11*, 129.
van Oss, C. J., Absolom, D. R., and Bronson, P. M. (1982b). Immunol. Commun. *11*, 139.
van Oss, C. J., Good, R. J., and Chaudhury, M. K. (1986a). J. Colloid Interface Sci. *111*, 378.
van Oss, C. J., Good, R. J., and Chaudhury, M. K. (1986b). J. Chromatog. *376*, 111.
van Oss, C. J., Gillman, C. F., and Neumann, A. W. (1975). *Phagocytic Engulfment and Cell Adhesiveness*, Marcel Dekker, New York, pp. 7–19, 84–99.
Visser, J. (1972). Adv. Colloid Interface Sci. *3*, 331.

17

Agglutination and Precipitation

Carel J. van Oss School of Medicine and School of Engineering and Applied Sciences, State University of New York at Buffalo, Buffalo, New York

I. AGGLUTINATION

A. Introduction

Agglutination, one of the oldest methods for demonstrating the occurrence of immune reactions in vitro, is the destabilization of relatively stable suspensions of antigenic particles by cross-linking them with antibodies directed to their antigenic determinants. Because destabilization of antigenic particles can readily be detected with rather small volumes (ca. 0.1 ml) of dilute particle suspensions and that relatively few antibody molecules suffice to achieve destabilization, agglutination is an uncommonly sensitive method for detecting quite low concentrations of antibody (as low as a few nanograms per milliliter). The sizes of antigenic particles, used in agglutination, may range from a few nm to about 10 μm in diameter. Antigenic particles may be cells (bacteria, red blood cells, etc.) displaying their native antigenic determinants, or inert particles (cells as well as synthetic carriers) to which antigenic molecules have been adsorbed or covalently attached.

B. Mechanisms of Visualization

Sedimentation Rate

According to Stokes' law, the force resisting sedimentation of a particle in a liquid medium is proportional to its radius, while the force inducing its sedimentation is proportional to the radius cubed. The net force causing a particle's sedimentation thus is proportional to the square of its radius. Therefore (other parameters being equal), when

clumping causes an increase in particle size by a factor X, the clumps will sediment X^2 times faster than the initial monodisperse particles. For instance, single human erythrocytes, suspended in saline water, sediment at a rate of about 1 cm/hr at ambient gravity, while clumps of agglutinated erythrocytes, comprising an average of 30 cross-linked cells, sediment 1 cm in about 6 min under the same conditions. Thus the simple visual observation of a 10-fold increase in sedimentation rate of a red cell suspension in a test tube is indicative of significant agglutination (van Oss, 1979). The same holds true for accelerated sedimentation in a centrifugal field, but as most cells (especially large cells such as erythrocytes) sediment rather quickly, even at relatively low g forces, the use of centrifugation in assessing agglutination focuses more on the scrutiny of the physical properties of the completely sedimented packed agglutinate than on the actual sedimentation rate (which is difficult to measure under conventional circumstances).

Properties of Sedimented Agglutinates

Once cells (agglutinated and otherwise) have been sedimented by centrifugation, there are two ways of recognizing agglutination: by the difference in adherence to the rounded bottom of the test tube, and by the difference in dispersability that may be observed while attempting to resuspend the sedimented cells (e.g., by shaking).

Nonagglutinated, monodisperse cells or particles pack very tightly when forced to the bottom of a test tube by a centrifugal field; nonagglutinated cells are therefore deposited in a small, round, sharply delineated button at the bottom of the tube. Agglutinated cells, on the other hand, form large, open network structures of many cells attached to each other at a few points only; such large "flocs" cannot be packed tightly and thus tend to be deposited on, and adhere to, the entire hemispherical inner surface of the bottom of the tube. Sedimented agglutinates thus appear as wide and thinly spread layers, often with jagged edges, covering much of the surface of the bottom of the tube, in sharp contrast with the small, round buttons formed by nonagglutinated cells.

For further verification, or in those cases where too few cells are available for visual inspection of the deposited cells with the naked eye, resuspension of the cells may be attempted (e.g., by vigorous shaking of the test tubes). Nonagglutinated cells can be completely redispersed in this manner, while agglutinates are indispersable and remain present as large flocs. With large amounts of cells that phenomenon is visible with the naked eye; however, when relatively few cells are present in each tube, one may revert to microscopic inspection of the contents of each tube.

Agglutination on Flat Surfaces

Agglutination can also be performed on flat plates (made of glass, plastic, or even cardboard). By that technique agglutination is recognized visually by the coarse graininess of agglutinated cell clumps, in contrast to the smooth aspect of monodisperse cell suspensions.

Agglutination in Gels

Agglutination of very small particles (e.g., sonicated fragments of erythrocyte stromata) can even be effected in gels, by double diffusion of such very small antigenic particles

versus a specific antiserum (Milgrom and Loza, 1967). This modification of agglutination is much akin to immune *precipitation* by double diffusion in gels (see Sec. II.C).

Automated Agglutination Methods

To quantitate the results of many hemagglutination determinations in quick sequence, the agglutinated cells can be separated from the nonagglutinated cell suspensions; the latter may be hemolyzed and the hemolysate conducted through a spectrophotometer which can record, via the absorptivity at 420-nm wavelength, the proportion of nonagglutinated cells, thus expressing, by difference from the total amount of cells used, a measure of the number of cells that have been agglutinated, for example, as a function of the concentration of antibody to be tested (Greenwalt and Steane, 1973).

For the quantitation of antibody (or antigen) by passive latex agglutination, the number of nonagglutinated particles (which are all close to the same size) may be automatically determined by light scattering, at such (relatively small) forward light-scattering angles as will best assure the exclusive counting of small particles, of the size of nonagglutinated particles only (Masson et al., 1981).

C. Mechanisms for Reducing Intercellular Distances

Hemagglutination with IgM and IgG-Class Antibodies

Since the early days of blood transfusion, *hemagglutination*, agglutination of erythrocytes with various antierythrocyte (usually blood group) antibodies, has been one of the most important analytical tools in immunohematology and blood banking. It has long been realized (see, e.g., Mollison, 1972) that with IgM-class antibodies, due to their size as well as to the availability of 10 antibody sites (Edberg et al., 1972) disposed at diametrical distances of about 300 Å (Dorrington and Mihaesco, 1970), hemagglutination is much more readily achieved than with antibodies of the IgG class, which have just two antibody sites, which are maximally about 120 Å apart (see, e.g., Valentine and Green, 1967; Labaw and Davies, 1971).

However, as IgG-class antibodies are the most prevalent among blood group antibodies as well as among other specificities, much effort and ingenuity has been devoted to modifications of the environment and properties of erythrocytes, to facilitate hemagglutination with IgG. With some IgG-class blood group antibodies (e.g., anti-A and anti-B, however, hemagglutination usually is readily achieved (Mollison, 1972); the reasons for this are discussed in the following sections.

The DLVO Theory and the Secondary Minimum

With the DLVO theory (named after its combined originators, Derjaguin and Landau, and Verwey and Overbeek; see, e.g., Verwey and Overbeek, 1948), it has been eluciated that with stable suspensions of particles or cells, spontaneous clumping of particles (due to the van der Waals attraction) is avoided when the electrostatic repulsion between particles, caused by their (usually negative) surface potential, is so strong that the particles are prevented from approaching each other closely enough for the van der Waals attraction to prevail. That distance, which can be calculated (van Oss and Absolom, 1983) when the particles' or cells' surface potential (measured by electrophoresis; see, e.g., Seaman and Brooks, 1979; van Oss and Fike, 1979), as well as their van der Waals attraction energy

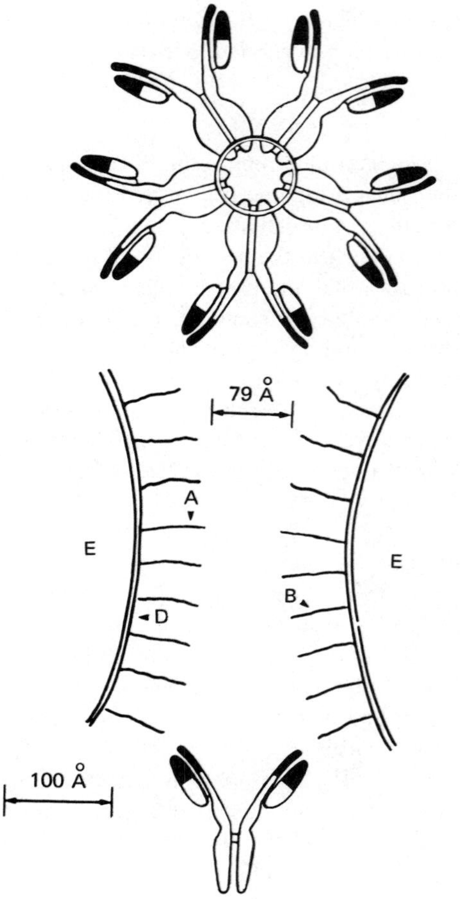

Figure 1 Diagram of the minimum distance of approach of two normal human erythrocytes, with an IgG and an IgM molecule drawn on the same scale. The curvature of the erythrocyte surfaces has been drawn vastly exaggerated, considering the scale: on this scale the diameter of an erythrocyte would be about 33 ft (≈10 m). E indicates the interior of the erythrocytes; A, B, and D indicate sites of A, B, and D (Rh_0) antigenic determinants. The closest approach between two unsensitized erythrocytes is ≈79 Å between the extremities of the sialoglycoprotein surfaces (drawn here as simple strands), which may be estimated to extend ≈50 Å from each cell membrane's lipid bilayer. [From C. J. van Oss and D. R. Absolom (1983), Vox Sang. *44*, 183; S. Karger, Basel.

(measured, e.g., via interfacial tensions; see Neumann et al., 1980), are known, is situated near the "secondary minimum" of intercellular attraction. For human erythrocytes that secondary minimum is at about 79 Å (van Oss and Absolom, 1983). which means that two erythrocytes can approach each other to within about 79 Å, counted from the extremities of their surface sialoglycoproteins. Thus (Fig. 2) the distance between two A or B antigenic determinants on different red cells is about 79 Å and

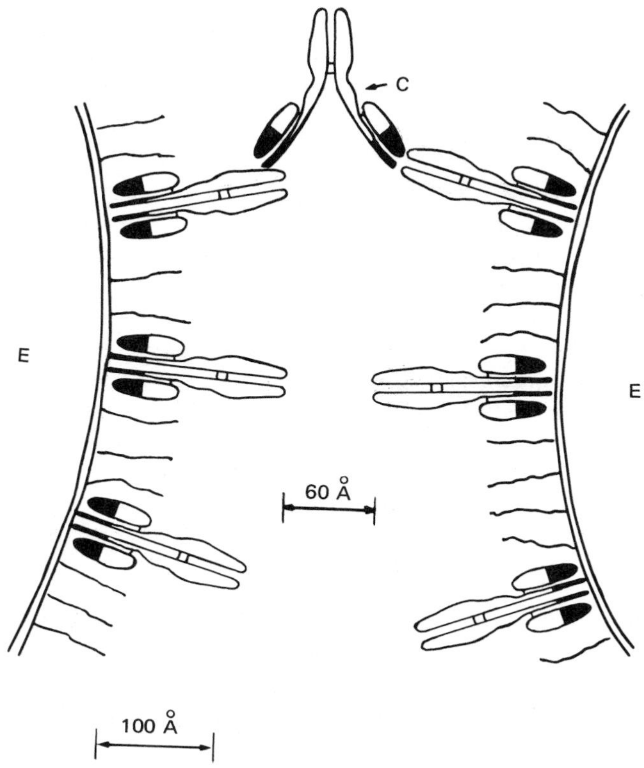

Figure 2 Diagram of the minimum distance of approach of two human erythrocytes, monogamously sensitized with IgG-class anti-D (Rh_0) antibodies. The curvature of the erythrocyte surfaces has been drawn vastly exaggerated, considering the scale. The closest approach between sensitized cells is ≈60 Å (from Fc tail to Fc tail). The extremities of two opposing IgG Fc tails obviously can easily be cross-linked by another (rabbit-anti-human IgG, or "Coombs") IgG-class molecule, here indicated by C. [From C. J. van Oss and D. R. Absolom (1983), Vox Sang. *44*, 183; S. Karger, Basel.]

can easily be bridged by IgG-class antibodies. However, the distance between two D (Rh_0) antigenic determinants, which are situated near the phospholipid part of the red cell membrane, is of the order of 2 × 50 Å + 79 Å = 179 Å (assuming that the sialoglycoprotein strands extend about 50 Å away from the actual cell membrane), which cannot be cross-linked by IgG-class antibodies, although with IgM-class antibodies that cross-linking is still readily achieved (see Fig. 2).

D (Rh_0)-positive erythrocytes, when "monogamously" sensitized with anti-D (Rh_0) antibodies of the IgG class, can easily be cross-linked by means of IgG-class (rabbit) anti-human IgG antibodies (antiglobulin or Coombs test; see Coombs et al., 1945) for two reasons: (1) The length of the sialoglycoprotein strands no longer matters, because the Fc tails of the monogamously bound IgG molecules protrude beyond these strands, even though their Fab sites are bound close to the "bottom" of these strands (see Fig. 2); and

(2) due to the higher surface tension *and* the lower ζ potential of red cells sensitized with IgG-class antibodies (van Oss and Absolom, 1983), the minimum distance of approach between cells (measured between the extremities of the Fc tails) has been reduced from about 79 Å to about 60 Å; see Fig. 2.

Influence of Physicochemical Factors on Hemagglutination

When properly analyzed, virtually all procedures used to facilitate hemagglutination can be shown to have as their major effect the shortening of the minimum distance between suspended erythrocytes (van Oss et al., 1978). Some of the more obvious methods are: forceful packing of cells by centrifugation; polymer bridging, with neutral (dextran) or positively charged polymers (polybrene), or even with euglobulins that are normally adsorbed on red cells (e.g., IgM), by insolubilizing them at low ionic strength. Among other methods that cause a decrease in intercellular distance are: diminishing the thickness of the pericellular ionic double layer by strong increases in the extracellular colloid-osmotic pressure brought about by the admixture of, for example, high concentrations of serum albumin; lowering the erythrocytes' surface potential (ζ potential) by treatment with proteolytic (e.g., papain, bromelin) or other enzymes (e.g., neuraminidase), which also tends to increase the cells' hydrophobicity, and thus decreases the distance to which they can approach each other.

If one causes the normally smooth biconcave erythrocyte surface to become spiculated, the multiple protuberances of a small radius of curvature can approach the surfaces of other cells much more closely, for purely steric reasons, as well as because of the fact that the tips of such spicules locally have a much diminished effective surface charge, which also decreases the intercellular distance at many points (van Oss et al., 1978). In addition, when spiculation causes a decrease in the total surface area with which two opposing cells can closely approach each other, their repulsive energy also decreases proportionally (van Oss and Absolom, 1983). Spiculation can be brought about by treatment with neutral polymers (such as dextran); it also occurs directly when erythrocytes react specifically with anti-A or anti-B antibodies (even of the IgG class) and even with blood group A-specific lectins, but not with anti-D antibodies (Salsbury and Clarke, 1967; van Oss and Mohn, 1970).

One other major approach to cross-linking erythrocytes with specific IgG-class antibodies is by lengthening the reach of the IgG molecules, rather than shortening the distance between erythrocytes by the antiglobulin or Coombs test (Coombs et al., 1945; discussed above).

Quantitative hemagglutination with an excess of radioactive antibodies can be used to determine the number of antigenic determinants of a given blood group specificity per red cell. Thus homozygous (DD, Rh_0-positive) human red cells have been shown to comprise 10,300 D determinants, while heterozygous (Dd) red cells contain 6400 D determinants (Masouredis, 1960). On A-positive cells the number of antigenic determinants is much higher: 1,000,000 per A_1 erythrocyte. The much greater number of A than of D determinants per erythrocyte may well be one of the major reasons why the interaction of anti-A antibodies causes spiculation in red cells, while the interaction of anti-D antibodies does not affect their smooth biconcave shape (Salsbury and Clarke, 1967; van Oss and Mohn, 1970).

D. Types of Agglutination

Inhibition of Hemagglutination

Inhibition of hemagglutination is used to measure the degree to which dissolved antigen can bind antibody, as shown by the inhibition of hemagglutination of cells coated with the same antigen. Inhibition of hemagglutination can also be used to characterize antibodies to viruses that can agglutinate various species of erythrocytes (e.g., antibodies to measles, rubella, influenza viruses).

Bacterial Agglutination

Of other agglutination methods, one of the oldest is bacterial agglutination, used for the demonstration of the presence of antibacterial antibodies in sera.

Passive Agglutination

To this point mainly active hemagglutination has been discussed, as used in blood grouping work. Hemagglutination can also be used as passive hemagglutination, with antigens that are not an integral part of the red cells, but may be adsorbed onto red cells that serve solely as carrier particles. Soluble polysaccharide antigens usually easily adsorb directly onto the red cell surface. However, protein antigens will adsorb onto red cells only after the latter have undergone treatment with, for example, 0.005% tannic acid (10 min at 37°C) or 1.65% glutaraldehyde (several hours at room temperature). Protein antigens and haptens can also be covalently coupled to red cells by diazotization and other chemical attachment procedures (Jou et al., 1983), which can make passive hemagglutination and inhibition of (passive) hemagglutination one of the most sensitive methods for the characterization of small amounts of antibody (Adler and Adler, 1980); the method has proven especially useful for the characterization of monoclonal antibodies (see Chap. 8). With erythrocytes coupled to protein antigens or haptens, one can of course also characterize antibodies as well as antibody-forming cells via passive complement-mediated hemolysis (in tubes as well as in plaque assays) (Adler and Adler, 1980; see also Chap. 23).

Passive agglutination with polystyrene latex and other particles has become an important tool in clinical tests, as well as in research, for the characterization of antigens as well as antibodies, either of which can be aspecifically but firmly adsorbed onto the rather hydrophobic surfaces of latex particles or covalently bound to chemically modified lattices (see, e.g., Masson et al., 1981). The first passive latex agglutination test was developed by Singer (1961) for the demonstration of rheumatoid factor (IgM-class antibody to slightly altered or aggregated IgG). In that method human IgG is aspecifically adsorbed onto the surface of polystyrene latex particles (0.8 μm or 0.25 μm in diameter), which then can be agglutinated by sera that contain rheumatoid factor. Latex agglutination is also used to demonstrate the presence of chorionic gonadotropic hormone (usually as an inhibition of agglutination test) in pregnancy urine, and also, for example, in tests for C-reactive protein, antithyroglobulin, and antistreptolysin O antibodies (van Oss, 1979), as well as in the determination of levels of circulating immune complexes, IgE, α-fetoprotein, and so on (Masson et al., 1981).

Agglutination Involving Two Cell Types

Mixed agglutination is a technique usually applied to cultured cell monolayers, for the characterization of antigens on the cultured cells, by the intermediary of antibodies bound to these antigens. Erythrocytes, sensitized with antierythrocyte antibodies, are used as indicators of the presence of sensitized cultured cells, by being cross-linked to them with anti-immunoglobulin antibodies, much like an antiglobulin or Coombs procedure (Coombs et al., 1945; see Sec. I.C), arranged in two flat layers of different opposing cells, as in a sandwich.

Coagglutination is a procedure to characterize bacteria sensitized with type-specific antibodies of the IgG class, by cross-linking them with *Staphylococcus aureus*, strain Cowan I, which carries protein A. Via protein A, which binds to the Fc moieties of IgG (in human IgG, to IgG1, IgG2, IgG4), the sensitized bacteria are then coagglutinated with formaldehyde and heat-treated staphylococci (Kronvall, 1973). When the type-specific antibodies are not of the IgG class, coagglutination is not possible (Hovanec et al., 1980), but in the case of IgM, direct agglutination is then practicable. Coagglutination is used for the characterization and/or typing of, for example, *Hemophilus influenzae, Streptococcus pneumoniae, Neisseria meningitidis*, and various groups of streptococci. Coagglutination with protein A-bearing staphylococci also can be used to demonstrate sensitization with IgG-class antibodies of erythrocytes and other mammalian cells.

Closely related to coagglutination as well as to mixed agglutination are the various *rosetting* techniques. These have been developed primarily for the characterization of different kinds of lymphocytes. For instance, sheep erythrocytes (E) are prone to attach themselves to human T lymphocytes, and to arrange themselves into a circular formation around such lymphocytes, reminiscent of a flower. Such *rosettes* are not only useful for detecting human T cells, but also for removing T cells from lymphocyte preparations,

Table 1 Some of the Principal Modes of Leukocyte Rosetting with Erythrocytes (E)

Type of rosetting	Type or treatment of erythrocytes	Types of leukocytes rosetted
E	Sheep E	T cells; 50% of NK cells (low-affinity E rosettes)
EAγ	E, sensitized with IgG-class antibodies	B cells, some T cells and NK cells; granulocytes, monocytes, macrophages
EAμ	E, sensitized with IgM-class	Some B and T cells
EAC	E, sensitized and treated with complement; ox erythrocytes often preferred, to avoid binding T cells with nonsensitized patches (see sheep E, above)	B cells, some non-B cells

making use of the fact that T lymphocytes rosetted with sheep erythrocytes (E rosettes) sediment much faster than the non-E-rosetted non-T lymphocytes. Table 1 summarizes a number of leukocyte-rosetting approaches. Protein A-bearing staphylococci, latex particles coated with IgG, and B lymphocytes have been used to characterize Rh (D) positive erythrocytes sensitized with IgG-class anti-Rh (D) antibodies (EA; see Table 1) (Loren et al., 1982).

There is no significant qualitative or quantitative difference between *rosetting* and *coagglutination*, except that in *rosetting* one usually employs erythrocytes, while staphylococci are more typically used in *coagglutination*.

II. PRECIPITATION

A. Introduction

Immune precipitation is the formation of insoluble complexes as a result of the specific interactions between dissolved antigen molecules with the corresponding antibody molecules in solution. Usually, the largest amounts of precipitate are obtained when soluble antigens and antibodies are present in approximately equal concentrations. As a method of determination of the concentration, or just of the presence of antibody, precipitation therefore requires considerable amounts of antigen; the method usually does not allow the detection of less than microgram quantities of antibody; that is, it is about 1000 times less sensitive than agglutination. Contrary to agglutination, divalent IgG (and not dekavalent IgM) is the immunoglobulin with the strongest precipitating power.

A major breakthrough that caused the diversification of immune precipitation into a variety of different and powerful analytical methods was the development of precipitation in gels. This approach gave rise to methods permitting, for example, the distinction of small differences between antigenic sites, and the characterization of 100 or more different blood serum proteins.

B. Mechanism

The solubility of nonpolar polymers in a given solvent seems on the whole (although there are various exceptions) to be favored when the value of the surface tension of the solvent is fairly close to that of the polymer (see Hildebrand and Scott, 1950), that is, when the polymer is readily "wetted" by the solvent or, more generally speaking, when the free energy of interaction between polymer molecules immersed in a liquid is small; for example, ΔF_{131} should usually have a smaller absolute value than -3 to -4 ergs/cm^2 to assure solubility; $\Delta F_{131} = -2\gamma_{13}$, where γ_{13} is the interfacial tension between polymer (1) and liquid (3). In aqueous solutions of (bio)polymers this means that biopolymers (e.g., proteins) will remain dissolved as long as their hydrophilic moieties are facing the liquid while their more hydrophobic aspects remain turned inward.

Distortion of the tertiary structure of proteins [antigen (Ag) as well as antibody (Ab)] in solution due to Ag–Ab interaction tends to expose more of the hydrophobic moieties of both Ag and Ab to the interface with the liquid, which leads to decreased solubility. Also, proteins (i.e., some Ag and all Ab) are amphoteric polyelectrolytes, which owe part of their hydrophilicity (and thus part of their solubility in water) to their ionized or charged moieties. These can be positively as well as negatively charged; proteins are most soluble when either sign of charge prevails, that is, when they are nega-

tively charged (at high pH) or positively charged (at low pH). When they are neutral, at the pH of their isoelectric point, proteins are least soluble, as under these conditions protein-binding due to charge interactions is most favored (see Chap. 13, Sec. II.A).

Other parameters being equal, polymers of the highest molecular weight will be the least soluble and a high degree of cross-linking may give rise only to swelling rather than to complete solution (see, e.g., Hildebrand and Scott, 1950).

When immune complexes beyond a critical size are formed, their solubility in water decreases rapidly. That critical size in most cases is between molecular weight 10^6 and 10^7. Large complexes of that type are increasingly prone to associate further with each other [yielding particles of about 5 μm in diameter, which subsequently combine in larger clumps; see Ritchie (1978), Chap. 2], both through specific rearrangements between antibody-active sites and antigenic determinants and via aspecific protein–protein interactions, partly of an electrostatic but probably mainly of a van der Waals ("hydrophobic") nature (see above), through binding between nonpolar moieties (van Oss, 1979; van Oss et al., 1980). It may take many hours before immune precipitate particles attain their ultimate size. The fact that immune complexes consist in an important part of immunoglobulins (especially IgG), which are among the less soluble blood serum proteins, contributes to their low solubility.

C. Conditions Favoring Precipitation

Various devices are used to enhance the precipitation of otherwise just marginally soluble Ag–Ab complexes. *Dehydration*, by the addition of salts [e.g., $(NH_4)_2SO_4$] in high concentration (Farr, 1958), or dehydration combined with polymer pore exclusion with polyethylene glycol (PEG), usually of a molecular weight of about 6000 (PEG-6000), in concentrations of 3 to 4% (w/v) (Polson, 1977; Creighton et al., 1973), strongly diminishes the solubility of Ag–Ab complexes. With DNA–anti-DNA complexes, precipitation with PEG is much to be preferred over $(NH_4)_2SO_4$, as high ionic strengths tend to dissociate the medium- and low-affinity complexes of that type (de Groot et al., 1980; Smeenk and Aarden, 1980; see Chap. 13, Sec. II.B).

Other conditions favoring Ag–Ab precipitation are: adjusting the pH as closely as possible to the *isoelectric point* of the complex (a pH that is usually intermediary between the isoelectric point of IgG (i.e., ca. 7) and that of the Ag, depending also on the Ag/Ab ratio); lowering the *temperature* (see Chap. 16); lowering the *ionic strength* (except in some instances where chicken Abs are used; with some of these Abs ionic strengths of 1.5 instead of 0.15 favor precipitate formation).

D. Precipitation in Tubes

Ag/Ab Ratios and Nonstoichiometry

Immune precipitation in aqueous media in test tubes is one of the oldest in vitro methods for obtaining quantitative data on the amount of specific antibody present in an antiserum. From precipitation in tubes it became clear that only at an "optimal" Ag/Ab ratio does a maximum amount of precipitate ensue; at slight Ab *or* Ag excess, much less precipitate is obtained, and at great Ab *or* Ag excess, no visible precipitate develops at all, although in the latter case the formation of soluble complexes can be detected (e.g., by analytical ultracentrifugation). At the "optimal" Ag/Ab ratio no dissolved Ag *or* Ab can

be detected in the supernatant liquid by ordinary means. [Due to the fact that the Ag-Ab reaction is reversible (see Chap. 16), one can detect the presence of very small amounts of free Ag and Ab on the supernatant even at "optimal" Ag/Ab ratios (e.g., by means of radioiodinated reagents; see van Oss et al., 1982b)]. One important conclusion emerges from these observations: *The Ag–Ab reaction is not stoichiometric, but can take place at a wide range of Ag/Ab ratios.* From Ag-Ab precipitates obtained at "optimal" ratios one can only obtain the valency ratios of Ag and Ab by fairly gross approximation; it is only at Ag excess that one can determine the exact valency of the Ab (Edberg et al., 1972), while the exact valency of the Ag is obtainable only at Ab excess.

Automated Precipitation Methods

Contrary to automated agglutination and hemagglutination methods, where the non-agglutinated particles or cells are quantitated (see Sec. I.B), in automated immune precipitation (AIP) methods only the amount of precipitated material is measured. That amount of precipitated material is measured by light scattering (either at angles around 90°, or at fairly sharp forward-scattering angles) (nephelometry) (Ritchie, 1978). Great care should be taken in AIP to avoid working under conditions of an excess of either Ab or Ag (to obviate decreased precipitate formation under those conditions; see Sec. II.C), by meticulous preliminary titrations of the Ab reagents as well as of the various Ag controls (Ritchie, 1978).

E. Double-Diffusion Precipitation in Gels

Cause of Line Formation

Slightly more than a third of a century ago, Ouchterlony (1968) developed the technique in which solutions of Ag and Ab are deposited in separate wells, punched in an agar gel slab, and allowed to diffuse toward each other. At their place of encounter, a sharp, usually somewhat curved Ag-Ab precipitate line is formed. That precipitate line actually is a precipitate membrane, or barrier, seen from above. The salient property of that precipitate membrane or barrier is its specific impermeability for the dissolved Ag and Ab molecules situated on either side of it (van Oss and Heck, 1961). Ag and Ab molecules unrelated to the Ag and Ab that formed the precipitate can freely pass that precipitate; this is why precipitate lines formed with identical Ag-Ab systems fuse, whereas such lines made by two unrelated Ag-Ab systems cross (van Oss and Heck, 1961; van Oss, 1979); see Fig. 4. The specific impermeability of the precipitate barrier formed by a given Ag-Ab system persists only as long as approximately equivalent amounts of Ag *and* of Ab remain present in solution, each on its own side of that barrier. This requirement furnishes the explanation for the specific impermeability, which is based on the fact that the barrier is *self-repairing*, that is, as soon as a hole is formed in it by accident, some of the soluble Ag present on one side of it will penetrate that hole, but then immediately will encounter an equivalent amount of the soluble Ab present on the other side of the barrier, and forms a precipitate with it, which plugs the hole (van Oss and Heck, 1961).

Place of First Formation of Precipitate Lines

To understand properly the behavior, formation, evolution, and decay of Ag-Ab precipitate lines formed by double diffusion in gels, it is indispensable to treat the first forma-

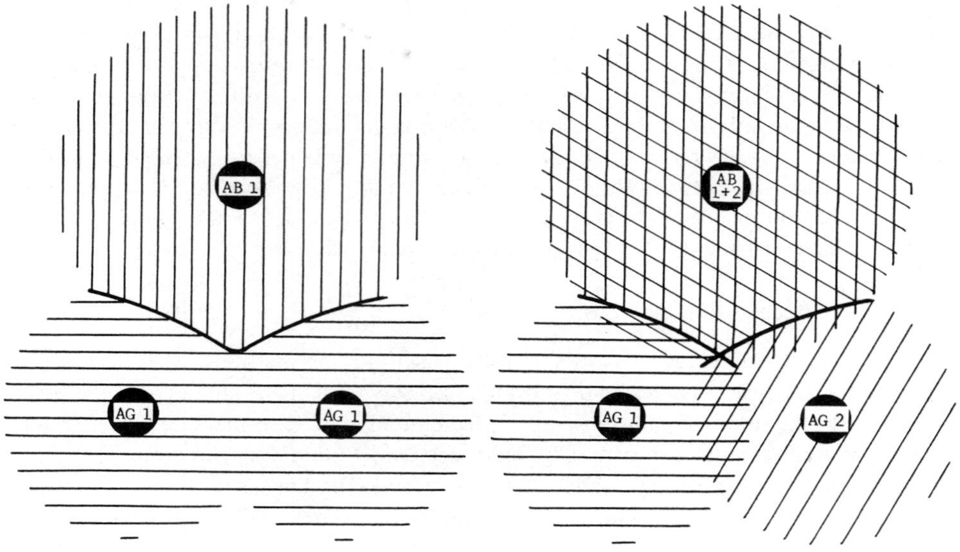

Figure 3 Diagrams of the fusing (left) and crossing (right) of precipitate lines in double immunodiffusion. (Left) When two identical Ags (AG1), deposited in two different wells, can both precipitate with an Ab (AB1) from the third well, fusing must occur on account of the specific impermeability of the precipitate line formed. (Right) When two different Ags (AG1 and AG2), deposited in two separate wells, each precipitates with its own Ab only (AB1 and AB2, respectively, mixed together in the third well, on top), the precipitate lines cross, as each is part of an entirely different system, in which each forms its own self-repairing, specifically impermeable precipitate barrier, which has nothing in common with the other.

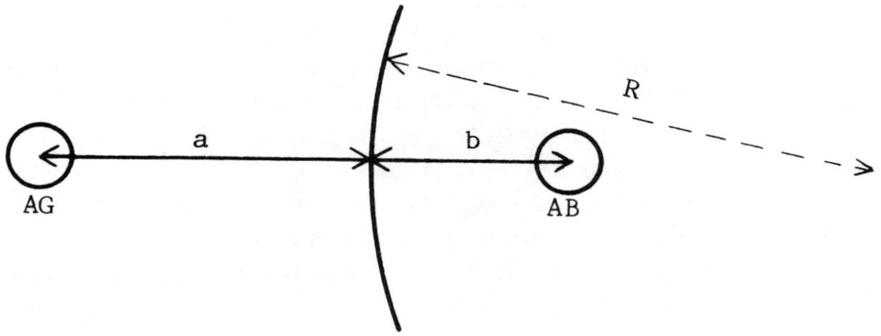

Figure 4 Circle-shaped precipitate line obtained by double-diffusion precipitation of an Ag with an Ab, such that $\sqrt{D_{Ag}}/\sqrt{D_{Ab}} = a/b$. The radius of the circle $R = ab/(a - b)$ [Eq. (3)].

tion of precipitate lines as an entirely separate phenomenon from their later evolution, subsequent to their initial formation.

Ag–Ab precipitate-forming systems, being essentially nonstoichiometric (see Sec. II.D), are *complex-forming systems* (as are precipitates or complex coacervates formed, e.g., by the interaction of anionic with cationic surfactants). Such systems tend to start precipitating when their concentrations are at equivalence (E) and have reached a certain *minimum value*, which is contrary to the precipitation of most noncomplex-forming systems, which occur once a *minimum solubility product* is exceeded. It can be proven fairly easily that owing to the fact that Ag–Ab precipitation occurs when

$$C_{Ag} = C_{Ab} = E \tag{1}$$

the *place* where that precipitate first starts in double diffusion is largely *independent* of the starting amounts (M_{Ag} and M_{Ab}) deposited in the respective wells (van Oss, 1968). The place where the Ag–Ab precipitate first forms depends only on the distance (a + b) between the wells and on the diffusion coefficients of Ag and Ab, such that

$$\frac{a}{b} = \sqrt{\frac{D_{Ag}}{D_{Ab}}} \tag{2}$$

As every point on the precipitate line of first formation has to obey that rule [Eq. (2)], the line of first formation, when the starting points are point-shaped (or small cylindrical) wells, must have the shape of a *circle*, with

$$R = \frac{ab}{a - b} \tag{3}$$

(see Fig. 4). When $D_{Ag} = D_{Ab}$ (which occurs, e.g., in the case of IgG–anti-IgG systems), a = b and R = ∞, resulting in a straight line perpendicular to the line connecting the Ag and Ab wells, and equidistant to these wells.

A point-shaped well versus a linear trough yields parabola-shaped precipitate lines, and two linear troughs give rise to linear precipitate lines (see the section "Immunoelectrophoresis," below).

Equation (2) can be used for the determination of the diffusion coefficient of a given Ag, if the diffusion coefficient of the corresponding Ab (or vice versa) is known. The best procedure for such a determination is shown in Fig. 5, in which two perpendicular rectilinear troughs are filled with Ag and Ab, respectively. The precipitate line obtained forms an angle such that

$$\tan \alpha = \sqrt{\frac{D_{Ag}}{D_{Ab}}} \tag{2a}$$

(see Allison and Humphrey, 1960). By this procedure one has the unique possibility of determining the diffusion coefficient of an Ag (and thus frequently its molecular weight; see Chap. 13, Sec. X.D) without actually purifying it, provided that one has some means (e.g., by an enzymatic color reaction, or by an immunological reaction of identity) to identify the line formed by the Ag. In the application of this method, it is useful to know that for IgG, $D_{IgG} \approx 4 \times 10^{-7}$ cm²/sec.

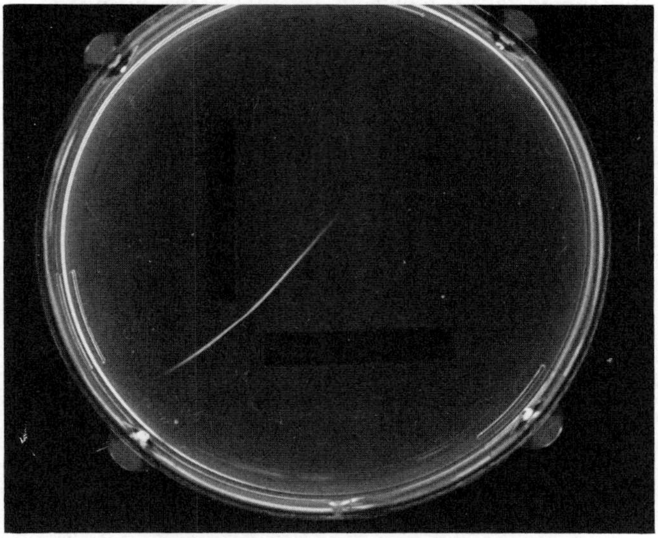

Figure 5 Two perpendicular troughs filled with, respectively, Ag (bovine serum albumin, or BSA) horizontally and Ab (goat anti-BSA antiserum) vertically. The angle α the precipitate line makes with the horizontal trough is characterized by $\tan \alpha = \sqrt{D_{Ag}/D_{Ab}}$ [Eq. (2a)]. As Ab is of the IgG class, its diffusion coefficient is known (i.e., $D_{Ab} = 4 \times 10^{-7}$ cm²/sec). From D_{Ab} and $\tan \alpha$, D_{Ag} can thus be derived via Eq. (2a). In this case, for $\tan \alpha \approx 1.22$, $D_{Ag} \approx 6 \times 10^{-7}$ cm²/sec, which corresponds to the value of the diffusion coefficient of BSA.

Further Evolution of Precipitate Lines

When a precipitate line is formed when Ag and Ab are present at equivalent concentrations in their respective wells, the precipitate lines they form will indefinitely remain in the same place where it was first formed, and it will remain thin. However, if one of the reagents (A) was present in its starting well in excess, it will ultimately be able to break down the barrier (when the other reagent begins to be exhausted), cross it, and form a new one farther down, in the direction of the reagent present in the lower concentration (B), with which it will react in a place closer to the starting well of B, where some B is still present in solution. Thus the precipitate may do *one* of three things (van Oss and Heck, 1961):

1. It will *thicken* in the direction of the reagent (B) initially present in the lowest concentration.
2. It will form new precipitate lines in the direction of the reagent (B) initially present in the lowest concentration; instead of one thick line, *multiple lines* will form.
3. Excess reagent A will dissolve the line of first formation, and form a new precipitate line closer to B; thus the precipitate line will appear to *move* in the direction of the reagent (B) present in the lowest concentration.

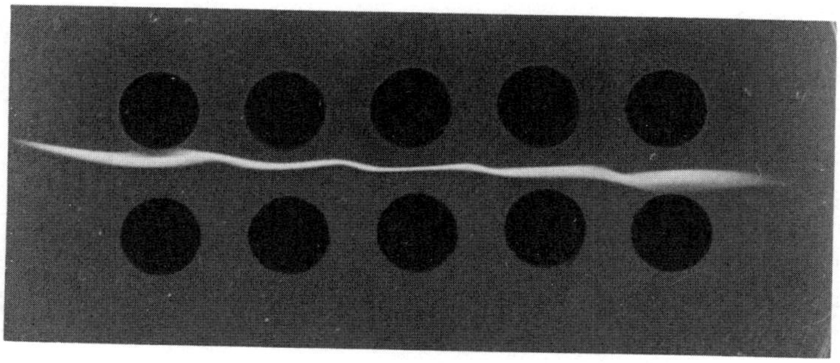

Figure 6 Titration of bovine serum albumin (BSA) with goat anti-BSA antiserum. (Top wells) Goat anti-BSA (diluted 1/2); bottom wells (left to right) 0.5% 0.25%, 0.125%, 0.0625%, and 0.0312% BSA. The equivalence point (point of optimal Ag/Ab ratio) is at 0.125% BSA.

The visible evolution of the place and/or thickness of the precipitate line when one of the reagents is present in excess, in contrast to the line's continuing immobility and thinness when Ag and Ab are present at equivalent concentrations, makes it possible to do a simple determination of the equivalence ratio of a given Ag–Ab system by a double-diffusion titration (van Oss, 1968; van Oss and Heck, 1961; Ouchterlony, 1968, p. 32); see Fig. 6.

Immunoelectrophoresis

As is often the case with two *different methods*, applied successively and *in directions perpendicular* to each other, the resolution of immunoelectrophoresis is a decimal order of magnitude greater than that of electrophoresis alone. Immunoelectrophoresis consists of the electrophoretic separation of an antigen mixture in the first dimension, followed by the characterization of each antigen by double-diffusion precipitation in a direction perpendicular to the electrophoretic step (Grabar and Burtin, 1964; Ouchterlony, 1968). The property of nonidentical antigens to form crossing precipitate lines upon interaction with their corresponding antibodies (even when the antigens still partly overlap after a less than total electrophoretic separation) allows the characterization of overlapping antigens (van Oss, 1979; Ouchterlony, 1968). The shape of the precipitate lines permits the indication of monoclonality (a parabola-shaped line, e.g., serum albumin) or polyclonality (a fairly straight line, e.g., IgG) of the separated fractions (see the section "Place of First Formation of Precipitate Lines," above). For this and other reasons, immunoelectrophoresis is especially useful for detecting monoclonal gammopathies. Due to an electrophoretic dilution effect (compounded by possible electrophoretic heterogeneities of given fractions), immunoelectrophoresis is less sensitive than ordinary double-diffusion immunoprecipitation tests, although with the help of radioactive Ag or Ab, by means of autoradiography, extremely small amounts of given Ag or Ab can still be demonstrated by radioimmunoelectrophoresis.

Nondiffusion-Driven Ag and Ab Transport Mechanisms for Immunoprecipitation

Counterelectrophoresis (also referred to as crossed-over electrophoresis, electrosyneresis, immunoelectroosmophoresis, or immunoosmophoresis) brings Ag and Ab together by means of electrophoresis. This is, of course, possible only at a pH that is intermediate between the isoelectric pHs of Ag and Ab; in other words, Ag and Ab must have different isoelectric points for this method to be applicable (van Oss, 1979). The method is faster (precipitate visible in a few hours) and more sensitive (by a factor of about 3 to 4) than double-diffusion precipitation. Contrary to the implication of some of the names proposed for this method (see above), electroosmotic flow plays *no* role in bringing Ag and Ab together.

Rheophoresis, or immunorheophoresis, is a method for bringing Ag and Ab together by hydrodynamic transport engendered by the evaporation of solvent through a gap in the cover (on top of the gel chamber) situated exactly above the place (between Ag and Ab) in the gel where the precipitate is to occur. Excess solvent should be provided at both extremities of the gel, to furnish liquid for capillary transport through the gel's pores. This method is also relatively fast (precipitate visible in 2 to 4 hr) and more sensitive (by a factor of 3 to 4) than double diffusion, and can even be used when Ag and Ab have close or identical isoelectric points (van Oss, 1979; van Oss and Bronson, 1969).

F. Single-Migration Precipitation in Gels

Radial Immunodiffusion

In single-migration methods, only one of the components (usually the Ag) migrates or diffuses into a gel that contains a homogeneous, low concentration of the other component (usually the Ab). The simplest example of this type of method is single diffusion, or radial immunodiffusion (Mancini et al., 1965), in which various concentrations of an Ag solution are deposited in wells punched in an Ag-containing gel. The initial concentrations of Ag deposited in the wells must be higher than the Ab concentration of the bulk of the gel for any Ag–Ab precipitate to become visible beyond the confines of the wells (see the section "Further Evolution of Precipitate Lines," above). After the precipitate rings around a well has reached its maximum size, the surface areas of that ring (or its diameter squared) are proportional to the amount (and thus to the concentration) of the Ag deposited in that well. Plots of Ag concentration versus precipitate ring diameter squared therefore yield straight lines. This method represents a reliable but slow (24 to 48 hr) method for measuring the concentration of plasma proteins and other Ags.

Electrophoretic Migration Methods in Antibody-Containing Gels

Instead of waiting for the diffusion of an Ag into an Ab-containing gel to run its (slow) course, the process can be accelerated by *electrophoresing* the Ag into the Ab-containing gel until all growth of the precipitate lines (now rocket-shaped) has stopped (Laurell, 1965). The final surface areas of the rockets (or their heights, as in first approximation the rockets may be likened to triangles with a unit baseline) are proportional to the Ag concentrations. This method, which is somewhat more accurate than radial immunodiffusion (see above), is also about a decimal order of magnitude faster.

Bidimensional electrophoresis in Ab-containing gels (or crossed immunoelectrophoresis) is comparable to immunoelectrophoresis in which the double immunodiffusion

step has been replaced by electrophoresis (at right angles to the first electrophoresis step) into an antiserum-containing gel. With complex Ag mixtures, this gives rise to a forest of rockets; the surface area under each of these is roughly proportional to the concentration of the corresponding Ag originally present in the mixture (see, e.g., van Oss, 1979).

Affinity Electrophoresis

Affinity electrophoresis uses the degree of retardation of the electrophoretic mobility of an Ag in a gel column in which the Ab is bound to (or admixed with) the gel for obtaining equilibrium constants (see Chap. 16). Various ways of monitoring the rate of advance of Ag in the gel column can be used, including the rate of advance of an Ag–Ab precipitate, in the cases where such a precipitate is formed. However, as the degree of retardation of the Ag-Ab precipitate does not necessarily accurately reflect the degree of retardation of the electrophoretic mobility of the bulk of the Ag (under precipitating as well as nonprecipitating conditions), affinity electrophoresis measured by the progression of the Ag–Ab precipitate front contains the risk of yielding erroneous equilibrium constants (van Oss et al., 1982a,b; see below).

Affinity Diffusion

If one can use affinity electrophoresis to measure the equilibrium constants of Ag–Ab interaction (see Chap. 16), one may also use affinity *diffusion* for the same purpose, in which case one measures the decrease in the diffusion coefficient of the Ag, in the Ab-containing gel, as a function of Ab concentration (van Oss et al., 1982a). However, as in affinity electrophoresis (see above), if the *progression of the Ag–Ab precipitate* front is measured for the purpose of deriving the dissociation constant of the Ag–Ab interaction, there is a serious risk of obtaining erroneous results. The reason for this is that in this manner one tends to measure the dissociation constant of exactly those precipitate-forming reagents with the strongest propensity to dissociation (van Oss et al., 1982a,b).

Oudin noted in 1949 (Oudin, 1971) that, as in molecular diffusion, the rate of progression of Ag–Ab precipitate fronts by single diffusion in Ab-containing gel tubes (with a supply of excess Ag in solution on top of the gel tubes) is proportional to the square root of time. However, it was not realized until recently that the rate of progression of such Ag–Ab precipitate fronts is considerably faster than the rate corresponding to the diffusion coefficient of the Ag in free solution (van Oss et al., 1982a; see also the data furnished by Oudin, 1971). The explanation of this apparent paradox is relatively simple: the molecular diffusion coefficient of, for example, albumin as measured by visible or ultraviolet light absorption, is obtained via the progression of the inflection point of the first derivative of the absorption versus distance curve with time. However, the progression of albumin in an anti-albumin-containing gel, as judged by the advancing precipitate front, reflects only the diffusivity of the fastest few Ag molecules that can just attain the solubility product necessary to achieve precipitation. These few fast Ag molecules tend to be many times faster than the molecules diffusing with the average of the bulk of the Ag (van Oss et al., 1982a).

REFERENCES

Adler, F. L., and Adler, L. T. (1980). Methods Enzymol. *70*, 455.
Allison, A. C., and Humphrey, J. H. (1960). Immunology *3*, 95.

Coombs, R. R. A., Mourant, A. E., and Race, R. R. (1945). Br. J. Expl. Pathol. *26*, 255.
Creighton, W. D., Lambert, P. H., and Miescher, P. A. (1973). J. Immunol. *111*, 1219.
de Groot, E. R., Lamers, M. C., Aarden, L. A., Smeenk, R. J. T., and van Oss, C. J. (1980). Immunol. Commun. *9*, 515.
Dorrington, K. J., and Mihaesco, C. (1970). Immunochemistry *7*, 651.
Edberg, S. C., Bronson, P. M., and van Oss, C. J. (1972). Immunochemistry *9*, 273.
Farr, R. S. (1958). J. Infect. Dis. *102*, 239.
Grabar, P., and Burtin, P., Eds. (1964). *Immunoelectrophoretic Analysis*. Elsevier, New York.
Greenwalt, T. J., and Steane, E. A. (1973). Br. J. Haematol. *25*, 227.
Hildebrand, J. H., and Scott, R. L. (1950). *The Solubility of Non-electrolytes*. Reinhold, New York, p. 346.
Hovanec, D. L., Absolom, D. R., van Oss, C. J., and Gorzynski, E. A. (1980). J. Clin. Microbiol. *12*, 608.
Jou, Y. H., Mazzaferro, P. K., Mayers, G. L., and Bankert, R. B. (1983). Methods Enzymol. *7X*, in press.
Kronvall, G. (1973). J. Med. Microbiol. *6*, 187.
Labaw, L. W., and Davies, D. R. (1971). J. Biochem. *246*, 3760.
Laurell, C. B. (1965). Anal. Biochem. *10*, 358.
Loren, A. B., Matsuo, T., Charman, D., and Yokoyama, M. M. (1982). Transfusion *22*, 194.
Mancini, G., Carbonara, A. O., and Heremans, J. F. (1965). Immunochemistry *2*, 235.
Masouredis, S. P. (1960). J. Clin. Invest. *39*, 1450.
Masson, P. L., Cambiaso, C. L., Collet-Cassart, D., Magnusson, C. G. M., Richards, C. B., and Sindic, C. J. M. (1981). Methods Enzymol. *74*, 106.
Milgrom, F., and Loza, U. (1967). J. Immunol. *98*, 102.
Mollison, P. L. (1972). *Blood Transfusion in Clinical Medicine*, 5th ed. Blackwell, Oxford.
Neumann, A. W., Absolom, D. R., Francis, D. W., and van Oss, C. J. (1980). Sep. Purif. Methods *9*, 69.
Ouchterlony, Ö. (1968). *Handbook of Immunodiffusion and Immunoelectrophoresis*. Ann Arbor Science Publishers, Ann Arbor, Mich.
Oudin, J. (1971). Methods Immunol. Immunochem. *3*, 118.
Polson, A. (1977). Prep. Biochem. *7*, 129.
Ritchie, R. F., Ed. (1978). *Automated Immunoanalysis*, Parts I and II. Marcel Dekker, New York.
Salsbury, A. J., and Clarke, J. A. (1967). Rev. Fr. Etud. Clin. Biol. *12*, 981.
Seaman, G. V. F., and Brooks, D. E. (1979). In *Electrokinetic Separation Methods* (P. G. Righetti, C. J. van Oss, and J. W. Vanderhoff, Eds.). Elsevier, Amsterdam, p. 95.
Singer, J. M. (1961). Am. J. Med. *31*, 766.
Smeenk, R. J. T., and Aarden, L. A. (1980). J. Immunol. Methods *39*, 165.
Valentine, R. C., and Green, N. M. (1967). J. Mol. Biol. *27*, 615.
van Oss, C. J. (1968). J. Colloid Interface Sci. *27*, 684.
van Oss, C. J. (1979). In *Principles of Immunology* (N. R. Rose, F. Milgrom, and C. J. van Oss, Eds.). Macmillan, New York, p. 80.
van Oss, C. J., and Absolom, D. R. (1983). Vox Sang. *44*, 183.
van Oss, C. J., and Bronson, P. M. (1969). Immunochemistry *6*, 775.
van Oss, C. J., and Fike, R. M. (1979). In *Electrokinetic Separation Methods* (P. G. Righetti, C. J. van Oss, and J. W. Vanderhoff, Eds.). Elsevier, Amsterdam, p. 111.
van Oss, C. J., and Heck, Y. S. L. (1961). Z. Immunol. Forsch. *122*, 44.
van Oss, C. J., and Mohn, J. F. (1970). Vox Sang. *19*, 432.

van Oss, C. J., Mohn, J. F., and Cunningham, R. K. (1978). Vox Sang. *34*, 351.
van Oss, C. J., Absolom, D. R., and Neumann, A. W. (1980). Colloid Polym. Sci. *258*, 424.
van Oss, C. J., Bronson, P. M., and Absolom, D. R. (1982a). Immunol. Commun. *11*, 129.
Verwey, E. J. W., and Overbeek, J. T. G. (1948). *Theory of the Stability of Lyophobic Colloids*, Elsevier, Amsterdam.

18

Circulating Immune Complexes

Boris Albini, A. M. Fagundus, and Adrian O. Vladutiu State University of New York at Buffalo, Buffalo, New York

I. INTRODUCTION

Recently, concentrated efforts have been made to establish reliable assays for the detection of circulating immune complexes (CICs). It seemed advisable to make such test procedures available since an increasing number of human diseases became associated with immune complex (IC)-mediated pathology (Williams, 1980; Table 1). In addition, until a few years ago, IC-mediated pathology was viewed primarily as induced by tissue deposition of CICs.

The sheer number of available tests (see Tables 2 to 4) suggests that an "optimal" or definitive test procedure has not yet been described. Difficulties in establishing assays for the detection of CICs arise from the following basic problems:

1. Difficulties in proving convincingly a causative link or, at least, a strong association between the presence of CICs and disease
2. Heterogeneity of CICs
3. Technical problems with the assays

The first point made, the problem of establishing a causative link between the presence of CICs and disease, cannot be discussed here at any length. Interested readers are referred to other reviews (Albini and Penner, 1982; Couser and Salant, 1980; Dixon, 1963; Germutn and Rodriguez, 1973; Kano and Milgrom, 1980; Penner and Albini, 1982; Williams, 1980). It must suffice to stress the fact that even when ICs have been implicated convincingly in the pathogenesis of a disease, CICs may not be involved; more and more evidence is accumulating making in situ formation of ICs as feasible a hypothesis for the explanation of IC-mediated pathology as tissue deposition of CICs. Unfortunately, it seems at the moment extremely difficult or even impossible to determine which of these

Table 1 Selected Diseases Associated with Immune Complexes

Systemic lupus erythematosus (SLE)
Rheumatoid arthritis (RA)
Mixed connective tissue disease
Sjögren syndrome
Felty syndrome
Glomerulonephritides
Systemic sclerosis
Primary biliary cirrhosis
Alcoholic liver cirrhosis
Retinal vasculitis
IC-mediated pathology in infectious diseases:
Bacteria
Viruses
Parasites

two modes of IC-mediated pathology is at work in individual human diseases. Therefore, in recognition of our lack of understanding of the mechanisms involved in the induction of almost all IC-associated human diseases, we may consider CICs simply as "markers" of specific diseases. It is often useful to rely on "association" of certain test results with disease processes as long as pathogenetic, causal links have not been firmly established. A number of extensive clinical studies are available on the presence of CICs in many of the IC-associated diseases (rev. Williams, 1980).

It is obvious that the presence of CICs can never, by itself, be a disease-specific diagnostic finding. There are an ever-increasing number of diseases which are associated to some degree with IC-mediated pathology, as seen in Table 1; in addition, CICs may be present in the wake of "trivial," often clinically latent infections and possibly also in other conditions. To make determinations of CICs more specific, it seems essential to define some of the characteristics of CICs seen in certain diseases. Here, first but very important steps have been made. The complement-fixation ability of CICs may be used to define somewhat better CICs associated with some diseases; for example, Furst et al. (1981) have determined that CICs associated with lung cancer do not bind complement. Association of CICs containing immunoglobulins of certain classes may be used to define "subtypes" of CICs; it has been shown that CICs found in liver diseases may contain IgA, and they usually occur in patients with alcoholic liver disease (Penner et al., 1982), or may contain IGM, as usually seen in patients with primary biliary cirrhosis (Gupta et al., 1978). It is even better, but much more complicated, to define the antigen involved in CICs; this not only may have implications for the understanding of the pathogenetic mechanisms involved in individual disease entities, but may, in the future, make determination of CICs truly disease specific (e.g., Deliere et al., 1978; Penner et al., 1982).

Since we have to assume a number of antigens and antibodies to be involved in the whole list of IC-mediated or associated diseases, we should expect CICs to be heterogeneous in their characteristics. Thus we just mention that patients with long carcinoma seem to have noncomplement-fixing CIC; most CICs found in systemic lupus erythematosus (SLE) fix complement. But even in the same patient, CICs may be composed of subpopulations differing in their characteristics and constituents. It seems

necessary to undertake extensive clinical and experimental studies to get some insight into this heterogeneity. Only with such knowledge can a meaningful comparison of results obtained with different assays be made.

As most of the assays used for the determination of CICs are rather complex, it is no surprise to find many technical difficulties associated with them. In many cases, it is cumbersome to standardize all components of an assay; thus we may use cells, as in the Raji cell assay, which may change significantly in their surface characteristics during culture; or we may use antisera, which may change in antibody reactivity during storage; or we may use standards to express our test values, which are only remotely and not always reproducibly comparable to the CICs we presume we test for. In addition, serum components may lead to false positive or false negative results in many of the assays. Here we ought to analyze, characterize, and thus define and understand all components and interferences of our test systems. Much ongoing research is concerned with these aspects of CIC assays. Despite all of these difficulties, it seems that assays for the detection of CIC do contribute significantly and positively to research, clinical diagnosis, and monitoring of patients (Albini and Penner, 1982). In research, assays for CICs are an important addition to other methods of demonstration and analysis of IC (e.g., Penner and Albini, 1982), and therefore may contribute to the understanding of pathogenetic mechanisms of many diseases. Alternatively, data from CIC tests may contribute to diagnosis and thus, despite not being disease specific in their present form, may be of the same importance for the diagnosis as, for example, an increased white blood cell count. Finally, determination of CICs may be used to monitor the progression of a disease and the efficiency of therapy. It is clear that tests for the determination of CICs are still in their developmental stage; with the necessary critical approach and accumulation of clinical and experimental data, they may be expected to develop over the next years into reliable and useful laboratory tools. Assays for CICs may be grouped by their mechanisms into those based on physicochemical and immunochemical aspects and those based on the biological reactivities of ICs, which differ both in type and in degree from those of noncomplexed antigen or antibody.

Several points should be borne in mind when discussing soluble ICs. As an effect of the nonstoichiometric nature of ICs and of the law of mass action, any procedure that involves the separation of the immune complexes from a mixture causes a change in equilibrium conditions, and will lead to a shift in the equilibrium between complexed and noncomplexed components, resulting in dissociation of some ICs and the establishment of a new equilibrium (see Chap. 16). Complexes with antibodies of low avidity will be affected preferentially. This may be of importance to the further manipulations of the assay or isolation procedure. One must consider this possibility when interpreting results gained through such methods as gel filtration and centrifugation.

Soluble ICs formed under conditions of antigen excess are heterogeneous in respect to molecular ratio of antigen to antibody, to the extent of lattice structure, and therefore to molecular size. An assay or isolation scheme will not affect equally the entire range of antigen-excess immune complexes, and different procedures may optimally detect different categories of immune complexes. As a result, there is often poor correlation between assays for ICs, which are based on different physicochemical or biological interactions of ICs.

Assays for ICs either rely on the specificity of the antigen in the complex or may be "antigen independent." A large number of the latter category of assays has been developed, as this category is of greatest clinical potential; in the overwhelming majority of cases, the antigens involved in IC-induced pathology in humans are not known.

Antigen-independent methodologies usually cannot distinguish between nonspecifically aggregated immunoglobulin and antigen-aggregated antibody. One must handle samples carefully to prevent nonspecific aggregation of immunoglobulin, avoiding such practices as repeated freezings and thawings or heating samples to high temperatures.

II. RADIOIMMUNOASSAYS FOR THE DETECTION OF CICs*

Radioimmunoassays (RIAs) are often employed for detection and "quantitation" of ICs and are in fact assays using reagents labeled with radionuclides. Among the advantages of these assays are the high sensitivity, common for RIA, and the fact that in these assays the radioactivity—as counts per minute (cpm)—is measured which is independent of temperature, physical form of the testing medium, and/or presence of other bystander substances in the serum. Among the disadvantages of RIA are the health hazards posed by radioactivity, especially during labeling (although with proper precaution this can be minimized) and the short half-life of ^{125}I, the most common radionuclide used for labeling. This latter factor is responsible for the relatively short shelf life of most labeled reagents (about 1 to 2 months). Furthermore, special equipment (a scintillation spectrometer) is needed for obtaining test results. The radiolabeling can be substituted with enzyme labeling, which yields a longer shelf life. The enzymatic activity can be measured with the usually available equipment. Unfortunately, enzyme activity is influenced by various activators and inhibitors, which may be present in biological test material.

In all RIA for measuring IC, a protein is radiolabeled with ^{125}I and will bind to one of the components of the IC, particularly the immunoglobulin or the complement (for complement-fixing IC). Tests for determining specific ICs can use labeled reagents that bind to the antigen component of ICs (see Sec. III.A), but these assays are of rather limited applicability.

Labeled anti-IgG (anti-IgA or anti-IgM), protein A, ^{125}I-labeled rheumatoid factor, or Clq, all of which bind to the Fc region of immunoglobulins, or anti-Clq, anti-C3, or conglutinin, have been used in various RIAs for determining ICs. Some of these assays are not "competitive binding" RIAs (i.e., not RIAs as in the original description by Yalow and Berson, 1966) but immunoradiometric assays (IRAs). Although there are several differences between "classical" RIA and IRA, they both are generally termed "RIA" because a radiolabeled tracer is used in all of them.

All proteins mentioned above are easily labeled with ^{125}I. Several methods of labeling are available. In general, high specific activity is not needed since the sensitivity of these assays is several orders of magnitude lower than that of most conventional RIA. Although the chloramine T method is the most commonly used method for labeling immunoglobulins, chloramine T can damage the antibodies. This is of special concern when affinity-purified antibodies are used, because these antibodies have already been subjected to conditions that can decrease their potency (e.g., nonphysiological pH). Therefore, milder methods (e.g., the lactoperoxidase technique; Yonemasu and Stroud, 1971) are more convenient to label Clq, conglutinin, or anti-IgG.

Protein A, a substance extracted from certain strains of *Staphylococcus aureus* (e.g., Cowan I strain) has been isolated in a very pure form and is commercially available (Forsgren, 1970). Protein A binds rapidly to the Fc region of IgG from many species (e.g., human IgG_1, G_2, and G_4, rabbit, guinea pig IgG, and to a lesser extent mouse IgG_{2a}, G_{2b}, G_3). Apparently, there are at least two sites for binding protein A on the IgG

*See also Chap. 19.

molecules. Protein A also binds to some IgM and even IgA molecules. Protein A allows a shorter incubation time than is necessary when anti-IgG antibody is used. In our hands, it is less affected by the chloramine T method of radiolabeling, and ^{125}I-labeled protein A is stable for 3 months or even longer after labeling. Furthermore, protein A as a pure substance is easy to quantitate. It can serve as a universal reagent for various RIAs; that is, it allows for detection of ICs in sera or body fluids of several species using the same reagent (see also Chaps. 5 and 13).

Some RIAs for measuring ICs are performed in solution and various additional methods are used to separate the tracer (^{125}I-labeled substance) bound to ICs from the free tracer. This separation may be accomplished by various concentrations of polyethylene glycol (PEG, 2.5 or 3.5%) or protein A-containing dry particles of *Staphylococcus.*

Other assays use a "solid phase" (e.g., plastic tubes or soft plastic U-shaped microtiter wells), on which ICs are first bound via complement or immunoglobulin component and subsequently react with the tracer. In some assays, cells are used instead of the solid phase (e.g., Raji cells, mouse leukemia cells, platelets, erythrocytes, or spermatozoa), on which ICs bind via Fc or C3 receptors on the cells. After reacting with the tracer, the cells are easily separated from the fluid phase by centrifugation and the radioactivity of the sediment is measured.

Because of their increased sensitivity as compared to other assay techniques (e.g., precipitation in gel), all RIAs (and enzyme-linked immunoassays; see Chap. 20) for IC determination not only detect very small amounts of IC but make attempts at quantitation feasible. To this end, the assays require the establishment of a "standard curve," as in any RIA. Using the standard curves, the concentrations of IC in the test can be estimated. In sharp contrast to RIAs for defined substances, in which the standard curve is constructed with known amounts of the pure ligand, ICs have such a variability with respect to composition and size that it is impossible to find a substance similar to all ICs. Therefore, aggregated IgG has been most commonly used to simulate IC since such aggregates share some of the biological effects of IC (e.g., binding and activation of complement). Aggregated IgGs were used as a model of ICs for development of assays to detect ICs, and for assessment of the sensitivity of these assays. The size and concentration of IgG aggregates influences their solubility and their reactivity with Fc and complement receptors. Usually, IgG aggregates are obtained by heating or alkali treatment. The duration of heating (10 to 40 min at 61 or 63°C) has some influence on the degree of aggregation. Even under identical conditions, the degree of aggregation varies, and freshly prepared aggregates of IgG are unstable and will aggregate further as assessed by sucrose density-gradient ultracentrifugation. Aggregates containing 46 to 80 molecules of IgG per aggregate appear to be most stable as far as complement activation is concerned. At higher concentrations of IgG, the aggregates rapidly build up into very large polymers, with a paucity of aggregates in the intermediate size (around 19S); in low concentrations, IgG aggregates poorly. It seems that alkali aggregation is more efficient in producing aggregates of intermediate size. Since the aggregates of IgG have a tendency to increase in size with storage, their use is limited to a few weeks. To avoid changes in size, which will give spurious results from one assay to another, aggregated IgG should be stored in aliquots at the lowest temperature available (at least -70°C and preferably in liquid nitrogen), and each aliquot should be thawed only once. To assure a relative homogeneity of aggregates, they can be filtered through Ultragel Ac, A34, or Sephadex G-200 or purified by ultracentrifugation in sucrose density gradient. Before aggregation, IgG is obtained from Cohn fraction II with further fractionation on a DEAE-52 cellulose column or AffiGel Blue

(Biorad). Ultracentrifugation at 100,000g will eliminate already aggregated IgG and determination of protein ($A^{1/0}_{280\,nm}$ for IgG is 14.5) is performed by micro-Kjeldahl or Lowry techniques (see Chap. 13).

In many RIAs, the concentration reactivity curves (also called calibration or standard curves) obtained with various dilutions of aggregated IgG are not linear (see Chap. 19). Furthermore, the standard curves show nonparallelism with dilutions of sera containing IC, evidenced by the fact that the same serum containing IC gives different results at various dilutions (i.e., the higher the dilution, the higher the values of IC obtained). Aggregates of IgG of a given size behave differently on dilutions in three assays used for IC quantitation (CIq binding assay, Clq solid phase, and Raji cell RIA), and aggregates of different sizes behave differently in dilutions in any one assay. Therefore, serum samples should be tested at the lowest acceptable dilution (usually 1:4) and new batches of aggregated IgG should always be compared to old ones.

Since all RIAs for quantitation of IC use aggregated IgG as standards, it appears that all these assays are semiquantitative at best. They cannot be compared to each other, and for accurate results many test samples should be run simultaneously in the same assay. (For further details on RIA, see Chap. 19.)

Addition of protein (e.g., 0.5% serum albumin or gelatin) to IgG aggregates has been shown to stabilize them and to prolong their storage without influencing the accuracy of the tests (Vanes and Daha, 1979). ICs made in vitro (e.g., tetanus toxoid-anti-toxoid) can be used to quantitate ICs, and recently the World Health Organization has established some provisional "standards" of IC in lyophilized form with added stabilizers.

To avoid some of the pitfalls just mentioned, it has been suggested to express IC in other ways than in aggregated IgG equivalent, for example, an "IC index," determined by dividing the values obtained for test serum (percentage binding of a tracer) by the mean value of a large pool of sera from healthy individuals. It has also been suggested to express the amount of IC in numbers of standard deviations above the mean, or as a percentage of the same positive standard. The normal binding is determined by assaying 100 samples from healthy individuals, and in each subsequent assay only 10 samples from this pool need be tested.

III. ASSAY METHODS

A. Antigen-Specific Methods

Precipitation of IC with antibodies to immunoglobulin or to complement and subsequent demonstration of a specific antigen in the precipitate have been used in experimental models (for reviews see Penner and Albini, 1982; Theofilopoulos and Dixon, 1979; Williams, 1980). After injection of radiolabeled antigen, the presence of the label in thoroughly washed ammonium sulfate precipitates strongly suggests IC involving the radiolabeled antigen. DNA labeled with radioactive actinomycin D may be used for this purpose (Harbeck et al., 1973). Similarly, infectivity of mouse serum containing viral IC may drop substantially after precipitation. The latter method is not feasible, however, if excess antigen is present in the serum. Under such circumstances, preparatory procedures separating free antigen from IC (such as sucrose density gradient fractionation or gel filtration) should be performed prior to precipitation.

Alternatively, patients with malaria were given radiolabeled antibodies to malarial antigens (Houba et al., 1976). Solubility of the antibodies was significantly

Table 2 Selected Assays for the Detection of CIC: I. Tests Based on Physicochemical and Morphological Properties

Sucrose density gradient separation
Polyacrylamide gel electrophoresis (PAGE)
Analytical ultracentrifugation
Polyethylene glycol precipitation
Cryoprecipitability
Zone electrophoresis in agarose
(High-resolution electron microscopy)

decreased after interaction with patients' serum in vivo. In the great majority of cases, however, the antigens involved in IC-induced pathology in humans are not known.

B. Detection of Circulating Immune Complexes Based on Their Physicochemical Characteristics

Many of the physicochemical methods use standard biochemical isolation and purification techniques developed for protein chemistry; for instance, adaptations of purification methods used for enzyme-substrate complexes to isolation of the antigen–antibody complexes.

Table 2 lists a number of methods that have been used to detect CIC. They are based on the differences in physicochemical characteristics of IC as contrasted to their noncomplexed components.

One method used early for the detection of IC in biological fluids was the demonstration of large, quickly sedimenting peaks of protein in sera using sedimentation velocity ultracentrifugation in an analytical ultracentrifuge (see Chap. 13). There is little material in normal serum that sediments between 7S and 19S, and little that sediments more rapidly than 19S. In 1957, Franklin et al. found proteins of approximately 22S in sera of patients with rheumatoid arthritis. Using this approach, Kunkel et al. (1961) demonstrated large amounts of components with multiple sedimentation rates intermediate to 7S and 19S, and in some cases, components sedimenting between 22S and 28S in sera of patients with rheumatoid arthritis, hypergammaglobulinemic purpura, and certain other disorders. The isolated intermediate-size components were readily and completely dissociated in 6 M urea or low-pH glycine-HCl buffer to homogeneous 7S, the >22S components to 7S and 19S fractions. The material of abnormal sedimentation rates in sera in this study was interpreted to be IC composed of 7S and 19S rheumatoid factors with 7S immunoglobulin. As a method for detection of CIC, analytical ultracentrifugation has the advantages of being performed under mild conditions without the need for addition of specific reagents, and provides information on molecular size and composition of the complexes (Stanworth and Johns, 1977; see also Chap. 13). However, it is relatively insensitive, and can be used only when large amounts of CIC are present in the sample.

Another ultracentrifugation method that has been much used both for detection and for concentration and isolation of CIC for further analysis is sucrose density gradient ultracentrifugation. In this sedimentation velocity method, buffers containing different concentrations of sucrose are placed into a centrifuge tube to form a gradient of increasing sucrose concentration from the top of the tube to the bottom. The sample is applied

at the low end of the concentration gradient and serum components are sedimented through increasing concentrations of sucrose, which serve to minimize diffusion and to prevent mixing of the separated components at the end of the run. Sedimentation rates of components are obtained by comparison with the sedimentation rates of labeled known reagents. In a recent study, sucrose density gradients were combined with gradients of polyethylene glycol to isolate IC from synovial fluid for further analysis (Male and Roitt, 1979; see also Chap. 13).

Characterization by size may also be used in the separation and detection of CIC by gel filtration (see Chap. 13). In this technique, components of a mixture passed through a gel filtration column are differentially retarded as dictated by their size and shape while passing through interstices of gel beads or around the beads. Subject to the influence of molecular shape and size of components and the effective filtration range of the gel, components are eluted in accord with their molecular weights. Gel filtration in short columns of Sephacryl S-300, together with a radioimmunoassay for IgG, was used by Kilpatrick et al. (1980) to measure CIC in 40 sera from patients with a number of diseases and from normal controls. They found that the method had the advantages of requiring only small volumes of serum, of using commercially available and relatively stable materials, of great sensitivity due to the combination with a radioimmunoassay, and of easy adaptability to measurement of CIC containing classes of immunoglobulin other than IgG. The assay was limited in that only CIC of very high molecular weight were detectable.

CICs have also been isolated for measurement and analysis by methods based on their solubility characteristics. The prototype of a very useful assay was described by Farr in 1958. The original Farr technique involved the precipitation of albumin-containing IC from solution by "salting out" the complexes with ammonium sulfate. Saturated ammonium sulfate was added to the solution of CIC in amounts to reach a final concentration of one-half saturation of ammonium sulfate (approximately 2 M). At this concentration, immunoglobulin is insoluble, and the antibody-bound albumin in IC precipitated while nonbound albumin remained in solution. The presence of immune complexes in the original mixture was detected by assaying the precipitate for antigen. The basic technique, however, has been modified by the use of different means to precipitate the IC (e.g., coprecipitation with antiglobulins, precipitation of complexes in low concentrations of water-soluble polymers). It can be used for detection of antibody in IC by choosing conditions such that immunoglobulin in IC is precipitated under conditions in which uncomplexed immunoglobulin is soluble. Certain types of IC, however, (e.g., DNA–anti-DNA complexes of medium to low affinity), tend to become *solubilized* at high salt concentrations (see Chaps. 13 and 17); this does not occur when polyethylene glycol is used instead of ammonium sulfate (see below).

It has been observed that immune precipitation is enhanced in the presence of high-molecular-weight linear polysaccharides and water-soluble polymers such as polyethylene glycol (Polson et al., 1964). Serum proteins are precipitated by polyethylene glycol of molecular weight 6000 (PEG-6000) in an inverse relationship to their molecular weights, with larger proteins precipitating at lower concentrations of PEG (Polson, 1977). However, molecular weight is not the only criterion for precipitation, as several investigators (Digeon et al., 1977; Zubler et al., 1977) have reported the precipitation of small CICs at low PEG concentrations (2 to 5%), at which certain larger proteins were not precipitated (macroglobulin, MW 800,000; β-lipoprotein, MW 3,200,000). The exact nature of the interaction between polymer and protein is not known. It has been suggested that the polymer-protein association masks surface charges which govern the solubility of proteins. An

alternative hypothesis is that the polymer sterically excludes the proteins from the volume of solvent occupied by the polymer in solution. This brings large proteins closer to their solubility limits in the solvent, whereupon they precipitate (Polson, 1977). The protein precipitation is reversible, as the protein can be resolubilized by dialyzing out the PEG.

Creighton et al. (1973) described modifications of the Farr assay in which PEG-6000 was used to precipitate CIC from solution. With the use of 7.5% PEG it was possible to precipitate ICs as small as a trimer of IgG, while precipitation of noncomplexed IgG was low. Digeon et al. (1977) and others, in contrast, have found significant precipitation of noncomplexed IgG at 7% PEG, and have preferred to use PEG concentrations of 3% or 3.5% in order to avoid nonspecific precipitation of IgG and other serum proteins such as the complement components, which might interfere with the immune complex marker used to quantitate the assay.

While some investigators have found direct spectrophotometric measurement of the precipitates to be adequate for quantitation of IC-screening assays involving PEG, the technique is more often used in conjunction with other markers for CICs or their components, such as radial immunodiffusion, immunoelectrophoresis, or radioimmunoassay for immunoglobulin, or direct and indirect markers for aggregated IgG such as Clq, conglutinin, protein A, and rheumatoid factors.

Advantages of PEG-precipitation assays include the ease of the precipitation procedure, its reversibility, which is important for further analysis of the precipitated IC, its concentrating effect, and its low cost. Disadvantages are the contamination of the precipitates with noncomplexed proteins, which are insoluble under the chosen conditions; the insensitivity of the method unless labeled markers are used, which adds both to the cost and to the difficulty of the procedure; the possibility that precipitated IC might activate complement, which would interfere with the use of complement components as IC markers; false positives contributed by aggregated IgG in some assays; and the possibility of false negative results due to the interference by DNA or complexes containing DNA, which have been found by some investigators to interfere with PEG-induced precipitation of IC and nonspecifically aggregated proteins (Zubler and Lambert, 1977), although others have not reported this to be a problem (Macanovic and Lachmann, 1979).

Polyacrylamide gel electrophoresis (PAGE) has been used to detect high-molecular-weight fractions containing antigen and immunoglobulin. Furthermore, identification of IC has been attempted based on the net surface charge difference between IC and uncomplexed antigen and antibody. Fox et al. (1974) have used an ion-exchange cellulose to adsorb IC from sera of New Zealand Black mice, and more recently, Kelly et al. (1980) have proposed the use of zone electrophoresis in agarose. In studies comparing zone electrophoresis with Raji cell and Clq binding assays, an acceptable correlation was found. The authors conclude that this method is rapid, sensitive, and inexpensive, and suggest it for use in clinical laboratories.

IC can also be detected by inspection of preparations from serum with high-resolution electron microscopy. This is relatively easy with ICs containing particulate antigen (e.g., viral particles). In serum samples containing IC, electron micrographs show morphologically recognizable antigen and immunoglobulin molecules. Suggestive evidence for the presence of ICs containing small and morphologically uncharacteristic antigen may be obtained by demonstration of aggregated ferritin or peroxidase-labeled immunoglobulin (Fig. 1). The morphological demonstration of IC, however, is cumbersome and in no way a feasible approach for routine clinical needs. It is a highly interesting technique for research (see also Chap. 22).

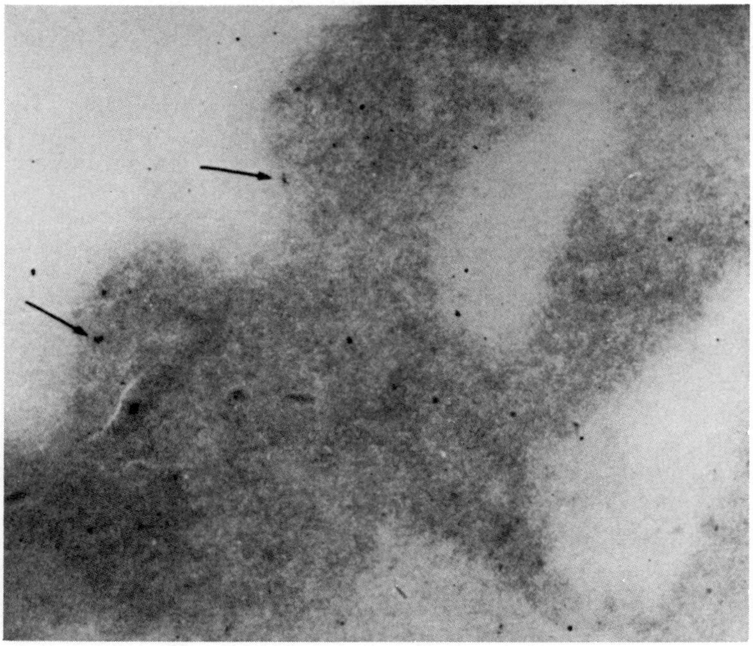

Figure 1 Ferritin (arrows) and rabbit antibodies to ferritin form immune complexes. The ferritin-immune complexes seem to be connected by masses of interacting "antiglobulins." × 101,600. (From Robinson and Schuffman, 1970).

C. Detection of CIC by Nonantigen-Specific Assays Based on Biological Reactivity

In the following, a few selected assays, which are performed most frequently or with which the authors are most familiar, will be discussed in some detail. Wherever possible radioimmunoassays (or ELISA) will be described, which usually allow for some degree of quantitation and are better suited than other methods to be performed in routine clinical laboratories. Most other available assay systems and pertinent literature may be found in Tables 2 to 4.

Techniques Based on Activation of Complement Components by CIC (see also Chap. 23)

IC may activate complement via the classical or the alternative pathways. The available tests using this biological property of IC vary in detecting either IC activating complement via one or the other pathways, or detecting IC irrespective of which pathway is used (Table 3). Obviously, all these procedures detect only IC capable of activating complement. For example, consumption of a known amount of complement by test sera is interpreted as an indication of the presence of IC in the microcomplement test (Mowbray et al., 1973). The amount of complement not consumed by the test serum is evaluated in an assay for lysis of IgM-antibody-sensitized sheep red blood cells. Test sera have to be heated (i.e., decomplemented) prior to the assay, which may induce formation of aggregates. This seems to be the major disadvantage of this otherwise sensitive assay detecting IC activating complement either via classical or alternative pathways.

Table 3 Selected Assays for the Detection of CLC: II. Tests Based on Biological Properties of CIC–A. Complement Activation

Complement component	Assay technique
Whole complement	Microcomplement test
Clq	Agarose Clq precipitation
	Precipitation with PEG
	Clq deviation
	Clq solid-phase RIA
C3	C3 precipitation
	EAC-rosette inhibition
	Solid-phase RIA with conglutinin
	Raji cell RIA
	Raji cell FITC microtest

Table 4 Selected Assays for the Detection of CIC: II. Tests Based on Biological Properties of CIC–B. Antibodies to Immunoglobulins and Cellular Fc Receptors

Reagent reacting with immunoglobulin of IC	Assay technique
Rheumatoid factors (RFs) (polyclonal or monoclonal)	Precipitation in agarose
	Inhibition of agglutination of IgG-coated latex particles
	Competitive inhibition assay employing RIA or ELISA
Low-avidity immune sera specific for human Igs	Inhibition of agglutination of IgG-coated latex particles
Anti-antibody	Agglutination inhibition assay
Cellular Fc receptor	Platelet aggregation
	L1210 murine leukemia RIA
	Inhibition of ADCC
	Inhibition of EA rosettes

Tests based on interaction of IC with Clq. IgG_1, IgG_3, to some extent IgG_2, and IgM may effectively bind Clq, and binding is significantly enhanced upon interaction with specific antigen or aggregation. This binding is reversible. Incorporation of exogenous Clq into ICs that already have fixed complement is possible. IgA, IgE, IgD, and IgG_4 do not interact with Clq, and ICs containing these immunoglobulins as antibody cannot be detected by the Clq assays. Clq has been purified from plasma by precipitation in the presence of chelating agents at low ionic strength (Yonemasu and Stroud, 1971).

Various polyanions or polycations, DNA, endotoxin, C-reactive protein, heparin, viruses, lipoproteins, and uric acid crystals bind to Clq. Binding of Clq is influenced by pH and ionic strength. It is considered that Clq recognizes preferentially large CIC (the Clq molecule has six recognition units). Although Clq may interact with monomeric IgG, a lattice involving three or more antibody molecules probably is required for strong binding leading to complement activation. Smaller CICs thus can be missed with Clq assays. Agnello and collaborators (1970) developed the original agarose Clq precipitation test. The test requires long incubation times and has very low sensitivity.

The *Clq binding* or *Clq fluid-phase* (Clq BA) *assay* uses radiolabeled Clq which binds to CIC (Nydegger et al., 1974). The CICs are precipitated with 2.5% PEG, which does not precipitate soluble Clq (or Clq bound to DNA) and the radioactivity of the precipitate is measured. The maximum ("100%") binding is determined by precipitating $[^{125}I]$ Clq in serum with trichloracetic acid. Initially, sera were heated to inactivate native complement; however, this may result in some aggregation of immunoglobulins or in decreasing the ability of IC to bind to Clq. Furthermore, some antigens are sensitive to heating and some IC may dissociate after heating. To circumvent these problems, ethylenediaminetetraacetic acid (EDTA) (0.13 M) was used to stabilize Clq and prevent incorporation of radiolabeled Clq into the endogenous, high-molecular-weight Clqrs complex. However, EDTA may release Clq from Cl in the test serum and the unlabeled Clq will compete with labeled Clq and decrease the sensitivity of the assay. Heparin can give false positive results, but protamine sulfate eliminates this effect. The sensitivity of the assay is 80 to 100 μg/ml equivalent aggregated IgG and the test has been shown by some researchers to correlate well with the Raji cell RIA (see Sec. II), although the latter seems to be more sensitive. Free DNA in serum inhibits binding of Clq to aggregated IgG and possibly to CIC. It was shown that the binding of plasma in this assay can be 90 to 400% higher than that of serum because of complexing of fibrinogen with labeled Clq. The test seems most sensitive for IC in antigen-antibody equivalence or antigen excess. The increase in size of aggregated IgG enhances the precipitation of $[^{125}I]$ Clq ("bonus" effect). As just mentioned, false positive results can be due to endotoxin ($>$100 μg/ml), heparin ($>$1 μg/ml), or DNA ($>$20 μg/ml). $Beta_2$-microglobulin and Clrs can also bind to Clq and give false positive results. The quality and purity of Clq is of utmost importance for the accuracy of the assay since Clq polymers (less soluble in PEG) or residual IgG (which will react with rheumatoid factor) give false positive results. Clq purity can be assessed by electron microscopy and its activity should be checked by latex agglutination. The IgG contamination should be determined by an RIA using radiolabeled IgG fraction of an anti-human IgG serum. Usually, labeled Clq is added to buffer containing 1% heat-inactivated normal serum, and is centrifuged for 40 min at 18,000*g* to eliminate the aggregated Clq.

Clq solid-phase assays use polystyrene tubes or plastic microtiter plates (wells have a superior surface/volume ratio) coated with purified Clq (Hay et al., 1976). Through Clq, IC are fixed on the solid-phase component and their amount is determined by measuring

the radioactivity of the tube (well) after addition of radiolabeled anti-IgG or radiolabeled protein A in excess. This assay is more sensitive than Clq binding assay (i.e., about 5 μg/ml of aggregated IgG is still detected). The serum is usually heated before testing, but EDTA can obviate the need for heating. The binding is not affected by monomeric IgG, but rheumatoid factors interfere with the results. The test is not influenced by heparin, DNA, or endotoxin and is little affected by suboptimal storage or handling conditions of the test sera.

Human Clq can be used for measuring CIC in other species (e.g., rat). The solid-phase Clq is most sensitive for IC with four times antigen in excess. The binding of CIC to Clq seems much greater at ionic strength lower than physiological; under these conditions, nonspecific absorption of Clq to the solid phase becomes a problem, as does nonspecific binding to other materials (DNA, some viruses, endotoxin, etc.). Adsorption of Clq to the solid phase varies with the brand of plastic used, and addition of Tween-20 to the buffer used for washing reduces aggregation and nonspecific adsorption in solid-phase systems. Clq, even unlabeled, cannot be stored for more than 2 months.

The *Clq deviation assay* is a fast and very sensitive assay (estimated sensitivity, 1 to 5 μg/ml aggregated IgG) for measuring CIC (Gabriel and Agnello, 1977). This is a competitive assay in which binding of radiolabeled Clq to a particulate substrate (e.g., sensitized sheep erythrocyte/anti-sheep IgG, IgG-coated latex particles, or IgG-coated Sepharose particles) is inhibited by CIC. First, the serum is incubated with $[^{125}I]$ Clq and then the binder (e.g., IgG-Sepharose) is added. Sera should be preheated and this can give false positive results. Endotoxin, DNA, aggregated immunoglobulins, rheumatoid factors, and so on, can also give false positive results, but increasing the ionic strength of the solution can obviate some of these problems. When using sensitized sheep erythrocytes, careful adjustment of the amount of antierythrocyte antibody is needed, so that "spontaneous" agglutination of red cells is not obtained and roughly 50% of the radiolabeled Clq is fixed by the sensitized red cells (the "binder"), much as in a "standard" RIA. Usually, the test material is centrifuged through sucrose and the percentage inhibition of uptake is calculated. As in the other assays using Clq, large IC (>19S) appear preferentially measured by Clq deviation assay.

Tests based on the activation of C3 by IC.

1. *Conglutinin RIA*. Conglutinin is a protein of high molecular weight, which occurs naturally in the sera of certain members of the family Bovidae, in contrast to immune conglutinin, which is found in most species (Lachmann, 1967). Conglutinin does not bind to native C3 but only to activation products of C3 of cow, horse, rabbit, guinea pig, humans, and perhaps mouse. It binds C3 fixed on IC, specifically C3bi, an early product of activation and inactivation of C3. Hence conglutinability appears at an early stage of the interaction of C3b with heated serum, where loss of C3 antigenicity is not yet detectable. C3b is degraded by C3b inactivator and β_1H. The formation of the intermediary product (formally called C3K) involves the cleavage of a peptide bond without dissociation of the resulting fragments, which may remain linked by other covalent bonds until they are cleaved by a heat-stable, trypsinlike serum enzyme. After prolonged storage of serum, the conglutinability decreases due to excessive exposure of C3 to cleaving enzymes. Thus conglutinin reactivity with IC is restricted to the first 60 to 400 min after the complement activation. Binding of conglutinin to C3bi is Ca^{2+} dependent, and chelators (and other substances, e.g., N-acetyl-d-glucosamine) can inhibit or even reverse the binding. Freezing and thawing also decrease the binding.

Several assays for measuring IC were devised with the use of conglutinin. These assays do not require terminal complement components and are influenced very little by 7S IgG. Conglutinin is fairly stable at 4°C. Conglutinin titers vary from 8 to 2000 and a number of cow sera have to be screened with horse erythrocytes and antibody to horse red cells to detect an adequately potent reagent. Sera with high conglutinin titers are heated for 60 min at 37°C. In a solid-phase RIA, purified conglutinin is used to coat plastic tubes or microtiter wells (Casali et al., 1977; Eisenberg et al., 1977). The IC are bound to the solid phase through their C3, and the amount of bound IgG (within IC) is quantitated with labeled anti-IgG antibodies or with radiolabeled or peroxidase-labeled protein A. Alternatively, radiolabeled antibodies to IgA or IgM could be used to detect IC containing IgA or IgM. The sensitivity of the assay is roughly 5 μg/ml of aggregated IgG equivalent.

In another RIA, radiolabeled conglutinin is used. This reacts first with IC and then is precipitated with PEG. The amount of IC precipitated is measured by determining the radioactivity of the precipitate. Conglutinin can also be used to isolate CICs (especially of large size, >19S), which activate complement by both classical and alternative pathways.

It seems that conglutinin reacts preferentially with large ICs, and conglutinin assays detect only a fraction of complement-fixing IC, because conglutinin is specific for C3bi, which, as mentioned before, has a short life. In our hands as well as in those of others, there was no good correlation between Raji cell RIA and conglutinin RIA for detecting CICs.

Since C3bi is destroyed by proteolytic enzymes, it is better to add enzyme inhibitors to the serum samples (e.g., phenylmethylsulfonylfluoride, 0.5 mM final concentration). The assay is not influenced by heparin, endotoxin, or DNA, but antigen released from its antibodies can still retain C3bi and bind to conglutinin. When antigen–antibody complexes were artificially formed in vitro, conglutinin assays were more sensitive for CIC in slight antigen excess.

2. *Raji cell assays*. Raji cells are EB virus-infected lymphoblastoid cells derived from a Burkitt lymphoma patient and kept in continuous culture. They have some B-cell characteristics but lack membrane-bound immunoglobulins. They have few low-affinity receptors for the Fc portion of IgG and many high-affinity receptors for C3b, C3d, Clq, and perhaps for intact C3 and other complement components. These receptors are uniformly distributed on the cell membrane. The binding of CIC to the Raji cells is effected primarily through C3b-C3d receptors and about 4×10^5 molecules of C3 are bound per cell. Although Clq receptors on Raji cells were considered of no practical importance for binding CIC, 30% of binding of the whole serum remained after depletion of C3, thus proving the Clq binding. The affinity for C3 components does not seem species specific and Raji cells have been used successfully to detect IC in mice, rats, and rabbits. Serum (usually diluted 1:4 to avoid binding through the Fc receptors) is incubated at 37°C, with gentle shaking, with a known number of viable Raji cells and, after washing, labeled anti-IgG or labeled protein A is reacted with the cells (Theofilopoulos et al., 1976). The amount of the labeled tracer bound to the Raji cells is related to the amount of IC fixed on the cells and is expressed in aggregated IgG equivalent per milliliter. The sensitivity of the assay is roughly 5 μg/ml of aggregated IgG equivalent, which is high when compared to other tests for IC measurement but low when compared to RIA for measuring drugs or hormones. To detect ICs containing other classes of immunoglobulin (e.g., IgA or IgM), one can use labeled anti-IgA or anti-IgM. Some of these "non-IgG" ICs activate complement predominantly or exclusively by the alternative pathway. For lowering the

background radioactivity, it is advisable to use affinity-purified antibodies. IgA- or IgM-containing ICs cannot be quantitated at all with the use of aggregated IgG since the amount of C3 bound on the latter complexes is not the same as that bound on IgA or IgM ICs.

It has been reported that Raji cell RIA best detects IC near antigen-antibody equivalence. Size determines the ability of IC to fix complement and hence to bind on the Raji cells. Binding of aggregated IgG to which various amounts of fresh serum were added was minimal at a low ratio of serum (i.e., complement) to antigen-antibody (<1:64); it increased sharply at intermediate ratios (1:32 to 1:4) and then decreased sharply at high ratios (>1:2). Care should be taken in counting the number and viability of Raji cells before the assay and after the incubation with test serum. The Raji cells should be checked periodically for their receptor characteristics and new cultures should be established when their growth is exceedingly high (e.g., more than 18×10^6 cells per 30 ml of culture). Reactivity of receptors for C3 and Fc of IgG varies in time, reflecting changes in density and/or avidity of surface receptors. Expression of receptors for C3 is maximal at 72 hr after changing the culture medium, at a time when Fc receptors have a minimal role in binding IC, although it may be difficult to block them completely.

Recently, glutaraldehyde-fixed Raji cells have been prepared and stored for longer periods of time, and this can assure more reproducible results. Raji cell RIA was also modified by addition of radiolabeled bovine serum albumin (BSA) (with ^{131}I) to correct for nonspecific entrapment. The test can be standardized better than any other solid-phase RIA. Furthermore, it is not influenced by heparin, DNA, or endotoxin and is only negligibly influenced by excess monomeric IgG or by storage and handling conditions of sera. Of particular importance for the reproducibility of the assay is finding the amount of radiolabeled protein A (or anti-IgG) needed for optimal results and the use of the same number of cells. Raji cells can be also used to isolate CIC, and labeled antibodies directed against the suspected antigens can be used to identify specific antigens in CIC.

Although Raji cell RIA has the advantage of using cells in continuous culture (i.e., relatively homogeneous, readily available cells), their maintenance is cumbersome. The test can give false positive results because of the presence of warm-reactive monomeric serum IgG antibodies with allo- or autospecificity and the presence of antibodies interacting with antigens released from nuclei or cytoplasms of dead cells. Sera from multiparous women, which often contain lymphocytotoxic antibodies, are negative in the Raji cell assay; however, it was shown that much of the IgG bound at 37°C to Raji cells from the sera of many patients with SLE are not IC but antibody to Raji cells. These antibodies did bind to cells after pronase digestion of Raji cells (which strips the complement receptors). Antilymphocytic antibodies were also found in viral infections, subacute bacterial endocarditis, and possibly in mixed connective tissue diseases, scleroderma, and Sjögren's syndrome (i.e., diseases in which there are often circulating IC). It appears that lymphocytotoxic antibodies of the IgG class bind better to the B cells (similar to Raji cells) and also bind at 37°C, whereas IgM lymphocytotoxic antibodies bind better at 4°C. As mentioned before, the quantitation of IC by Raji cell RIA is actually an approximation since the level of IC bound varies with the size and character of CIC.

The Raji Cell Assay with the Membrane Immunofluorescence Micromethod

Using defined immunofluorescent staining (Beutner et al., 1968) and the procedure described by Schauenstein et al. (1976), large numbers of patient sera can be tested in a

short time allowing for semiquantitative results: namely, the determination of end titers. The washing fluids (RPMI 1640 and BBS) should contain 0.7 mM NaN_3 to prevent capping. Experiments with in vitro-formed ICs containing human or rabbit antibodies suggest that the maximum sensitivity of the test is obtained with ICs at antigen–antibody equivalence or in slight antigen or antibody excess. As expected, antigen can be detected more easily in antigen-excess CIC, whereas immunoglobulin is better demonstrable in antibody-excess CIC.

If reasonably fresh sera (less than 18 months) are used, a titer of 4 and higher is considered "positive." The highest titers observed by us was 512. Normal blood donor populations gave a positive result in 6%, and among 22 healthy laboratory personnel, 1 had CIC (also confirmed in other CIC assay).

Decomplemented sera may be used after addition of guinea pig complement. The optimal complement concentration has to be established by titration. Excess of complement may lead to "solubilization" of CIC (Takahashi et al., 1980), and thus to false negative results. Still, the titers obtained for IC in sera after decomplementation and reconstruction of complement are lower for IC-containing preparations, and higher for IC-free-sera than found in fresh preparations. Under these conditions, only titers higher than 8 can be considered positive. Therefore, addition of complement to sera prior to Raji cell assay should be resorted to only as a "last measure."

With many patients' sera, prozonelike phenomena may be observed at serum dilutions of 1:2 and 1:4. In addition, heavy immune precipitates may blurr the reading of preparations at these low dilutions. The typical appearance of immunofluorescent staining of IC on Raji cells is shown in Fig. 2.

The presence of antibodies to lymphoid cell surface antigens in sera may lead to the presence of linear to granular immunoglobulin deposits on the cell. With antilymphocytic antibodies, however, Raji cells often show intracellular uptake of the FITC conjugate used. Results with sera of patients with SLE, rheumatoid arthritis, alcoholic liver diseases, primary biliary cirrhosis, membranous nephropathy, and leprosy are summarized in Table 1.

CIC involving antibodies of various immunoglobulin classes can be distinguished in Raji cell assays. ICs containing IgA were first described using the immunofluorescence micromethod (Penner et al., 1982) and then with the use of radiolabeled antisera. CICs containing IgM were detected in the sera of patients with leprosy and primary biliary

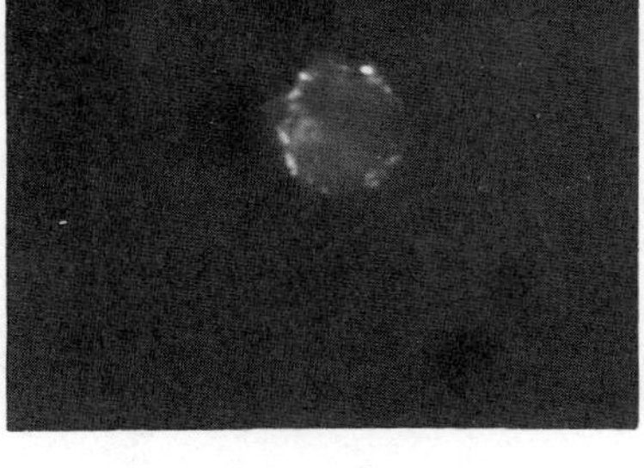

Figure 2 Immunofluorescence test for circulating immune complexes using Raji cells. Immune complexes bound to the cell surface are visualized by reaction with fluorescein-isothiocyanate-labeled antibody to human IgG. ×400.

cirrhosis (e.g., Kano et al., 1981). Antigens may be detected in IC using Raji cells, and fluorochrome conjugates antigen-specific antisera. Complement components cannot (or can only inconsistently) be detected in CIC after reaction with C receptors on Raji cells. Thus use of fluorochrome or radiolabeled antisera to complement components seems not feasible.

Techniques Based on the Interaction of Antibodies with Epitopes of Antigen-Complexed Immunoglobulin

Antibodies to immunoglobulin reacting preferentially with aggregated or antigen-complexed immunoglobulin molecules arise spontaneously under certain circumstances in humans and in animals. Among these antibodies, rheumatoid factor (RF) and anti-antibody (AA) are of special importance. Among RFs, monoclonal 19S reactants of the IgM class seem better to distinguish between monomeric and aggregated or complexed immunoglobulin. The multivalent, low-avidity rheumatoid factors preferentially react with immune aggregates because of the multiplicity of epitopes in such moieties. In addition, low-affinity IgM antibodies to human IgG (and other immunoglobulin classes) may be obtained by immunization of rabbits (Levinsky and Soothill, 1977). Both reagents (RFs and anti-immunoglobulins) have been used for the detection of IC, and in both tests the specificity (i.e., the capability of the procedure to differentiate between monomeric and aggregated or complexed immunoglobulins) depends primarily on the characteristics of the antiglobulin reagent. Quite different is the method based on the reactivity of AA with the Fab fragments of immunoglobulin (Kano et al., 1978), since this antiglobulin moiety reacts with neither monomeric nor heat-aggregated IgG (Kano and Milgrom, 1980). In addition to these tests, interaction of cell-bound Fc receptors and staphylococcal protein A may be used in tests for detection of IC.

RIAs with rheumatoid factor (monoclonal or polyclonal) are sensitive (1 to 25 μg/ml of aggregated IgG) and are reported to detect CIC as small as 8S. These assays are independent of the complement-fixing properties of the IC. Rheumatoid factor binding inhibition is a competitive-type RIA, similar in principle to the Clq inhibition assay (Davey and Korngold, 1981; Lurhuma et al., 1977). In this assay, radiolabeled monoclonal rheumatoid factor (mRF) is used as a tracer, which competes with IC for the binding sites of IgG-Sepharose. In another sensitive assay, polyclonal rheumatoid factor (pRF) is used as a binder and radiolabeled aggregated IgG competes with IC for the pRF binding sites. The pRF is precipitated with anti-IgM serum and the radioactivity of the precipitate is measured.

Monoclonal RF can also be used in a solid-phase assay. Microtiter wells are coated with RF, serum is added, and the amount of IC is determined by use of radiolabeled or peroxidase-labeled protein A, or alternatively, ^{125}I-labeled $F(ab')_2$ rabbit anti-Fab bound per well.

Other RIAs for measuring IC use staphylococcal protein A, which binds to Fc of IgG and some IgA and IgM molecules. In such an assay, serum is precipitated initially with 5% PEG in which only 5% of monomeric IgG (versus 82% of aggregated IgG) is precipitated. The dissolved precipitate is incubated with heat-killed, formalin-fixed staphylococci containing protein A. The incubation is carried out under conditions in which complexed IgG is preferentially bound and the monomeric IgG is competitively eluted by monomeric rabbit IgG. Residual IgG present on the bacteria is quantitated by incubation with ^{125}I-labeled rabbit anti-human IgG serum.

Assays using immune sera with antiglobulin activity. Low-avidity antibodies to human Igs raised in rabbits have been used to assess the presence or absence of IC in sera (Levinsky and Soothill, 1977). In this test, human immunoglobulins (IgG, IgA, and IgM) were used to coat latex particles; low-avidity IgM-class antibodies to human immunoglobulins raised in rabbits agglutinate the latex particles unless reacted previously with IC-containing test sera. This simple agglutination inhibition assay apparently is not influenced by monomeric human immunoglobulins, and electronic quantitation of unagglutinated latex particles should make automation feasible. EDTA and incubation with IgG-coated sigma cells circumvents interference by Clq and RF. It is obvious, however, that elimination of RF also removes IC containing RF. Using this assay, Levinsky and Soothill (1977) have demonstrated and characterized IC presumably containing IgA in patients with Henoch-Schönlein nephritis, and have documented a correlation between serum concentrations of IC and disease activity in a patient with diffuse proliferative glomerulonephritis.

Anti-antibody inhibition assay. Anti-antibodies (AAs) were first described in 1956 by Milgrom et al. These antibodies occur in the sera of some humans and rabbits. In humans, they are of the IgM class. They occur in relatively low frequency in the populations of healthy or hospitalized individuals tested thus far (1 out of 2000). AAs react with other antibodies after these have interacted with antigen. Aggregated gamma globulin reportedly does not react with AA. The anti-antibody combines with an antigenic determinant in the $F(ab')_2$ fragment of antibodies.

AAs are used in an agglutination-inhibition test for the detection of CIC. The anti-antibody is incubated first with the test serum; IC of the IgG class present in the serum will combine with AA. The indicator reaction in this consumption test is agglutination of human erythrocytes sensitized with "incomplete" antibodies to Rh-blood group antigens. Agglutination occurs only when the antibodies reacted with erythrocyte antigens are "bridged" by AA. Therefore, absence of agglutination indicated the presence of IC in a test serum, and positive agglutination indicated the absence of CIC. The degree of inhibition is expressed as titer (i.e., the highest dilution of test serum still giving a negative reaction in the agglutination indicator system). The antigenic specificities responsible for reaction with high-titered AA seem not to be allotypic determinants, despite some suggestive evidence that low-titered AA indeed may show allotypic specificities.

The AA inhibition assay is highly sensitive in the detection of CIC in antibody-antigen equivalence or in slight antigen excess. This test is unique among all assays described up to now in not showing false positive results with aggregated IgG (Kano et al., 1978).

RF present in test sera may aggregate sensitized erythrocytes. This may lead to false negative results in this assay. Pretreatment of test sera with 2-mercaptoethanol destroys RF. ICs consisting of IgM-class RF and IgG (now in the role of antigen) are not detected by the AA inhibition assay; IgG has to react as antibody in CIC to be detected in this assay. The assay is easy to perform and has high specificity for IC. The necessity for elaborate and extensive screening procedures for detecting sera containing high-titered AA limits, at least at present, the usefulness of this procedure for general clinical laboratories.

IV. CONCLUSIONS

To account for the differences in detectability of CIC with various tests and the complexity of CIC with different characteristics found in individual sera, a variety of assay

Table 5 Proposed "Battery" of Assays to Be Used in Routine Clinical Laboratories[a]

Assay	Remarks
1. Raji cell assay (RIA or FITC microassay)	Detects IC binding complement via alternative or classical pathways; preferentially IC near Ag/Ab equivalence; with appropriate antisera, IC containing IgG, IgA, and IgM may be distinguished or tested for putative antigens. Requires cell culture facilities, gamma counter, or immunofluorescence microscope; high false positive rate
	and/or
2. Conglutinin solid-phase RIA	Detects IC binding complement via alternative or classical pathways; requires gamma counter
	and
3. Anti-antibody agglutination inhibition assay	Highly specific, simple assay for IC containing IgG in Ab/Ag equivalence zone; limiting factor: availability of high-titered anti-antibodies
	and/or
4. Monoclonal rheumatoid factor competitive inhibition assay (RIA or ELISA)	Detects IC containing IgG; reported to be sensitive also for very low molecular weight IC; best results with monoclonal RF in RIA or ELISA: limiting factor; availability of monoclonal RF; requires gamma counter or spectrophotometer
	and/or
5. Low-avidity antibodies to immunoglobulin (agglutination inhibition assay)	Detects ICs of IgG, IgA, IgM, and IgE (?) classes; requires immunization of rabbits and production of low-avidity antibodies
	and
6. Clq solid-phase RIA	Detects IC binding complement via classical pathway; preferentially IC of high molecular weight; high false positive rate; requires gamma counter

[a]We are aware that the selection proposed here is subjective and influenced by our personal experience.

procedures should be used for optimal results (Table 5). Presently, assays for the detection of CIC are not disease specific; they may be useful in clinical diagnosis and in monitoring a patient's disease and its therapy. However, they have great potential for basic immunological and clinical immunopathological research.

REFERENCES

Agnello, V., Winchester, R. J., and Kunkel, H. G. (1970). Immunology *19*, 909.

Albini, B., and Penner, E. (1982). Diagnostic value of assays for the determination of circulating immune complexes. Behringwerke Mtlg.

Beutner, E. H., Sepulveda, M. R., and Barnett, E. V. (1968). WHO Bull. *39*, 587.

Casali, P., Bossus, A., Carpentier, N. A., and Lambert, P. H. (1977). Clin. Exp. Immunol. *29*, 342.

Couser, W. G., and Salant, D. J. (1980). Kidney Int. *17*, 1.

Creighton, W. D., Lambert, P. H., and Miescher, P. A. (1973). J. Immunol. *111*, 1219.

Davey, M. P., and Korngold, L. (1981). J. Immunol. Methods *44*, 87.

Deliere, M., Cambiaso, C. L., and Messon, P. L. (1978). Nature *272*, 632.

Digeon, M., Laver, M., Riza, J., and Bach, J. F. (1977). J. Immunol. Methods *16*, 165.

Dixon, F. J. (1963). Harvey Lect. *58*, 21.

Eisenberg, R. A., Theofilopoulos, A. N., and Dixon, F. J. (1977). J. Immunol. *118*, 1428.

Farr, R. S. (1958). J. Infect. Dis. *103*, 239.

Forsgren, A. (1970). Infect. Immun. *2*, 672.

Fox, A. E., Plfscia, O. J., and Mellors, R. C. (1974). Immunology *26*, 367.

Franklin, E. C. Holman, H. R., Müller-Eberhard, H. J., and Kunkel, H. G. (1957). J. Exp. Med. *105*, 425.

Furst, G., Fekete, B., Angyal, I., Jakab, A., Pal, A., Merety, K., Falus, A., Torok, K., Szegedi, G., Kavia, M., Puskas, E., Csesci-Nagy, M., Szabo, T., Lenkey, A., and Misz, M. J. (1981). J. Immunol. Methods *46*, 259.

Gabriel, A., and Agnello, V. (1977). J. Clin. Invest. *59*, 990.

Germuth, F. J., Jr., and Rodriguez, E. (1973). *Immunopathology of the Renal Glomerulus*. Little, Brown, Boston.

Gupta, R. C., Dickson, E. R., McDuffie, F. C., and Baggenstoss, A. H. (1978). Clin. Exp. Immunol. *34*, 19.

Harbeck, R. J., Bardana, E. J., Kohler, P. F., and Carr, R. I. (1973). J. Clin. Invest. *52*, 789.

Hay, F. C., Nineham, L. J., and Roitt, I. M. (1976). Clin. Exp. Immunol. *24*, 396.

Houba, V., Lambert, P. H., Voller, A., and Soyamwo, M. A. O. (1976). Clin. Immunol. Immunopathol. *6*, 112.

Kano, K., and Milgrom, F. (1980). Vox. Sang. *38*, 121.

Kano, K., Nishimaki, T., Palosuo, T., Loza, U., and Milgrom, F. (1978). Clin. Immunol. Immunopathol. *9*, 425.

Kano, K., Aranzazu, N., Nishimaki, T., Convit, J., Albini, B., and Milgrom, F. (1981). Int. Arch. Allergy *64*, 19.

Kelly, R. H., Scholl, M. A., Harvey, V. S., and Devemyi, A. G. (1980). Clin. Chem. *26*, 396.

Kilpatrick, D. C., Yap, P. L., and Irvine, W. J. (1980). J. Immunol. Methods *39*, 269.

Kunkel, H. G., Müller-Eberhard, H. J., Fudenberg, H. H., and Tomasi, T. B. (1961). J. Clin. Invest. *40*, 117.

Lachmann, P. J. (1967). Adv. Immunol. *6*, 2479.

Levinsky, R. J., and Soothill, J. F. (1977). Clin. Exp. Immunol. *29*, 428.

Lurhuma, A. Z., Ricconi, H., and Masson, P. L. (1977). Clin. Exp. Immunol. *128*, 49.

Macanović, M., and Lachmann, P. J. (1979). Clin. Exp. Immunol. *38*, 274.
Male, D., and Roitt, I. M. (1979). Mol. Immunol. *16*, 197.
Milgrom, F., Dubiski, S., and Woznicsko, G. (1956). Vox. Sang. *1*, 172.
Mowbray, J. F., Holborow, E. J., Hoffbrand, A. V., and Seah, P. P. (1973). Lancet *1*, 400.
Nydegger, U. E., Lambert, P. H., Gerber, H., and Miescher, P. A. (1974). J. Clin. Invest. *54*, 297.
Penner, E., and Albini, B. (1982). *Immunofluorescence Technology* (G. Wick, K. N. Trail, and K. Schauenstein, Eds.). Elsevier/North-Holland, Amsterdam, p. 349.
Penner, E., Albini, B., and Milgrom, F. (1978). Clin. Exp. Immunol. *34*, 28.
Penner, E., Goldenberg, H., Albini, B., Weiser, M. M., and Milgrom, F. (1982). Clin. Immunol. Immunopathol. *22*, 394.
Polson, A. (1977). Prep. Biochem. *7*, 129.
Polson, A., Potgieter, G. M., Largier, J. F., Mears, G. E. F., and Joubert, F. J. (1964). Biochim. Biophys. Acta *82*, 463.
Robinson, J. P., and Schultman, S. S. (1970). Immunology *20*, 883.
Schauenstein, K., Wick, G., and Kink, H. (1976). J. Immunol. Methods *10*, 143.
Stanworth, D. R., and Johns, P. (1977). Ann. Rheum. Dis. *36*, S12.
Takahashi, M., Takahashi, S., and Hirose, S. (1980). Prog. Allergy *27*, 134.
Theofilopoulos, A. N., and Dixon, F. J. (1979). Adv. Immunol. *28*, 89.
Theofilopoulos, A. N., Wilson, C. B., and Dixon, F. J. (1976). J. Clin. Invest. *57*, 169.
Vanes, L. A., and Daha, M. R. (1979). J. Immunol. Methods *31*, 11.
Williams, R. C., Jr. (1980). *Immune Complexes in Clinical and Experimental Medicine*. Harvard University, Cambridge, Mass.
Yalow, R. S., and Berson, S. A. (1966). Trans. N.Y. Acad. Sci. *28*, 1033.
Yonemasu, K., and Stroud, R. M. (1971). J. Immunol. *106*, 304.
Zubler, R. H., and Lambert, P. H. (1977). Ann. Rheum. Dis. *36*(1), S27.
Zubler, R. H., Perrin, L. H., Creighton, W. D., and Lambert, P. H. (1977). Ann. Rheum. Dis. *36*, S23.

19

Radioimmunoassay

Daniel J. McCormick and H. Edward Schmitz Mayo Clinic, Rochester, Minnesota

I. INTRODUCTION

Radioimmunoassay (RIA) was first introduced as a useful clinical technique by Yalow and Berson (1960). This method, which was an extension of basic procedures used by Farr (1958) to study antibody-antigen interactions, described the use of guinea pig antibodies to ^{131}I-labeled bovine insulin for the direct quantitation of plasma insulin levels. Since this initial introduction in the field of biological investigation, RIA has gained widespread use as an extremely sensitive method for the quantification of a great number of biologically active compounds.

RIA methods derive their applicability as an indispensable tool in the laboratory from two properties: the great sensitivity of isotope determinations and the high specificity of antibody-antigen interactions. Almost any compound to which an antibody can be produced may be measured by RIA down to the picogram (10^{-12} g) range. A list of substances measurable by this technique is given in Table 1. Although by no means exhaustive, this table illustrates the numerous variety of substances which can be accurately measured, thus demonstrating the great versatility and sensitivity of this method.

Application of RIA procedures have profoundly revolutionized the field of clinical endocrinology since many hormones can be detected and measured in serial blood samples taken at appropriate intervals. RIA has also found numerous uses in other clinical and experimental disciplines, such as cell biology, immunochemistry, tumor immunology, pharmaceutical research, and blood banking.

In this chapter we describe general procedures for the production of RIA antisera, methods most commonly used to radiolabel antigens, experimental conditions that influence the performance of the fluid-phase RIA, and recent developments in solid-phase RIA techniques.

Table 1 Substances Measurable by RIA Techniques

Substance	Sensitivity
I. Intracellular messengers	
1. cAMP, cGMP	5 fmol
2. Prostaglandins and metabolites	
a. $PGE_{1,2}$	1.0 pg
b. $PGF_{1\alpha,2\alpha}$	1.0 pg
c. $PGA_{1,2}$	1.0 pg
d. 15-Keto-$PGF_{2\alpha}$; 15-keto-PGE_2	500 pg
e. Thromboxane B_2	10 pg
II. Hypothalmic and pineal hormones	
1. Thyrotropin-releasing hormone (TRH)	5 pg
2. Gonadotropin-releasing hormone (GRH)	5 pg
3. Somatostatin (SRIF)	1.5 pg
4. Melatonin (MT)	50 pg
5. Substance P	0.5–1.0 fmol
6. Neurotens in (NT)	1–3 fmol
III. Pituitary hormones	
1. Folicle-stimulating hormone (FSH)	100 pg
2. Luteinizing hormone (LH)	100 pg
3. Prolactin (PRL)	50–150 pg
4. Growth hormone (GH)	50 pg
5. Adrenocorticotropic hormone (ACTH)	1.0 pg
6. Melanocyte-stimulating hormone (MSH)	50 pg
7. Lipotropic hormone (LPH)	1–5 pg
8. Thyrotropin (TSH)	1.0 pg
9. Oxytocin	1.0 pg
10. Vasopressin	0.1 pg
IV. Thyroid and parathyroid hormones	
1. Calcitonin (CT)	5 pg/ml
2. Thyroxine (T_4) and triiodothyronine (T_3)	100 pg
3. Parathyroid hormone (PTH)	1–5 pg
4. Thyroglobulin	
V. Renal hormones	
1. Erythropoietin	1.0 pg/ml
2. Vitamin D metabolites	
a. 25-Hydroxycalciferol (25-OHD)	250 pg
b. 1,25-Dihydroxycalciferol [1,25$(OH)_2$D]	20 pg
VI. Gastrointestinal hormones	
1. Gastrin (I and II)	1.0 pg
2. Secretin	10 pg/ml
3. Cholecystokinin (CCK)	5 pg/ml
4. Pancreozymin	5 pg/ml
5. Serotonin	100 pg
6. Gastric inhibitory polypeptide (GIH)	100 pg/ml
7. Vasoactive intestinal peptide (VIP)	1 pmol/liter
8. Motilin	100 pg/ml
9. Bombesin (gastrin releasing factor)	20 pmol/liter
10. Bile acids	5 pmol

Table 1 (continued)

Substance	Sensitivity
VII. Pancreatic hormones	
1. Insulin, proinsulin, C peptides	100 pg
2. Glucagon	5 pg
3. Human pancreatic peptide (HPP)	2–5 pg
VIII. Steroid hormones	
1. Estradiol	25 pg/ml
2. Estrone	25 pg/ml
3. Estriol	25 pg/ml
4. Urinary estriol glucuronide	250 pg
5. Progesterone	200 pg/ml
6. 20α-Dihydroprogesterone	100 pg/ml
7. Testosterone	10 pg
8. Dihydrotestosterone	10 pg
9. Aldosterole	2–3 pg
Aldolactone	6 pg
10. Deoxycorticosterone (DOC)	4 pg
11. Cortisol	5–30 pg
12. Cortisone	5–30 pg
13. Corticosterone	5–30 pg
14. Compound S	5–30 pg
15. Ecdysteroid (athropod molting hormone)	2.5 pmol
IX. Uterine and placental hormones	
1. Human chorionic gonadotropin (hCG)	50 pg
2. Human placental lactogen (hPL)	50 pg
3. Human chorionic thyrotropin (hCT)	0.5 pg/ml
4. Relaxin	30 pg
X. Vasoactive peptide hormones	
1. Renin	10 pg
2. Angiotensin I, II, and III	10 pg
3. Bradykinin	20 pg
XI. Growth factors	
1. Urogastrone-epidermal growth factor (EGF)	5–50 pg
2. Nerve growth factor (NGF)	0.5 ng/ml
3. Somatomedin A and C	100 pg
XII. Nonhormonal polypeptides	
1. Collagenase	10 pg
2. α-Fetoprotein	10 pg
3. α-1-Antitrypsin	100 pg
4. Rheumatoid factors	100 pg
XIII. Drugs	
1. Digitoxin	100 pg
2. Digoxin	100 pg
3. Morphine	100 pg
4. Ouabian	100 pg

Table 1 (continued)

Substance	Sensitivity
XIII. Drugs (continued)	
5. Antibiotics	
a. Kanamycin	100 pg
b. Gentamicin	100 pg
c. Amikacin	100 pg
d. Nebcin	100 pg
XIV. Other substances	
1. Folates (^{5}N-methyltetrahydrofolate)	1.0 pg
2. Anti-DNA antibodies	———
3. Carcinoembryonic antigen (CEA)	———
4. Fibrinopeptide A	1.0 pg
5. Hepatitis B antigen	0.1 pg
6. IgE-serum (RAST test)	1 pg
7. Intrinsic factor	———
8. Vitamin B_{12}	25 pg

II. PRODUCTION OF REFERENCE ANTISERA

Traditional methods can be used to prepare reference antisera to most macromolecular proteins, polysaccharides, polyamino acids, nucleic acids, and synthetic polymers of clinical or research importance (Garvey et al., 1977). Although such antigens usually contain more than one antigenic determinant and are thus capable of eliciting a heterospecific antibody response (Karush, 1962), hyperimmune reference antisera can be produced that contain a predominant species of high-affinity antibody for a single antigenic determinant (Steiner and Eisen, 1967). The avidity of such an antiserum approximates the affinity of the predominant antibody. On the other hand, severe antibody heterogenicity has also provided the impetus to develop new methods for the production of monospecific antibodies.

One development in recent years, for example, has been the extension of RIA methods in hapten-specific assays. Although small molecules of molecular weight less than 10,000 (i.e., haptens) do not usually elicit the production of antibodies that can be used in RIA procedures, the immunogenicity of these substances may be enhanced by covalently attaching them to large immunogenic carrier molecules such as bovine serum albumin or keyhole limpet hemocyanin. Antisera to these conjugated preparations will contain a mixed population of antibody specificities, but only those directed against the hapten will be involved in binding reactions with free hapten.

Several techniques used to prepare antigenic hapten-carrier conjugates with hapten molecules that possess functional carboxyl (R-COOH), amino (R-NH2), hydroxyl (R-OH), and carbonyl (R-C=O) groups have been described (Jaffe and Behrman, 1979). Further details regarding the preparation of purine and pyrimidine protein conjugates (Beiser et al., 1968), nucleoside-specific synthetic antigens (Sela and Ungar-Waron, 1968), antigenic vitamin and coenzyme derivatives (Jaton and Ungar-Waron, 1970), carbohydrate antigens (Gray, 1978), cyclic nucleosides (Steiner, 1974), steroids (Segre et al., 1975), and postaglandins (Gutierrez-Cernosek et al., 1975) have also been reported. With

larger hapten molecules (MW>5,000), novel synthetic approaches to obtain specific antibodies have been taken. In the case of parathyroid hormone (MW = 9,500), small peptides with amino acid sequences identical to those contained in several portions of the hormone were synthesized by Segre and co-workers (1975). Animals immunized with these peptide-carrier conjugates produced antibodies that were able to bind to defined areas of the hormone.

Most recent advances in RIA procedures involve the use of monoclonal antibodies whose specificities are directed toward a single antigenic determinant (for a review, see Haber, 1982). Such antibodies are homogeneous with respect to antibody specificity and affinity. They are therefore ideal immunoreagents in hapten-specific radioimmunoassays.

III. PREPARATION OF RADIOLABELED ANTIGENS

Antigens may be radioactively labeled for radioimmunoassay procedures by several selected methods involving the incorporation of different radioactive nucleides. The isotopes most commonly used are tritium (^3H), iodine-125 or iodine-131 (^{125}I, ^{131}I), and carbon-14 (^{14}C). The relatively short half-life of ^{131}I and the low specific activities of ^{14}C-labeled or ^3H-labeled antigens generally limit their practical use in radioimmunoassays. The various properties of the radioisotopes are given in Table 2.

The relative advantage of using radioiodine in antigen-labeling procedures may be seen from the following estimation. One atom of ^{131}I, substituted into an insulin molecule, will generate approximately 200 times as many radioactive disintegrations per unit time as would be produced if all 263 carbon atoms in the molecule were labeled with ^{14}C; about 600 atoms of ^3H must be incorporated into the same molecule to equal the specific activity produced by one atom of ^{131}I. Therefore, the radioisotope of choice for most RIA procedures, is either ^{131}I or ^{125}I. If the antigen to be labeled is stable over long periods of time, the longer half-life of ^{125}I is usually preferred in RIA procedures.

A. Labeling with Isotopes Other Than Iodine

Antigens and low-molecular-weight haptens that are labeled with tritium (^3H) are usually prepared by custom synthesis or purchased from commercial laboratories. Although such compounds are stable for very long periods of time, the low specific activity of these preparations limits their overall usefulness. Tritiated substances, however, are essentially identical to the unlabeled molecule being measured in RIA procedures. Tritium is also more effective for labeling certain compounds, such as steroid hormones or drugs, than other radioisotopes.

Table 2 Properties of Radioisotopes Used in RIA Procedures

	^3H	^{125}I	^{131}I	^{14}C
Half-life, $t^{1/2}$	12.3 yr	60 days	8 days	5730 yr
Radiation type	β^-	γ	γ, β^-	β^-
Average energy (MEV) per disintegration	0.0056	0.021	0.180	0.049
Specific activity	30 μCi/ milliatom	2200 Ci/ milliatom	16,000 Ci/ milliatom	62 mCi/ milliatom

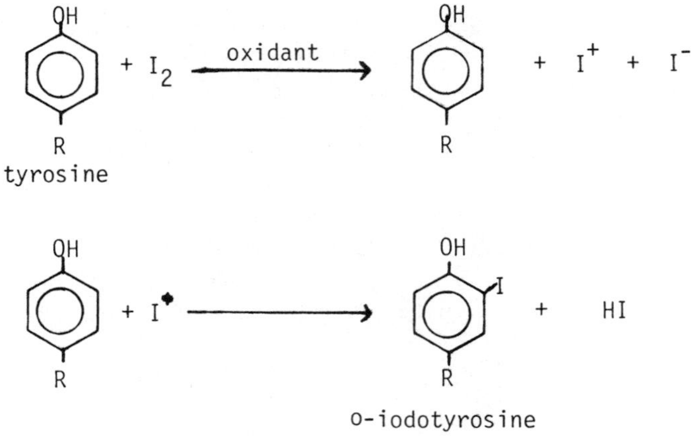

Figure 1 General reaction scheme for the iodination of the phenolic ring structure of tyrosine residues.

^{14}C nucleides may be introduced into antigens by use of the reagent [^{14}C]acetic anhydride. This reagent was used to prepare [^{14}C]acetyl bovine growth hormone preparations (Collipp et al., 1965). Labeled preparations of the hormone were reported to be biologically and antigenically similar to unlabeled bovine growth hormone material. The specificity of this reaction appears to involve the introduction of [^{14}C]acetyl groups at the α-amino terminals and lysine residues, with some reactivity of the reagent at serine, threonine, and tyrosine. As discussed previously, the low specific activity of ^{14}C limits its practical use in radioimmunoassay measurements.

B. Labeling with Radioiodine

The chemistry of radioiodination is essentially a substitution reaction of iodine into the phenolic ring structure of tyrosine, and to a lesser degree, some histidine residues. Under varying conditions, a reaction of iodine with sulfhydryl groups and tryptophan may occur. The reader is referred to a review by Hughes regarding the chemistry of iodination (1957). The conditions of the iodination substitution of tyrosine groups are selected so as to favor this reaction. This is usually achieved by using buffers of low ionic strength and a pH on the alkaline side of neutrality (pH 7.5). An illustration of the incorporation of iodide, produced by oxidation of iodine, into tyrosine groups of a protein or peptide is given in Fig. 1. A detailed study of conditions that influence the iodination reaction of proteins is discussed by Glover et al. (1967).

The introduction of ^{125}I or ^{131}I isotopes is usually carried out by either chemical (e.g., persulfates, iodates, nitrites, chloramine T), or enzymatic (e.g., lactoperoxidase) oxidation of iodine (I_2) to a reactive iodide form (I^- or I^+). In the case of the reaction described in Fig. 1, a maximum theoretical utilization of half of the available iodine can be expected. McFarlane (1958) introduced the use of iodine monochloride to facilitate a maximum incorporation of 100% of the available iodine, as seen in Fig. 2.

In recent years, three methods of radioiodination to high specific activities have been described: the conjugation-labeling method, enzymatic iodination with lactoperoxidase, and the chloramine T procedure of Hunter and Greenwood (1962). The last procedure is used extensively and will, therefore, be presented in detail.

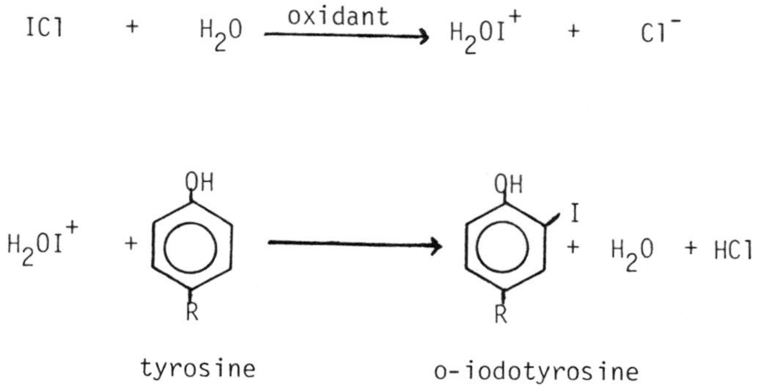

Figure 2 Iodination of tyrosine residues by iodine monochloride.

Chloramine T Labeling

Chloramine T is the sodium salt of N-monochloro-p-toluenesulfonamide. This reagent yields a slow release of hypochlorous acid, a mild oxidizing agent, in aqueous solutions. When chloramine T is added to an alkaline solution of iodine (usually as $Na^{125}I$) cationic iodine ($^{125}I^+$) is formed and spontaneously incorporated into tyrosine, as shown in Fig. 3. Iodination is carried out in aqueous solutions, under mild conditions (pH 7.5), with a very low concentration of reagents. Very small quantities of protein or peptides (10 to 25 μg) may be iodinated and subsequently separated from free iodide and other low-molecular-weight reactants by gel filtration on cross-linked dextrans (Sephadex) or agarose (BioGel).

For conventional iodination at pH 7.5, the percent incorporation of iodine is proportional to the concentration of chloramine T, the tyrosine content of the protein and the $\log_{10}$ of the protein concentration. It is important to note that all the tyrosine residues in a protein or peptide may not iodinate with equal incorporation, since the extent of iodine substitution using the chloramine T method may be dependent on the surface accessibility or exposure of the tyrosine groups.

When attempts are made to label antigens to high specific activities, certain problems may arise that can lead to a loss of immunoreactivity of the labeled antigen in RIA procedures. This loss of immunoreactivity may be due to one or more of the following causes: (1) alteration of antigenicity due to the substitution of iodine for hydogen in the structure of the proteins, (2) endogenous radiation damage to the antigen, (3) chemical

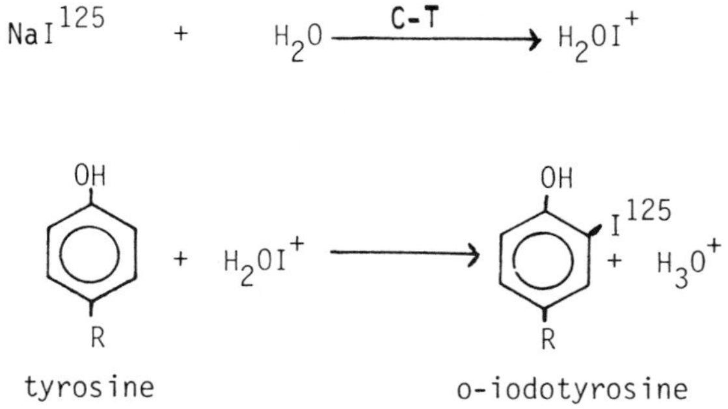

Figure 3 Iodination of tyrosine residues by the chloramine T (C-T) method.

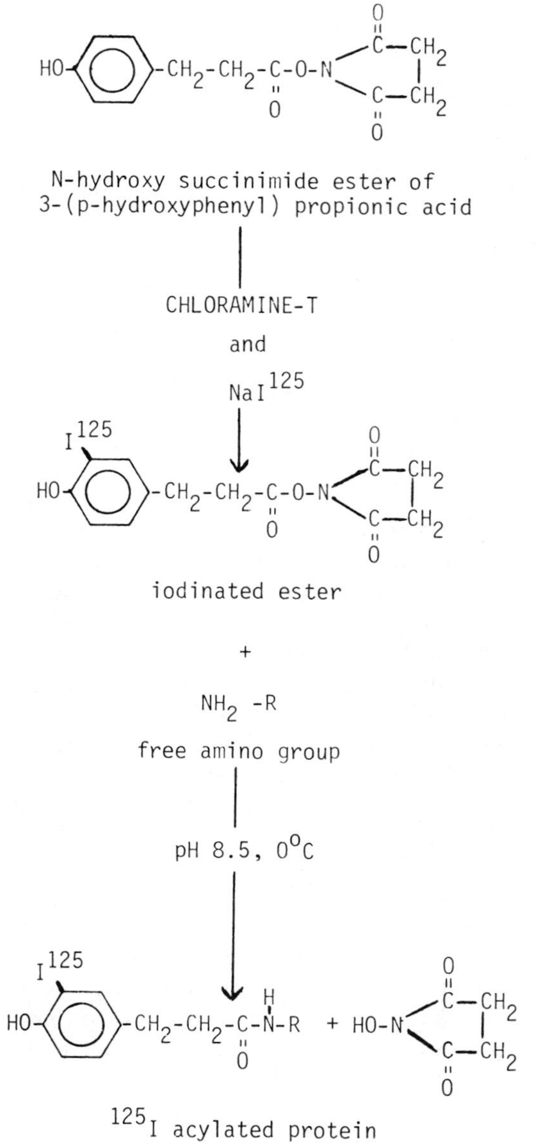

Figure 4 Reaction scheme for the conjugation labeling of iodine in proteins. The symbol NH_2-R represents free ϵ-amino groups contained in lysine side chains or free N-terminal α-amino groups of the protein.

damage to antigens caused by the reagents required to oxidize iodide to iodine, and (4) chemical damage to the antigen caused by noxious agents (i.e., polymerized iodine) present in the radiochemical reagents.

Conjugation Labeling

An alternative method, described as conjugation radioiodination, was developed by Bolton and Hunter (1973) to avoid iodination damage caused by the direct chloramine T

iodination procedure. In this method, radioiodine is first incorporated by the chloramine T method into the N-hydroxysuccinimide ester of 3(p-hydroxyphenyl) propionic acid. The iodinated acylating agent is then extracted into benzene and subsequently dried. A spontaneous condensation reaction and peptide linkage of the iodinated acyl group to amino groups (ϵ-NH_2 or α-NH_2) results when the dried iodinated reagent is added to an aqueous solution of protein or antigen. The reaction scheme for conjugation labeling is illustrated in Fig. 4.

The Bolton-Hunter method of iodination offers several advantages over the chloramine T procedure: (1) the [^{125}I] succinimide ester reagent is an excellent alternative when proteins lack tyrosine or when tyrosine iodination reduces immunoreactivity or biological activity of the antigen; (2) there is essentially no incorporation of iodine into histidine residues; and (3) the protein or antigen is not directly exposed to the oxidizing agents or noxious chemicals present in the radiolabeling reaction. A serious disadvantage of the conjugation-labeling procedure is the possible acylation of essential lysine residues that may be required in antigen–antibody binding.

The iodinated N-succinimdyl propionate ester was the first conjugation reagent to be used to prepare a great variety of labeled antigens for radioimmunoassay procedures (Langone, 1980). Similar reagents, however, have been synthesized from methyl p-hydroxybenzimidate or diazotized aniline and used to iodinate proteins, but not specifically for use in radioimmunoassays.

Peroxidase-Catalyzed Iodination

Peroxidase-catalyzed iodination of proteins with the enzyme lactoperoxidase has several advantages over the chemical methods used to iodinate proteins and antigens. Peroxidase iodination has fewer side reactions because the oxidant peroxide can be used in very low concentrations. Lactoperoxidase will iodinate only those tyrosine residues that can form an enzyme-substrate complex with the peroxidase (i.e., only those residues exposed on the three-dimensional surface of the protein). Iodination by this technique does not in general readily perturb the three-dimensional structure of the protein and hence the biological activities of the molecule will be preserved.

Peroxidase iodination procedures are usually easily controlled and the incorporation of iodine into the molecule may be readily followed by employing an iodine-sensitive electrode or spectrophotometric techniques. The chemistry of the reaction involves oxidation of iodide ($Na^{125}I$) to cationic iodine (I^+) by the enzyme in the presence of hydrogen peroxide (H_2O_2) followed by the spontaneous incorporation of iodine into tyrosine, and to a lesser extent, histidine residues. Removal of the labeled material from free ^{125}I and lactoperoxidase may be accomplished by gel filtration, dialysis, or ion exchange chromatography procedures.

IV. FLUID-PHASE RIA

In a fluid-phase RIA, antibody and radiolabeled antigen are allowed to interact in a liquid buffer system to form soluble antibody–antigen complexes that exist in a state of equilibrium with the free forms of the two reactants. The complexes are subsequently recovered from the system by techniques that do not disturb the equilibrium position of the system. The amount of radioactivity measured in these complexes is proportional to the amount of antigen bound to antibody. Results obtained in this manner can therefore be used to study antibody–antigen interactions or, when unlabeled antigen is used to inhibit the reaction, to quantify unknown concentrations of antigen in other solutions.

A. Standard Calibration Curve

In general, when a fluid-phase RIA is used to measure antigen concentrations, a standard calibration curve is produced by reacting a fixed amount of radiolabeled antigen (Ag*) and increasing amount of unlabeled antigen (Ag) with a limited amount of antigen-specific antibody (Ab) to form antibody–antigen complexes (either Ab•Ag or Ab•Ag*). Assuming that the reaction energies for labeled and unlabeled antigen are the same,

$$\mathrm{Ab} + \begin{cases} \mathrm{Ag} \rightleftharpoons \mathrm{Ab{\cdot}Ag} + \mathrm{Ag^*} \\ \mathrm{Ag^*} \rightleftharpoons \mathrm{Ab{\cdot}Ag^*} + \mathrm{Ag} \end{cases} \tag{1}$$

Eq. (1) shows that increasing the quantity of unlabeled antigen of known concentration results in a dose-dependent displacement of bound labeled antigen to the unbound fraction. Since this decrease in bound radioactive antigen is proportional to the amount of unlabeled antigen added, the concentration of the competing antigen in the sample can be determined from the standard calibration curve. This is accomplished by correlating that fraction of labeled antigen bound in the presence of sample antigen to that fraction bound in the presence of an equivalent amount of the standard antigen.

B. Optimization of the Assay

In a fluid-phase RIA, predetermined concentrations of the reference antiserum and antigen are used to optimize the sensitivity and maximize the range of precision of the assay. The antigen–antibody interaction, therefore, must be well defined in terms of the extent of the reaction (equilibrium constant) and the relative homogeneity of the reactants.

Sensitivity

Sensivitiy is the detection limit of the assay and has been defined (Ekins et al., 1968) as that concentration of antigen which can be distinguished from zero concentration with a stated degree of probability. Sensitivity is therefore related to the concentration of unlabeled antigen at which the fraction of bound labeled antigen changes as the unlabeled antigen concentration is increased. Sensitivity is also an expression of the titer, affinity, and specificity of the reference antibody used in the RIA. As such, it can be affected by possible differences in antibody affinity for unlabeled and labeled antigen, heterogeneity of the sample antigen, iodination or incubation damage to the labeled antigen, and the presence of cross-reactive antigens or interfering substances or conditions in the test sample, as well as separation artifacts generated by experimental technique. It is important, therefore, that the reference antigen and antisera be homogeneous and used at concentrations that will optimize the sensitivity and precision of the assay. These concentrations must be determined empirically for each standard antibody and antigen by means of preliminary titration studies.

Precision

The precision of an RIA can be defined as the reproducibility of the results obtained in that assay. It is a measure of the errors that arise during the performance of the assay. These errors can be classified as either determinate or indeterminate in origin. Determinate errors, such as those that may arise from the improper use of equipment, can be

measured and usually corrected. Indeterminate errors, which result from extending a system of measurement to its maximum, cannot be identified but fluctuate in a random manner that produces a scatter of results for replicate measurements. They are described by statistical methods and can be controlled to some extent by increasing the number of replicate samples in order that the experimental mean value for counts bound will fall, within a certain degree of probability, within a smaller predetermined range around the true mean value. Although detailed treatment of statistical methods used to evaluate the performance of a given RIA procedure is beyond the scope of this chapter (for details, see Rodbard and Catt, 1972), it is important to note that, within the detection limits of a given experimental design, there is an inverse relationship between the sensitivity of an RIA (increased by using a low concentration of reference antibody) and the precision and working range of the RIA (which are improved by increasing the concentration of the reference antibody). In practice, a reasonable degree of sensitivity and precision within an adequate working range is obtained by using that concentration of reference antibody that will bind approximately 50% of labeled antigen in the absence of unlabeled antigen.

Evaluations of Reaction Avidity and Heterogeneity

As stated previously, optimization of an RIA procedure requires that the interaction between antibody and antigen be well defined in terms of the equilibrium constant and heterogeneity index for that reaction. According to the law of mass action, the quantitative relationship between the two reactants at equilibrium can be expressed by the equation

$$K = \frac{k_1}{k_2} = \frac{[Ab \cdot Ag]}{[Ab]\ [Ag]} \tag{2}$$

where K is the equilibrium constant for the reaction, k_1 the association constant, k_2 the dissociation constant, Ab·Ag represents combined antibody-antigen complementation units, Ab represents free and unbound univalent antibody combining sites, and Ag represents free and unbound monovalent antigen determinants; see Chap. 16.

The magnitude of K, expressed in units of liters per mole, is a thermodynamic measure of the extent to which the reaction will proceed to form stable Ab·Ag complexes. K can also be defined in terms of the affinity expressed by a given antibody for a particular antigen. A high-affinity antibody therefore is one that forms a strong bond with its complementary antigen to give an Ab·Ag complex with a low tendency to dissociate. A low-affinity antibody, on the other hand, forms a complex that requires less energy to dissociate.

The affinity constant for an ideal antibody-antigen interaction can be determined by means of a direct titration study in which a constant amount of the antibody is reacted with radiolabeled antigen over a range of radiolabeled antigen concentrations (Farr, 1958). Bound radiolabeled antigen is measured in such an experiment. Furthermore, it is initially assumed that each reactant is homogeneous with respect to determinants and binding sites, that the number of determinants on each antigen molecule (antigen valence) is uniform, and that the number of binding sites on each antibody molecule (antibody valence) is the same. Thus for a monovalent antigen in antigen excess, Eq. (2) can be rearranged to the form

$$\frac{r}{C} = nK - rK \tag{3}$$

where $r = [Ag \cdot Ab]/[A]$ is the molar concentration of bound antigenic sites per molar concentration of total antibody molecules, [A] ; $c = [Ag]$ the molar concentration of free antigen molecules; and n the antibody valence. By plotting r/C against r, as shown in Fig. 1A, Chap. 16, a single curve called a Scatchard plot (Scatchard, 1949) is obtained whose slope defines the affinity constant for the reaction (see Chap. 16).

A Scatchard plot should be linear if equilibrium disturbances are avoided when bound antigen is separated from free antigen and if the initial assumptions regarding antibody and antigen homogeneity are good approximations of the true reactant conditions. An additional test of reactant homogeneity is afforded by a logarithmic variation (Nisonoff and Pressman, 1958a,b; Karush, 1962) of the Sips (1948) equation derived from Eq. (3):

$$\log \frac{r}{n - r} = a \log K + a \log C \tag{4}$$

where a is the index of heterogeneity. By plotting log [r/(n - r)] against log C, as illustrated in Fig. 1C, Chap. 16, one obtains the best linear fit and calculates from its slope the heterogeneity index. When the reactants are homogeneous, the index a = 1 (see Chap. 16). Such a degree of homogeneity may be expressed by a monoclonal antibody or by an antiserum prepared against a simple hapten of a monodeterminate and univalent antigen. Relative homogeneity may also be expressed by an antiserum to a multideterminant antigen if the antiserum consists of a predominant species of high-affinity antibody. On the other hand, a heterogeneity index considerably less than 1 may be encountered when a suboptimal immunization with a multideterminant antigen results in the production of a heterospecific antisera containing various amounts of mixed-affinity antibodies. Because a number of very complex equilibria and stoichiometries may exist under such conditions, the term *avidity*, rather than affinity (see Chap. 16), should be used to describe the overall interaction of this antiserum with its antigen. Furthermore, although various mathematical transformations may serve to correct this situation in terms of linearization of RIA standard curves, severe expression of heterogeneity warrants the rejection of such an antiserum as a potential reference reagent in most RIA studies.

Experimental Conditions

Other experimental conditions can also influence the performance of an RIA. Should the binding energy of the reference antibody be temperature dependent, for example, the sensitivity of the RIA can be controlled by incubating the reagents at that temperature with the lowest energy of reaction. Lower temperatures, however, often require prolonged equilibration times during which rigid adherence to the reaction temperature must be maintained to prevent a significant deterioration in assay precision.

Reaction volumes may affect the performance of fluid-phase RIAs. An increase in diluent volume can decrease the slope or the standard curve and shift it vertically so as to decrease the detection limit of the RIA (see Chap. 16). Therefore, reaction volumes are usually minimized at a fixed value so as to obtain optimal assay sensitivity and precision.

Other experimental conditions can be selected so as to favor the binding of antigen to antibody and to assure the stability of all the reagents used during subsequent incubation procedures. Fluid-phase RIAs are normally performed in aqueous isotonic buffers at a pH near 7.2. Suitable carrier proteins can be added to the diluent buffers to prevent serious losses of antigen or antibody by adsorption to glass or plastic vessels. This protein

Table 3 Separation Procedures Used in Radioimmunoassay

I.	Physicochemical methods (see also Chap. 13)
	Electrophoresis (cellulose acetate, 1% starch gel, 5% polyacrylamide)
	Molecular sieving (gel filtration, microfiltration, gel centrifugation, dialysis)
	Absorption (charcoal–dextran coated, hydroxyapatite, zirconyl phosphate gel, silicates)
	Ion exchange (Whatman DEAE[a] and CM[b] ion exchange paper and resins, Dowex, Amberlite, Biorad ion exchange resins)
	Fractional precipitation (NaCl: ethanol, dioxane, sodium sulfite, trichloroacetic acid, 4 M ammonium sulfate, 20% polyethylene glycol)
II.	Immunological methods
	Double- or second-antibody precipitation (postprecipitation; primary antibody preprecipitation; nonimmune globulin preprecipitation)
	Solid-phase antibody precipitation (adsorption of antibody to particles, tubes, disks, or covalent attachment to dextran)

[a]Diethylaminoethyl.
[b]Carboxymethyl.

carrier should not cross-react with the reference antisera. In addition, the specific activity and stability of the radiolabeled antigen should be monitored routinely since it has a limited useful life span and is subject to the destabilizing influences of its diluent.

C. Separation of Antibody-Bound Antigen from Free Antigen

In standard fluid-phase RIA procedures, the binding of labeled antigen with specific antibody is the only reaction that is measured. The measurement of this binding is feasible provided that a means is available for separating the antibody-antigen complex from unbound antigen. To achieve separation, procedures (either physicochemical or immunological) that select for difference in properties between the bound and unbound components are used (see Table 3).

The methods of separation to be considered must be determined on the basis of the following factors: (1) the efficiency of separating free or unbound Ag* from antibody bound Ag*, (2) the practicality of rapidly separating several hundreds of samples in a routine assay, (3) the volume of the fluid-phase system, (4) the disruptive effects that a particular separation procedure may have upon the antigen-antibody complex itself, and (5) the possible distortions of equilibrium positions between antigen and antibody during removal of the complexes from the system.

The physicochemical methods and immunological techniques that can be used for separating antibody-bound antigen from unbound antigen are described in the following sections.

Physicochemical Methods

In general, most antibody preparations contain proteins that are larger than the bound antigen, especially when the antigen is a hormone. The physicochemical separation of the

antibody-bound antigen complex from free antigen will therefore approximate the physicochemical behavior of the antibody molecule itself.

Separation by electrophoretic methods. The separation of bound from free antigen may be obtained by electrophoretic techniques on paper, cellulose acetate, starch, or polyacrylamide gels. Cellulose acetate electrophoresis is usually too expensive and tedious for routine use in RIA.

Electrophoresis on 1% starch or 5% polyacrylamide gels has been used to a larger extent. It offers a high resolving power between bound and free antigen components due to its ability to separate molecules on the basis of charge and size. In the case of polyacrylamide electrophoresis, the antibody-antigen complex enters the gel, but due to its large size and small net charge, it migrates slowly, while the free antigen moves with a more rapid migration. However, these electrophoretic procedures are generally too elaborate to be used routinely (see also Chap. 13).

Separation by molecular size differences. These techniques exploit the difference in size between the smaller free antigen and the antibody–antigen complex. Several of these methods are described below (see also Chap. 13).

Gel filtration (e.g., on Sephadex or BioGel) offers a routine method for the separation of free antigen from its antibody-bound form provided that the antigen has a molecular weight of less than 100,000. The disadvantages of this method are the nonspecific adsorption of free antigen to the dextran matrix and the possible disruption of the antigen–antibody complex in hapten-specific assays. Microfiltration through glass fiber filters supported by disposable microfilters has provided a rapid and efficient means of separating large numbers of samples in immunoassay procedures (Chalkley and Renshaw, 1980).

Ultracentrifugation techniques have been used in assays demonstrating the presence of insulin; however, this procedure is too cumbersome for repetitive work in routine assay systems. Another procedure, described as gel centrifugation through microcolumns of Sephadex G-25 (Larsen and Rasmussen, 1980), has found widespread use in assays requiring the separation of antibody-bound steroids from free steroid antigen. The practicality of this method allows a 95% efficient separation of over 100 samples in 1 hr with presumably no disruption of the steroid-antibody complex.

Adsorption systems. Adsorption systems consist of material that contains irregular surfaces and charged molecules. These materials are added to the incubates as powders or slurries in order to adsorb free but not bound antigen. The nonadsorbed antigen-bound complexes are then separated from the adsorbent by subsequent centrifugation.

Charcoal has been widely employed to separate antibody-bound antigen from nonbound antigen in several types of immunoassay systems (e.g., vitamin B_{12}, insulin, steroids, prostaglandins). The practical use of this agent has been reviewed in detail (Odell, 1980). In general, charcoal has irregularities or "in-pocketings" on its surface. The sizes of these "in-pocketings" varies so that the amount of surface area accessible to a given protein or antigen depends on the size, shape, and charge of the antigen molecule. Charcoal may be used effectively if the antigen is small relative to the antibody–antigen complex. For antigens equaling or exceeding the immunoglobulin in size, the size-charge differences between free antigen and antigen-bound antibody are too small for charcoal to be used effectively. The use of charcoal in a particular assay system requires considerable investigation, since this material is an avid adsorbent that will take up the bound as

well as the free antigen moieties under various conditions. Excellent separation systems have been devised, however, where the charcoal is first coated with dextran or a carrier protein (Herbert et al., 1965). In addition to charcoal, other adsorbents, such as hydroxyapatite and zirconyl phosphate, have been used successfully in immunoassay procedures.

In general, all adsorbent assay procedures must take into account the following variables: (1) the amount of adsorbent to be added to the incubates (determined by a dose-response adsorption curve), (2) the nature and amount of coating substance on the adsorbent, (3) the ionic species involved, (4) the concentration and pH of the buffer system to be used, and (5) the time course of the adsorption.

Separation by ion exchange. Ion exchange papers and resins (DE-52, DEAE-Sephacel, CM-Sephadex) have been used to separate free antigen moieties in incubates in the same way that adsorbents have been employed. However, this method has not been widely used because considerable experimentation is required before a specific ion exchange material may be used in a single immunoassay system.

Separation by fractional precipitation. The precipitation of proteins with organic solvents and salts was one of the earliest methods devised for the separation of free and bound antigen in RIA procedures. In these systems, the concentration of the precipitating agent is usually adjusted so as to precipitate the bound antigen while leaving the unbound fraction in solution. A variety of agents, such as ethanol, sodium chloride, dioxane, sodium sulfite, trichloracetic acid, ammonium sulfate, and polyethylene glycol, have been used successfully in fractional precipitation methods (see also Chap. 13). Highly nonpolar solvents have generally been avoided because of their dissociative effects upon antigen antibody complexes (see Chap. 16).

Ammonium sulfate (4 M solution) and polyethylene glycol (20% w/v) have been used extensively in several salt precipitation systems, and their use is more thoroughly discussed by Chard (1980); see Chap. 13. The use of salt precipitation methods in immunoassay procedures is desirable because of their speed, simplicity, reproducibility, and low cost, and the conditions that separate the two component systems are usually determined without much difficulty. The salting-out systems, however, tend to be unsuccessful when used to separate bound from free antigen at the nanogram to picogram per milliliter levels. This limitation appears to be due to the nonspecific trapping or absorption of the free antigen to the precipitate latice, which is designed to contain only the bound antigen fraction.

Immunological Methods

Antibody-bound antigen may be separated effectively from unbound antigen by immunological methods. These methods are widely utilized in radioimmunoassays because of their great applicability to more types of RIA than any other separation procedure devised. They can be used with any ligand or antigen that is able to be bound by an antibody. The two most common groups of immunological separation techniques currently used in RIA procedures are described in the following sections.

Double-antibody precipitation. The double-antibody method is a general method first described by Morgan and Lazarow (1962). This technique is based on the finding that most of the antigenic sites of an immunoglobulin molecule are distinct from its antigen binding sites. An antibody, therefore, may form a complex with antigen and then itself be complexed to a second antibody. In conventional radioimmunoassay procedures,

the primary antigen–antibody complex is too dilute to be precipitated. When a carrier nonimmune serum of the same species as the first antibody is added followed by the addition of a second antibody against the primary antibody, a sizable precipitation latice, including the labeled antigen bound to the first antibody, is developed. Separation of the bound component in the precipitate is then achieved by centrifugation.

A single experiment is generally sufficient to optimize a double-antibody system with a given second antiserum, and once optimized for one antigen, identical conditions may be suitable for a large number of other antigens provided that the first antibody is raised in the same species. Most double-antibody precipitations are complete by 16 hr at 4°C, and generally give excellent separations (95%) of bound labeled antigen from free antigen. In general, most double-antibody systems are able to handle any convenient volume of incubate, and they are well suited to the gentle, simple, and rapid separation of a very large number of samples. Double-antibody methods, therefore, are used to obtain rapid separations with excellent specificity, sensitivity, precision, and reproducibility. The reader is referred to the review by Hunter (1977) for exhaustive details on the double-antibody precipitation method.

The various methods of utilizing a second antibody for the purpose of antibody-bound antigen separation may be divided into three types of procedures: double-antibody postprecipitation (addition of second antibody after incubation of primary antibody with antigen in the assay mixture), primary antibody preprecipitation (in which a precipitate of the first and second antibodies is added to tubes containing antigen), and nonimmune globulin preprecipitation (in which a precipitate of the second antibody and nonimmune globulin of the same species as the primary antibody is added to assay tubes containing the primary antibody and antigen). Although the first procedure (postprecipitation) is the most common method, the last method combines the advantages of postprecipitation and double-antibody solid-phase precipitation and appears to be the procedure of choice.

Double-antibody solid-phase separation. In the double-antibody solid-phase technique developed by den Hollander and Schuurs (1971), the second antibody, raised to the primary antibody of the system, is linked to a solid support medium (Sepharose) and subsequently added as a suspension to an incubate containing primary antibody and antigen, as illustrated in Fig. 5. The specifics of this procedure are identical to the double-antibody postprecipitation method described above.

D. Data Presentation

RIA standard curves are most often plotted by means of least-squares linear regression techniques using untransformed or transformed data (Meinert and McHugh, 1968). As shown in Fig. 6, the quantity of bound radiolabeled antigen in an arithmetic plot is often represented as a percentage of the amount bound in the absence of unlabeled antigen. Only the linear part of such a calibration curve, however, can be used to estimate the unknown concentration of an antigen in a biological sample. To simplify what is often a curvilinear relationship to a linear one, arithmetic RIA data are often transformed so that the proportion of the variance of the dependent variable (fraction of labeled antigen bound) explained by the independent variable (unlabeled antigen concentration) is generally increased, while the distribution of replicate sample points around the regression line tends to become normal with a common variance.

Logarithmic transformations are frequently used to linearize RIA standard curves (Rodbard and Hutt, 1974). These transformations may be applied to the controlled

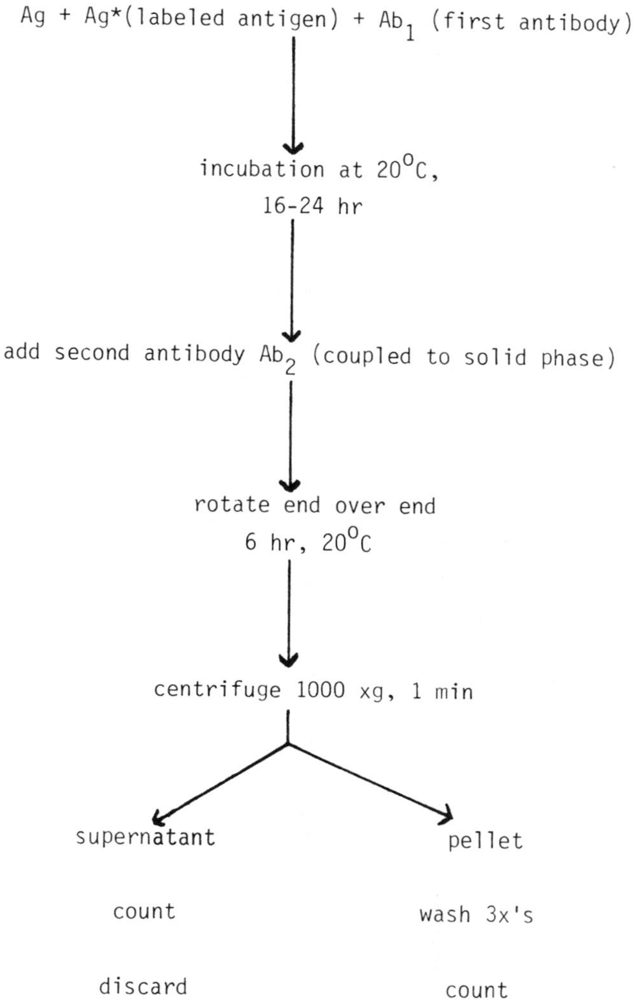

Figure 5 General scheme for the double-antibody solid-phase precipitation method. Illustrated is a competitive inhibition system for the quantitative determination of antigen. Labeled antigen (Ag*) and solid-phase coupled antibodies (Ab) are in a fixed concentration so that only 20 to 30% of the total labeled antigen is bound in the absence of unlabeled antigen. (From Twining and Atassi, 1979.)

values of unlabeled antigen concentrations as well as to the observed values for the corresponding bound radiolabeled antigen. Logarithmic transformation of the independent variable is effective when proportional changes in the independent variable produce linear responses in the dependent variable. Figure 6 illustrates how logarithmic transformation of the independent variable may affect the regression line. For computations, the unlabeled antigen concentrations are transformed into logarithms.

Logarithmic transformations of both variables are frequently used to linearize the standard curve when the antigen-antibody interaction is described by an experimental

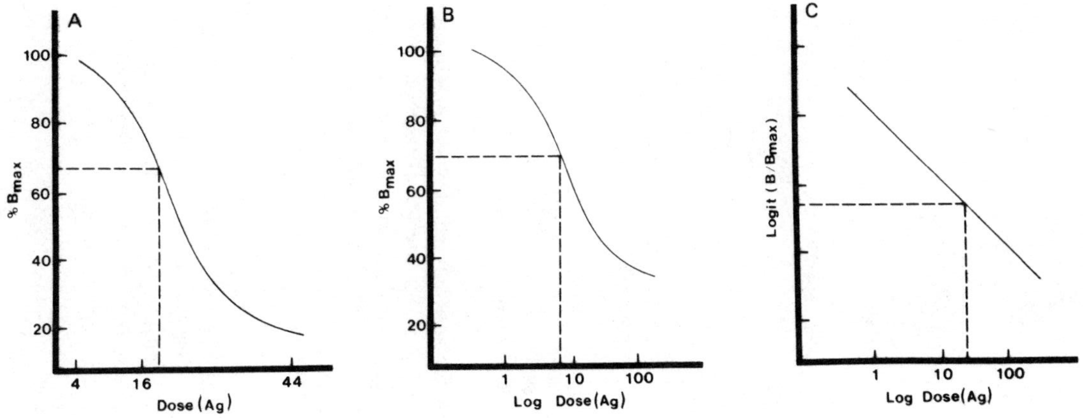

Figure 6 Idealized RIA standard inhibition curves showing progressive decreases in the amount of radiolabeled antigen bound, B, with increasing concentrations of unlabeled antigen, Ag. In (A) and (B), the amount of bound radiolabeled antigen is expressed as a percentage of that amount of radiolabeled antigen bound in the absence of Ag (B_{max}). In (A), Ag concentrations are plotted on an arithmetic scale. In (B), Ag concentrations are plotted on a logarithmic scale. The logit-log plot is illustrated in (C), where logit $(B/B_{max}) = \log_e [(B/B_{max})/(1 - B/B_{max})]$. The linear portions of these calibration curves can be used to quantify unknown concentrations of antigen in various sample preparations.

function. This situation may arise when the major binding species of the reference antisera is contaminated with significant amounts of low-affinity antibodies. The logit-log transformation, as shown in Fig. 6, often corrects for the nonlinear effects of antibody heterogeneity. Other weighted and unweighted data transformations have been described (Rodgers, 1974), and with the advent of small computers the use of these transformations is becoming more popular.

V. SOLID-PHASE RIA

Recent advances in solid-phase matrix technology has led to the development of extremely sensitive and convenient RIA procedures that can be used to measure unknown concentrations of antigen, to determine antibody-antigen equilibrium constants, and to detect as well as to characterize both serum antibodies produced by traditional methods and monoclonal antibodies secreted by hybridoma cell lines.

When a solid-phase method is used to quantify an antigen (Wide, 1971), the primary antibody is covalently linked to a solid support (such as cellulose, cross-linked dextran, or cross-linked agarose) that is not soluble in assay buffers or wash reagents. Increasing amounts of the unlabeled antigen are incubated with the immobilized antibody in the presence of a constant amount of radiolabeled antigen until a state of equilibrium has been attained. The solid and liquid phases are then separated, and the amounts of radioactivity in each phase are measured. The amount detected on the solid support corresponds to the amount of radiolabeled antigen bound to the immobilized antibody. The amount in the liquid phase corresponds to that of free radiolabeled antigen. Because unlabeled antigen interferes with antibody binding of labeled antigen in a manner that is inversely proportional to the concentrations of unlabeled antigen, a calibration curve may be constructed that can be used to determine the concentrations of antigen in various sample preparations.

Solid-phase RIAs are also used to determine affinity constants for antigen-antibody interactions. This is accomplished by titrating a constant amount of the solid-phase antibody with decreasing amounts of the fluid-phase radiolabeled antigen over a detectable range of antigen concentrations. Once the reaction is complete and the two phases have been separated, the fractions of bound and free antigen can be determined and the results evaluated by means of a Scatchard analysis to obtain a binding constant (see Chap. 16). This equilibrium constant, however, may be substantially different from one obtained using strictly fluid-phase techniques. The solid matrix may interfere with the binding of antigen to the antibody. Chemical modifications introduced into the antibody when it is conjugated to the solid support may also alter the equilibrium constant of the reaction. The double-antibody solid-phase technique introduced by den Hollander and Schuurs (1971), however, can be used to circumvent such distortions because it allows for the formation of antibody-antigen complexes in a single fluid phase. A solid-phase secondary antibody, specific for the primary antibody, is then used to separate antibody-bound antigen from free antigen with a degree of rapidity common to most solid-phase techniques.

Another recent advance in the use of solid-phase RIA technology has been the development of the quantitative immunoadsorbent titration method established by Twining and Atassi (1979). In this procedure, antigens and small synthetic peptides comprising regions of a specific antigenic determinant are covalently coupled to a Sepharase

Table 4 Typical Binding Experiment of ^{125}I-Labeled Antibodies with Immunoadsorbents of Myoglobin and Its Synthetic Sites[a]

Immunoadsorbent	Antibody bound	
	cpm	Percent
Mb	5810	100
Site 1	1485	25.6
Site 2	1260	21.7
Site 3	990	17.0
Site 4	1020	17.6
Site 5	1040	17.9
Control peptide 1–6	0	0
Control peptide 121–127	0	0
Total for sites 1–5	5770	99.8

[a]A fixed amount of ^{125}I-labeled antibodies to myoglobin (Mb) was titrated with increasing amounts of Mb-Sepharose or peptide-Sepharose. The results represent the maximum (plateau) amounts of antibodies bound by an immunoadsorbent.
Source: Twining and Atassi (1979).

solid-phase matrix by CN-Br activation. This method has been used to study the binding of radiolabeled antibody to sperm-whale myoglobin and its five synthetic antigenic sites (Twining and Atassi, 1979), to hen egg white lysozyme and its three discontinuous antigenic regions (Atassi, 1978; Atassi and Lee, 1978), the five synthetic antigenic sites of bovine serum albumin (Atassi et al., 1979), and the antigenic regions of the α and β chains of human hemoglobin (Kazim and Atassi, 1982; McCormick and Atassi, 1983a; Atassi and McCormick, 1983). Also, the interaction of haptoglobin with human hemoglobin and its subunits and the precise localization of the haptoglobin binding site on the α chain (Kazim and Atassi, 1981; McCormick and Atassi, 1983b) has been recently investigated using the quantitative immunoadsorbent titration method.

The immunoadsorbent titration procedure offers a variation to other solid-phase RIA methods by allowing the direct coupling of antigen, rather than antibody, to the solid-phase matrix. The ability to couple antigens and synthetic peptides to adsorbents provides a simple means for quantitating the amount of antibodies in a given serum directed toward a specific antigenic region (see Table 4). Quantitative immunoadsorption also allows for the direct determination of the maximum amount of antibody (plateau value) that can bind with a given antigen adsorbent, and for the precise localization of antigenic regions by the direct binding of antibody to peptide adsorbents comprising these regions.

Briefly, the immunoadsorbent titration procedure involves the use of a fixed amount of ^{125}I-labeled antibody that is added to varying amounts of protein or peptide adsorbents. This titration of labeled antibody with the adsorbents establishes binding curves (Fig. 7) that plateau rapidly for protein molecules such as myoglobin or hemoglobin. Synthetic peptides, however, comprising various antigenic regions of the whole molecule, plateau at a much slower rate, and usually require higher volumes of adsorbents to reach plateau values of antibody binding. In addition, the net binding of antibody to peptide adsorbents is generally lower than the values observed for binding of antibody to protein adsorbents.

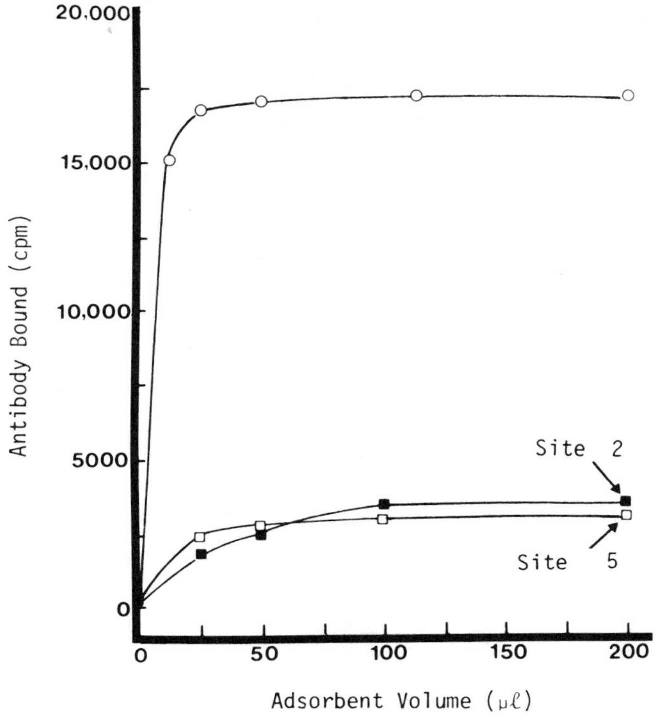

Figure 7 Example of immunoadsorbent titration studies using a fixed amount (3.6×10^5 cpm) of ^{125}I-labeled goat anti-Mb IgG (○) Mb; (□) site 5 peptide 145–151; (■) site 2 peptide 56–62. (Adapted from Twining and Atassi, 1979.)

Assay conditions for immunoadsorbent titration studies include the incubation of the antibody and the adsorbents at room temperature in isotonic buffers [e.g., phosphate buffered saline (PBS), pH 7 to 7.5] containing a carrier protein [e.g., 0.1% bovine serum albumine (BSA)] to minimize nonspecific adsorption of antibody to the vessel walls. Other conditions, including the determination of the amounts of antigen or peptide to be used for coupling, coupling parameters, antibody concentrations, and the potential limitations of the titration system, have been described (Twining and Atassi, 1979).

Another solid-phase method that is growing in popularity takes advantage of the fact that proteins and other macromolecules tend to stick spontaneously to the inner well surfaces of polyvinyl chloride microtiter plates (see Tsu and Herzenberg, 1980). This absorption is essentially irreversible in the buffered salt solutions most commonly used as aqueous-based solvents for polar protein molecules. The amount of protein that is adsorbed depends on its initial concentration, its inherent ability to bind to plastic, and on the amount of time that it is allowed to adsorb to the plastic surface at a given temperature. After the wells have been coated with the protein, however, they are usually incubated with a high concentration of a noncross-reacting protein in order to prevent nonspecific binding of subsequent assay reagents. Residual blocking protein is then removed by repeated washings with an appropriate wash reagent such as phosphate buffered saline at pH 7.2

If a reference RIA antiserum has been adsorbed onto the plate wells, solutions containing a constant amount of radiolabeled antigen and increasing amounts of unlabeled antigen can be added to the wells to initiate a binding competition reaction with the solid-phase antibody. When the reaction is complete, the wells are washed with buffer to remove residual unbound label. They are then separated and the amount of radiolabeled antigen bound to the solid-phase antibody is measured in a gamma scintillation counter. These results are used to make a standard calibration curve for determining antigen concentrations in other solutions.

If an antigen has been used to coat the plate wells, both serum and hybridoma antibodies can be detected and further characterized using modifications of the solid-phase plate binding assay. This is accomplished by incubating the solid-phase antigen with appropriate dilutions of the antibody sample using predetermined assay conditions that maximize specific antibody binding. A radiolabeled secondary antibody, specific for the primary antibody, is then used to detect binding of the primary antibody to the solid-phase antigen. If the secondary antibody is allotype or isotype specific, the primary antibody can also be classified with respect to these determinants.

Additional variations in the solid-phase plate binding assay have been used to calculate monoclonal antibody affinity constants (Frankel and Gerhard, 1979), to confirm the preselected submolecular binding specificities of monoclonal antibodies by using carrier-free synthetic peptides as immunogens (Schmitz et al., 1982), to characterize hybridoma antibodies to cellular surface determinants (Williams, 1977), to detect trace amounts of reaginic antibodies (Liu et al., 1980), and to determine antibody binding to synthetic peptides comprising antigenic sites of protein molecules such as influenza virus (Atassi and Webster, 1983). Other protocols continue to be developed to meet specific research and clinical needs.

REFERENCES

Atassi, M. Z. (1978). Immunochemistry *15*, 909.

Atassi, M. Z., and Lee, C. L. (1978). Biochem. J. *171*, 429.

Atassi, M. Z., and McCormick, D. J. (1983). Fed. Proc. *42*, 704.

Atassi, M. Z., and Webster, R. G. (1983). Proc. Natl. Acad. Sci. USA *80*, in press.

Atassi, M. Z., Sakata, S., and Kazim, A. L. (1979). Biochem. J. *179*, 327.

Beiser, S. M., Tanenbaum, S. W., and Erlanger, B. F. (1968). Methods Enzymol. *12B*, 889.

Bolton, A. E., and Hunter, W. M. (1973). Biochem. J. *133*, 529.

Chalkley, S. R. C., and Renshaw, A. (1980). Methods Enzymol. *70A*, 305.

Chard, T. (1980). Methods Enzymol. *70A*, 280.

Collipp, P. J., Kaplan, S. A., and Boyle, D. C., and Schimizic, C. S. N. (1965). J. Biol. Chem. *240*, 143.

den Hollander, F. C., and Schuurs, A. H. (1971). In *Radioimmunoassay Methods* (K. E. Kirkham and W. M. Hunter, Eds.). Williams & Wilkins, Baltimore, pp. 419–426.

Ekins, R. P., Newman, G. B., and O'Riordan, J. L. H. (1968). In *Radioisotopes in Medicine: In Vitro Studies* (R. L. Hagar, F. A. Gaswit, and B. E. P. Murphy, Eds.). U.S. Atomic Energy Commission, Oak Ridge, Tenn., p. 59.

Farr, R. S. (1958). J. Infect. Dis. *103*, 239.

Frankel, M. E., and Gerhard, W. (1979). Mol. Immunol. *16*, 101.

Garvey, J. S., Cremer, N. E., and Susdorf, D. H. (1977). In *Methods in Immunology.* W. A. Benjamin, Reading, Mass., pp. 7–38.

Glover, J. S., Salter, D. N., and Shepherd, B. P. (1967). Biochem. J. *103*, 120.

Gray, G. R. (1978). Methods Enzymol. *50*, 155.
Gutierrez-Cernosek, R. M., Levine, L., and Gjika, H. (1975). Methods Enzymol. *35*, 287.
Haber, E. (1982). In *Monoclonal Antibodies in Clinical Medicine* (A. J. McMichael and J. W. Fabre, Eds.). Academic Press, New York, pp. 477–500.
Herbert, V., Lau, K. S., Gottlieb, C. W., and Bleicher, S. J. (1965). J. Clin. Endocrinol. Metab. *25*, 1375.
Hughes, W. L. (1957). Ann. N.Y. Acad. Sci. *70*, 3.
Hunter, W. M. (1977). In *Handbook of Experimental Immunology, Immunochemistry*, Vol. 1 (D. M. Weir, Ed.). Blackwell, Oxford, pp. 14.1–14.40.
Hunter, W. M., and Greenwood, F. C. (1962). Nature (Lond.) *194*, 495.
Jaffe, B. M., and Behrman, H. R. (1979). *Methods of Hormone Radioimmunoassay*. Academic Press, New York, pp. 981–1003.
Jaton, J. C., and Ungar-Waron, H. (1970). Methods Enzymol. *18A*, 609; *18B*, 735.
Karush, F. (1962). Adv. Immunol. *2*, 1.
Kazim, A. L., and Atassi, M. Z. (1981). Biochem. J. *197*, 507.
Kazim, A. L., and Atassi, M. Z. (1982). Biochem. J. *203*, 201.
Langone, J. J. (1980). Methods Enzymol. *70A*, 226.
Larsen, J., and Rasmussen, W. F. (1980). Methods Enzymol. *70A*, 315.
Liu, F. T., Bohn, J. W., Ferry, E. L., Yamamoto, H., Molinardo, C. A., Sherman, L. A., Klinman, N. R., and Katz, D. H. (1980). J. Immunol. *124*, 2728.
McCormick, D. J., and Atassi, M. Z. (1983a). Fed. Proc. *42*, 704.
McCormick, D. J., and Atassi, M. Z. (1983b). Fed. Proc. *42*, 2001.
McFarlane, A. S. (1958). Nature (Lond.) *182*, 53.
Meinert, C. L., and McHugh, R. H. (1968). Math. Biosci. *2*, 319.
Morgan, C. R., and Lazarow, A. (1962). Proc. Soc. Exp. Biol. Med. *110*, pp. 29–32.
Nisonoff, A., and Pressman, D. (1958a). J. Immunol. *80*, 417.
Nisonoff, A., and Pressman, D. (1958b). J. Immunol. *81*, pp. 126.
Odell, W. D. (1980). Methods Enzymol. *70A*, 274.
Rodbard, D., and Catt, K. J. (1972). J. Steroid Biochem. *3*, 255.
Rodbard, D., and Hutt, D. M. (1974). In *Symposium on Radioimmunoassay and Related Procedures in Clinical Medicine and Research*. International Atomic Energy Agency, Vienna/Unipub, New York, p. 165.
Rodgers, R. C. (1974). *Radioimmunoassay Theory for Health Professionals*. Hewlett-Packard, Loveland, Colo.
Scatchard, G. (1949). Ann. N.Y. Acad. Sci. *51*, 660.
Schmitz, H. E., Atassi, H., and Atassi, M. Z. (1982). Mol. Immunol. *19*, 1699.
Segre, G. V., Tregar, G. W., and Potts, J. T. (1975). Methods Enzymol. *37*, 38.
Sela, M., and Ungar-Waron, H. (1968). Methods Enzymol. *12B*, 900.
Sips, R. (1948). J. Chem Phys. *16*, 490.
Steiner, L. A. (1974). Methods Enzymol. *38*, 96.
Steiner, L. A., and Eisen, H. N. (1967). J. Exp. Med. *126*, 1161.
Tsu, T. T., and Herzenberg, L. A. (1980). In *Selected Methods in Cellular Immunology* (B. B. Mishell and S. M. Shiigi, Eds.). pp. 373–397.
Twining, S. S., and Atassi, M. Z. (1979). J. Immunol. Methods *30*, 139.
Wide, L. (1971). In *Radioimmunoassay Methods* (K. E. Kirkham and W. M. Hunter, Eds.). Williams & Wilkins, Baltimore, pp. 405–412.
Williams, A. F. (1977). Contemp. Top. Mol. Immunol. *6*, 83.
Yalow, R. S., and Berson, S. A. (1960). J. Clin. Invest. *39*, 1157.

20

Enzyme Immunoassays

Kenneth W. Walls Centers for Disease Control, Public Health Service, U.S. Department of Health and Human Services, Atlanta, Georgia

I. INTRODUCTION

Although relatively new, enzyme immunoassays (EIAs) have had an important impact on the immunodiagnosis of infectious diseases. Preceded by two highly effective procedures, radioimmunoassays (RIAs) and fluorescent immunoassays (FIAs) EIA was intended to combine the best features of each—the sensitivity of RIA and the convenience of FIA. First described by Van Weeman and Schuurs (1971), and almost simultaneously by Engvall and Perlmann (1971), ELISA, the enzyme-linked immunosorbent assay, began a new era of immunoassay procedures. Although ELISA originally described the detection of antibody, modifications of the procedure have become numerous and now involve homogeneous and nonhomogeneous assays, direct and indirect assays, inhibition assays, assays combining RIA and EIA or FIA and EIA, and assays for antigen, antibody, complexes, or metabolites. Consequently, ELISA can no longer be used as a generic term and should be superseded by EIA-enzyme immunoassays. ELISA will still have priority in those discussions concerned with modifications of the original procedure designed to measure antibody.

For the detection of antibody, heterogeneous EIA, like its predecessors, depends on the fixation of antigen to a solid phase; reaction with patient's serum; detection with an enzyme-conjugated antiglobulin; and measurement with the appropriate substrate which is converted to a measurable product or its degradation products, which, in turn, cause a measurable reaction. Each of these stages is a complex variable, a description of which could entail an entire chapter but will be condensed here to its pertinent facts.

In many infectious diseases the detection of antigen is perhaps more critical for diagnosis than is the measurement of antibody, which may only reflect antibody persisting from an infection months or even years earlier. EIA offers a practical system for

measuring antigen fixed to tissue or circulating in body fluids. The methods vary little from those used in antibody detection except that they are used in reversed order. For use in body fluids, unlabeled anti-antigen is fixed to the solid phase, followed by the suspect fluid and then with labeled anti-antigen. For detection of antigen in tissue, only labeled anti-antigen is required, but the substrate must be reduced to an insoluble product which can usually be visualized by light or electron micrscopy.

Homogeneous assays, by definition, are those in which all reactants remain suspended throughout the measurement without a physical separation stage. Their use is limited, therefore, to competitive inhibition or stearic interference with the enzyme label. Because of these limitations, homogeneous EIA has been restricted to the measurement of metabolites and antimicrobial agents and as such, has been used extensively in clinical chemistry.

Because of their sensitivity, specificity and reactivity, enzyme immunoassays have been extensively exploited as aids to clinical diagnosis and for research applications. There is an excellent monograph on EIA, edited by E. T. Maggio (1980). One cannot assume the simplicity of performance of EIA, however, since these assays combine all of the inherent problems associated with enzymes, immunology, and the chemistry and physics of heterogeneous systems. In establishing a procedure, each variable must be taken into consideration.

II. HETEROGENEOUS EIA

By far the most commonly used EIA procedure is the heterogeneous assay for antibody. These assays are characterized by the necessity to separate reacted from nonreacted analyte. Generally speaking, this has been accomplished by adsorbing one ligand to a solid phase such as tubes, plates, slides, beads, or paper disks. The nonreacted analyte can then be removed easily by washing, without the laborious process of centrifugation. First introduced in 1971 as ELISA, the enzyme-linked immunosorbent assay, in the ensuing decade this procedure has seen a rapid acceptance and introduction into virtually every area of infectious diseases.

Adapted from RIA procedures, ELISA has proven to be equal or superior to RIA in every way. Numerous comparative studies have shown sensitivity and specificity to be equivalent; ELISA reagents are more stable, nonhazardous, and easier to use; and at least semiquantitative results can be obtained with visual reading of the results and no specialized equipment. With the addition of spectrophotometric measurement, the precision and sensitivity of RIA is realized.

There has been considerable debate concerning the relative sensitivities of RIA and EIA. Widely divergent results have been reported, but as seen in Table 1, Wolters et al. (1976) have shown EIA sensitivity equivalent to RIA in detecting hepatitis B surface antigen. Engvall et al. (1971), Dray et al. (1975), and others have all clearly demonstrated that EIA can measure at least in the picogram level of antigen or antibody. In fact, by combining the two methods into an ultrasensitive enzymatic radioimmunoassay, Harris et al. (1979) have been able to measure 10^{-16} g of cholera toxin. Undoubtedly, the specific reagents involved affect the ability to detect minute quantities, but the evidence is clear that properly performed and standardized methods permit exquisite sensitivity of EIA.

Potentially, the specificity of ELISA should be somewhat less than RIA since many sera contain enzyme inhibitors or free enzyme. In most cases these agents have little effect on the test since all serum components not taking part in the test are thoroughly

Table 1 Titration of Two Different HB_SAg-Positive Sera of Subtype ay

	RIA ratio		EIA ratio	
Dilution factor	1	2	1	2
2^{11}	6.0	4.4	12.0	9.9
2^{12}	3.3	3.3	5.5	6.7
2^{13}	2.0	1.9	3.8	3.1
2^{14}	2.3	1.6	1.5	1.6
2^{15}	1.6	1.4	1.4	1.2
2^{6}	2.8	3.4	9.2	13.0
2^{7}	2.6	2.5	5.3	5.6
2^{8}	1.8	2.2	2.3	2.6
2^{9}	1.2	1.5	2.2	2.0
2^{10}	0.9	1.3	1.2	1.2
2^{11}	1.1	1.3	0.9	0.5

Source: Wolters et al. (1976).

washed from the system prior to the addition of conjugate and substrate. As a result, the major nonspecificity factor is the same as in RIA, which is "background" due to nonspecific adsorption of conjugate to the solid phase. Frequently, this can be reduced or eliminated by the use of a surfactant such as Tween-20 and a nonspecific protein such as bovine serum albumin. Properly adjusted, the EIA systems appear to show a specificity equivalent to other procedures.

A. ELISA Procedure

For illustration and as a guide to the succeeding discussion of the individual components, a brief outline of the ELISA microtitration procedure used in our laboratory for determining serum antibodies follows. Schematically, this is illustrated in Fig. 1(A). To maintain generalities, specific reagent identity and dilution have been purposefully omitted.

1. Polystyrene microtitration plates are appropriately labeled.
2. 50 μl of antigen, optimally diluted in pH 9.6 carbonate buffer, is added to each well.
3. Plates are covered and incubated for 2 hr in a 37°C water bath and then overnight at 4°C. Sealed with cellophane tape, they can remain usable at 4°C up to 6 months.
4. Antigen solution is removed from the plates, and wells are filled with phosphate buffered saline, pH 7.2, containing 0.5% Tween-20 (PBS-Tween). Plates are allowed to stand at room temperature for 3 min.
5. Repeat step 4 twice more.
6. 50 μl of PBS-Tween is added to each well.
7. Using 50 μl diluter loops, serum is added to the first well of each row. Twofold dilutions are then made directly in the plates.
8. Plates are covered and incubated for 30 min in a 37°C water bath.
9. Plates are washed as in steps 4 and 5.
10. 50 μl of horseradish peroxidase-labeled goat anti-human IgG, optimally diluted in PBS-Tween, is added to each well.
11. Plates are covered and incubated for 30 min in a 37°C water bath.

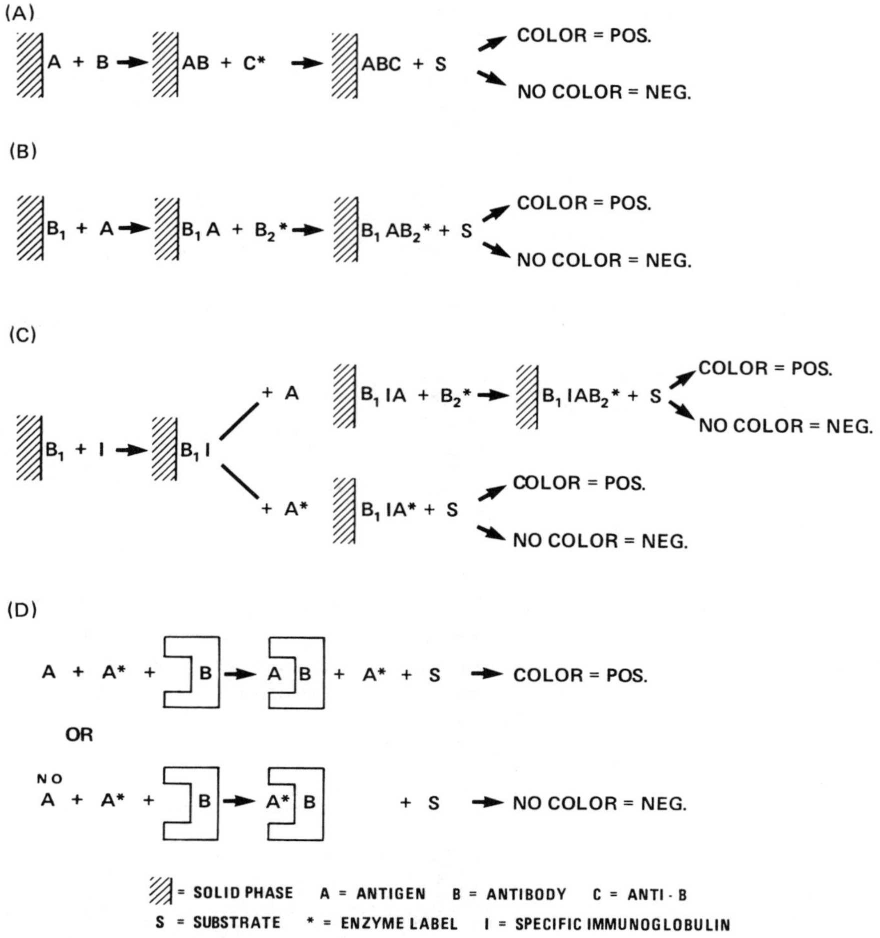

Figure 1 Schematic diagrams for EIA methods.

12. Plates are washed as in steps 4 and 5.
13. 200 μl of 1.0 M 0-phenylenediamine in pH 4.5 acetate buffer with 0.3% H_2O_2 is added to each well.
14. Plates are incubated for 30 min at room temperature in the dark.
15. Color reaction is stopped by adding 25 μl of 8 N H_2SO_4 to each well.
16. Results are read spectrophotometrically at 490 nM. If plates are protected from drying, color reaction is stable for at least 2 days after the addition of acid.

B. Solid Phase

A great deal has been written about the appropriate solid phase to be used in ELISA. Beads, disks, tubes, and plates have all been used successfully. Except for the disks, all of the systems rely on a styrene or vinyl surface to adsorb the ligand. Pesce et al. (1977) and

Herrmann and Collins (1976) have described the adsorption of protein to styrene surfaces and indicate their usefulness and the problems involved. Unfortunately, all styrenes, even those from the same manufacturer, are not alike. As a consequence, in the most popular systems, which use microtitration plates, selection is a very subjective decision, in most cases relying on precedent procedures or personal preference. Few absolute data are available to document the advantages or disadvantages of any styrene. A variety of special polystyrene plates have been produced by several manufacturers. Each has its special characteristics which must be evaluated in any new system.

Polyvinyl is considerably more adsorptive for some proteins than polystyrene and has been used in a number of applications. It is, however, plagued by excessive background due to nonspecific adsorption. In addition, the optical qualities of polyvinyl microtitration plates are considerably inferior to the special polystyrene plates. These disadvantages have resulted in restricting the primary use of polyvinyl plates to those systems in which maximum adsorption is required due to the properties of the ligand being used and in which visual determination of reactivity is adequate. To date, no plate—styrene or polyvinyl—stands out as the primary solid-phase agent.

Criteria that must be considered when evaluating a solid-phase agent are:

1. The proper adsorption of the primary ligand. In general, proteins adsorb adequately to polystyrene and frequently excessively to polyvinyl. Polysaccharides adsorb quite variably to styrene and usually adequately to vinyl. Some authors have reported success in increasing adsorption of polysaccharides to polyvinyl by linkage through poly-L-lysine. Few authors have found it necessary or have suggested the advantages of covalent bonding of the primary ligand.

2. The amount of background that will be present and the amount that can be accepted. In some systems using either impure antigens or very weak antigens, a large amount of reagent must be adsorbed, which may result in nonspecific background. However, the specific reaction may be sufficiently strong to allow the background to be acceptable by simply subtracting the background from specific reactivity to provide the final quantitation. In other systems, the specific reactions may be weak and cannot be clearly separated from background. A variety of techniques have been used to eliminate the background and, again, differ with the system used. Virtually all procedures have used a surfactant, usually Tween-20, in all reagents except the buffer in which the first ligand is suspended for attaching to the solid phase. In addition, some procedures call for reacting the plate with 1% bovine serum albumin (BSA) after washing out the excess of the primary sensitizing reagent. Still other procedures suggest including 1% BSA in all the diluents in the procedure other than the original sensitizing step.

3. Specific performance requirements of the procedure. Microtitration plates are most useful in those procedures requiring serial dilutions. Microtitration loops or multichannel pipettes can be used to make the dilutions directly in the plates. Plates with removable wells are available from several manufacturers and permit custom arranging when a serum is being tested quantitatively against a variety of antigens. Beads, both metal and plastic, and cellulose acetate disks allow complete flexibility in that each well or vial is an independent system which can be handled in concert with others or in complete independence. In using these systems, however, allowance must be made for special washing and transfer mechanisms. Reading will also be dependent on either visual interpretation or removal and separate measurement of the color of the supernate. In some cases, an insoluble substrate has been used which affixes to the cellulose disk, and reactivity is determined by visual comparison of the disk after washing.

A number of investigators have reported an "edge effect" when working with microtitration plates. Surprisingly, both positive and negative effects have been described. When seen, it is recognized that the wells on the outer rim of the plate react differently from the rest. So marked was this in some studies that the outer wells were not used. Numerous suggestions have been made as to possible causes, and the only one that seems to have been accepted is the differential temperature that the outer wells experience. These wells, obviously more exposed, would warm faster in an incubator and cool faster when removed. Whatever the cause, edge effect has never been a problem in our laboratories.

C. Antigen

One of the advantages of ELISA over othei immunoassays is the flexibility in the type of antigen used in the test. Any component that can be attached to the solid phase can be used. Successful reports have included whole organisms, disrupted organisms, metabolic products, toxins, highly fractionated proteins and carbohydrates, antibiotics, drugs, and serum proteins. As previously stated, procedural modifications may be necessary to assure fixation of some of the agents to the solid phase. It is also apparent that both sensitivity and specificity depend on the purity of the antigen used.

Most procedures call for fixation of the antigen in an alkaline pH, about 9.6. In some cases, because of the lability of the antigen, more modest pH is used and may only be as high as 7.8 or 8.0. After fixation, the plates can be stored with the antigen solution remaining in them, the antigen solution can be removed and the plate stored dry, or the antigen solution can be allowed to dry in the plate then stored. Depending on the stability of the antigen, all three methods have been used successfully. It has been reported that sensitivity is increased with the latter method, in which antigen is dried in the plate.

The amount of antigen necessary to sensitize a plate must be determined by a chessboard titration, but a number of reports agree that about 5 μg/ml of a protein antigen is an optimal concentration. Surprisingly, it has also been reported that at this concentration the solution can be retrieved and used to sensitize at least two more times.

D. Washing

Although few published data are available, it is an accepted concept that perhaps one of the most important steps in assuring reproducibility in ELISA is correct washing. Whether done manually or by automated instruments, erroneous results directly related to improper washing are obvious. Particularly in microtitration plates, drops tend to cling and air bubbles tend to form in the plate wells if a special effort is not made to exchange reagents. Removal cannot be accomplished by simple dumping of a plate. The plate must be *shaken vigorously*, and preferably some system of aspirating used. Wells must be filled by direct injection; when using other methods of filling, bubbles that might be trapped must be manually removed. Most of the commercial instruments available fill wells very well. Disks and beads are handled in very special ways and in general allow more complete washing than plates. Two concepts of washing have been proposed, and each is successful. Early reports suggested a minimum number of washes, each accompanied by a prolonged soaking period. Recently, multiple rapid washes with more-or-less violent exchanges and no soaking period have been used. The rationale of the latter method is that more loosely bound nonspecific reactions will be mechanically broken, whereas firmly bound specific reactions would not. Soaking might not make this differentiation.

E. Conjugate

Today, for most purposes, commercial conjugates are of very high quality. The antibodies are generally heavy chain and/or light chain specific; most are affinity chromatograph purified and have exceptionally high reactivity levels, permitting economical use. They may be obtained with general reactivity against the whole globulin fraction of serum or may have exquisite sensitivity by using the $F(ab)_2$ fraction of an antiserum against a specific immunoglobulin group (kappa, lambda, mu, etc.).

A variety of enzymes have been used as labels, but the most common have been alkaline phosphatase (Aptase) and horseradish peroxidase (HRP). There are a variety of reasons for the choice of enzyme, but most have little relevance and in general reflect personal preference. Initially, proponents and opponents of Aptase and HRP used three criteria for differentiation: cost, stability, and carcinogenicity. Although Aptase, at present, is more expensive, as the demand increases, newer production methods will undoubtedly overcome this problem. HRP is unquestionably more stable, which is a distinct advantage in some research areas where the conjugate might be exposed to some rather harsh treatment; but in terms of shelf life, the conjugates can be stored in such a way that both phosphatase and peroxidase have virtually unlimited shelf life. Initially, carcinogenic chromogens were used for HRP, but newer chromogens have few or no carcinogenic or mutagenic properties.

Of more importance is the specific metabolic rate of each enzyme. Since sensitivity is directly related to the ability to detect a reaction, the greater the amount of reaction per unit of time, the smaller number of enzyme molecules necessary for measurement. HRP is somewhat more efficient than Aptase and with the proper selection of substrates can be improved. Neither, however, approaches some of the other enzymes, which, up to now, have been impractical to bind to antibodies for use in immunoassays. Adequate for the state of the art, alkaline phosphatase and horseradish peroxidase enzymes will continue to be used for some time, but as pressure builds to convert to kinetic systems, turnover rate will become the limiting factor, and more efficient enzymes will have to be used.

One of the problems associated with labeled immunoassays, including ELISA, RIA, and FIA, is the restriction that the antiserum containing the label must be species specific. In hospitals and clinical laboratories where only human samples are tested, this creates no problem. In research and in veterinary clinics where a variety of animals are evaluated, a different conjugate must be used for each species. Some techniques have been suggested to overcome this problem but have not been readily accepted. One such system suggested the inclusion in the primary reaction of complement and with the indicator system then using a labeled anticomplement. Although this overcame the species dependency, the inherent problems associated with complement and the extra procedural step decreased its attractiveness.

F. Substrate

The choice of substrates and chromogens for measuring the reaction is somewhat more objective than the selection of enzyme. With few exceptions, para-nitrophenylphosphate (p-NPP) has been the exclusive substrate used in procedures involving alkaline phosphatase. This substrate is easily prepared, is water soluble, is nonhazardous, and converts to a yellow color when reduced. Initially, the yellow color was the impetus for the introduction of horseradish peroxidase. Unless the results were read spectrophotometrically, weak reactions were difficult to discern.

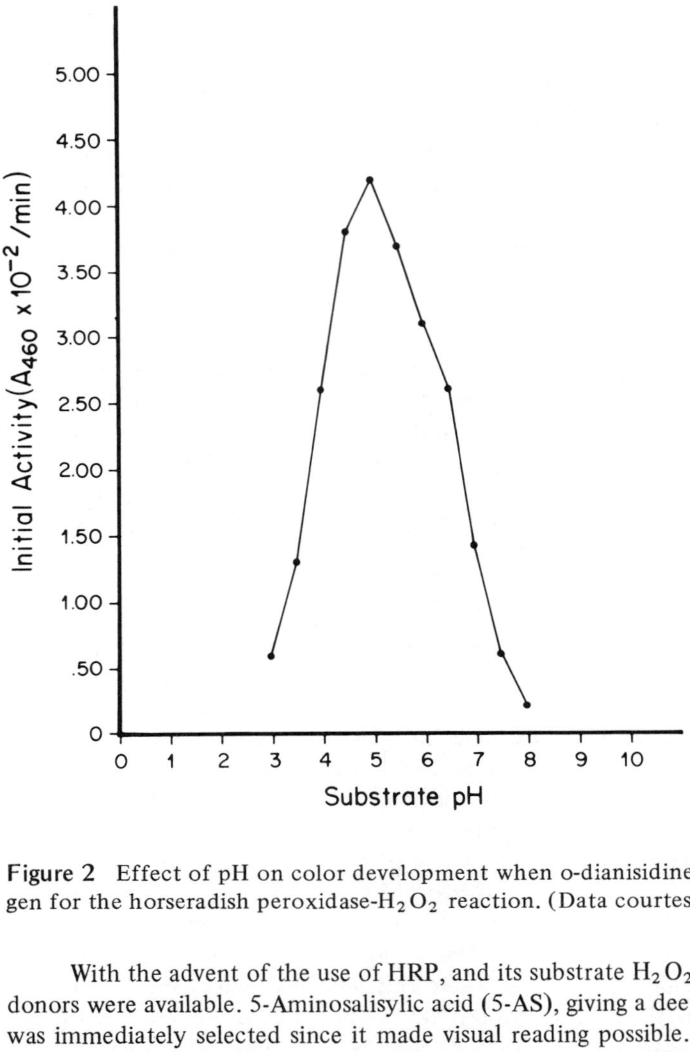

Figure 2 Effect of pH on color development when o-dianisidine is used as the chromogen for the horseradish peroxidase-H_2O_2 reaction. (Data courtesy of V. C. W. Tsang.)

With the advent of the use of HRP, and its substrate H_2O_2, a choice of hydrogen donors were available. 5-Aminosalisylic acid (5-AS), giving a deep purple-brown reaction, was immediately selected since it made visual reading possible. Its disadvantages were soon recognized and alternatives pursued. 5-AS is carcinogenic; it is a slow converter, requiring more enzyme activity to cause color production; and it spontaneously converts in the presence of light (Bullock and Walls, 1977). Two other compounds, o-phenylenediamine (OPD) and ABTS [2,2′-azino-di(3-ethylbenzthiazolinesulfonic acid)], are now commonly used. Both produce sufficient color to permit visual estimation of the reaction, and both are about 10 times more efficient hydrogen donors than is 5-AS. Other substrates are selected for specific functions: molecular size, solubility, or color.

Regardless of which enzyme and which substrate one chooses, pH optimum is critical. Figure 2 illustrates the narrow range of maximum color development when HRP is measured with o-dianisidine. A variation of only 0.5 pH unit from the optimum will create measurable decreases in activity. The selection of the substrate reaction pH must be a compromise between the optimum for enzymatic action and the optimum for substrate hydrogen donation. Although Aptase and HRP function best near neutrality, when using colored substrates this changes considerably. Alkaline phosphatase with p-NPP functions best near neutrality, while horseradish peroxidase with OPD reacts at pH 4.5.

Both enzymes are completely inhibited at very high or low pH levels. When using HRP with OPD or ABTS, the reactions are virtually stabilized by the addition of H_2SO_4 to drop the pH. Because of the color inhibition, Aptase with p-NPP is better stopped by the addition of NaOH to raise the pH.

G. Standardization

Unfortunately, one of the most serious problems in the performance of ELISA, and the most difficult to solve, is the lack of standardization. There are no guidelines to suggest what should be accepted as proper reactivity, what specificities should be required of the conjugates, which enzyme-substrate systems are appropriate, or how to read and interpret results. Although serving as a basic framework, the original ELISA procedure has been modified so extensively by a variety of individuals that it has become virtually unrecognizable. Even though most methodology changes are small, their effects are often marked (e.g., shorter or longer incubation times frequently alter sensitivity considerably). As a result, reagent standardization becomes almost impossible when there is no technical standardization. The broad acceptance of EIA as a viable alternative to many other methods in a variety of clinical areas has led to total chaos in standardization. Although some efforts have been made to prepare an international standard conjugate and in some areas international standard sera are available, there still remains a strong need for international agreement on reagents and techniques.

III. APPLICATIONS OF EIA

A. Antibody Determinations

In spite of the procedural variations that have been introduced, the principle of EIA remains the ability to measure antigen or antibody. A review of the literature reveals that EIA can efficiently measure antibody to any antigen, provided that the antigen can be immobilized sufficiently to be detected. Tests have been described for detecting antibody to more than 20 viruses, 30 bacteria, 12 parasites, and 4 fungi. In addition, toxins, serum proteins, and antibodies have all been measured by EIA methods. ELISA parameters were first described for the detection of immunoglobulins and thereafter for *Salmonella*. The obvious progression was the rapid comparison of ELISA to existing serologic methods for a variety of bacteria. Of more significance, however, was the introduction of tests for new disease agents or those in which tests were inadequate or unavailable.

Tests have now been described, for example, for *Chlamydia* (Schachter, 1978), *Legionella* (Farshy et al., 1978), group B streptococci (Rote et al., 1980), meningococcus (Beuvery et al., 1982), *Brucella* (Boraker et al., 1981), *Hemophilus* (Dahlberg, 1981), and *Salmonella* (Carlson et al., 1972). For some of these, tests had been available, but they were inadequate, needing greater sensitivity or the ability to detect class specific antibody. For others such as *Chlamydia* and *Legionella*, no adequate tests were available.

In virology, ELISA has, for the most part, served to increase sensitivity or to replace RIA with a more convenient system. Although tests for most viruses of medical importance have at least been evaluated, significant progress has specifically been made in rotaviruses (Yolken et al., 1977), rubella (Voller and Bidwell, 1975), cytomegalovirus (Castellano et al., 1977), herpesvirus (Friedman and Kimmel, 1982), influenza (Hammond et al., 1980), norwalk (Greenberg et al., 1978), and respiratory syncytial viruses (Hendry and McIntosh, 1982). In Tables 2 and 3 we see the comparative reactivity of EIA versus hemagglutination inhibition for rubella and indirect hemagglutination for

Table 2 Comparison of Micro-ELISA and Hemagglutination Inhibition Tests for Rubella Antibody

	Micro-ELISA results		
Hemagglutination inhibition titers	Number of negative sera (E400 ⩽ 0.02)	Number of positive sera (E400 > 0.2)	Mean ELISA value (E400)
⩽8	26	5	0.15
16 and 32	2	18	0.34
64 and 128	0	72	0.40
256 and 512	0	49	0.49
⩾1024	0	16	0.71

Source: Voller and Bidwell (1975).

cytomegalovirus (CMV). In each case, the newer ELISA procedure was equal to or superior to the conventionally accepted procedure. ELISA has become a major tool in the investigation of rotavirus enteritis in children; it has made possible more unified serodiagnostic programs for rubella, CMV, and herpesvirus; and it has greatly improved the sensitivity and specificity of tests for the other virus infections for which it has been introduced.

Serodiagnosis of visceral larva migrans, a serious parasitic infection caused by *Toxocara canis* in young children, was extremely difficult until the development of an ELISA procedure by Cypess et al. (1977). Specific diagnosis can now be made to differentiate this disease from other life-threatening diseases and appropriate therapy initiated. Under the impetus mainly from Voller et al. (1977), most parasitic infections can now be tested by ELISA. Laboratories previously using as many as six different serologic procedures can now rely upon modifications of one.

Although a number of fungal diseases have been evaluated, most of the efforts in mycology have centered around the identification of antigen and antibody to *candida*. Tests are now available for the detection of circulating mannans and to the antibodies associated with them. Tests have been described for *Histoplasma, Blastomyces, Cryptococcus*, and *Coccidioides* and will surely be refined to replace the presently used precipitin and complement fixation methods.

B. Antigen Detection

As mentioned earlier, antigen detection is as important in most infectious diseases as in antibody detection. An extensive discussion of EIA techniques for antigen detection has been published by Yolken (1982). As seen in Fig. 1(B), heterogeneous EIA for antigen detection differs from antibody detection only by the addition of an initial capture antibody. Antibody 1 against the antigen is affixed to the solid phase; the antigen is captured from the clinical specimens; an enzyme-labeled antibody 2 against the antigen detects its presence; and with the appropriate substrate chromogen the reactions can be quantitated.

Three major complications not seen in ELISA must be considered. First, antibody 1 must be directed against the appropriate antigen to be measured. In some cases monoclonal antibody has been highly efficient as the capture antibody when looking for antigens such as specific toxins. In other cases the precise circulating antigen in an infectious disease may not be known, and monoclonal antibodies have proven to be too specific and

Table 3 Results of Enzyme-Linked Immunosorbent Assay (ELISA) and Indirect HA (IHA) Test for Antibody to Cytomegalovirus in 30 Sera

Serum	Absorbance (400 nm) of 1:100 dilution of serum	ELISA titer	IHA titer
1	3.0550	⩾ 10,000	1,024
2	2.4295	⩾ 10,000	⩾ 2,048
3	2.3640	⩾ 10,000	⩾ 2,048
4	2.1050	5,000	1,024
5	2.0995	1,000	1,024
6	1.9370	1,000	2,048
7	1.8075	5,000	⩾ 2,048
8	1.3025	1,000	128
9	1.2775	1,000	512
10	1.1745	1,000	128
11	0.9445	1,000	512
12	0.9260	1,000	64
13	0.8865	500	32
14	0.8040	500	64
15	0.6275	500	16
16	0.4725	100	16
17	0.1045	< 50	8
18	0.0945	< 50	< 4
19	0.0940	< 50	< 4
20	0.0885	< 50	< 4
21	0.0745	< 50	< 4
22	0.0730	< 50	< 4
23	0.0710	< 50	< 4
24	0.0660	< 50	< 4
25	0.0655	< 50	< 4
26	0.0510	< 50	< 4
27	0.0350	< 50	< 4
28	0.0220	< 50	< 4
29	0.0180	< 50	< 4
30	0.0025	< 50	< 4
Known positive	0.9920	1,000	128
Known negative	0.0665	< 50	< 4

Source: Castellano et al. (1977).

will not always detect the antigen available at that moment. In these cases a polyclonal antiserum or an antiserum produced by combining monoclonal antibodies has been useful.

Second, antibody 1 should not nonspecifically bind globulin from the patient sample or from antibody 2. A number of investigators have noted that globulins will nonspecifically bind through the Fc portion of antibody. Binding of these from the patient's sample may in many cases inhibit the specific binding of antigen. More seriously, nonspecific binding of globulins from antibody 2 produces a false positive.

Finally, antibody 1 and antibody 2 must be sufficiently similar, or dissimilar, to avoid interspecies cross-reactions. Such reactions occur as a result of blood group substances or Forrsman-type antibodies. Antisera prepared in totally unrelated species may avoid these problems but, of course, require the preparation of two antisera. Many investigators have fractionated a portion of a serum for Fab_2 to use as antibody 1 while labeling another portion of the same serum with enzyme to use as antibody 2.

EIA for antigen detection has been used extensively in virology, where direct staining in tissue with enzyme-labeled antiserum can identify viral particles or inclusions while heterogeneous EIA can identify viruses in virtually any body fluid. The tests have been used, for example, to identify respiratory syncytial virus in nasal secretions (Hendry and McIntosh, 1982), influenza A and adenovirus in nasal secretions and throat swabs (Harmon and Pawlik, 1982), hepatitis B surface antigen in serum (Wolters et al., 1976), and coxsackievirus from stool (Yolken and Torsch, 1981). Tests are described for all major respiro- and enteroviruses or their antigens. Specificity of the procedures is sufficient for grouping or typing isolates such as influenza A and B (Murphy et al., 1981) or herpes type I or type II (Mills et al., 1978). Although similar tests have been evaluated in bacteriology, emphasis has been more on the detection and identification of antigens and toxins than on microbial identification. M protein in staphylococcus (Russell et al., 1976), pneumococcus polysaccharides (Pedersen and Henrichsen, 1982), and meningococcal antigens (Sippel et al., 1978) in serum have all been effectively measured. In addition, identification, and in some cases quantitation, of bacterial toxins in clinical samples can be useful. Tests have been designed to detect *Escherichia coli* enterotoxins (Stiffler-Rosenberg and Fey, 1978), diphtheria toxins (Svenson and Larsen, 1977), and staphylococcal enterotoxins (Saunders and Bartlett, 1977).

Many parasitic infections have both chronic and acute phases. Detection of antigen, particularly specific life-cycle antigens, can be of immeasurable importance in diagnosis. Detection of circulating antigens in toxoplasmosis has been described by several investigators (Van Knappen and Panggabean, 1977; Araujo and Remington, 1980) and is frequently used as the indicator of currently active infection. The particular life-cycle stage of schistosomes—eggs, cercarial, or adult—can indicate the activity of a disease characterized by chronicity. Similarly, tests to detect antigen of *Trypanosoma cruzi* (Araujo, 1982), filarial worms (Karavodin and Ash, 1981), and protozoa (Taylor et al., 1981) have been evaluated. A commercial test for detecting amoeba antigens in stool samples proved to be three to five times more sensitive than conventional stool examinations.

Of the variety of fungal agents that have been evaluated in EIA tests, most progress has been reported on detecting, identifying, and typing the mannans from *Candida albicans* (Segal et al., 1979). Important in establishing the clinical significance of the yeast, considerable emphasis has been placed on identifying and characterizing these antigens. Additionally, tests for measuring of cryptococcal antigens and H and M antigens in histoplasmosis have been reported (Sharma et al., 1982).

In addition to detecting infectious agents, tests have been designed for the detection of specific serum proteins, hormones, and drugs. Early reports on techniques for carcinoembryonic antigen and α-fetoprotein were superior to standard methods used currently. Since then, tests have appeared for T_3 (triiodothyronine), T_4 (thyroxine), thyroglobulin, gentamycin, and other aminoglycosides, and many hormones and serum proteins.

C. Measuring Specific Class Antibody

Initially, EIA methods were patterned after RIA and FIA tests in which specific class antibody was detected by simply substituting enzyme-labeled antiserum against the class to be measured into the conventional ELISA procedure. All other parameters remained essentially unchanged. Particularly in the measurement of IgM antibody, several problems were encountered: False positives resulted from the presence of rheumatoid factor, IgM antibody may be blocked by IgG antibody, and the test is frequently undersensitive. As seen in Fig 1(C), two new modifications have been introduced which have proved to be highly efficient. First described by Duermeyer et al. (1978), anti-μ antibodies (B_1) are affixed to the solid phase, and the patient's IgM is captured by this antibody. Then, as seen in the upper diagram, the antigen for the disease in question is added and finally enzyme-labeled anti-antigen (B_2*). Schmitz et al. (1980) modified the procedure as seen in the lower diagram. Anti-μ (B_1) is affixed and the patient's IgM captured. Then enzyme-labeled antigen (A*) is reacted. The advantages of these techniques are readily apparent. The capture system effectively removes the IgM from the IgG, thus preventing any blocking action by the latter; all IgG is washed from the reaction, preventing any possible rheumatoid activity; and finally, since only IgM antibodies are present to bind with the antigen sites, the test is much more sensitive and reactive than older techniques. Two major precautions are necessary for proper test performance. Specific affinity-chromatographed heavy chain specific antiserum *must* be used to coat the solid phase, since even minor contamination will result in false positives. In using the Schmitz method, highly purified antigens should be used for labeling, since extraneous labeled material can cause interference and wastes enzyme. We have also found that an enzyme label to antigen ratio of 4:1, which is heavier than earlier reports, gives higher specificity and reactivity with lower background readings.

Still in its infancy, the technique has shown great promise. To date, the Duermeyer modification (Duermeyer et al., 1978) has most frequently been used and shown to measure IgM antibodies to hepatitis A and B core antigen (Roggendorf et al., 1980), *Toxoplasma gondii* (Naot and Remington, 1980), and tick-borne encephalitis virus (Heinz

Table 4 Comparative Reactivity Between R-EIA and IGM-IIF (*Staphylococcus* A Adsorbed) in 122 Serum Samples from 32 Patients Involved in an Outbreak of Toxoplasmosis and 80 Control Samples from Normal Individuals

Adsorbed IgM-IIF titer	Number of sera	R-EIA mean titer	Titer ratio R-EIA/IIF
<16 (controls	80	<4	–
<16 (patients)	9	110.6	>6.9
16	5	371.5	23.2
64	12	937.8	14.7
256	24	2,412.9	9.4
1,024	55	12,816.0	12.5
4,096	16	31,068.6	7.6
16,384	1	66,461.0	4.1

et al., 1981). The Schmitz method has been used for cytomegalovirus (Schmitz et al., 1980) and toxoplasmosis (Franco et al., 1981). In each case, IgM-specific antibodies have been detected to a higher level. Table 4 shows the results we found when the Schmitz method, which we refer to as R-EIA, is compared to the conventional IgM indirect immunofluorescence test. Note that patients negative by IIF are positive by R-EIA. In addition, R-EIA is up to 23 times more reactive than IIF. Recent reports using the ELISA methods have shown IgM antibodies for *Toxoplasma* to persist for at least 1 year, much longer than previously understood. Classical concepts for diagnostic purposes had suggested that IgM antibodies persist for only 3 to 6 months and thus could be used as an indicator of early infection. The capture IgM systems obviously alter this theory and change the significance of the presence of IgM antibody.

Although the procedure should be readily adaptable for other class antibodies, this has not yet been explored. Since IgG antibodies occur in greater abundance in patient's serum and may be in response to a variety of antigenic stimuli, a capture system for IgG would probably be unrewarding, since possibly a very small percentage of captured antibody would be directed against the specific antigen or the test. The other classes that occur naturally in much lower concentrations should be more appropriate for the procedure.

D. Other Modifications

To obtain maximum reactivity, a variety of enhancement substrates have been suggested. Chemiluminescence has been the most frequently investigated system. In these tests, a light-emitting substrate, luminol, is activated by enzymatic action and can be measured by fairly simple photometers or modification or a scintillation counter. The resulting sensitivity is nearly 100 times greater than that of colored FIA reactions. The major drawback of this methodology is that the luminescence is nearly instantaneous and noncumulative. Consequently, readings must be read almost simultaneously with substrate addition and are not a multiplying reaction as are the color reactions.

Fluorescent substrates such as umbelliferone have been used to produce a fluorescent reaction. Bound umbelliferone, such as β-galactosylumbelliferone, fluoresces at a different wavelength than the unbound molecule. Consequently, enzymatic action by β-galactosi-

Table 5 Comparison of ELISA and Ultrasensitive Enzymatic Radioimmunoassay (Useria) for Cholera Toxin

Cholera toxin		ELISA[a]	USERIA[b]		
g	M	(1000 min)	10 min	100 min	1000 min
10^{-10}	10^{-15}	1.61 ± 0.14	759 ± 8	3882 ± 36	8799 ± 63
10^{-11}	10^{-16}	0.77 ± 0.09	189 ± 12	483 ± 18	2205 ± 66
10^{-12}	10^{-17}	0.26 ± 0.05	24 ± 6	186 ± 12	696 ± 36
10^{-13}	10^{-18}	0.08 ± 0.03	18 ± 6	126 ± 9	567 ± 33
10^{-14}	10^{-19}	0.0	15 ± 6	93 ± 9	408 ± 27
10^{-15}	10^{-20}	0.0	6 ± 6	57 ± 6	291 ± 24
10^{-16}	10^{-21}	0.0	0	9 ± 6	93 ± 21
10^{-17}	10^{-22}	0.0	0	0	3 ± 15

[a]Values are expressed as A_{405nm} units.
[b]Values are expressed as dpm (× 10^{-2}) of [^{3}H]adenosine produced during 10, 100, and 1000 min of incubation.
Source: Harris et al. (1979).

dase releases free umbelliferone and a measurable fluorescence. This is quantitative and cumulative and when measured in a fluorometer is far more sensitive than color reactions.

One highly sensitive system has been described in which EIA and RIA are combined by Harris et al. (1979) in a technique they named USERIA. Tritium-labeled adenosine monophosphate is used as the substrate for alkaline phosphatase. The labeled adenosine is released and can be separated from the monophosphate salt and counted in a scintillation counter. Table 5 demonstrates the extreme sensitivity of the procedure, in which endotoxin is detected at 10^{-16} M.

IV. HOMOGENEOUS EIA

As described in Fig. 1(D), this method is usually an inhibition test, since by definition the reactants are not separated for measurement. The test can be done as a one- or two-step

Table 6 Homogeneous Enzyme Immunoassays

Substance	Enzyme label	Sensivity range of commercially available assays[a] (μg/ml)
Morphine	Lysozyme	⩾0.5
Methadone		⩾0.5
Barbiturate		⩾2.0
Amphetamine		⩾2.0
Benzoyl ecgonine (cocaine metabolite)		⩾1.6
Benzodiazapine		⩾0.7
Propoxyphene		⩾2.0
Tyroxine	MDH	0.02–0.2
Cannabinoids		⩾0.05
Lidocaine	G6PDH	1–12
Phencyclidine		⩾0.025
Theophylline		2.5–40
Digoxin		0.0005–0.006
Cortisol		0.02–0.5
Phenytoin		2.5–30
Phenobarbital		5–80
Primidone		2.5–20
Carbamazepine		2–20
Ethosuximide		10–150
Gentamicin		1–16
Quinidine		1–8
Procainamide		1–16
N-Acetylprocainamide		1–16
Methotrexate		0.09–0.9
Tobramycin		1–26

[a]Performance data available from Syva Company, Palo Alto, California.
Source: Maggio (1980).

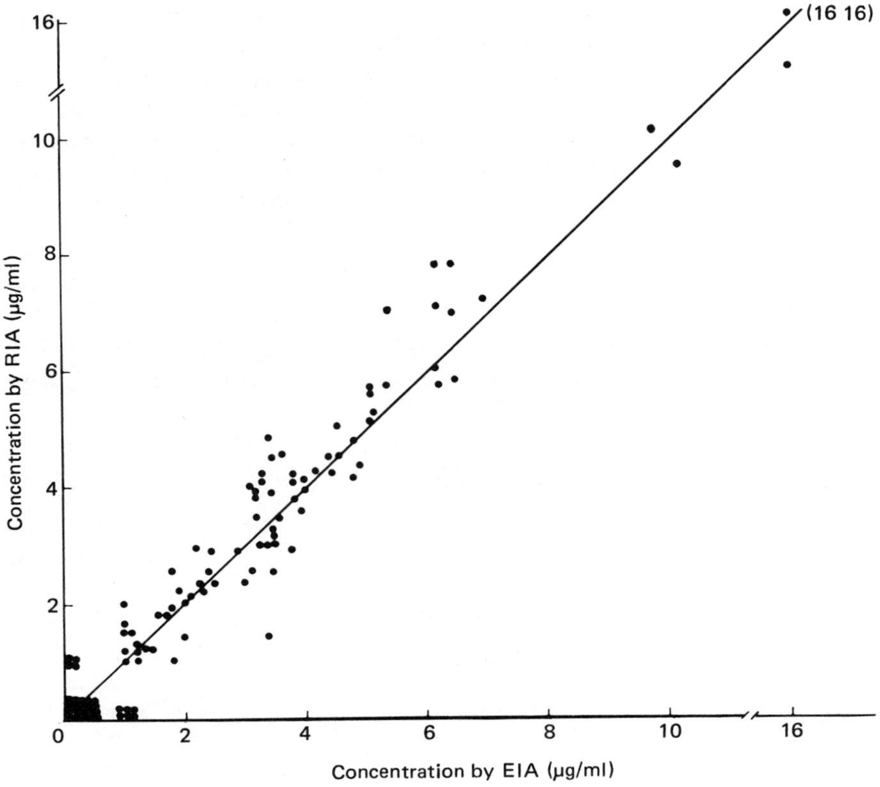

Figure 3 Assays of gentamycin-containing sera using the EIA and RIA methods. (From Phaneuf et al., 1980.)

method. In the two-step test, unknown sample (A) is mixed with antibodies against the specific substance being measured (B). After suitable incubation, a known limited amount of labeled antigen is added (A*) and incubated. Substrate is added and reacts proportionately to the amount of free labeled antigen. It should be noted that for the homogeneous reaction to be measured, there must be inhibition of the enzyme label of the bound labeled antigen. Although several physical phenoma occur, the most important in this case appears to be stearic hinderance. When the labeled antigen reacts with antibody, substrate is stearically prevented from interacting with the enzyme. Reaction of the unlabeled antigen with the antibody prevents reaction with the labeled, leaving free enzyme to react with the substrate.

This is critical to the usefulness of the procedure since it obviously restricts its use to the detection of antigen or haptens which are considerably smaller than the antibody molecule and can thus be stearically blocked. As a consequence, the method has been used almost exclusively for detecting hormones, metabolites, serum proteins, and antibiotic levels in body fluids.

As perhaps the most commercially utilized FIA methodology, EMIT, enzyme multiplied immunoassay technique (Syva Company, Palo Alto, California),* has been successfully used for the detection of drugs, antibiotics, and cannabinoids. Table 6 is a partial listing of substances that have been detected in blood, saliva, or urine by this homogeneous assay. The procedure is simple, very rapid, and agrees closely with other accepted procedures for measuring these substances. In Fig. 3, comparing EMIT and RIA for the detection of gentamycin, the two methods appear to be quite comparable in sensitivity. EMIT, however, is much simpler, much faster, and uses readily available equipment. The major problem of EMIT lies in its simplicity. Because there is no separation phase, one must contend with interfering substances and endogenous enzyme in the sample.

Although a variety of modifications have been designed to utilize homogeneous assays for high-molecular-weight compounds, none have yet proven to be practical. This is an area that will undoubtedly be pursued because of the major advantages of homogeneous assays. Since there is no solid phase involved, there is no limitation to the number of antibody molecules that can be introduced into the system, thus making the detection system a nonlimiting portion of the reaction and thereby increasing the test sensitivity.

V. KINETIC ASSAYS

Highest sensitivity and accuracy of measuring enzyme activity is by kinetic assay. Most EIA procedures are time limited; that is, enzyme and substrate are allowed to react for a predetermined length of time, and then a reagent is added to stop the enzymatic reaction. This introduces two errors in the determination of relative activity: one is not always successful in completely terminating the reactions, and enzymatic reactions are rarely linear beyond the initial few minutes of reaction. As a result, even reactions actually terminated have far exceeded the linear time period and the final results are not linearly related to the analyte present. To avoid these discrepancies from actual linearity, kinetic measurements must be made.

EMIT is measured on a kinetic system in that optical density (OD) readings are made at T_1 and T_2, and the OD related to standard values to obtain analyte concentration. Since no special equipment is normally used, this kinetic measurement must be somewhat inaccurate, but has been shown to have excellent sensitivity.

Recently, a heterogeneous procedure referred to as K-ELISA was described (Tsang et al., 1980). Using a computer-assisted reaction-rate analyzer, precise timing and reactive slope analysis permitted extreme sensitivity and reproducibility of results. In this procedure, the kinetic reading made possible a method in which all reagents were in extreme excess except the unknown analyte, which became the rate-limiting reagent. Equally effective for both antigen or antibody detection, one of its major applications appears to be its ability to detect even minor variations in antigen preparations. In analyzing the purity of any series of antigens, one needs only a specific antiserum, and "antigen units"

*Use of trade names and commercial sources is for identification only and does not constitute endorsement by the Public Health Service or the U.S. Department of Health and Human Services.

can be determined for each and, with appropriate control antisera, even minor cross-reactions can be identified.

One of the more exciting innovations in EIA, it promises to become one of the more important. Its precision and linearity make possible uses not attainable before. Although readings on a standard rate analyzer are too slow for large-scale antibody measurements in clinical laboratories, feasibility has already been shown in a microtitration system. Instrumentation capable of handling the processing is already available. Only adaptation to the precise methodology remains before micro-K-ELISA can become a reality.

VI. CONCLUSION

We have discussed only the EIA procedures that have been utilized to date to detect antigen and antibodies in body fluids. Other modifications are available for detecting antigens or infectious agents in tissue. Enzyme labels are used for light and electron microscopy. Cell surface antigens can be measured by EIA. A variety of immunologic phenomena and histochemical measurements are monitored by EIA. The uses are as limitless as the imagination. The beauty of the system lies in the inherent multiplication of reaction by the enzyme. Although never realized, theoretically a single enzyme molecule can be measured if given sufficient time to convert enough substrate to be measured. Consequently, in theory, monomolecular reactions can be detected. Will we ever reach that goal?

REFERENCES

Araujo, F. G. (1982). Ann. Trop. Med. Parasitol. *76*, 25.

Araujo, F. G., and Remington, J. S. (1980). J. Infect. Dis. *141*, 144.

Beuvery, E. C., Leussink, A. B., Van Delft, R. W., Tiesjema, R. H., and Nagel, J. (1982). Infect. Immun. *37*, 539.

Boraker, D. K., Stinebring, W. R., and Kunkel, J. R. (1981). J. Clin. Microbiol. *14*, 396.

Bullock, S. L., and Walls, K. W. (1977). J. Infect. Dis. *136*, S729.

Carlsson, H. E., Lindberg, A. A., and Hammarstrom, S. (1972). Infect. Immun. *6*, 703.

Castellano, G. A., Hazzard, G. T., Madden, D. L., and Sever, J. L. (1977). J. Infect. Dis. *136*, S337.

Cypess, R. H., Karol, M. H., Zidian, J. L., Glickman, L. T., and Gitlin, D. (1977). J. Infect. Dis. *135*, 633.

Dahlberg, T. (1981). J. Clin. Microbiol. *13*, 1080.

Dray, F., Andrieu, J. M., and Renaud, F. (1975). Biochim. Biophys. Acta *403*, 131.

Duermeyer, W., Wielaard, F., and Van Der Veen, J. (1978). J. Med. Virol. *4*, 25.

Engvall, E., and Perlmann, P. (1971). Immunochemistry *8*, 871.

Engvall, E., Johsson, K., and Perlmann, P. (1971). Biochim. Biophys. Acta *251*, 427.

Farshy, C. E., Klein, G. C., and Feeley, J. C. (1978). J. Clin. Microbiol. *7*, 327.

Franco, E. L., Walls, K. W., and Sulzer, A. J. (1981). J. Clin. Microbiol. *13*, 859.

Friedman, M. G., and Kimmel, N. (1982). Infect. Immun. *37*, 374.

Greenberg, H. B., Wyatt, R. G., Valdesuso, J., Kalica, A. R., London, W. T., Chanock, R. M., and Kapikian, A. Z. (1978). J. Med. Virol. *11*, 546.

Hammond, G. W., Smith, S. J., and Noble, G. R. (1980). J. Infect. Dis. *141*(5), 644.

Harmon, M. W., and Pawlik, K. M. (1982). J. Clin. Microbiol. *15*, 5.

Harris, C. C., Yolken, R. H., Krokan, H., and Hsu, I. C. (1979). Proc. Natl. Acad. Sci. USA *76*, 5336.

Heinz, F. X., Roggendorf, M., Hofmann, H., Kunz, C., and Deinhardt, F. (1981). J. Clin. Microbiol. *14*, 141.

Hendry, R. M., and McIntosh, K. (1982). J. Clin. Microbiol. *16*, 324.

Herrmann, J. E., and Collins, M. F. (1976). J. Immunol. Methods *10*, 363.

Karavodin, L. M., and Ash, L. R. (1981). Infect. Immun. *34*, 105.

Maggio, E. T. (1980). *Enzyme Immunoassay* (E. T. Maggio, Ed.). CRC Press, Boca Raton, Fla.

Mills, K. W., Gerlach, E. H., Bell, J. W., Farkas, M. E., and Taylor, R. J. (1978). J. Clin. Microbiol. *7*, 73.

Murphy, B. R., Phelan, M. A., Nelzon, D. L., Yarchoan, R., Tierney, E. L., Alling, D., and Chanock, R. M. (1981). J. Clin. Microbiol. *13*, 554.

Naot, Y., and Remington, J. S. (1980). J. Infect. Dis. *142*, 757.

Pedersen, F. K., and Henrichsen, J. (1982). J. Clin. Microbiol. *15*, 372.

Pesce, A. J., Ford, D. J., Gaizutis, M., and Pollak, V. E. (1977). Biochim. Biophys. Acta *492*, 399.

Phaneuf, D., Francke, E., and Neu, H. C. (1980). J. Clin. Microbiol. *11*, 266.

Roggendorf, M., Frosner, G. G., Deinhardt, F., and Scheid, R. (1980). J. Med. Virol. *5*, 47.

Rote, N. S., Taylor, N. L., Shigeoka, A. O., Scott, J. R., and Hill, H. R. (1980). Infect. Immun. *27*, 118.

Russell, H., Facklam, R. R., and Edwards, L. R. (1976). J. Clin. Microbiol. *3*, 501.

Saunders, G. C., and Bartlett, M. L. (1977). Appl. Environ. Microbiol. *34*, 518.

Schachter, J. (1978). N. Engl. J. Med. *298*, 428.

Schmitz, H., Von Deimling, U., and Flehmig, B. (1980). J. Gen. Virol. *50*, 59.

Segal, E., Berg, R. A., Pizzo, P. A., and Bennett, J. E. (1979). J. Clin. Microbiol. *10*, 116.

Sharma, A., Johri, B. N., and Shriniwas, A. (1982). J. Immunol. Methods *50*, 115.

Sippel, J. E., Mamay, H. K., Weiss, E., Joseph, S. W., and Beasley, W. J. (1978). J. Clin. Microbiol. *7*, 372.

Stiffler-Rosenberg, G., and Fey, H. (1978). J. Clin. Microbiol. *8*, 473.

Svenson, S. B., and Larsen, K. (1977). J. Immunol. Methods *17*, 249.

Taylor, D. W., Kim, K. J., Munoz, P. A., Evans, C. B., and Asofsky, R. (1981). Infect. Immun. *32*, 563.

Tsang, V. C. W., Wilson, B. C., and Maddison, S. E. (1980). *26*, 1255.

Ullman, E. F., and Maggio, E. T. (1980). In *Enzyme Immunoassay* (E. T. Maggio, Ed.). CRC Press, Boca Raton, Fla., pp. 105–134.

Van Knappen, F., and Panggabean, S. O. (1977). J. Clin. Microbiol. *6*, 545.

Van Weeman, B. K., and Schuurs, A. H. W. M. (1971). FEBS Lett. *15*, 232.

Voller, A., and Bidwell, D. E. (1975). Br. J. Exp. Pathol. *56*, 338.

Voller, A., Bidwell, D. E., Bartlett, A., and Edwards, R. (1977). Trans. R. Soc. Trop. Med. Hyg. *71*, 431.

Wolters, G., Kuijpers, L., Kacaki, J., and Schuurs, A. H. W. M. (1976). *29*, 873.

Yolken, R. H. (1982). Rev. Infect. Dis. *4*, 35.

Yolken, R. H., and Torsch, V. M. (1981). Infect. Immun. *31*, 742.

Yolken, R. H., Kim, H. W., Clem, T., Wyatt, R. G., Kalica, A. R., Chanock, R. M., and Kapikian, A. Z. (1977). Lancet *2*, 263.

21

Immunofluorescence: Principles and Procedures

Boris Albini State University of New York at Buffalo, Buffalo, New York

Konrad Schauenstein and Georg Wick Institute of General and Experimental Pathology, University of Innsbruck Medical School, Innsbruck, Austria

I. INTRODUCTION

Immunofluorescent staining techniques are methods used to visualize immune reactions by labeling one of the immune reactants with a fluorochrome. This reactant is called *conjugate*, as it consists of an immunologically reactive component (e.g., antibody, antigen) and a *marker*, a molecule capable of emitting fluorescent light. Immunofluorescence (IF) combines the specificity of an immune reactant with visibility caused by fluorescence of the marker molecule (Beutner et al., 1979; Goldman, 1968; Nairn, 1976; Polak and Van Noorden, 1983).

IF constitutes only one category of fluorescent tracing techniques. These include use of fluorochrome-labeled dextrans or microspheres for permeability studies and use of autofluorescence of electrolytes (Ca, Mg, S, P), steroids, lipids, proteins, amino acids, enzymes, drugs (barbiturates, salicylates, tetracyclines, quinidine, LSD), and some metabolites (porphyrins, ketones, ammonia, etc.) in clinical pathology and public health. On the other hand, IF is one of the indirect methods for demonstration of antigen-antibody reactions. In contrast to simple precipitation tests, in which the reaction product of antibody binding to antigen is visualized as such, in indirect immunological assays, a marker is used to visualize otherwise undetectable immunological reactions. This marker may be a dye, as is the case in IF; an erythrocyte, as is the case in passive hemagglutination; or a more complex entity, as, for example, the indicator system in complement consumption assays.

IF staining techniques may be used for tracer studies in vivo or in vitro and are useful in both immunohistology and serology. In immunohistology, the presence of immunoreactants in tissues or cells and the presence of inflammatory mediators may be tested. IF techniques are only one methodology available to accomplish localization of antigen-antibody reactions in tissues or cells (Andres et al., 1978; Sternberger, 1979).

Table 1 Methods in Immunohistology

	Light microscopy	Electron microscopy
Immunofluorescence	××	0[a]
Immunoenzyme techniques (e.g., immunoperoxidase techniques)	×	××
Autoradiography with immune reactants	××	××
Ferritin immunohistology	(×)	××
Heavy metals (e.g., gold) immunohistology	(×)	××
Haemadsorption techniques	××	0

[a]May be used in scanning electron microscopy.

A number of other approaches are possible and a list of some other immunohistological methods is given in Table 1. Whereas immunofluorescence seems the most widely used and standardized of these methods, optimal immunohistological evaluation can be achieved only by the rational and critical application and selection of available methodologies. In serology, both antigens and antibodies may be screened using IF techniques. Often, solid-phase techniques are used employing tissue as antigenic substrate. Sensitivity and specificity of IF used in serology may be comparable to those of hemagglutination assays, radioimmunoassay (RIA), or enzyme-linked immunosorbent assay (ELISA).

In the early 1930s it was shown that various chemical groups can be attached to antibody or antigen without appreciable loss of immunological reactivity. Azodyes were used by Heidelberger and his colleagues (1933) to label egg albumin. The labeled antigen was used to quantitate colorimetrically precipitating antibodies reacting with the antigen. Marrack (1934) used the same dye to color antibodies to thyroid bacilli. Such dye-antibody or dye-antigen conjugates were useful in serology, but the color was too weak to make them detectable under the microscope. To circumvent this limitation, Coons and collaborators in 1941 investigated the use of a fluorescent compound (i.e., anthracene) as a marker. Coons was able to observe "pneumococci type 3" stained blue with the appropriate fluorochrome-antibody conjugate in ultraviolet light. However, because of the blue autofluorescence of tissues elicited by ultraviolet light, the anthracene conjugate could not be applied to studies involving tissues. It was only 1 year later that Coons and his associates (1942) described the coupling of fluorescein-4-isocyanate to antibodies. The resulting conjugates gave a yellow-green fluorescence readily differentiable from the autofluorescence of tissue (Fig. 1). Since then, IF staining procedures have been modified and greatly improved (Beutner et al., 1968; Goldman, 1968; Nairn, 1976). The quality of the optical hardware has contributed significantly to this development. The introduction of epiillumination equipment by Ploem (1967) significantly expanded the application and increased the sensitivity and resolution of IF. The use of lasers promises to ensure well-characterized and steady light sources for IF technology. Automatization of reading of test samples and combination of high-resolution optics and computers are opening new horizons for fast and reliable clinical testing and comprehensive image analysis (Ploem, 1982). Biological reagents used in IF tests have been characterized to some extent, and reproducibility and reliability of IF procedures have been improved in a quest for "defined immunofluorescence." A World Health Organization subcommittee on standardization of immunofluorescence provides a framework for comparison and "calibration" of laboratory procedures and values.

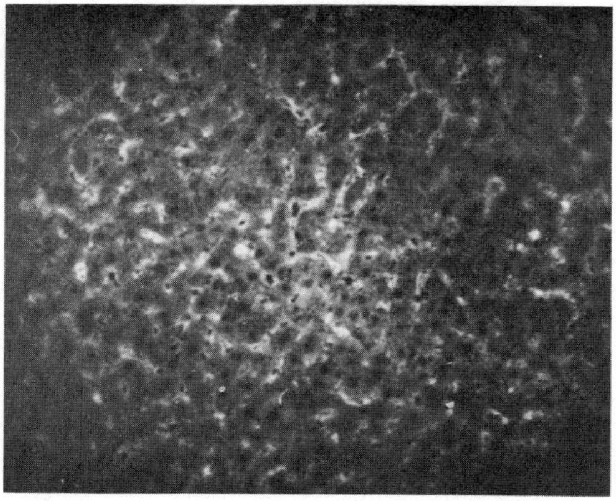

Figure 1 Photomicrograph through fluorescence-microscope of the liver of a mouse moribund with pneumococcal 3 infection. The antigen in the liver stained with fluorescein-carbamido-antipneumococcal 3 rabbit serum. Black areas are blood vessels; gray areas are the blue-gray fluorescence of formalin-fixed hepatic cells; white areas and fine white lines represent the green fluorescence of fluorescein-antibody and indicate the location of pneumococcal 3 antigenic material. 20 × apochromatic Zeiss objective, 10 × ocular; in the ocular Wratten filters 2A and 8 (K2). Leitz 1 × microphotographic apparatus. Eastman film Super Ortho Press. Exposure: 20 min. (From Coons et al., 1942.)

IF not only makes it feasible to identify the site of immune reactants in tissues or cells but also to quantitate the immune reactants using fluorometry. Specificity of the reaction in IF is conveyed by antibodies (to detect antigen) or antigens (used to detect antibodies). Lectins coupled to fluorochromes also may be used, in analogy to "classical" IF. Lectin fluorescence is not based on immunological reactions, but has a very high specificity and is used widely in immunological research. The high specificity of lectins for sugar moieties opens new perspectives for histochemistry.

The principles underlying immunological specificity are discussed extensively in Chaps. 1, 2, and 16 of this volume. Suffice it here to mention that immunological reagents used in IF have to be pure and in high concentration (or "titer") and their biological reactivity should survive conjugation procedures and exposure to intensive high-energy light.

II. FLUORESCENCE

Absorption of energy by a molecule or atom may cause the emission of photons. This emission of light by an "excited" atom or molecule is termed *luminescence* and probably was consciously observed first by Monardes in 1565. Energy-exciting atoms or molecules may be of various types. Inducing luminescence requires absorption of energy quanta by an atom or molecule; this changes the internal structure of the atom or molecule and may ultimately lead to emission of energy quanta in the form of photons (Fig. 2). In the context of our subject, we are mainly concerned with photoluminescence, that is, luminescence induced by the absorption of photons. The basic mechanisms of absorption and

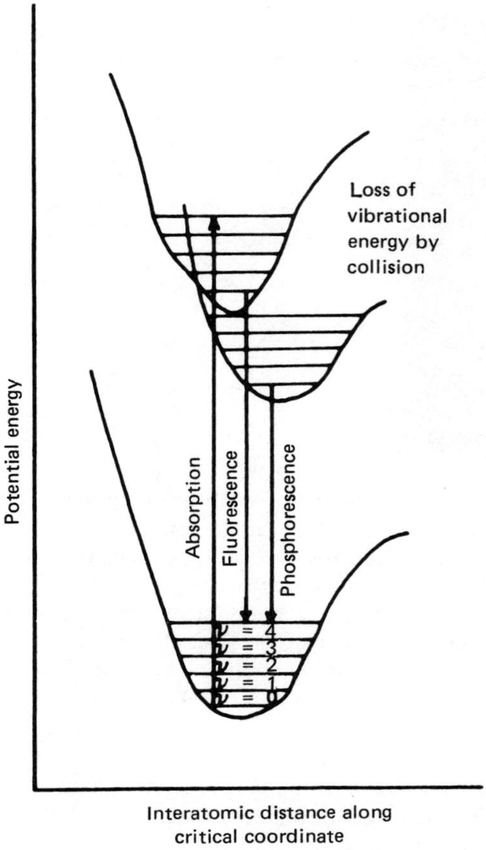

Figure 2 Schematic energy-level diagram for a diatomic molecule. Transition of electrons from ground to excited electronic states. (From Guilbault, 1973.)

emission of photons in the course of photoluminescence were studied comprehensively by G. G. Stokes in the 1850s. Photoluminescence may manifest itself as fluorescence,* phosphorescence, or "delayed fluorescence" (Guilbault, 1973; Udenfriend, 1969).

Light, a form of electromagnetic radiation, formally is described either as a wave function or as consisting of particles, the photons. Using the wave functions, light is defined by λ, the wavelength, and ν, the frequency. These two variables are interdependent according to

$$\nu = \frac{c}{\lambda} \tag{1}$$

with c being the velocity of light (i.e., 3×10^8 m/sec).

*The term *fluorescence* was coined by Stokes. He was studying a mineral giving impressive blue-white fluorescence; the name of the mineral happened to be "fluorspar."

Light encountering matter may either pass through it or be absorbed. When absorption occurs, energy is transferred to the absorbing atom or molecule in discrete units, the quanta. Quanta and energy are related according to

$$E = h\nu = \frac{hc}{\lambda} \qquad (2)$$

with E being the energy and h being Planck's constant (i.e., 6.62×10^{-27} erg/sec). Equation (2) makes it obvious that E is inversely related to λ: the higher the energy of light, the shorter its wavelength.

The Bohr model of the atom depicts a nucleus in which protons and neutrons, and thus most of the atomic mass, are concentrated. Negatively charged elementary particles of low mass, the electrons, orbit the nucleus. In an atom not experiencing influx of energy, the electrons will orbit in "ground electronic states." Absorption of energy, however, may lift an electron into a higher energy level. In a very simplified model, this transition may be depicted as shown in Fig. 2. In this case, an energy quantum is absorbed by an atom, placing one electron into a higher-energy state. Subsequently, the electron "falls down" into its original ground state, releasing another quantum of energy, possibly in the form of light. In molecules containing more than one atom, emitted light is always of a larger wavelength than the absorbed light.

Every molecule has a number of closely spaced energy levels. As in the case of an atom, the absorption of a quantum of light may induce the transition of an electron from a lower to a higher energy level. The distribution of energy levels in a molecule varies according to the interatomic distance of its atoms. Energy levels vibrate according to the state of the molecule, as depicted in Fig. 2. When a molecule is at the lowest vibrational level (i.e., v_0) absorption most likely will occur at the mean vibrational position, because this place is occupied for the longest time. Absorption at higher vibrational levels, in contrast, probably occurs at the two extremes of the vibration. A curve similar to the ground energy level may be drawn for the excited states. The minima of these two curves, however, do not coincide. After absorption of a quantum of light energy by a diatomic molecule in the lowest vibrational level of the ground state, the electron is transferred to the excited state. The electron changes from the lowest vibrational level of the ground state to one of the highest vibrational levels of the excited state. This transition usually takes 10^{-15} sec.

The electron may spend some 10^{-4} sec in the excited state. During this time, energy is dissipated as the electron falls through lower vibrational levels of the excited state. Then, in the case of fluorescence, the electron jumps back to one of the higher vibrational levels of the ground energy state; the energy freed during this process is emitted in the form of an energy quantum of light. As can be seen by comparing the length of the arrows "Absorption" and "Fluorescence" in (Fig. 2), the emitted energy is lower than the absorbed energy, and therefore the wavelength of the emitted light will be longer than that of absorbed light. During the absorption and emission processes, the interatomic distance in the molecule remains constant because the transition of electrons occurs much faster than movements of molecular atoms. This is known as the *Franck-Condon principle*. Absorption of light is highly specific for various molecules. Radiation of a particular energy is absorbed only by a characteristic atomic or molecular structure.

In fluorescence, transition of π electrons are most important. Absorption of light quanta by π electrons occurs in the visible and near-ultraviolet light ranges. The σ electrons, which also participate in bonding, may be involved in light absorptions of higher

energy (i.e., in the high-energy ultraviolet range). The transition of π electrons from a ground to an excited energy state may be depicted as $\pi \rightarrow \pi^*$. Electrons which are not involved in bonding may be lifted to the next free π orbital by relatively low energy quanta. The latter type of transition (which often may interfere with fluorescence, i.e., $n \rightarrow \pi^*$) has a much longer lifetime of the excited state than that of $\pi \rightarrow \pi^*$ transitions.

Transition from ground energy levels may occur to the excited singlet electron state (S*) or to the excited triplet state (T*). S* and T* differ in the spin of the electrons. By definition, electrons have a spin equal to ±1/2. Molecules in the ground state have an even number of electrons with + and -spins. Each electron with a spin of 1/2 is matched by another electron of a spin of -1/2. The orbital angular momentum or "multiplicity" (M) is related to spin(s) by

$$M = 2s + 1 \tag{3}$$

M is 1 when all electrons are paired. This characterizes the *singlet* electronic state. Two unpaired electrons (i.e., two electrons with the same spin) result in M = 3. This electronic state is termed the *triplet state*. According to Hund's rule, triplet states are in general of lower energy than the equivalent singlet states. Transitions from singlet to triplet states and vice versa may occur and are known as *intersystem crossings*.

Possible consequences resulting from absorption of light quanta are summarized in the *Jablonski diagram* shown in Fig. 3. Absorption of light quantum may lead to fluorescence as described earlier. Alternatively, the acquired energy may be lost by collision deactivation or quenching, a process in which no radiation is emitted (*radiationless deactivation*). Furthermore, intersystem crossing may occur: The energy level of excited singlet

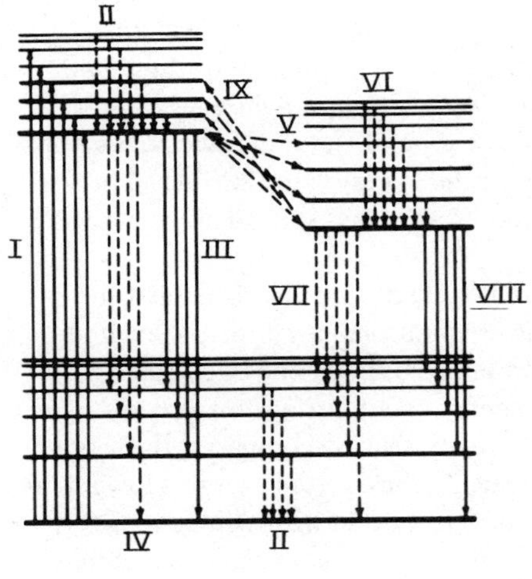

Figure 3 Energy-level diagram of a typical organic molecule inducing only ground singlet, first excited singlet, and its corresponding triplet state. Solid lines indicate radiational transitions and dashed lines indicate nonradiational transitions. I, Absorption; II, radiationless vibrational deactivation; III, fluorescence; IV, quenching of excited singlet state; V, radiationless intersystem crossing; VI, radiationless vibrational deactivation; VII, quenching of first triplet state; VIII, phosphorescence; IX, radiationless intersystem crossing. (From Udenfriend, 1969.)

state is abandoned and the triplet state is reached, a process in which no radiation is produced; the electron may drop to the ground energy levels from the excited triplet state, emitting phosphorescence. Phosphorescence is characterized by the *afterglow effect*, the continuation of light emission after removal of the exciting light source. The transition times in phosphorescence are much longer than in fluorescence (10^{-4} to 10 sec). More likely than phosphorescence, quenching of triplet states may occur, that is, radiationless deactivation, or *collision deactivation*. If additional energy is absorbed, triplet states may revert to singlet states. In this case, the newly reached singlet state may be abandoned, giving rise to fluorescence. The transition time of this fluorescence obviously is longer than that of fluorescence not involving passage through a triplet state; it is therefore termed *delayed* fluorescence. Direct transitions of electrons from ground to triplet states are extremely rare.

Absorption of light quanta may have still other consequences. The absorbed energy may be transferred from one molecule to another. This process is called *sensibilized* fluorescence. Alternatively, the excited molecule may collide with other molecules, or the absorbed energy may be set free as heat. Rarely, photodissociation may result when the absorbed light energy is sufficient to increase the interatomic distance significantly and the molecule breaks up. Similarly improbable is ionization by photoprocesses. Finally, the excited molecule may undergo intramolecular changes or associate with other molecules.

The efficiency of a fluorescent conditions is defined by the fluorescence quantum yield. The fluorescence quantum yield (Φ) is defined by

$$\Phi = \frac{\text{number of quanta emitted}}{\text{number of quanta absorbed}} \qquad (4)$$

Nonfluorescent molecules obviously have $\Phi = 0$, and the greater the fluorescence efficiency of a compound, the higher the value of Φ. Φ decreases with increasing temperature; changes in fluorescence intensity are roughly 1% per 1°C; increased temperature lowers fluorescence efficiency because it favors dissipation of energy by vibrational energy transitions. Quantum yields of some compounds are given in Table 2.

The fluorescence lifetime τ is in the range of 10^{-9} sec for most fluorescent organic compounds (Table 2). The probability P that a given molecule is still excited at time t after absorption is

$$P_t = e^{-t/\tau} \qquad (5)$$

Table 2 Characteristics of Some Fluorochromes (at 21–25°C)

	Quantum yield	Lifetime, τ (nsec)
Anthracene (in hexane)	0.33	5.75
Fluorescein (in 0.1 N NaOH)	0.92	4.6
Rhodamine B (in ethanol)	0.97	–
GO 6 (in water)	–	5.8
Eosin (in 0.1 N NaOH)	0.19	1.7
Uranyl acetate (in water)	0.04	–

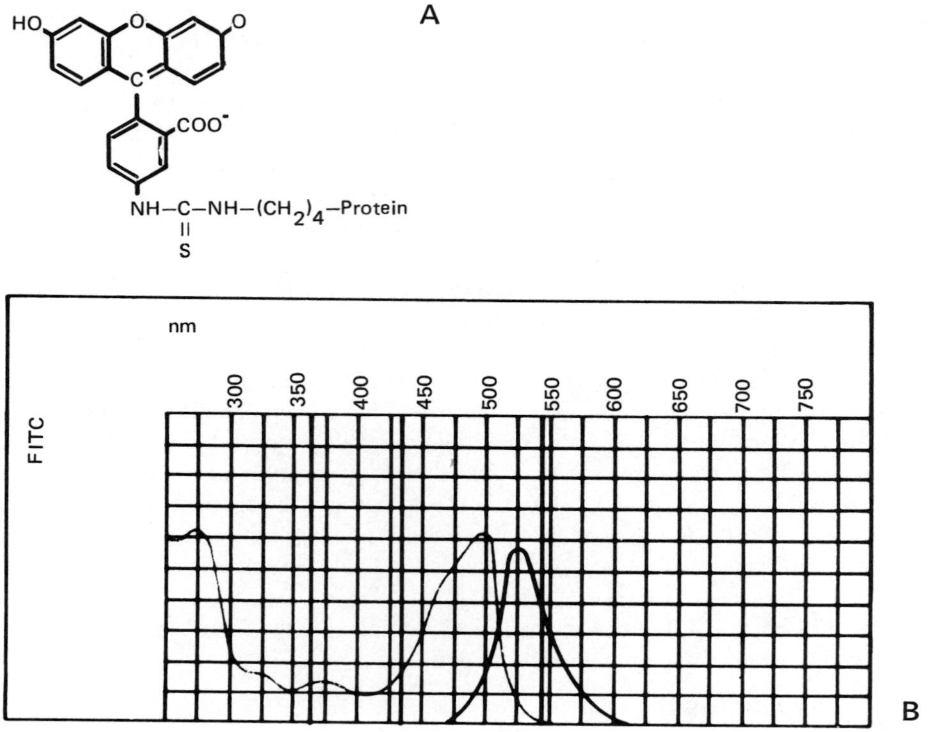

Figure 4 Fluoresceinisothiocyanate (FITC). (A) molecular structure. (From Nairn, 1976). (B) absorption (gray) and emission (black) spectra. The dark vertical lines indicate maxima of energy output by high-pressure mercury vapor lamps. (From Wick et al., 1976.)

Fluorescence intensity is related to fluorescence lifetime according to

$$I = I_0 e^{-t/\tau} \tag{6}$$

where I is the fluorescence at time t and I_0 is the maximum fluorescence during excitation. Table 2 lists τ values for some fluorescent compounds.

Each fluorescent molecule has characteristic excitation and emission spectra. Excitation and emission spectra of fluorescein and rhodamine are shown in Figs. 4 and 5. As a general rule, the longest wavelength peak of an excitation spectrum should be used to induce fluorescence. Selection of the lowest excitation energy minimizes the risks of molecular decomposition and of changes in interatomic distance, dipole moment, and configuration. This reduces the extent of processes that may interfere with fluorescence. Both the shape of the emission spectrum and the quantum yield are *independent* of the wavelength used for excitation. Obviously, when the exciting radiation has a wavelength differing from the excitation maxima, the amount of energy absorbed will be suboptimal, as will be the energy emitted; Φ, however, will be the same as observed with optimal excitation wavelengths. Theoretically, emission spectra of fluorochromes should be mirror images of their absorption spectra. Deviations from this principle indicate the occurrence of "scatter" (Rayleigh-Tyndall or Raman scatter) or the presence of impurities.

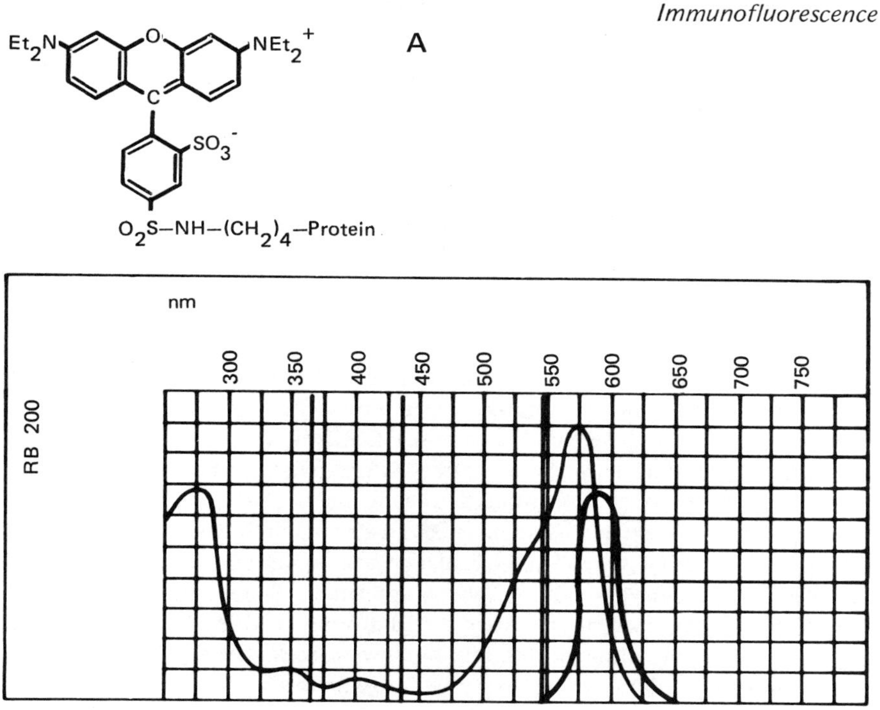

B

Figure 5 Lissamine rhodamine B (RB 200). (A) molecular structure. (From Nairn, 1976.) (B) absorption (gray) and emission (black) spectra. The dark vertical lines indicate maxima of energy output by high-pressure mercury vapor lamps. (From Wick et al., 1976.)

In fluorescence, the wavelength of emitted energy and the wavelength of exciting energy are related according to the

$$\text{"Stokes' shift"} = 10^7 \times \left(\frac{1}{\lambda_{ex}} - \frac{1}{\lambda_{em}} \right) \quad (7)$$

where λ_{ex} and λ_{em} are the maximum wavelengths for excitation and emission, respectively. The value for Stokes' shift is a physical constant characteristic for individual luminescent molecular species. Stokes' shift values vary with changes of the molecular conformation, and therefore depend also on the pH. The shift in wavelength between excitation and emission maxima may range from 10 to 280 nm.

Over a range of rather low concentration of fluorescent compounds, fluorescence intensity remains roughly linear and may be estimated using an expression analogous to Beer's law in spectrophotometry, namely

$$F = K\psi I_0 \epsilon bc \quad (8)$$

with F being the fluorescence intensity, ϵ the molar absorptivity, b the length of the light path through the solution or "cell" containing the fluorochrome, and c the molar concentration.* At high concentrations, quenching becomes significant and may lead to a net

*For higher concentrations of fluorochromes, use the equation

$$F = \psi I_0 (1 - e^{-\epsilon bc}) \quad (9)$$

decrease of fluorescence intensity, and scattering may reach significant levels. Thus the linear relationship of concentration of fluorochrome to fluorescence intensity is seen only when less than 5% of the exciting radiation is absorbed.

Use of fluorescence has some advantages over use of absorption as markers for the presence of a well-defined reactant. The advantages of fluorescence in both photometric and histologic evaluation are primarily its high sensitivity and specificity. Fluorometric methods in general may be up to 1000 times more sensitive than spectrophotometry. Similarly, the spectral specificity of fluorescent dyes is greater than that of simple dyes: first, only a few of the molecules that absorb light are fluorescent; and second, two variables define fluorochromes (excitation and emission spectra) compared to only one for simple dyes. Thus two compounds that may absorb at the same wavelength may have two readily distinguishable emission spectra, or vice versa.

Quenching due to photochemical decomposition is likely to occur with very high energy (i.e., short-wave ultraviolet) light irradiation. Alternatively, quenching may be caused by excessive concentration of fluorochrome, as already mentioned. When absorbance of a medium containing fluorescent molecules becomes greater than 5%, light is not evenly distributed along the light path. Fluorescent molecules near the light source will absorb so much light that molecules farther away from the light source will not be reached by an optimum of light energy. Fluorescence will not be distributed uniformly. Ultimately, no light may be able to pass the first "rows" of molecules, causing the *inner-cell effect*. This concentration quenching may contribute to the "prozone" phenomena seen sometimes with immunofluorescence. In addition, the energy transfer between molecules increases with the sixth power of the concentration and becomes significant at concentrations of 10^{-3} M or above. Similarly, high concentrations of fluorescent molecules favor dimer or polymer formation; these compounds have a longer emission wavelength than the monomer, and this emission easily may be missed ("*excimer quenching*"). Oxygen present in solutions is a potent fluorescence quencher (10^{-3} M oxygen may educe fluorescence by as much as 20%). Finally, trace amounts of iodide and nitrogen oxides may decrease fluorochrome emission. It is obvious that the presence of highly absorbing compounds (e.g., dichromate) may influence the intensity of fluorescence by competing with the fluorochrome for light energy.

The best way to minimize quenching is:

1. To use the lowest-energy excitation maximum of the fluorochrome.
2. To use low concentrations of fluorochrome or fluorochrome-conjugated immune reactants.
3. To avoid deleterious substances (e.g., oxygen, iodide, dichromate) in all reagents, media, or glassware.
4. To interpret or record the results of immunofluorescence immediately after excitation; therefore, when not observing the preparations, slides should not be left with the light beam focused on the sections.
5. To read slides as soon as possible after finishing the preparation, although storage of slides is feasible (in the dark at 4°C) for some days.

III. FLUOROCHROMES

With a fluorochrome to be used in IF, fluorescence should predominate over radiationless decay and photochemical reactions. To assure this, the following premises with regard to the molecular structure should be fulfilled:

Table 3 Molecular structure and Fluorescence

	Quantum yield	λ_{abs}	λ_{em}	Color of fluorescence
Benzene	0.11	205	278	Ultraviolet
Naphthalene	0.29	286	321	Ultraviolet
Anthracene	0.46	365	400	Blue
Tetracene	0.60	390	480	Green
Pentacene	0.52	580	640	Red

1. The spin-allowed electronic absorption transition of lowest energy should be very intense.
2. The energy of the lowest spin-allowed absorption transition should be low.
3. The electron absorbing the energy should not be *strongly* involved in bonding (π but not σ electrons).
4. The structure and functional groups of the molecule should not promote radiationless processes.

Aromatic compounds usually exhibit strong fluorescence because of the availability of π electrons: $\pi \rightarrow \pi^*$ transitions are strongly favored in aromatic compounds. Other highly conjugated compounds (e.g., carotenes) may be fluorescent, also because of the availability of $\pi \rightarrow \pi^*$ transitions. Heterocyclic compounds and molecules containing carbonyl groups may also fluoresce because of $n \rightarrow \pi^*$ transitions (see Sec. II). These compounds have, however, a weaker fluorescence than aromatic hydrocarbons or heteroaromatics.

By increasing the degree of conjugation, aromatic compounds are made more intensely fluorescent; concomitantly, there is a shift toward longer wavelengths in the emission maxima (Table 3). Similarly, addition of polar groups increases the light absorption.

Fused aromatic rings seem to be essential for many fluorescent dyes. Thus, fluorescein, which possesses fused aromatic rings, does fluoresce, but the analogous compound lacking fused rings, phenolphthalein, does not. This occurs also with rosamine and malachite green, respectively. Fluorescein and rosamine also show rigidity, planarity, and extended conjugated bond system. Oxygen bridges enhance all of these characteristics. The importance of rigidity for fluorescence is illustrated by some azodyes which become fluorescent only after chelation with metal ions. Planarity is important, as shown by the fact that only plane transisomers of thioindigodyes fluoresce.

IV. COUPLING OF FLUOROCHROMES TO PROTEINS (CONJUGATION)

Tagging of proteins with fluorochromes has to be stable (i.e., covalent bonding), and the process of coupling the two compounds has to have as little effect as possible on the immunological reactivity of the protein and on the luminescence of the fluorochrome. Modifications of fluorochromes necessary for attachment to immunological reagents may diminish or abolish the fluorescence; therefore, many fluorochromes used successfully as fluorescent histological stains are *not* useful in immunohistology.

Fluorochromes to be used in immunofluorescence should fulfill the following criteria, first proposed by Chadwick et al. (1965) and by Pearse (1968).

1. A chemical group should be available on the fluorochrome which makes stable covalent bonding with proteins possible without destroying the fluorescent potential of the fluorochrome.
2. Unconjugated fluorochrome should be separable from the conjugate.
3. The quantum yield of the dye should be high and not be impaired by conjugation.
4. The conjugation of fluorochrome to immune reactant should not diminish the immunological specificity and reactivity of the protein.
5. The wavelength of the fluorescence of the conjugate should differ significantly from autofluorescence of the tissue.
6. The conjugate should be stable under normal storage conditions.
7. The conjugation procedure for a fluorochrome should be simple and efficient.

The main classes of chemically reactive groups used to couple fluorochromes to proteins are (1) diazonium salts, (2) acid chlorides, and of most practical importance, (3) isothiocyanates (or isocyanates). Sometimes, such groups are already present in a native fluorochrome; more often, these reactive groups have to be introduced into fluorochromes by synthesis (Nairn, 1976).

1. Diazonium salts have been used for the conjugation of dyes to immunologically reactive proteins since Heidelberger's and Landsteiner's classical studies. Fluorochromes are coupled to aromatic amino acids. The diazonium salt may then be prepared using nitrous acid at 0°C following the reaction

$$RNH_2 + HNO + HCl \rightarrow RN_2 \cdot Cl + 2H_2O \tag{10}$$

R being the aromatic amino acid derivative of the fluorochrome. Subsequently, the diazonium salt is conjugated with the protein (e.g., immunoglobulin) in alkaline solution. Immunological reactants retain much of their biological activity when low ratios of fluorochrome to protein are used. Diazonium salts preferentially react with tyrosin and histidine (phenolic group) tryptophan (indole group), imino, and aliphatic amino groups of the proteins. Diazotized nuclear fast red conjugated to proteins provides a red-fluorescent conjugate. Unfortunately, the fluorochrome is chemically not well defined, precluding wide use of this conjugate.

2. Acid chloride groups are easily introduced into fluorochromes containing sulfonic acid groups by treatment with phosphorus pentachloride:

$$RSO_3H + PCl_5 \rightarrow RSO_2Cl + POCl_3 + HCl \tag{11}$$

The sulfonyl chloride, which is stable in acids or stored dried under anhydrous conditions, can be extracted with organic solvents (e.g., acetone) or, when insoluble in water, may be separated by filtration from an aqueous solution. Fluorescence is not appreciably altered by the acid chloride group, and conjugation to protein occurs readily. Acid chloride derivatives of fluorochromes react with histidine, tyrosin, the thiol group of cystein, and the ε-amino group of lysine residues. Although sulfofluorescein and lissamine rhodamine B have been coupled to immunological reagents via acid chloride groups and used successfully in immunofluorescent assays, their use is rather limited as a consequence of the better performance of isothiocyanate derivates of the same dyes. 1-Dimethylaminophthalene-5-sulfonic acid (DANS), fluolite C, and lissamine flavine FFS give a blue to blue-green fluorescence and therefore usually are not suitable for work with tissue that shows autofluorescence in the same spectral region.

3. The fluorochromes used most commonly and with best results in immunofluorescence are isothiocyanates of fluorescein and rhodamine. Conjugation of proteins with

phenylisocyanate was reported in 1933 by Hopkins and Wormall, and used by Creech and Jones to conjugate isocyanate derivatives of aromatic hydrocarbons to proteins. Coons used this method in 1942 to couple fluorescein to antibody. The conjugate gave a bright green fluorescence, allowing for good contrast with the background of tissue autofluorescence (Fig. 1). Isothiocyanates, however, are much more stable than isocyanates, and are almost exclusively used nowadays. The reaction used to introduce an isothiocyanate group into a fluorochrome is the following:

$$RNH_2 + CSCl_2 \rightarrow RN{=}C{=}S + 2HCl \qquad (12)$$

There are two isomers (Isomer I and II) of FITC. It is best used as crystallized isomer I, which can be stored for at least 3 years at 4°C without deterioration. Light should be kept out of the fluorochrome and its container as should any water (store with desiccators). Similarly, tetramethylrhodamine is used as an isothiocyanate for conjugation with protein (TMRITC; sometimes also TRITC) TMRITC crystalline isomer R seems to perform better than the amorphous compound. Both FITC and TMRITC, once dissolved, should be used immediately. The purity of the preparations may be assayed using paper or thin layer chromatography; isothiocyanate content may be estimated by infrared absorption.

The most reactive group of the protein in the conjugation reaction with FITC or TRMITC is the ϵ-amino group of lysine residues, but reactions with thiol groups also seem probable. Sodium azide should not be used to preserve antisera or antigenic preparations prior to the conjugation step, as it interferes with coupling of fluorochromes to proteins.

A large number of protocols have been proposed for the actual conjugation procedure and for further information the interested reader is referred to the publications cited in the reference section (in particular, Goldman, 1968; Johnson et al., 1978; Nairn, 1976; Wick et al., 1976). During the conjugation procedure, the following general guidelines have to be observed:

1. The fluorochromes should be added slowly, drop by drop, to the solution containing the protein, to avoid "local concentration" effects.
2. Continuous and efficient but not too vigorous stirring during the conjugation procedure should allow uniform contact between fluorochrome and protein.
3. The pH has to be maintained well in the alkaline region, optimally between 9.0 and 9.5. The ratio of fluorochrome to protein in the reaction mixture roughly determines the effective number of fluorochrome molecules per protein molecule (the molar F/P ratio of the conjugate). Salt concentration and pH are important variables influencing the outcome of the conjugation procedure.

At 0 to 2°C, 50% of the reaction is completed after 2 hr and the conjugation of proteins to FITC reaches a plateau after 24 hr; at room temperature, the conjugation is virtually complete after 30 min, and similar behavior can be expected in TMRITC conjugation to protein. Acid chloride derivatives of fluorochrome usually react faster with proteins; lissamine rhodamine sulfonyl chloride reacts with protein in 30 min at 0 to 2°C, and conjugation of DANS to protein is completed after 6 hr under standard conditions. Usually, it is advisable to proceed with the conjugation at 0 to 2°C because of the better preservation of the immunological reactivity at this temperature.

V. PURIFICATION OF CONJUGATES

Upon termination of the conjugation reaction, the reaction mixture contains a number of undesired components: organic solvents, buffer salts, unreacted fluorescent material (UFM), and inefficiently reacted protein. Organ solvents and buffer salts—and some of the UFM—may be removed by dialysis against borate saline or phosphate buffered saline. Prolonged dialysis is usually required to remove UFM efficiently because fluorochromes, especially impurities found in some of the commercial preparations, are strongly adsorbed to serum proteins, especially to albumin. Additional purification procedures have, therefore, been developed, among them absorption with tissue powder, repeated protein precipitation, and extraction with charcoal. However, the best available methods for conjugate purification are gel filtration, fractionation on DEAE-cellulose, or electrophoresis in polyvinyl chloride (PVC) or starch. These methods allow for complete removal of UFM with no or only minimal loss of protein or immunological reactivity of proteins.

Fractionation on DEAE-cellulose, based on its ion exchange properties, allows for removal of UFM and of excessively labeled and therefore strongly acidic molecular species of the conjugate. The latter components are the most notorious causes for nonspecific (predominantly coulombic) interactions with positively charged groups of tissue sections. PVC electrophoresis allows for elimination of UFM and fractionation of conjugate molecular species according to varying ratios of fluorochrome to protein. Bands with different F/P ratios differ in color and are easily cut out of the PVC carrier. In addition to obtaining conjugates with very narrowly defined F/P ratios, the electrophoresis allows for elimination of insufficiently labeled or unconjugated protein. This protein may compete with the conjugate for binding sites and thus inhibit the desired immune reaction and therefore the staining of the preparation.

Immunologically reactive fluorochrome-protein conjugates should be aliquoted (0.05- to 0.2-ml aliquots seem to be most practical) and stored in plastic or glass vials at -30°C. Alternatively, conjugates may be lyophilized and stored as dry powder at 0 to 4°C. Prolonged storage may lead to enhanced formation of aggregates. Conjugates should therefore be centrifuged at top speed of table centrifuges to remove large aggregates which may precipitate on tissue, slides, or around cells in the preparation to be stained. Merthiolate and sodium azide may be used to preserve conjugates against bacterial contamination.

VI. CHARACTERIZATION OF FITC AND TRMITC CONJUGATES AND DETERMINATION OF OPTIMAL WORKING DILUTIONS

Fluorochrome-protein conjugates should be characterized with regard to a number of properties (Beutner et al., 1968; Nairn, 1976; Wick et al., 1976). Specificity may be ascertained in double diffusion in gel (Ouchterlony) assays and in immunoelectrophoresis or by testing the conjugate against antigen to antibody in any other suitable immunological procedure. The optimal test system for specificity of a fluorochrome-protein conjugate is immunofluorescence itself. Thus fluorochrome-conjugated antibodies to immunoglobulin classes and subclasses may be tested against myeloma cells with known specificity. Similarly, antibodies to microorganisms and multicellular parasites may be tested for specificity in IF assays.

Purified antigens may also be coupled to agarose or polymeric dextrans (Sepharose). This *defined antigenic substrate system* (DASS) allows for probing for specificity of a

conjugate in immunofluorescence using photometric evaluation. Many of the DASS preparations may be lyophilized without loss of activity and stored over a long period of time with stable antigenic reactivity (Knapp, 1982). Similar preparations have been obtained using antigen cross-linked by glutaraldehyde.

Sensitivity of a purified conjugate depends both on the titer and the avidity of the conjugated antibody (or on concentration or accessibility of antigenic determinants of conjugated antigen) and on the number of fluorochrome molecules attached to the protein molecule. The antibody reactivity may be determined using antigen-antibody precipitation in the equivalence zone (see Sec. I) as described by Heidelberger. This procedure, however, is not economical as a routine characterization method for conjugates. A rough estimate of antibody reactivity of a conjugate may be obtained using simple double diffusion in gel against a solution of antigen (1 mg/ml). The conjugate is serially diluted and the titers are transcribed as *unitage*: thus a conjugate giving a visible precipitation reaction up to a dilution of 1:8 is considered as having 8 "precipitation units." A more accurate estimate of antibody content may be obtained by a reversed radial immunodiffusion procedure.

The amount of whole protein is determined to assess the F/P (i.e., fluorochrome to protein) ratio and to evaluate the presence of "ballast" protein. An antiserum usually contains 0.8 to 3% antibody out of measurable protein, and its gamma globulin fractions 3 to 15%. Average preparations contain about 50 to 200 mg per 100 ml. The amount of total protein decreases from 6 to 7 g/ml for the crude antiserum to 1.2 to 1.4 per 100 ml in the gamma globulin fraction. Obviously, antibody concentration may increase impressively in relation to total protein after affinity chromatography. A conjugate must contain a high enough titer of antibody reactivity (or of antigenic determinants) to allow detection of its corresponding immune reactant.

Molar F/P ratios are determined measuring the concentrations of the fluorochrome and protein by spectrophotometry. TMRITC is measured at $\lambda = 550$ nm, FITC at $\lambda = 495$ nm, and protein at $\lambda = 280$ nm. For determination of FITC, a standard of fluorescein diamine (FDA) is used. The ratio of weights of fluorochrome (F) and protein (P) are then transformed into molar ratios using the molecular weights (MwF or MwP). Thus computation of F/P ratios follows the equation

$$\text{F/P molar} = \frac{\dfrac{\text{F}}{\text{MwF}}}{\dfrac{\text{P}}{\text{MwP}}} \tag{13}$$

For example for FITC-conjugated IgG, this is

$$\text{F/P molar} = 0.411\,\frac{\text{F}}{\text{P}} \tag{14}$$

The optimal working dilution (WD) of conjugates has to be established for each individual test system. For direct staining procedures (see Sec. IX), simple titration of the conjugate allows for determination of dilutions giving no background staining and optimal desired staining intensity. The second-highest dilution of conjugate giving satisfactory desired staining should be used.

For direct staining procedures (see Sec. IX), a chessboard titration with a known positive antiserum should be performed as described by Beutner et al. in 1968. The protocol and results obtained in a chessboard titration for a human anti-IgG conjugate in

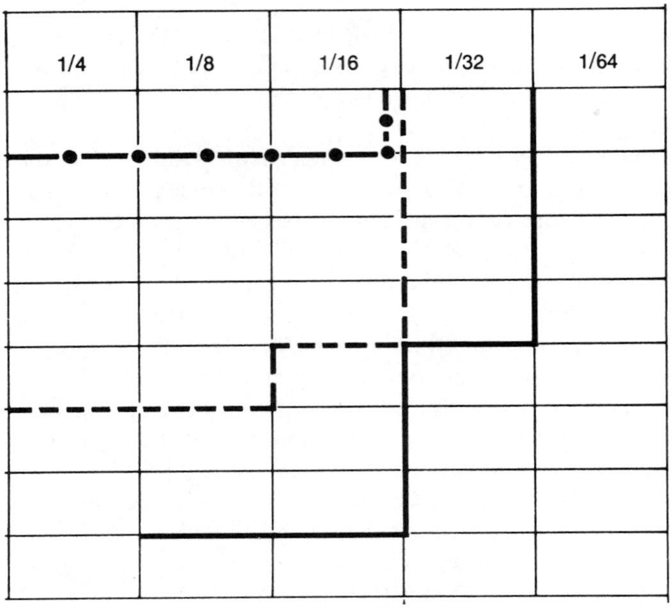

Figure 6 Chessboard titration. Decreasing concentrations of a FITC-conjugated antiserum to human IgG are plotted horizontally (the values given represent precipitation units). Decreasing (top to bottom) concentrations of a patient's serum containing antinuclear factors are plotted vertically. Full line, FITC conjugate with F/P = 4.2; dashed line, FITC conjugate with F/P = 1.4; dashed-dotted line, FITC conjugate with F/P = 0.7. The PEP is the same for all these conjugates; PT varies with different F/P ratios.

an antinuclear antibody system on chicken erythrocyte nuclei are shown in Fig. 6. Dilutions of conjugate are given in units determined in double diffusion in gel assays (from left to right); a known antiserum to nuclear antigens is diluted from 1:40 to 1:5120 (from top to bottom). The titers obtained with the antiserum remain the same over a broad range of conjugate dilutions. This titer is termed *plateau titer* (PT). The titer drops for at least two titer steps at a certain dilution of conjugate (usually in the range 1/8 to 1/32 unit). This conjugate dilution is termed *plateau end point* (PEP). *PEP depends on antibody concentration of the conjugate and PT on its F/P ratio*. The background staining seen with normal human serum disappears at higher conjugate dilutions. The optimal dilution is the second to the left from PEP (Fig. 6), not the dilution directly at PEP because minor technical errors could push the actual dilution of the conjugate to the right of the PEP.

The quality of the conjugate has a paramount influence on the ratio of desired specific staining (DSS) to undesired and nonspecific staining (NSS). In an attempt to express this relation, the following approximation may be used:

$$\frac{\text{DSS}}{\text{NSS}} \hat{=} k \frac{\text{Ab/P}}{\text{F/P}} \hat{=} k \cdot \text{Ab/F} \qquad (15)$$

in which F is the concentration of fluorochrome, P the concentration of protein, Ab the concentration of specific antibody, and k a coefficient determined by quality of antigenic

substrate, optical system used, and similar rather poorly defined parameters. The equation suggests that increased DSS/NSS ratios may be associated with increased specificity (Ab/P) and decreased sensitivity (F/P) of the conjugate.

VII. THE OPTICAL SYSTEM

A schematic representation of the optical system used for transmitted illumination and epiillumination fluorescence microscopy is shown in Fig. 7. In transmission immunofluorescence microscopy, the light passes an exciter filter and is focused on the slide containing the preparation under study by a condenser; light emitted during fluorescence enters the objective lens, passes a barrier filter, and reaches the observer's eye via the ocular. With *epiillumination*, after passing an exciter filter, the light passes a dichroic mirror and is directed to the specimen by an objective lens; the emitted light passes the objective lens again, passes through the mirror and a barrier filter, and reaches the observer's eye through the ocular. Epiillumination offers a number of advantages over transmission microscopy: (1) the thickness of the preparation under scrutiny is of no importance, as the light does not have to pass through the preparation as in transmission microscopy; (2) the area of the preparation that is exposed to exciting light is kept roughly limited to the viewing field, which minimizes the quenching of the preparation, especially when high magnification is used; (3) any inner-cell effect (see Sec. II) is avoided; and (4) no separate condenser lenses are necessary.

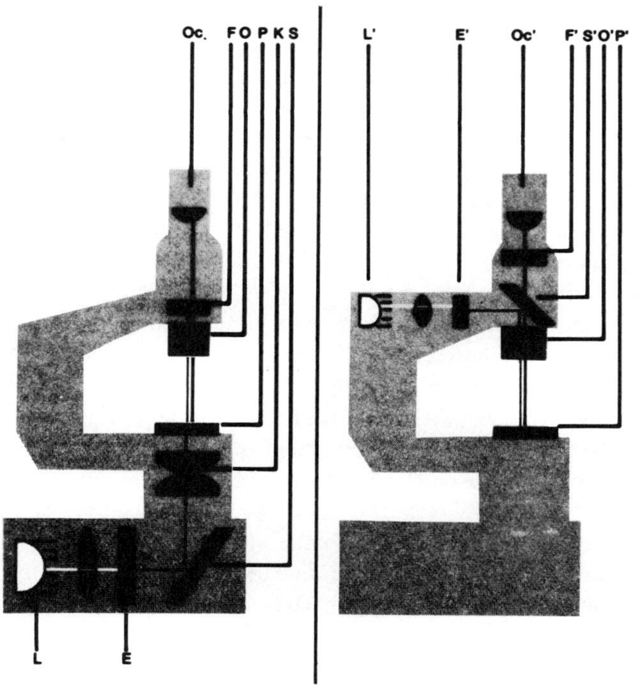

Figure 7 Immunofluorescence microscopes. (Left) Transmitted light; (right) epiillumination. L, L′, lamp; E, E′, exciter filter; S, S′, mirror; K, K′, condensor; P, P′, preparation to be examined; O, O′, objective; F, F′, barrier filter; Oc, Oc′, ocular. (From Wick et al., 1976.)

A. The Light Source

Three types of lamps may be used for fluorescence microscopy:

1. Low-voltage (tungsten) halogen lamps
2. Gas discharge lamps
3. Lasers

Low-voltage halogen lamps usually are of the iodine-quartz type and have an output of 60 to 100 W. They have a continuous spectrum and are useful for routine IF examinations. They produce no ultraviolet (UV) light.

High pressure (up to 70 atm) mercury vapor lamps provide powerful ultraviolet blue and green light in a discontinuous spectrum. The brightness of these lamps reaches 30,000 stilbs (or candles/cm^2); 50- to 200-W lamps are widely used in immunofluorescence. The main emission peaks of the most popular 200-W lamp are at 365, 405, 435, and 546 nm, with the full spectrum encompassing 280 to 600 nm. The intensity of emitted light varies during operation over short time periods, especially when the lamp has been operated for 80 to 100 hr. With age, the brightness of the lamp diminishes significantly.

The xenon lamp has a relatively stable output of light and may be used for operating time up to 1000 hr. Its disadvantages include powerful emission at long wavelengths and the high price of the lamp and the housing.

Lasers used as light sources have the advantages of high-intensity monochromatic and monophasic light of constant energy (Wick et al. , 1976). For immunofluorescence the most suitable type is the argon ion laser (Schauenstein et al., 1982), which is tunable to emit two bands at wavelengths (488 and 514 nm), corresponding to the excitation maxima of fluorescein and rhodamin dyes, respectively.

Experimental work with argon lasers as an incident light source for immunofluorescence microscopes has revealed methodological advantages and interesting theoretical findings: (1) Using a laser model with an output of 10 mW/sec, about 1000 times higher fluorescence intensities can be obtained than with conventional light sources (e.g., high-pressure mercury vapor bulbs). Thus for systems with negligible background (e.g., cell membrane fluorescence) this kind of illumination may drastically raise the sensitivity of the method. (2) Recovery of fluorescence intensity after short-time bleaching has been observed with short-pulse laser excitations of fluorescein dyes after dark intervals of sufficient length [Fig. 8(C)]. (3) For the systematic analysis of the bleaching characteristics of FITC during excitation, laser illumination has been particularly useful. As this process is extremely sensitive to physicochemical changes in the environment of the excited dye molecule, it has been used successfully to correlate certain bleaching characteristics with the nature of binding of antibody-FITC conjugates. Figure 8(A) shows bleaching curves of an anti-rabbit Ig conjugate adsorbed to Sephadex G-25 beads coated with bovine serum albumin (BSA). The specific binding of an anti-BSA conjugate to the same bead preparation resulted in a marked inhibition of FITC bleaching, as seen in Fig. 8(B). In the future, this might provide a valuable tool for identifying nonspecific staining effects in immunofluorescence.

B. Exciter Filters

The light sources available for immunofluorescence, with the exception of lasers, emit energy of a spectrum quite different from the absorption spectra of the fluorochromes used. To avoid undesired light, which may have untoward effects on the fluorochrome,

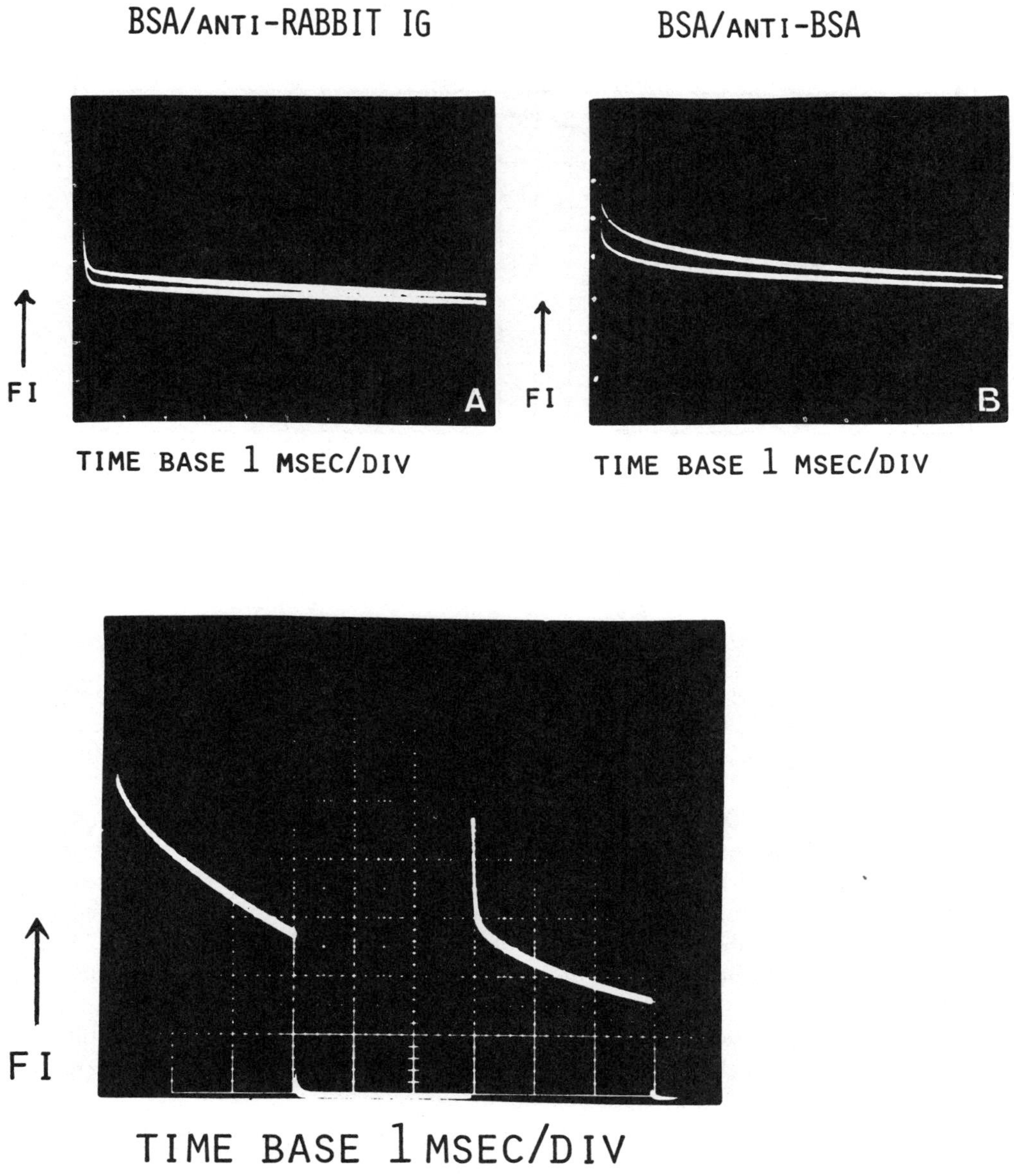

Figure 8 (A) Discrimination of nonspecific immunofluorescence staining by bleaching analysis. A marked inhibition of the initial bleaching is observed with FITC-antibody conjugate (anti-BSA) specifically bound to insoluble carriers (Sephadex G-25) coated with BSA (B) as compared to the bleaching characteristics of an anti-rabbit Ig conjugate nonspecifically adsorbed to the same antigenic substrate (A) (FI, fluorescence intensity). (C) Bleaching and recovery of FITC fluorescence under laser excitation. An FITC-labeled Sephadex G-25 bead was exposed to two pulses of laser light (488 nm) each of 3-msec duration and 3 msec apart. Eighty percent of the initial fluorescence intensity was recovered rapidly.

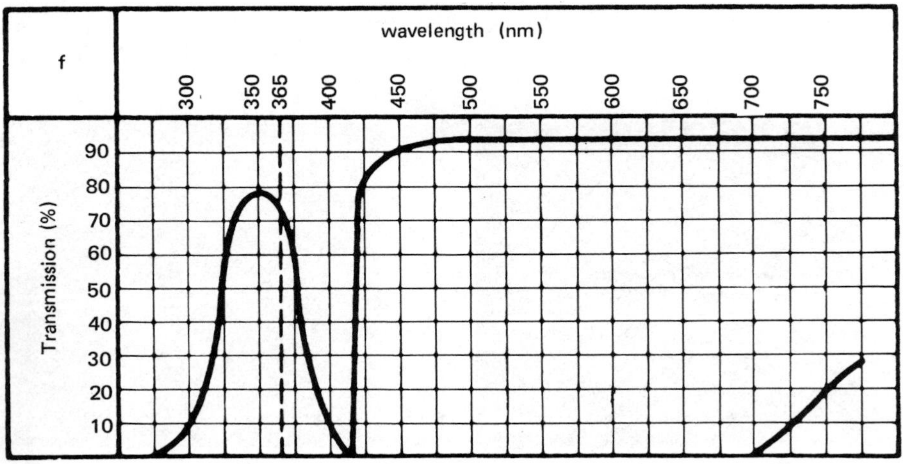

Figure 9 Transmittance curves: 1.5 UG1 and barrier filter KV418. [From F. Herzog et al. (1973), J. Immunol. Methods *3*, 211.]

on extent of autofluorescence, and on background qualities, exciter filters are placed in the pathway of the light as depicted in Fig. 7.

Two basic types of performance may be obtained from filters: (1) filters may transmit only ultraviolet light and stop all visible illumination and heat, or (2) filters may transmit only a band of the visible spectrum around the absorption maximum of a fluorochrome.

An example of a filter for ultraviolet transmission is the UG1, whose spectrum of transmitted light is depicted in Fig. 9. This filter is suitable for dark-field work. An advantage in using UV light for excitation of FITC may be good visibility of tissue structures caused by intensive blue autofluorescence. Filters of the second type, giving a "band" centering around an absorption maximum of individual fluorochromes may give stronger light intensities (broadband filters), or may reduce interfering, unnecessary light (narrow-band filters).

Two basic makes of filters are used: (1) colored glasses and (2) interference filters. The first category is cheaper, the second better in selectivity. The following criteria should be considered when selecting exciter filters:

1. Which fluorochrome is used
2. Whether dark- or bright-ground work is done
3. Whether high efficiency or high specificity is required
4. Whether demonstration of tissue structures is of paramount importance
5. Whether rapid fading (quenching) of fluorescence is acceptable (not so for microphography)

A comparison of the intensity and quenching of filters used for FITC-fluorescence is given in Fig. 10.

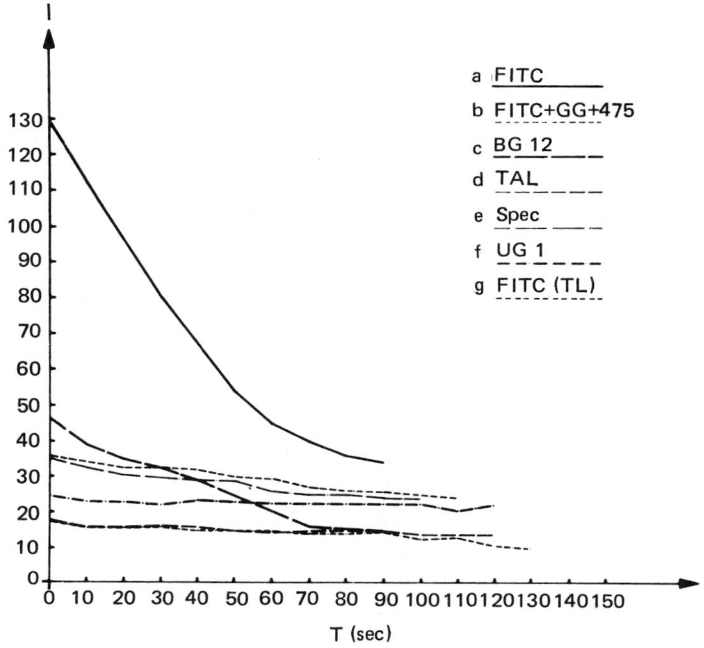

Figure 10 Fading of fluorescence. I, intensity in relative photometric units; a–g, various exciter filters. [From F. Herzog et al. (1973), J. Immunol. Methods *3*, 211.]

C. Condensers

In transmitted-light immunofluorescence, dark-ground condensers are used almost exclusively. The best and most versatile performance of dark-ground condensers is obtained by the use of torus lenses. The torus lens makes it possible to use the whole width of the available light bundle and still prevents almost any light from passing directly into the objective. This type of condenser requires immersion fluid to be placed between the condenser and the slide holding the preparation. Immersion oil has the refractory index most comparable to the glass, but is difficult to remove. Use of the only slightly less refractory glycerol allows easy removal because of its water solubility. Obviously, immersion fluids used have to be free of fluorescent contaminants. All dark-field condensers should have a high numerical aperture and good UV transmission.

D. Dichroic Mirrors

In epiillumination, proper channeling of exciting and emitted light is accomplished by use of mirrors with transmission and reflection properties that change according to the wavelength of the incident light. This is necessary because the incident and emitted light paths cross each other and travel for some time along identical pathways. Such mirrors are produced by applying extremely thin layers of metals onto glass. Modern epiillumination devices have a dichroic mirror allowing work with both fluorescein and rhodamine derivates. It has two transmission and two reflection maxima at the appropriate places in the spectra of the two fluorochromes.

E. Barrier Filters

To assure complete darkness of the background and to avoid damage to the observer's eye when UV light is used, any light used for excitation has to be filtered out before reaching the observer's eye. This is accomplished by appropriate barrier filters (Fig. 9).

VIII. PREPARATION OF ANTIGENS (OR ANTIBODIES) AS SUBSTRATES FOR IMMUNOFLUORESCENCE

In immunohistology, tissue preparation has to fulfill three basic requirements (Sternberger, 1979): (1) antigenicity (or antibody reactivity) should be preserved; (2) the tissue should be amenable to cutting; and (3) antigen (or antibody) should be prevented from being washed out of the tissue. Preservation of antigen or antibody properties obviously is essential. The best way to preserve most of the antigenicity of tissue and still allow for sectioning of the specimen is the use of freezing and cutting in a cryostat. Unfixed frozen sections can be used successfully in many immunofluorescence tests. With water-soluble antigens, mild fixation may be necessary to prevent loss of antigen into the washing fluids. It is useful to cut tissue specimens to obtain thin sections and thus to avoid scattering of the image-forming light. This is especially true for transmission microscopy. Only with very thin tissue (e.g., serosal membranes) and the use of epiillumination may unsectioned tissue be used. The sections should be thinner than the thickness of a cell (i.e., 2 to 6 μm). Very thin sections, 1 μm and below, are extremely well suited for work at high resolution (magnifications of 600 times and above), since they allow for excellent localization, especially when used in conjunction with phase-contrast microscopy. However, it must be kept in mind that with the use of thinner sections, the sensitivity of the procedure is also decreased. The freezing of specimens is most conveniently accomplished by immersion in liquid nitrogen or in a dry ice-acetone mixture. The frozen tissue section is placed on a slide or a coverslip. It may be stored at -80°C or, for a short time, at -30°C. To avoid dehydration of the specimen and to facilitate uniform freezing, the specimen may be embedded in OCT (optimal control temperature fluid). Frozen tissue sections may then be fixed or used unfixed.

Fixatives are agents that cross-link proteins or, less frequently, carbohydrates. At the same time, fixation prevents leakage of water-soluble antigens. Unfortunately, fixation also entails more or less extensive deterioration of antigenicity. Formaldehyde or paraformaldehyde may be used in low concentration (1% at 4°C) for immunofluorescent work, and acetone, alcohols, and ether may be beneficial in some instances. Sometimes, specimens fixed in higher concentrations of formalin at room temperature for prolonged periods of time and embedded in paraffin may be used in immunofluorescence, often only after treatment of the fixed section with trypsin or pronase prior to staining. Carcinoembryonic antigen can be detected in such preparations without any pretreatment; immunoglobulins, fibrin, and hepatitis B antigen only after digestion with trypsin; complement usually cannot be detected even after this procedure. Formaldehyde and glutaraldehyde increase both the autofluorescence of tissues and specific staining. Dimethylsuperimidate, a bifunctional reagent reacting with the ϵ-amino groups of lysine, may leave many antigens unaltered; a fixative containing lysine, paraformaldehyde, and periodate has been proposed to cross-link sialic acid residues. Unfortunately, the latter fixatives have not yet been explored extensively for IF. Incubation of tissue in a detergent (e.g., Triton X-100 as a 0.3% solution added to the antibody or conjugate

preparation) may facilitate penetration in relatively dense tissue components (e.g., nerve fibers). Usually, a compromise between preservation of tissue structure and antigenic reactivity, as well as prevention of loss of antigen during washing, has to be found. Viable cells may be used untreated in suspensions in IF of surface components (*membrane IF*) or may be fixed, especially to allow interaction of conjugates or antisera with intracellular components.

IX. STAINING PROCEDURES

The principle of applying a fluorochrome-tagged and thus detectable immunoreactant to the detection of tissue-bound or tissue-reactive antigen or antibody is used in several variations (Beutner et al., 1968; Goldman, 1968; Johnson et al., 1978; Nairn, 1976; Wick et al., 1976). Each has its own field of application. The tissue sections or cells have to be prevented from drying out during the staining procedure.

1. In *direct IF* staining (DIF, Fig. 11), an antigen (or antibody) in tissue or on cells reacts with antibody (or antigen) coupled to a fluorochrome. Twenty to thirty minutes of incubation with the conjugate at room temperature or 30 to 45 min at 0 to 4°C is sufficient for cell suspensions or tissue sections of 2 to 6 μm thick. For unsectioned preparations or thicker sections, the incubation time has to be increased. After incubation, slides are washed in buffered saline [usually phosphate buffered saline (PBS), 0.1 to 0.15 M, at pH 7.2 to 7.4 is used]. After 30 to 45 min of washing, the slides are wiped off around the tissue specimen, and coverslipped using PBS-buffered glycerol. Slides are usually read immediately upon completion of the staining procedure, but may be kept at 4°C for some days without significant loss of fluorescence activity. Cell suspensions are treated similarly. Direct tests are used in immunohistology, for example, in the evaluation of tissue biopsies.

2. *Indirect immunofluorescence* tests (IIF) include an additional step of incubation with a "primary" antibody (Fig. 11). Incubation with tissue or cells follows the outline given for direct staining. This technique may have advantages over the direct test: (a) it may be more sensitive, increasing the number of conjugate molecules binding to reactive sites in the tissue; (b) it may make it possible to use one conjugate (e.g., an antibody to human IgG coupled to FITC) for the detection of a variety of antigens by using appropriate unconjugated primary antibodies; and (c) it allows for the use of immunofluorescence in serology. In this case, a known antigenic substrate is used, bestowing specificity on the test; the conjugate may then detect any immunoglobulin binding to this substrate; thus patients' sera may be assayed for the presence of antibody to a number of foreign and "self" antigens. A possible disadvantage of IIF may be a decrease in specificity.

3. Another variant of the immunofluorescence test is the *sandwich* technique (SIF). This technique is used almost exclusively to detect specific antibody in plasma cells. First, an antigen is incubated with the tissue sections; after washing, the conjugate specific for the antigen is applied.

4. *Complement immunofluorescence* (CIF) makes use of the complement fixation abilities of certain antibodies to increase sensitivity of the staining procedure. In vivo-bound, complement-activating immunoglobulins may be visualized by incubating the tissue with complement and, after washing, with antibody to complement components conjugated to a fluorochrome (Fig. 11). Similarly, complement binding antibody in serum may be detected in a manner analogous to IIF. This type of IF also allows for evaluation of the complement binding capacity of antibodies bound in vivo or present in serum or other body fluids.

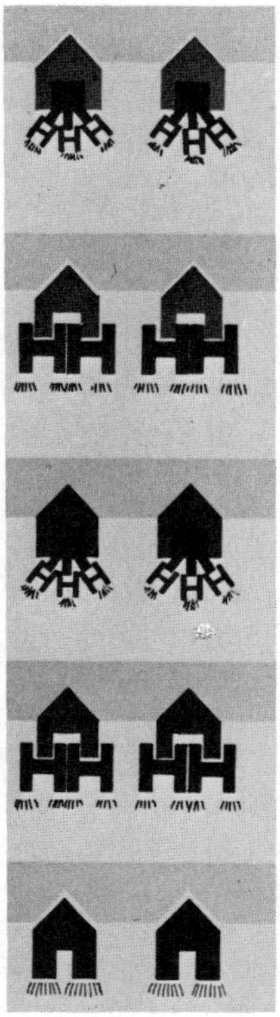

Figure 11 Immunofluorescence staining procedures. Top three rows, direct staining. First row, fluorochrome-conjugated antibodies react with tissue antigens; second row, fluorochrome-conjugated antiserum reacts with immune complexes in tissues; third row, fluorochrome-conjugated antibodies to complement react with complement deposited in tissue. Fourth row, indirect staining; an unconjugated antibody reacts with tissue antigens; in a second step, a fluorochrome-conjugated antiserum to the immunoglobulin class of the unconjugated antibody is applied. Fifth row, complement immunofluorescence; an unconjugated antibody reacts with tissue antigens; subsequently, complement is added; in a third step, a fluorochrome-conjugated antiserum to complement is applied. (Modified from Wick et al., 1976.)

5. Only recently developed, *biotin immunofluorescence* (BIF) makes use of the extremely high affinity of avidin, a basic glycoprotein from egg white, for biotin (Heggeness and Ash, 1977). The affinity constant in avidin-biotin interactions is 10^{15} per mole. Biotin is coupled to an immunologically reactive protein (antibody or antigen) and avidin to a fluorochrome. Biotinization has very little effect on the biological properties of proteins. Because many biotin molecules can be coupled to one protein moiety, biotin immunofluorescence ensures increased sensitivity compared to classical immunofluorescence. Avidin-coupled fluorochrome also produces much less background than do antibody-fluorochrome conjugates. Biotin-avidin systems have been used most extensively in the study of cell surface antigens. An indirect BIF also is possible using biotinylated antibody, evidin, and fluorochrome-labeled biotin.

6. Another important addition to the arsenal of IF is the *hapten-sandwich* technique (HIF; Lamm et al., 1972). It is a variant of the indirect staining techniques. The primary antibody is linked covalently to a hapten. Subsequently, antibodies to this hapten conjugated to a fluorochrome are used for visualization of the antibody-antigen reaction. This method has the advantage of augmentation of sensitivity by linking many hapten moieties to one antibody molecule, which then can react with a high number of fluorochrome-labeled antihapten antibodies. Using two or more distinct haptens, two or more antibodies of distinct specificities but raised in the same species may be used in a double staining procedure. This is of special interest in work with monoclonal antibodies of murine origin.

Some researchers prefer to use fluorochrome-labeled staphylococcal protein A (SPA) in lieu of fluorochrome-conjugated antibodies to immunoglobulins. This molecule has strong affinity for the Fc portion of some human and animal immunoglobulins. SPA can be obtained commercially as a purified, well-standardized, and homogeneous preparation. However, in some test systems, the sensitivity of SPA conjugates may be inferior to that of conventional immunofluorescence with antibodies.

X. SPECIAL APPLICATIONS OF IMMUNOFLUORESCENCE

Double staining (DS) techniques in IF enable the observer to detect simultaneously two immune reactants in a specimen (Goldman, 1968; Nairn, 1976). This may be achieved by labeling one immunologically reactive moiety with FITC and another with TRMITC. DS may be accomplished either by sequential or simultaneous application of the two conjugates.

DS may also be achieved by *sequential* techniques using two conjugates containing the same fluorochrome. Thus a FITC-conjugated antibody to one antigen may first be applied to tissues and the staining pattern recorded; subsequently, the fluorescence is quenched by prolonged exposure, the specimens are incubated with another FITC conjugated antibody of differing specificity, and the staining patterns are compared to those obtained with the first conjugate.

Membrane IF (MIF) is widely used for the detection of cell surface markers (Goldman, 1968; Nairn, 1976; Wick et al., 1976). The procedure follows guidelines for tissue IF. Living cells bar entrance to antibodies, and therefore allow only for interaction of conjugates with surface membrane moieties. Dead cells are easily penetrated by antibody and therefore also by conjugate; they show, in MIF, a cloudy staining of the entire cell, whereas living cells show linear, granular, or "cap" staining. Early in staining, a linear pattern is usually observed, which changes during the incubation time into a granular pattern. Subsequently, the granules may converge on one side of the cell and form a "cap." The formation of a cap is energy dependent and may be prevented by inhibitors of cell metabolism [e.g., by sodium azide (NaH_3)]. Cells are usually stained in suspension, washed and recovered by centrifugation, and dried under a fan onto slides. Intracellular antigens or antibody may be detected in cells fixed in alcohol, alcohol-ether, or acetone.

Photometric equipment in conjunction with appropriate standards allows for *quantitation of immunofluorescence* staining (Nairn, 1976; Wick et al., 1976). Homogeneous antigenic substrates, a well-defined conjugate, and a steadily burning light source are essential in such attempts. Photometric evaluation of tests has been used in attempts to automatize routine clinical tests (Ten Veen and Feltkamp, 1969).

Fluorescence-activated cell sorts (FACSs) allow for rapid analysis and separation of cells in suspension using simultaneous measurements of size (by evaluation of light scatter) and fluroescence of cells (Bonner et al., 1972). Cells are forced to pass, one by one, between a laser light source and highly sensitive light detectors. Computers coupled to the detectors allow for presentation of analytical profiles of cell subpopulations. Modern FACSs can differentiate two fluorochromes in addition to cell size. This procedure allows for rapid analysis or purification: between 10^7 and 10^9 cells may be processed per hour. The separation procedure does not impair cell viability. The FACS method is of utmost importance in the rapidly progressing field of cell surface markers and definition of cell population and subpopulations, especially of B and T lymphocytes. Monoclonal antibodies have opened new perspectives for the use of FACS. Space limitation precludes exhaustive discussion of this important immunological tool; interested readers are referred to specialized reviews.

Interestingly, FITC-labeled antibodies may be used in *scanning electron microscopy* because they have high differential ionization potentials compared to those of surrounding tissue (cited in Sternberger, 1979). This property is caused by the aromatic rings of FITC. Aromatic compounds possess a much lower ionization potential than that of aliphatic moieties. As images formed in the scanning electron microscope result from release of secondary electrons, FITC-labeled antibodies will induce, at the site of their reaction in the tissue or cell, a brighter image than that of the surrounding unstained structures.

XI. CONTROL PROCEDURES IN IMMUNOFLUORESCENCE

Immunofluorescence is a rather complex method. Proper functioning of the optical system, optimal environmental conditions for fluorescence emission, use of adequate conjugates, optimal conditions of the antigenic substrate and awareness and experience of the observer are essential. Therefore, a number of rigorous controls is necessary to assure proper performance and reading of tests. The appropriate controls for IF will depend on the test system and questions asked. Obviously, defined immunofluorescence requires well-serviced and well-kept optical instruments, characterization of the conjugates and careful sectioning and staining techniques. The following controls should be performed to ensure immunological specificity and proper staining:

1. To control for the specificity of the antibody conjugate, tissues should be exposed to immunologically unrelated conjugates, preferentially including conjugated preimmune serum from the same animal in which the specific antibody has been raised.
2. In the indirect test, normal serum from the same species and antiserum unrelated in specificity to the one used in tests should be applied instead of the test serum.
3. Conjugates (or primary antisera in indirect tests) should be absorbed with the presumed specific antigen and retested after absorption.
4. Conjugates (or primary antisera) should be tested on tissue known or assumed to lack the specific antigen.
5. Conjugates (or primary antisera) should be preabsorbed with the tissues tested.
6. In indirect immunofluorescent staining techniques, controls should be run with the conjugate alone [i.e., omitting the antiserum ("saline control")].
7. Blocking tests should be performed by using unlabeled antibodies prior to conjugates. It has to be remembered, however, that in many systems complete blocking may not be achieved.

Table 4 Routine Tests Using Immunofluorescence (Selection)

A. Serology (indirect immunofluorescence tests)
  1. *Infectious diseases*: antibodies to:
     - *Bacillus anthracis*
     - *Legionella pneumophila*
     - *Brucella* spp.
     - *Treponema pallidum*
     - *Mycoplasma pneumoniae*
     - *Cryprococcus neoformans*
     - Ricketsiae
     - Chlamydiae
     - Cytomegalo virus
     - Varicella-zoster virus
     - Epstein-Barr virus
     - Arboviruses
     - Rabies virus
     - Arenaviruses
     - *Leishmania* spp.
     - *Pneumocystis carinii*
     - *Toxoplasma gondii*
     - Plasmodia
     - Helminthic antigens
  2. *Diseases with autoimmune phenoma*: antibodies to:
     - Thyroglobulin
     - Microsomes of thyroid cells
     - Thyroid cell surface
     - Adrenocortical antigens
     - Parathyroid antigens
     - Islet cells (Langerhans islets)
     - Spermatozoa
     - Ovarian cells of zona granulosa and theca
     - Ova
     - Zona pellucida ovarii
     - Skeletal muscle antigens
     - Acetylcholine receptors
     - Heart muscle
     - Glomerular basement membrane of kidney
     - Tubular basement membrane of kidney
     - Smooth muscles
     - Oral mucosa
     - Gastric parietal cells
     - Reticulin
     - Intercellular antigen of skin and mucosae
     - Basement membrane zone of skin
     - Nuclear components (e.g., DNA, histones, "nucleoprotein")
     - Cytoplasmic components (e.g., "R_0" antigen)
     - Lymphocytes

B. Use of immunofluorescence with cells in suspension
  1. *Infectious diseases*:
     - Immunoglobulin on enterobacteriaceae in urine
     - *Vibrio cholerae*—microorganisms in feces
     - Herpes simplex antigens in cells from the cerebrospinal fluid

Table 4 (continued)

B. (Continued)
   2. *Immune complex-mediated diseases*:
      Circulating immune complexes in the Raji cell assay
      Circulating immune complexes in the phagocytosis assay
   3. *Derrangements of the immune system*:
      Definition of B- and T-cell subpopulations (usually using FACS)

C. Use of immunofluorescence with tissue sections
   1. *Infectious diseases*: tissue-bound antibodies to:
      Meningococcal capsules
      Gonococci
      B. anthracis
      L. pneumophila
      T. pallidum
      Herpes simplex (skin, brain)
      Hepatitis B core antigen
      Rabies virus
   2. *Diseases with autoimmune phenomena and/or immune complexes*: immune deposits (usually immunoglobulin, antigen, complement) in:
      Skin and mucosae
      Kidney
      Testis
      Ovary
      Stomach
      Jejunum
      Rectum
      Colon
      Muscle

8. Competitive binding assays are similar to blocking experiments in their principles but consist of simultaneous application of unlabeled and labeled antibodies. This should lead to a significant decrease in staining if compared to incubation with conjugate alone.
9. Sensitivity of the tests should be ascertained by including positive standard preparations in the test run. Such preparations are now available for some test systems (e.g., reference preparations of antinuclear antibodies available from the World Health Organization).
10. Photometric equipment should be calibrated using stable fluorochrome preparations.

Fading in the light beam under the microscope is a characteristic of fluorochromes. Absence of fading suggests nonspecific (usually auto-) fluorescence.

XII. APPLICATIONS OF IMMUNOFLUORESCENCE

IF is widely used for research and diagnostic purposes (Nairn, 1976; Sternberger, 1979; Wick et al., 1982). Selected routine tests are summarized in Table 4. Selected patterns obtained with IF are seen in Fig. 12.

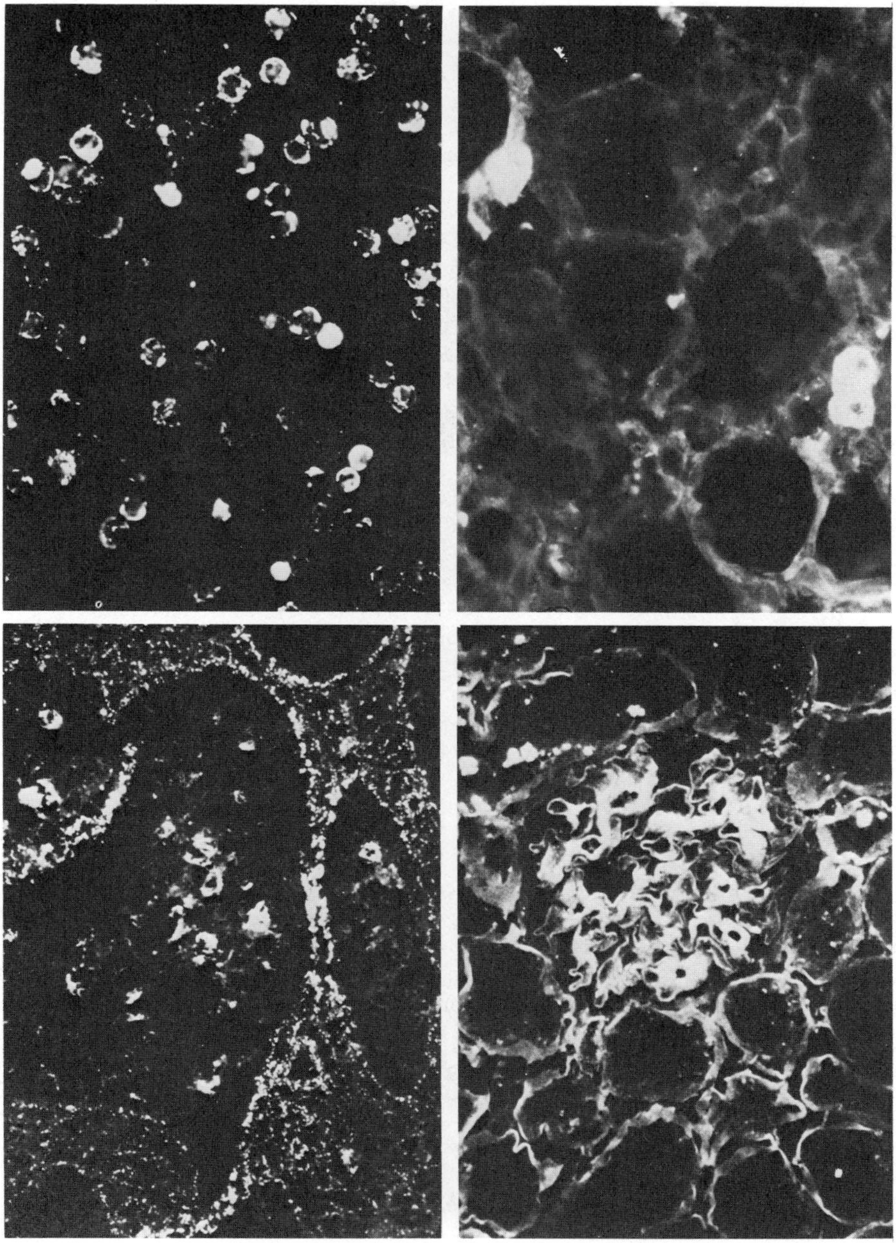

Figure 12 Patterns of immunofluorescence. (Upper row, left) Immunofluorescent staining of cell membrane-associated reactants (Raji cell assay for circulating immune complexes with FITC-conjugated antiserum to human IgG). (Upper row, right) Immunofluorescent staining of cytoplasm (plasma cells). (Lower row, left) granular staining suggesting immune complex deposits (tissue section from a spleen of a rabbit with chronic serum sickness indiced by administration of bovine serum albumin, BSA; reaction of tissue with FITC-conjugated antibodies to BSA). (Lower row, right) Linear staining of kidney glomerular and tubular basement membranes and patchy staining of glomerular mesangium (tissue section from a kidney of a rat injected with mercury chloride developing antibodies to basement membranes and mesangial matrix; reaction of tissue-bound antibodies with FITC-conjugated antiserum to rat IgG). Magnifications, from upper left to lower right: × 250; × 400; × 250; × 250.

REFERENCES

Andres, G. A., Hsu, K. C., and Seegal, B. C. (1978). In *Handbook of Experimental Immunology*, 3rd ed. (D. M. Weir, Ed.). Blackwell, Oxford, p. 37.1.

Beutner, E. H., Sepulveda, M. R., and Barnett, E. V. (1968). WHO Bull. *39*, 587.

Beutner, E. H., Nisengard, R. J., and Kumar, V. (1979). In *Immunopathology of the Skin*, 2nd ed. (E. H. Beutner, T. P. Cherzelski, and S. F. Bean, Eds.). Wiley, New York, p. 29.

Bonner, W. A., Hulett, H. R., Sweet, R. G., and Herzenberg, L. A. (1972). Rev. Sci. Instrum. *43*, 404.

Chadwick, C. S., McEntegart, M. G., and Nairn, R. C. (1965). Immunology *1*, 315.

Coons, A. H., Creech, H. J., and Jones, R. N. (1941). Proc. Soc. Exp. Biol. *47*, 200.

Coons, A. H., Creech, H. J., Jones, R. N., and Berliner, E. (1942). J. Immunol. *45*, 159.

Goldman, M. (1968). *Fluorescent Antibody Methods*. Academic Press, New York.

Guilbault, G. G. (1973). *Practical Fluorescence*. Marcel Dekker, New York.

Heggeness, M. H., and Ash, J. F. (1977). Biologist *73*, 783.

Heidelberger, M., Kendall, F. E., and Soo Hoo, C. M. (1933). J. Exp. Med. *58*, 137.

Johnson, G. D., Holborow, E. J., Derling, J. (1978). In *Handbook of Experimental Immunology*, 3rd ed. (D. M. Weir, Ed.). Blackwell, Oxford, p. 15.1.

Knapp, W. (1982). In *Immunofluorescence Technology* (G. Wick, K. N. Trail, and K. Schauenstein, Eds.). Elsevier, Amsterdam, p. 11.

Lamm, M. E., Koo, G. C., Stockpole, C. W., and Hemerling, V. (1972). Proc. Natl. Acad. Sci. USA *69*, 2732.

Marrack, J. (1934). Nature (Lond.) *133*, 292.

Nairn, R. C., Ed. (1976). *Fluorescent Protein Tracing*, 4th ed. Churchill Livingstone, Edinburgh.

Pearse, A. G. E. (1968). *Histochemistry–Theoretical and Applied*, 3rd ed., Vol. 1. Churchill, London.

Ploem, J. S. (1967). Z. Wiss. Mikrosk. *68*, 129.

Ploem, J. S. (1982). In *Immunofluorescence Technology* (G. Wick, R. N. Trail, and K. Schauenstein, Eds.). Elsevier, Amsterdam, p. 75.

Schauenstein, K., Böck, G., and Wick, G. (1982). In *Immunofluorescence Technology* (G. Wick, K. N. Trail, and K. Schauenstein, Eds.). Elsevier, Amsterdam, p. 27.

Schiffer, M. S., Michael, A. F., Kim, Y., and Fish, A. J. (1981). Lab. Invest. *44*, 234.

Sternberger, L. A. (1979). *Immunocytochemistry*, 2nd ed. Wiley, New York.

Ten Veen, J. H. and Feltkamp, T. E. W. (1969). Clin. Exp. Immunol. *5*, 673.

Udenfriend, S. (1969). *Fluorescence Assay in Biology and Medicine*, Vol. 2. Academic Press, New York.

Wick, G., Baudner, S., and Herzog, F. (1976). *Immunofluorescence*. Medizinsch Verlagsanstalt, Marburg an der Lahn.

Wick, G., Trail, K. N., and Schauenstein, K., Eds. (1982). *Immunofluorescence Technology*. Elsevier, Amsterdam.

22

Immunoelectron Microscopy

Matthew A. Gonda NCI-Frederick Cancer Research Facility, Frederick, Maryland

I. INTRODUCTION

The potential of employing fluorescein-labeled antibodies to localize antigens at the light microscopic level was first conceived and employed by Coons and colleagues (1942). They showed that such conjugates would combine with specific antigen present in a tissue section and that the bound antibody could be visualized in the ultraviolet microscope through the emission of fluorescence. With the advent of transmission electron microscopes (TEM), and their greater resolving power and improved methods for the visualization of the fine structure of cells, this concept was expanded to permit the localization of antigens at the ultrastructural level.

In the TEM, the specific antigen-antibody precipitin reaction was localized by the coupling of an electron-dense marker to the specific antibody replacing the fluorochromes, rhodamine, or fluorescein. Singer (1959) was the first to develop a method for conjugating an electron-dense marker, ferritin, to antibodies without concomitant loss of their binding capacity. Rifkind et al. (1960) were the first to employ such conjugates for the immunologic detection of antigens by electron microscopy, hence the term *immunoelectron microscopy*. Since these early developments, many antibody methods have been devised to demonstrate antigen sites on cells and in tissues through the antigen–antibody precipitin reaction. Some of the systems use electron-dense markers such as ferritin as originally proposed (Singer, 1959), while others have modified this technique using markers which are morphologically recognizable or enzymes and their reaction products which are rendered electron dense (Nakane and Pierce, 1966; Sternberger, 1974; Farr and Nakane, 1981). Some investigators have relied on the functional properties of unlabeled antibodies and their interaction with other facets of the immune system (i.e., complement) to demonstrate the presence of a specific antigen.

There has been a rapid explosion in microscope technology during the past two decades which has extended the classical two-dimensional ultrastructural studies performed with the TEM to include three-dimensional imaging of cells and tissues with the development of the scanning electron microscope (SEM). With its introduction, methods to study the antigenic composition of cell surfaces have followed as extensions of those marker systems used in TEM. Singularly or in combination, these two methods of microscopic analysis have proved to be an extremely important technique for the study of cells, cellular organelles, and the molecular organization of cell structures when used in conjunction with immunologic probes.

As the diversity of methods used in immunoelectron microscopy are too great to adequately present detailed protocols for each method, this chapter presents basic information on methodology, considerations in choosing a labeling system, and problems that may be encountered. With the use of numerous micrographs it is intended to be a pictorial essay for those aspiring to study antigen–antibody interactions by electron microscopy. It is hoped that by covering a wide variety of techniques and their application, this chapter will provide a better insight into immunoelectron microscopy for the investigator than could be obtained by a "cookbook"-style reading.

II. MARKERS FOR IMMUNOELECTRON MICROSCOPY

The choice of a marker system for immunoelectron microscopic detection of a particular antigen depends on many variables, some of which are discussed below. The first variable one must consider is the mode of imaging. Does the particular study lend itself to TEM, or SEM, or both? When one performs localization experiments with TEM in mind, a suitable choice would be a marker which is electron dense, such as ferritin (Singer, 1959; Rifkind et al., 1960) or colloidal gold (Horisberger and Rosset, 1977; Geoghegan and Ackerman, 1977; Horisberger, 1981). Alternatively, a marker with a definitive shape, such as hemocyanin (Karnovsky et al., 1972; Nemanic, 1979; Gonda et al., 1979 to 1981) or southern mosaic bean virus (Hammerling et al., 1969), would suffice. If the particular study deals with SEM localization, where electron density plays less of a role because metallic conductive coatings are applied for imaging with secondary electrons in contrast to transmitted electrons with TEM, the markers must be large enough to be easily resolved and have a distinct shape and uniform size so that they may be discerned from topographical cell surface features.

The variety of markers are quite diverse, ranging from the most often used biological macromolecule, hemocyanin (Carter, 1978; Gonda et al., 1979; Nemanic, 1979; Gonda et al., 1980, 1981), to plant and animal viruses (Kumon, 1976). In addition, various synthetic polymeric latices (Molday, 1976; Molday and Maher, 1980) and colloidal particles (Horisberger and Rosset, 1977; Peters et al., 1978; Horisberger, 1981) have been developed which serve as excellent immunological markers. Quite often, correlative techniques employing both TEM and SEM localization are required and, therefore, a marker resolvable by both transmitted and secondary electrons must be selected (Carter, 1978; Gonda et al., 1979; Nemanic, 1979; Gonda et al., 1980, 1981).

Another important variable that must be considered is spatial resolution. To determine by TEM or SEM whether a particular cell is positive or negative for a particular antigen in a similar fashion as fluorescence microscopy is used, spatial resolution of the marker is not a prerequisite and a large marker is often desirable for the easy detection and quantitation of the target cell. However, to detect a particular viral or cell surface

Table 1 Markers for Immunoelectron Microscopy

Marker	Approximate size (nm)	Characteristic shape	Basis of detection		Instrument application	
			Morphology	Electron density[a]	TEM[a]	SEM[a]
Ferritin	14	Spherical	Marginal	Yes	Yes	Marginal
Peroxidase reaction product	–	Amorphous particles	Marginal	Yes	Yes	Marginal
Hemocyanin	35 × 50	Cylindrical	Yes	Marginal	Yes	Yes
Southern mosaic bean virus	30	Spherical	Yes	Marginal	Yes	N.D.
Bushy stunt virus	30	Spherical	Yes	Marginal	Yes	Yes
Tobacco mosaic virus	15 × 300	Rod	Yes	Marginal	Yes	Yes
Bacteriophage T4	100 (head) 220 (head-tail)	Hexagonal head with tail	Yes	Marginal	Yes	Yes
Colloidal gold	5–150	Spherical or oval	Marginal	Yes	Yes	Yes
Silica spheres	7–25	Spherical	Yes	N.D.	N.D.	Yes
Latex microspheres	30–340	Spherical	Yes	Marginal	Yes	Yes

[a]N.D., not determined.

antigen in relation to the virus or a virus-specific topographical feature, it is advised to use a marker that is small enough to be detectable and yet which does not obscure the cell surface feature. This is particularly the case with SEM, where morphological recognition of cell surface features are not as clear cut as with TEM and immunological recognition of a particular topographical feature (i.e., viruses) bears the greater burden of proof.

In addition to the foregoing points regarding the selection of a marker for immunoelectron microscopy, several other features of markers for TEM or SEM are important. The marker should show little tendency for nonspecific adsorption. It should have a uniform size, distinct shape, and be capable of being reproducibly synthesized and/or purified. Finally, the marker should be stable and amenable to all preparatory procedures used during labeling and processing for electron microscopy. A list of markers that have been most widely used in TEM and SEM, and their relative size, shape, electron density, and instrument application (Andres et al., 1978; Molday and Maher, 1980), are shown in Table 1.

III. DIRECT AND INDIRECT LABELING METHODS

There are two basic approaches for antigen localization, direct and indirect assays in immunoelectron microscopy. A schematic representation of these techniques is shown in Fig. 1. In the direct test, marker is conjugated directly to the specific antibody which is applied directly to the substrate (Singer, 1959; Andres et al., 1978). An alternative approach for the direct test would use a single hybrid antibody with dual specificity to bridge the specific antigen being localized and the marker (Hammerling et al., 1969).

In the indirect test, the unlabeled antibody (specific antibody) is applied directly to the substrate. The bound antibody is visualized by treatment with a marker-conjugated anti-immunoglobulin (secondary antibody) which was raised to IgG of the species in which the specific antibody was made. There are two additional indirect methods that do not involve conjugation of antibody to marker. The first, the hapten-sandwich method (Carter, 1978; Nemanic, 1979) involves three steps. Specific antibody modified with hapten is reacted with substrate followed by bridging antihapten antibody. A hapten modified marker is then bound and used for visualization of the specific antibody–antigen interaction. The second, the unlabeled antibody bridge method (Gonda et al., 1979; Gonda et al., 1980, 1981) involves four steps. Unlabeled antibody (specific antibody) is first reacted with the substrate. An unlabeled bridging anti-immunoglobulin which was raised to the specific antibody is then bound. A third antibody to a marker which was made in the same species as specific antibody is then applied, being bound by the free Fab portion of the divalent bridging (anti-immunoglobulin) antibody. The marker is then applied being bound by the free Fab arms of the marker-specific antibody (third antibody).

Although all of the foregoing approaches are applicable to TEM or SEM, some have particular advantages and disadvantages. The direct test is advantageous in that only a single antiserum need be raised, conjugated, and used for labeling. This method gives excellent spatial resolution when small markers are used. Several disadvantages of the direct test become readily apparent. Despite its simplicity the single conjugated antiserum can only be used for that particular antigen being localized and additional antisera would have to be made and conjugated for each antigen being localized. When few antigen sites

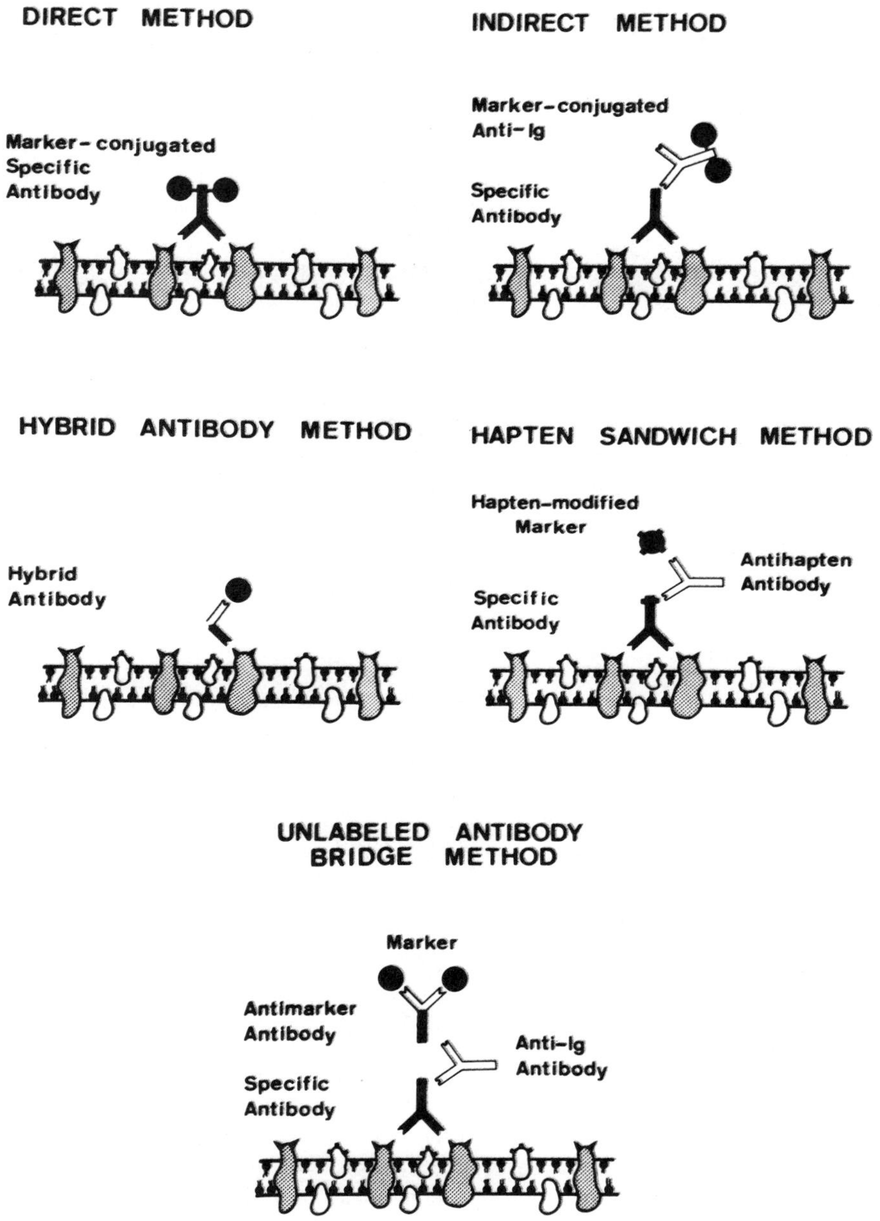

Figure 1 Schematic representation of the direct and indirect immunological labeling techniques used to localize cellular antigens for immunoelectron microscopy. Although the illustrations depict labeling of the plasma membrane, some of the methods are applicable to labeling cytoplasmic and nucleic acid associated antigens.

are available or the specific antiserum has a low titer, the direct test is somewhat inefficient. The indirect test is advantageous in that only a single anti-immunoglobulin need be raised and conjugated. The conjugated anti-immunoglobulin can be used with any specific antibody from the particular species to which it was raised.

IgG is the most abundant and easily prepared class of immunoglobulin. Since most antibodies in hyperimmune sera used as specific antibody are of the IgG class, whose light chains confer cross-reactivity with other immunoglobulin classes, labeled anti-IgG is most commonly used. IgG possesses numerous antigenic determinants (epitopes) that increase the number of sites available to be bound by the anti-IgG marker. The technique thus amplifies the precipitin reaction by binding many markers and making the location of the bound specific antibody more visible either when few antigen sites are available or the specific antiserum has a low titer.

The unlabeled antibody bridge method shares some of the features of the indirect test where secondary antibody is conjugated to marker, in that it is an amplification method in which conjugation of marker is not needed. The disadvantage of the unlabeled antibody bridge is that multiple antisera need to be made, in contrast to the indirect method. However, once they are made, they can be used with any specific antibody from the particular species to which the anti-immunoglobulin was made. The hapten-sandwich technique is also a bridging method with amplification properties but involves extensive conjugation steps of hapten to specific antibody and marker for reaction with anti-hapten antibody.

IV. UNLABELED AND LABELED ANTIBODY METHODS

Labeled antibodies for direct and indirect methods have been prepared for almost all the markers presented in Table 1 by coupling the electron-dense or morphologically recognizable marker to the specific antibody (in the case of the direct method) or anti-immunoglobulin (in the case of the indirect method) antibody. Examples of bifunctional-coupling reagents used to produce labeled antibodies are glutaraldehyde, metaxylylene diisocyanate, toluene-2,4-diisocyanate, cyanogen bromide, carbodiimide, and p,p′-difluoro-M-M′-dinitrodiphenyl sulfone, to name a few (Molday, 1976; Andres et al., 1978). Electrostatic coupling has also been achieved reproducibly for negatively charged sols such as colloidal gold (Geoghegan and Ackerman, 1977; Horisberger and Rosset, 1977, Geoghegan et al., 1980; Horisberger, 1981).

The unlabeled antibody bridge has been used successfully with ferritin, hemocyanin, and enzymes such as peroxidase (Sternberger, 1974; Andres et al., 1978; Gonda et al., 1979 to 1981; Willingham, 1980). As described above, this marker system uses no chemical coupling reagents to link the marker to the precipitin site. The technique's success rests solely on the ability of the divalent antibody in the bridging step to couple the primary (specific) antibody to the marking antibody that binds the visual marker.

Hammerling et al. (1969) have devised a technique using ferritin or southern bean mosaic virus as the marker in a direct test without using any conjugated antibody. The technique involves the assembly of an antibody with dual specificity; one is against the marker, the other is against the particular antigen of interest. This is accomplished by cleavage of the Fc receptor followed by reduction of disulfide bonds, which yields a monovalent Fab fragment. When antigen-specific and marker-specific antibodies treated as described above are mixed in equimolar concentrations in the presence of a mild oxidant, the Fab monomers reconstitute into $F(ab)_2$ dimers, some of which are hybrids with dual specificity (see Chap. 7).

There are various other unlabeled antibody methods that are less sophisticated than the methods described above, but are nonetheless direct, easy to perform, and efficient. Some of these methods are described in the following section together with labeled and unlabeled antibody methods previously discussed.

V. LABELING OF CELL SURFACE ANTIGENS

Surface antigens can be detected and localized by the use of both labeled and unlabeled antibodies. Because of their size, antibodies do not readily penetrate live or well-fixed cells; thus the precipitin reaction is compartmentalized to the exterior of the plasma membrane. Based on this property and the importance of organizational changes in plasma membranes during normal and pathologic states, the labeling of cell surface antigens has been the most widely exploited and successfully used method in immunoelectron microscopy. The interaction of antibody–antigen can be studied with both the TEM and SEM using appropriate markers and can range from simple binding of antibody to antigen to agglutination. Some of the applications of immunoelectron microscopy to cell surface labeling are discussed.

A. Transmission Electron Microscopic Studies

The immunoferritin technique is the most widely used in cell surface labeling studies. It has been used in the detection of viral, bacterial, and cellular antigens as well as surface receptors and immunoglobulins (Fig. 2). Because of its relatively small size and electron density, ferritin is easily recognized and permits good spatial resolution when small entities such as viruses are being labeled (Singer, 1959; Rifkind et al., 1960; Andres et al., 1978; Willingham, 1980). The ferritin marker has been used in direct, indirect, hybrid-antibody, and unlabeled antibody bridge methods.

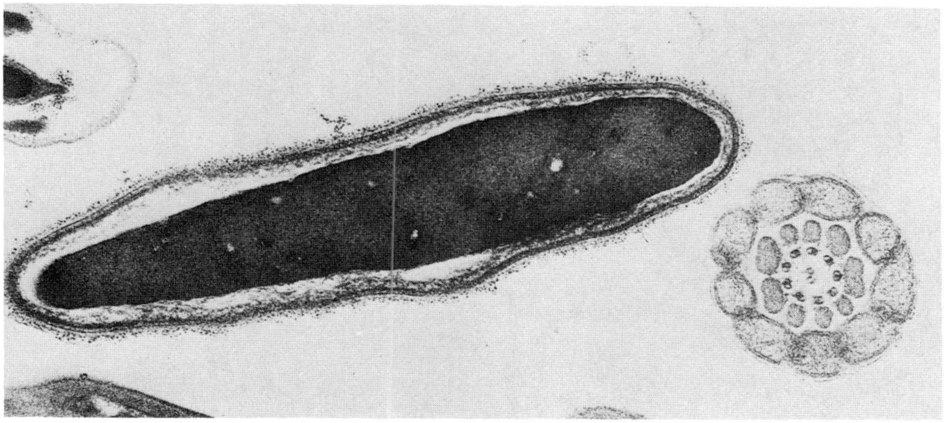

Figure 2 Electron micrograph showing the results of an experiment in which sperms from a normal rabbit were first treated with rabbit antiacrosomal antibodies and then incubated with ferritin-conjugated goat anti-rabbit immunoglobulin. Ferritin is localized on the cell membrane of the sperm head but not on the sperm tail (lower right corner). Scale bar = 0.5 μm. [From P. E. Bigazzi et al. (1976), J. Exp. Med. *143*, 382.]

The immunoperoxidase technique (Nakane and Pierce, 1966; Sternberger, 1974; Farr and Nakane, 1981) has also been used in studies of membrane and surface immunoglobulins (Fig. 3). It has a greater sensitivity when compared to the immunoferritin technique because of the coupling of an enzyme (peroxidase) whose reaction product can be amplified at the deposition site of the antigen–antibody precipitin reaction. However, because of the amorphous nature of the reaction product, good spatial resolution can only be achieved at very high dilution of the specific antiserum. The immunoperoxidase technique has been used in direct, indirect, and the unlabeled antibody bridge methods.

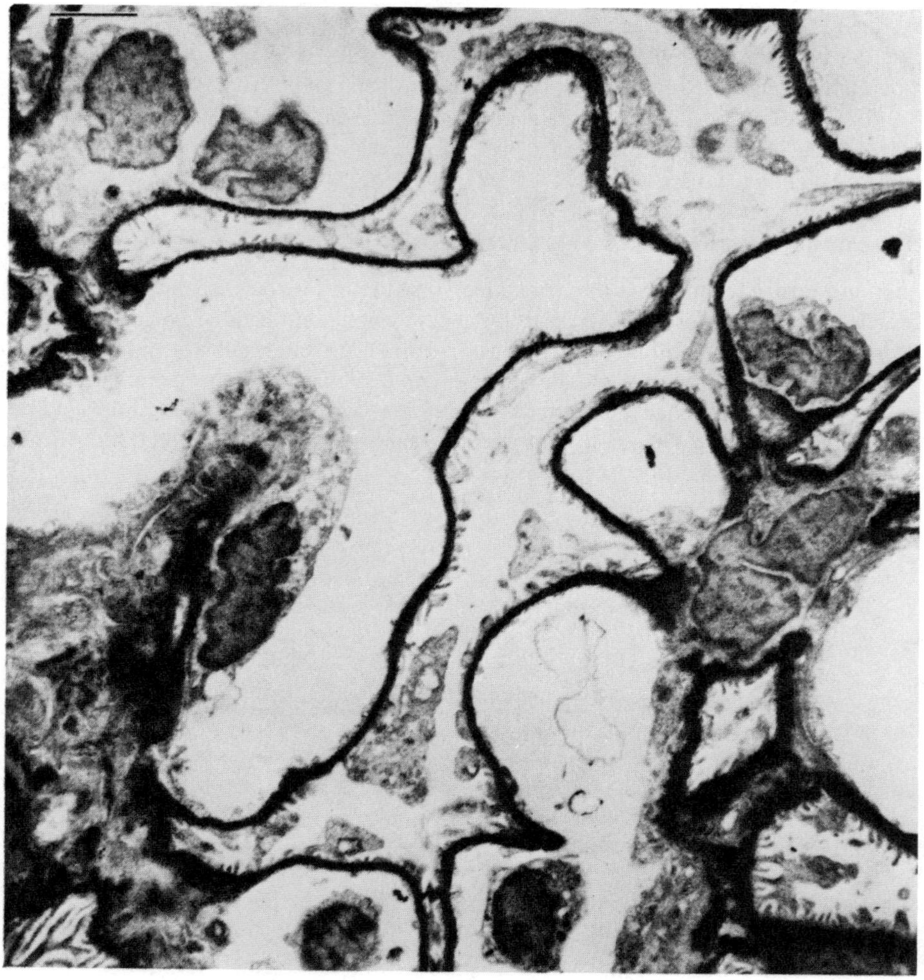

Figure 3 Kidney section of a rat which had been rendered nephritic by the injection of rabbit anti-rat glomerular basement membrane serum. The tissue was first incubated with sheep anti-rabbit globulin serum, followed by rabbit antiperoxidase and peroxidase. The reaction products are localized in the glomerular basement membrane. Scale bar = 2.0 μm. (Courtesy of G. A. Andres.)

The reaction product is visible in the light microscope and the immunoperoxidase technique is therefore applicable to correlative studies at the light and electron microscopic levels without any further modifications.

A recent innovation in immunoelectron microscopy was the introduction of the colloidal gold technique (Horisberger and Rosset, 1977; Geoghegan and Ackerman, 1977; Geoghegan et al., 1980; Horisberger, 1981). Monodisperse colloidal gold can be synthesized relatively easily by the reduction of auric chloride in a size range of 5 to 150 nm. The successful preparation and coupling of gold markers depend only on a few stringent conditions, as reviewed by Geoghegan et al. (1980) and Horisberger (1981). Stable antibody-gold complexes generally form at pH values near the isoelectric point of the immunoglobulin. The colloidal gold labeling system rivals the ferritin technique in spatial resolution when small (5 to 25 nm) gold sols are prepared and used in marking experiments (Fig. 4). The gold sol, electrostatically coupled to immunoglobulin, can be used in the direct or indirect methods.

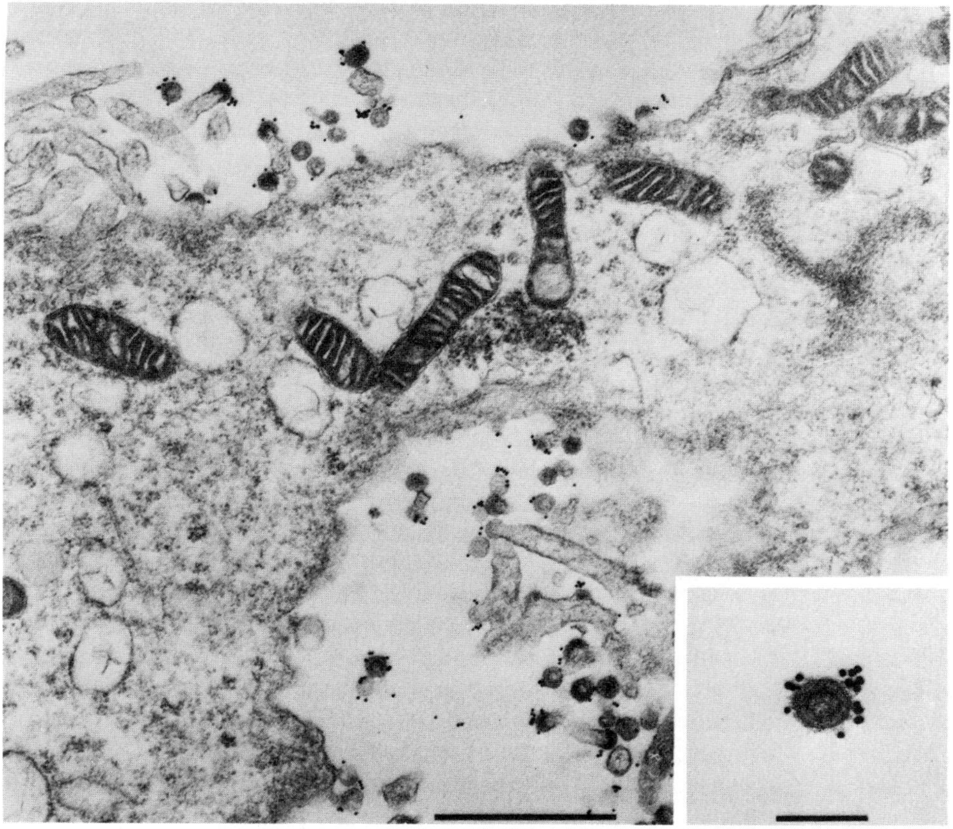

Figure 4 Labeling of mouse mammary tumor virus with mouse monoclonal antibodies (IgG) to the major envelope glycoprotein, gp52, followed by colloidal gold (18 to 20 nm)-labeled goat anti-mouse IgG purified by affinity chromatography. Virus particles containing gp52 antigen are labeled by the gold-coupled antibodies. Scale bar = 1.0 μm; inset scale bar = 0.2 μm.

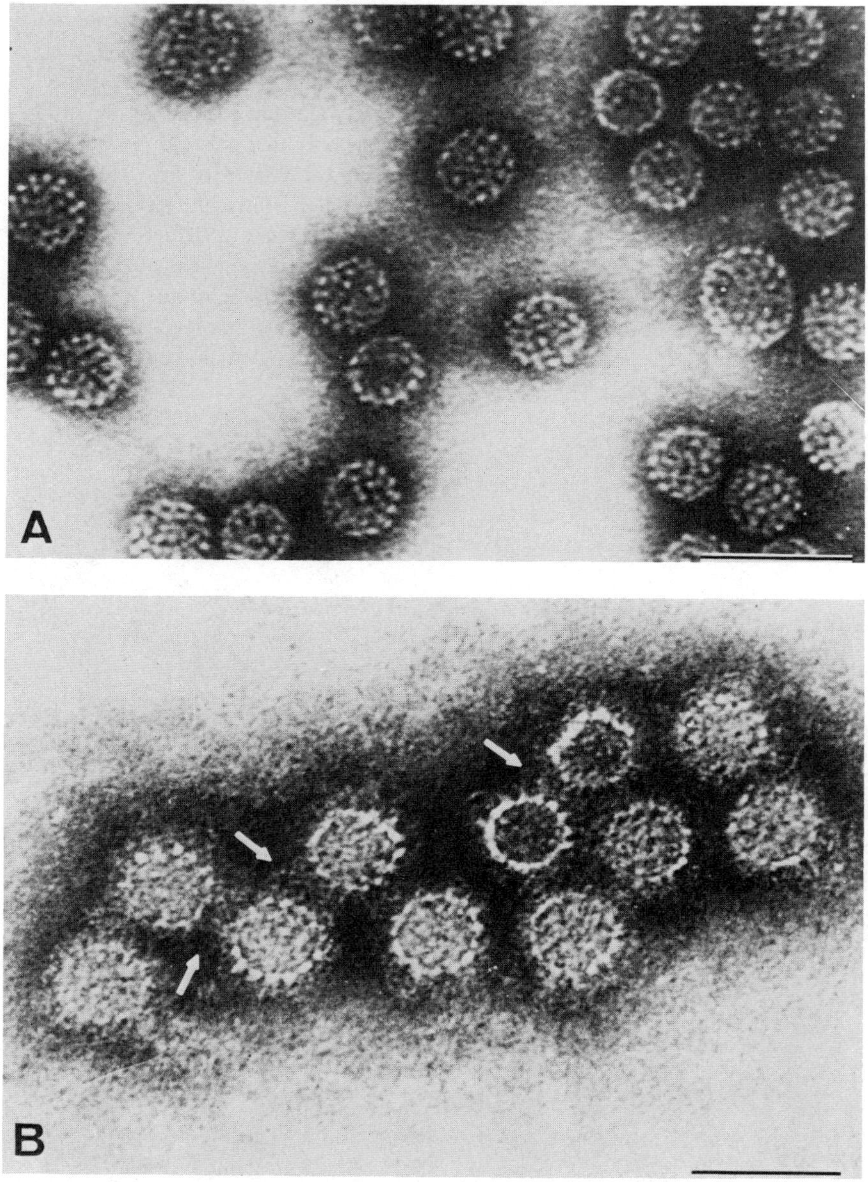

Figure 5 Phosphotungstic acid negative staining showing agglutination of human papilloma virus from Verruca vulgaris (common wart). (A) Wart virus incubated with normal nonreactive serum. The virus particles are not agglutinated. (B) Wart virus incubated with unlabeled anti-Verruca vulgaris virus antibodies. The particles are agglutinated as indicated by the cross-linking of antibodies between virus particles (→). Note the presence of hairpin (loops) coatings on the perimeter of the virus. Scale bar = 0.1 μm. [From S. Oroszlan and M. A. Rich (1964), Science *146*, 531.]

Unlabeled antibody methods have been extensively used in immunoelectron microscopic detection of surface antigens. Figure 5 demonstrates the use of specific antibody to viral proteins to cross-link multivalent viral envelope antigens. This leads to agglutination and precipitation, which demonstrates the presence of an antibody–antigen interaction. Unlabeled antibodies have been used in conjunction with the full complement system. Negatively stained preparations in the electron microscope show "pits" or "craters" which correspond to individual sites of complement activation where destruction of the mucopeptide layer occurs (Fig. 6). The unlabeled antibody hemocyanin bridge (Gonda et al., 1979 to 1981) has been used extensively to label virus particles and viral antigens on infected cells (Fig. 7). The unconjugated marker is bound in the last step of a series of antigen–antibody precipitin reactions bridging specific antigen and marker. Hemocyanin is morphologically recognizable in TEM and SEM. Furthermore, when fluorescein is conjugated to bridging antibodies, the technique can also be used in fluorescence microscopy.

Quite often the resolving power of SEM (3 to 10 nm) is not great enough to give good spatial resolution of small markers such as ferritin and colloidal gold. Carbon-platinum replicas of cell surfaces are amenable to high-resolution studies in the TEM and

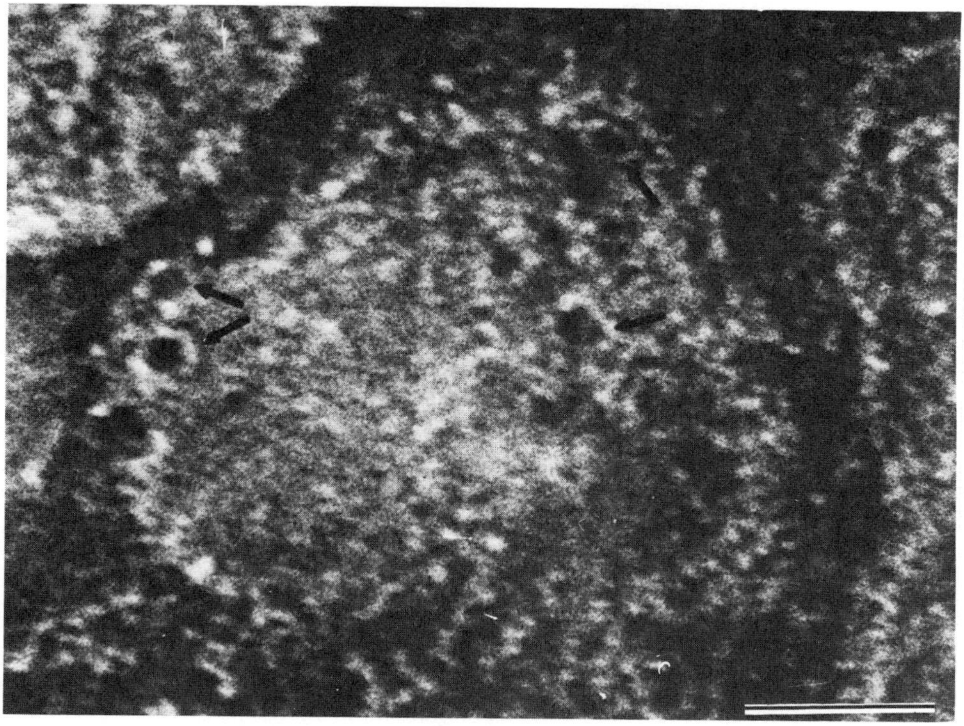

Figure 6 Electron micrograph of "pit- or craterlike" lesions seen on the surface of Moloney murine leukemia virus after incubation with antibody to surface virus antigens and complement as seen by negative staining with phosphotungstic acid. The arrows (→) point to the craters, which measure about 10 nm. Scale bar = 0.05 μm. (Courtesy of M. B. A. Oldstone.)

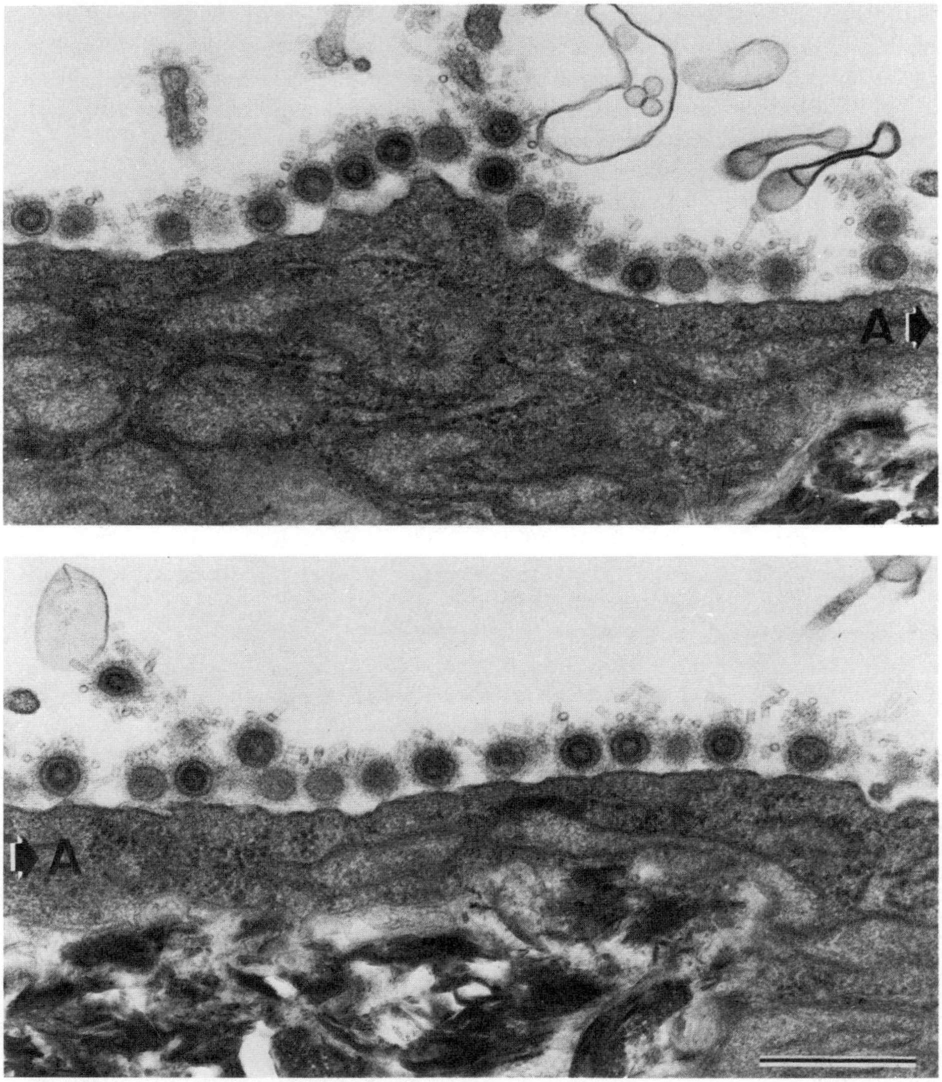

Figure 7 Mouse mammary tumor virus-infected cells labeled by the unlabeled antibody hemocyanin bridge technique. Cells were first incubated with monoclonal anti-mouse mammary tumor virus gp52 envelope protein (IgG) followed by rabbit anti-mouse (IgG), mouse antihemocyanin (IgG), and the marker, hemocyanin. The section is through a marginal region of cytoplasm. Virions in all stages of morphogenesis are labeled with hemocyanin. Lettered arrows (→) indicate the junction of two sequential micrographs taken of this region. Intense labeling of virus particles and cell membrane is observed. Scale bar = 0.5 μm. (From Gonda et al., 1981.)

the application of ferritin and colloidal gold labeling systems has been utilized to localize cell surface antigens while still giving the appearance of a quasi-three-dimensional image (Pinto Da Silva et al., 1981). These carbon-platinum replicas, in which the electron-dense marker is left in the replica after acid digestion of organic materials, have been used in the

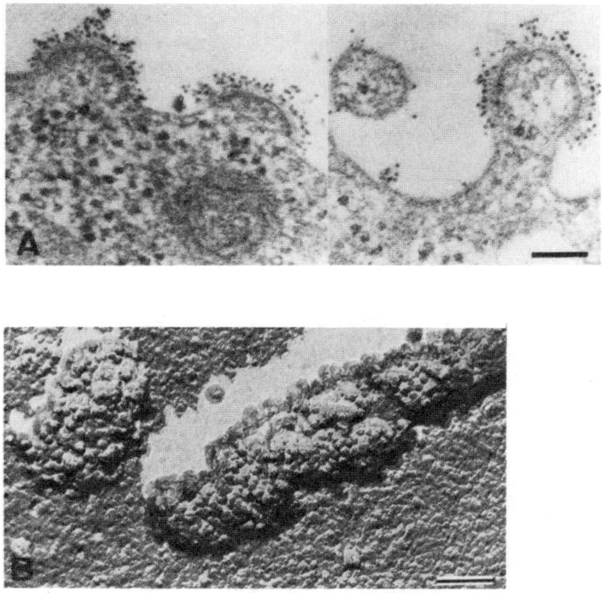

Figure 8 Detection of measles virus-specific antigen by thin section electron microscopy and replica immunocytology. (A) Measles virus-infected HeLa cells incubated with rabbit antimeasles antibodies followed by ferritin-conjugated goat anti-rabbit (IgG). Measles virus antigen is marked by a halo of ferritin molecules over the virus-specific structures. (B) Replica immunocytochemistry with measles virus-infected HeLa cells labeled as in (A) but the marker is demonstrated in transmission electron micrograph of a cell surface replica. Scale bar = 0.1 μm. (Courtesy of K. Mannweiler.)

immunocytochemical study of virus-specific antigen in measles virus-infected cells (Figs. 8 and 9).

B. Scanning Electron Microscopic Studies

Immunoelectron microscopy has recently been extended to the SEM, where the ability to display the surface membrane of large numbers of cells at high resolution (3 to 10 nm) has permitted simultaneous morphologic studies of the three-dimensional structure of cells and immunologic studies of cell surface antigen composition. The images observed are quite dramatic, especially when cell surface labeling techniques are also incorporated. The choice of a marking system again depends on the specific application. Various biologic and synthetic markers have been used successfully to identify specific cell types and to map and quantitate the topographical distribution of surface antigens and receptors.

Virus and viral cell surface antigens have been mapped by the unlabeled antibody hemocyanin bridge technique using antibodies to mouse mammary tumor virus on chronically infected cells (Gonda et al., 1979 to 1981). In these studies, both monoclonal antibodies to the major envelope glycoprotein (gp52) and polyvalent sera to whole virus were used (Fig. 10). The labeling patterns with either antibody were the same. Hemocyanin's small size and distinct shape give good spatial resolution in virus studies and it is morphologically recognizable in both the TEM and SEM (compare Figs. 7 and 10).

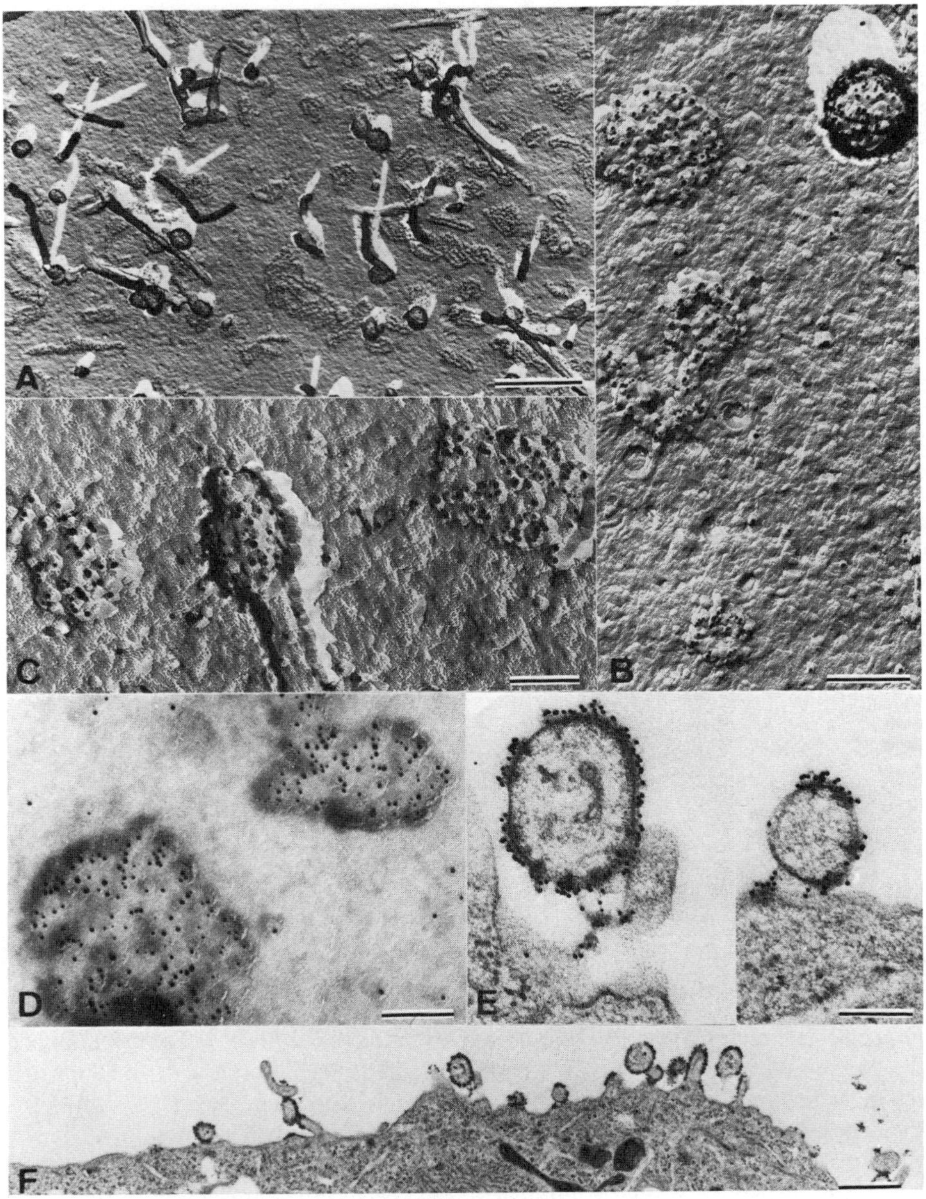

Figure 9 Detection of measles virus-specific antigen by thin section and replica immunocytology. (A–D) Surface replica of measles virus-infected HeLa cells labeled for virus-specific surface antigens with rabbit antimeasles virus serum followed by *Staphylococcus* protein A-coated colloidal gold. Gold particles are almost exclusively located on virus-specific structures. (E, F) Ultrathin sections of embedded cells derived from the same preparation as in (A–D). Scale bar = 1.0 μm in (A); 0.2 μm in (B); 0.1 μm in (C–E); and 1.0 μm in (F). [From K. Mannweiler et al. (1982), J. Microsc.]

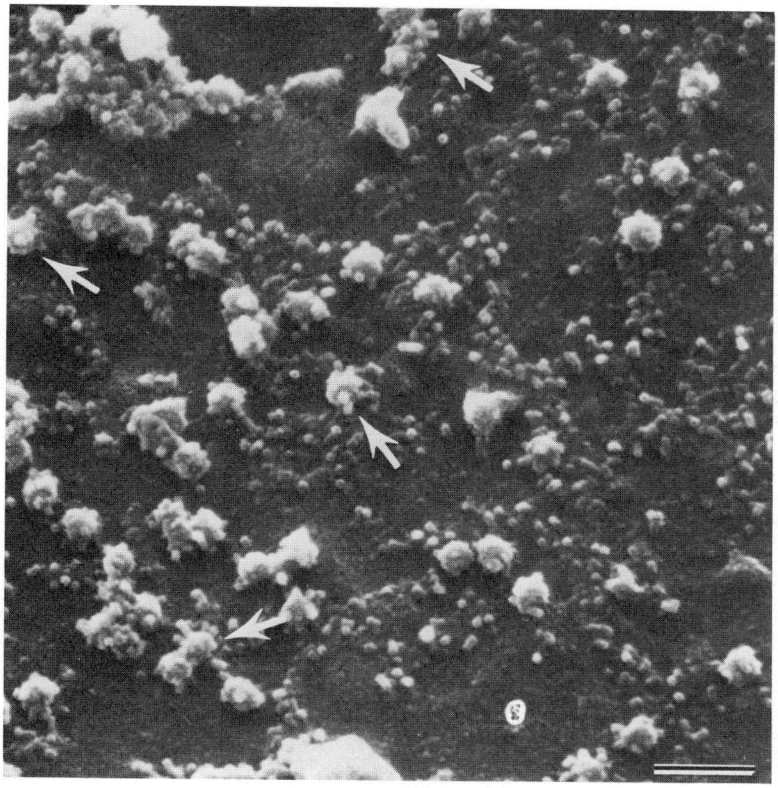

Figure 10 Labeling of mouse mammary tumor virus using specific polyvalent antiserum to whole virus and the unlabeled antibody hemocyanin bridge. Virus particles (→) and discrete patches of cell membrane are labeled. Scale bar = 0.5 μm. (From Gonda, 1979.)

Bacteria and plant viruses have also been used to detect cell surface antigens (Kumon, 1976). T4, in particular, has been used to study influenza virus antigens on infected cells (Fig. 11). Its hexagonal head and tail assembly make it an excellent morphologic marker of the intermediate resolution range. This feature makes it feasible to survey samples at relatively low magnifications, while scanning at high magnification to display the site specificity. However, because of its head size and shape, it is not useful in detecting entities of its own size range or smaller by SEM, but it is amenable to virus or cell surface studies by TEM.

Biological molecules such as peroxidase and their reaction products are not as amenable to detection by SEM because of their amorphous shape. Although ferritin is very useful in TEM, it is only of marginal use in the SEM because it is difficult to distinguish with certainty ferritin-conjugated antibody from the rough cell surface often encountered in labeling experiments, even when the finest metallic coatings and the best equipped high-resolution SEM are used. The limiting step for most small markers in SEM is not their size but rather the size of antibodies (ca. 25 nm) used in the labeling

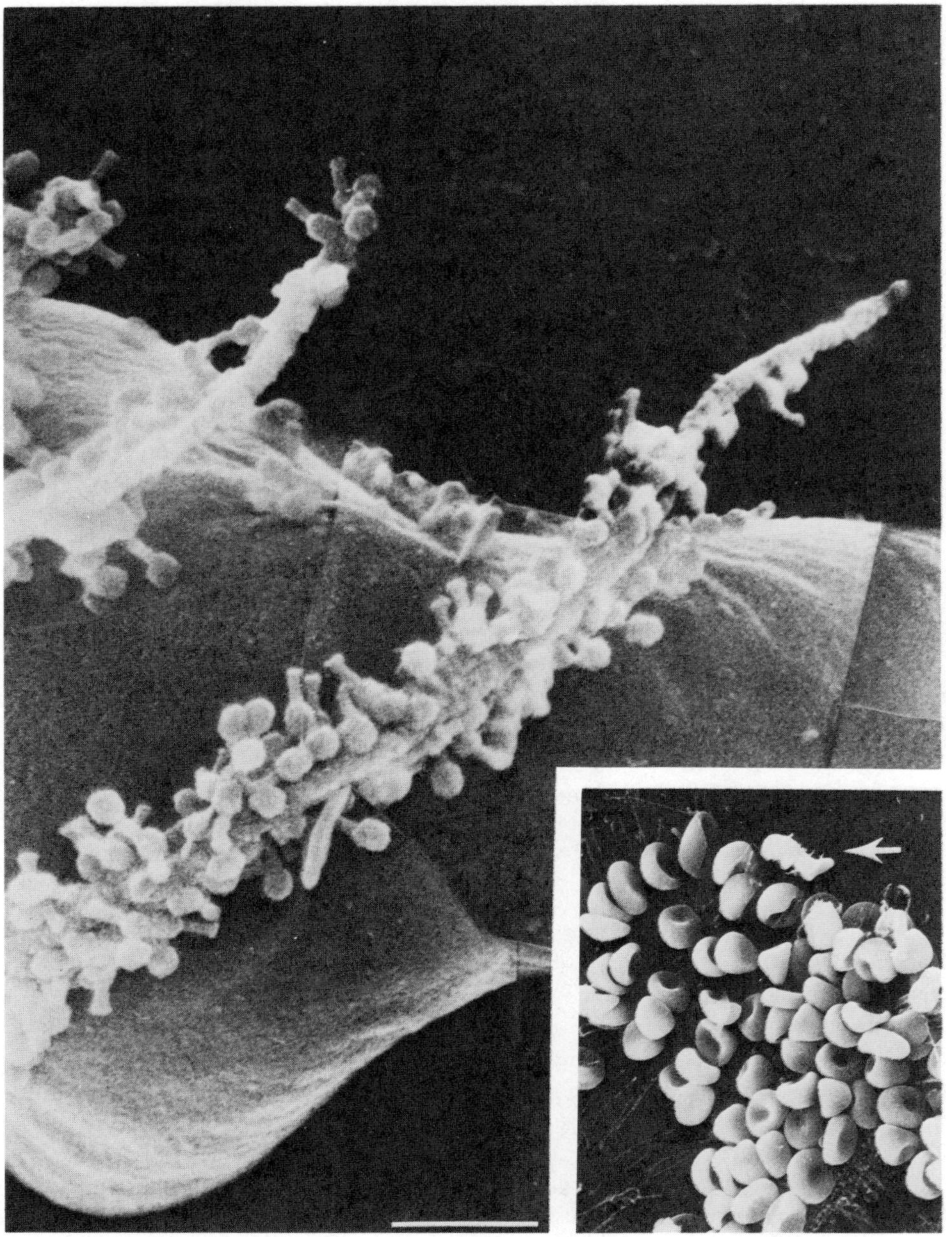

Figure 11 High-magnification micrograph of the cell protrusions of an influenza virus-infected cell and adsorbed RBC, indicated by the arrow in the inset. Cells have been indirectly labeled with influenza antiserum followed by T4 anti-immunoglobulin after hemadsorption. Scale bar = 0.5 μm. (From H. Kumon, 1976.)

experiments. The usefulness of a morphologically recognizable marker for SEM would therefore depend on it having a size greater than the size of the antibody. This is not the case for TEM, where much smaller electron-dense markers are used (Gonda et al., 1979).

Synthetic immunolatex spheres have been used quite extensively. They range in size from 30 to 340 nm and are of a spherical shape (Molday, 1976; Molday and Mayer, 1980). They have been used in detecting cell surface antigens on human red blood cells (Fig. 12), identifying mouse splenic lymphocytes (B cells) bearing immunoglobulin receptors (Fig. 13), and studying the movement of receptors in the plane of lymphocyte cell membranes as manifested by the formation of patches (capping) when the receptors are labeled with divalent antibodies (Fig. 14). The smaller immunolatexes in the 30-nm-diameter range are particularly useful for mapping cell surface antigens, whereas larger spheres are more applicable in typing cells. Latex spheres have also been used in TEM studies of viral antigen expression (Gonda et al., 1979).

Colloidal gold particles are quite suitable for direct SEM visualization of cell surface antigens by means of adsorbed immunoglobulins. Chemically fixed *Candida utilis* yeast cells have been labeled with a total anti-*C. utilis* antiserum electrostatically coupled to 60-nm colloidal gold particles using direct labeling procedures [Fig. 15(A)]. The marking

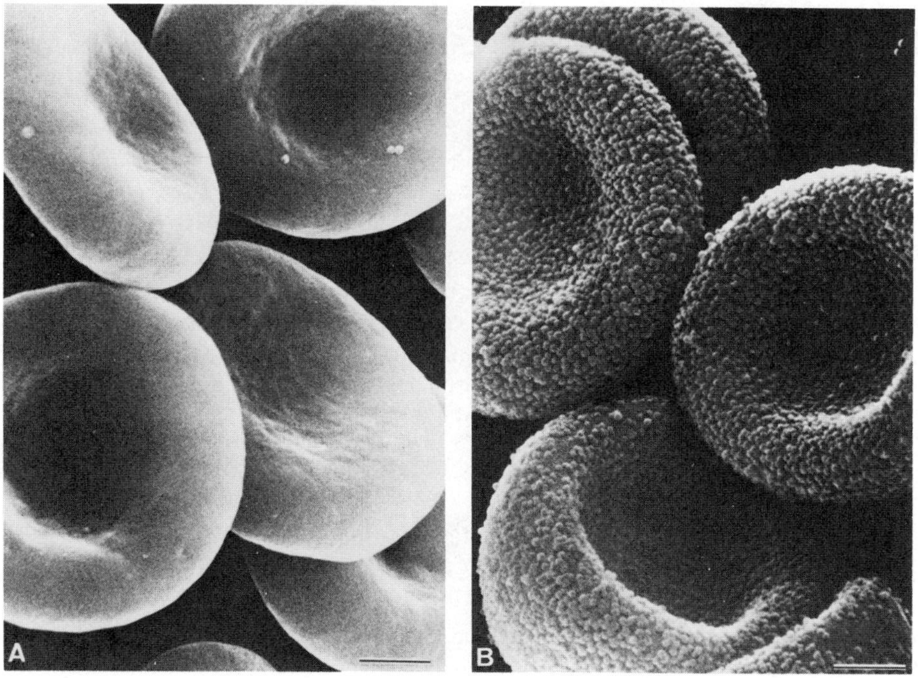

Figure 12 (A) Human RBC treated with normal rabbit serum and goat anti-rabbit microspheres. No labeling is observed. (B) Human RBC treated with rabbit anti-human RBC antiserum followed by goat anti-rabbit microspheres. Scale bar = 1.0 μm. (Courtesy of R. S. Molday.)

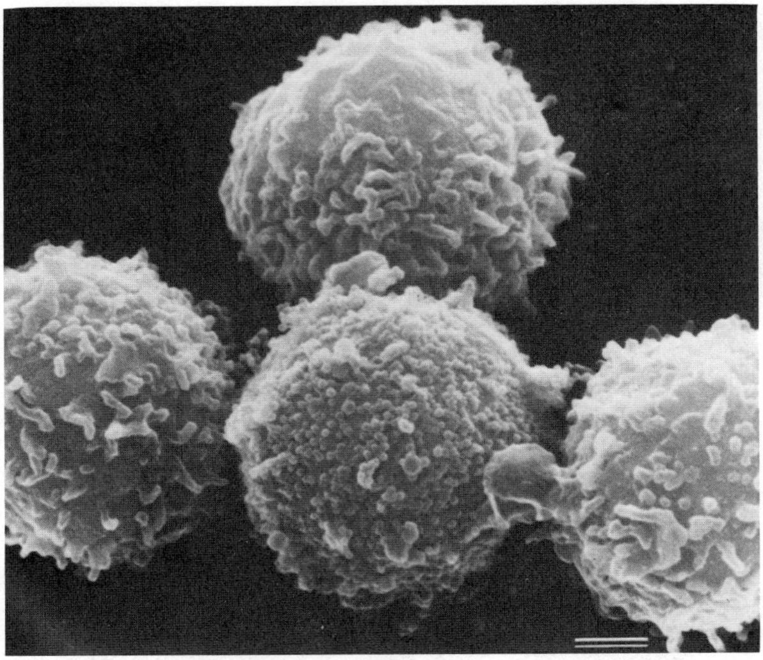

Figure 13 Mouse splenic lymphocytes indirectly labeled for surface Ig with rabbit anti-mouse IgG followed by goat anti-rabbit microsphere conjugates. Only the center cell is labeled. Scale bar = 1.0 μm. (Courtesy of R. S. Molday.)

is random but quite homogeneous over the whole cell wall surface. Practically no label is observed when *Saccharomyces cerevisiae*, an antigenically distinct yeast, is labeled with the same antiserum [Fig. 15(B)]. With colloidal gold particles in this size range no metallic coatings were required for visualization of the labeled sites. Compared to organic markers, the advantage of a heavy metal label is mainly its strong emission of secondary electrons, which are easily distinguished from the lower emission of secondary electrons from the yeast cell wall.

VI. LABELING OF CYTOPLASMIC ANTIGENS

Labeling of intracytoplasmic antigens poses several problems not encountered in cell surface labeling studies. To internalize the specific antibody, the cells must be damaged. To avoid morphological degeneration of cellular fine structure during this process, it is necessary to fix the tissue. The greater the degree of fixation, however, the more difficult it is for antibody to penetrate the cytoplasmic matrix. In addition, those chemical fixatives that confer the best ultrastructural preservation (e.g., glutaraldehyde) most often reduce the reactivity of specific antibody for its antigen due to complex changes in antigen structure. Some compromise in this process of preservation, which also allows penetration, has been studied in detail by Willingham (1980). By the use of a new primary fixative, ethyldimethylaminopropyl carbodiimide-glutaraldehyde-Tris combined with the

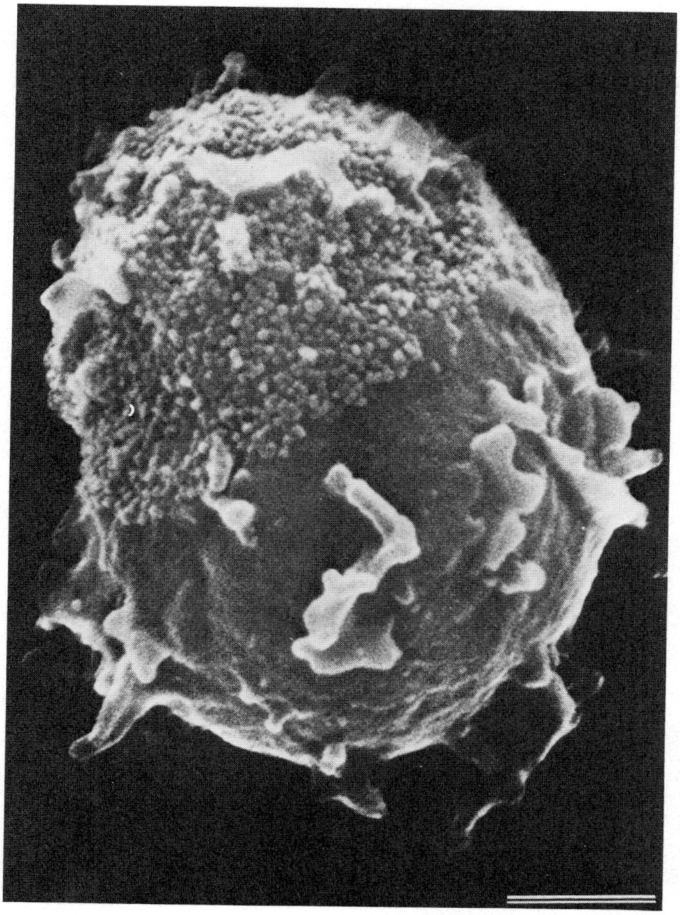

Figure 14 Cap on unfixed mouse splenic lymphocytes indirectly labeled for surface Ig using rabbit anti-mouse Ig antiserum followed by goat anti-rabbit Ig antibody-latex conjugate. Scale bar = 1.0 μm. (Courtesy of R. S. Molday.)

use of saponin (EGS) for membrane permeablization, ultrastructural morphology is preserved, yet the cytoplasmic matrix is permeable to immunoglobulin molecules. This allows preembedding localization of intracellular protein antigens in cultured cells.

Using the unlabeled antibody ferritin bridge with specific antibodies to actin and the EGS procedure, actin was localized in microvilli of mouse Swiss 3T3-4 and L929 cells (Fig. 16) as well as surface ruffles, microfilament bundles, microfilament mat, and leading lamellae. In a similar fashion, antibodies to brain tubulin were used to localize tubulin in cultured Swiss 3T3-4 fibroblasts (Fig. 17). The unlabeled antibody ferritin bridge readily allows the penetration of all components in the bridge because of the smaller size of each of the individual components in comparison to antibody-marker conjugates in which the marker is relatively large.

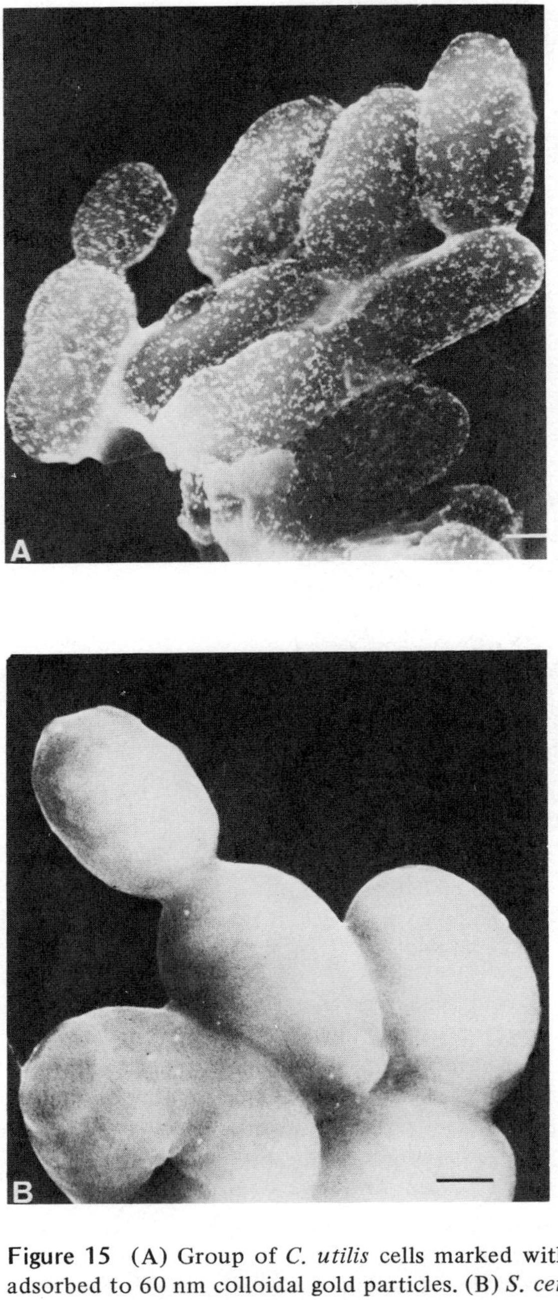

Figure 15 (A) Group of *C. utilis* cells marked with the total anti-*C. utilis* antiserum adsorbed to 60 nm colloidal gold particles. (B) *S. cerevisiae* cells incubated with the total anti-*C. utilis* antiserum. Practically no label is observed. Scale bar A = 1.0 μm; B = 1.0 μm. [From M. Horisberger (1975), Experimentia *31*, 1147.]

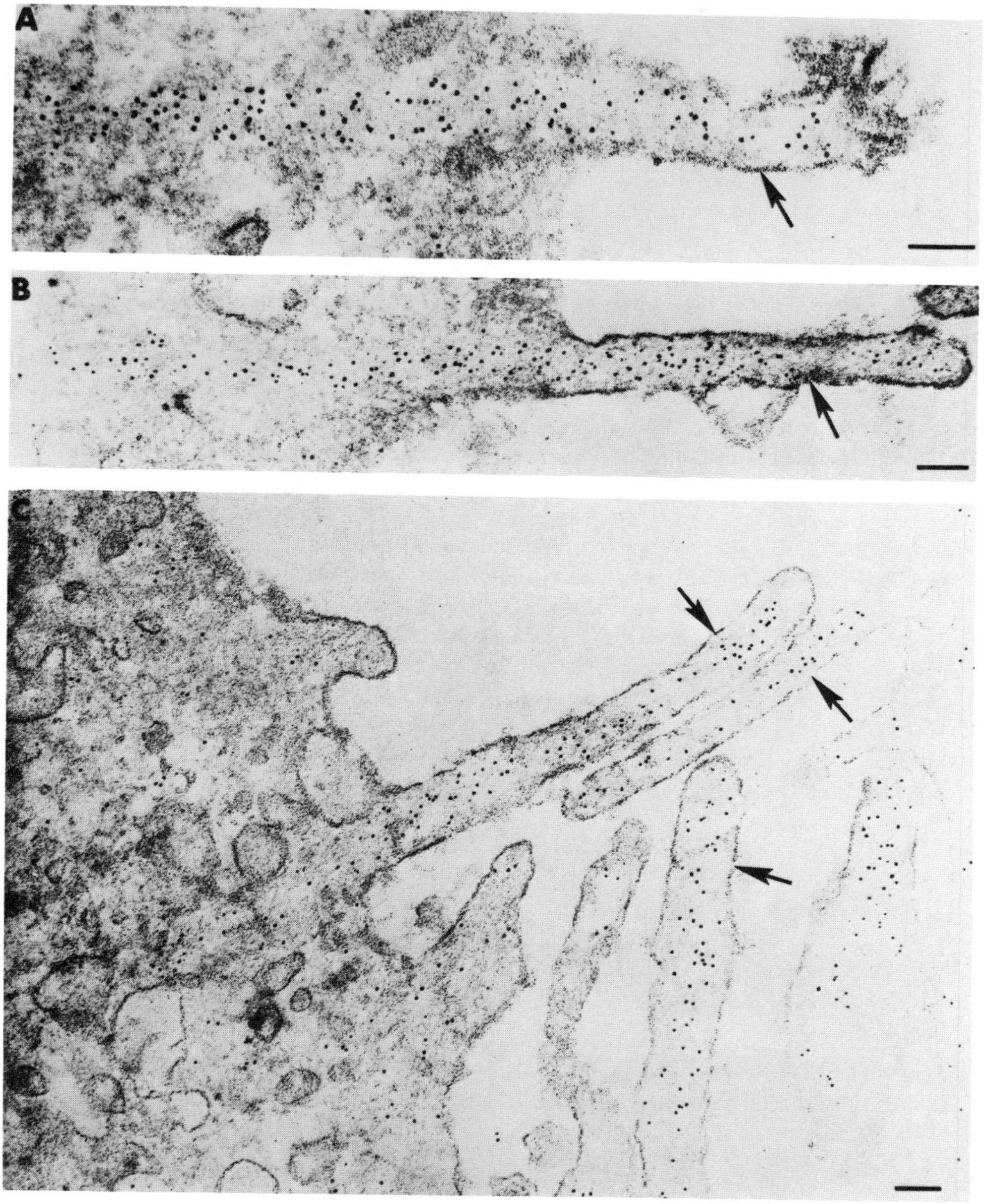

Figure 16 Localization of actin in microvilli of L929 cells. (A), (B), and (C) show different views of surface microvilli (→) on L929 which show extensive actin localization in their interior with extensions of localization into the cell interior. Scale bar = 0.1 μm. [From M. C. Willingham et al. (1981), J. Histochem. Cytochem. *29*, 17.]

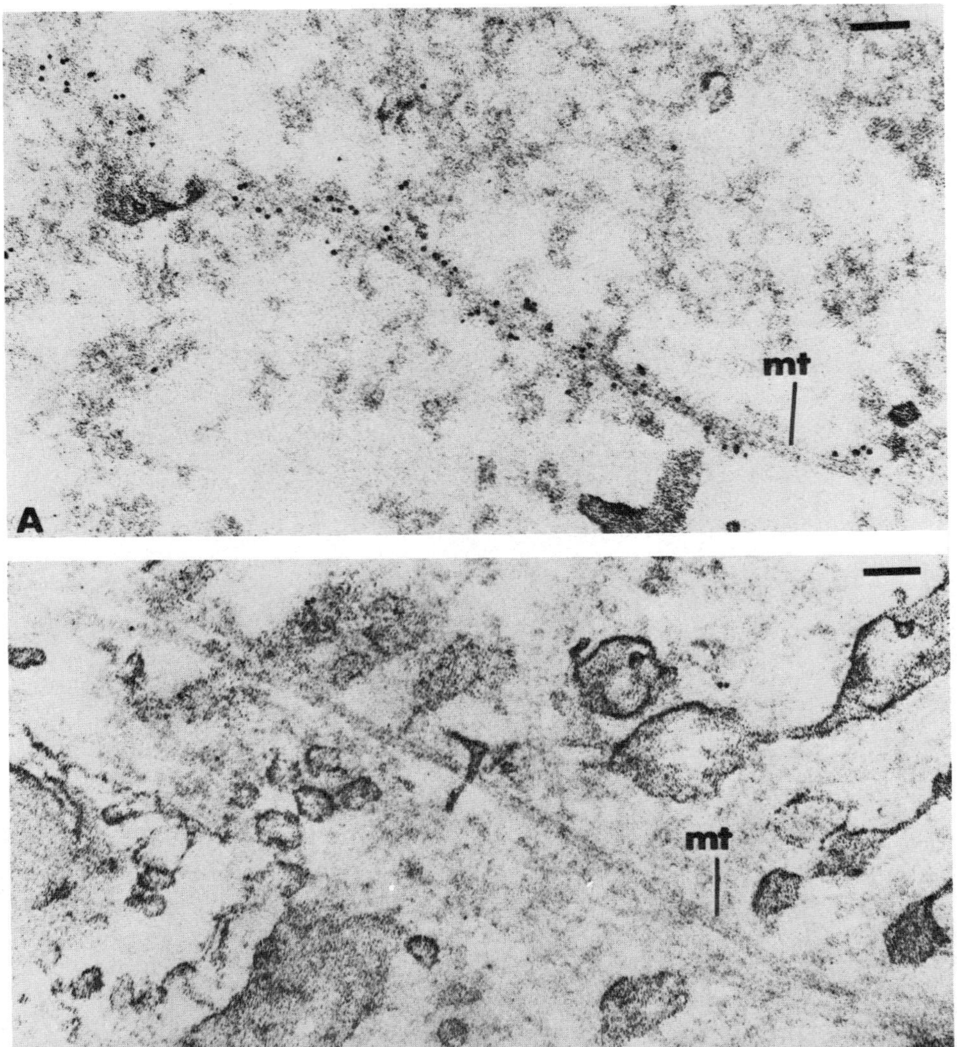

Figure 17 Ultrastructural localization of antitubulin antibody on microtubules. (A) Antitubulin; (B) normal globulin control (mt = microtubules). Scale bar = 0.1 μm. [From M. C. Willingham et al. (1980), J. Histochem. Cytochem. *28*, 453.]

Another approach to localizing intracytoplasmic antigen is to perform the labeling postembedding on thin sections (Horisberger, 1981). Mannan was localized on thin sections of *C. utilis* with homologous antimannan antibodies electrostatically coupled to colloidal gold. Mannan was primarily found in cell walls and plasmalemma (Fig. 18). The colloidal gold marking system is amenable to double-labeling experiments using different probes, each coupled to a different-size gold particle. *C. utilis* has been simultaneously

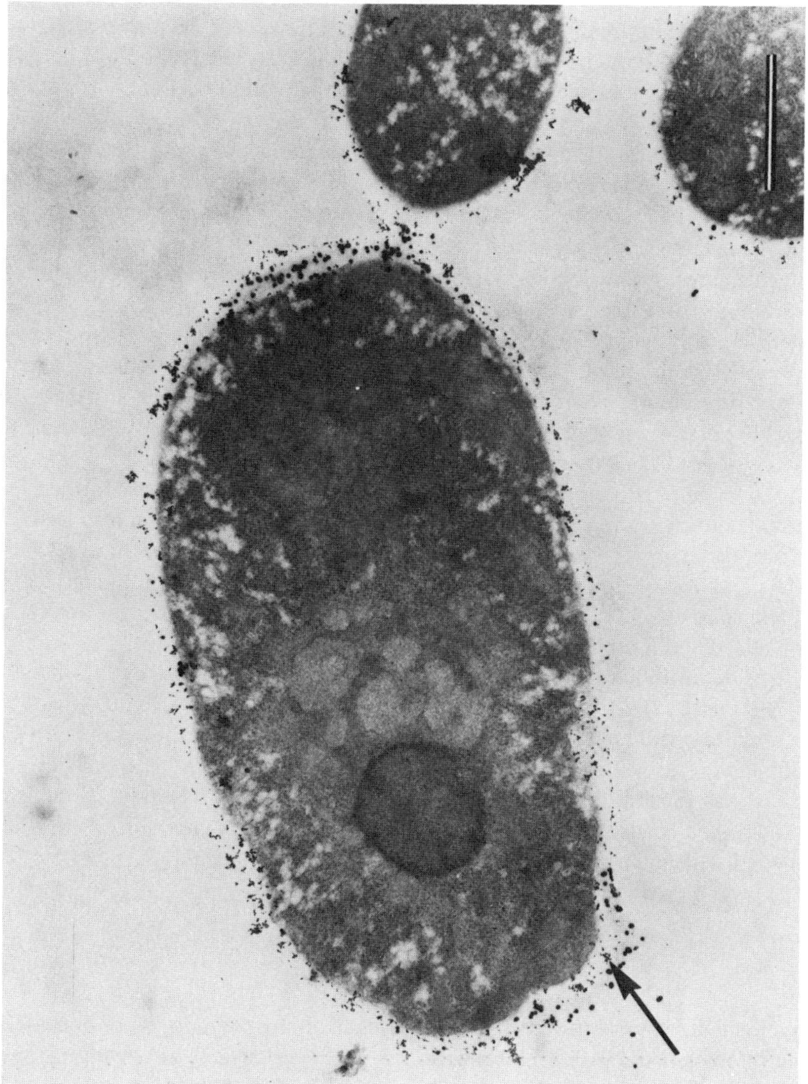

Figure 18 *C. utilis* thin sections marked simultaneously with antimannan antibody coupled-colloidal gold (5 nm) and wheat germ agglutinin-colloidal gold (26 nm). The arrow (→) indicates presence of mannan in bud scar. Scale bar = 1.0 μm. [From M. Horisberger and M. Vonlanthen (1977), Arch. Microbiol. *115*, 1.]

labeled using wheat germ agglutinin, a plant lectin that binds to chitin, and antimannan antibodies. Chitin was found primarily in bud scars but was also localized with mannan in the cell wall and on the plasmalemma (Fig. 18).

VII. LABELING OF NUCLEIC ACIDS AND ASSOCIATED PROTEINS

Studies of nucleic acid structure and the functional role of nucleic acid associated proteins have been aided by the use of electron microscope analysis. The technology has been transferred from methodologies typically used in the labeling of cellular antigens through immunoelectron microscopy.

Single-stranded DNA can be detected with precision and accuracy through the use of anti-DNA-binding protein antibody. These types of studies are important in localizing the position of sequences in DNA/DNA or RNA/DNA heteroduplexes where single-stranded DNA is difficult to distinguish from duplexed DNA. Partially single-stranded duplexed molecules are reacted sequentially with DNA-binding protein (specific for single-stranded DNA) from *Escherichia coli* and monovalent (Fab) antibodies to this protein before mounting for electron microscopy. The protein binds to single-stranded regions and causes them to appear considerably thicker than double-stranded regions after reaction with unlabeled monovalent antibody (Fig. 19). Ferritin-conjugated antibodies in an indirect test have also been used to study the origin of replication of SV40 DNA at the binding site of T antigen (Fig. 20).

Electron microscopic determinations of protein assembly have been performed for ribosomal proteins during transcription of ribosomal DNA. The method employs antibodies (IgG) to ribosomal proteins that are reacted during active transcription of ribosomal DNA. The location of bound IgG is detected with anti-immunoglobulin conjugated-polymethacrylate spheres that can be observed as electron-dense reaction cores on RNA transcripts in nucleic acid mounts (Fig. 21).

Antinucleoside antibodies highly specific for purine or pyrimidine determinant groups on single-stranded DNA have been developed and have been used to investigate the molecular architecture of metaphase chromosomes. Antinucleoside antibodies to 5-methylcytodine were used to label human metaphase chromosomes. The unlabeled specific antibody was localized by the indirect immunoperoxidase technique in which peroxidase was coupled to an anti-immunoglobulin. Discrete areas of the spread chromosomes were labeled with the electron-dense peroxidase reaction product (Fig. 22).

VIII. POLYCLONAL VERSUS MONOCLONAL ANTISERA AS IMMUNOSPECIFIC PROBES

Although polyclonal serological probes for immunoelectron microscopy are of considerable use in immunodiagnostic studies, localization studies have often yielded ambiguous results due to the presence of nonspecifically reacting antibodies. The antibodies were generated either by contamination in the original inoculum used to produce the polyclonal antiserum or preformed nonspecific antibodies present in animals used to generate the antiserum. To reduce the nonspecific antibodies and obtain a relatively pure population of reactive monospecific antibody, affinity column purification steps or extensive adsorption would have to be employed (see Chap. 13).

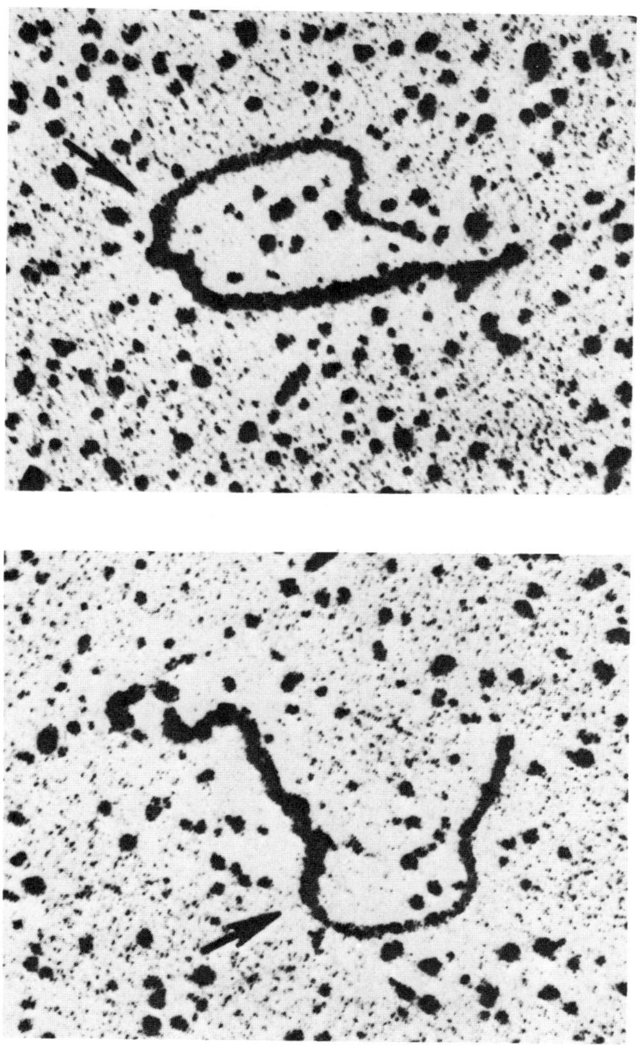

Figure 19 Electron micrographs of an *Eco* RI linear SV40 L strand hybridized to the smaller SV40 *Eco* RI/*Hpa* II restriction fragment followed by treatment with DNA binding protein from *E. coli* and Fab fragment from anti-DNA binding protein IgG. The single-stranded DNA segment can be distinguished from double-stranded DNA (→) since it is thicker and has greater contrast. [From S. I. Reed and J. C. Alwine (1977), Cell *11*, 523.]

Köhler and Milstein (1975) developed the technique of somatic cell hybridization to make continuous cultures of hybrid B lymphocytes-myeloma cells (hybridomas) producing monoclonal antibodies of predefined specificity (see Chap. 8). The use of monoclonal antibodies has distinct advantages over polyclonal sera in immunoelectron microscopic studies: (1) they are reactive to a single antigenic determinant, thus allowing

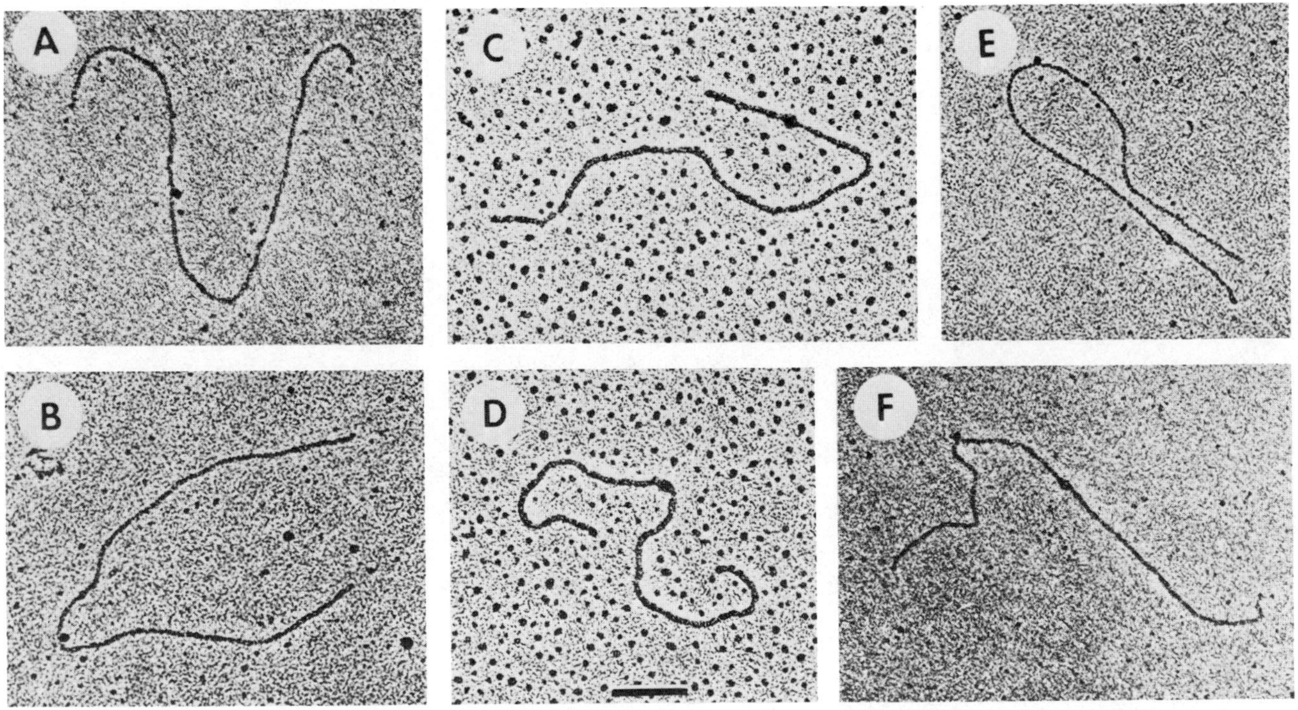

Figure 20 Electron micrographs of SV40 DNA prepared in various ways. (A, B) SV40(I) DNA incubated with T antigen, hamster anti-T IgG, and ferritin-conjugated anti-hamster IgG and cleaved with *Eco* RI endonuclease. (C, D) SV40(I) DNA treated with *E. coli* DNA unwinding protein and cleaved with *Eco* RI endonuclease. (E, F) Same as (A, B) except that the sample was treated with *E. coli* DNA unwinding protein before cleavage with *Eco* RI endonuclease. Scale bar = 0.2 μm. [From S. I. Reed et al. (1975), Proc. Natl. Acad. Sci. USA *72*, 1605.]

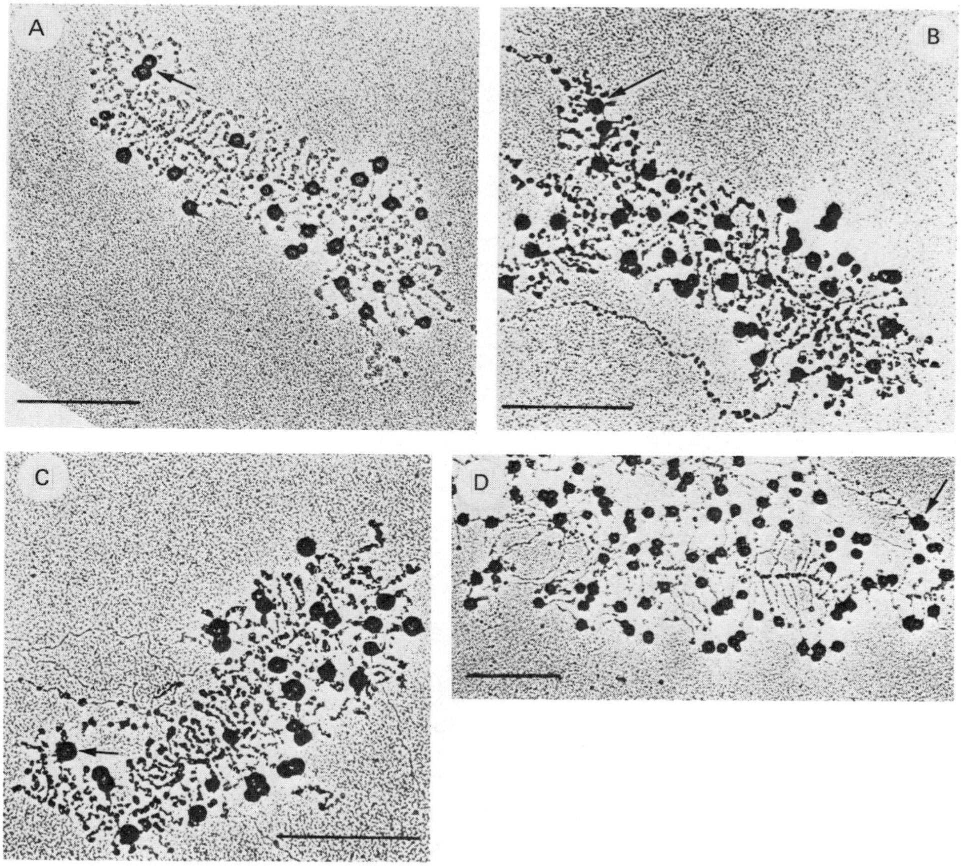

Figure 21 Gallery of ribosomal transcriptional units (rTU) that had been allowed to react with anti-S14 (ribosomal protein) IgG. The first antibody detected on a rTU is denoted by the arrow (→). Samples (A)–(C) were treated with the primary antibody at 1, 5, and 10 μg/ml, respectively, followed by polymethacrylate sphere-conjugated anti-immunoglobulin; sample (D) was treated with 50 μg/ml. Scale bar = 0.5 μm. [From W. Y. Chooi and K. R. Leiby (1981), Proc. Natl. Acad. Sci. USA *78*, 4823.]

greater resolution and specificity in cell surface labeling experiments than heretofore available with polyvalent sera; (2) they provide an infinite source of homogeneous antibodies since the hybridoma cells have been immortalized and the resultant cell lines can be stored for future use; and (3) not only are the antibodies pure biological labels of a homogeneous nature, but they can be produced in large quantities in tissue culture or as an ascitic tumor in syngeneic mice. Thus the tedious procedures of repeated inoculations, bleedings, and purifying antibodies and the variability of the polyvalent reagent are avoided.

Monoclonal antibodies are very useful in resolving antigenic differences which exist between closely related proteins and can be made with a narrow specificity (see Chap. 8). They can also be made broadly reactive and used for detecting antigenic determinants

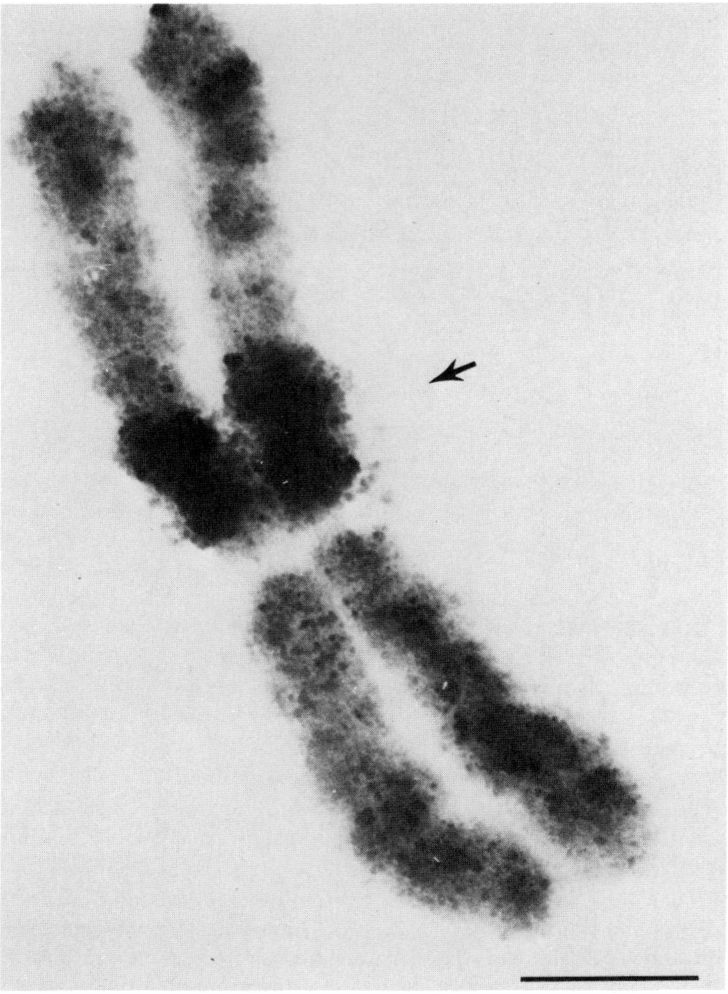

Figure 22 Electron micrograph of chromosome 1 stained by indirect immunoperoxidase procedure to localize 5-methylcytidine antibodies. The arrow indicates electron-dense staining of chromosome. Scale bar = 1.0 μm. [From B. W. Lubit et al. (1976), Cell *9*, 503.]

that have been conserved, thus resembling polyvalent sera. Because of the narrow specificity of most monoclonals, care should be taken in the interpretation of negative results by immunoelectron microscopy and these studies should be supplemented with other immunochemical tests. Immunoelectron microscopy results with cell surface labeling in particular may be considered as more consistent with the in vivo antibody response since the target antigens are still native with regard to three-dimensional structure and topographical position (Gonda et al., 1980, 1981). Although a greater amount of initial work is required for obtaining hybridomas to produce monoclonal antibodies in contrast to polyvalent sera, the consistent quality and homogeneous nature of the reagent certainly outweigh the inconsistencies inherent in the preparation of multiple polyvalent sera that vary from animal to animal and even from bleeding to bleeding.

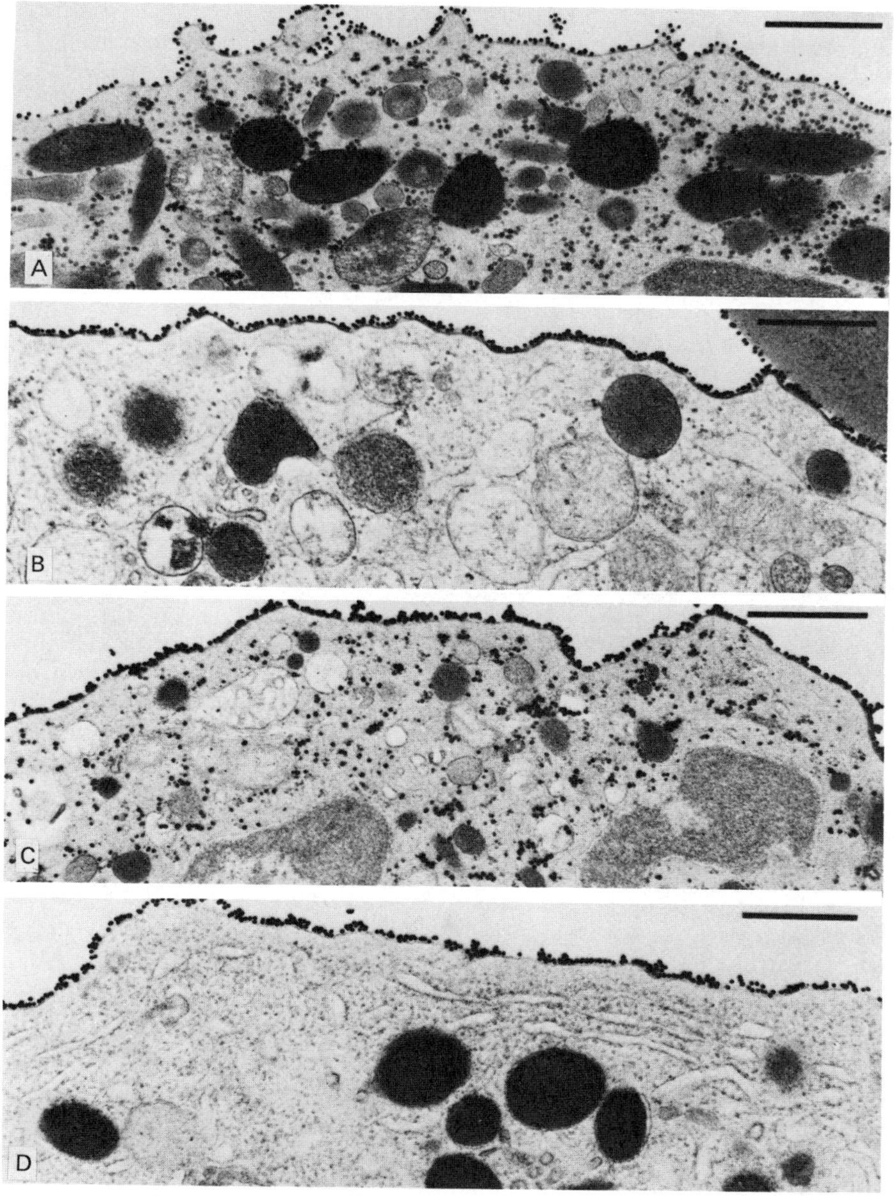

Figure 23 (A–D) Human and guinea pig bone marrow cells labeled by the wheat germ agglutinin-ovomucoid-coupled colloidal gold method. (A) Segmented neutrophil from human bone marrow shows heavy wheat germ agglutinin surface labeling. (B) Neutrophilic promyelocyte from human bone marrow shows heavy wheat germ agglutinin surface labeling. (C) Segmented neutrophil from guinea pig bone marrow exhibits heavier surface wheat germ agglutinin than human cells in (A). (D) Neutrophilic promyelocyte from guinea pig bone marrow is heavily labeled but appears to have slightly less wheat germ agglutinin binding than the segmented neutrophil in (C). Scale bar = 0.5 μm. [From G. A. Ackerman (1979), Anat. Rec. *195*, 641.]

IX. USE OF LECTINS AS PROBES OF ANTIGEN SITES

Plant lectins have been extensively used in localization studies of biological membranes. They are reactive to surface glycoproteins with different lectins having different specificities. Their action in selecting a particular receptor is similar to antibody. Several known lectins are: concanavalin A, soybean agglutinin, wheat germ agglutinin, and peanut lectin.

Labeled and unlabeled marker methods have been used to detect lectin receptors on cells. Hemocyanin, which contains a glycoprotein in its outer structure, can be used to mark indirectly the specific lectin receptor by being applied to the cells after they have been treated with lectin. The lectin is capable of bridging receptor and marker. The

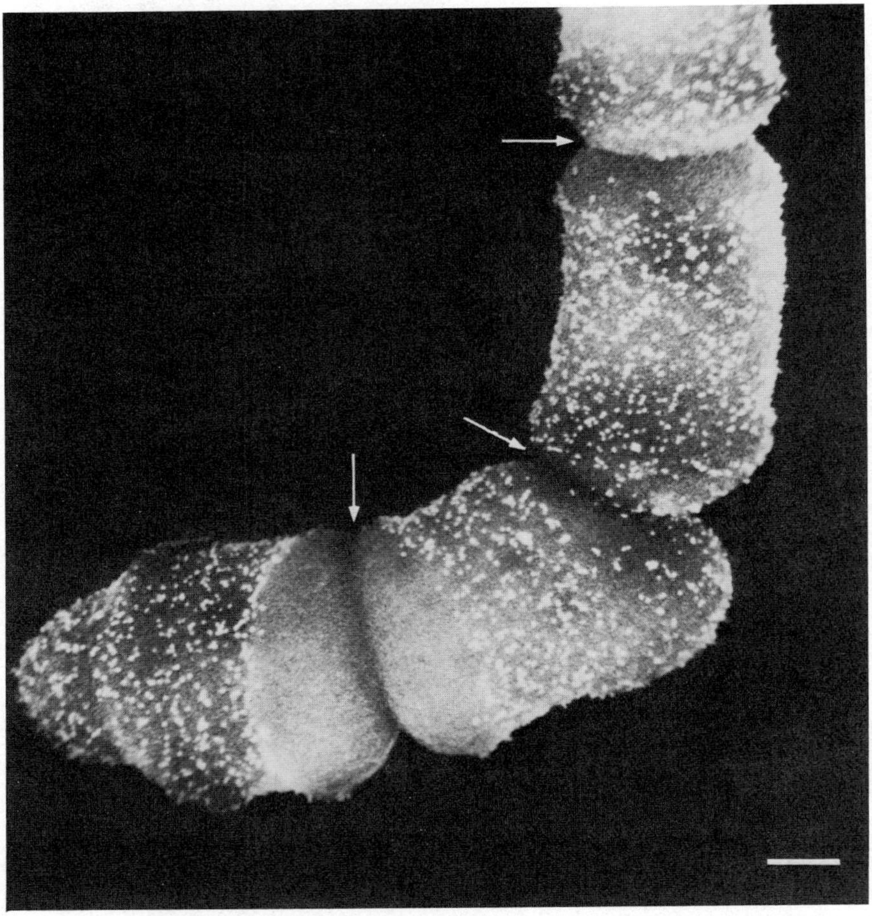

Figure 24 Localization of galactomannan on the surface of *Schizosaccharomyces pombe* by the sandwich-gold method. The gold marker indicating the presence of galactomannan was found mainly on walls growing by extension. Walls established by division were practically not labeled (→). Scale bar = 1.0 μm. [From M. Horisberger et al. (1978), Arch. Microbiol. *119*, 107.]

distribution of wheat germ agglutinin binding sites on normal human and guinea pig bone marrow cells has also been studied in an indirect method using colloidal gold. Ovomucoid-coupled colloidal gold particles (Geoghegan and Ackerman, 1977) were reacted with human or guinea pig bone marrow cells that had previously been treated with wheat germ agglutinin. The lectin bridged the cell surface receptor and the ovomucoid-coupled colloidal gold which was visualized by TEM (Fig. 23). The wheat germ agglutinin is believed to bind selectively to N-acetylglucosamine residues. Colloidal gold has been used in SEM studies to localize α-galactomannan receptors in *Schizosaccharomyces pombe* by a sandwich method using the lectin of *Bandeiraea simplicifolia*. The receptors were evenly

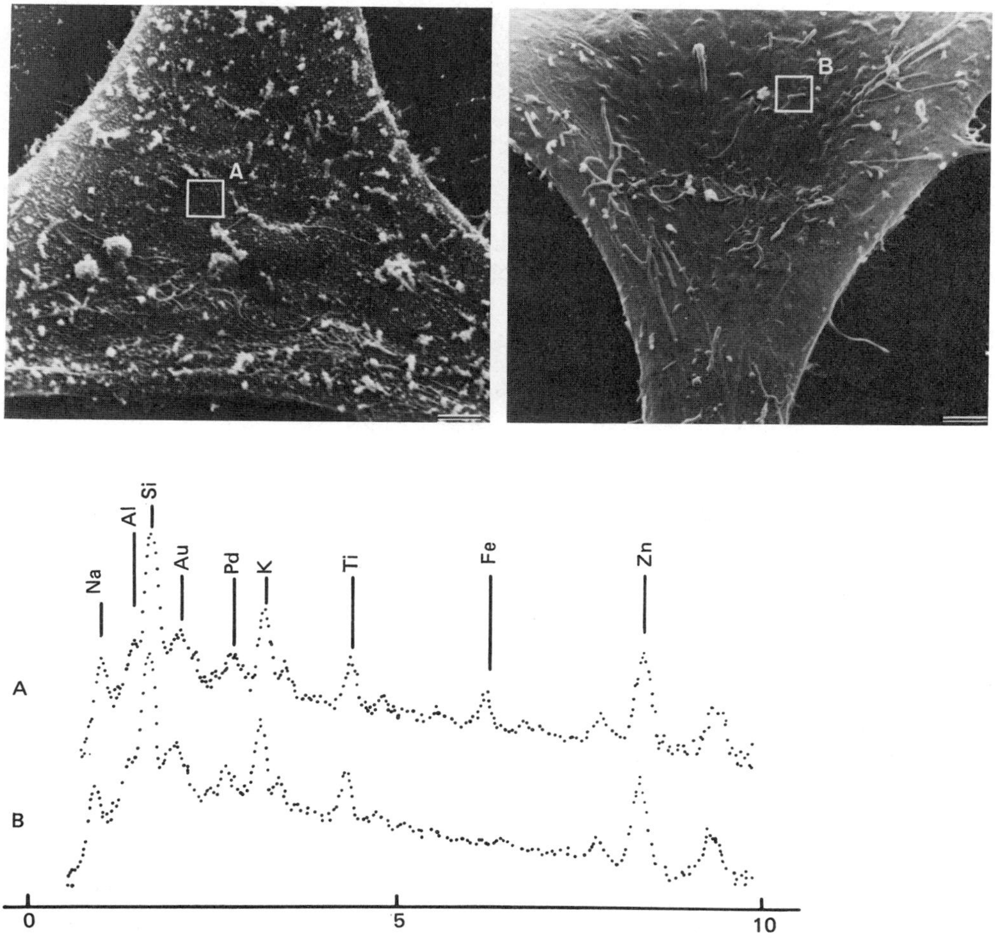

Figure 25 Mouse 3T3 fibroblast labeled for *Ricinus communis* agglutinin receptors with iron microspheres in the absence (A) and presence (B) of the inhibitor galactose. At the bottom is an x-ray spectrum of the boxed areas in (A) and (B). An iron (Fe) peak appears only in the iron-microsphere-labeled cell. Other elements are derived from the glass coverslip, support stub, and gold-palladium cell coating. Scale bar = 1.0 μm. [From R. S. Molday (1981), Biomedical Research *2*, 23.]

distributed over the cell surface except at areas of bud scars (Fig. 24). Lectins have also been conjugated to iron-containing latex microspheres and used in cell surface studies of lectin receptors on mouse 3T3 fibroblasts. The iron content enables one to detect these markers on cell surfaces under the SEM using x-ray analysis and secondary electron imaging (Fig. 25).

X. CONCLUSIONS

This chapter on immunoelectron microscopy has covered basic methodology, considerations in choosing a marker and labeling system, and problems that may be encountered with a particular approach. A variety of specific applications have been presented to demonstrate the feasibility and range of immunoelectron microscopy in the study of antigen–antibody interactions by TEM and SEM. In addition to using electron-dense or morphologically recognizable inert particles, autoradiography has also become a valuable tool in biological research for the localization of antigen using radioactive isotopes bound to antibody. The technique is applicable to TEM, SEM, and light microscopy.

Before any successful labeling experiments can be carried out the labeling reagents should be well characterized. Antibody specificity and titer must be determined. Control experiments must be included and convincing demonstration that the labeling is specific is necessary. This can be accomplished by blocking experiments, or by using positive and negative controls, or both (Nemanic, 1979; Gonda et al., 1981).

While immunoelectron microscopy in the strictest terms implies that antibodies are used, this chapter has also presented the use of lectins as probes of cell surface moieties. Molecules such as *Staphylococcus* protein A have also been used in indirect methods to couple the Fc receptor of antibody to the marker (see Fig. 9). In addition, antibody fragments (i.e., Fab monomers) have been conjugated to markers for use in mapping cell surface and intracytoplasmic antigens. With the advent of monoclonal antibodies, we can expect to continue to see a greater use of these highly specific probes in immunoelectron microscopy and the resolution of many problems that were not readily feasible with polyclonal sera.

REFERENCES

Andres, G. A., Hsu, K. C., and Seegal, B. C. (1978). In *Handbook of Experimental Immunology*, Vol 2: *Cellular Immunology* (D. M. Weir, Ed.), Blackwell, Oxford, Chap. 37.

Carter, D. P. (1978). In *Principles and Techniques of Scanning Electron Microscopy*, Vol. 6 (M. A. Hayat, Ed.). Van Nostrand Reinhold, New York, pp. 317–329.

Coons, A. H., Creech, H. J., and Jones, R. N. (1942). J. Immunol. *45*, 259.

Farr, A. G., and Nakane, P. K. (1981). J. Immunol. Methods *47*, 129.

Geoghegan, W. D., and Ackerman, G. A. (1977). J. Histochem. Cytochem. *25*, 1187.

Geoghegan, W. D., Ambegaonkar, S., and Calvanico, N. J. (1980). J. Immunol. Methods *34*, 11.

Gonda, M. A., Gilden, R. V., and Hsu, K. C. (1979). Scan. Electron Microsc. *3*, 583.

Gonda, M. A., Arthur, L. O., Massey, R. J., Elser, J. E., Schochetman, G., and Hsu, K. C. (1980). Scan. Electron Microsc. *3*, 13.

Gonda, M. A., Benton, C. V., Massey, R. J., and Schultz, A. M. (1981). Scan. Electron Microsc. *2*, 45.

Hammerling, H., Aoki, T., Wood, H. A., Old, L. J., Boyse, E. A., and de Harven, E. (1969). Nature (Lond.) *223*, 1158.

Horisberger, M. (1981). Scan. Electron Microsc. *2*, 9.

Horisberger, M., and Rosset, J. (1977). J. Histochem. Cytochem. *25*, 295.

Karnovsky, M. J., Unanue, E. R., and Leventhal, M. (1972). J. Exp. Med. *136*, 907.

Köhler, G., and Milstein, C (1975). Nature (Lond.) *256*, 495.

Kumon, H. (1976). Virology *74*, 93.

Molday, R. S. (1976). In *Principles and Techniques of Scanning Electron Microscopy*, Vol. 5: *Biological Applications* (M. A. Hayat, Ed.). Van Nostrand Reinhold, New York, pp. 53–77.

Molday, R. S., and Maher, P. (1980). Histochem. J. *12*, 273.

Nakane, P., and Pierce, G. P. (1966). J. Histochem. Cytochem. *14*, 929.

Nemanic, M. K. (1979). Scan. Electron Microsc. *3*, 537.

Peters, K., Rutter, G., Gschwender, N. H., and Haller, W. (1978). J. Cell Biol. *78*, 309.

Pinto Da Silva, P., Kachar, B., Torrisi, M. R., Brown, C., and Parkison, C. (1981). Science *213*, 230.

Rifkind, R. A., Hsu, K. C., Morgan, C., Seegal, B. C., Knox, A. W., and Rose, H. M. (1960). Nature (Lond.) *187*, 1094.

Singer, S. J. (1959). Nature (Lond.) *183*, 1523.

Sternberger, L. A. (1974). In *Foundations of Immunology Series* (A. Olser and L. Weiss, Eds.). Prentice-Hall, Englewood Cliffs, N.J., pp. 1–171.

Willingham, M. C. (1980). Histochem. J. *12*, 419.

23

Structure and Function of Complement

Douglas T. Fearon Harvard Medical School, Boston, Massachusetts

I. INTRODUCTION

The complement system is comprised of 18 proteins which are present in highest concentration in plasma (Table 1). The primary function of this system is the mediation of host defense against microbial infection. This object is achieved during activation of complement by the elaboration of peptides, larger protein fragments, and multimolecular complexes that opsonize and lyse the activating target; induce chemotactic, secretory, and metabolic responses of leukocytes; and alter vascular permeability. Taken together, these activities constitute an inflammatory response.

The complement system is the most complex of the several protein activation systems in blood, but this complexity can be reduced by considering the system to be comprised of three functional divisions: two pathways for activation, the classical and alternative pathways, and a common effector sequence to which the activating pathways are directed and from which are derived many of the biological activities of complement (Fig. 1). This chapter presents the molecular reactions of the three functional divisions of complement, a consideration of the biologic functions of complement and their relation to specific receptors on leukocytes, and the consequences of inherited deficiencies of complement proteins.

The nomenclature of the complement proteins reflects the order in which the two activating pathways were discovered rather than considerations of relative importance or of phylogenetic precedence. Indeed, the alternative pathway may have evolved before the classical pathway and may be more critical for host defense against infection, characteristics that are not suggested by its name. Components of the classical pathway and the effector sequence are designated by the letter C and a number, which indicates the sequence of interaction of the proteins (except for C2, C3, and C4, which were named

Table 1 Proteins of the Complement System

Name	Former designation	Molecular weight	Serum concentration (μg/ml)
C1q	—	400,000	70
C1r	—	95,000	35
C1s	—	85,000	35
C4	—	180,000	400
C2	—	117,000	25
C3	—	185,000	1500
C5	—	200,000	85
C6	—	128,000	75
C7	—	121,000	55
C8	—	153,000	55
C9	—	80,000	200
B	C3 proactivator Glycine-rich β-Glycoprotein	95,000	250
D	C3 proactivator convertase	25,000	2
P	Properdin	160,000	25
$C\bar{1}$ inhibitor	—	105,000	180
C4 binding protein	—	1.2×10^6	250
H	β1H	150,000	400
I	KAF C3b inactivator	90,000	50

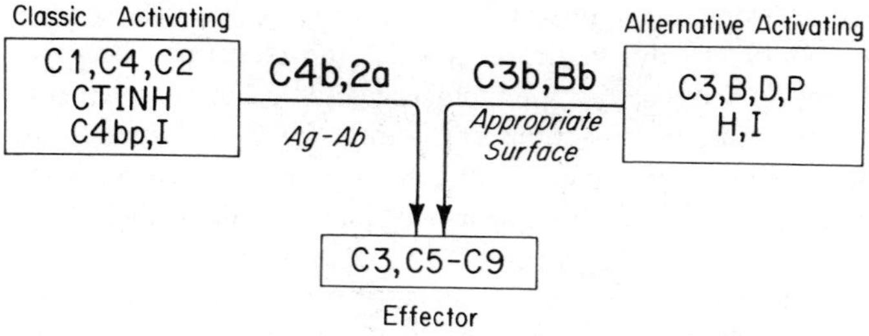

Figure 1 The three functional divisions of complement and their constituent proteins. The classical activating pathway forms its C3 convertase, C4b,2a, following interaction with antigen–antibody (Ag–Ab) complexes, and the alternative activating pathway generates the amplification C3 convertase, C3b,Bb, upon interaction with certain cell surfaces. Both enzymes have identical substrate specificities and are capable of activating the effector sequence.

before their position in the reaction sequence was known). Proteins of the alternative pathway are signified by capital letters, except for C3, which retains its classical pathway designation. Fragments derived by proteolysis of proteins are denoted by lowercase letters, as in C4a, C4b, and Ba, Bb. Fragments lacking activity in the reaction sequence have the letter i, as in iC3b and Bbi. An overbar indicates the enzymatically active form of a component, as in $\overline{\text{Clr}}$ and $\overline{\text{Cls}}$, when this activity cannot be indicate by the method for denoting fragments.

II. CLASSICAL PATHWAY OF COMPLEMENT ACTIVATION

The classical pathway of complement activation has six components: C1, C4, and C2, which interact to form an enzyme that activates the effector sequence, and C$\overline{1}$ inhibitor (C$\overline{1}$ INH), C4 binding protein (C4-bp), and C4b/C3b inactivator (I), which limit the magnitude of activation of C3–C9 by this pathway. The pathway is initiated when antibody, which has been aggregated by interaction with antigen, binds C1. This component becomes an active protease which hydrolyzes C4 and C2, whose major cleavage fragments form a bimolecular complex, C4b,2a, which is the classical pathway C3 convertase or C3-activating enzyme (Fig. 2).

C1 is a complex of the proteins, Clq, Clr, and Cls, which are held together by calcium in the molar ratio 1Clq:2Clr:2Cls (Ziccardi and Cooper, 1977). Each component has a distinct function: Clq binds to the Fc region of immunoglobulin to initiate the activation reaction; Clr is a proenzyme that is converted to a proteolytic form, $\overline{\text{Clr}}$, following binding of Clq to antibody; and Cls is a proenzyme that is activated by $\overline{\text{Clr}}$ to a form, $\overline{\text{Cls}}$, that is capable of hydrolyzing C4 and C2. When visualized by electron microscopy, the Clq molecule appears to have six globular regions connected by fibrils to a central stemlike region. The globular portions of the protein have no unusual primary structural features, while the fibrillar and stemlike regions resemble collagen in possessing a high content of hydroxyproline and hydroxylysine and a repeating X-Y-Gly sequence

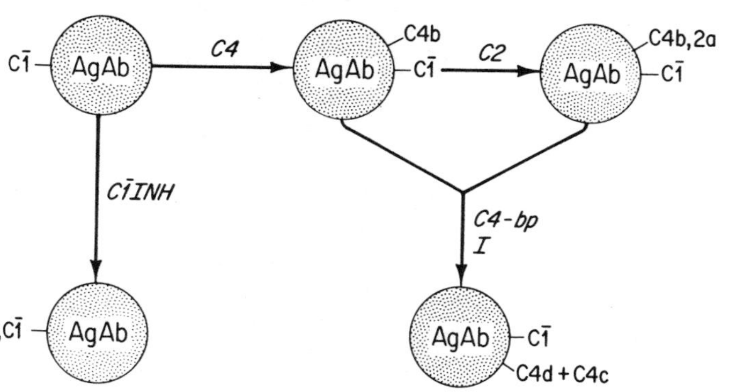

Figure 2 Molecular reactions of the classical pathway of complement activation. An antigen–antibody complex (AgAb) binds and activates C1. The activated first component, C$\overline{1}$, sequentially cleaves C4 and C2 whose major fragments form the classical C3 convertase, C4b,2a. Those reactions are regulated by C$\overline{1}$ INH, which inhibits the C$\overline{1}$r and C$\overline{1}$s subcomponents of C$\overline{1}$, and by C4-bp and I, which inactivate C4b by cleavage into the C4c and C4d fragments.

(Porter, 1977). The globular regions contain the sites that mediate binding of Clq to immunoglobulin and have a specificity for the Cμ4 and Cγ2 domains of IgM and IgG, respectively, Although Clq has a valence of 6 for antibody, binding of only two of the globular regions is necessary and sufficient for activation of C1. The stemlike region of Clq carries the calcium-dependent attachment sites for dimers of Clr and Cls, and is resistant to treatment with pepsin but is sensitive to proteolysis by collagenase, which is additional evidence for its collagenlike structure.

The mechanism of activation of Clr subsequent to binding of Clq to immunoglobulin is not entirely understood. Currently held hypotheses state that a conformational change induced in Clq by binding of at least two globular regions causes Clr to enter a transitional state in which its proteolytic site is "exposed." By an intra- or intermolecular peptide-bond cleavage, Clr, which is comprised of a single polypeptide chain, is converted to $\overline{\mathrm{Clr}}$, which has two polypeptides linked by disulfide bond(s). The serine protease site on the light chain of $\overline{\mathrm{Clr}}$ then hydrolyzes Cls at a single site, converting it to $\overline{\mathrm{Cls}}$, which also carries a serine protease site on the smaller of two, disulfide-linked polypeptides. $\overline{\mathrm{Cls}}$ is the enzyme responsible for activation of C4 and C2.

C4 is comprised of three disulfide-linked polypeptides, α, β, and γ, and the α chain contains an internal thiol ester formed between cysteine and glutamic acid residues (Harrison et al., 1982). On cleavage of the α chain near its amino terminus by $\overline{\mathrm{Cls}}$, the C4a peptide is released and the thiol ester on the major C4b fragment is apparently hydrolyzed to yield a free sulfhydryl and a highly reactive acyl group of the glutamyl residue. This acyl group may form an amide bond with amino groups or an ester linkage with hydroxyl groups of constituents on the cell surface or the antigen–antibody complex near the site of C$\overline{1}$. The C4b cannot bind to sites distant from the locus of its generation because the reactive acyl group reacts rapidly with H_2O to yield the stable carboxyl group, which has no covalent binding function.

The bound C4b forms a magnesium-dependent reversible complex with C2, which is then activated by cleavage with $\overline{\mathrm{Cls}}$ (Müller-Eberhard et al., 1967). The larger C2a fragment remains complexed to C4b and carries the serine protease site with specificity for C3 and C5. This proteolytic activity of C2a is expressed only when the protein is complexed with C4b. Thus the intact classical pathway C3 convertase, C4b,2a, is firmly anchored through its C4b subunit to a site immediately adjacent to the antigen–antibody–C$\overline{1}$ complex.

Three regulatory proteins limit the formation and function of C4b,2a. C$\overline{1}$ INH forms a covalent complex with $\overline{\mathrm{Cls}}$ (Harpel and Cooper, 1975) and blocks activation of C4 and C2. C$\overline{1}$ INH can also inhibit $\overline{\mathrm{Clr}}$ and may retard autoactivation of Clr. C4-bp binds to C4b and prevents uptake of C2 or displaces C2a (Gigli et al., 1979). Binding of C4-bp to C4b also leads to cleavage of the latter protein by I at two sites to yield the fragments C4c and C4d. C4d contains the covalent binding site of C4b and remains bound to the immune complex, but has no function in complement activation.

III. ALTERNATIVE PATHWAY OF COMPLEMENT ACTIVATION

The human alternative pathway of complement activation differs from the classical pathway in two general respects. First, it can be activated by certain bacteria, fungi, yeast, heterologous mammalian cells, and virally infected homologous cell in the absence of specific antibody. This property of the alternative pathway indicates that it can provide the nonimmune individual with a means for resistance to microbial infection, whereas the

classical activating pathway usually operates only after an individual has produced specific antibody. Second, formation of an effective C3 convertase by the alternative pathway depends on a positive feedback reaction that involves C3b, the product of C3 cleavage; no positive feedback reaction exists in the classical activating pathway. Six proteins constitute the alternative pathway (Table 1) and there are two types of C3-cleaving enzymes, the "priming" C3 convertase, which has the composition C3,Bb, and the "amplification" C3 convertase, C3b,Bb. The latter enzyme mediates the positive feedback reaction by utilizing C3b, the product of C3 cleavage, for its formation, and the former enzyme generates small amounts of C3b that are necessary to initiate the amplification reaction (Fearon, 1979).

Although the priming C3 convertase has not been isolated, it apparently is slowly and continually being formed in the fluid phase by the interaction of C3, B, D, and P. C3 consists of two polypeptide chains, and the α chain, like that of C4, contains an intrachain thiol ester between cysteine and glutamyl residues (Tack et al., 1980). This cyclic structure is thought to hydrolyze spontaneously at a slow rate, causing a conformational change to occur that yields a form of C3 which has an affinity for B (Pangburn a and Müller-Eberhard, 1980). B is a proenzyme which is activated when cleaved into its Ba and Bb fragments by D. The serine protease site of B resides on the Bb fragment, is specific for C3 and C5, and is expressed only when Bb is complexed to the altered C3 or C3b. D is a serine protease that is in its active state in plasma and only cleaves B that is bound to altered C3 or the C3b. Thus the proposed reaction sequence for formation of the priming C3 convertase is the following: spontaneous hydrolysis of the thiol ester in C3, reversible magnesium-dependent binding of B to altered C3, and cleavage of bound B by D to form C3,Bb. This C3 convertase then cleaves native C3 to generate C3a and C3b, and the latter fragment participates in the amplification phase of the alternative pathway. P enhances this reaction, by retarding the decay of C3,Bb.

The thiol ester within the α chain of C3, like that of C4, is the basis for the covalent binding, or "fixing," function of the protein (Law and Levine, 1977). Immediately following the cleavage of the C3a peptide from the α chain, the thiol ester becomes exposed in a manner that permits it to engage in a transesterification reaction. If C3b randomly collides with a carbohydrate presenting hydroxyl groups, an ester bond is formed between the glutamyl portion of the thiol ester, while the cysteinyl constituent accepts a proton to acquire a free sulfhydryl. If a carbohydrate is not encountered within milliseconds after cleavage of the α chain of C3, the reactive acyl group derived from the thiol ester reacts with H_2O and the C3b loses its covalent binding function (Fig. 3). Since cell surfaces present abundant amounts of carbohydrate in the forms of glycoproteins, glycolipids, and polysaccharides, this mechanism for C3b attachment is suitable for the fixation of this component to foreign and autologous cells.

The bound C3b then initiates formation of the amplification C3 convertase, C3b,Bb. C3b, in the presence of 0.5 mM magnesium, reversibly binds B with a K_a value of approximately 2×10^5 M^{-1}. The complexed B is activated by cleavage with D into its Bb and Ba fragments to form C3b,Bb (Fig. 4). This C3 convertase is labile and loses its C3-cleaving activity when Bb dissociates irreversibly from C3b in a time- and temperature-dependent manner. Although C3b can reform a new C3 convertase by interaction with additional B and D, the proteolytic site on Bb becomes "buried" upon dissociation of the protein from C3b and is termed Bbi. The decay of the amplification C3 convertase can be modified by P, which binds to C3b and retards the rate of dissociation of Bb. For example, the half-life of C3b,Bb at 30°C is increased from 4 min in the absence of P to as

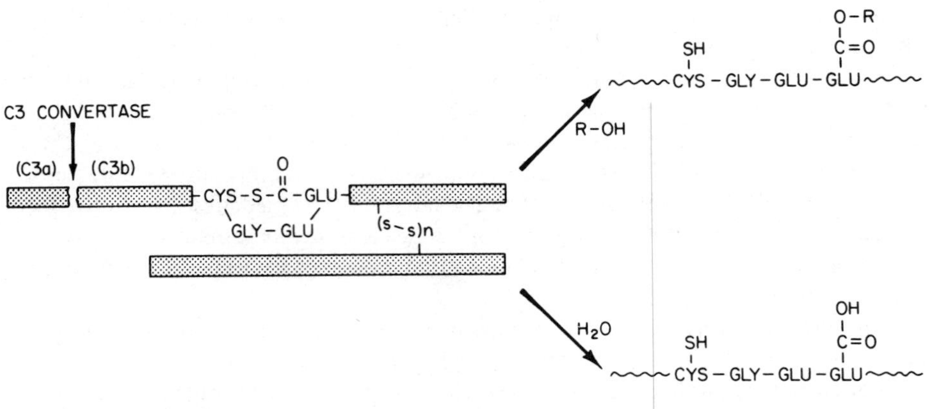

Figure 3 Transesterification reaction of C3b. The internal thiol ester within the α chain of C3 hydrolyzes when the C3a peptide is released by the action of a C3 convertase. The reactive acyl of the glutamyl residue then either forms an ester linkages with a cell surface constituent containing a hydroxyl, R-OH, or reacts with H_2O, which abolishes the covalent binding function of C3b.

much as 30 min in the presence of the protein. This activity of P greatly augments C3 cleavage by the alternative pathway (Fearon, 1979).

The C3b that is generated by the priming C3 convertase does not automatically lead to amplified C3 cleavage because of the actions of the two regulatory proteins. H and I (Fig. 4). H binds to C3b and prevents uptake of B or even causes the dissociation of B or Bb that has already bound, thus preventing the formation and function of the amplification C3 convertase (Weiler et al., 1976; Whaley and Ruddy, 1976). I is a serine protease

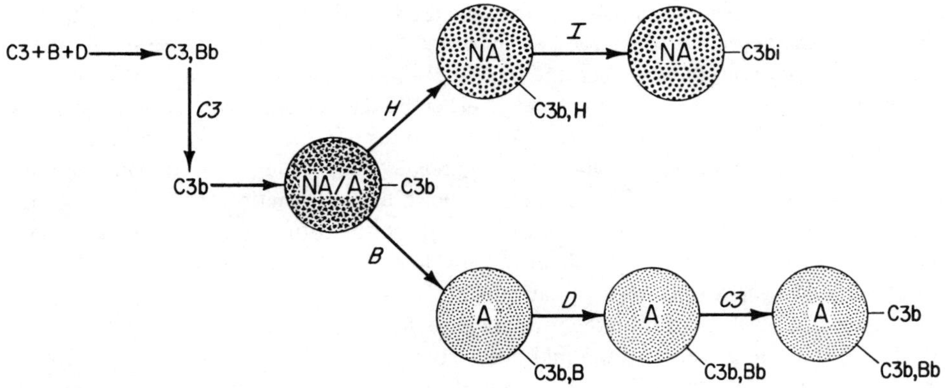

Figure 4 Alternative pathway of complement activation. C3b, which is slowly and continuously generated in the fluid phase by a priming C3 convertase, C3,Bb, may attach to bystander cells regardless of their activating function. If the cell is an activator (A), the amplification C3 convertase, C3b,Bb, is formed and catalyzes the cleavage and deposition of additional molecules of C3b. If C3b is initially bound to a nonactivator (NA), binding of H is facilitated and the C3b is converted to inactive iC3b by I. P, which is not represented in this figure, augments the activating reaction by stabilizing C3b,Bb.

that coverts C3b to iC3b, an inactive form of the protein, by cleaving site(s) in the α chain. This action of I is dependent on the simultaneous presence of H and it is presumed that the latter protein reversibly alters the conformation of C3b to expose the site(s) that is susceptible to I. Removal of either of these proteins from serum causes rapid cleavage of C3 and B to occur because of the impaired regulation of the amplification C3 convertase. There is no plasma protein inhibitor of D, which, as already noted, resides in plasma in its active form; however, B is not cleaved until it is complexed with C3b, so that H and I indirectly regulate D activity by their actions on C3b.

Activation of the alternative pathway is synonymous with impaired regulation of the amplification C3 convertase (Fearon and Austen, 1977; Fearon, 1979) and the regulatory step that is modified by certain cell surfaces is the binding of H to C3b (Kazatchkine et al., 1979). C3b that is in the fluid phase or affixed to the surface of a nonactivator of the pathway binds H with a K_a value of approximately 10^7 M^{-1}, almost 100-fold greater than the affinity with which it binds B. Since B and H are present in plasma at equimolar concentrations, $\sim 3 \times 10^{-6}$ M, binding of H to C3b is likely to occur, and the C3b can then be inactivated by I. In contrast, C3b that is affixed to the surface of an activator of the alternative pathway, such as zymosan, has a lower affinity for H, binding it with a K_a in the range of 10^5 M^{-1}, which is similar to the K_a value with which it binds B. In this circumstance, B can effectively compete with H, and formation of C3b,Bb will occur. The C3 convertase can then cleave C3 to deposit additional molecules of C3b, which continue the amplification reaction.

A cell surface constituent that regulates the affinity of membrane-associated C3b for H is sialic acid, which is present in glycoproteins, glycolipids, and polysaccharides of cells. Naturally occurring activators of the human alternative pathway, such as zymosan, an insoluble derivative of yeast cell walls, and rabbit erythrocytes, have absent and diminished amounts of sialic acid on their respective surfaces. Enzymatic removal or chemical modification of sialic acid on a nonactivator, the sheep erythrocyte, converts this cell into an activator of the pathway by decreasing the affinity of membrane-bound C3b for H (Fearon, 1979). The capacity of the alternative complement pathway to respond to cells that are relatively deficient in surface sialic acid may relate to its role in natural resistance to infection, since most bacteria and all plants lack sialic acid. Interesting in this regard is the observation that some bacterial species which have capsular sialic acid, such as type III, group b *Streptococcus*, group B and group C *Neisseria meningitidis*, and K1 *Eschericia coli*, are pathogenic for nonimmune individuals.

In a model system, the coupling of heparin glycosaminoglycan to zymosan suppressed the capacity of this microbial material to activate the human alternative pathway by facilitating the regulatory action of H and I on particle-bound C3b. Heparin and a closely related mucopolysaccharide, heparin sulfate, can be specifically bound to nucleated cells, and nucleated cells in culture bear heparin-related polysaccharides. These sulfated glycosaminoglycans may represent another family of cell surface constituents that modulate activation of the alternative pathway.

In addition to providing the nonimmune individual with a means for host defense, the alternative pathway can utilize elements of acquired immunity. Activation of the alternative pathway is augmented by specific antibody to particulate activators, as observed for bacteria, such as *Streptococcus pneumoniae* and group B, type III *Streptococcus*, for zymosan, and for mammalian cells such as rabbit and mouse erythrocytes (E) and HeLa cells infected with measles virus. In all of these studies, the $F(ab')_2$ fragment of IgG has been as effective as the intact molecule, indicating that the Clq-binding Fc

fragment is not required. Studies of mechanism have demonstrated that the presence of specific IgG increased the rates but not the final magnitude of consumption of C3 and uptake of C3b during activation by zymosan of human serum. IgG-sensitized HeLa cells that were infected with measles virus also deposited C3b more rapidly than nonsensitized cells during incubation with purified C3, B, D, P, H, and I. Rabbit IgG anti-mouse E increased the efficiency of deposition of C3b during the cleavage of C3 by cell-bound amplification C3 convertase sites, C3b,Bb. The augmented deposition of C3b, without enhanced cleavage of C3, resulted in formation of greater numbers of C5 convertase sites and more effective cytolysis of the IgG-sensitized mouse E. IgG anti-mouse E did not stabilize C3b,Bb sites to the decay accelerating action of H, and thus its apparent mechanism of action on activating particles was distinct from the regulatory effects of membrane sialic acid. In addition to augmentation of alternative pathway activation by cells already capable of initiating C3 consumption and C3b deposition by this pathway, guinea pig antibody may convert a nonactivating sheep erythrocyte into an activator of the altnerative pathway. IgG_1 but not IgG_2 is active in this reaction, suggesting that a structural feature of the antibody subclass is critical. Human IgG directed to a sialic acid-containing antigen of the capsule of group B, type III, *Streptococcus* converts this bacterium to an activator of the human alternative pathway. Thus several mechanisms for the potentiating activity of antibody may be envisioned, which include alteration of the target cell surface, perhaps by covering sialic acid residues, and provision of additional covalent binding sites for C3b on the antibody molecules or on the cell surface itself.

An unusual mechanism of antibody-mediated activation of the alternative pathway occurs in some patients with membranoproliferative glomerulonephritis and in most patients with partial lipodystrophy with or without glomerulonephritis. The sera of these individuals exhibit a serum complement profile indicative of alternative pathway activation (i.e., low levels of C3 and normal concentrations of C1, C4, and C2). These sera also contain an IgG auto-antibody (Davis et al., 1977), termed C3 nephritic factor, that is specific for antigenic determinants expressed by the amplification C3 convertase. One molecule of nephritic factor binds per C3b,Bb, creating a stable trimolecular enzyme that is resistant to H-mediated dissociation of the Bb subunit (Daha et al., 1977).

IV. EFFECTOR COMPLEMENT SEQUENCE: ASSEMBLY OF THE MEMBRANE ATTACK COMPLEX

The C4b,2a and C3b,Bb enzymes are identical in their substrate specificity, as they both cleave C3 at the arg 74-Ser 78 bond of the α chain to generate the C3a and C3b fragments. When this C3b covalently attaches to a site on the cell or immune complex that is adjacent to the C3 convertase, the C4b,2a,3b complex acquires the ability to cleave C5 and initiate the interaction between the components of the effector sequence C5–C9 that form the membrane attack complex. The C3b is thought to function by interacting with C5 and appropriately "modifying" it so that proteolysis may occur.

Both C5 convertases, C4b,2a,3b and C3b,Bb, C3b, cleave C5 at the arg 74–75 bond in the α chain of this protein, to generate the C5b and the glycopeptide, C5a. The C5b fragment then forms a firm, noncovalent complex with C6, which reversibly binds to C3b on the cell surface. Binding of C7 to C5b,6 creates a trimolecular complex which leaves C3b and inserts into the lipid bilayer of the target membrane. If the C5b,6,7 complex does not encounter a membrane shortly after formation, it loses membrane-associating activity, presumably because a conformational rearrangement in the complex causes a

folding in of the hydrophobic domains necessary for membrane insertion. C8 then binds to the complex to create a "small" transmembrane channel that allows slow leakage of ions. Binding of C9 to the membrane-bound C5b-8 induces formation of a more stable transmembrane channel, and the effective pore diameter is increased as the multiplicity of C9 per C5b-8 complex approaches 5:1. The effective diameter of stable channels formed on sheep erythrocytes by guinea pig C5b-9 were in three ranges: 0.5 to 0.8 nm, 0.9 to 3 nm, and 3 nm up to at least 5.5 nm. One mechanism by which C9 promotes the formation of larger channels may be the creation of dimers of C5b-8 complexes.

The location and chemical composition of the hydrophilic passage across the lipid bilayer membrane is probably a protein channel located within the interior of the $C5b\text{-}(9)_n$ complex. Thus the surface of the interior of the complex would be comprised of hydrophilic amino acid residues and possibly carbohydrates, whereas the exterior surface of the complex that is exposed to the phospholipids of the bilayer would have hydrophobic amino acids. This model of the transmembrane channel of $C5b\text{-}(9)_n$ is supported by electron microscopic studies of complexes which have been extracted from sheep erythrocytes with detergent and incorporated into unilamellar liposomes. The structures appear as hollow, thin-walled cylinders of approximately 10 nm internal diameter and of 15 nm height which penetrate the bilayer. The outer portion of the cylinder terminates in a 3- to 5-nm annulus of 10 nm diameter that projects above the membrane and represents the classical electron microscopic lesion seen on cells that have been reacted with complement (Mayer et al., 1981).

Another mechanism by which these amphipathic $(C5b\text{-}9)_n$ complexes can alter the integrity of membranes is by their detergentlike effect. When large numbers of the complexes are formed on membranes, their capacity to bind phospholipid leads to a gross disruption of the bilayer rather than formation of discrete channels. Membrane damage by this mechanism may be important for lysis of viruses whose envelope membranes do not function as osmotic barriers.

V. BIOLOGIC ACTIVITIES OF COMPLEMENT

Many of the biologic effects of the complement system occur following the interaction of cleavage fragments of components with specific receptors on cells that mediate the inflammatory and immune responses. Listed in Table 2 are the products of complement reactions which interact with cells, the cell types involved, and the consequences of these interactions. These responses are of three types: exocytic, chemotactic, and endocytic.

Exocytic reactions are elicited by the cleavage peptides, C4a, C3a, and C5a, and are mediated by mast cells, basophils, and granulocytes. Mast cells and basophils have specific membrane receptors for C3a, C5a, and probably C4a. C4a probably binds to the same receptor as does C3a, since these peptides exhibit cross-tachyphylaxis, while the C5a receptor is distinct from that for C3a. Binding of the peptides to these cells induces secretion of histamine, which accounts for their historical name of "anaphylatoxins." The order of potency for inducing histamine-dependent wheal and flare reactions in human skin is C5a > C3a > C4a, with C5a being active at a dose of 10^{-10} to 10^{-12} mol, C3a at 10^{-8} to 10^{-10} mol, and C4a at 10^{-6} to 10^{-8} mol. Some investigators contend that all spasmogenic activity of C3a and C5a is dependent on histamine release, while others believe that these peptides may have an additional direct smooth muscle contractile activity. Secretagogue activity is abolished or greatly diminished when the C-terminal arginine is removed from each human peptide by a serum carboxypeptidase. The desArg

Table 2 Complement Receptors

Component or Fragment	Cell type bearing receptor	Response
Clq	Neutrophil, monocyte B lymphocyte	?
C4a	Mast cell	Histamine secretion
C4b	? Same as C3b	? Same as C3b
C3a	Mast cell, basophil Eosinophil Neutrophil	Histamine secretion Enzyme secretion ?
C3b	Erythrocyte Neutrophil, eosinophil, monocyte Macrophage, mast cell B and T lymphocytes Glomerular podocyte	Adherence Adherence and enhanced phagocytosis; absorptive endocytosis Above effects and phagocytosis ? ?
iC3b	Neutrophil, monocyte, NK cell	Adherence and phagocytosis
C3d	B lymphocytes	?
C5a	Mast cell basophil Neutrophil, eosinophil, monocyte	Histamine secretion Chemotaxis, enzyme secretion, vascular adherence, oxidative metabolism

form of human C5a may retain some anaphylactic activity at high concentrations, and chemical removal of the oligosaccharide chain increases this residual spasmogenic activity (Hugli, 1981).

C5a induces a chemotactic response by binding to receptors on enutrophils (Chenoweth and Hugli, 1978), eosinophils, and monocytes that may not be identical to C5a receptors on mast cells and basophils. C5a desArg retains a capacity to bind to this receptor and is the factor that is primarily or entirely responsible for the chemotactic activity found in whole serum after complement activation. This activity of purified C5a desArg is greatly augmented by a serum factor that is apparently not derived from the complement system. An additional response of granulocytes to C5a or C5a desArg is increased adherence to themselves or to other cell types, such as endothelial cells. This phenomenon may help the granulocytes adhere to the substratum over which they are crawling and promote their egress from the intravascular space to the interstitial site at which complement activation is occurring. When complement activation occurs intravascularly, as may occur during hemodialysis or cardiopulmonary bypass because of the alternative pathway-activating capacity of the dialysis and oxygenating membranes, sequestration of granulocytes in the pulmonary vascular bed is observed, presumably as a consequence of the increased "stickiness" of the cells. C5a, in higher concentrations than are necessary for chemotaxis or in the presence of cytochalasin B, also can induce lysosomal enzyme release from neutrophils and monocytes.

Systemic activation of complement in vivo leads to an immediate, transient decrease in the number of circulating granulocytes secondary to their trapping in the microvasculature in response to C5a. This response is followed in 30 to 60 min by a rise above baseline levels of the peripheral blood granulocyte count which is thought to be mediated by C3e, a 10,000-molecular weight peptide derived from the α chain of C3. C3e, which can be generated from C3 or C3b by prolonged treatment with trypsin, releases granulocytes from perfused rat femurs. The proteases that are involved in production of this leukocytosis-inducing factor in vivo are not known.

The components that promote endocytic reactions are C4 and C3, the two complement proteins that covalently attach to the targets of complement activation. A particle or immune complex that bears C3b binds to certain cells (Table 2) because of the presence on these cells of C3b receptors. Particles bearing larger amounts of C4b also bind to C3b receptor-bearing cells. The C3b receptor of human peripheral blood erythrocytes, granulocytes, monocytes, and B lymphocytes is a glycoprotein comprised of a single polypeptide chain having an apparent molecular weight of 205,000 (Fearon, 1980). The C3b receptor that is found on glomerular podocytes and on 5 to 15% of peripheral blood T cells has been shown to be antigenically similar to that found on other cells, but structural analyses have not been performed.

Functions of the C3b receptor are incompletely understood and probably are related to the cell type being investigated. C3b receptors of granulocytes, monocytes, and macrophages promote the phagocytosis of C3b-coated particles, but the role of the receptor may vary according to the particle or the state of the cell. Sheep erythrocytes bearing C3b adhere to but are not ingested by human neutrophils and monocytes; the C3b receptor-dependent adherence reaction does greatly facilitate phagocytosis of erythrocytes sensitized with suboptimal amounts of IgG (Mantovani et al., 1972). If the sheep erythrocyte surface is altered by removal of sialic acid, adherence will be accompanied by phagocytosis. Examples of altered C3b receptor function are found with human monocytes cultured for 1 week and with mouse peritoneal macrophages exposed to a T-cell product (Griffin and Griffin, 1979). These cells exhibit a capacity to ingest C3b-coated sheep erythrocytes. In summary, these studies have suggested that C3b receptors have a critical role in the phagocytosis of particles that have activated complement so as to be coated with C3b; this role may be limited to the adherence phase of phagocytosis or may be involved with the phagocytic event itself.

C3b receptors on neutrophils and monocytes have been shown to mediate adsorptive pinocytosis, which probably involves different cytoskeletal elements from those participating in phagocytosis. Soluble ligand, either in the form of rabbit $F(ab')_2$ anti-C3b receptor or as cross-linked C3b, that is bound to these cells is internalized at 37°C within 7 min by a mechanism involving the formation of coated pits and coated vesicles. Much of the internalized ligand is rapidly delivered to lysosomes, where degradation may occur. This function of C3b receptors may be involved in the clearance of C3b-bearing immune complexes.

A role for C3b receptors on B lymphocytes in their development into antibody-secreting cells and into memory B cells has been suggested. In these experiments, C3 was depleted from animals or from in vitro culture systems in order to observe these impaired B-cell functions, so that the participation of the C3b receptor was only indirectly examined. A further complication of the in vivo experiments in which C3 was depleted by proteolysis into its C3b and C3a fragments is the recent finding that C3a or related fragments may have immunosuppressive activity (Hobbs et al., 1982). Thus no firm

conclusions as to the role of the C3b receptor in these phenomena can be made. Another possible function for C3b receptors was found when lymphocytes which had been incubated with fluid phase C3b secreted a lymphokine that was chemotactic for monocytes. The cellular source of this chemotactic mediator is unclear because all B cells and 5 to 15% of peripheral blood T cells have C3b receptors. Although it is evident that C3b receptors are probably participating in regulation of lymphocyte functions, a clear definition of their role is lacking.

The roles of C3b receptors on two cell types not considered part of the immune system, erythrocytes and glomerular podocytes, are also incompletely understood. Erythrocytes bear only 300 to 700 C3b receptors per cell, far less than the 10,000 to 30,000 receptors per cell found on granulocytes, monocytes, and B lymphocytes (Fearon, 1980). However, because of the great numbers of erythrocytes in peripheral blood compared to those other cell types, approximately 85% of the receptors in the vascular compartment are on erythrocytes. They may function to bind temporarily C3b-bearing complexes and particles for subsequent delivery to the reticuloendothelial system, and the finding of diminished numbers of C3b receptors on erythrocytes of patients with systemic lupus erythematosus may indicate an impairment of this delivery function. The glomerular podocyte, which projects foot processes over the extracapillary side of the glomerular basement membrane, appears to have large numbers of C3b receptors when observed by immunofluorescence. Although the function of these receptors has not been elucidated, they have been shown to be absent in patients having severe proliferative glomerulonephritis of systemic lupus erythematosus.

Receptors for two fragments of C3b, iC3b and C3d, are also present on the erythrocytes and leukocytes. Cleavage of C3b by I in the presence of H genrates iC3b; tryptic hydrolysis of iC3b yields C3d, a 35,000-molecular weight fragment of the α chain, and C3c, which contains intact β chain and residual α chain. Both iC3b and C3d remain covalently attached to complement-activating target because the thiol ester sequence resides in the C3d region. Receptors for iC3b appear to reside on neutrophils, monocytes, and NK cells C3b and they may have functions similar to those of C3b receptors, although studies are relatively incomplete. C3d receptors in man have been found only on B lymphocytes and their biologic role is unknown (Lambris et al., 1981). A receptor for H which can mediate secretion of I also has been identified on leukocytes.

VI. INHERITED ABNORMALITIES OF COMPLEMENT

Genetically controlled deficiencies have been described for many of the components of the complement system (Table 3) (Colten et al., 1978; Glass et al., 1982). Deficiency of one component, C3, consistently leads to marked impairment of host defenses, manifested in vivo by recurrent pyrogenic and gram-negative infections and impaired leukocyte mobilization, and in vitro by absent serum bactericidal, chemotactic, and opsonic enhancing activity. Deficiency of an inhibitor protein, I, allows uncontrolled activation of the alternative complement pathway with secondary C3 and B deficiencies. I deficiency therefore produces clinical manifestations similar to those of C3 deficiency. Deficiencies of the later complement components C5, C6, C7, and C8 are associated with recurrent neisserial infections with bacteremia, suggesting that the cytolytic action of complement has a critical role in host defense against these organisms. Deficiency of $C\overline{1}$ inhibitor is associated with the syndrome of hereditary angioedema involving the skin, gastrointestinal tract, or upper respiratory tract. This autosomal dominant disease is

Table 3 Inherited Abnormalities of Complement

Protein	Kindreds	Associated diseases	Homozygotes without disease
C1q	2	Lupuslike syndromes, glomerulonephritis, sepsis, skin disease	1
C1r	2	Lupuslike syndromes, glomerulonephritis, discoid lupus	2
C1s	1	Systemic lupus erythematosus	
C4	2	Systemic lupus erythematosus	
C2	Many	Systemic lupus erythematosus, discoid lupus, dermatomyositis, Henoch-Schönlein purpura, inflammatory bowel disease	Yes
C3	5	Recurrent bacterial infections, nephritis	
C5	2	Recurrent bacterial infections (*Neisseria*), systemic lupus erythematosus	1
C6	9	Recurrent bacterial infections (*Neisseria*)	3
C7	8	Recurrent bacterial infections (*Neisseria*), Raynaud's phenomenon with sclerodactyly, systemic lupus erythematosus	4
C8	5	Recurrent bacterial infections (*Neisseria*), systemic lupus	3
C9	3	None	3
P	1	Meningitis	
$C\bar{1}$ INH	Many	Hereditary angioedema, glomerulonephritis	
I	3	Recurrent bacterial infections	

infrequent, but not rare. Correct diagnosis may avert needless abdominal surgery or death by asphyxiation because effective pharmacologic intervention is now available in the form of attenuated androgens that stimulate $C\bar{1}$ INH production.

The apparent increased frequency of diseases such as SLE, systemic lupus erythematosus, Henoch–Schönlein purpura, or polymyositis in individuals with complement deficiencies would not have been predicted from the postulated role of complement as a mediator of inflammation in these rheumatic diseases. If anything, a protective effect might have been expected. With the exception of single C5-, C7-, and C8-deficient

individuals, all instances of association between rheumatic diseases and complements deficiency have involved proteins of the classic activating pathway. There is even a report of systemic lupus erythematosus occurring in patients with hypercatabolism of C4 and C2 due to $C\bar{1}$ inhibitor deficiency. Since C1 and C4 have been demonstrated to increase the efficiency of antibody in the neutralization of certain viruses, the most direct explanation for the association would involve failure of neutralization of the etiologic agent. An attractive alternative would invoke failure of clearance of antigen–antibody complexes irrespective of the nature of the antigen. In animal models the failure to clear viral antibody has been shown to be important in the pathogenesis of immune complex disease. The latter view would argue that while the alternative pathway is essential for host defence, the adaptive immunity augmentation afforded by the classic activating pathway is less important than the need to clear the products of the immune response.

The location of the genes controlling the C2, B, and C4 polymorphisms in the major histocompatibility complex affords yet another explanation. In the mouse and guinea pig, genes controlling immune responsiveness (Ir) to certain synthetic antigens are located in or near the histocompatibility complex. A similar linkage of Ir genes to histocompatibility complexes has been demonstrated in rhesus monkeys and seems likely in humans. The products of the HLA-D locus are most likely to function as Ir genes, and it is to this locus that the four complement loci are most closely linked. Association between rheumatic disease and complement deficiencies may result from linkage disequilibrium between a silent or null allele for the complement component and an allele of an Ir gene locus. The association between Ir genes and disease may be primary and that with the complement allele secondary. There is some evidence for this. C2 deficiency in homozygous and in heterozygous form occurs in increased frequency in patients with systemic lupus erythematosus, as does HLA-DR2. HLA-DR2 is in linkage disequilibrium with C2 deficiency, and more patients with systemic lupus erythematosus have HLA-DR2 than C2 deficiency. The greater frequency of HLA-DR2 in this disease and the marginal effect of heterozygous C2 deficiency on complement function suggest that the important association with respect to pathogenesis may be with the HLA-D region allel rather than with the complement component allele. On the other hand, rheumatic diseases are also associated with component deficiencies that are not coded for in the major histocompatibility region, namely, Clr, Cls, and C1 INH. These associations may represent selection bias resulting from the type of population screened for complement polymorphism. Alternatively, the association may indicate that impaired clearance alone may be sufficient to increase disease susceptibility.

VII. CONCLUSIONS

The complement system is comprised of 18 proteins and has two pathways for activating a common effector pathway. The classical activating pathway is initiated by antigen–antibody complexes and the alternative pathway is directly activated by cell surfaces but can be enhanced by specific antibody. Both pathways lead to formation of C3 and C5 convertase, which have identical substrate specificities. Most of the biologic activities of complement are mediated by the products of C3 and C5–C9. These activities include direct cytolysis of the target, increased local vascular permeability secondary to stimulation of mast cell secretion, adherence and chemataxis of granulocytes and monocytes, and opsonization of the target. Genetic deficiencies of individual components of complement have emphasized the role of this system in host defense. These abnormalities have

also raised the possibility that complement may clear the products of an immune response and serve to protect the individual from autoimmune disease.

REFERENCES

Chenoweth, D. E., and Hugli, T. E. (1978). Proc. Natl. Acad. Sci. USA *75*, 3943.

Colten, H. R., Alper, C. A., and Rosen, F. S. (1978). In *Clinical Aspects of the Complement System* (W. Opferkuch, K. Rother, and D. R. Schultz, Eds.). G. Thieme, Stuttgart, pp. 200–206.

Daha, M. R., Austen, K. F., and Fearon, D. T. (1977). J. Immunol. *119*, 812.

Davis, A. E., III, Ziegler, J. B., Gelfand, E. W., Rosen, F. S., and Alper, C. A. (1977). Proc. Natl. Acad. Sci. USA *74*, 3980.

Fearon, D. T. (1979). CRC Crit. Rev. Immunol. *1*, 1.

Fearon, D. T. (1980). J. Exp. Med. *152*, 20.

Fearon, D. T., and Austen, K. F. (1977). Proc. Natl. Acad. Sci. USA *74*, 1683.

Gigli, I., Fujita, T., and Nussenzweig, V. (1979). Proc. Natl. Acad. Sci. USA *76*, 6596.

Glass, D. N., Fearon, D. T., and Austen, K. F. (1982). In *The Metabolic Basis of Inherited Disease* (J. B. Standbury, Ed.). McGraw-Hill, New York, pp. 1934–1955.

Griffin, J. A., and Griffin, F. M., Jr. (1979). J. Exp. Med. *150*, 653.

Harpel, P., and Cooper, N. R. (1975). J. Clin. Invest. *55*, 593.

Harrison, R. A., Thomas, M. L., and Tack, B. F. (1981). Proc. Natl. Acad. Sci. USA, 7388.

Hobbs, M. V., Feldbush, T. L., Needleman, B. W., and Weiler, J. M. (1982). J. Immunol. *128*, 1470.

Hugli, T. E. (1981). CRC Crit. Rev. Immunol. *1*, 321.

Kazatchkine, M. D., Fearon, D. T., and Austen, K. F. (1979). J. Immunol. *122*, 75.

Lambris, J. D., Dobson, N. J., and Ross, G. D. (1981). Proc. Natl. Acad. Sci. USA *78*, 1828.

Law, S. K., and Levine, R. P. (1977). Proc. Natl. Acad. Sci. USA *74*, 2701.

Mantovani, B., Rabinovitch, M., and Nussenzweig, V. (1972). J. Exp. Med. *135*, 780.

Mayer, M. M., Michaels, D. W., Ramm, L. E., Shin, M. L., Whitlow, M. B., and Willoughy, J. B. (1981). CRC Crit. Rev. Immunol. *2*, 133.

Müller-Eberhard, H. J., Polley, M. J., and Calcott, M. A. (1967). J. Exp. Med. *125*, 359.

Pangburn, M. K., and Müller-Eberhard, H. J. (1980). J. Exp. Med. *152*, 1102.

Porter, R. R. (1977). Fed. Proc. *36*, 2191.

Tack, B. F., Harrison, R. A., Janatova, J., Thomas, M., and Prahl, J. W. (1980). Proc. Natl. Acad. Sci. USA *73*, 3268.

Weiler, J. M., Daha, M. R., Austen, K. F., and Fearon, D. T. (1976). Proc. Natl. Acad. Sci. USA *73*, 3268.

Whaley, K., and Ruddy, S. (1976). J. Exp. Med. *144*, 1147.

Ziccardi, R. J., and Cooper, N. R. (1977). J. Immunol. *118*, 2047.

24

Lymphocyte Interactions

Charles A. Janeway, Jr. Yale University School of Medicine and Howard Hughes Medical Institute, New Haven, Connecticut

I. INTRODUCTION

The division of immune responses into humoral and cellular responses refers only to the effector phase of the response: all immune responses involve cells, in particular lymphocytes, in their induction and regulation. Furthermore, with rare exceptions, all cellular immune responses involve interactions between various subpopulations of lymphocytes, and between lymphocytes and various types of accessory cells. Therefore, to understand the generation and regulation of all immunological responses, one must characterize the lymphocytes involved in the response and their interactions with other cells. This chapter examines this problem from two points of view. The individual cellular components of the immune system are described first, followed by a description of their interactions with other cells leading to immune responses.

Such a description will necessarily reflect large areas in which the available information is not complete. One particularly difficult problem is to reconcile findings made in different centers using different reagents and completely different experimental systems. The fact that this resolution cannot be made at the present time reflects both the relatively incomplete nature of our knowledge and the great diversity of the immune system and of the cells that comprise it. A second problem is that the relative importance of each type of cell and interaction in the generation of a particular response in vivo must await a unified description of all lymphocyte sets and their interactions. Only with this information in hand can an integrated approach to in vivo immune responses be attempted.

II. DEFINITION OF LYMPHOCYTE SUBPOPULATIONS

Lymphocytes make up a heterogeneous family of small mononuclear cells found in the blood, in lymphoid tissues, and in small amounts of virtually all tissues of the body. They

For a brief discussion of the connection between Chaps. 9 and 24–28, see the appendix.

are by nature migratory, as befits their primary function, surveillance of the body for foreign (i.e., non-self) materials. The resting, mature lymphocyte has a characteristic morphology: a rounded nucleus with a thin rim of cytoplasm containing very little protein synthetic activity. Until the pioneering studies of Gowans, lymphocytes had no known functions. Gowans and McGregor (1965) and subsequent workers have shown that lymphocytes individually and collectively express the cardinal properties of all immune responses: specificity, inducibility, memory, regulation, and amplification. Thus, the study of immune responses is of necessity the study of lymphocytes.

Because of their morphological homogeneity, the discrimination of lymphocyte subpopulations has relied on other distinctions: localization in specific tissues, physical separation, and differential rates and sites of maturation have all been employed. However, the most powerful means of defining lymphocyte subpopulations has involved the use of cell surface antigenic determinants detected with antibodies. Since particular constellations of these antigens are found reproducibly on cells expressing particular differentiated functions, these antigens are referred to as *cell surface differentiation antigens*. The profile of cell surface differentiation antigens borne by a cell correlates with function and is used to assign both function and cell surface antigen phenotype to that cell. Besides function and cell surface antigen phenotype, three other properties of lymphocytes have been useful in their description: specificity for foreign (or nominal) antigen, specificity for self antigens (also known as restriction specificity, see below), and state of activation.

A. Cell Surface Differentiation Antigens

A very large number of differentiation antigens on lymphocytes have been defined (for a review, see McKenzie and Potter, 1979). For the present chapter, only those that have been of greatest importance are described. (See also Chap. 28.)

Immunoglobulin

Cell surface immunoglobulin (Ig) is the most universal B-lymphocyte differention antigen. Ig serves the B lymphocyte as its antigen-specific receptor, and a modified form of the Ig with the same specificity is secreted when a B cell is activated by antigen plus accessory signals, usually delivered by T cells.

Thy-1

Thy-1 is an alloantigen, that is, an antigen found in different forms in different individuals of the same species: antibodies to the different allelic forms of Thy-1 (alloantibodies) specifically react with this molecule. Thy-1 is found on all T lymphocytes, and is not found on B lymphocytes.

Ly Antigens

Differentiation antigens found only on lymphocytes are designated Ly. Lyb antigens are found on B lymphocytes but not on T lymphocytes, whereas Lyt antigens are found on T lymphocytes but not on B lymphocytes. The recognized Lyb and Lyt antigens are alloantigens. Of particular importance in defining T-lymphocyte interactions are the Lyt antigens Lyt-1 and Lyt-2, which have been shown to be expressed by distinct subpopulations of T lymphocytes having distinct functions.

Ia Antigens

The region of the major histocompatibility complex (MHC) involved in genetic control of responses to specific antigens (Ir genes) encodes cell surface antigens (Ia antigens) expressed on subpopulations of lymphocytes. Products of the I-A and I-E subregions are expressed on B lymphocytes and specialized antigen presenting cells, while products of the I-J subregion and perhaps other subregions as well are expressed on subpopulations of T lymphocytes, especially those involved in the regulation of the immune response.

B. Functions

A great many different functions of various lymphocytes have been described. In this section a very abbreviated list of commonly measured functions is given: in a later section, subsets of cells having functions described below and defined by specific cell surface antigens are listed as a guide. These functions are divided into effector functions, regulatory functions, and proliferation.

Effector Functions

Antibody production. All antibody-secreting cells are members of the B-cell lineage. The only known function of B lymphocytes is to secrete antibody; T lymphocytes do not secrete antibody.

Delayed-type hypersensitivity. DTH is a complex inflammatory reaction initiated by lymphocytes interacting with antigen in cutaneous tissues. It is initiated by specific T lymphocytes, and is under T-lymphocyte-mediated regulation.

Cytotoxicity. A lymphocyte that can cause the death of another cell is said to be cyto- (from cell) toxic (poisonous).

Lymphokine release. Many lymphocyte functions are mediated by the release of a myriad of nonantigen-specific biologically active molecules. These are referred to collectively as lymphokines, since they are released by lymphocytes and have activating or inactivating effects on other cells (see Chap. 30).

Regulatory Functions

Helper function. Any lymphocyte that does not itself mediate an effector function, but which augments the effector activity generated by another set of lymphocytes can be referred to as a helper lymphocyte. The term *inducer* has also been used to describe such cells.

Suppression. Any lymphocyte whose influence leads directly or indirectly to a decrease in the effector function of another lymphocyte is said to express suppressor function.

Contrasuppression. A lymphocyte that inactivates the suppressive activity of other lymphocytes is referred to as contrasuppressive; such cells must be clearly distinguished from helper cells.

Transduction. A lymphocyte that accepts an inducing signal from one cell and differentiates into or causes the differentiation of another lymphocyte now performing an effector or regulatory function will be referred to here as a transducer cell; such cells have no established name in the literature.

Proliferation

Proliferation of lymphocytes is often referred to as a lymphocyte function; this is not strictly true. Proliferation should be thought of as just that, a reaction of lymphocytes to signals leading to an increase in their number, which is a central and important feature of many immunological responses. However, the function of the proliferating cells is not proliferation per se but rather one of the differentiated functions listed above. Proliferation serves to increase the number of cells having that function.

C. Specificity

Specificity for foreign (or nominal) antigen is a feature of all lymphocytes. Each lymphocyte bears a clonally distributed receptor capable of precisely recognizing a particular foreign antigenic determinant. In the case of B lymphocytes, this function is served by the cell surface immunoglobulin, which expresses products of a single Igh-V and Igl-V gene, giving it a unique specificity. In the case of T lymphocytes, the nature of this molecule is not yet known (see Chap. 28). Specificity for foreign antigen is expressed largely independently of function; thus lymphocytes having a variety of distinct functions may recognize the same antigenic determinant.

D. Restriction

T lymphocytes are specialized to recognize antigen in the context of self specificities (see Chaps. 25 to 28, and below). This is ordinarily demonstrated as a failure to recognize the same foreign antigen in the context of cells from another member of the same species, and has been termed *restriction* for this reason. However, it is clear that this restriction reflects the ability of the T lymphocyte to recognize self determinants, and will be referred to in this chapter as self recognition, or recognition of self antigenic determinants. The best characterized of the self determinants recognized are those encoded in the MHC. However, evidence from several systems suggests that products of the Igh gene complex also serve as self specificities recognized by T lymphocytes. Like the receptors for foreign antigens, these self recognition units are also clonally distributed on T cells, either as an integral part of the receptor for the foreign antigen or as an independent combining site on the same or a different polypeptide chain. Because T-lymphocyte function and the self specificity recognized by such cells are coordinately expressed, one can further characterize cells on the basis of their self specificity.

E. Activation

The biological activity of a lymphocyte is determined by its specificity, self recognition units, function, cell surface antigen phenotype, and by its state of activation. The four lymphocyte activation states that are well characterized are:

1. Virgin or resting lymphocytes, which have no biological activity since they have not been activated by antigen.
2. Activated lymphocytes, which, upon encounter with their specific foreign antigen, proliferate and differentiate into lymphocytes actively mediating their effector or regulatory function; it would appear that lymphocytes are committed to function prior to deliberate activation.

3. Memory lymphocytes, which are lymphocytes that have matured as a result of encounter with antigen, but are presently not activated, due to the removal of antigen from the system. Such memory lymphocytes are more sensitive to antigenic stimulation than are virgin lymphocytes, and show other evidence of their differentiated state.
4. Inactivated lymphocytes, which are lymphocytes whose activity has been actively suppressed by encounter with antigen and/or accessory suppressive signals. Lymphocyte inactivation plays an important role in self-non-self discrimination and in regulation of all immune responses. The measured functional activity of a population of lymphocytes depends on the activation states of its various components.

III. LYMPHOCYTE SUBPOPULATIONS

Before describing lymphocyte interactions, it is necessary to designate each of the subpopulations according to the characteristics listed in the preceding section. However, specificity for nominal antigen and state of activation are not, in general, markers of subpopulations of lymphocytes but rather of individual members of a population; although exceptions to this statement do exist, they are beyond the scope of this chapter (see Chap. 1). Since the cell surface antigen profile and function are useful in defining these subpopulations, these will provide the structure of this section.

A. B Lymphocytes

All B lymphocytes have the ability to synthesize immunoglobulin of a particular specificity, defined by the VH and VL genes rearranged in these cells. Their only known function is the secretion of immunoglobulin. B lymphocytes also express Ia antigens encoded in the I-A and I-E subregions of the MHC, in common with antigen presenting cells. Various subpopulations of B cells have been defined. Thus far, the most useful dissection has been with the X-linked immunodeficiency gene of the CBA/N mouse; F1 male progeny of CBA/N females express this deficiency and are unable to respond to certain thymus-independent classes of antigen. These cells also lack the mature B-cell differentiation antigens Lyb3 and Lyb5.

B. Antigen Presenting Cells

Antigen presenting cells *are not lymphocytes*. They are included here because they appear to play a central role in the induction of an immune response. These cells are distinguished as non-T (Thy-1$^-$), non-B (Ig$^-$) cells from lymphoid organs that bear MHC-encoded molecules from the same genes in I-A and I-E subregions as do B cells. Their specialized function appears to be the "presentation" of antigen to certain classes of T lymphocytes. Antigen presenting cells (APCs) may also bear acceptor sites for certain lymphocyte-derived immunoregulatory molecules, and may play a central role in the exchange of information between lymphocyte subpopulations (analogous to a bulletin board).

C. T Lymphocytes

T lymphocytes appear to be more heterogeneous than B lymphocytes: certainly, more subpopulations have been defined thus far. In this section, these will be listed according

to names that connote function, and their cell surface markers defined. In Sec. IV their interactions and the molecules employed by these cells are described; see also Chap. 25.

T-Helper Cells (Th)

All T-helper cells thus far defined are Lyt-1^+, 2^-. At least three distinct sets of these cells appear to exist:

1. *ThMHC.* These cells recognize antigen in association with self Ia molecules. They are activated by APCs bearing foreign antigen in the context of self Ia molecules, and they help B cells capable of binding antigen to their Ig receptor and also bearing the appropriate Ia molecules.
2. *ThTRF.* These cells also recognize antigen in association with self Ia molecules. However, they appear to mediate help by releasing nonspecific helper factors, including T-cell growth factor or interleukin 2. Such ThTRF (T-cell replacing factor) are involved in helping certain B-cell responses in vitro and perhaps in vivo (this has not been conclusively demonstrated), and are not distinguishable from the helper cells required for optimal generation of cytotoxic effector cells.
3. *ThIg.* These cells recognize antigen independently of MHC-encoded molecules, but bear recognition units for self Ig determinants. They interact with B cells via the B cell's Ig receptor, and require activation by antigen to exert their effect.

Delayed-Type Hypersensitivity (DTH) Effector Cells (DTHE)

These cells are Lyt-1^+, 2^-. They recognize antigen in association with self MHC-encoded Ia antigens, and release a family of lymphokines that together elicit the classical DTH reaction. One cell may be able to release all of these factors. It is not clear whether such cells are distinct from ThMHC or ThTRF. They may be distinguishable by a particular propensity to home to cutaneous sites. However, cells deriving from a single lymphocyte have been shown to mediate helper function (probably TRF secretion) and DTH in vivo (Bianchi et al., 1981).

Suppressor Effector T Cells

All suppressor effector T cells thus far identified have the cell surface antigens Lyt-1^-, 2^+. There is controversy surrounding the expression of I-J-subregion-encoded molecules on such cells. In the most careful studies performed thus far, those T cells whose function is to inactivate Lyl helper T cells are I-J^-. The assignment of I-J to suppressor T cells may identify a separable subset of suppressive cells. Ts appear to recognize antigen in the free form, independent of either MHC- or Igh-encoded determinants, and Ts can bind antigen directly.

Inducers of Suppression (Is)

The effective generation of Ly2 Ts cells requires a signal from an Lyt-1^+,2^-, I-J^+ T cell that has no demonstrable helper function for B cells. This cell is antigen specific in inducing suppression. While its recognition of antigen appears to be independent of MHC- or Igh-linked genes, its ability to communicate with its target requires matching of genes mapping to the Igh-V segment of chromosome 12.

Contrasuppressor Effector Cells (CSE)

The action of Ts cells on Th cells can be interferred with by a subset of Lyt-1^+,2^-, I-J^+ cells termed contrasuppressors by Gershon et al. (1981). These cells are distinct in cell surface antigens and function from all other described T-cell sets. They are not Th because Th do not bear I-J, and they differ both in function and in the I-J determinants expressed from Is cells. The molecules they recognize on their targets, and their specificity for antigen, are at present uncharacterized.

Inducers of Contrasuppression (CSI)

These cells induce their target to become effectors of contrasuppression. These CSI are Lyt-1^-,2^+ and bear the same I-J determinants as the CSE. The specificity of these cells has been defined only in terms of their cell-free products (see below).

"Supersuppressor" T Cells (Tss)

Tada and Okumura (1979) have defined very potent suppressor T cells that are Lyt-1^-,2^+, I-J^+. These cells are activated by hyperimmunization in vivo. I will term them Tss to distinguish them from the Ts cells defined above as I-J^-. They are also distinguished in terms of their requirements for activation.

Inducers of "Supersuppressors" (Iss)

Tss are activated by an inducer cell that has been shown to bind to free antigen and to be Lyt-1^-,2^+, I-J^+.

Suppressors of DTH

Benacerraf and his group have defined a series of at least three cells that are involved in suppression of DTH. The relationship of these cells to the cells described by the groups of Cantor and Gershon and of Tada is not clear at the present time. The cells have been termed Ts1, Ts2, and Ts3 to indicate the order in which they become activated following antigen administration.

1. *Ts1*. Ts1 cells are activated by tolerance-inducing regimens of antigen administration. These cells are Lyt-1^+,2^-, bind antigen, bear an idiotype associated with antibodies binding the same antigen, and are I-J^+. They or their product, TsF1 (see below), activate Ts2 cells. This interaction is not limited by genes mapping in the MHC or linked to Igh-V. However, a "pseudorestriction" by Igh-V-linked genes is observed, since Ts2 cells are anti-idiotypic, and act only on cells bearing the appropriate Igh-V controlled idiotype. Thus Ts1 can activate Ts2 from any strain, but the Ts2 must in turn act on the strain of origin of the Ts1 cells or factor.

2. *Ts2*. Ts2 cells are activated by Ts1 cells or TsF1. They are anti-idiotypic and do not bind antigen. They are Lyt-$1^?$,2^+ and I-J^+. They are also restricted in their interaction with their target cells, Ts3, by the I-J borne by the target cell. The Ts3 cell must bear the same Igh-V linked genes as the Ts1 that activated the Ts2, but it need not be Igh-V matched with the Ts2 donor. Hence communication in this system appears to be of the type idiotype → anti-idiotype → idiotype, and the second step is both anti-idiotypic and MHC (I-J) restricted.

3. *Ts3*. The Ts2 cell is not an effector of suppression, but rather activates the Ts3 cell. Pre-Ts3 cells are found among cells activated by antigen immunization. This cell is idiotypic, I-J^+, Lyt-$1^?$,2^+. Its priming requires an immunogenic form of antigen presented to the Ts3 precursor on cells matched at I-J. Thus this cell is antigen-specific, idiotype positive, and MHC (I-J) restricted.

Transducer T Cells

The activation of many different cell types involves an inducing signal delivered to a functionally inactive cell with the cell surface antigens Lyt-1^+,2^+. These cells have thus far been characterized primarily by the lack of generation of the activated cells when the inducers are cultured in the absence of Ly12 cells. Thus little is known of their specificity for antigen or the self specificities they may recognize. Nevertheless, at least three distinct sets of these cells have been defined:

1. *Transducers of suppression (Trans-s).* Unlike the majority of Ly12 cells, these cells are I-J^+. They are activated by Is cells to either activate or differentiate into Ts cells.
2. *Transducers of contrasuppression (Trans-C).* These cells are also I-J^+, but bear I-J determinants distinct from those borne by the Is and Trans-s cells.
3. *Transducers of "supersuppression" (Trans-ss).* Again, little is known of these cells, but they are activated by Iss cells and either activate or differentiate into Tss cells. They bear the same I-J determinants as the Tss cells.

IV. LYMPHOCYTE INTERACTIONS

In this section, interactions between parts of cells are described in terms of the measurable result of the interaction and in terms of the genetic constraints placed on the interaction.

A. Antigen Presenting Cells → ThMHC, DTHE, ThTRF

Antigen presenting cells, while still poorly defined, appear to have several characteristics critical for activation of those T cells involved in the activation of immune responses: they interact with antigen in a unique and poorly characterized way, apparently rapidly acquiring and only slowly losing antigens with which they come into contact. They are richly endowed with Ia antigens. They secrete a factor, lymphocyte activating factor or interleukin 1, whose role is still poorly characterized, but which is required for secretion of IL2. The primary outcome of this interaction is the induction of proliferation and differentiation by Lyl T cells that recognize antigen in the context of self Ia molecules. Thus this interaction is critical in the initiation of virtually all immune responses. T cells, once primed, can interact with APCs again to secrete a variety of lymphokines, including IL2, TRF, and the lymphokines involved in DTH reactions. Whether the same cell is involved in both types of response, primary activation and secondary lymphokine secretion, is not known.

B. Th → B Cell Interaction

B cells are not normally effective in presenting antigen to Th cells in such a way that the Th cells proliferate and differentiate. However, they do present antigen very effectively to Th cells in the sense that very small doses of antigen, bounnd specifically to the Ig

receptor on the B cell, will associate with self Ia in sufficient quantity to lead to the proliferation and differentiation to antibody secretion of the antigen binding B cell in the presence of specific Th cells. Higher doses of antigen will lead to polyclonal B cell proliferation and antibody secretion (Jones and Janeway, 1981). There are two requirements for this interaction: the B cell must bind an antigenic determinant that is physically linked to the antigenic determinant recognized by the Th cell, and the Th cell must bear a recognition unit for the Ia antigen expressed by the B cell.

A second type of Th → B cell interaction occurs in some in vitro systems and may occur in vivo. The Th cell is stimulated by antigen on an APC (in association with self Ia determinants) and releases a nonspecific helper factor(s) called *T-cell replacing factor* (TRF). For the B cell to receive this factor, it must have been triggered by antigens or perhaps by B-cell mitogens, in some cases requiring specific T-cell help. The relative importance of this pathway of B-cell activation in a response depends on the experimental system.

A third type of Th → B interaction involves ThIg cells. As presently defined, these cells will act only on B cells already activated by ThMHC plus antigen, and selectively activate B cells bearing particular Ig molecules for which the ThIg are specific (isotype, allotype, or idiotype). There is some controversy surrounding the specificity of these cells for antigen and for MHC. This probably results from the existence of two distinct Th cells, both with Ig specificity, one of which is also antigen specific but does not recognize self MHC (Bottomly, 1981) and the other of which is specific for self Ig in association with self MHC-encoded determinants (Jørgensen et al., 1981) (and should therefore be termed a ThMHC). The relative importance of these two types of cell is unknown.

C. Ts → B Cell Interaction

Certain types of Ts cells can clearly act directly on B cells, as shown in systems in which B-cell growth and secretory activity are T independent. This is shown most clearly in experiments in which the secretion of myeloma proteins by B cells is suppressed by anti-idiotypic T cells (Abbas et al., 1980). Suppression has also been observed in plaque assays by antigen-specific suppressor T cells acting on B cells in the presence of antigen and preventing antibody secretion (Warren and Davie, 1977).

D. Ts → Th Interaction

Most suppressor T cells appear to act on other T cells, particularly Th cells. The first example was actually the inactivation of a ThIg cell by allotype-specific Ly2 Ts cells. More recently, Fresno et al. (1981) have used cloned suppressor T cells to inactive purified Th cells, and similar results have been obtained by Gershon et al. (1981). This interaction appears to be antigen specific and to function optimally in the presence of antigen.

E. Is → Ts Interation

Inducers of suppression activate Ly12 cells to become or to further activate Ly2 cells to become active T-suppressor cells. The interaction between the Is cell and its acceptor involves recognition of polymorphic determinants the genes for which are linked to the Igh-V gene complex. However, this recognition event appears not to involve antigen specificity, and therefore apparently is not an idiotype-anti-idiotype interaction. The precise nature of the determinants involved is not known.

F. Ics → CSE Interaction

It is not known what activates the Ics, but it is involved in the generation of contrasuppressor effector function via an Ly12 cell.

G. Iss → Tss Interaction

Iss cells activate Ly12 cells to become Tss cells, which will then suppress responses in a carrier-specific fashion, probably via inactivation of T-helper cells.

H. CSE → Th Interaction

In the presence of active suppression, contrasuppressor effector cells will protect the T-helper cells from suppression. More important, the residual helper activity is now quite resistant to further suppressive signals. Whether this applies to all T-helper cells or only to certain subsets, that is, whether contrasuppressor cells amplify the outgrowth of suppressor resistant helper cells or render sensitive helper cells resistant is not known.

V. MOLECULAR MEDIATORS OF CELLULAR INTERACTIONS

The study of cell interaction has progressed in the past dozen years from in vivo cell mixing experiments to isolation and chemical characterization of specific factors that can mimic the in vitro activity of many of the cells. These have been described in a wide variety of systems, and no one system has had all of its known interactions mimicked by factors. Similarly, there are factors that affect immune responses whose cellular equivalent has not yet been clearly defined. Finally, there are also many different nonspecific factors, whose role may be the final mediation of an interaction that is initially determined by cell–cell contact, and whose normal in vivo role is to carry intercellular messages over very limited distances. These lymphokines are described in Chap. 30. Here, certain relatively clearly defined molecules that may mediate such interactions will be described, acknowledging that many others have been reported that are not listed here.

A. Genetically Restricted Factor (GRF)

This factor, originally described by Erb et al. (1976), appears to be a complex of antigen and self Ia molecules released by antigen presenting cells and capable of interacting with and activating T-helper cells specific for that antigen and recognizing that MHC-encoded determinant. Similar factors have more recently been shown to bind to clones and hybrids of specific, MHC-restricted T-helper cells; see Chaps. 25, 26, and 27.

B. Helper Factor

A large number of different factors have been isolated that will amplify an immune response. Most of these do not accurately reflect the biological activity of any of the described helper cells and have been assayed in complex cell mixtures. These factors may be involved in the activation of helper or contrasuppressor T cells rather than in B-cell activation. Recently, Cantor and his colleagues have isolated an antigen binding molecule from a clone of T-helper cells that mimics the antigen-specific, Ia-recognizing behavior of the parent clone. The structure of this molecule is not known. A similar result has been published by Lonai et al. (1981).

C. Suppressor Factors

A wealth of different factors involved in the generation and/or effecting of suppression have been isolated, and many are clearly distinct, one from another, both chemically, antigenically, and functionally. These have been reviewed by Germain and Benacerraf (1980).

Suppressor Effector Factor

This factor acts on an Lyl T-helper cell and causes inactivation of the Lyl T-helper cell; it is the molecular product of a suppressor effector cell. Fresno et al. (1981) have cloned this cell and have isolated its molecular product. Their results, together with those of Yamauchi et al. (1981), suggest that this product is of molecular weight 70,000 and binds to antigen. In the presence of antigen, it is specifically suppressive of T-helper cells. Its antigen binding moiety is separable from its suppressive moiety by proteolytic cleavage (Fresno et al., 1982). It is not I-J positive, but in Gershon's experiments, such molecules are MHC restricted in their ability to suppress Lyl-helper cells. This MHC restriction can be overcome by adding I-J^+ molecules derived from mice identical at I-J with the donor of the suppressor effector factor. Thus this factor, while able to be purified to apparent homogeneity, is complex in its final active form.

Suppressor Inducer Factor

Gershon's group (Yamauchi et al., 1982) have defined a two-chain product that is derived from I-J^+ Lyl T cells that mimics the biological effects of the inducer of feedback suppression. This complex of molecules consists of two chains. It is not known if both are produced by the same cell. One chain binds antigen. It associates with an I-J^+ chain that is also the chain that bears a recognition unit for an Igh-V-linked molecule borne by the target Lyl^+2^+ transducer cell. Whether these two chains physically associate is at present a moot point, since they are readily separable by physical means under physiological conditions.

TsFl

Benacerraf and colleagues have defined a factor, TsF1, that mimics the biological and immunological properties of Tsl cells. It is I-J^+, antigen binding, and idiotype positive. It activates Ts2 cells in vivo or in vitro. The molecular characteristics of TsF1 have not been elucidated except by Krupen et al. (1982), who find that this molecule has a molecular weight of 25,000. They find that such molecules are highly hydrophobic. They have succeeded in in vitro translating messenger RNA for this product, and the biological activity is retained by the unglycosylated form, as are its immunochemical properties (Wieder et al., 1982). There is no evidence for a two-chain structure for this molecule at present, but its assay involves addition to whole spleen cells, so it may activate the production of or associate with other moieties in the cultures that are required for its full biological potential.

TsF2

TsF2 mimics the biological activity of Ts2 cells. It is I-J^+, anti-idiotypic, and is I-J restricted in its action on its target cell, the Ts3 cell. Little is known of its molecular structure.

TsF

Taniguchi et al. (1980) have extensively characterized TsF, which mimics the action of inducers of supersuppression. This molecule can be obtained from a T-cell hybrid source. In its cell-associated form, it is a two chain molecule readily falling apart on Sepharose columns of antibodies. One chain bears I-J determinants, the other bears antigen specificity. The latter chain carries antigenic determinants encoded in or near the Igh-C region that are *not* part of conventional Ig constant region determinants, analogous to the Igh-C-linked T-cell receptor antibodies of Owen and Spurll (1981). In the cell-free form, secreted by the hybrid cell line, the two chains are disulfide bonded. This molecule has been produced by in vitro translation of two separable mRNA species (Taniguchi et al., 1982), and requires both translates for any biological activity to be generated. The I-J- and Igh-C-linked antigens are retained after in vitro translation of mRNA.

VI. CONCLUSIONS

All immune responses involve interactions between distinct sets of lymphocytes. These interactions form the subject matter of cellular immunology. It is clear that they are highly complex. The present chapter describes a number of moderately well characterized cell interactions, and lists some of the better characterized molecular products that mediate some of these interactions. It is clearly going to be very difficult to describe an in vivo immune response definitively, because of the complexity of the system and of its components. However, the technique of cell cloning, the availability of monoclonal antibodies, and the promise of molecular genetic analysis of the molecules involved may allow us ultimately to analyze these interactions at a molecular level. Already, some of the products have been produced in cell-free systems in vitro. The difficulty will be in analyzing the products obtained in terms of their mode of action on their target cells, and in integrating this information into a working understanding of the genesis of an immune response. Use of cloned molecular products, and inhibition of immunoregulatory function with highly specific monoclonal antibodies in in vitro and in vivo immune responses, may allow a fuller comprehension of the role of each class of interaction on the induction or regulation of diverse immune responses.

REFERENCES

Abbas, A. K., Burakoff, S. J., Gefter, M. L., and Green, M. I. (1980). J. Exp. Med. *152*, 968.

Bianchi, A. T. J., Hooijkaas, H., Benner, R., Tees, R., Nordin, A. A., and Schreier, M. H. (1981). Nature (Lond.) *290*, 62.

Bottomly, K. (1981). In *Immunoglobulin Idiotypes* (C. Janeway, E. Sercarz, and H. Wigzell, Eds.). Academic Press, New York, pp. 517–532.

Erb, P., Feldman, M., and Hogg, N. (1976). Eur. J. Immunol. *6*, 365.

Fresno, M. L., McVay-Boudreau, L., Nabel, G., and Cantor, H. (1981). J. Exp. Med. *153*, 1260.

Fresno, M., McVay-Boudreau, L., and Cantor, H. (1982). J. Exp. Med. *155*, 981.

Germain, R., and Benacerraf, B. (1980). Springer Sem. Immunopathol. *3*, 93.

Gershon, R. K., Eardley, D. D., Durum, S., Green, D. R., Shen, F.-W., Yamauchi, K., Cantor, H., and Murphy, D. B. (1981). J. Exp. Med. *153*, 1533.

Gowans, J. L., and McGregor, D. D. (1965). Prog. Allergy *9*, 1.

Green, D. R., Gershon, R. K., and Eardly, D. D. (1981). Proc. Natl. Acad. Sci. USA *78*, 3819.

Jones, B., and Janeway, C. A., Jr. (1981). Nature (Lond.) *292*, 547.

Jϕrgensen, T., Boegen, B., and Hannestad, K. (1981). In *Immunoglobulin Idiotypes* (C. Janeway, E. Sercarz, and H. Witzell, Eds.). Academic Press, New York, pp. 573–580.

Krupen, K., Araneo, B. A., Brink, L., Kapp, J. A., Stein, S., Wieder, K. J., and Webb, D. R. (1982). Proc. Natl. Acad. Sci. USA *79*, 1254.

Lonai, P., Bitton, S., Savelkoul, H. F. J., Puri, J., and Hammerling, G. J. (1981). J. Exp. Med. *154*, 1910.

McKenzie, I. F. C., and Potter, T. (1979). Adv. Immunol. *27*, 179.

Owen, F. L., and Spurll, G. M. (1981). In *Immunoglobulin Idiotypes* (C. Janeway, E. Sercarz, and H. Wigzell, Eds.). Academic Press, New York, pp. 419–428.

Puri, J., and Lonai, P. (1980). Eur. J. Immunol. *10*, 273.

Tada, T., and Okumura, K. (1979). Adv. Immunol. *28*, 1.

Taniguchi, M., Takei, I., and Tada, T. (1980). Nature (Lond.) *283*, 227.

Taniguchi, M., Tokuhisa, T., Kanno, M., Yaoita, Y., Shimizu, A., and Honjo, T. (1982). Nature. *298*, 172.

Warren, R. W., and Davie, J. M. (1977). J. Exp. Med. *146*, 1627.

Wieder, K., Araneo, B. A., Kapp, J. A., and Webb, D. R. (1982). Proc. Natl. Acad. Sci. USA, in press.

Yamauchi, K., Murphy, D. B., Cantor, H., and Gershon, R. K. (1981). Eur. J. Immunol. *11*, 913.

Yamauchi, K., Chao, N., Murphy, D. B., and Gershon, R. K. (1982). J. Exp. Med. *155*, 655.

25

Genetic Regulation of Immune Response

Patrick M. Long and Chella S. David Mayo Medical School, Rochester, Minnesota

I. HISTORICAL PERSPECTIVE

The history of the H-2 system dates back to the early 1900s with the discovery of transplantable tumors in mice. Cancer biologists recognized tumors as being unique in that they grew best in the animal in which they arose. This uniqueness and individuality of tumors limited the time they could be studied, because they could not be transplanted from one animal to another with great success. Thus the tumor was eventually lost because of its inability to grow in other mice. In 1903, Jensen propagated a mouse tumor for 19 generations. The animals he used were from a closed colony that had achieved a high degree of inbreeding, because mating occurred only among relatives. When this tumor was transplanted into mice that were not from his inbred colony, it failed to grow. Other workers confirmed these observation of Jensen using other stocks of mice. This led Little to formulate a genetic theory for tumor transplantation. He proposed that susceptibility to a tumor transplant was governed by several dominant genes. Little and Tyzzer confirmed this theory, which stimulated the development of inbred mouse strains.

The genetic theory of tumor transplantation explained why tumors grew in some strains of mice, but did not explain why they failed to grow in others. In 1933, Haldane proposed an immunologic mechanism for tumor rejection. Tumors arising in tissue retained the alloantigens of the host. Alloantigens are antigens possessed by some members of the species but not by other members of that species. They are controlled by allelic genes. Tumors were rejected because of these alloantigens on their surface, which elicit an immune response when injected into an inbred strain lacking them. Gorer was the first to detect blood group antigens in mice. He was able to distinguish four blood groups, which he termed I, II, III, and IV. While working to demonstrate if a relationship

For a brief discussion of the connection between Chaps. 9 and 24–28, see the appendix.

existed between his blood group genes and the genes for tumor susceptibility, Gorer noted that an antibody to the group II antigen appeared following tumor regression. The presence of antigen II on the tumor cells was confirmed by absorption experiments. In 1938, Gorer formulated an immunologic theory of tumor transplantation. "Normal and neoplastic tissues contain iso-antigenic factors [alloantigens] which are genetically determined. Iso-antigenic factors present in the grafted tissue and absent in the host are capable of eliciting a response which results in the destruction of the graft."

It was not until 1943 that the immunologic basis of skin graft rejection was demonstrated by Gibson and Medawar. Medawar and his colleagues demonstrated that second-set skin graft rejection was accelerated, specific, and systemic. Medawar felt that these three characteristics proved that allograft rejection was indeed immunologic in nature. By means of adoptive transfer experiments, Mitchison in 1954 provided evidence that this reaction was not antibody mediated. Mitchison could transfer tissue immunity only with lymphocytes from a sensitized animal and not with its serum. These experiments established a cellular basis for graft rejection. Snell in 1948 termed the antigens responsible for tissue compatibility, *histocompatibility antigens*, and the genes coding for these antigens, *histocompatibility genes* or *H genes*. He recognized that in conventional inbred strains of mice, the role of an individual histocompatibility locus cannot be assessed because any two inbred strains differ at several histocompatibility loci. He then set out to establish strains of mice that differed in only one histocompatibility locus—congenic strains. Using these congenic strains in collaboration with Gorer, Snell verified the identity of a gene, confirming tumor resistance with the gene coding for antigen II. This histocompatibility gene was designated *H-2* and was found to be linked to the gene for fused tail, which was assigned to linkage group IX, later identified as chromosome 17.

The complexity of the *H-2* locus began to emerge as techniques to detect them were refined. Different *H-2* alleles were seen to control an array of antigens, some of which were limited to one allele, whereas others were shared by several alleles. The genetic complexity of the *H-2* locus was seen by Snell when he noted that the F_1 of $H\text{-}2^d \times H\text{-}2^k$ accepted tumors of $H\text{-}2^a$ origin, a characteristic absent from either parental strain. This phenomenon (later termed the *DK effect*) was explained by assuming that the *H-2* locus is composed of two components, d and k, and that the $H\text{-}2^a$ allele is a genetic recombinant between these components. The first intra-H-2 recombinants were soon detected by histogenic and serological methods. Their analysis revealed the *H-2* locus to be a complex of several regions separable by genetic crossing over. The *H-2* locus is much more polymorphic than any other known mouse locus. Because of the genetic complexity and polymorphism of the *H-2* locus, it has become known as the major histocompatibility complex (MHC) of the mouse (Fig. 1).

The MHC of the mouse has been shown to code for at least three classes of molecules. The class I molecules fall into two divisions. The classical transplantation antigens are present on all cells and are coded for by genes in the K and D (L, R) regions of the MHC. Hematopoietic differentiation antigens are coded for by genes in the Qa and TL regions and are expressed on subsets of lymphoid cells. Both the transplantation antigens and the hematopoietic differentiation antigens are transmembrane proteins composed of two chains, a 45,000-molecular weight polypeptide encoded by the MHC and a noncovalently associated 12,000-molecular weight polypeptide, β_2 microglobulin, encoded by chromosome 2. Class II molecules are encoded for by the I region of the MHC and are involved in interactions among T cells, B cells, and macrophages which regulate the immune response to many antigens. The I-A and I-E molecules are composed of a 32,000- to 34,000-molecular weight α chain (A_α or E_α) and a 25,000- to 28,000-molecular weight

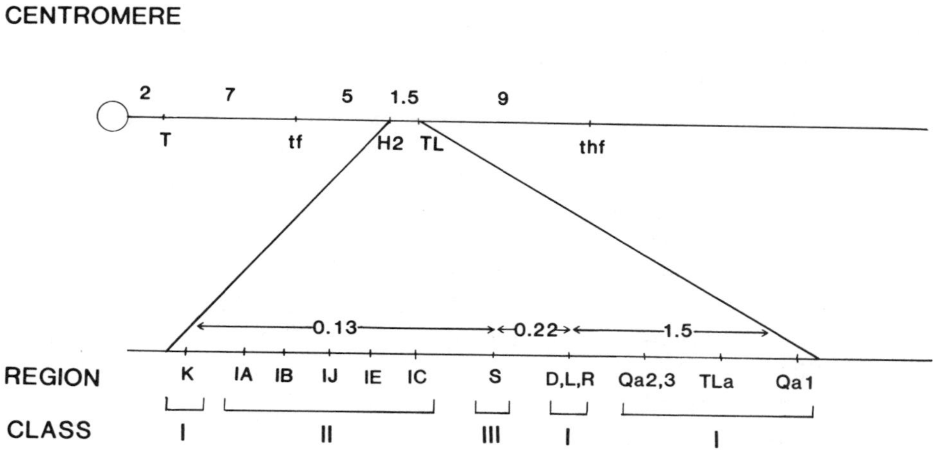

Figure 1 Genetic map of *H-2* complex.

β chain (Aβ or E_{β}). Genes in the I-J subregion encode for immune-associated (Ia) antigens present on both T-suppressor cells and soluble suppressor factors. Table 1 lists the major functions associated with the I subregions. The class III genes encode for several components of the complement system.

The principle of a genetic basis for inherent resistances to diseases was explored by Webster during the 1920s to 1940s. At this time he noted that in a stock of outbred Swiss

Table 1 Functions Associated with I Subregions

Function	Subregion				
	A	B	J	E	C
Tissue expression					
B cells	+	-	-	+	-
T cells	+	-	+	+	+
Macrophages	+	-	-	+	-
Langerhans cells	+	-	-	+	-
Neutrophils	+	-	-	+	-
Eosinophils	+	-	-	+	-
Peyer's patch	+	-	-	+	
Tumor cells	+	-		+	-
Serum	+	-	-	+	-
Ir genes	+	+	-	+	+
Is genes	+	-	-	-	-
T-cell helper factors	+	-	+	-	-
Antigen-specific suppressor factors	-	-	+	-	-
Macrophage helper factors	+	-	-	-	-
T-cell receptors	+	-	+	-	+
B-cell receptors	+	-	-	-	-
Antigen presentation	+	+	+	+	+
Antigen recognition	+	+	+	+	+

mice there was variation in their susceptibility to infection by *Bacillus enteriditis*. By selective inbreeding techniques, he was able to develop two lines of mice: bacteria susceptible, termed BS, and bacteria resistant, termed BR. Further investigation with Louping ill virus indicated that there was a dissimilarity in the BS and BR lines with respect to infection with this virus. While some of the BS mice were resistant to virus infection, it was observed that essentially all of the BR mice were susceptible to infection with Louping ill virus. These observations gave rise to a series of inbreeding programs which resulted in the establishment of four strains of mice. Mice of the BS strains that were resistant to virus infection were inbred, resulting in a new substrain that was susceptible to *B. enteriditis* infection but resistant to Louping ill virus, and were thus designated BSVR. The BR mice that were susceptible to the virus infection were inbred to produce the BRVS subline. As the BSVR subline was inbred, occasionally up to 50% of the animals of some litters were susceptible to infection with Louping ill virus. These susceptible mice from the BSVR strain were inbred to give rise to the BSVS strain of mouse. Thus mice of the BSVS strain were highly susceptible to infection with both bacteria and virus, whereas mice of the BSVR strain were susceptible to bacterial infection but resistant to viral infection. When the BRVS and BSVR strains were mated, the resulting F_1 hybrid population were resistant to both bacterial and viral infections. These F_1 hybrid mice were inbred to produce the BRVR strain of mouse.

Additional studies demonstrated that the BRVR and BSVR strains were resistant to other viral infections, such as St. Louis encephalitis virus and Russian spring and summer encephalitis viruses, in addition to the Louping ill virus. The BRVR and BRVS mouse strains were also resistant to infection with *Salmonella* bacteria. Genetic studies disclosed that the VR trait and BR trait segregated independently as simple dominant mendelian genes. Webster thus demonstrated that resistance or susceptibility to bacterial and viral infections was not due to previous contact with virus or bacteria but, on the contrary, resided in the genetic makeup of the mouse. These genetic resistance traits were capable of being passed on to subsequent offspring prior to any contact with the infectious agents. Subsequent H-2 typing of the strains has shown that BRVR is H-2^k ($K^kI^kS^kD^k$), BSVS is H-2^{t5} ($K^sI^sS^dD^d$), and BSVR is H-2^{a3} ($K^kI^kS^dD^d$). Thus, as a consequence of selection for virus resistance, the K^k gene has been fixed in the BRVR and BSVR strains, while selection for bacterial resistance has fixed the D^d gene in the BSVS and BSVR strains. The basis for future studies on the genetics of resistance and susceptibility and subsequent identification of immune response (Ir) genes was thus formulated by these early studies of Webster.

It had been noted by immunologists for years that there were certain individuals who were poor responders to a given antigen whereas others responded vigorously to the same antigen challenge. This was true not only of noninfectious antigens but of infectious agents as well. Thus certain individuals of outbred populations were found to be more susceptible to diseases than others. During an outbreak of influenza only a limited number of the population become ill. Some people are always coming down with a cold, whereas others are never sick a day in their lives. Not only is this seen with infectious agents such as viruses and bacteria, but with noninfectious agents as well. Not everyone exposed to the same carcinogen succumbs with cancer. Although not all of these susceptibilities or histories of familial diseases can be associated with the MHC, several alleles of the human major histocompatibility complex (HLA) occur with high frequent in certain diseases, and this is probably related to the role of the MHC in immune reactions. These immune reactions can be an abnormally active response, such as in autoimmune reactions

and inflammatory diseases, or a feeble response toward a particular pathogenic agent, be it viral, bacterial, or tumor antigen.

The heterogeneity of determinants presented by most antigens discouraged the genetic investigations of immune responses. However, with the advent of the use of synthetic polypeptides as antigens, the field of immunogenetics began to expand rapidly. These polypeptides were composed of a limited number of L-amino acids and thus were restricted in their heterogeneity. Other antigens that have proven to be extremely valuable in the identification of *Ir* genes include the use of weak native proteins such as the minor histocompatibility antigens as well as strong native proteins such as bovine serum albumin (BSA) given in a limited dose range. These three varieties of antigens, the synthetic polypeptides, weak native antigens, and limited doses of strong native antigens, all have the common feature of presenting to the recipients' immune system antigens of limited heterogeneity. Such restricted antigens have led to the identification of specific immune response genes in numerous species. The first specific immune response gene identified was in the guinea pig with the pioneering work of Benacerraf and his co-workers (Levine et al., 1963). Using the synthetic polypeptide poly-L-lysine (PLL) as an antigen, it was found that of the two inbred strains of guinea pigs, strain 2 and strain 13, only strain 2 could respond to DNP-poly-L-lysine; strain 13 could not, although a cross between strain 2 and strain 13 guinea pigs resulted in F_1 animals that were responders. The work in the guinea pig led to the identification of genetic control of responsiveness to other synthetic polypeptide antigens, such as poly-L-arginine, random copolymers of L-glutamic acid and L-alanine (GA), L-glutamic acid and L-tyrosine (GT), and L-glutamic acid and L-lysine (GL) (Bluestein et al., 1971). Strain 2 guinea pigs are high responders to PLL and GA, while strain 13 is a high responder to GT. Using randomly bred Hartley guinea pigs, the genes regulating GA and PLL responsiveness were shown to be linked. Nonetheless, a small percentage of these Hartley guinea pigs respond to GA but not to PLL, and vice versa. Thus the response to GA and PLL can be separated as a result of crossover, indicating the existence of two separate genes. At this time, the *Ir* genes in the guinea pig were not recognized as being associated with the major histocompatibility complex of the guinea pig. It was the work of McDevitt and his co-workers that linked the immune response genes with the major histocompatibility complex and mapped such *Ir* genes in the I region using the mouse as a model. Earlier work by Pinchuch and Maurer had shown that homopolymers and linear copolymers of two amino acids (GL, GA, or GT) were not immunogenic in mice. Introduction of the third amino acid to these copolymers resulted in an immunogenic complex; however, there were strain response differences identified in studies using these antigens, suggesting the existence of *Ir* genes in mouse. McDevitt and Sela (1965) established the existence of *Ir* gene-controlled responses in inbred mice to the synthetic polypeptide antigens (T,G)-A-L, (H,G)-A-L, and (P,G)-A-L. These antigens are composed of a poly-L-lysine backbone with side chains consisting of specific terminal amino acids of either tyrosine and glutamic acid, histidine and glutamic acid, or phenylalanine and glutamic acid attached to poly-DL-alanine. Through linkage studies using backcross populations, it was recognized that the responsiveness to the aforementioned antigens were associated with the *H-2* genotype of the mouse (McDevitt and Chinitz, 1969) (Table 2). The availability of congenic resistant mouse strains and strains with recombinant events within the *H-2* complex allowed for the mapping of the *Ir-1* locus controlling responsiveness to these antigens to a unique region of the mouse *H-2* complex designated the I region (McDevitt et al., 1972). The H-2I region was mapped between the H-2K and S regions using H-2 recombinant strains.

Table 2 Responsiveness to Synthetic Polypeptides Based on H-2 Haplotype

H-2 Haplotype	Synthetic antigen			
	GLØ	(T,G)-A-L	(H-G)-A-L	(P,G)-A-L
a	L[a]	L	H[b]	H
b	L	H	L	H
d	H	I[c]	I	H
f	L	L	L	H
j	H	L	I	H
k	L	L	H	H
p	H	L	L	H
q	H	L	L	H
r	H	L	–[d]	–
s	L	L	L	L

[a]Low responsiveness.
[b]High responsiveness.
[c]Intermediate responsiveness.
[d]Not determined.

At this time, the *Ir* genes were postulated to code for the antigen receptors on thymus-derived lymphocytes by Benacerraf and McDevitt (1972). The general properties of the MHC-linked *Ir* genes which led to this hypothesis were the fact that *Ir* genes controlled recognition of carrier molecules on an antigen which is a function of thymus-derived cells. Second, while these genes controlled both humoral and cellular responses, the control of antibody production resulted as a need for T-cell help in triggering antibody synthesis from B cells. Finally, only T-dependent antigens were found to be under *Ir* gene control. T-independent antigens that activate B cells directly were not under MHC-linked *Ir* gene control.

With this in mind, several laboratories attempted to produce antiserum against these putative receptors on T cells. Production of alloantisera to I-region products was possible because of the availability of several pairs of recombinant mice that were identical at the H-2K and H-2D region and differed only in the I and S region. These strains were the A.TH ($K^sI^sS^sD^d$) and A.TL ($K^sI^kS^kD^d$) (David and Schreffler, 1972) pair as well as the B10.AQR ($K^qI^kS^dD^d$) and B10.T(6R) ($K^qI^qS^sD^d$) pair (Klein et al., 1970). Reciprocal immunization of these pairs of mice produced antisera that reacted with cell membrane alloantigens controlled by the I region (referred to as Ia antigens). It was immediately recognized that this antiserum was reacting with the subpopulation of lymphocytes because it failed to kill 100% of normal lymph node cells or spleen cells. Because *Ir* gene products were expected to be primarily on T cells, it was important to distinguish between the T-cell and B-cell reactivity of these alloantisera. Studies found that predominant activity was with B lymphocytes (Shreffler and David, 1975). Because no direct proof was found that the reaction was with the *Ir* gene product, the designation immune response associated or Ia antigen was given. Biochemical analysis with anti-Ia sera showed that the Ia determinants on B cells were glycotproteins composed of two polypeptide chains, termed α and β, with molecular weights of 33,000 and 28,000, respectively.

Subsequent studies have demonstrated Ia antigen expression on subsets of T cells as well as on concanavalin A (ConA)-stimulated T cells by both cytotoxicity and absorption studies. Immunoprecipitation and immunofluorescent studies have located Ia antigens on

macrophages as well. Thus, depending on the method of study, Ia antigen expression can be found on T cells and B cells as well as on the macrophage. Ia antigens have also been identified on cells other than those of the T- and B-cell lineage. They have been found on Langerhans cells, neutrophils, eosinophils, and sperm cells.

II. NON-MHC-LINKED IMMUNE RESPONSE GENES

Although the most thoroughly studied immune response genes are those linked to the MHC, it must be pointed out that several non-MHC-linked immune response genes have been identified. In mice, these non-MHC-linked immune response genes influence the resistance or susceptibility to infection from DNA, RNA, or tumor viruses, as well as responses to H-2 alloantigens, minor histocompatibility antigens, erythrocyte alloantigens, and certain native proteins, such as bovine serum albumin, hemoglobin, and acetylcholine receptor, to mention a few. Gasser and co-workers have studied the genetic control of the immune response to the mouse erythrocyte alloantigen Ea-1. The response to this antigen is controlled by a major gene (*Ir-2* on chromosome 2) as well as additional modifying genetic factors, one linked to the MHC and a third not linked to chromosome 2. An unusual characteristic of *Ir-2* gene function is that unresponsiveness is dominant. Biozzi and collaborators also examined multiple genetic factors influencing the immune response. Starting with randomly bred Swiss mice, they developed two lines of mice selected for their ability to generate high or low titers of hemagglutinating antibody to sheep and pigeon red blood cells. By selective breeding based on such antibody production, a clear separation of high- and low-responder lines occurred. This separation of high- and low-responder lines occurred in 16 generations and indicated multiple gene control of the response (10 to 18 genes), one of which is in the *H-2* complex. An important observation in these studies was the discovery that selection for high responsiveness to the red blood cell antigens coselected for high responsiveness for antigens not used in the selection process. Thus the high-responder strain was also a high responder to unrelated antigens such as the *Salmonella typhi* O and H antigens and hen ovalbumin. It has subsequently been observed that the high-responder strain BZ.H is hyperresponsive to a multitude of antigens.

A sex-linked *Ir* locus has been found in the CBA/N strain. The antibody response to type III pneumococcal polysaccharide, double-stranded RNA, and denatured DNA is very low in this strain. F_1 males from crosses involving female CBA/N with males from high-responder strains are low responders, whereas female F_1's are high responders. These results indicate the involvement of genes linked to the X chromosome in the immune response to the antigens noted above.

Immune response genes have been reported in association with the mouse heavy chain constant region locus, designated *Ig-1*. These include the responses to α-1,3-dextran, the H-2 specificity H-2.32, and the naturally occurring antibodies to chicken erythrocytes found in normal mouse serum.

III. HOW Ia MOLECULES INFLUENCE THE IMMUNE RESPONSE

Several lines of evidence have been developed over the past several years which have implicated the role of I-region products in cell interactions involving the immune response. These cell interactions include T cell/macrophage interaction, T cell/B cell interaction, and T cell/T cell interaction. Shevach and Rosenthal (1973) made the initial observation

in the guinea pig for the involvement of Ir genes in macrophage/T cell inteaction. They investigated the role of the macrophage in regulation of the immune response against DNP-GT and DNP-GL in strain 2, strain 13, or F_1 between strains 2 and 13 guinea pigs. Strain 2 is a responder to DNP-GL. T-cells from (2 × 13) F_1 guinea pigs were primed to both antigens in vivo, and then were presented with antigen on the surface of macrophages in vitro. Macrophages of strain 2 or F_1 origin could present DNP-GL to the primed F_1 T cells, whereas macrophages of strain 13 origin could not. Similarly, macrophages of strain 13 of F_1 origin could present DNP-GT, whereas macrophages of strain 2 failed to do so. These findings led to the conclusion that immune response genes or at least some immune response gene products are expressed on macrophage cell membranes. The inability of the macrophage in the experiments above to present antigens to primed T cells was due to a defect at the macrophage level. This conclusion has recently been challenged by Klein and his co-workers (Ishii et al., 1982). They proposed that Ir genes in these F_1 experiments are not expressed at the macrophage level, but rather at the T-cell level. The defect in the nonresponder animal is due to a deletion in the ability of their own T cells to recognize the antigen in question in the context of self Ia. They have shown in a mouse system that nonresponder T cells can respond when presented antigens by allogeneic macrophages of either high- or low-responder strains. They fail to respond only when presented antigen in association with their own MHC antigens. The authors have developed a unique system for using allogenic macrophages to present antigens to the T cell. T cells were first cultured with allogeneic antigen presenting cells, and the alloreactive T cells that proliferate in response to the alloantigen on the antigen presenting cells are destroyed by treatment with 5-bromo-2-deoxyuridine and light on day 3. The T cells remaining after this procedure are then cultured for 7 days with allogeneic antigen presenting cells and an antigen, the response to which is known to be controlled by immune response genes. A secondary proliferative response is then induced with the same antigen and variety of antigen presenting cells obtained from several mouse strains. In all allogeneic combinations tested, they obtained a proliferative response regardless of the source of either the T cells or the antigen presenting cells. Thus T cells from nonresponders could respond to the antigen presented on either responder or nonresponder macrophages, and T cells from responder animals could respond to antigens presented on either type of antigen presenting cells. The only time that nonresponder T cells failed to respond was when the antigen was presented on the T cells' own antigen presenting cells. These results suggest that immune response gene-controlled defects are associated with holes in the T-cell repertoire recognizing self MHC antigens.

The expression of immune response genes on B cells has been demonstrated through experimental studies using much the same methodology as that employed in the identification of *Ir* genes on macrophages. Using F_1 T cells obtained from crosses between responder and nonresponder mice, it was found that such T cells, when primed to a carrier under *Ir* gene control such as GLT, could cooperate only with B cells of F_1 and responder parental origin and would not cooperate with B cells of the nonresponder parent in response to DNP conjugated to GLT. The results point out that immune response genes are expressed on B cells and their action is concerned with successful T/B-cell cooperation involved in carrier recognition.

The expression of immune response genes on T cells has been less well established. As mentioned previously, alloantiserum directed to I-region determinants reacted primarily with B cells and subpopulations of macrophages. However, several studies have shown that at least some Ia specificities are on T cells. Cortisone-resistant T cells as well

as mitogen-stimulated T cells have been shown to express Ia specificities. Ia antigens are not easily detected on normal thymocytes, suggesting that T cells require some stimulating signal before expression of Ia specificities. Immunoprecipitation studies using ConA blasts internally labeled with tridiated leucine have shown that most Ia specificities are present on such activated T cells. The presence of Ia antigens coded for by the I-J region have been shown to be expressed on both-T-suppressor cells as well as soluble suppressor factors.

Antigen-specific T-suppressor cells have been identified in several antigenic systems. In the mouse, one of the best studied was the response to GT. All mouse strains are nonresponders to GT. However, they are capable of making an anti-GT antibody response if the antigen is first coupled to an immunogenetic carrier such as methylated bovine serum albumin. Thus the B cells of these nonresponder animals are not defective. Further studies have proven that GT could be presented by nonresponder macrophages. Therefore, the antigen presenting cell was not defective. When animals were preimmunized with soluble antigen, that is, GT alone, T-suppressor cells are generated which inhibit a specific GT response upon challenge with GT coupled to methylated bovine serum albumin. Further work has demonstrated that such T-suppressor cells could be generated in only some nonresponding strains. For this reason, T-suppressor cell generation could not be a general explanation for nonresponsiveness in all of the strains tested. The genes controlling the generation of GT T-suppressor cells have been mapped to the I region of the MHC. Several lines of indirect evidence have identified Ia antigens on T-helper cells. Tausig has identified antigen-specific molecules capable of replacing T cells in T-dependent responses. These molecules interact with receptor sites on B cells, leading to a successful antibody response in the presence of antigen. Both the B-cell activator molecule and the T-cell factor express Ia antigen determinants.

IV. Ia MOLECULES ARE THE Ir GENE PRODUCT

Several lines of evidence have been developed over the past few years which point to the idea that Ia antigens are indeed the *Ir* gene products and are responsible for the genetic regulation of the immune response. Mapping studies, antibody blocking studies using both conventional alloantisera to Ia and monoclonal anti-Ia antibodies, cloned lines of antigen-specific T cells, and recent studies with mice bearing an Ia-region mutation have all identified Ia antigens as the true products of *Ir* genes. The first indication that this was the case came from blocking studies using alloantisera directed to MHC antigens to inhibit T-cell proliferation assays in vitro (Shevach et al., 1972; Meo et al., 1975). It was shown that antiserum directed to either the K or the D product of the MHC could not inhibit while antisera containing activity against I-region-encoded determinants could. The continuous presence of anti-Ia antibodies inhibited the T-cell proliferative response to GLT, (T,G)-A-L, GAL, lactate dehydrogenase H_4, staphylococcal nuclease, and the IgA myeloma protein TEPC-15. The response to all of these antigens had previously been shown to be under MHC control. In F_1's made between high and low responders to the antigens above, only anti-Ia directed toward antigens of the responder inhibited T-cell proliferation. This indicated that neither cell death nor nonspecific blocking but rather blockade of antigen stimulation at the cell surface was the cause of the inhibition. The investigations concluded that Ia antigens are indeed determinants on *Ir* gene products. In addition, anti-Ia antibodies with specificities to certain I subregions only inhibited responses toward antigens which were controlled by that subregion. The response to GLϕ

Table 3 Reactivity of T-Cell Clones Specific for GAT Derived from $(B6A)F_1$ Mice[a]

Antigen presenting cells	Group 1	Group 2	Group 3
A	-	+	-
B6	+	-	-
A + B6[b]	+	+	-
(B6A) F_1	+	+	+

[a]+/- represent positive or negative responses as incorporation of $[^3H]$TdR in the presence of 1×10^6 irradiated antigen presenting cells and 200 μg/ml GAT.
[b]Strain A and B6 cells were mixed 1:1 for source of antigen presenting cells.
Source: Summarized from Kimoto and Fathman (1980).

(a random linear polymer with 55% glutamate, 35% L-lysine, and 10% phenylalanine) is controlled by the I-E subregion. Ia.7 is a determinant coded for by this subregion. An anti-Ia.7 antibody inhibited the response to GLϕ, but not the response to antigens that were under I-A subregion control (Schwartz et al., 1976). The results of the blocking studies have been supported recently through the use of monoclonal anti-Ia antibodies, thus ruling out the possibility of contaminating antibodies in the conventional antisera being responsible for inhibition. The availability of cloned T cell lines specific for a given antigen has provided more supportive evidence for the identity of Ia molecules as *Ir* gene products. It has recently become possible to maintain long-term murine T-cell cultures which proliferate in response to soluble antigens. The response of these cloned T cells has been shown to be MHC restricted to the specific antigen seen in the context of Ia-compatible antigen presenting cells. Using GAT (a random linear polymer composed of 60% L-glutamate, 30% L-alanine, and 10% tyrosine) as antigen, several antigen reactive T-cell clones have been derived from (B6 × A) F_1 mice (Kimoto and Fathman, 1980). These clones could be classified into three categories based on the recognition of antigen in association with various antigen presenting cells (Table 3). One group could be stimulated by antigen presented by the B6 parent or (B6 × A) F_1 antigen presenting cells. The second group recognized antigen presented by the A parent or (B6 × A) F_1 cells. The third group could be stimulated by antigen only when presented by (B6 × A) F_1 cells, indicating that there are unique hybrid antigen presenting determinants on F_1 cells which are not present on either parent and that such hybrid determinants can present antigen to T cells. Mapping studies in the system above have shown that these unique F_1 hybrid determinants are controlled by the I-A subregion (Table 4). These hybrid Ia molecules are derived by combinatorial association of α and β chains, resulting in at least four I-A antigens being expressed on F_1's, due to cis- and trans-complementation. In the (B6 × A) F_1, one Ia molecule is composed of $A_\alpha^k A_\beta^k$ from the A parent, a second Ia molecule is composed of $A_\alpha^b A_\beta^b$ from the B6 parent, and two hybrid Ia molecules arise from trans-complementation ($A_\alpha^k A_\beta^b$; $A_\alpha^b A_\beta^k$).

In attempts to confirm the I-A subregion origin of the hybrid determinants above, use was made of the mutant B6.C-H-2^{bm12} (Kimoto et al., 1982). This mutation is a spontaneous gain-loss type which arose in the B6 strain. The mutation has been mapped to the I-A^b subregion. Biochemical studies indicate a lack of or a greatly reduced quantity of I-A^b product on the cell surface. More refined biochemical analysis suggests that the mutation has occurred in the A_β chain of the bm12. Using (bm12 × A) F_1 cells as antigen presenting cells to antigen reactive T-cell clones, it was seen that certain clones which can be stimulated with antigen when present on (B6 × A) F_1 antigen presenting cells could

Table 4 Mapping of F_1 Hybrid Determinants to the I-A Subregion

Antigen presenting cells	MHC Haplotype K	A	E	S	D	Proliferation after GAT stimulation
A	k	k	k	d	d	–
B6	b	b	b	b	b	–
A + B6	k/b	k/b	k/b	k/b	k/b	–
B10.A(4R)	k	k	b	b	b	–
(B6A) F_1	b/k	b/k	b/k	b/k	b/k	+
[B10.A(4R) × B6] F_1	k/b	k/b	b/b	b/b	b/b	–
[B10.A(4R) × B10.A] F_1	k/k	k/k	b/k	b/d	b/d	+
(B10.MBR × A.AL) F_1	b/k	k/k	k/k	k/k	q/d	–

not recognize antigen presented by (bm12 × A) F_1 cells. This supports the idea that a normal I-A^b-region product is not present on the (bm12 × A) F_1 cells. This would arise due to the lack of trans-complementation due to the mutation in the A_β chain of the bm12 mutant. Thus, using the bm12 mutant, there are serological, biochemical, as well as functional data which strongly support the idea that Ia antigens are the restricting elements in immune response.

V. MHC RESTRICTIONS

An abundance of information has been accumulated in the last decade regarding the central role of the MHC antigens in cellular interactions of the immune response. Evidence points to the concept that T cells recognize antigen only in association with MHC antigens. This has led to the concept of MHC associative antigen recognition by T cells. The results of studies involving virus- or hapten-modified cells as well as the responses to minor transplantation antigens point to the fact that cytotoxic responses to such antigens are MHC restricted. That is, the cytotoxic lymphocytes (CTLs) generated recognize the foreign antigen (virus, hapten, or minor histocompatibility antigens) only when it is presented in the context of the MHC antigens used to initiate the response. In the case of cytotoxic responses the restricting elements have been mapped to the H-2K and H-2D regions of the *H-2* complex. In the activation of T cells for the generation of helper function in antibody formation the restricting element has been localized in the I region. It is important to note here that two of the major classes of T cells are restricted by different regions of the MHC. That is, the T-helper cells are restricted by I-region determinants, while T-effector cells are restricted by K/D-region determinants.

To demonstrate MHC restriction of CTL responses, the antigens initially used included trinitrophenyl-modified target cells (Shearer), virus-infected cells (Zinkernagel and Dougherty), or cells bearing minor histocompatibility differences (Bevan) (see Zinkernagel and Dougherty, 1979 for a review). When animals were immunized with the antigens above, CTLs were generated that would lyse only antigen-bearing target cells which were of the same H-2 haplotype as the host. Mapping studies employing

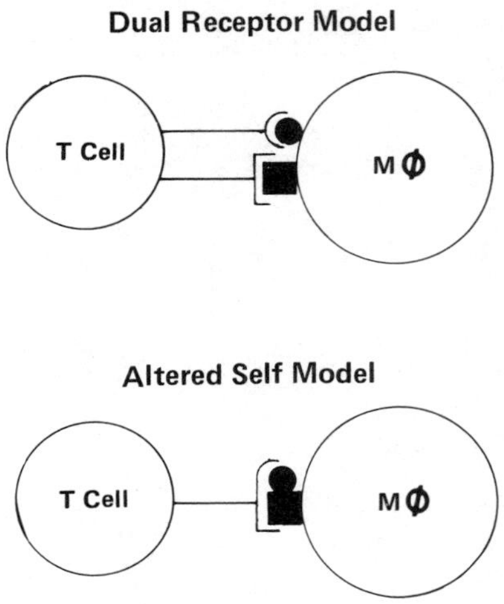

Figure 2 Models for associative recognition of antigen by T cells. ■, MHC restricting element; ●, specific antigen.

appropriate H-2 recombinants mapped the restricting elements to the K and/or D region of the H-2. While both K and D regions were shown to restrict CTL effector function, one region may dominate depending on the antigen or the H-2 haplotype in question. The restriction was shown to be in the generation of the effector cells and not an event in the effector phase of the response by studies in F_1 animals. Effector cells from F_1 animals immunized with antigen presented on cells of one parent lyse only antigen bearing cells of the parent used to immunize and are not capable of lysing antigen bearing cells of the other parent. By studying appropriate F_1 into parent cell transfer experiments, it was demonstrated that the environment in which the cells differentiate gave rise to the restricting MHC phenomena. The F_1 bone marrow cells differentiating in one parent became restricted to the MHC antigens of that parent. This differentiation phase takes place in the thymus and is influenced by self-MHC gene products.

The mechanisms by which MHC-restricted CTLs recognized targets is still under scrutiny. The possibility of either single recognition of an altered self determinant composed of a combinatorial determinant of foreign antigen and self-MHC antigen is one possibility. A second possibility is that of the dual receptor model where the effector cells has two separate receptors on the surface, one for the foreign antigen and one for self (Fig. 2).

VI. CLONAL DISSECTION OF THE IMMUNE RESPONSE

Through the use of T-cell clones reactive to sperm whale myoglobin, Atassi and colleagues have recently examined in detail the mechanisms of T-cell recognition of antigen (Infante et al., 1981). Sperm whale myoglobin has been studied extensively and its amino acid sequence, three-dimensional structure, and antigenic determinants recognized by antibody are well known. Myoglobin is a helical molecule of 153 amino acids which upon cleavage with cyanogen bromide yields three peptides comprising residues 1 to 55, 56 to 131, and 132 to 153 (Fig. 3). Upon immunization with myoglobin, serum antibodies are

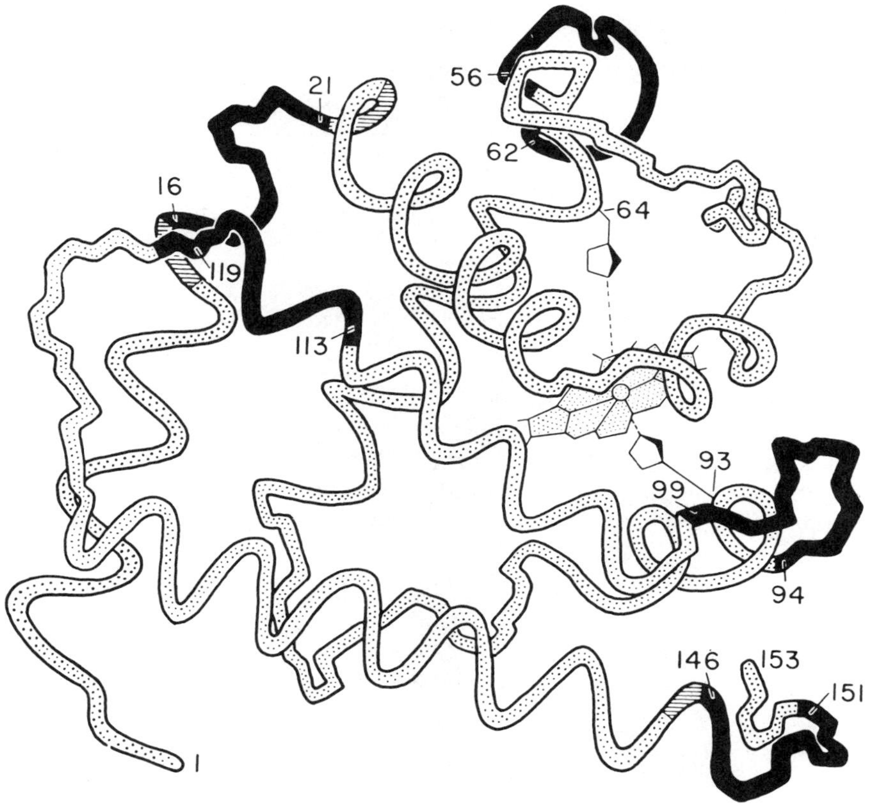

Figure 3 Structure of myoglobin molecule with five major antigenic sites. [From M. Z. Atassi, (1975), Immunochemistry *12*, 423.]

obtained which are reactive with five major antigenic sites. These consist of residues 16 to 21, 56 to 62, 94 to 99, 113 to 119, and 146 to 151. T-cell clones have now been isolated from mice which demonstrate specific I-region restriction and activity to the three cyanogen bromide polypeptides of myoglobin (Table 5). Fragment 1 contains antigenic site 1, fragment 2 contains antigenic sites 2 to 4, and fragment 3 contains antigenic site 5. One isolated T-cell clone was shown to react only to fragment 2, not to fragments 1 and 3. This response was also shown to be restrictive to specific I-A-region molecules. A second clone, responding to fragment 1, was shown to be restrictive to combinatorial Ia determinants composed of $E^b_\beta E^k_\alpha$ chains. Another clone reacting to fragment 1 was identified but was restricted to the I-A region of the mouse MHC. Finally, a third clone reactive to fragment 1, but restricted to recognition of antigen in association with I-E molecules was identified. These results were consistent with determinant selection theories stating that antigen-specific T cells recognize antigen in association with distinct epitopes on Ia molecules. Shevach has recently come to the same conclusions using T-cell colonies obtained from antigen-primed guinea pigs (Clark et al., 1982). Using monoclonal anti-Ia antibodies to guinea pig Ia antigens in conjunction with antigen-specific T-cell colonies, they have shown that the proliferative responses could be dramatically inhibited by only

Table 5 Myoglobin-Reactive Long-Term T-Cell Clones Dervied from $(B6A)F_1$ Mice

Clone	H-2 Restriction molecule	Myoglobin	Fragment recognition 1	2	3
26.17	I-A	+	–	+	–
45.49[a]	I-E	+	+	–	–
28.12	I-A	+	+	–	–
69[b]	I-E	+	+	–	–

[a]45.49 responds to fragment 1 presented on $E_\beta^b E_\alpha^k$ molecule.

[b]69 responds to fragment 1 presented on $E_\beta^k E_\alpha^k$ molecule.

a few of the monoclonal antibodies. Thus the T-cell colonies were recognizing antigen in association with different Ia epitopes. The findings are agreeable with the theories that T cells recognize neoantigenic determinants composed of the Ia of the antigen determinants composed of the Ia of the antigen presenting cell and antigen, or they recognize self-Ia antigens and foreign antigens by two separate receptors (Fig. 4).

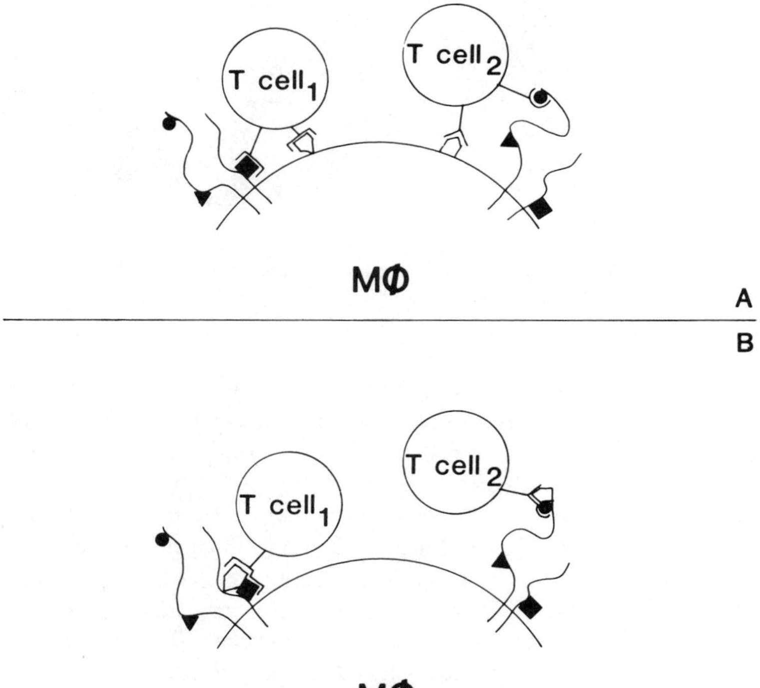

Figure 4 Determinant selection model for presentation of distinct antigenic determinants by different epitopes of I-region molecules. ●, ▲, ■ represent different Ia epitope on I-region molecules. ⌂ represents the specific antigen. Different determinants of the antigen are presented to the T cells by various restricting epitopes on the I-region molecule. (A) Dual receptor model; (B) alerted self model.

VII. SUBPOPULATIONS OF CELLS INVOLVED IN THE IMMUNE RESPONSE

Lymphocytes can be distinguished by their cell surface membrane antigens. The major division into T cells or B cells can be made based on the expression of the Thy-1 antigen by T cells and surface immunoglobulins by B cells; see Chap. 24. The T cells can be further divided into various subpopulations based on their expression of various Ly, Qa, or Ia antigens. There appears to be a correlation between the Ly phenotype expressed and T-cell function. The most thoroughly studied Ly antigens are the Lyt-1, Lyt-2, and Lyt-3. All thymocytes express Lyt-1, 2, and 3. About 50% of peripheral T cells express the Lyt-1, 2, and 3 antigens, while 5 to 10% are Lyt-1^-23^+. The peripheral Tcells expressing Ly-1, 2, and 3 are precursors of Lyt-1^+ and Lyt-2^+,3^+ cells and regulate the development of T-helper and T-suppressor cells. T-helper cells are Lyt-1^+23^-, while T-effector cells are Lyt-1^-23^+. T-suppressor cells are also Lyt-1^-23^+; however, unlike T-effector cells, they express Ia antigens. It has recently been shown that T-helper cells are composed of at least two subsets of Lyt-1^+ cells; one is Qa^- and a second is Qa^+ and functions during the induction of T-suppressor cells. It is generally accepted that T cells do not recognize antigen alone, but only in association with self-MHC products. The T-helper cells (Lyt-1^+) recognize antigens in association with I-region MHC molecules, while the T-effector cells (Lyt-2^+,3^+) recognize antigens in association with K and D molecules. Recently, this correlation has been questioned. It appears that T help can be found in T cells responding to K or D alloantigens and is mediated by Lyt-1^+,2^+. Thus it might be said that the Ly phenotype correlates better with the function of the T cells than to the MHC product that it recognizes. Further studies on this point are warranted before such a conclusion can be fully accepted.

VIII. ROLE OF MHC GENES IN PARASITIC DISEASE, AUTOIMMUNE DISEASES, AND TUMOR RESISTANCE

A role for the MHC has been implicated in several autoimmune types of diseases (see also Chaps. 26 and 27). These include thyroiditis, allergic encephalomyelitis, myasthenia gravis, and collagen-induced arthritis. In a normal individual, B cells are present which are reactive to self determinants, but they fail to respond due to the absence of T-helper cells for self which have been deleted during tolerance induction, or they are actively suppressed by T-suppressor cells. A disruption in T-suppressor-cell activity or in induction of T-helper-cell activity in these systems can lead to autoimmune phenomena. Any alteration in self antigen due to such agents as bacteria, viruses, drugs, or degradation can lead to the generation of T-helper-cell activity and thus autoimmune phenomena.

In an experimental autoimmune model of myasthenia gravis in mice, Lennon and David have demonstrated that the lymphocyte proliferative response to torpedo acetylcholine receptor, antibody responses to mouse muscle acetylcholine receptor, and susceptibility to experimental allergic myasthenia gravis are all controlled by genes mapping to the I-A subregion of the *H-2* complex (Christados et al., 1981). Disease can be induced by intradermal injection of acetylcholine receptor (AChR) emulsified in complete Freund's adjuvant into susceptible strains. It is characterized in susceptible strains by cellular and humoral immune responses to AChR, complexing of anti-AChR antibody in muscles, and impairment of neuromuscular transmission. H-2^b strains are high responders to AChR, whereas mice of the H-2^k haplotype are low responders. The cellular proliferative responses could be blocked by anti-Ia alloantisera. A mutation in the A_β chain of the B6

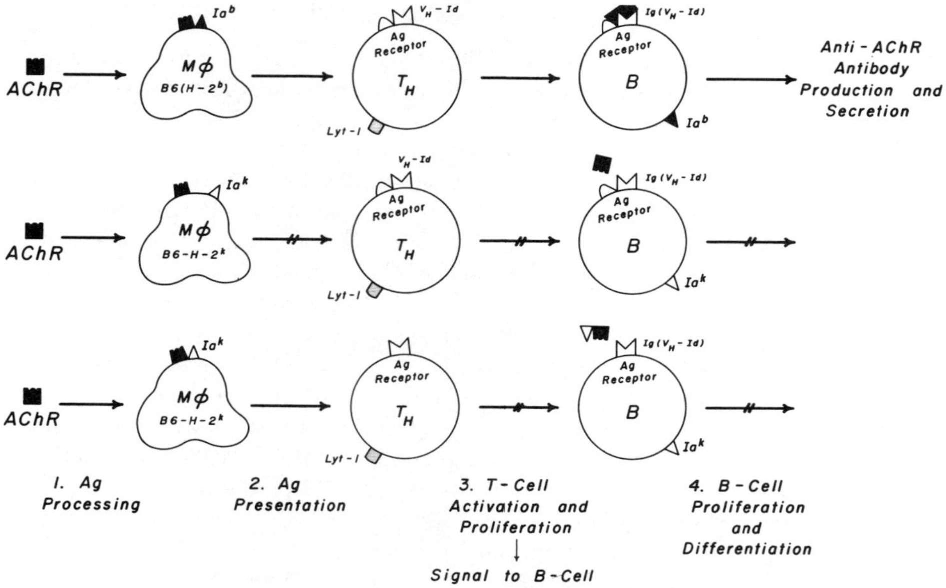

Figure 5 Role of Ia molecules in the induction of autoimmunity to AChR. The immunologic outcome after introduction of AChR in immunogenic form to the lymphoid system of MHC congenic mice of B6 ot B10 background is dependent on genes at the I-A subregion of the H-2 complex. AChR is taken up and processed by macrophages. In order to be recognized by T-helper cells, AChR must become associated with the I-A gene product expressed on macrophages. Ia molecules of B6 macrophage (Ia^b) provide optimal presentation of AChR for stimulation of T-helper cells (Lyt 1^+) to proliferate. AChR also binds to B cells, some of which recognize autoantigenic determinants of AChR. To trigger proliferation and differentiation on the B cells into an anti-AChR antibody-producing cell, a second signal is required from the stimulated T-helper cell. A state of antigen-specific unresponsiveness occurs when the I-A gene product is Ia^k (e.g., B6-H-2^k). Neither T-helper cells nor B cells are stimulated. This unresponsive state could result from either failure of AChR to associate with the macrophage's Ia^k molecule, which would prevent appropriate presentation of AChR to the T-helper cell, or failure of the T-helper cell to recognize AChR in association with Ia^k. (From Christadoss et al., 1981.)

(B6.C-H-2^{bm12}) strain can convert this high-responsive strain into a low responder. The cellular responses were shown to be dependent on both Lyt-1^+23^- cells as well as adherent cells. The data are indicative of a macrophage-associated Ia molecule which functions by presenting the acetylcholine receptor to T-helper cells, and thereby initiates B-cell lymphocyte differentiation into antiacetylcholine receptor antibody-forming cells (Fig. 5). In both high- and low-responder strains AChR is phagocytized and processed by macrophages. T-helper cells recognize the AChR in association with the Ia molecule expressed on the macrophage. If the macrophage is a high-responder haplotype, presentation occurs, resulting in the stimulation of T-helper cells (Lyt-1^+23^-). These T-helper cells trigger the proliferation and differentiation of antigen-reactive B cells into anti-AChR antibody-producing cells. Some of the stimulated B cells recognize autoantigenic

determinants of AChR. In nonresponder haplotypes neither T-helper cells nor B cells are stimulated. It is proposed that the AChR either failed to associate with Ia molecules of nonresponder macrophages, resulting in lack of presentation to T-helper cells, or the combination of AChR in association with nonresponder Ia molecules is not recognized by their own T-helper cells. The indication of Ia molecules in susceptibility to experimental autoimmune diseases is again seen in the autoimmune thyroiditis mouse model with the addition of D-region immune response gene affects.

Recent studies from David's laboratory have linked the induction of type II collagen-induced arthritis in mice to the I region of the MHC (Wooley et al., 1981). Upon intradermal injection into mice of native type II collagen of chick origin in complete Freund's adjuvant, arthritis is induced in susceptible strains. The arthritis is portrayed by an initial redness and swelling of one or more limbs followed by gross joint deformation and loss of joint mobility. Only mice of the H-2^q haplotype were susceptible to disease (B10.Q, B10.G, DBA/1J). Using recombinant strains of mice, the susceptibility was localized to the I^q locus. F_1 progeny from crosses of susceptible by resistant strains were susceptible to disease. Mouse strains could be classified as high, intermediate, or low responders by their anti-type II antibody level measured in a radioimmune assay. All susceptible strains were found to be high antibody producers. It was postulated that Ia antigen on the macrophage of the I^q strains can effectively present type II chick collagen to initiate an antiself response. This antiself response could be the result of a specific response against type II collagen determinants expressed within the joint, or alternatively the immune response of I^q haplotype mice to chick collagen may cross-react with autologous mouse collagen, whereas the response of non-I^q haplotype does not cross-react. The immune response to collagen could also be directed toward a common antigenic determinant in all mouse strains, with the response of the I^q haplotype being higher and more aggressive, resulting in damage to joint cartilage and thus disease.

The genetics of the host-parasite relationship has been examined for genetic control of the host susceptibility to infection. Perhaps one of the more thoroughly examined systems was the *Trichinella spiralis* studies performed by Wassom et al. (1979). Using independent haplotypes on the B10 background, they found that the s and q haplotypes were resistant to parasite infection, whereas the b and u haplotypes were intermediate in susceptibility. The k and f haplotypes, on the other hand, were found to be susceptible to *Trichinella* infection. Mapping studies with recombinant strains have placed the genes that influence the susceptibility to infection in the K end and possibly in the D end of the *H-2* complex. Non-*H-2* genes were also shown to affect susceptibility to a certain extent. In vitro T-cell proliferation assays using an extract of *T. spiralis* muscle larvae as antigen to prime in vivo and challenge in culture have recently been used to map the proliferative response in resistant strains to the I region.

A role for the major histocompatibility complex in virally induced leukemias has been discovered by Lilly (1981). He has demonstrated that there is an *H-2*-linked susceptibility gene for both exogenous and endogenous murine leukemia viruses. The susceptibility to gross murine leukemia virus was seen to be associated with H-2^k and H-2^d haplotypes. The immune response to virus-controlled antigens expressed on leukemia cells mapped to the K/I end for the gross murine leukemia virus, but to the D end for the Friend murine leukemia virus. The major factor in resistance to Friend and Gross leukemogenesis is tumor-specific cytotoxic lymphocytes. These cytotoxic lymphocytes are specific for a cell surface viral antigen that exhibits MHC restriction. For the Friend murine leukemia virus, the CTLs were shown to be restricted to the D end in the H-2^b

haplotype. This suggested that the D^b antigen could associate with Friend murine leukemia virus antigen molecule on the cell surface. This was shown to be the case when antibody to the D^b of the target cell was able to block the cytotoxic lymphocyte and capping of the cell surface Friend viral antigen cocapped the D^b molecule but not the K^b molecules.

IX. CURRENT THEORIES OF Ir GENE FUNCTION

Several theories concerning immune response gene function have been synthesized over the years since the discovery of the genetic control of the immune response. No single theory has been put forth that can accommodate all of the experimental observations made on *Ir* gene function. The theories must account for the exquisite specificity evident in the immune response, the lack of immune responses to certain antigens, the problem of dual recognition of antigen plus self-MHC molecules, the cell type in which the immune response genes are expressed, and the apparent small number of I-region genes available in the genome, yet the ability to respond to an almost limitless array of antigens. It was recognized early that the T cell does not see antigen alone, but rather in association with major histocompatibility products. This led to the proposal that the T-cell recognition molecule had dual specificity. This can occur by either a two-receptor model, one being for self, the other for antigen, or a one-receptor model where the receptor is specific for altered self. However, in competition experiments the observation was made that T-cell function could not be inhibited by the addition of either free antigen or free MHC molecules. This shed doubt on the validity of a two-receptor molecule for the T-cell recognition structure. Experimental evidence has been put forth which shows that at least part of the T-cell receptor is coded for by a gene linked to the *Igh* locus which has been termed V_T (specific for antigen) (Fig. 6). In a gene system similar to that of the V and C

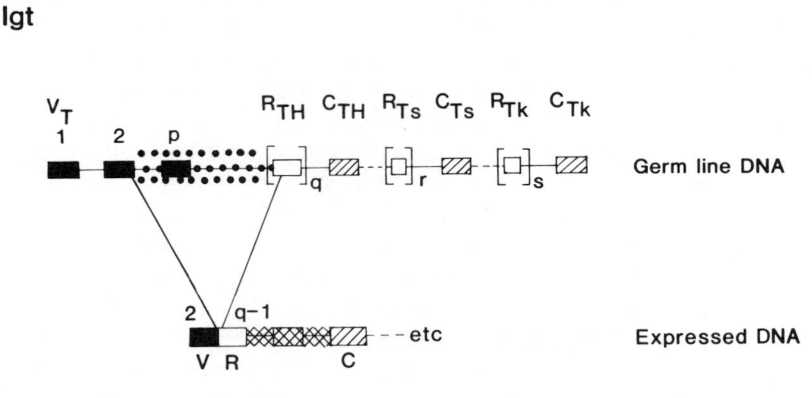

Figure 6 Proposed arrangement of gene segments at the *Igt* locus. Prior to gene expression, V- and R-segment rearrangement analogous to V-D-J joining at *Igh* is postulated. V_T, multiple family of genes specific for antigen; *R*, a set of MHC restricting genes functioning as joining elements (in a manner analogous to *J* genes) in expression of *VT* and *CT* genes. Each corresponds to a T-cell function: R_{Th} (helper), R_{Ts} (suppressor), and R_{Tk} (cytotoxic). *C*, a set of genes for constant portion of molecule, each corresponds to a T-cell function: C_{Th} (helper), C_{Ts} (suppressor), and C_{Tk} (cytotoxic). (From Williamson, 1982.)

regions of the *Ig* chain, a second component of the T receptor, termed R (specific for the MHC molecule), has been proposed. Rearrangement of the V_T and R segments would be analogous to V-D-J joining for immunoglobulin genes. Thus the T-cell receptor would be one recognition unit with specificity for both antigen and MHC elements. Recent studies with H-2-restricted T-cell hybridomas have been performed which verify the tight linkage of the V_T and R structures. Double hybrids were produced from T cells each specific for a distinct antigen in association with different H-2 molecules. In the hybrids formed, no mixing of specificity was found. That is, the restriction of antigen in association with a given MHC element cosegregated.

Blocking studies with alloantiserum and monoclonal antibodies have proven that Ia antigens are *Ir* gene products. Identical mapping of Ia antigens and *Ir* gene function to the same regions of the mouse genome as well as functional and biochemical studies on the Ia mutant mouse strain bm12 substantiate this conclusion. How, then, can just a few I-region molecules give rise to a system that can recognize such a numerous quantity of different antigens? Perhaps more I-region gene products are present but have not been detected due to their inability to cause alloantibody formation. Or, alternatively, they have been undetected due to being present in small quantities on the cell surface. Another explanation is that of free combinatorial association of existing Ia molecules. With the four Ia molecules (A_α, A_β, E_β, E_α) identified as being involved in *Ir* gene function, you could generate an antigen recognition system by free recombinatorial association, particularly in a heterozygous animal (four for I-A and two for I-E). If the A- and E-region molecules could also associate, 16 molecules would be possible. It has also been shown by blocking studies that different antigenic determinants are presented in association with different determinants on I-region molecules. Therefore, each I-region molecule could react with several different antigens, thus leading to considerable diversity. The site on an antigen being recognized by Ia molecules are apparently small in size, on the order of four to five amino acids based on the immunogenicity of insulin A and B chains. There are, however, no direct data on the binding of antigen by Ia molecules.

Reports of Lafuse et al. (1982) have indicated the existence of multiple Ia molecules in both the I-A and I-E regions. Using well-defined monoclonal anti-Ia antibodies they have evidence for the existence of at least two separate E_β genes and two separate E_α genes (Fig. 7). By sequential immunoprecipitation with 13-4 (anti-Ia.7 determinant on E_α) and 17-3-3 (anti-Ia.22 determinant on E_β) monoclonal antibodies, it was found that 17-3-3 only partially removed the I-E molecules immunoprecipitated by 13-4. In addition, 13-4 only partially removed the I-E molecules immunoprecipitated by 17-3-3. The results suggest a minimum of three I-E molecules, one expressing both 13-4 and 17-3-3 determinants, one expressing only 13-4 determinants, and a third expressing 17-3-3 determinants. The existence of a fourth I-E molecule expressing neither 13-4 nor 17-3-3 determinants is theoretically possible, but there is no evidence at present for its existence (Fig. 7). Isoelectric focusing gels suggest the existence of perhaps as many as five E_β chains precipitated by 13-4 and 17-3-3 monoclonal antibodies. If a comparable number of E_α chains existed and there was free association between all E_α and E_β chains, 25 separate I-E molecules could be formed. If it were also possible to create 25 separate I-A molecules due to multiple A_α and A_β chains, a total of 50 distinct molecules capable of interacting with antigen would exist. If each could react with several different antigens (determinant selection), one begins to generate the diversity necessary to explain the specificity seen in the immune response.

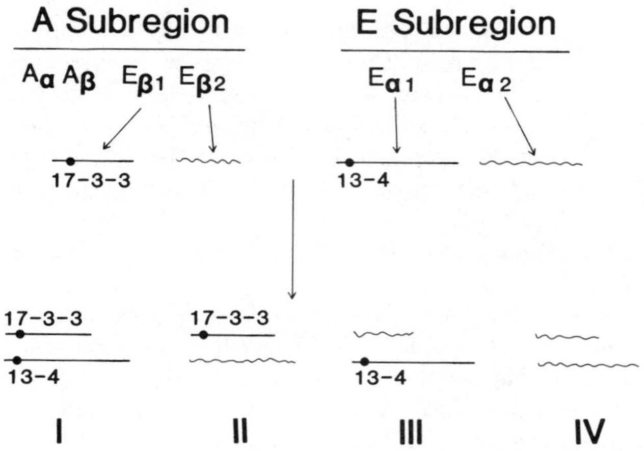

Figure 7 Model of multiple I-E molecules as defined by monoclonal antibodies 17-3-3 and 13-4. Strains used to describe these molecules are B10.A ($E^k_\beta E^k_\alpha$), B10.A(3R) ($E^b_\beta E^k_\alpha$), B.10A(5R) ($E^b_\beta E^k_\alpha$), B10.S(9R) ($E^s_\beta E^k_\alpha$), and B10.RIII ($E^r_\beta E^r_\alpha$). (From Lafuse et al., 1982.)

Based on most reported evidence, the *Ir* genes are expressed at the level of the macrophage and B cells, where they are involved in determinant selection (see Sec. III for the lack of *Ir* genes in macrophages). Macrophages present antigen in association with MHC products to T cells. When the T cell binds antigen, the antigen presenting cell is induced to give a differentiation signal such as lymphokines, which activate the T cell. The primary function of Ia molecules may be in the presentation of antigen at the macrophage level, and a secondary function may be the binding of recognition of the complex self plus antigen by T cells. Defects in the immune response genes can be accounted for in several ways. For some combination of antigen with MHC product, there may be no T-cell receptor present that can recognize it. Such a hole in the T-cell repertoire would arise during tolerance induction in the thymus or bone marrow. If a foreign antigen plus MHC mimic a self antigen plus MHC, the animal would have not T cells capable of recognizing it because such a cell would have been deleted during tolerance induction. T cells must learn to recognize self in their environment. Thus negative selection would eliminate any cells recognizing self antigen in association with self MHC. The possibility also exists that some MHC alleles cannot associate with a given antigenic determinant.

The genetic control of the immune response is a complex biological phenomenon. The crucial role of the MHC has been established beyond any doubt. The immune response genes control responses to bacterial infections, viral infections, and parasitic infections, as well as challenge by alloantigens and foreign proteins. A balanced interaction between macrophage, B cell, and T cell and their subsets leads to a successful immune response. The rapidly expanding technology regarding monoclonal antibody and antigen-reactive T-cell clones will prove to be invaluable in further dissection of the intricacies of the immune response.

REFERENCES

Benacerraf, B., and McDevitt, H. O. (1972). Science *175*, 273.
Bluestein, H. G., Green, I., and Benacerraf, B. (1971). J. Exp. Med. *134*, 1538.

Christadoss, P., Lennon, V. A., Krco, C. J., Lambert, E. H., and David, C. S. (1981). N.Y. Acad. Sci. *377*, 258.

Clark, R. B., Chiba, J., Zweig, S. E., and Shevach, E. M. (1982). Nature (Lond.) *295*, 412.

David, C. S., and Shreffler, D. C. (1972). Tissue Antigens *2*, 241.

Infante, A. J., Atassi, M. Z., and Fathman, C. G. (1981). J. Exp. Med. *154*, 1342.

Ishii, N. Nagy, Z. A., and Klein, J. (1982). Nature (Lond.) *295*, 531.

Kimoto, M., and Fathman, C. G. (1980). J. Exp. Med. *152*, 759.

Kimoto, M., Beck, B., Shigeta, M., and Fathman, C. G. (1982). In *Ia Antigens* (C. S. David and S. Ferrone, Eds.). CRC Press, Boca Raton, Fla.

Klein, J., Klein, D., and Shreffler, D. C. (1970). Transplantation *10*, 309.

Lafuse, W. P., Corser, P. S., and David, C. S. (1982). Immunogenetics *15*, 365.

Levine, B. B., Ojida, A., and Benacerraf, B. (1963). Nature (Lond.) *200*, 544.

Lilly, F. (1981). In *Immunobiology of the Major Histocompatibility Complex*, 7th Int. Convoc. Immunol. (M. B. Zaleski, C. J. Abeyounis, and K. Kano, Eds.). S. Karger, Basel, p. 347.

McDevitt, H. O., and Chinitz, A. (1969). Science *163*, 1207.

McDevitt, H. O., and Sela, A. (1965). J. Exp. Med. *122*, 517.

McDevitt, H. O., Deak, B. D., Shreffler, D. C., Klein, J., Stimpfling, J. H., and Snell, G. D. (1972). J. Exp. Med. *135*, 1259.

Meo, T., David, C. S., Rijnbeck, A. M., Nabholz, M., Miggiano, V. C., and Shreffler, D. C. (1975). Transplant. Proc. *7*(Suppl. 1), 127.

Schwartz, R. H., David, C. S., Sachs, D. H., and Paul, W. E. (1976). J. Immunol. *117*, 531.

Shevach, E. M., and Rosenthal, A. S. (1973). J. Exp. Med. *138*, 1213.

Shevach, E. M., Paul, W. E., and Green, I. (1972). J. Exp. Med. *136*, 1207.

Shreffler, D. C., and David, C. S. (1975). Adv. Immunol. *20*, 125.

Wassom, D. L., David, C. S., and Gleich, G. J. (1979). Immunogenetics *9*, 491.

Williamson, A. K. (1982). Immunol. Today *3*, 68.

Wooley, P. H., Luthra, H. S., Stuart, J. M., and David, C. S. (1981). J. Exp. Med. *154*, 688.

Zinkernagel, R. M., and Doherty, P. C. (1979). Adv. Immunol. *27*, 51.

26

The HLA System

Takeo Juji The Tokyo Women's Medical College, Tokyo, Japan

Kyoichi Kano The Institute of Medical Science, University of Tokyo, Tokyo, Japan

I. INTRODUCTION

Unlike the situation in the discovery of the murine MHC that was described in Chap. 25 (see also Chap. 27), the human major histocompatibility complex (MHC), the HLA system, was discovered in the 1950s during studies on leukocyte antibodies in patients with various blood dyscrasia. In 1952, Dausset and Nenna found leukocyte alloantibodies in sera of leukopenic patients. Subsequently, Dausset (Dausset, 1958), using sera of patients with multiple transfusions, discovered in 1958 the first HLA antigen, Mac (A2), by means of leukoagglutination. For this contribution, Dausset received the Nobel prize in 1980.

Shortly after Dausset's discovery, Payne (Payne, 1958) and van Rood (van Rood, 1963) independently found similar leukogglutinins in sera of multiparous women. Because of the complexity of the system and technical difficulties in detecting these alloantigens, progress in establishment of the HLA system was slowed down for the next several years. In 1964, Terasaki and his associates (Terasaki and McClelland, 1964) established an elegant test, a microdroplet-lymphocytotoxicity test (LCT), for detection of the alloantigens. Investigators working in the field immediately appreciated the remarkable reproducibility of the test and its practicality (i.e., use of a minimal amount of reagents and a small number of test lymphocytes). For these reasons, this test has been adopted as the standard procedure in the subsequent international workshops.

In the history of the HLA system, it is quite unique and should be emphasized that international collaborative efforts in the form of periodic workshops starting with the first one organized by Amos in 1964 have been devoted to the establishment of this highly polymorphic system. At present, more than 90 HLA alleles belonging to five separate but closely linked loci at the short arm of the sixth chromosomes were officially recognized after the 8th International Workshop, which was held in 1980.

For a brief discussion of the connection between Chaps. 9 and 24–28, see the appendix.

II. GENETIC ASPECTS OF THE HLA SYSTEM

A cluster of genes coding for major histocompatibility antigens is called the *major histocompatibility complex* (MHC). Every mammalian species thus far studied has been shown to possess one MHC. The formal evidence for the HLA system being the MHC of humans was provided by experimental skin grafting between ABO blood group-matched siblings. Mean survival time of these skin grafts showed a clear-cut bimodal distribution, 10 days and 22 days. Grafts showing longer survival time turned out to be those from HLA-identical siblings and the remaining grafts with shorter survival time from HLA-nonidentical siblings (Snell et al., 1977).

The HLA complex was mapped on the short arm of the sixth chromosome based on the result obtained from family studies and studies on human-mouse somatic cell hybridization (Fig. 1). It was shown that those hybrid clones retaining human sixth chromosome express HLA antigens, whereas those clones that lost the chromosome do not. Another approach has also confirmed the presence of HLA gene complex on the sixth chromosome: Lymphoblastoid cells with known HLA phenotype were exposed to gamma radiation, and variant cells that had lost HLA antigens were selected by incubation of the cells with corresponding cytotoxic HLA antibodies and complement. Karyotype analysis of these variant cells without expression of the HLA antigens revealed morphological abnormalities in the short arm of the sixth chromosome (Kavathas et al., 1980). Recent

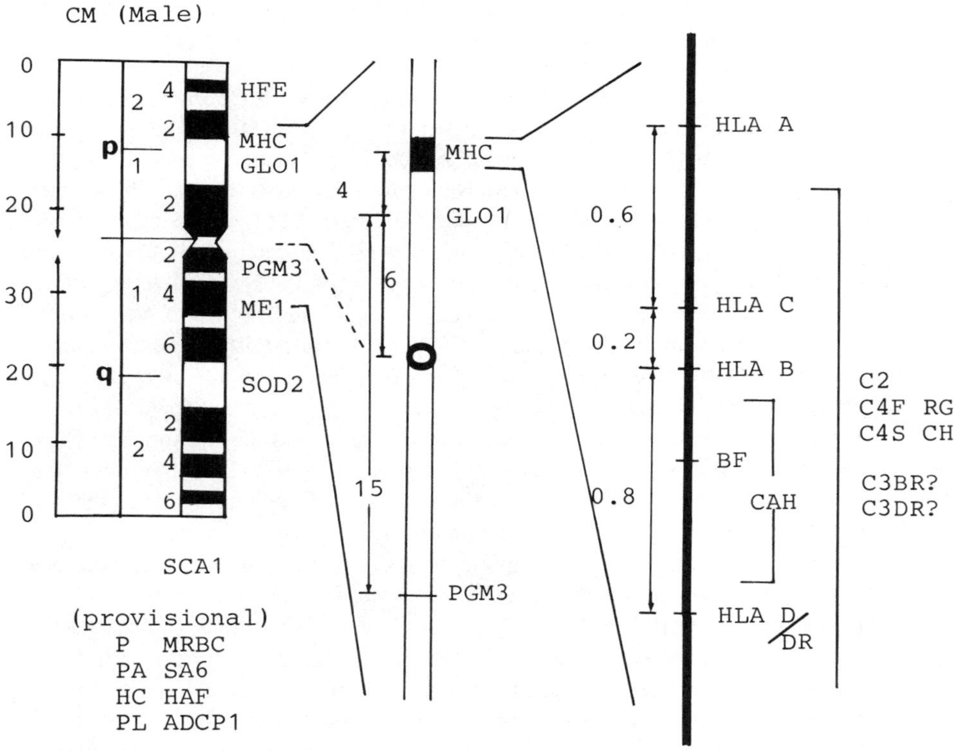

Figure 1 Gene loci on human chromosome 6.

Table 1 HLA Serotyping of a Family[a]

	Reactions with antisera to HLA							
	A1	A2	A3	A9	B5	B7	B8	B12
Father	–	+	–	–	+	+	–	–
Mother	+	–	+	–	–	+	+	–
Children								
I	+	+	–	–	–	+	+	–
II	–	+	+	–	+	+	–	–
III	–	+	+	–	+	+	–	–
IV	+	+	–	–	+	–	+	–
V	–	+	+	–	–	+	–	–

[a]Parental haplotypes: father, *A2-B5, A2-B7;* mother, *A1-B8, A3-B7.*

cytogenetic studies clearly demonstrated that the HLA complex forms a linkage group with phosphoglucomutase-3 (PGM-3), glyoxalase (GLO), urine pepsinogen-5 (pg-5), and allotypes of complement components, C2, C4, and factor B. As shown in Fig. 1, the order of HLA loci *A, C, B*, and *D* (*DR*) on the chromosome was determined on the basis of accumulated data on recombinants between these loci.

HLA antigens are inherited as simple mendelian condominant traits; there are no recessive genes or amorphogenes in the HLA system. Because of rather low recombinant frequencies (see Fig. 1), in a typical two-generation family, an HLA antigen controlled by a gene at one locus segregates together with other antigens controlled by genes at the remaining loci. The combination of alleles at the segment of a single HLA chromosome is called an *HLA haplotype*. HLA genotype of a single individual consists of two haplotypes, one from the father and the other from the mother. Therefore, in a two-generation family of four HLA haplotypes, two paternal and two maternal haplotypes exist and they can be deduced from phenotypes of the family members. Table 1 shows such an example; since the father and all children but the mother possess A2 antigen, the father is homozygous for A2 and, therefore, paternal haplotypes would be *A2-B5* and *A2-B7*. The mother's A1 and B8 antigens are inherited together by children I and IV and A3 and B7 antigens by children II, III, and V. Therefore, the maternal haplotypes are *A1-B8* and *A3-B7*.

It is, however, sometimes possible to identify the most likely haplotypes of an individual in a given population without family studies, since some haplotypes in a given population appear more frequently than others. The tendency of two alleles at closely linked loci appearing together more frequently than the expected value based on gene frequency of each of the alleles is called *linkage disequilibrium*. Such association of certain alleles between HLA-*A* and *B* and HLA-*B* and *DR* loci is characteristic for each ethnic group. For example, *A1-B8, A3-B7, B7-DR2*, and *B8-DR3* are very common haplotypes in Caucasians, whereas *Aw33-Bw44, Aw24-B7*, and *Bw52-DR2* are common haplotypes in Japanese. On the other hand, certain haplotypes of HLA-*B* and *C* loci such as *B15-Cw3* and *Bw35-Cw4* are frequently observed in all ethnic groups thus far studied.

These linkage disequilibria of the HLA system are supposed to have originated from the common ancestry of the human beings and preserved for a long period of time, presumably 50,000 to 100,000 years in each ethnic group (Bodmer, 1978).

III. DETECTION OF HLA ANTIGENS

A. HLA-*A*, *B*, and *C* Locus Antigens (Class I Antigens)

HLA antigens in this class are demonstrated serologically by monospecific alloantisera obtained primarily from multiparous women. These alloantibodies are raised by feto-material immunization. As mentioned, LCT established by Terasaki and his associates is now most widely used. This procedure is employed as the standard test for HLA-*A*, *B*, and *C* typing in most transplant centers all over the world. Briefly, 2000 lymphocytes separated from peripheral blood are incubated with 1 μl of antiserum and 5 μl of fresh rabbit serum as a source of complement. If the antibodies react with corresponding antigens on the lymphocytes, the lymphocytes are killed in the presence of the complement. After the killed lymphocytes are stained with eosin, the reactions with each of the HLA antisera are examined under a phase-contrast microscope and the percentage of killed cells is determined.

B. HLA-DR Antigens (Class II Antigens)

The modification of the LCT is necessary to detect HLA-*DR* antigens which are predominantly expressed on B lymphocytes. To prepare a B-lymphocyte-enriched suspension, a method with nylon fibers is now widely employed. When peripheral blood lymphocytes are passed through a straw packed with nylon fibers, T lymphocytes and "null" cells of the lymphocytes do not adhere to nylon fibers, whereas B lymphocytes do. After washing fibers, the straw is squeezed and a B-lymphocyte-enriched preparation consisting of more than 90% of B cells is obtained. To demonstrate *DR* antigens, naturally occurring nonspecific B-cell antibodylike factors should be carefully eliminated from both the antisera and the rabbit complement. Furthermore, each antiserum has to be preabsorbed with pooled platelets to remove contaminated HLA-*A*, *B*, and *C* antibodies. The fresh rabbit serum (complement) should also be carefully selected for the absence of cytotoxic factors against unsensitized B lymphocytes.

C. HLA-*D* Antigens (Class II Antigens)

T lymphocytes are transformed into lymphoblasts and proliferate, when lymphocytes of two persons with different *D* locus antigens are cultivated together for 5 to 7 days (mixed lymphocyte culture, MLC), whereas no such transformation takes place when the lymphocytes of each person are cultured separately or lymphocytes of two identical twins are cultured together under the same experimental conditions. If [^{3}H] thymidine is added to the culture at the end stage of MLC, it is incorporated into the DNA of transformed and proliferating T lymphocytes, and the amounts of thymidine incorporated correspond to the degree of responses to allogeneic *D* locus antigens.

For typing of a person for HLA *D* locus antigens by means of MLC, homozygous typing cells (HTC) must be employed. These HTC are the lymphocytes originating from some persons of cousin marriages. HTC are pretreated with mitomycin C, or irradiation. Such treated cells are not able to respond in MLC, but are able to stimulate the untreated lymphocytes from donors with different type of HLA-*D* locus antigens. If an HTC, for example Dw1/Dw1 cells, failed to stimulate a person's lymphocytes of unknown HLA *D* type, this person must possess at least one Dw1 allele. In this way, by looking for negative MLC with a series of HTC, the *D* type of the person can be determined.

IV. SPECIFICITIES OF HLA SYSTEM

The complexity of the HLA system was fully realized soon after its discovery by demonstration of a series of antigens which were apparently coded for by multiple alleles at a single locus. As mentioned previously, HLA complex consists of at least four separate but closely linked loci with multiple alleles. This makes the HLA system the most polymorphic system known in humans. Table 2 lists presently recognized HLA specificities. The "w" stands for International Workshops nomenclature and specificities with or without "w" are those accepted officially by the World Health Organization (WHO) HLA nomenclature committee.

A. HLA-*A*, *B*, and *C* Locus Specificities

The series of specificities coded for by alleles at the locus *A* were formerly called the first sublocus, or LA series antigens. At present, 20 alleles are officially accepted by the WHO HLA nomenclature committee. Aw23 and Aw24 are split specificities of A9. In other words, A9 specificity includes two subspecificities, Aw23 and Aw24. A25 and A26 are split antigens of A10, and Aw19 includes A29, Aw30, Aw31, Aw32, and Aw33 antigens.

The series of antigens coded for alleles at the locus *B* are more polymorphic than that of *A* locus, and 40 alleles are officially designated. Formerly, they were called the *second sublocus* or *Four series antigens*. The following antigens are serologically subdivided into two split antigens: B5(bw51 and Bw52), B12(Bw44 and Bw45), B15(Bw62 and Bw63), Bw16(Bw38 and Bw39), B17(Bw57 and Bw58), B21(Bw49 and Bw50), Bw22(Bw55 and Bw56), and B40(bw60 and Bw61).

The immunogenicity of the series of antigens coded for by alleles at the locus *C* are rather weak, and only eight alleles of this locus have been identified thus far. In platlet transfusion, incompatibilities at this locus are sometimes disregarded because of their relatively low immunogenicity.

B. HLA-*D* Locus Specificities

This series of specificities were officially accepted after the 6th International Histocompatibility Workshop in 1975. At present, a total of 12 alleles (Dw1 to Dw12) are officially defined (Table 2). As described previously, these antigens are determined by means of HTC typing. Although no significant response in MLC between lymphocytes of HLA-*D* identical siblings has been observed, there have been claims of difficulties in HTC typing. This is presumably due to the presence of antigens other than HLA-D locus antigens which are involved in MLC. In fact, two HTCs defining the same *D* locus antigens often stimulate each other in MLC.

C. HLA-*DR* Locus Specificities

This series of antigens, which are serologically detectable on B but not on T lymphocytes, were officially designated as *DR* antigens. *DR* stands for "*D* related," since most of the *DR* antigens show strong associations with the corresponding HLA-*D* specificities and no recombinant was discovered between *D* and *DR* loci. The seven specificities of HLA-*DR* alleles (*DR*1 to *DR*7) were designated by the same antigen numbers of the corresponding HLA-*D* alleles at the 7th International Histocompatibility Workshop in 1977. However, the three newly accepted alleles (*DR*w8, 9, and 10) at the 8th Workshop were designated

Table 2 WHO-Recognized HLA Specificities (1980)

HLA- A locus	HLA-B locus		HLA-C locus	HLA-D locus	HLA-DR locus
A1	B5	Bw46	Cw1	Dw1	DR1
A2	B7	Bw47	Cw2	Dw2	DR2
A3	B8	Bw48	Cw3	Dw3	DR3
A9	B12	Bw49 (Bw21)	Cw4	Dw4	DR4
A10	B13	Bw50 (Bw21)	Cw5	Dw5	DR5
A11	B14	Bw51	Cw6	Dw6	DRw6
Aw19	B15	Bw52	Cw7	Dw7	DR7
Aw23 (A9)	Bw16	Bw53	Cw8	Dw8	DRw8
Aw24 (A9)	B17	Bw54		Dw9	DRw9
A25 (A10)	B18	Bw55 (Bw22)		Dw10	DRw10
A26 (A10)	Bw21	Bw56 (Bw22)		Dw11	
A28	Bw22	Bw57 (B17)		Dw12	
A29 (Aw19)	B27	Bw58 (B17)			
Aw30 (Aw19)	Bw35	Bw59			
Aw31 (Aw19)	B37	Bw60 (B40)			
Aw32 (Aw19)	Bw38 (Bw16)	Bw61 (B40)			
Aw33 (Aw19)	Bw39 (Bw16)	Bw62(B15)			
Aw34	B40	Bw63(B15)			
Aw36	Bw41				
Aw43	Bw42	Bw4			
	Bw44 (B12)	Bw6			
	Bw45 (B12)				

independently from the antigen numbers of HLA-*D* alleles. The phenotypic frequency of each HLA-*DR* antigen is diverse in various ethnic groups, similar to the situation of HLA antigens of other loci. For example, the frequencies of HLA-*DR*3, *DR*7 and *DR*w10 are extremely low in Japanese and *DR*w9 is quite rare in Caucasians.

It should be mentioned that these 10 *DR* alleles represent approximately 75 to 80% of the gene pool, and the data of the 8th Workshop did not seem to fit the Hardy-Weinberg equilibrium. Presumably, this was due to the presence of an excess number of DR null cells (DR-/-) and difficulties to define some of the specificities, such as *DR*w6. These facts clearly indicate that the present status of *DR* series is still far from complete and is hampered by technical difficulties, including poor reproducibility.

Supertypic cross-reacting specificities of HLA-*DR* antigens MT-1, MT-2, and MT-3 were recently recognized. MT-1 includes specificities of *DR*1, *DR*2, *DR*w10, and parts of *DR*w6 and *DR*w8. MT-2 includes specificities of *DR*3, *DR*4, *DR*5, *DR*w6, and *DR*w8, and MT-3 *DR*4, *DR*7, and *DR*w9. An hypothesis was recently advanced that MT antigens are controlled by genes at two separate loci, *MT "A"* for MT-1 and *MT "B"* for MT-2 and MT-3. However, immunochemical analyses suggest that *DR* determinants may reside on different molecules from those carrying MT antigens. There are, however, no recombinants between HLA-*DR* and MT genes.

The secondary B-cell (SB) antigens are recently defined by means of the primed lymphocyte typing (PLT). They are apparently a new series of antigens controlled by genes on the HLA region. The PLT test is based on the observation that a very rapid proliferation of "primed" lymphocytes occurs if the primed cells are restimulated with cells carrying the same antigens as the priming cells used in the initial MLC. Studies of recombinant families suggest that the SB antigens are encoded by genes that map between HLA-*DR* and GLO.

More recently, a new antigenic system was described which might represent the human counterpart of the T1a system in the mouse. Selected sera of multiparous women reacted with a panel of PHA-stimulated T lymphocytes and the reaction patterns of these sera appeared to be independent from any of those given by antisera to HLA-*A, B, C*, or *DR* antigens.

Evidence for the presence of the second locus for B-lymphocyte alloantigens independent from the *DR* series has also been accumulating, although it has not yet been accepted officially. Antibodies to these antigens are not reactive with monocytes in the cytotoxicity test, whereas antisera to *DR* antigens and MT "B" antigens usually give positive reactions with monocytes.

V. GENE PRODUCTS OF THE HLA COMPLEX

Gene products of the HLA system are expressed as cell surface antigens of nucleated cells, except allotypic expression of complement components. According to the membrane model by Singer and Nicholson, HLA antigenic molecules are believed to "float" or to "swim" in the double layer of fatty acids and phospholipids. The hydrophobic portion of solubilized HLA antigenic molecules is supposed to be immersed in the lipid double layer of the cell membrane.

It is generally accepted that HLA antigenic molecules controlled by genes at the HLA-*A, B,* and *C* loci (class I) are expressed on practically all nucleated cells, analogous to H-2K and H-2D molecules. However, the concentrations of these HLA antigens are different in some cells and tissues than in others. Although murine (class I) H-2K or H-2D

molecules are expressed on the erythrocytes at a detectable level and are traditionally used for determination of H-2 types of mice, the amount of HLA antigen on human erythrocytes is extremely small, if any, and therefore it is impossible to use erythrocytes for HLA typing. Gene products of I-A and I-E subregions (class II or Ia antigens) of the mouse and human analog, HLA-*DR* antigens, show a similar restriction in their tissue and cell distribution. HLA-*DR* antigens are found on B lymphocytes; monocytes or peripheral blood; macrophages in tissues, endothelial cells, epidermal cells, but not on platelets; unstimulated T lymphocytes; granulocytes; and erythrocytes.

A. Solubilization of HLA Antigens

To study the chemical structure of the HLA antigens, it is necessary to isolate antigenic molecules from the membrane in sufficient quantity. Although many procedures have been employed to extract HLA antigens from the membrane, two methods are commonly used to solubilize HLA antigens: detergent treatment and digestion with papain. Cultured lymphoid cell lines are most conveniently utilized as a source of the antigens, since the yield of HLA antigens from the cell lines is several times more than that obtained from peripheral blood lymphocytes. In the first method, membrane components are treated with nonionic detergents. A group of nonionic detergents (Lubrol PX, Triton X-67, Tween-40, and NP-40) have been evaluated for their efficiency to extract each molecular component from different cell types. Recently, the NP-40 detergent is widely used in conjunction with electrofocusing to isolate the active fragment of HLA molecules.

In the second method, the pellet and supernatant of homogenized lymphoid cells are treated with papain. It has been demonstrated that papain digests a certain part of the polypeptide of HLA heavy chains of class I molecules and that the larger pieces of cleaved heavy chains are solubilized.

More recently, plant lectins (see Chap. 15) were used for the isolation of alloantigens. Lens culinaris phytohemagglutinin (LCH) has specific binding activity to glycoproteins because of its antibodylike activity against sugar residues on glycoproteins. Glycoproteins, including HLA antigens, are retained in the LCH-coupled Sepharose 4B column, and are eluted with α-methyl mannoside. Most of the HLA activity added can be recovered in the eluted fraction.

For further purification, the solubilized HLA antigenic molecules are then reacted with corresponding alloantisera or monoclonal antibodies, and the antigen-antibody complexes are precipitated by a heteroimmune serum to human immunoglobulins, or protein A. After washing the precipitate, HLA molecules are released from the precipitate by boiling in sodium dodecyl sulfate (SDS), and the molecules are identified by SDS-polyacrylamide gel electrophoresis (SDS-PAGE). As illustrated in Fig. 2, in detergent-solubilized preparations, an intermolecular S-S bridge is formed between half-cysteines of two heavy chains, and a dimer of two heavy chains with two light chains is obtained. Papain digestion of a heavy chain with a molecular weight of 45,000 yields two pieces of 34,000 and 12,000 molecular weight. The smaller piece, of 12,000 molecular weight, contains a hydrophobic part of polypeptide, and this hydrophobic part keeps a heavy chain stabilized in its structure immersed in the double lipid layer of the cell membrane and, therefore, this fragment is designated as FM (a membrane fragment). The larger piece, of 34,000 molecular weight, is called FH (a heavy fragment) and binds with the light chain noncovalently. The FH molecule with a single chain is called FS (a soluble fragment). In fact, direct digestion of the membrane by papain yields FS molecules (see Fig. 2).

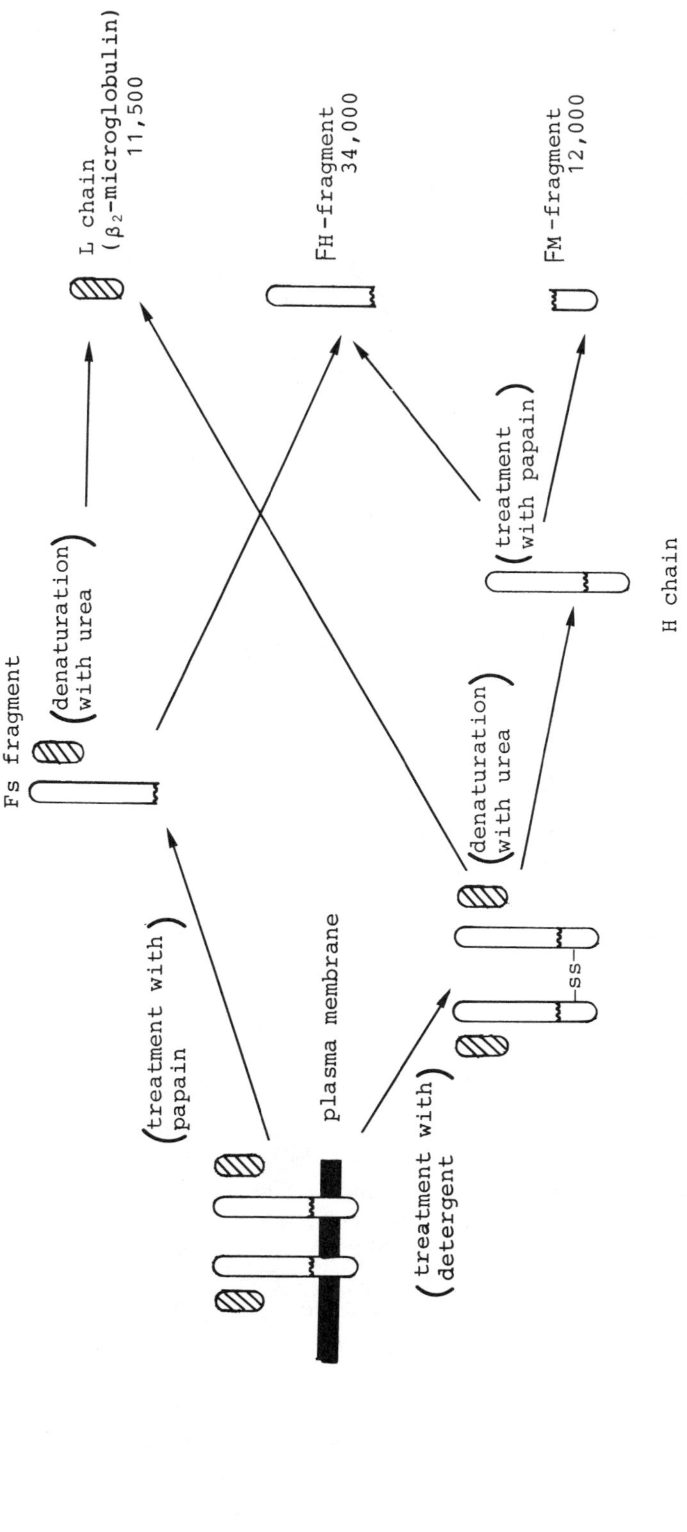

Figure 2 Cleaved fragments of HLA-A, B, C molecules.

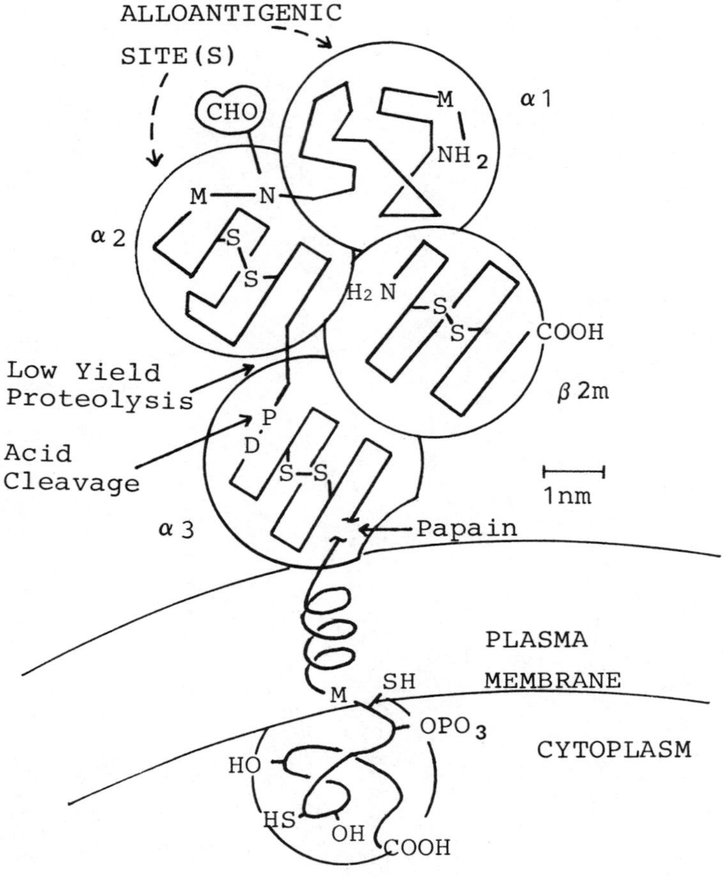

A MODEL OF HLA-B7

Figure 3 Schematic representation of the structure of an HLA-A or -B antigen. [From J. L. Strominger (1981), *Immunology 80: Progress in Immunology IV*, Vol. 2 (M. Fougereau and J. Dausset, Eds.), Academic Press, New York, p. 542.]

B. Biochemical Structure of HLA Antigens

As shown in Fig. 3, it is now generally accepted that the detergent-solubilized HLA-*A, B,* and *C* (class I) molecules are composed of noncovalently linked polypeptide chains with molecular weights of 45,000 (heavy chain) and 11,500 (β_2 microglobulin, β_2M), even though most HLA antigenic molecules isolated with detergent NP40 are eluated in fractions with a molecular weight of about 140,000 by Sephadex gel filtration. After being mildly denatured, molecules in these fractions were split into fractions with smaller molecular weights of about 90,000, 45,000, and 11,500.

The light chain with a molecular weight of 11,500 contains two half-cysteines that form an intramolecular loop. This molecule dissociated from HLA heavy chain was originally identified by heteroimmune sera. It was revealed later that this molecule was serologically identical with a protein that was also present in serum: β_2M which does not

carry any determinants for allogenic specificities. However, it plays quite an important role in the occurrence of HLA antigenic molecules on the cell membrane. Cells cannot express HLA antigenic molecules if they have no β_2M molecules on their cell membrane. For instance, Daudi cells (a lymphoblastoid cell line) do not express both H and L chains of (class I) HLA molecules. However, hybrid cells produced by fusion of Daudi with any other cells carrying normal β_2M exhibit all the HLA antigens that are identified in the donor of the Daudi cells. A β_2M molecule can reversibly associate and dissociate with heavy chains of HLA molecules on the cell surface. This holds true even between chains of different species. If murine cells are incubated with human β_2M, the human β_2M associates physically with heavy chains of murine H-2 molecules on the cell membrane. Interestingly, analysis of the amino acid sequence of β_2M revealed an unexpected high homology to certain C domains of immunoglobulins.

Murine B lymphocytes express two main classes of Ia (class II) molecules, I-A and I-E (or I-A/E). Each of the Ia molecules consists of an α chain and β chain, with molecular weights of 34,000 and 29,000, respectively, and is not associated with β_2M. Both chains are transmembrane glycoproteins. The HLA-*DR* (class II) molecules are glycotproteins in

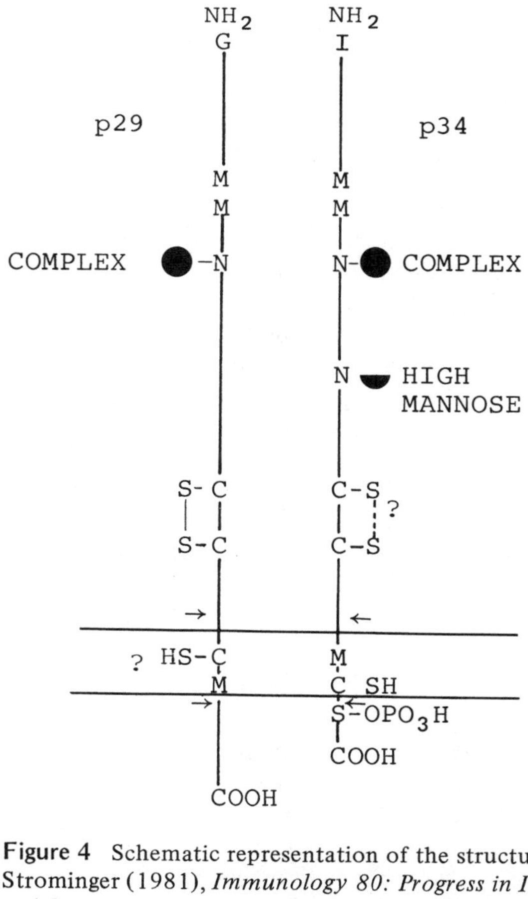

Figure 4 Schematic representation of the structure of an HLA-DR antigen. [From J. L. Strominger (1981), *Immunology 80: Progress in Immunology IV*, Vol. 2 (M. Fougereau and J. Dausset, Eds.), Academic Press, New York, p. 550.]

nature and are composed of an α chain and a β chain with molecular weights of about 34,000 and 29,000, respectively. Both α and β chains have the transmembrane structure in the lipid double layer of cell membranes, as illustrated by Fig. 4.

C. Amino Acid Sequence of HLA Antigenic Molecule

Recently, the complete primary structure of HLA-B7 heavy chain and 97% of the amino sequence of HLA-A2 have been determined by Strominger et al., as shown in Fig. 5. These data have been compared with data accumulated for the murine H-2K and H-2D molecules by Nathenson et al. The heavy chain consists of five regions. The first three regions are outside the cell membrane. The fourth region is the transmembranous segment of 25 amino acid residues. The fifth region, of 31 residues, is phosphorylated and resides in the cytoplasm. The cysteines at positions 101 and 164 in the second region form the first disulfide loop and the cysteines at positions 203 and 259 form the second loop. In the first region, the carbohydrate was attached to asparagine at position 86. The sequences of the third region are very highly conserved and have been shown to be homologous to sequences of constant domains of immunoglobulins and β_2M. These data suggest that both immunoglobulins and HLA antigens might have evolved from the same ancient, probably membrane-bound defense molecules. In contrast, the highly variable areas are between positions 65 and 80 in the first region and between positions 105 and 115 in the second region. These areas are considered to be exposed to the outside of the molecule and represent the alloantigenic determinants.

D. Density of HLA Antigens in the Cell Cycle

The progression of cells through the cell cycle is associated with an orderly series of events, some of which control the expression of cell surface molecules. The cell cycle consists of pre- and postsynthetic gaps (G1 and G2), the DNA-synthetic period (S), the mitotic period (M), and the stationary phase (G_0). The relative amount of cell surface HLA molecules per cell and the relative mean cell volume were determined by the FACS instrument with fluorescein-labeled monoclonal antibodies to HLA antigens and the Coulter counter. The amounts of HLA-*A* and *B* locus antigens increase in accordance with cell volume expansion in the early part of the cell cycle, but there is no further increase in the amounts between the late and the premitotic stages, although the volume expands again rapidly. On the other hand, the amounts of HLA-*DR* antigens increase as the cell volume increases. As shown in Fig. 6, the density of HLA-*DR* antigens increases rapidly at the premitotic stage, whereas fluctuations of the density are small at the premitotic stage, whereas fluctuations of the density are small from G_0 to the end of the S phase. On the other hand, the density of HLA-*A*- and *B*-locus antigens does not change significantly throughout the entire cell cycle.

E. Function of Human MHC Gene Products

Recent studies on murine MHC gene products have established that MHC gene products play fundamental roles in both humoral and cellular immune responses. Murine Ia (class II) molecules control the magnitude and character of humoral immune responses to certain synthetic polypeptides, some foreign proteins, and minor histocompatibility antigens; and they restrict the cell collaboration in antibody formation against certain T-dependent antigens. Induction of cytotoxic T cells, as well as their action in the

```
                 1         2         3         4         5         6         7         8
HLA-B7  GSHSMRYFYTSVSRPGRGEPRFISVGYVDDTQFVRFDSDAASPREEPRAPWIEQEGPEYWDRNTQIYKAQAQTDRESLRN

HLA-A2  --------F--------------A------------------Z-M--------------------KV--H-H-V-VD-GT

                           1         1         1         1         1         1         1
                 9         0         1         2         3         4         5         6
HLA-B7  LRGYYNQSEAGSHTLQSMYGCDVGPDGRLLRGHDQYAYDGKDYIALNEDLRSWTAADTAAQITQRKWEAAREAEQRRAYL

HLA-A2  ----------------R-------S-W-F---YH------------K----------M-----KH------V---L----

                 1         1         1         2         2         2         2         2
                 7         8         9         0         1         2         3         4
HLA-B7  EGECVEWLRRYLENGKDKLERADPPKTHVTHHPISDHEATLRCWALGFYPAEITLTWQRDGEDQTQDTELVETRPAGDRT

HLA-A2  --T----------QNRET-----T----H---AV----------------------------------------------

                 2         2         2
                 5         6         7
HLA-B7  FEKWAAVVVPSGEEQRYTCHVQHEGLPKPLT

HLA-A2  ------------Q------------------
```

Figure 5 Complete amino acid sequence of papain-solubilized HLA-B7 and its comparison with 97% of the sequence of HLA-A_2. [From J. L. Strominger (1981), *Immunology 80: Progress in Immunology IV*, Vol. 2 (M. Fougereau and J. Dausset, Eds.), Academic Press, New York, p. 544.]

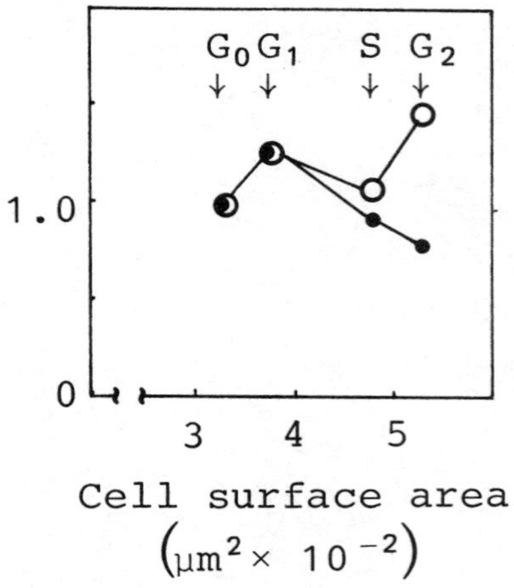

Figure 6 Fluctuation of antigen density in the WI-L2 cell cycle. The antigen density values were obtained by dividing the mean fluorescence (representing the average antibody binding to surface antigens) by cell surface area at respective cell cycle phases. These values were normalized relative to the values obtained at G_0 (set equal to 1) and are plotted against the mean cell surface area of the respective phase. ●, HLA-A,B antigens; ○, HLA-DR antigens. [From S. Sarkar, M. C. Glassy, S. Ferrone, and O. V. Jones (1980), Proc. Natl. Acad. Sci. USA 77(12), 7297.]

effector phase, are restricted by *K* or *D* (class I) molecules. These MHC-related phenomena are explained by dual recognition by lymphocytes of foreign antigens in the context of self-MHC molecules.

A similar type of dual recognition has been observed in humans. For example, lymphocytes from a HLA-A2-positive female patient transplanted with bone marrow cells of a HLA-identical brother showed cytotoxic activities against all the lymphocytes obtained from the HLA-A2-positive male but not those of the female. This observation was interpreted as follows. The patient's cytotoxic lymphocytes were generated against HY antigen of the HLA-identical donor upon transplantation and male target cells were killed when they shared the HLA-A2 antigen with the cytotoxic lymphocytes.

A possible role for MHC gene products in immune surveillance against malignancies has also been speculated. Evidence has already been presented for the presence of alien major and minor histocompatibility antigens in malignant cells of experimental animals (Kano, 1981).

VI. CLINICAL SIGNIFICANCE OF THE HLA SYSTEM

A. Transplantation

As mentioned previously, formal evidence that the HLA antigens are the major histocompatibility (transplantation) antigens has been obtained by experimental skin grafting between ABO blood group-matched siblings. The relevance of HLA-*A* and *B* matching of donor and recipient in clinical transplantation has been well documented in renal transplantation from related donors; grafts from ABO-compatible, HLA-identical siblings have a 3-year survival rate of over 90%. However, the effects of *A* and *B* matching on the

survival of renal grafts from unrelated donors are not clear. Recent statistical analyses have demonstrated beyond any doubt that the number of pretransplant blood transfusions influences significantly the survival of renal grafts from unrelated donors (Snell et al., 1976).

In bone marrow transplantation, HLA matching influences not only graft survival but also the occurrence and severity of graft versus host reaction. Therefore, bone marrow transplantation has thus far been restricted to HLA-idential siblings. Some information on HLA matching in cardiac and liver transplantation is available, but in neither instance has any apparent influence of the matching emerged.

B. Transfusion

It has been reported that HLA antibodies of the recipient sometimes elicit unpleasant reactions, such as fever, in blood transfusions with HLA-incompatible platelets and white blood cells. The cause of the reactions seems to be due to the presence of HLA antibodies in the recipient's circulation. The most frustrating problem encountered following multiple platelet transfusions is the development of alloimmunization and refractoriness of the immune response to random donor platelets. This refractoriness results primarily from alloimmunization against HLA antigens, especially HLA-*A* and *B* antigens, since alloimmunized recipients frequently respond successfully to platelet transfusions from HLA-matched donors. However, it is not easy in many situations to find HLA-compatible platelet donors, because in recent years the number of siblings in a family has been getting smaller and HLA antigens are extremely polymorphic. Regional blood centers are now establishing registration of prospective platelet donors who have been typed for HLA-*A* and *B* antigens. To cover the demands of most patients in a region, at least several thousand donors should be available (Duquesnoy et al., 1977).

C. Disease Associations

Many interesting data have been accumulated regarding the associations of various diseases with certain HLA antigens. Most of the available reports have been collected in the International Workshops to form the HLA and Disease Registry (Sveigaard et al., 1980). The most striking association is that between ankylosing spondylitis and HLA-B27 in every ethnic group thus far examined. Especially in Japanese, B27 is extremely rare in healthy controls (less than 0.3%); the relative risk of this disease is only 1 in over several hundred among B-27-positive Japanese. On the other hand, in cases of diseases associated with HLA-B8 in Caucasians, no such association has been confirmed in the Japanese population under exactly the same criteria of diagnosis of the diseases (Snell et al., 1976). Some autoimmune diseases, however, have been shown to be associated with the same alleles of HLA-*DR* locus among Caucasians and Japanese. These data suggest that the linkage disequilibria between the genes influencing susceptibility to these autoimmune diseases and the alleles of HLA-*DR* may have been preserved for a long period in each ethnic group from common ancestry (Bodmer, 1978).

Studies on HLA and disease association have contributed to mapping of some disease-related genes within or close to the HLA and also to our understanding of the heterogeneity of some diseases, each of which otherwise has been considered as a single entity. Congenital adrenal hyperplasia due to 21-hydroxylase deficiency is an example for mapping of the disease-related gene within the HLA. This disease is probably due to the lack of this enzyme. The gene responsible for susceptibility to the disease has been

mapped between HLA *B* and *DR* loci, closer to the *B* locus, and therefore the gene coding for the enzyme probably belongs to the HLA complex.

The first example of separation of a disease into two subgroups through HLA association studies would be psoriasis; psoriasis vulgaris but not pustular psoriasis is associated with HLA, which has confirmed the clinical distinction between these two types of psoriasis.

REFERENCES

Bodmer, W. F. (1978). Br. Med. Bull. *45*(3), 213.

Dausset, J. (1958). Acta Haematol. (Basel) *20*, 156.

Duquesnoy, R. J., Fillip, D. J., Rodey, G. E., and Aster, R. H. (1977). Am. J. Hematol. *2*, 219.

Kano, K. (1981). In *Immunobiology of the Major Histocompatibility Complex*, 7th Int. Convoc. Immunol. (M. B. Zaleski, C. J. Abeyounis, and K. Kano, Eds.). S. Karger, Basel, p. 353-361.

Kavathas, P., Bach, F. H., and De Mars, R. (1980). Proc. Natl. Acad. Sci. USA *77*, 4251.

Payne, R., and Rolfs, M. R. (1958). J. Clin. Invest. *37*, 1756.

Snell, G. D., Dausset, J., and Nathenson, S. (1976). In *Histocompatibility*. Academic Press, New York, p. 181-247.

Svejgaard, A., Morling, N., Platz, P., Ryder, L. P., and Thomsen, M. (1981). In *Immunology 80: Progress in Immunology IV*, Vol. 2 (M. Fougereau and J. Dausset, Eds.). Academic Press, New York, p. 530.

Terasaki, P. I., and McClelland, J. D. (1964). Nature (Lond.) *204*, 998.

van Rood, J. J., and van Leeuwen, A. (1963). J. Clin. Invest. *42*, 1382.

27

Immunochemistry of Murine Major Histocompatibility Antigens

Marek B. Zaleski School of Medicine, State University of New York at Buffalo, Buffalo, New York

I. INTRODUCTION

Over 40 years have passed since the discovery by Gorer of the first component of the major histocompatibility complex (MHC) of mice (see also Chap. 25). During the subsequent years a constant evolution of the concepts and ideas has occurred and is still taking place today. It is precisely because of this constant evolution that any attempt to present the facts and speculations concerning the MHC carries an inherent danger of either oversimplifying or confusing the issues. Nevertheless, one cannot avoid such an attempt since there is little, if any, doubt that the MHC has a profound effect on a wide variety of immunologic phenomena. With this in mind, it will be attempted in this chapter to present a basic knowledge concerning the structure and function of the murine MHC (i.e., the *H-2* complex). The choice of the murine MHC is dictated by the fact that it has been more thoroughly studied than the MHC of any other species, and today represents the reference point for studies of the MHC in other animals.

II. GENETICS OF THE *H-2* COMPLEX

A. Structure of the *H-2* Complex

The ever-changing concepts concerning the genetic structure of the *H-2* complex best exemplify the constant evolution mentioned above. The original single *H-2* locus was replaced by two closely linked loci, *H-2K* and *H-2D*, only to be replaced again by the series of regions and subregions comprising the *H-2* complex or haplotype. Indeed, not too long ago the *H-2* complex was believed, at least by some, to be a cluster of several tens, if not several hundreds, of loci. The steadily increasing number of loci assigned to the *H-2* complex was a reflection of the idea that any phenotypic trait or phenomenon

For a brief discussion of the connection between Chaps. 9 and 24–28, see the appendix.

associated with the H-2 phenotype must be determined by a discrete gene within the *H-2* complex even if the molecular product of such a gene cannot be demonstrated. Thus, as the number of the H-2-associated phenomena reported increased, so did the number of genes. Recently, however, a new trend has been gaining impetus. According to this trend, the *H-2* complex consists of a relatively limited number of loci and their respective genes. Each gene determines a demonstrable molecular product that may be responsible for a wide variety of the phenomena underlying the multiple traits associated with the H-2 phenotypes. It is more than likely that, in reality, the truth is somewhere in between the two concepts, which are represented schematically in Table 1. Whichever concept ultimately prevails, certain facts seem to be well established and these facts constitute the framework of this chapter.

The *H-2* complex consists of a cluster of closely linked genes located on the seventeenth chromosome (IX linkage group). The size of the complex is believed to be 1.5 cM, but this may be an underestimate since genes in its vicinity are known to suppress crossing over within the *H-2* complex. It was accepted that the *H-2K* and *H-2D* (or *H-2L*) loci constitute centromeric and telomeric limits of the complex, respectively (for references, see Götze, 1977; Klein, 1975; Snell et al., 1976). However, these limits have been established arbitrarily and should not be considered as fixed. Indeed, there are some indications that genes beyond the telomeric limit may be structurally and functionally closely related to those encompassed by the *H-2* complex. A similar, albeit weaker case could be made for the genes on the centromeric side of the complex.

On the basis of the molecular structure and functional properties of the products, the genes comprising the *H-2* complex can be assigned to three classes: class I (*H-2K, H-2D, H-2L*), class II (A_α, A_β, A_e, and E_α), and class III (*Ss, Slp*) (Klein et al., 1981). Classification of other genes defined only functionally will remain the subject of debate until their structural products are demonstrated and isolated.

From the genetic point of view the *H-2* complex has two unique properties (for reviews, see Fougereau and Dausset, 1980; Klein et al., 1981; Zaleski et al., 1981). The genes belonging to class I and class II display extraordinary polymorphism matched only by that of some immunoglobulin (*Ig*) genes. At present class I (*H-2K* and *H-2D*) genes have 30 and 38 alleles and at the class II (A_α-A_β and A_e-E_α) loci 21 and 6 alleles are known. The low polymorphism of molecules determined by A_e and E_α genes may be due to the fact that at the E_α locus, the silent (amorphic) allele is quite common. These actual figures permit theoretical estimates of 50 to 100 and 20 to 60 alleles of class I and class II genes, respectively. All of these alleles seem to occur in more or less equal frequencies (3 to 6%) in a global population of mice. Gene cloning experiments suggest that there may be several A_β and A_e genes.

Recently, it has been suggested that the phenotypic polymorphism of class I genes could be accounted for by the simultaneous presence of multiple copies of a given class I gene similar to multiple copies of *Ig* genes. Under normal conditions only one copy would be expressed, while all others would be repressed by undefined regulatory genes. However, in some situations (e.g., neoplastic transformation) the normal copy could be repressed and the other copy derepressed. The product of such a derepressed copy, corresponding to an allelic variant, is often referred to as an *alien* H-2 antigen. This interesting concept, based on the actual finding of such an alien antigen, still requires experimental confirmation.

A second striking genetic feature of the *H-2* complex is its tendency for heterozygosity. The heterozygosity reaches the level of over 90% for class I loci. This is in sharp contrast to the common finding of only 10 to 20% heterozygosity for loci of various

Table 1 Two Versions of the Genetic Composition of the *H-2* Complex

Traditional version										
Regions	*K*			*I*			*S*		*D*	
Subregions		*A*	*B*	*J*	*E*	*C*				
Loci (genes)[a]	*H-2K*	*H-2A*				*H-2C*			*H-2D, H-2L*	
	Lad	*Lad*			*Lad*	*Lad*			*Ss*	*Lad*
		Ia-1	*Ia-2*	*Ia-4*	*Ia-5*	*Ia-3*			*Slp*	
	Ir(?)	*Ir-1A*	*Ir-1B*		*Ir*(?)	*Ir-1C*				*Ir*(?)
	Is					*Is*				
	CI									
Centromere ○ ———										——— Telomere
New version										
Genes	*H-2K*	A_α, A_β, A_e	?	?	E_α	?	*Ss*	*Slp*	*H-2D*	*H-2L*
Products	K	$A_\alpha, A_\beta, E_\beta$ (Ia-1)	?	?	E_α (Ia-5)	?	Ss (C4)	Slp	D	L
Class	I	II			II		III	III	I	I

[a]For definitions of genes, see the text.

other genetic systems. These two features are extraordinarily pronounced in mice, but may also be observed to a lesser degree for the MHC in other species studied thus far.

B. Origin and Evolution of the *H-2* Complex

Striking similarities in the genetic structure as well as in the chemistry and function of the products of the MHC genes in different species strongly suggests that there was a common evolutionary pathway leading to the development of the MHC (for reviews and references, see Fougereau and Dausset, 1980; Hildeman et al., 1981; Klein, 1975). Unfortunately, tracing the origin and history of the MHC is based mostly on guesswork and the piecing together of rather fragmentary data. Thus only speculations, often biased by the researcher's own philosophy, can be made at this time. Nevertheless, it seems that even a brief outline of the concept of *H-2* evolution should contribute invaluably to the considerations dealing with the biologic significance and function of the MHC (Table 2).

It is proposed that the MHC evolved from an archetypal gene consisting of about 300 bases and determining a product consisting of about 100 amino acid residues. This product might have played an essential role in cell–cell interactions in the primitive organisms. This hypothetical archetypal gene, by duplication and translocation, gave rise to three independent but related primitive genes which assumed various specialized functions in more highly developed organisms. The *T* gene evolved as a system of the polymorphic genes determining the differentiation antigens affecting embryonal morphogenesis. The primitive receptor gene *R* became a gene for B-cell receptor and ultimately evolved into the *Ig* genes as well as gene for the still elusive T-cell receptors.

Finally, the primitive MHC gene, by a series of duplications, or even triplications, reversion, and probably deletions, generated the MHC as we know it today. Independent mutations of the duplicated primitive genes resulted in two classes of genes (I and II) and subsequently their polymorphism. Recently, it was suggested that diversification of MHC to several loci and their polymorphism antedates speciation.

An obvious conclusion from the concept outlined above is that the MHC and Ig systems have a common origin and therefore should be functionally related. Indeed, this conclusion seems to gain credence in light of recent developments in immunology and molecular genetics (Weiss et al., 1983). Second, the extreme polymorphism of both systems cannot be considered an incidental phenomenon but rather the result of selective pressure to generate and maintain the polymorphism. This conclusion is supported by finding extensive polymorphism of both systems in virtually all species studied thus far.

III. MOLECULAR PRODUCTS OF THE *H-2* COMPLEX

A. Class I Molecules

A given *H-2* haplotype determines at least three distinct class I molecules, designated K, D, and L (for reviews, see Fudenberg et al., 1978; Götze, 1977; Klein, 1975; McKenzie and Potter, 1979). Since class I alleles are codominant, the *H-2* heterozygous mouse may have as many as six different class I molecules. In fact, recently data have been reported indicating that some *H-2* haplotypes may determine additional class I molecules (K1, K2, D1, D2, M, R), probably due to the duplication of some class I genes. However, these observations and their significance must await further confirmation and extension (for a review, see Zaleski et al., 1981). The class I molecules are ubiquitously distributed in virtually all cells but at different concentrations. It is suggested that the concentration of class I

Table 2 Evolution of the MHC[a]

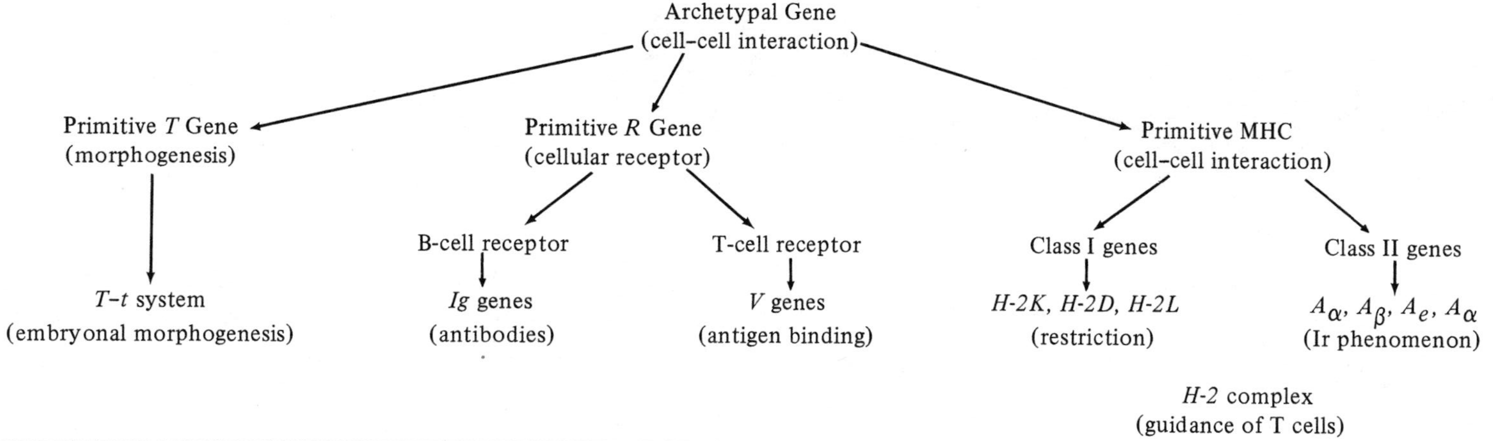

[a]Functions are given in parentheses).

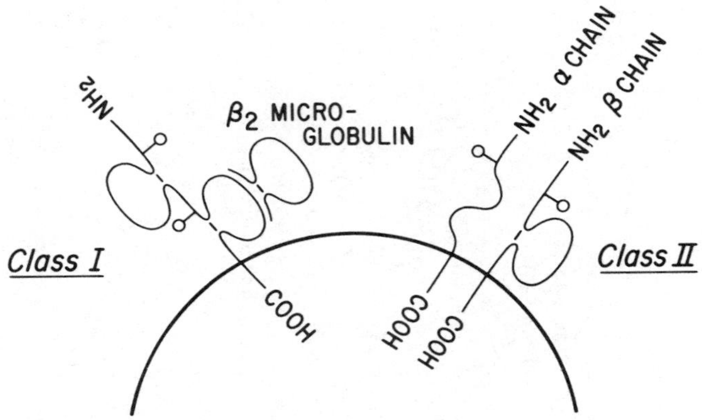

Figure 1 Schematic representation of two classes of murine MHC molecules on the surface of a cell.

molecules decreases with the progress of cellular differentiation. Indeed, relatively undifferentiated hemopoietic or lymphoid cells have a high concentration of class I molecules, whereas highly differentiated neurons, muscle fibers, or erythrocytes have a very low concentration of these molecules. The class I molecules have homogeneous distribution on the cell surface. At the cellular and subcellular levels, class I molecules represent transmembrane moieties (Fig. 1), with the larger portion exposed on the outer surface of the cell membrane. This particular relationship between the molecules and the membrane is responsible for the variability in the reported determination of molecular weight, depending on the method of solubilization employed. It is presently accepted that an average class I molecule has a molecular weight of 45,000. When present on the cell surface the class I molecule is noncovalently associated with a single molecule of β_2 microglobulin (see Chap. 9) in the ratio 1:1.

Biochemistry of the Class I Molecules

Class I molecules are glycoproteins consisting of two sugar moieties and a single polypeptide chain (for reviews, see Dorf, 1981; Ploegh et al., 1981). The sugar moieties (of molecular weight 3300 each) are attached to aspargine residues at positions 86 and 176. The currently proposed composition of the sugar moiety is as follows:

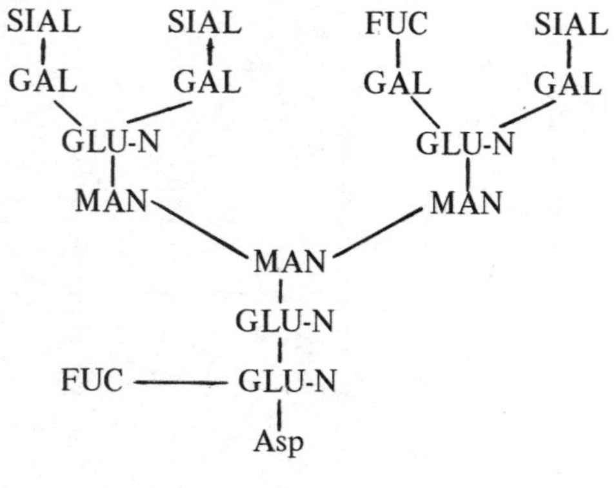

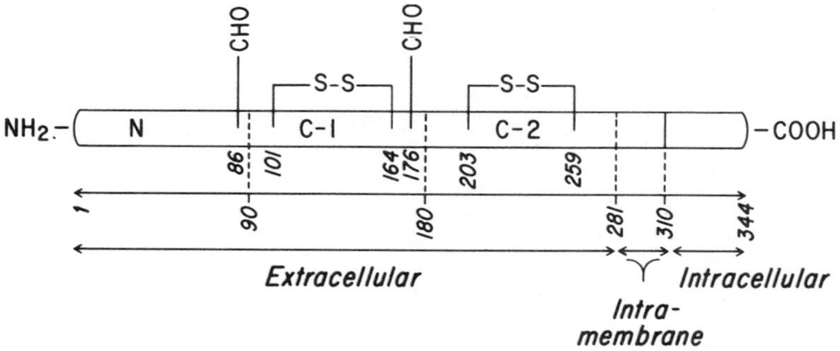

Figure 2 Structure of the class I molecule. N, C-1, and C-2, distinct regions; CHO, oligosaccharide residue; S-S, the disulfide bridge forming the intrachain loop.

In this scheme the abbreviations are: SIAL, sialic acid; GAL, galactose; GLU-N, glucosamine; MAN, Mannose; FUC, fucose; and Asp, aspargine.

The polypeptide chain (Fig. 2) consists of about 344 amino acids and can be divided into three parts. The *intracellular part* extends into the cytoplasm and contains 39 amino acids, half of which are polar. Characteristically, this part abounds in serine and the portion closest to the cell membrane has a sequence of the basic amino acids arginine and lysine, which presumably participate in binding to cellular phospholipids. This binding, as well as the presence of a cysteine that can form a disulfide bridge and a serine that can be phosphorylated, assures the proper anchoring of the molecule. The *intramembrane part* is located within the lipid layer of the cell membrane. It consists of 24 hydrophobic amino acids that are remarkably preserved in various allelic molecules but not in class I molecules of other species. The *extracellular part*, the largest segment of the molecule, consists of 281 amino acids with the NH_2 terminus located peripherally. In murine molecules three regions, N, C-1, and C-2, can be distinguished. The most peripheral region, N, extends from residue 1 to residue 90; the C-1 region is defined by residues 91 to 180, and the C-2 region by residues 181 to 281. The two C regions form intrachain loops by disulfide bridges between cysteines in positions 101 and 164, and 203 and 259. Interestingly, the C-2 loop has significant homology with the heavy chain domains of immunoglobulins. The full amino acid sequence has been determined for several allelic class I molecules, allowing some degree of speculation as to the homology between allelic molecules, between different class I molecules of a given *H-2* haplotype, and between molecules of different species. However, since the spatial arrangement of these molecules known, conclusions concerning their antigenic properties must be accepted with some reservations. Nevertheless, there is consensus that the degree of homology between allelic molecules is greater than that between different molecules determined by the same haplotype. Still, the homology between the different molecules is greater than the homology between the corresponding molecules from different species. The intramembrane part is the least variable within the species. The extracellular portion of allelic molecules contains about 10 amino acid substitutions per 100 residues. Substitutions are distributed evenly along the molecule, but a few clusters of substitutions were found, especially in the vicinity of residues 61 to 83 and 95 to 99. The intracellular segment is the most divergent when different molecules are compared.

Table 3 Summary of Known Mutations of the Class I Genes

Gene	Allele	Number of mutations	Symbols
H-2K	*b*	19	bm1–bm11, bm16–bm23
	d	2	dm4, dm5
	f	1	fm1
	k	1	km1
H-2D	*b*	2	bm13, bm14
	d	3	dm1, dm3, dm6
	f	1	fm2
H-2L	*d*	2	dm1, dm2

Molecular Basis of Mutations of Class I Genes

Even before primary structure of the class I molecules was determined, a search for mutations was undertaken (for reviews, see Dorf, 1981; Zaleski et al., 1981). As a result, 23 mutants of K molecules, 6 mutants of D molecules, and 2 mutants of L molecules were reported (Table 3). Phenotypically, most of these mutations were of the gain-loss type, as judged by immunologic properties of the mutant molecules (see below). At present the precise nature and type of amino acid substitutions is known for some of the mutations (Table 4). In all instances the mutant molecule has either one or at the most three amino acid substitutions. Interestingly, the substitutions are located at the extracellular segments of the molecules that were found to be most variable (i.e., between positions 65 and 80 and in the vicinity of residues 110 and 175). Furthermore, in several independent mutations some substitutions occurred at the same position (bm5, bm6, bm7, bm9, bm16), suggesting that there may be a common mechanism underlying susceptibility to the alterations at this particular segment of the molecule.

The initial concept that most, if not all, amino acid substitutions result from the point mutations became recently hardly tenable. Notably, to account for proven substitutions in some of the mutants one would have to propose a substitution of as many as nine DNA bases—a very unlikely event indeed. Recent reports on the *H-2* gene cloning (Weiss et al., 1983; Melief, 1983) strongly suggest that some mutations may be a result of a gene conversion phenomenon. Such a phenomenon would involve unidirectional or bidirectional exchange of small segments of the DNA between two related genes or pseudogenes indeed, in some haplotypes, over 30 class I-like genes and pseudogenes were identified. However, the strongest support for the concept of gene conversion comes from an observation that the amino acid sequences characteristic of a given mutant were identified in the same location in the nonmutated class I molecules.

Biosynthesis of the Class I Molecules

Data concerning the intracellular synthesis of the class I molecules have only recently been reported (for references, see Ploegh et al., 1981). The transcription of the class I gene results in over 1600 bases long RNA that consists of transcript of 8 exons separated by 7 introns. The latter are removed during processing that results in translatable mRNA (Steinmetz et al., 1981). After the onset of translation the synthesized polypeptide is passed to the endoplasmic reticulum, due to the N-terminal extension of the chain.

Table 4 Summary of Known Amino Acid Substitutions in Mutants of Class I Molecules

		Amino acid	
Mutant	Position	Wild type	Mutant
bm1, bm2	152	Glu	Ala
	155	Arg	Tyr
	156	Leu	Tyr
bm3	77	Asp	?
	89	Lys	?
bm5	116	Tyr	Phe
bm6, bm7	116	Tyr	Phe
	121	Cys	Arg
	22	Tyr	Phe
bm8	23	Met	Ile
	24	Glu	Ser
bm9	116	Tyr	Phe
	121	Cys	Arg
bm10	165	Val	Met
	173	Lys	Glu
bm11	77	Asp	?
bm16	116	Tyr	Phe

As the translation progresses the attachment of sugar moiety and the association with β_2 microglobulin takes place inside the endoplasmic reticulum. The latter process and its timing are of crucial significance. It was shown that preventing association with β_2 microglobulin or the unavailability of β_2 microglobulin precludes the expression of class I molecules on the cell surface. The process of transporting the molecule from reticulum to membrane and its subsequent anchoring into the membrane takes up to 60 min. As the molecule becomes inserted into the membrane a modification of the sugar moiety occurs, but the biological significance of this process remains unknown.

Immunologic Properties of the Class I Molecules

The original discovery of class I molecules was possible because of their immunogenicity (i.e., ability to elicit a specific immune response). Such a response can be detected by either *histogenetic* or *serologic* methods (for reviews and references, see Klein, 1975; McKenzie and Potter, 1979; Snell et al., 1976; Zaleski et al., 1983). The histogenetic methods detect a cell-mediated response and the one most commonly employed is based on the fact that such a response results in the rapid rejection of an allogeneic skin graft. The other methods belonging to this category are merely extensions or modifications of allograft rejection [(graft versus host reaction (GVHR), cell-mediated lymphocytotoxicity

(CML)]. The major drawback of all of these methods is their qualitative or only semi-quantitative nature. The serologic methods detect the antibody-mediated response. The alloantibodies elicited by class I molecules are directed to specific determinants. There are more than 100 such determinants, commonly referred to as *specificities*. Their exact amino acid composition or localization along a polypeptide chain remains unknown. In fact, the search for hypervariable segments of the chain, carrying these specificities, has yet to yield clear-cut results. A particular class I molecule has 10 to 15 distinct specificities. Among them, one, called *private*, is unique for a given allelic molecule, whereas the remaining specificities, called *public*, are shared by different allelic molecules as well as the molecules determined by different class I genes. The antibodies specific for the private specificities can be used not only to detect the presence of corresponding molecules but also to isolate such a molecule in a preliminary (affinity chromatography) step of the biochemical analysis (see Chap. 13). The relationship between the antigenic determinants responsible for cell-medicated responses and those eliciting antibody responses is not yet clear, nor is the significance of these antigenic properties for the biological function of the class I molecules known (see below).

B. Class II Molecules

Our knowledge of class II molecules is significantly less extensive than that of class I molecules (for reviews, see Ferrone and David, 1982; Klein et al., 1981; Zaleski et al., 1983). A primary reason for the paucity of data stems from the relatively recent (1973) discovery of these molecules. A given *H-2* haplotype usually determines four polypeptides which assemble into two class II molecules, also referred to as Ia molecules. Since phenotypic expression of the class II genes is codominant, an *H-2* heterozygous mouse will have four discrete class II molecules, none associated with β_2 microglobulin. There are two demonstrable (Ia-1 and Ia-3/5) molecules and two presumptive (Ia-2 and Ia-4) molecules, corresponding to the still undetectable products of the genes at *I-B* and *I-J* subregions. The molecules are selectively expressed on some cells (lymphoid, epithelial, macrophages) but are absent or undetectable on others (e.g., muscle fibers, erythrocytes, cells of liver and kidney). Whether the concentration of class II molecules is related to the degree of cellular differentiation is not yet established. However, the increased expression of Ia molecules after mitogen treatment of lymphocytes is strongly suggestive in this regard. On the cellular level the class II molecules are transmembrane entities, although the exact relationship of extracellular, intramembrane, and intracellular segments is undefined (Fig. 1).

Biochemistry of Class II Molecules

The two demonstrable class II molecules are glycoproteins, each consisting of two polypeptides: α of approximate molecular weight 33,000 and β of approximate molecular weight 28,000 (for reviews, see Ferrone and David, 1982; Dorf, 1981; Zaleski et al., 1983). There are some apparent differences in molecular weight of different A_α and E_α or A_β and A_e, as well as in the molecular weights of allelic variants of a given chain. At this time, there is no satisfactory explanation for these differences. Both chains carry oligosaccharide residues; however, their number and location on the chains are unknown. The mechanism underlying the assembly of the two polypeptides into a single Ia moiety is also unknown. However, the available data suggest that disulfide bridging is not involved,

although such a bond is possible chemically. Presumably, the β chain forms an internal disulfide loop. Complete amino acid sequencing of some Ia molecules has been performed. Preliminary sequence data seemed to indicate little, if any, homology between N termini class I and class II molecules. Also, there was no apparent homology between different α or different β chains. However, a clear homology of more proximal sections of these molecules was recently reported. From these studies it appears that of the two polypeptides, the α chain is less variable. In fact, no amino acid differences were found in preliminary sequencing of allelic E_α chains, but allelic A_α chains are clearly variable. The β chain displays great variability between different *H-2* haplotypes. This variability appears to be especially pronounced within three discrete stretches between residues 2 to 13, 21 to 28, and 68 to 80 of allelic A_β chains. A similar situation exists in allelic A_e (E_β) chains. Very little is known about the biosynthesis of class II molecules. Preliminary data indicate that each chain is synthesized on a separate mRNA and the initial product has a small (about 20 amino acids) extension which is cleaved by the time the chain is inserted into the membrane.

Mutant of the Class II Gene

At this time only one mutant of class II molecules is known. The mutant call $H\text{-}2^{bm12}$ affected the β chain of the Ia-1 molecule. Phenotypically, the mutation resulted in a deficient expression of the molecule as well as in some changes in its antigenic properties (see below). The only other *H-2* haplotype ($H\text{-}2^{g2}$) determining a class II molecule suspected of having mutated has been derived by intra-*H-2* crossing over. With this in mind it may be argued that the altered class II molecule, probably the β chain of Ia-3/5, resulted from intragenic crossing over rather than from base substitution.

Immunogenic Properties of the Class II Molecules

The original definition of the class II loci and their presumptive products was based on the association between *I*-region incompatibility and stimulation of blast transformation and proliferation of lymphocytes in vitro (for reviews, see Ferrone and David, 1982; Götze, 1977; Klein, 1975). The antigenic determinants responsible for the phenomena were originally called lymphocyte activating determinants (Lad), determined by corresponding *Lad* genes. Soon thereafter it was discovered that the genes at the *I* region also determine the molecules that can induce the production of cytotoxic antibodies in an allogeneic recipient. The molecules were called *I-region associated* (Ia) and their putative genes were correspondingly termed *Ia*. In addition, genes controlling immune responsiveness (*Ir*) and genes determining antigens responsible for allograft rejection (*H-2I*) were assigned to the *I* region. In the last part of this chapter it will be argued that the *Lad, Ia, Ir,* and *H-2I* genes are identical with the A_α, A_β, A_e, and E_α genes which encode the polypeptides of class II molecules. There is a general consensus that class II molecules can elicit both antibody-mediated and cell-mediated responses.

The antibody-mediated response is directed against 50 discrete determinants, analogous to the specificities described above for the class I molecules. However, among these determinants, two types can be distinguished (see Ferrone and David, 1982; Zaleski et al., 1981). An *ordinary allotypic* specificity is detectable in the assembled Ia molecule as well as in the α or β chain forming such a molecule. The *combinatorial specificities* are detectable in the complete molecule but not in either of the chains. There are two

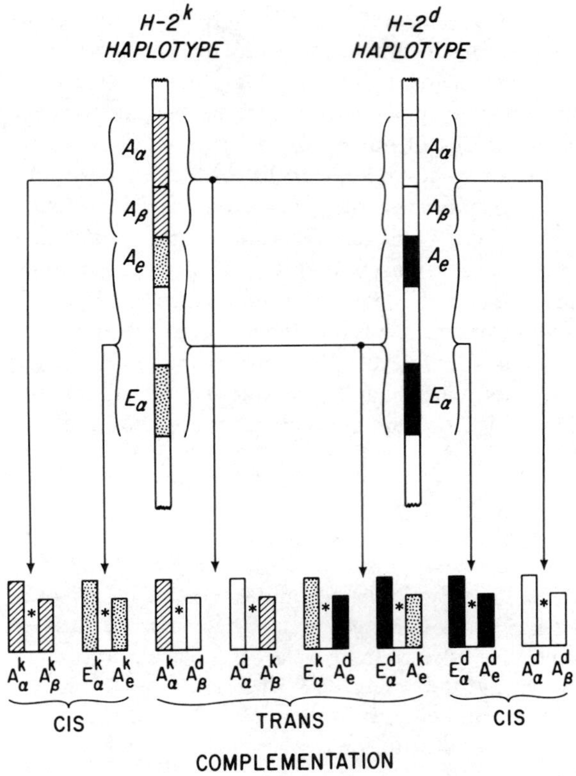

Figure 3 Schematic representation of the genetic determination of the class II molecules in the (H-2^k × H-2^d F_1 hybrid. The asterisks represent combinatorial or hybrid specificities formed as the result of the combining of two polypeptides into dimer.

concepts that attempt to account for the formation of combinatorial specificities. According to one concept, the specificity present in the chain is unmasked when the chain assembles into a molecule. Alternatively, conformational changes occurring during the assembly of the chains might result in the formation of the combinatorial specificities. In either model the assembling of two chains is a prerequisite for the detection of such a specificity, and thus it may be said that the two chains or their genes complement each other. Such a complementation may take place between A_α and A_β or A_e and E_α genes of the same *H-2* haplotype (*cis* complementation) or between corresponding α and β genes from two different *H-2* haplotypes (*trans* complementation) in the heterozygote. The combinatorial specificity resulting from *trans* complementation is referred to as a hybrid or F_1 specificity and is obviously absent in either parental animal. An important aspect of the phenomenon of combinatorial specificities is the fact that it increases the variability of class II molecules without increasing the number of discrete genes (Fig. 3). The distribution of the various specificities among the two chains and their contribution to the ability to induce a cell-mediated response remain unknown.

C. Class III Molecules of the *H-2* Complex

There are two class III molecules determined by any *H-2* haplotype (for reviews, see Dorf, 1981; Götze, 1977; Klein, 1975; Snell et al., 1976). The serum substance (Ss) and sex-limited protein (Slp) are determined by corresponding *Ss* and *Slp* genes in the *S* region. The two genes are closely linked to each other and thus far no crossing-over between them has been found. In fact, it is believed that the two genes evolved by duplication of a single ancestral gene. Both genes have at least two alleles each, Ss^l or Ss^h and Slp^a or Slp^0. However, recent observations strongly suggest that each gene may have a wide spectrum of alleles. The allelic variations are detectable phenotypically as a low (Ss^l) or high (Ss^h) serum concentration of Ss and the presence (Slp^a) or absence (Slp^0) of Slp. In addition, allotypic structural specificities have been found in the Ss molecule. In contrast to class I and class II molecules, class III molecules are present predominantly in plasma or body fluids and it is only by passive absorption that they become associated with the cell membrane of erythrocytes. Because of their association with plasma rather than with cell membrane, the class III molecules do not induce a cell-mediated response and are not considered to be histocompatibility antigens.

Each class III molecule is synthesized in macrophages and liver cells as a single moiety of molecular weight 200,000. This molecule is then cleaved into three chains: α (MW 100,000), β (MW 75,000), and γ (MW 35,000). The synthesis is influenced by sex hormones, with males displaying a higher level of Ss and the presence of Slp, whereas females have a lower level of Ss and, with few exceptions, lack Slp. Whether the hormonal effect is exerted directly on the structural gene or via regulatory genes is a subject of debate. The two class III molecules, although in many respects similar, differ from each other in their tryptic peptide map.

IV. FUNCTIONAL CONSIDERATIONS

A. H-2 Molecules as Histocompatibility Antigens

The discovery and early history of the *H-2* complex and its products (class I molecules) was prompted by the demonstration of its role in the acceptance or rejection of allografts of neoplastic and normal cells, tissues, and organs (for reviews, see Fudenberg et al., 1978; Hildeman et al., 1981; Klein, 1975; Snell et al., 1976). As a rule the incompatibility of the donor and responder H-2 antigens resulted in the rejection of the graft within 3 weeks (usually about 12 days). Conversely, compatibility permitted the prolonged and in some instances even indefinite survival of the graft. Thus it appeared that the acquisition of detailed knowledge of genotypic determination and phenotypic expression of H-2 molecules is of vital importance for progress in tissue transplantation. However, it was always suspected that the involvement in graft rejection, a human-made phenomenon, cannot be the true biological role of the MHC. That the MHC must be of immense biologic importance became obvious when the complex was detected in every species studied. While basic studies in mice served as a framework for subsequent studies in other species, including humans (see Chap. 26), the true biological importance of the MHC has been realized only recently. The discoveries of two phenomena (Table 5) were instrumental in this realization.

Table 5 Comparison of Ir Phenomenon and MHC Restriction

Feature	Ir Phenomenon	MHC Restriction
Genes	*Ir* = *Ia* (class II)	*H-2K*, *H-2D*, and *H-2L* (class I)
Antigens	Synthetic polypeptides Heterologous proteins Alloantigens	Haptens Viruses Alloantigens
Response	Antibody-mediated	Cell-mediated
Cells affected	T_h macrophages	T_c
Postulated mechanism	Associative recognition of antigen + class II molecules	Associative recognition of antigen + class I molecules

B. Ir Phenomenon

Although it is beyond the scope of this chapter to discuss all aspects and details of the Ir phenomenon, a brief description is indispensable for understanding the function of the MHC and its products (for reviews, see Blanden, 1981; Ferrone and David, 1982; Dorf, 1981; Klein et al., 1981; Zaleski et al., 1983; Zinkernagel and Doherty, 1979). The term *Ir phenomenon* (or *Ir control*) refers to the effect of the responder's genes on the magnitude of the immune response to some antigens, or, strictly speaking, antigenic determinants. Among outbred animals good and poor responders are distinguished, while among inbred animals good and poor responder strains are distinguished.

The genes responsible for the effect are referred to as *immune response* (Ir) genes. Early studies were consistent with the notion that a given *Ir* gene affects the response to one antigen (determinant). Accordingly, each time a response to a given antigen was found to be under genetic control, a distinct *Ir* gene was postulated. Such a postulated *Ir* gene had to have at least two alleles: one determining a good response and the other determining a poor response to the antigen in question. Since most, but not all, putative *Ir* genes were mapped to the *I-A, I-B, I-E* and/or *I-C* subregions of the *H-2* complex, it was considered that this segment of the complex contains a large number of discrete genes. However, in spite of an extensive search, no molecular product for any presumed *Ir* gene was found. In fact, the class II molecules appear to be the only molecular products which undoubtedly are determined by the genes in the *I* region.

The antigens that elicit an immune response subject to Ir control belong to at least three categories: synthetic polypeptides, heterologous proteins, and alloantigens. Regardless of the category of the antigen, the response affected by the Ir phenomenon is thymus dependent and therefore, involves T-helper (T_h) cells. Indeed, it was established that Ir control influences the generation of the specific T_h cells and, presumably by doing so, exerts its effect predominantly on the secondary antibody response, while the primary response often remains unaffected.

The cellular and molecular mechanisms underlying the Ir phenomenon have attracted a great deal of interest (for references, see Dorf, 1981; Klein et al., 1981; Zaleski et al., 1983). The original concept, which proposed that *Ir* genes determine antigen-specific receptors, did not find support in the experimental data. On the one side,

such receptors eluded attempts at isolation, while on the other hand, no evidence could be obtained that any such receptor is determined by the genes within the *H-2* complex. The decisive blow to the receptor concept was delivered when it was demonstrated that the Ir effect is mediated by macrophages, a rather unexpected finding if the receptor concept was correct. An alternative concept of associative recognition proposed that to be recognized and to trigger a good response, an antigen must be presented in association with class II molecules. Such an association, however, would be possible only between certain antigens and certain allelic class II molecules. Thus while some antigens would be capable of associating with class II molecules of a given individual, permitting a good response, other antigens would not form the association, resulting in a poor or no response. It then follows that to assure a good response to a wide spectrum of antigens, both individuals and populations should possess a large repertoire of different class II molecules. This is achieved at the level of the individual by the multiplicity of class II molecules and combinatorial determinants and at the level of the population by polymorphism, heterozygosity, and the formation of hybrid determinants (see Ferrone and David, 1982; Klein et al., 1981; Zaleski et al., 1981). An important implication of the associative recognition concept is that class II genes and molecules would be identical with the *Ir* genes and their products. Experiments in which antibodies to class II molecules of a good responder abolished a good response provided convincing evidence of the identity of class II molecules and the products of *Ir* genes.

C. MHC Restriction

The second important phenomenon shown to be intimately associated with MHC is referred to as *H-2 restriction* or, more generally, *MHC restriction* (for reviews, see Hildeman et al., 1981; Zaleski et al., 1981; Zinkernagel and Doherty, 1979). Similar to the Ir phenomenon, MHC restriction represents the effect of the responder's genes on the magnitude of the immune response to certain antigens. Specifically, a good response is achieved only when an antigen during the stimulatory and effectoral phases is presented by the cells carrying the alleles, which are identical with the alleles of the responder.

The genes responsible for the restriction were mapped to the K and D regions of the *H-2* complex. It was demonstrated unequivocally that the products of the *H-2K, H-2D,* and *H-2L* genes (i.e., class I molecules) exert the restrictive effect. The most convincing evidence in this regard is that mutations of these genes result in pronounced changes of the restriction specificity.

The antigens eliciting the immune responses which are restricted belong to three categories: chemical haptens, viruses, and cell surface alloantigens. A common denominator for these antigens is that they are associated with the cell surface and elicit a response mediated by T-cytotoxic (T_c) cells. T_c cells have double specificity and in order to display their cytotoxic activity against target cells, they must recognize on target cells the specific antigen and the restricting molecule, both of which must be identical with those that originally induced the T_c cells.

To account for the double specificity of T_c cells and the restriction phenomenon, a concept of associative recognition has been proposed. According to this concept, an antigen to be recognized and to trigger the generation of T_c cells, must be presented in association with class I molecules. The association would be possible between certain antigen and certain allelic class I molecules in much the same way as an antigen and

class II molecule might associate in the case of the Ir phenomenon. The unavailability of a class I molecule capable of associating with a given antigen would result in the inability to recognize such an antigen, and a poor response or no response would ensue. Again, the widest possible spectrum of class I molecules achieved by multiplicity of class I loci, polymorphism, and heterozygosity assures the recognition of many different antigens by an individual and a population.

D. Unifying Concept of MHC Function

The two phenomena above described—Ir control and MHC restriction—were originally considered as being entirely different. However, it now appears quite obvious that the phenomena are closely related. The belief in the relatedness of the two phenomena stems from the idea that the primary, if not an exclusive function of the *H-2* complex is the guidance of various subpopulations of T cells through the process of differentiating between self and nonself molecules (for reviews, see Klein et al., 1981; Klein, 1982; Zaleski et al., 1981). The self, represented by class I and class II molecules, would provide the framework or reference point for T cells. The nonself, an antigen, would be recognized effectively only if it is associated with a self molecule of the responder. The mechanism of such an association remains obscure, but considering the variability of antigens and of self molecules, it is plausible to consider that more than one mechanism may be operational. Implicit in the concept of associative recognition is that the ability to recognize broad spectrum of antigens is facilitated by the availability of a large number of self molecules. The variability of self molecules assures that most or all antigens will find in an individual, or in a population, suitable self molecules with which they may form an association. The association of nonself with self is often viewed as an alteration of self and thus the response elicited would be aimed at the elimination of the altered self rather than the antigen per se. Development and presence of two classes of major histocompatibility molecules coincides with the two basic types of immune response, which require the stimulation of two different T-cell subpopulations and result in two different effector mechanisms. The cell-mediated response involves the generation of T_c cells and results in the destruction of target cells carrying an altered self composed of class I molecule(s) and antigen. The antibody-mediated response, in turn, encompasses the generation of T_h cells by an altered self composed of class II molecule(s) and antigen and results in the stimulation of B cells. This stimulation leads to the production of antibodies, which eliminate that antigen. According to this interpretation, it could be predicted that the interaction of the various cells participating in a given immune response should be restricted by the self component (for a review and references, see Goldberger, 1980). Indeed, responder-stimulator, effector-target, T cell/macrophage, and T cell/B cell interactions are restricted by self-MHC molecules. The outlined concept inevitably leads to the conclusion that many traits and their putative genes assigned to the *H-2* complex in reality reflect the pleiotropism of a relatively few genes (Klein et al., 1981). Thus graft rejection (*H* genes), mixed lymphocyte reaction (*Lad* genes), cell interaction (*CI* genes), and immune responsiveness (*Ir* genes) are actually phenotypic representations of various aspects of the same basic function of the *H-2* complex accomplished by either class I or class II genes. Antigen-specific immunosuppression mediated by T-suppressor (T_s) cells is an additional MHC-associated phenomenon that falls within this scheme by postulating that certain antigens in association with self molecules, especially belonging to class II, may simultaneously induce T_h and T_s cells, the latter

interfering with the function of the former. The explanation of antigen-specific suppression prior to the concept of associative recognition required postulation of specific *immune suppression* (*Is*) genes.

As attractive in its simplicity as the unifying concept of structure and function of the *H-2* complex may be, it remains in the sphere of hypothesis. There are numerous phenomena that have not yet been comfortably accommodated. Although a detailed discussion of the function of class III molecules is beyond the scope of this chapter, a brief comment is warranted. The Ss and Slp molecules are structurally very similar, although not identical, but only Ss molecules have the C4 activity. There are two lines of reasoning concerning the possible relationship of class III molecules with the other two classes. First, it is speculated that since in most species complement genes are found within or near the MHC, such genes are an integral part of the complex. To strengthen this argument it is further speculated that the complement cascade evolved from a primitive system involved in cell–cell interaction. On the other hand, some authors consider the MHC-linked complement genes as accidental immigrants to the MHC via random insertion events.

E. The Receptors

The pivotal issue concerning the function of the MHC in associative recognition is the nature and origin of the T-cell receptors, which are vested with the ability to recognize the self + nonself complex. Because the recognition of the complex requires dual specificity, the receptors would also have to have dual specificity. The presence of the receptors is suggested but as yet not definitively demonstrated. The speculations, however, are in great abundance. Essentially two types of hypothesis concerning the nature and origin of the receptors may be distinguished (for reviews, see Goldberger, 1980; Zinkernagel and Doherty, 1979).

The hypothesis of antigen-specific receptor being determined by MHC genes has been already mentioned in the discussion of the Ir phenomenon. This type of hypothesis simply postulates that from the preexisting and stable pool of receptor genes different subsets of T cells express only one. The good responder would have a specific subset for a given antigen, whereas the poor responder would not. This concept is irreconcilable with findings that a poor responder may be converted to a good responder (i.e., generate a new receptor) when the MHC molecules of such a good responder are present during ontogenesis of the T cells.

The alternative hypothesis in the simplest terms proposes that from a pool of germ line receptor genes a repertoire of genes for different antigens develops via mutations. The receptor genes are independent of the MHC and are possibly related to the *V* genes of immunoglobulins. In one form of the hypothesis a single germ line produces two types of receptors. One receptor is specific for MHC molecules and if the receptor happens to be specific for self-MHC molecules, further mutation becomes precluded. The second receptor ultimately becomes specific for the antigen. The good responder has T cells that carry simultaneously receptors for a given antigen and for self, whereas the poor responder has cells that carry the receptor for an antigen but not for self. A variant of this hypothesis is that only receptors for various self determinants are formed and such receptors recognize both self and cross-reactive antigens. The cells with receptors only for antigens (nonself) are eliminated. In the other form it is proposed that there are two pools of germ lines: one for receptor for self and a second for the antigen. Whereas genes

of the first pool are precluded from mutation, genes of the second pool can mutate and thereby produce a limited repertoire of receptors for different antigens.

In either case a crucial point of any receptor hypothesis is that the two receptors—for self and for antigen—must be expressed on the same T cell and must somehow be spatially related. There is not yet any real experimental proof or disproof of the preclusion models. However, experiments showing that T cells specific for various unrelated antigens are restricted by the same segment of the self molecule strongly suggest that receptors for those antigens might have developed from the same germ line as the receptors for self.

An issue that is equally cloudy is the actual structure of the receptor (for references, see Goldberger, 1980). The isolation and purification of the receptors from the T cells still eludes researchers. Two alternative concepts are presently being debated. According to some investigators, the receptor would contain a variable portion of an immunoglobulin associated with a constant portion distinct from any known Ig. Alternatively, a variable portion would be associated with a nonimmunoglobulin constant molecule. The origin of such a molecule remains enigmatic, although some consider the possibility that it is modified β chain of class II molecule. In either concept the variable portion seems to carry idiotypic specificities, but it is not clear whether it contains both framework and hypervariable segments or only the latter.

REFERENCES

Blanden, R. V. (1981). Immunol. Today *1*, 33.

Dorf, M., Ed. (1981). *The Role of the Major Histocompatibility Complex in Immunobiology*. Garland Press, New York.

Ferrone, S., and David, C. S., Eds. (1981). *Ia Antigens*. CRC Press, Boca Raton, Fla.

Fougereau, M., and Dausset, J., Eds. (1981). *Immunology 80: Progress in Immunology IV*. Academic Press, New York.

Fudenberg, H. H., Pink, J. R. L., Wang, An-Chuan, and Douglas, S. D. (1978). *Basic Immunogenetics*. Oxford University Press, New York.

Goldberger, R. F., Ed. (1980). *Biological Regulation and Development*. Plenum, New York.

Götze, D., Ed. (1977). *The Major Histocompatibility System in Man and Animals*. Springer-Verlag, New York.

Hildeman, W. H., Clark, E. A., and Raison, R. L. (1981). *Comprehensive Immunogenetics*. Elsevier, New York.

Klein, J. (1975). *Biology of the Mouse Histocompatibility-2 Complex*. Springer-Verlag, New York.

Klein, J. (1982). *Immunology: Science of Self-Nonself Discrimination*. John Wiley & Sons, New York.

Klein, J., Juretić, A., Baxevanis, C. N., and Nagy, Z. A. (1981). Nature (Lond.) *291*, 455.

McKenzie, I. F. C., and Potter, T. (1979). Adv. Immunol. *27*, 179.

Melief, C. (1983). Immunol. Today *4*, 58.

Ploegh, H. L., Orr, H. T., and Strominger, J. L. (1981). Cell *24*, 287.

Snell, G. D., Dausset, J., and Nathenson, S. (1976). *Histocompatibility*. Academic Press, New York.

Steinmetz, M., Moore, K. W., Frelinger, J. G., Sher, B. T., Shen, F., Boyse, E. A., and Hood, L. (1981). Cell. *25*, 683.

Weiss, E. H., Mellor, A., Golden, L., Fahrner, K., Simpson, E., Hurst, J., and Flavell, R. A. (1983). Nature *301*, 671.

Zaleski, M. B., Abeyounis, C. J., and Kano, K. Eds. (1981). *Immunobiology of the Major Histocompatibility Complex*. S. Karger, Basel.

Zaleski, M. B., Dubiski, S., Niles, E. G., and Cunningham, R. K. (1983). *Immunogenetics*. Pitman, Marshfield, Mass.

Zinkernagel, R. M., and Doherty, P. C. (1979). Adv. Immunol. *27*, 51.

28

Lymphocyte Antigens and Receptors

Regina R. Skelly* Merck Sharp and Dohme Research Laboratories, Rahway, New Jersey

Terry A. Potter Albert Einstein College of Medicine, Bronx, New York

Aftab Ahmed U S. Naval Medical Research Unit 3, FPO, New York

I. INTRODUCTION

The immune system is a dynamic array of cellular and molecular interactions that are determined by recognition events at the level of the cell membrane. An immune response, resulting in antibody production, phagocytosis, or cytolysis, is thus initiated by the cellular recognition of a foreign or exogenous entity, with subsequent interactions occurring through the recognition of appropriate cell surface molecules on subsets of these cells; the products of these activated cells can then interact with other cells through appropriate recognition molecules.

The cell membrane consists of different types of molecules: structural molecules required to maintain the integrity of the cell membrane, molecules that facilitate the transport of ions and nutrients through the cell membrane, enzyme molecules, virus proteins, and receptor molecules for hormones, factors, and antigens. All of these molecules are potentially identifiable by unique antigenic determinants.

The main focus of this chapter will be the description of lymphocyte surface molecules defined by serological techniques and/or a receptor function. Operationally, receptors are cell surface determinants whose biological and/or biochemical interactions with a substrate have been defined. In contrast, for most of the lymphocyte antigens no function has been assigned; however, some of the these molecules have been useful in defining unique populations of cells. Needless to say, the presence of a unique antigen on a subset of cells without a defined activity implies that it is also functionally important, although its molecular and biochemical interactions have not been deciphered.

*Current affiliation: U.S. Naval Medical Research Unit 3, FPO New York.
For a brief discussion of the connection between Chaps. 9 and 24–28, see the appendix.

The majority of our current information about cell surface antigens and receptors has been derived by the use of genetically identical inbred lines of mice. Based on these murine studies, and the assumption that the major functional subsets of lymphocytes have been phylogenetically conserved, immunologists are now able to describe various subsets of lymphocytes in the human. Thus, while the initial analysis in the mouse has proved invaluable, these studies are now being extended to the human. Where appropriate, the human candidates of the antigens defined by mouse work will be described in this chapter, although the major emphasis is on murine cell surface antigens and receptors.

The description of the majority of cell surface antigens/receptors has been facilitated by the use of allo-, hetero-, or xenoantisera. A major disadvantage to the use of these reagents is the possibility of the presence of more than one antibody. This potential problem is now alleviated by hybridoma technology through which monospecific antibodies are produced by somatic cell hybridization. For the most part, antibodies are produced that, in the presence of complement, will lyse the appropriate cell; noncytotoxic antibodies are detected by absorption, by the ability of antibody-coated red blood cells to form rosettes, by the ability of antibody to bind cells as monitored with fluorescein-labeled developing reagents and fluorescence microscopy or flow microcytofluorometry, or by the ability of antibody to precipitate cell surface antigens. In addition, the methodology of affinity chromatography has improved the purification of both antibodies and antigens, reciprocally.

Recent advances in the field of molecular biology are having profound influences on membrane biology research. These techniques involve the identification, isolation, and cloning of DNA for specific gene products. Thus it is now possible using complementary DNA (cDNA) to probe specific subsets of lymphoid cells for the number of DNA copies of a particular gene, to isolate normal as well as variant genes, and to insert new DNA for specific proteins into the genome. At present, the focus of gene cloning research in the immunological field is centered around the genes for immunoglobulin heavy and light chains and major histocompatibility (MHC) antigens. Recently, the genes for the H-2L antigen have been cloned and the DNA inserted into mouse fibroblasts which normally do not express H-2L. These fibroblasts now express H-2L on their cell surface and the antigen is functional as defined by its role in viral cytotoxicity assays (Goodenow et al., 1982; Woodward et al., 1982).

II. ANTIGENS AND RECEPTORS ON BOTH T AND B LYMPHOCYTES

The criteria used to differentiate antigens as unique include strain distributions among inbred, congenic, and recombinant-inbred lines of mice, backcross analyses, the identification of multiple structural alleles, and the ontogenetic appearance of unique antigens. The functional status of positively selected cells is also assessed. In the latter case, however, the interpretation of the data is dependent on the assumption that appropriate accessory cells have not been manipulated, resulting in the loss of their functional reactivity. More detailed descriptions of and references for all these lymphocyte determinants can be found in recent reviews of non-MHC murine alloantigens by Ahmed and Smith (1982), McKenzie and Potter (1979), and McKenzie et al. (1982); only the more recent references, not cited in these reviews, will be mentioned in this chapter where appropriate.

A. Histocompatibility Antigens

The histocompatibility loci genes encode for and regulate the expression of many cell surface and serum antigens. The major histocompatibility loci, as described in Chaps. 27 and 28 in more detail, encode for three major classes of antigens. In the mouse, class I antigens, encoded for by *H-2D, H-2K,* and *H-2L* loci, are present on virtually all cells. Class I antigens in the human are classified as HLA-A, HLA-B, and HLA-C. These antigens are the determinants responsible for allograft rejection and are thus considered the major recognition determinants that serve to distinguish self from nonself. Class II antigens, typified by H-2I-region gene products in the mouse, and HLA-DR, SB, MB, and DC antigens in the human, have a more restricted tissue distribution than do class I antigens, and will be described later as antigens on B cells or T cells. Class III antigens are regulatory and structural proteins that include some of the serum complement proteins which are encoded for by genes with the MHC of both mouse and human.

The relative amount and/or density of class I histocompatibility antigens vary among different cell types. For example, murine thymocytes express much less H-2K, -D, or -L antigens that do spleen or lymph node cells. Whether this difference is due to the number of structural gene copies, regulatory gene differences, or metabolic turnover rates in distinct cell types is not fully appreciated. However, with the availability of cDNA probes to examine the *H-2* loci at the DNA level, these questions can soon be answered (Festenstein and Schmidt, 1981).

B. Qa Antigens

The five Qa antigens are encoded for by a series of closely linked genes located on chromosome 17 in the Tla region, linked to *H-31* and *H-32* and distal to the *H-2* complex in the mouse. Each of the Qa loci possesses a null allele, that is, an allele for which no complementary cell surface product has been identified (Flaherty, 1979). These antigens are structurally similar to the class I histocompatibility antigens and are predominantly expressed on T cells, but are also present on myeloid cells and on a small percentage of B cells. The characteristics of each of these alloantigens are listed in Table 1. Recently, a hybridoma antibody identifying a human T-cell subset and natural killer cell has been described that detects an alloantigen having properties similar to the murine Qa alloantigens (Chun et al., 1981).

C. Lymphocyte Antigens (Ly)

A number of antigens have been described that present predominantly, although not exclusively, on subpopulations of murine lymphocytes. These antigens are currently being designated numerically, based on their chronological isolation and description. Lyt-1, 2, and 3, found primarily on T cells, and Lyb alloantigens, found on B cells, will be discussed separately. Table 2 summarizes the remaining murine alloantigens and their properties. To date, no correlations have been made between these murine antigens and human cell surface antigens.

D. Hormone Receptors

The interaction of pharmacologically active hormones with lymphoid cells may occur through a unique functional receptor for the hormone or through recognition of the

Table 1 Biochemical, Genetic, and Functional Characteristics of Murine QA Antigens

Alloantigen	Identification	Number of alleles	Chromosome	Linkage	Molecular weight and composition
Qa-1	Alloantisera	3	17	*TL*	45,000 and 14,000; β_2 microglobulin
Qa-2	Alloantisera, monoclonal Ab	2	17	*H-2, TL, H-31, H-32, H-2T, Qa-3, Qat-4, Qat-5*	40,000–45,000 and 14,000; β_2 microglobulin
Qa-3	Alloantisera	2	17	*H-2, TL, H-31, H-32, H-2T Qa-2, Qat-4, Qat-5*	45,000 and 14,000; β_2 microglobulin
Qat-4	Monoclonal Ab	2	17	*H-2, TL, H-31, H-32, H-2T, Qa-2, Qa-3, Qat-5*	–
Qat-5	Monoclonal Ab	2	17	*H-2, TL, H-31, H-2T, Qat-2, Qat-3, Qat-4*	–

	Distribution									
	T-cell populations[a]						B-cell populations[a]			
Alloantigen	T:	Helper	Suppressor	Cytotoxic	MLR[b]	Mitogen	B:	AB[c]	Mitogen	Other
Qa-1	±		+	–	–	+	±	+	–	
Qa-2	+				+	+	±	+	+	Bone marrow, Myeloid progenitors
Qa-3	+				±	+	±	±	±	Natural killer cells
Qat-4	±						±		±	Null cells, Natural killer cells
Qat-5	±		±							Null cells, Natural killer cells

[a]+, all cells; ±, subpopulation.
[b]MLR, mixed lymphocyte reaction.
[c]AB, antibody producing.

Table 2 Biochemical, Genetic, and Functional Characteristics of Murine Ly Alloantigens

Alloantigen	Identification	Number of alleles	Chromosome	Linkage	Molecular weight
Ly-4[a]	Alloantisera, Monoclonal Ab	2	2	*H-13*, *Ly-m11* *H-3*, β_2 microglobulin, *Ir-2*	45,000
Ly-5	Alloantisera	2			200,000, glycoproteins
Ly-6[b]	Alloantisera, Monoclonal Ab	2	19	*Ly-8*, *Th-B*, *Thy-1*, *ALA-1*	18,000–22,000
Ly-7[c]	Alloantisera, Monoclonal Ab	1			
Ly-8	Alloantisera	1	19	*Ly-6*, *Th-B*	
Ly-9[d]	Alloantisera, xenoantisera, monoclonal Ab	2	1	*Ly-M*, *Mls*, *H-25*, *Ly-m20*, *Ly-17*	100,000, glycoprotein
Ly-10[e]	Alloantisera, monoclonal Ab	1	19	*Ly-1*	
Ly-11	Alloantisera, monoclonal Ab	2			18,000–22,000

Ly-12	Alloantisera	1			
Ly-13	Alloantisera	1			
Ly-14	Alloantisera	1	7	*Albinism*	73,000
Ly-15	Alloantisera, monoclonal Ab	2			>100,000
Ly-16	Alloantisera	2			
Ly-17[f]	Alloantisera	1		*Ly-M, Lym-20, Mls, H-25*	
Ly-18	Monoclonal Ab	1			
Ly-m19[g]	Monoclonal Ab	1	4	*Lyb-2, Lyb-4, Lyb-6, Mup,* LPS receptor	
Ly-m20	Monoclonal Ab	1	1	*Mls, H-25, Ly-M, Ly-9*	
Ly-m21[h]	Monoclonal Ab	1	7	Albinism	

Table continues

Table 2 (continued)

	Distribution										
	T-cell populations[i]						B-cell populations[i]				
Alloantigen	T:	Helper	Suppressor	Cytotoxic	MLR responder	Mitogen	B:	Ab-producing	Mitogen responder	MLR stimulator	Other
Ly-4[a]	±						+	±	+	±	Eosinophils, macrophages, neutrophils
Ly-5	+	+	+	+		+	+		+		Macrophages, eosinophils, neutrophils, mastocytomas, thymomas, B-cell lymphomas, myelopoietic cells, natural killer cells
Ly-6[b]	+	–	+	+	+	+	+	±	+	–	Macrophages, neutrophils, bone marrow cells, kidney, liver, brain
Ly-7[c]	+	+	–	+	+	+	±	±	+	–	
Ly-8	±						±	±			
Ly-9[d]	+	+	+				+				Brain, liver
Ly-10[e]	±						±				Bone marrow cells, progenitors, progenitors for granulocytes and megakaryocytes

Ly-11	±	±	Kidney
Ly-12	±	±	
Ly-13	+	+	
Ly-14	+	+	Brain, kidney, liver, erythrocytes
Ly-15	+	+	Liver, null cells
Ly-16	±		Brain
Ly-17[f]	–	±	Liver
Ly-m18	±	±	Stem cell tumor lines, bone marrow cells, kidney, brain, liver, granulopoietic cells
Ly-m19[g]	±	±	
Ly-m20	±	±	
Ly-m21			Increased expression T leukemia cells

[a]Ly-4: polymorphism of β_2 microglobulin? Alloantibody response to Ly-4 regulated by H-2D-region genes.
[b]Ly-6: Probably identical to ALA-1, REN-1, and DAG, closely linked to antigen defined by H9/25 (Takei, 1982).
[c]Ly-7: Alloantibody response to Ly-7 regulated by H-2I, E/C-region genes.
[d]Ly-9: Identical to T-100 and Lgp100.
[e]Ly-10: Appears on immature T cells.
[f]Ly-17: Found predominantly on immature B cells.
[g]Ly-m19: Absent from nonlymphoid tissues.
[h]Kennard and Meruelo (1982).
[i]+, all cells; ±, subpopulation.

antigenic determinants on the hormone. The fine distinction between the pharmacological versus antigenic nature of the receptor resides in the nature of the biochemical events that occur subsequent to hormone binding. In this section the emphasis is on defining the pharmacological agents that can indeed bind to lymphocytes and evoke hormone-specific responses.

Histamine

As a vasoactive hormone, histamine is involved in a number of biologically diverse functions. When released from mast cells, histamine contributes to allergic and anaphylactic reactions and inflammation; histamine is also involved in gastric acid secretion, neurotransmission, and rapid tissue proliferation, especially in wound healing and some experimental tumors. Biochemically, histamine binding activates cyclic AMP and alterations in intracellular cyclic AMP have profound regulatory influences on immune function. Histamine binding has been demonstrated on both murine B and T cells as well as on human lymphocytes. The function of histamine-mediated activation of lymphocytes awaits biochemical analysis using purified subsets of cells; in the murine system, the histamine receptor identifies subpopulations of T-suppressor cells, precursors of cytotoxic T cells, and cortisone-resistant thymocytes.

Insulin

Both murine and human lymphocytes acquire receptors for insulin upon activation, particularly in mixed lymphocyte reactions; the appearance of the insulin receptors is an additional assessment of cellular activation and/or correlates with biochemical requirements induced by activation. Insulin can also bind to macrophages and in this capacity functions to initiate specific immune responses as measured by subsequent T-cell proliferation and/or antibody formation by B cells.

Glucocorticosteroid

Receptors for glucocorticosteroids have been detected on T cells from rat, mouse, and human, and on human leukemia cells. The receptor appears heterogeneous on different subpopulations of human T cells and there is no apparent correlation between the number of receptors exposed on the cell surface and the sensitivity of that cell to the glucocorticosteroid; however, the number of receptors on cells may be related to the state of immunological maturation (Ranelletti et al., 1981).

Acetylcholine

Acetylcholine receptors are found predominantly at neuromuscular junctions. In the autoimmune disease myasthenia gravis, antibodies to this receptor are found both in circulation and in the thymus. Recent studies have detected an acetylcholine receptorlike antigen on the surface of human thymus cells using antibody. However, the receptor has not yet been biochemically purified from lymphoid cells and compared with the receptor isolated from neuromuscular junctions. The exposure of T cells, but not B cells, to acetylcholine results in the depolarization of the membrane. These findings suggest that T lymphocytes possess a receptor for acetylcholine, although the function of the receptor and its relationship to immunological communication has not been established (Shapiro and Strom, 1980).

E. Lymphokines

Lymphokines are nonimmunoglobulin effector molecules produced by lymphoid cells and are now being demonstrated to exert profound regulatory influences on lymphoid cells. Among these molecules are the interferons and the interleukins, which are nonpolymorphic within a species. Both of these classes of molecules can be absorbed by different lymphoid cells, strongly suggesting that specific receptors for these molecules exist. Three interleukins (IL) have been identified. IL-1 is produced by macrophages and monocytes and augments the proliferative responses of mitogen- and antigen-activated T cells. IL-2 is produced by T cells and promotes T-cell growth. IL-3 is biochemically distinct from IL-1 and IL-2, and is produced by mitogen-activated spleen cell cultures. IL-3 activates mature T-helper cells, as measured by enhanced enzyme activities, and also promotes the growth of mast cells. As mentioned above, the activities of these interleukins can be absorbed by lymphoid cell preparations, indicative that B and/or T cells are targets of their biological activity by virtue of a specific receptor, and the receptors are now being isolated and characterized.

Interferons are biologically active proteins that are described by their ability to inhibit viral replication nonspecifically, and by their cellular origin. These molecules are produced and secreted by both lymphoid and nonlymphoid cells. Type I interferon is produced by virus-induced lymphocytes (α-interferon) or fibroblasts (β interferon). Type II interferon (γ interferon) is produced by LPS-activated B cells or antigen-stimulated T cells.

Biochemically, type II interferon possesses two components, with molecular weights of 90,000 and 40,000. Both type I and II interferons have immunoregulatory effects on both T and B cells, although receptors for these molecules have not been isolated. Type I interferon increases the expression of H-2K and H-2D antigens, but not Ia, on cell surfaces, as well as augmenting the number of antigen binding lymphocytes. Type II interferon enhances suboptimal mitogen or antigen activation of lymphocytes. The activities of the interferons are reviewed elsehwere (Stewart, 1981).

III. T-CELL ANTIGENS

The antigens described in this section define subsets of T lymphocytes, yet no functional activities for these surface molecules have been defined. The biochemical, genetic, and functional characteristics of the mouse and human antigens are summarized in Tables 3 and 4.

A. Thy-1

The Thy-1 antigen was the first serologically defined alloantigen that distinguished a major subpopulation of murine lymphocytes. This antigen is also present on peripheral nerve tissues, brain tissues, fibroblasts, epidermal cells, and fetal skeletal muscle but is absent from immunoglobulin-positive B cells. While the original alloantisera also possessed other antibodies against T-lymphocyte antigens, the availability of monoclonal anti-Thy-1 antibodies now confirms that the Thy-1 antigen is present on all T lymphocytes, although the density of Thy-1 does vary among different T-cell subsets. Functionally, the elimination of Thy-1-positive cells results in a complete loss of all known T-cell responses. The role of the Thy-1 antigen, has not been defined specifically, although it is speculated that molecules whose expression is regulated during differentiation may be involved in the control of cell–cell interactions.

Table 3 Biochemical, Genetic, and Functional Characteristics of Antigens That Identify T-Cell Subsets

Alloantigen	Identification	Number of alleles	Chromosome	Linkage	Molecular weight and composition	Function	Human equivalent
Thy-1	Xenoantisera, alloantisera, monoclonal Ab	2	9	*Ly-6*, *Ly-8*, *Th-B*	20,000–30,000; glycoprotein, glycolipid	Present on all mature and immature T cells	
TL	Alloantisera, monoclonal Ab	5	17	*H-2*, *Qa*	45,000 and 14,000; β_2 microglobulin	Present only on thymus cells and some leukemias	Leu-6, OKT6, NA1/34
Lyt-1	Alloantisera, monoclonal Ab	2	19	*Ly-10*	67,000; single glycoprotein chain	Helper, amplifier, regulatory functions	Leu-1, OKT1
Lyt-2	Alloantisera, monoclonal Ab	2	6	*Lyt-3*	68,000–75,000; glycoprotein dimers of 30,000 and 35,000	Suppressor, cytotoxic, regulatory functions	Leu-2a,2b; OKT8
Lyt-3	Alloantisera, monoclonal Ab	2	6	*Lyt-2*	35,000; glycoprotein	Suppressor, cytotoxic, regulatory functions	

Table 4 Human T-Cell Surface Antigens and Receptors

Designation	Function(s)	Molecular weight and composition	Expression
OKT1, Leu-1, $T1_A$		69,000	All mature peripheral T cells, medullary thymocytes
OKT3, Leu-4, $T3_A$		20,000	All T cells, cortical thymocytes mitogenic for T cells
OKT4, Leu-3a, Leu-3b, $T4_A$	Helper/Inducer T cells for class II antigen cytotoxic cells	62,000	Subpopulation of peripheral T cells, majority of thymocytes
OKT5, OKT8, Leu-2a, Leu-2b, $T8_A$	Suppressor cells and class I antigen cytotoxic cells	76,000, dimer of 30,000 and 32,000	Subpopulation of peripheral T cells
OKT6		49,000	Cortical thymocytes, homologous to murine TL
OKT9	Transferrin receptor	190,000 and 94,000; dimer	Subpopulation thymocytes, activated T cells
$OKT10_7$		37,000	Majority of thymocytes, T cells activated
OKT11, 9.6, Leu-5, T11	Sheep erythrocyte (E-Rosette) receptor	55,000	All T cells and thymocytes

The Thy-1 antigen first appears on the rapidly dividing cortisone-sensitive cells in the cortex of the murine thymus between days 15 and 16 of gestation. These cortical-thymic cells have a high density of Thy-1, whereas in the medular-thymic areas, where immunocompetent T cells are located, the density of Thy-1 is lower. The T cells from the thymic medulla migrate to peripheral circulation and organs, where these cells acquire immunocompetence, through post-thymic maturation. In the absence of a thymus, nude mice fail to develop immunocompetent and functional T cells. A small percentage of spleen cells from nude mice possess a low density of Thy-1, suggesting that T-cell precursors possess Thy-1 and may be analogous to the low-density Thy-1-positive cells present in adult bone marrow; the latter cells, termed *prothymocytes*, differentiate in vitro into phenotypically mature T cells by the addition of various thymic hormones. Receptors for either thymic hormones or thymic polypeptides have not been isolated and/or characterized.

B. Thymus-Leukemia Antigens (TL)

The TL alloantigens are present only on some leukemias and immunologically incompetent, immature, Thy-1-positive thymocytes. The TL-positive, cortisone-sensitive medullar thymocytes are those capable of differentiating into terminal deoxytransferase-positive and Lyt-1,2,3-positive, functional T cells.

Biochemically, the TL antigens are similar to H-2K, H-2D, or H-2L antigens (Flaherty, 1979). The TL antigen exists in either dimeric or tetrameric conformations of heavy and light chains; the light chain is nonconvalently associated with the heavy chains and is β_2 microglobulin. There appears to be an inverse relation between the expression of TL antigens and H-2D antigens on thymic cells, which may suggest phenotypic regulation of expression of these antigens during differentiation of thymus cells. In addition to H-2D and TL, other alloantigens encoded for in this region, Qa, Ia, and C4, may also be under the same regulatory influences with regard to their expression or the density of their expression; this regulation may reflect the number of gene copies for that antigen in a given cell type, which can be evaluated using the appropriate cDNA probe.

C. Lymphocyte Antigens on T Cells (Lyt)

Ly-1 (Lyt-1)

This alloantigen is expressed predominantly on T cells but has been detected on Thy-1-negative germinal center cells and primary follicles of lymph node and spleen, and on certain B-cell lymphomas (Lanier et al., 1981). The presence of Ly-1 on murine T cells has not yet been correlated with a discrete function. Ly-1 is present on all T-cell subsets, although the density of this antigen varies. The presence of Ly-1 (high density) on murine T cells is associated with helper and/or amplifier functions, graft rejection and delayed-type hypersensitivity, and regulation through helper and suppressor activities. The interactions of these cells in regulatory circuits are discussed in detail in other chapters.

Ontogenetically, hematopoietic stem cells migrate from the yolk sac to the bone marrow to the thymus; Ly-1,2,3-positive cells arise in the cortex of the thymus and are present on both cortisone-sensitive and cortisone-resistant cells (thymocytes). Since the phenotype of the cortisone-resistant thymocytes is that seen in the peripheral tissues, it has been suggested that this population of cells migrates out of the thymus for post-thymic maturation. Alternative pathways of differentiation are also feasible; in the spleen, Ly-1,2,3-positive cells appear by 1 week of age, with subsequent increases in the frequency of Ly-1- and Ly-2-3-positive cells, reaching adult levels by 10 weeks of age.

Monoclonal antibodies to defined subpopulations of cells offer an improved therapeutic potential when compared with hetero- or alloantisera. Hence the in vivo administration of anti-Thy-1 alloantisera has not resulted in a generalized immunosuppression of the T-cell compartment, perhaps due to the high density of Thy-1 on T cells and other cell types or due to the rapid clearance and/or proteolytic degradation of the antibodies in vivo. In contrast, an antilymphocyte heteroantisera does result in generalized immunosuppression, although anti-idiotypic and antiallotypic antibodies develop. Recent studies using monoclonal mouse anti-Ly-1 in vivo have demonstrated T-cell specific suppression of all immune functions studied, including allograft rejection, T-cell-dependent antibody production, delayed-type hypersensitivity, and cytotoxic T-cell responses; a concomitant decrease in Ly-1-positive cells in the spleen and lymph node was also documented. In primate systems, the administration of mouse anti-human Leu-1 (Ly-1 equivalent)

prolonged renal allograft survival in monkeys. The mechanism of action of anti-Ly-1 through either cytotoxic elimination and/or antigenic modulation of an important cell surface receptor function remains to be determined (Michaelides et al., 1981).

Ly-2 (Lyt-2)

Ly-2 is genetically linked to the locus for *Ly-3* and the antigen is coexpressed on T cells with Ly-3; thus antibodies to either Ly-2 or Ly-3 will functionally block both of these antigens on the cell surface. However, the biochemical data indicate that by sequential immunoprecipitation, Ly-2 and Ly-3 are unique molecules and not different epitopes. T cells possessing Ly-2 and Ly-3 also express a low density of Ly-1. Functional Ly-2/3-positive cells differentiate post-thymically from Ly-1,2/3-positive cells.

Although the function of the Ly-2 antigen has not been determined, the presence of this antigen has been a useful marker for the identification of a major T-cell subset. The activities of the mature cell population include both suppressor and regulatory functions. The Ly-2/3-positive cells are cytotoxic effector cells for both allogenic and viral determinants, suppressor cells in antigen-specific, concanavalin A (ConA)-induced and allotype suppression systems, and secrete antigen-specific suppressor factors. The cells with the latter function have been fused with thymoma cells, cloned, and now exist as permanent cell lines, secreting their antigen-specific suppressor factors.

Ly-3 (Lyt-3)

As discussed above, Ly-3 and Ly-2 are coexpressed on T cells. The biochemical and functional relationships between this molecule and the T-cell subset it defines are currently unknown.

D. I-Region Antigens

The I region of the *H-2* gene complex encodes for Ia molecules that are associated with immune responses and the interactions of different cell types during an immune response. Although the Ia molecule is closely associated with immune function, the precise biochemical events that occur subsequent to immune-recognition interactions are not currently known. Ia molecules are expressed predominantly on B cells and macrophages; however, activated T cells and functionally differentiated T cells do express Ia (Natali et al., 1982; Lafuse and David, 1982). The biochemistry of the Ia molecule is considered in more detail in Sec. V.

Ia molecules, in particular molecules encoded for by the I-J subregion, are expressed on antigen-specific suppressor T cells. These determinants are intimately associated with the idiotypic determinants for antigen on these cells. Furthermore, antigen-specific suppressor factors, secreted by cloned T cells, possess I-J determinants as well as the idiotypic determinants for antigen (Okuda et al., 1980).

Ia molecules encoded for by the I-A and I-E/C subregions are present on mitogen- and alloantigen-activated T-cell blasts. The expression of these determinants has not yet been correlated with biochemical function in either human or murine systems.

T-cell recognition of antigens presented by macrophages is associated with recognition of appropriate Ia determinants. Ia antigens expressed on B cells are potent stimulators of T-cell proliferation in autologous as well as allogeneic mixed lymphocyte reactions. The mechanism of T-cell recognition of self- and allo-Ia molecules has not been deciphered.

IV. T-CELL RECEPTORS

In this section the functional and biochemical role of T-cell surface receptors will be discussed, with emphasis on antigen-specific and factor-specific associations. Surface receptors on human T cells have been included in Table 4.

A. Antigen Receptors (Idiotype)

Immune responses to most antigens are regulated by T lymphocytes; however, little is known about the T-cell recognition structure for the antigen. The evidence is clear that B cells recognize antigen through immunoglobulin present on the cell surface, specifically through the antigen binding site (idiotype). The initial studies of the antigen receptor on T cells were confused by the data suggesting that this receptor was genetically linked to the major histocompatibility locus (*I-J*) rather than to the genes coding for heavy or light chains of immunoglobulin. Recent data, utilizing techniques of somatic cell hybridization, suggest that on regulatory T-suppressor cells, both I-J and idiotypic determinants are associated with the receptor for antigen (Okuda et al., 1980).

Biochemically, the T-cell receptor for antigen possesses the variable region of the heavy chain and the idiotype of the immunoglobulin molecule. The in vivo administration of anti-idiotypic antibodies to naive animals results in the production of idiotype-specific antibodies in the absence of the specific antigen (Bluestone et al., 1981). These data strongly suggest that idiotypic determinants are expressed on T cells and that the binding of this molecule results in the functional activation of the T cell. Attempts to achieve the same result using different antigen systems have resulted in suppression, possibly due to the selective activation of T-suppressor cells rather than T-helper-cell subpopulations.

Surface immunoglobulin from T cells has not been isolated by current biochemical methodology; this may be due to the lack of the appropriate reagents or due to the low affinity and/or low number of Ig-like, antigen-specific receptors on T cells. Thus, in the case of the idiotype receptor for antigen on T cells, as well as all other receptor molecules, methods to detect very small numbers of antigenic sites on cells are very much needed; the regulation of lymphoid cell populations and their interactions may then be very dependent on the density of these surface molecules in relation to other surface molecules.

B. Fc Receptors

T cells have been shown to have receptors for the constant (Fc) portion of the immunoglobulin molecule. The binding of the Fc molecule is not species specific. Separate Fc receptors for all immunoglobulin isotypes have been identified on various cell types, particularly on macrophages. Fc receptors for IgM and IgA have been detected on human T cells, but the IgG isotype specificities of Fc receptors on human or mouse T cells have not been established.

Isolated as a class of receptors, Fc receptors consist of phospho-, lipo-, and glycoproteins with a molecular weight of 68,000. The biochemistry of isotype-specific Fc receptors has not been determined.

The function of the Fc receptor as expressed on T cells has not been determined, although Fc-positive T cells are ConA and phytohemagglutinin (PHA) mitogen responsive, while Fc-negative T cells respond only to PHA. Cytotoxic T cells possess the Fc receptor, although the precursor of these effectors are Fc-receptor negative.

C. Factor Receptors

A number of factors with unique characteristics have been described and although these factors may be bound by distinct cell types, the relationship between recognition and cell activation has not been determined.

I-A-Region Factors

Factors that bear I-region determinants have been described to have both enhancing and suppressing functions, some of which are antigenically and genetically restricted and others that are nonspecific. Antigen-pulsed macrophages produce genetically restricted helper factors that bind to subpopulations of T cells. T-suppressor cells produce suppressor factors that bear both I-J and idiotypic determinants and act on other Ly-2-positive T cells. The action of these factors on T-suppressor cells is not genetically restricted, but is antigen specific. A genetically restricted suppressor factor has been described that also bears I-J, antigen, and idiotypic determinants; this factor binds to nylon wool-adherent Ly-1,2,3-positive T cells. Genetically restricted helper factors are also antigen specific and bind to T cells via an *I-A*-region encoded molecule.

Non-I-A-Region Factor

ConA-activated T cells produce a 67,000-molecular weight glycoprotein that suppresses in vitro human immune responses. This factor is also absorbed by mouse spleen and thymus cells, suggesting the presence of receptors for this molecule. However, the relevant site of action may actually be the macrophage that is then activated to produce suppressor factors. The ConA-activated factor is neither genetically nor antigenically restricted and may in fact be one of the interleukin molecules.

D. Lectin Receptors

The ability of lectins to bind to cell surfaces is discussed elsewhere in this book. The specificity of the lectins for particular sugar residues has been well characterized; however, the biochemistry of the events that occur subsequent to receptor-ligand interaction and resulting in cellular proliferation has not been examined.

Several lectins, such as peanut agglutinin (PNA), are useful in identifying subsets of lymphocytes; high-density PNA-positive T cells are located primarily in the cortex of the thymus, suggesting that this lectin is a marker for immature T cells and germinal center cells. Neuraminidase treatment renders all T cells positive for PNA, indicating that the lectin binding site may be present on all T-cell surfaces, although its expression, as measured by density, may vary among the different subpopulations. PNA binds to membrane glycoproteins with molecular weights of 130,000 to 150,000 on human T-lymphoblastoid cell lines. The fine specificity of the binding is therefore dependent on the sensitivity of the assay used to detect surface receptors for PNA and other sugar binding sites.

V. B-CELL ANTIGENS

A. Lyb Antigens

Antigens that are restricted to the murine B-lymphocyte lineage are referred to as Lyb antigens and have been reviewed recently (Ahmed et al., 1982). The biochemistry, genetics, and functions of these antigens are shown in Table 5.

Table 5 Biochemical, Genetic, and Functional Characteristics of Antigens That Identify B-Cell Subsets

Alloantigen	Identification	Number of alleles	Chromosome	Linkage	Molecular weight and composition	Functional characteristics
Lyb-2	Alloantisera, monoclonal Ab	3	4	*Lyb-4*, *Lyb-6*, LPS receptor, *Mup*, *Ly-m19*		Expressed on: colony-forming B-cell precursors, antibody-secreting B cells
Lyb-3	Alloantisera,	1	–		68,000	Differentiation antigen identifying mature B cells; antisera augments in vivo antibody response to sub-optimal doses of antigens
Lyb-4	Alloantisera	2	4	*Lyb-2*, *Lyb-6*, LPS receptor, *Mup*, *Ly-m19*	44,000; structurally similar to H-2K or H-2D, without β_2 microglobulin	Antisera blocks Mls stimulatory cells; ontogenetically, appears during first week of life
Lyb-5	Alloantisera	2	–			Present on mature B cells with low sIgM/sIgD ratio, B cells responsive to type II T-cell-independent antigens
Lyb-6	Alloantisera	1	4			Antisera stimulates proliferation of B cells
Lyb-7	Alloantisera	2	12	Ig heavy chain		Antisera blocks type II T-cell-independent antigen responses in vitro
14G8	Monoclonal Ab	1	–			Immature B cell, cells responsive to anti-IgM stimulation

Interestingly, the genes for several of these antigens are linked together, forming a cluster of loci coding for cell surface antigens. The functional significance of gene clusters for cell surface molecules and regulatory proteins is currently unknown. However, it is becoming readily apparent that many such clusters exist and the molecular biology in the future will be instrumental in determining whether these arrays are fortuitous or coordinated for the expressions of cellular differentiation antigens. Analogous clusters of genes for other cell surface antigens are eluded to in the tables describing features of cell surface antigens; clusters of genes outside the MHC locus in the human have not yet been described (Tada et al., 1981a).

Lyb-1 (Ly-4)

This alloantigen was originally thought to be present only on B cells; however, when T cells were shown to possess this alloantigen as well, it was redisgnated Ly-4 (see Sec. III.C). Together with the genes coding H-3 and H-13 histocompatibility antigens, Ir-2, and polymorphisms of C5 and β_2 microglobulin, these loci on chromosome 2 of the mouse manifest a striking similarity to the major histocompatibility locus on chromosome 17. Similar loci have not yet been identified in the human. The relationship of these clusters of genes to the cell surface expression of these antigens, either evolutionarily or functionally, needs further investigation.

Lyb-2, Lyb-4, and Lyb-6

The functions of these antigens, present on B cells, have not been described. The genes coding for these three B-cell alloantigens are closely linked to each other on chromosome 4. These loci exist as a cluster of genes that also include genes for the LPS receptor, present on both B cells and macrophages, Ly-m19, present on B and T cells, and the major urinary proteins (MUP).

Lyb-3, Lyb-5, and Lyb-7

These alloantigens are defined by structural alleles that are present on all strains of mice except CBA/N. The CBA/N mice possess an X-linked immune defect gene (*Xid*) that results in these mice lacking a mature subpopulation of B lymphocytes (Scher, 1981). The biochemical nature of the *Xid* gene mutation has not been defined although this strain has been very useful in the delineation of B-cell alloantigens, especially Lyb-3, Lyb-Lyb-5, Lyb-7, and 14G8.

The Lyb-3 alloantigen is a differentiation antigen, present on mature B cells. Antibody to this antigen augments antibody responses to suboptimal doses of antigens, suggesting that this antigen is important in the amplification of activation signals to B cells.

The Lyb-5 alloantigen also defines a subpopulation of mature B lymphocytes, which are absent in CBA/N mice. Lyb-5 is present on complement-receptor-positive, immunoglobulin-positive, and Mls-positive B cells with a low density of surface Ia and IgM and a low-to-intermediate density of sIgD. Ontogenetically, Lyb-5 appears at 2 weeks of age and reaches adult levels by 6 weeks of age. Functionally, the B cells defined by Lyb-5 respond to type II T-cell-independent antigens, typified by polysaccharide antigens. Lyb-5-bearing B cells are not genetically restricted by MHC antigens in their cooperation with T cells to produce antibody; in contrast, the Lyb-5-negative B cells, typified by immature B cells from CBA/N mice, are MHC restricted (Singer et al., 1981).

The biochemical role of this antigen in discriminating these B-cell subsets has not yet been described, nor have potential regulatory effects of this alloantigen been described.

Lyb-7 is an alloantigen that is functionally described by the ability of the antisera to it to inhibit antibody formation to type II T-cell-independent antigens in in vitro cultures of spleen cells; antibody responses to type I T-cell-independent antigens and T-cell-dependent antigens are not affected. The biochemical basis for this inhibition is not known, although the target cell is presumably the mature, Lyb-5-positive subset of B cells.

14G8

This alloantigen is defined by a monoclonal antibody that reacts with the immature B cells and B cells from the CBA/N mice. Neither the biochemistry of this antigen nor the function of this molecule have been characterized. However, this antigen may prove useful in identifying immature B-cell subpopulations that are responsive to stimulation by anti-IgM (Kung et al., 1982).

B. Minor Lymphocyte-Stimulatory Locus (Mls) Antigens

Mls Antigens

These antigens were originally described in strains of mice that were H-2 identical, yet were capable of generating mixed lymphocyte reactions between each other; four alleles have been described for these determinants: a, b, c, and d, where only the b allele is nonstimulatory. The *Mls* locus is part of a cluster of genes that code for cell surface antigens, analogous to the I-region encoded-stimulatory antigens and Lyb alloantigens (Tada et al., 1981b). The biochemistry of the Mls determinants is currently unknown and therefore a structural comparison with H-2 encoded determinants is not possible.

The ontogenetic appearance of Mls determinants after immunoglobulin and complement receptors, and the absence of Mls determinants from Lyb-5-negative B cells, indicates that these determinants are alloantigens of mature B cells. The function of these determinants is not fully appreciated except that they serve to activate T cells to proliferate and secrete lymphokines.

LyM-1

The LyM antigen is also a stimulatory antigen present on B cells. The genes encoding for this alloantigen are closely linked to the *Mls* locus, but separable by recombination. These loci are also closely linked to *Ly-m9, Ly-17*, and *Ly-m20* on chromosome 1, forming another gene cluster (Kimura et al., 1981). By strain analysis of the cytotoxicity effected by this antisera and complement, and comparison with the known Mls types, three patterns of reactivity are apparent, suggesting that three alleles exist for the LyM alloantigens system. Functionally, the LyM-1-positive spleen cells are those that respond to LPS-induced mitogenic stimulation and are antibody-producing cells. The biochemical composition of the LyM-1 molecule is unknown.

C. Immune-Associated Antigens

The immune-associated (Ia) antigens are a series of cell surface molecules encoded for by genes of the I region of the major histocompatibility locus in mice and by genes of the

DR, SB, MB, and DC regions in humans. The murine I region is divided into five subregions, I-A, I-B, I-J, I-C, and I-E (reviewed by Lafuse and David, 1982). Biochemically, Ia antigens are glycotproteins consisting of three nondisulfide- and noncovalently linked polypeptide chains: the α chain (MW 30,000), the β chain (MW 27,000), and a third invariant polypeptide chain. The invariant chain has been identified by two-dimensional gel electrophoresis and appears to be nonpolymorphic; neither its gene location nor its functional relationship to the α and β polypeptides has thus been identified. The genetic mapping of private Ia specificities of the α and β chains to the subregions of the I region has been done using recombinant strains of mice; the biochemistry of the molecule has been studied primarily by immunoprecipitation. These analyses have indicated that polypeptides coded for by different subregions combine to form the cell surface Ia molecule in F_1 hybrid mice (Auchincloss et al., 1981); the appearance of different I-subregion products on different cell types has suggested that the function of the Ia molecule is related to the specificity of the subregion peptide associations. For example, the predominant Ia specificity expressed on T cells maps to the I-J subregion and the T cells expressing these Ia molecules are functionally involved in suppressive interactions with B cells. In addition, an Ia specificity mapping to the I-A subregion (Ia.w39) is associated only with a mature subpopulation of B cells, suggesting that this Ia molecule is a differentiation antigen (Huber et al., 1981). Thus the relationship between the expression of different I-region determinants and cell function needs to be evaluated at the receptor level.

As mentioned earlier, Ia antigens possess a wide tissue distribution. In addition to B cells, macrophages, and activated T cells, Ia antigens have been detected on epidermal cells, eosinophils, neutrophils, endothelial cells, Langerhans cells, dendritic cells, and sperm in both mouse and human.

Functions that have been ascribed to the I-region gene products include immune responses to specific antigens and mixed lymphocytes reactivity. The use of products of hybridoma technology has resulted in studies that indicate that immune responses to antigens are regulated by specific Ia epitopes; however, the disadvantage to these antibody blocking experiments is that the antibody may sterically interfere with other receptor functions necessary for immune function. Thus there is evidence to support I-region associations with immune responses to particular antigens, but the biological activities of recognition, presentation, or amplification have not been biochemically analyzed.

T cells recognize allogeneic and autologous Ia determinants on B cells and respond by proliferation. Thus Ia determinants may serve as a powerful recognition molecule for discrimination of self versus nonself. The human DR, MB, DC, and SB antigens are equivalent to I-region-controlled antigens, although several of these antigens are coded for outside the human MHC, in contrast to the murine MHC. A number of immune responses in the human now appear to correlate with a particular DR antigen, similar to those immune associations seen in the mouse.

VI. B-CELL RECEPTORS

A. Immunoglobulin

The immunoglobulin (Ig) molecule on the cell surface of B lymphocytes not only serves as an antigen to identify this subpopulation of lymphocytes, but also serves to recognize and bind antigen specifically. The major surface immunoglobulins are sIgM and sIgD,

which share the same idiotype and differ slightly from their serum counterparts in molecular weight and carbohydrate content. Ontogenetically, murine sIgM is first detected in the cytoplasm of fetal liver cells and subsequently appears on the cell surface before birth. The density of sIgM on these cells is very high and upon further differentiation and maturation, decreases. Immature B cells are therefore characterized by a high density of sIgM. IgD first appears on the cell surface after birth, at which time the density of sIgM decreases.

With the acquisition of sIgD, functional changes are also observed. Immature B cells are very easily rendered unresponsive to antigens, or tolerized, whereas adult B cells are resistant to antigen-induced tolerance. Thus the presence of sIgD may facilitate B-cell activation and resist tolerance induction. The sIgM of immature B cells is very easily modulated, as demonstrated by the irreversible capping of these determinants with anti-Ig reagents; in contrast, sIg of mature B cells is much more difficult to modulate by capping, but is reversible and followed by reexpression of sIg.

Functionally, sIg is the receptor for antigen on B cells. Both sIgM and sIgD share the same idiotype for antigen on B cells. Both sIgM and sIgD share the same idiotype for antigen and the precise role that each of these isotypes plays in the activation of the B cell to secrete antibody is not fully appreciated.

The antibody secreted by B cells shares the idiotype of its cell-surface receptors. As the committed B cell differentiates, the isotype of the antibody changes. T cells have been demonstrated to contribute to and/or regulate this isotype switch, yet this biochemical mechanism for the switching process is not understood (Mongini and Paul, 1981; Rosenberg et al., 1981). At the DNA level, the genes for the heavy chain classes of Ig are closely linked on chromosome 12 in the 5′ to 3′ order IgM-IgD-IgG3-IgG2a-IgG1-IgE-IgA. The isotype switching is associated with a deletion of constant regions to the left of the genes being transcribed and is therefore both directional and irreversible. Thus the isotype switch is a differentiation step wherein the selective loss of sIgM and sIgD is followed by the cell surface expression of the isotype of the secreted Ig (Adams, 1980).

An additional role of sIg is the recognition of T cells. Antigen-activated T cells express antigen binding idiotypic determinants comparable to those expressed by the antigen-specific B cell. Indeed, T cells express a functional preference for B cells sharing the same idiotype. During the course of an immune response, a network of idiotype and anti-idiotype antibodies is established which potentially regulates B-cell antibody production to the original antigen. The profound modulatory capacity by the idiotype network is demonstrated by the elimination of idiotype-specific clones of B or T cells in mice treated from birth with anti-idiotype antibody (Kelsoe et al., 1980). The structure of Ig molecules, both secreted and membrane bound, has been discussed in more detail in Chap. 7.

B. Fc Receptors*

The Fc portion of the immunoglobulin molecule dictates a number of isotype-specific effector functions, best exemplified by the ability of certain Ig isotypes to activate the classical pathway of complement. As mentioned previously, Fc receptors are present predominantly on B cells and macrophages, but also on many other cells, including T cells, null cells, mast cells, basophilic and other granulocytes; isotype-specific Fc receptors have been identified for all classes and subclasses of immunoglobulin. IgE-Fc receptors have been well characterized on mast cells and the IgG-Fc receptors present on

*See also Chap. 29.

macrophage cell lines have been isolated and characterized. The Fc receptors on B cells have not been characterized biochemically, but a number of functional studies using Fc-receptor ligand interactions have been done to describe their functional significance to the B cell (Unkeless et al., 1981).

The interaction of Fc receptors with purified Fc fragments results in proliferation. When B lymphocytes are activated in vitro by either anti-IgM or anti-IgD antibody in the presence of physiologically relevant concentrations of monomeric or aggregated IgG, specific ligand-Fc receptor associations occur. Whenever sIgM is involved in cell activation, the Fc receptor, in the presence of monomeric IgG, is also involved in the response; in contrast, sIgD-mediated activation of B cells is associated with Fc receptor interactions with antigen-antibody complexed IgG Fc fragments. Thus Fc receptors appear to be involved in a number of activation-signlaing events, although the precise biochemical mechanisms involved in this signaling remain undefined (Dickler et al., 1982).

C. Complement (C) Receptors*

B cells, macrophages, and null cells all possess at least two distinct receptors for the C3-derived fragments. T cells also possess a C3 receptor, but it is present in lower density than is observed on B cells. B cells from both the mouse and human possess these receptors, and the reagents for typing these receptors usually detect both the C3b-C4d (CR-1) and C3d-C3bi (CR-2) receptors. CR-1s are also present on human natural killer cells and lymphoblastoid cell lines. C receptors are cell surface antigens of mature B cells as defined ontogenetically, functionally, and by their absence from the CBA/N mice (Lindsten and Andersson, 1981; Schier et al., 1981).

C-receptor-positive B cells define a subset of B cells that respond to T-cell-independent (type II) antigens and respond well to anti-Ig-induced proliferation; the C-receptor-negative B cells do not respond to these stimuli. Both of these B-cell subsets respond to T-cell-dependent antigens.

The use of agents that deplete the complement system, such as cobra venom factor, have been used both in vivo and in vitro in efforts to study the influence of the complement system on the immune systems. Unfortunately, most of these studies have not resulted in a definition of the precise function of this system, owing primarily to the inhibitory activities of other antibodies and/or enzymes.

D. Receptors for β1H Globulins†

The complement system is composed of a series of serum regulatory proteins, ultimately resulting in either antibody-dependent cellular cytotoxicity or the clearance of immune complexes. One of the regulatory proteins of the complement cascade is β1H globulin (β1H). β1H is an essential cofactor for the cleavage of fluid-phase C3b by C3b inactivator and is a potentiator of C3bi-mediated cleavage of surface-bound C3b. A receptor for β1H has been demonstrated on the surface of human bone marrow and lymphoblastoid cell lines (Lambris et al., 1980).

When β1H is bound to the cell surface, endogenously synthesized C3b inactivator is released; when CR-1 is absent from or blocked on these cells, C3b (on C3b-coated red blood cells) is cleaved into C3d-C3bi; these red cells then form rosettes with the cell, through the CR-2 receptor. Thus, in vitro, the β1H receptor influences the binding of C3b to cell surfaces; its role in regulation of the immune response in vivo is not yet fully appreciated.

*See Chap. 23.
†The official nomenclature for β1H now is H; see Chap. 23, pp. 512–518.

E. Mitogen Receptors

LPS Receptor

Lipopolysaccharides (LPSs) are major components of the outer cell wall of gram-negative bacteria that display a number of potent immunopathological effects (Morrison and Ryan, 1979). LPSs, particularly the lipid A moiety, for example, are able to stimulate proliferation of B lymphocytes but not T lymphocytes. One structural allele for the receptor for lipid A has been identified using a xenoantisera. This antisera confirmed that the LPS receptor is absent from T cells but is present on B cells, bone marrow cells, Peyer's patch cells, fibroblasts, and macrophages. The gene coding for the receptor is located on chromosome 4 and is linked to the *Lyb-2, Lyb-4, Lyb-6* gene cluster. An analysis of the frequency of LPS receptors suggests that, analogous to the immunoglobulin molecules, the expression of this receptor may be allelically excluded.

LPSs are potent adjuvants of antigen-specific antibody synthesis, although the biochemistry of the receptor-ligand interaction to achieve this activity is not understood. LPSs are also able, in the absence of antigen, to stimulate nonspecifically antibody production by B cells. Cell surface membrane capping studies suggest that in some but not all cells, the LPS-receptor can cocap with IgM, Fc, and IgD receptors. These data suggest that receptor-mediated activation of the cell may require the involvement of more than one specific receptor. Bacterial products also contribute to the expansion and differentiation of the antigen-specific clones of B cells (Morrison and Ryan, 1979).

Dextran Receptor

Dextran is a B-cell activator of both proliferation and specific antibody synthesis. The mitogen receptor was identified using an anti-idiotype antiserum, and this receptor was demonstrated to be distinct from the antigen-specific immunoglobulin receptor on the B cell. The ontogenetic, genetic, and biochemical characterizations of the dextran receptor have not been reported.

Poly I:Poly C Receptor (PI:C)

Polyinosinic-polycytidilic acid (PI:C) provides a proliferative signal to murine B cells and its receptor has been isolated. PI:C receptors are present on some but not all LPS-receptor-positive cells, indicating that these receptors are distinct and occur on overlapping cell populations. The genetics, ontogeny, and biochemistry of this receptor are not yet established; however, it is known that PI:C can also stimulate production of IL-1 by macrophages.

F. Lectin Receptors: Peanut Agglutinin (PNA)

Receptors for PNA are found on both T and B cells, as mentioned previously. By rosetting and fluorescence assays, PNA is present at a very low density on splenic B cells and in higher concentrations on germinal center cells in the Peyer's patches and lymph node. The glycoprotein receptors for PNA on human B-lymphoblastoid cell lines have molecules weights of 83,000 and 150,000. In contrast to murine T cells, in which the PNA receptor is masked as the cell differentiates, the B-cell PNA receptor is gradually lost as pre-B cells mature to B cells and plasma cells (Newman and Boss, 1980).

G. Lymphokine Receptors

B-Cell Growth Factor (BCGF)

This lymphokine is biochemically distinct from IL-2 and appears to stimulate B-cell growth, especially among B-cell lines. The activity of BCGF can be absorbed by B cells, indicating that a receptor for this molecule exists on B cells. The proliferation induced by BCGF is synergistic with that induced by IL-2.

T-Cell Replacing Factor (TRF)

This lymphokine acts on B cells that have been clonally expanded after antigen stimulation and induces antibody secretion. This lymphokine is therefore distinct from BCGF since the responsive cell is different. The interrelationship of IL-2 with TRF is currently unknown. These factors were originally described by their ability to replace nonspecifically the need for T cells in in vitro antibody synthesis.

Allogeneic Effect Factor (AEF)

This lymphokine possesses activities similar to those of TRF, although it is produced in a different culture system. AEF and TRF also appear distinct based on a difference in the ontogenetic appearance of receptors for these lymphokines or B cells and the differential expression of receptors on B cells from CBA/N mice (Takatsu et al., 1981). Definitive biochemical analysis of these lymphokines and their receptors has not been determined.

VII. CONCLUSIONS

This chapter has attempted to summarize relevent information concerning receptor-ligand interactions of T and B lymphocytes. It is apparent that many lymphoid cell surface antigens have been described, but the analysis of the molecular, biological, and biochemical events that occur subsequent to receptor-ligand interactions needs more evaluation. Cell surface antigens as currently defined by antibodies may represent unique recognition determinants for cell-cell communication, or allelic variants of enzymes, substrates, or polymorphic structural proteins. However, their expression requires analysis at the level of both the protein and the DNA. Insights into both the structural and functional regulation imposed by clusters of genes for cell surface antigens as well as the biochemical activity of these molecules will provide the basis for the determination of the communication network among lymphoid cells as well as other nonlymphoid cells which share the same determinants. The presence of gene clusters for cell surface antigens (at least six such clusters have been identified in the mouse) provide unique opportunities for the molecular biologist to study the organization, expression, and function of unique classes of cell surface molecules and the genes that encode these molecules.

ACKNOWLEDGMENTS

We thank E. Frees and D. Sloan for their editorial assistance. The opinions contained herein are the private ones of the authors and are not to be construed as official or as reflecting the views of the Navy Department. This work was supported by the Naval Medical Research and Development Command, Bethesda, Md., under Work Unit #3M161102BS10.AI.430.

REFERENCES

Adams, J. M. (1980). Immunol. Today *1*, 10.

Ahmed, A., and Smith, A. H. (1982). Crit. Rev. Immunol. *3*, 331.

Ahmed, A., Kessler, S., Subbarao, B., and Humphreys, R. E. (1983). In *Differentiation Antigens of Lymphocytes* (N. L. Warner, Ed.). Academic Press, New York, in press.

Auchincloss, Jr., H., Ozato, K., and Sachs, D. H. (1981). J. Immunol. *127*, 1839.

Bluestone, J. A., Sharrow, S. O., Epstein, S. L., Ozato, K., and Sachs, D. H. (1981). Nature (Lond.) *291*, 233.

Chun, M., Fernandes, G., and Hoffmann, M. K. (1981). J. Immunol. *126*, 331.

Dickler, H. B., Kubicek, M. T., and Finkelman, F. D. (1982). J. Immunol. *128*, 1271.

Festenstein, H., and Schmidt, W. (1981). Immunol. Rev. *60*, 85.

Flaherty, L. (1979). In *Role of the MHC in Immunobiology* (M. Dorf, Ed.). Garland, New York, pp. 33–55.

Goodenow, R. S., McMillan, M., Nicolson, M., Sher, B. T., Eakle, K., Davidson, N., and Hood, L. (1982). Nature (Lond.) *300*, 231.

Huber, B. T., Jones, P. P., and Thorley-Lawson, D. A. (1981). Proc. Natl. Acad. Sci. (USA) *78*, 4525.

Kelsoe, G., Reth, M., and Rajewsky, K. (1980). Immunol. Rev. *52*, 75.

Kennard, J., and Meruelo, D. (1982). Immunogenetics *15*, 239.

Kimura, S., Tada, N., Liu, Y., and Hämmerling, U. (1981). Immunogenetics *13*, 547.

Kung, J. T., Sharrow, S. O., Ahmed, A., Habbersett, R., Scher, I., and Paul, W. E. (1982). J. Immunol. *128*, 2049.

Lafuse, W. P., and David, C. S. (1982). In *Ia Antigens* (S. Ferrone and C. S. David, Eds.). CRC Press, Boca Raton, Fla., pp. 105–125.

Lambris, J. D., Dobson, N. J., and Ross, G. D. (1980). J. Exp. Med. *152*, 1625.

Lanier, L. L., Warner, N. L., Ledbetter, J. A., and Herzenberg, L. A. (1981). J. Exp. Med. *153*, 998.

Lindsten, T., and Andersson, B. (1981). Cell. Immunol. *61*, 386.

McKenzie, I. F. C., and Potter, T. (1979). Adv. Immunol. *27*, 179.

McKenzie, I. F. C., Hogarth, P. M., and Potter, T. (1983). In *Differentiation Antigens of Lymphocytes* (N. L. Warner, Ed.). Academic Press, New York, in press.

Michaelides, M., Hogarth, P. M., and McKenzie, I. F. C. (1981). Eur. J. Immunol. *11*, 1005.

Michaelson, J. (1982). In press.

Mongini, P. K. A., and Paul, W. E. (1981). In *B Lymphocytes in the Immune Response* (N. Klinman, D. E. Mosier, I. Scher, and E. S. Vitetta, Eds.). Elsevier/North-Holland, New York, pp. 369–376.

Morrison, D. C., and Ryan, J. L. (1979). Adv. Immunol. *28*, 294.

Natali, P. G., Russo, C., Ng, A. K., Giocomini, P., Indiveri, F., Pelligrino, M. A., and Ferrone, S. (1982). In press.

Newman, R. A., and Boss, M. A. (1980). Immunology *40*, 193.

Okuda, K., Minami, M., Ju, S.-T., and Dorf, M. E. (1980). Proc. Natl. Acad. Sci USA *73*, 4557.

Ranelletti, F. O., Piantelli, M., Lacobelli, S., Musiani, P., Longo, P., Lauriola, L., and Marchetti, P. (1981). J. Immunol. *127*, 849.

Rosenberg, Y. J., Lieberman, R., and Asofsky, R. (1981). In *B Lymphocyte in the Immune Response* (N. Klinman, D. E. Mosier, I. Scher, E. S. Vitetta, Eds.). Elsevier/North-Holland, New York, pp. 385–392.

Sato, H., Kimura, S., and Itakura, K. (1981). J. Immunogenet. *8*, 27.

Scher, I. (1981). Crit. Rev. Immunol. *1*, 287.

Schier, R. D., Kao, M. Y., Hattori, M., and Moorhead, J. W. (1981). Cell. Immunol. *62*, 324.

Shapiro, H. M., and Strom, T. B. (1981). Proc. Natl. Acad. Sci (USA) *77*, 4317.

Singer, A., Morrisey, P. J., Hathcock, K. S., Ahmed, A., Scher, I., and Hodes, R. J. (1981). J. Exp. Med. *154*, 267.

Stewart, W. E. (1981). *The Interferon System.* Springer-Verlag, New York.

Tada, N., Kimura, S., and Hämmerling (1981a). In *Monoclonal Antibodies and T-Cell Hybridomas* (G. J. Hämmerling, U. Hämmerling, and F. Kearney, Eds.). Elsevier/ North-Holland, New York, pp. 38–44.

Tada, N., Kimura, S., Liu, Y., Taylor, B. A., and Hämmerling, U. (1981b). Immunogenetics *13*, 539.

Takatsu, K., Sano, Y., Tonita, S., Hashimoto, N., and Hamaoka, T. (1981). Nature (Lond.) *292*, 360.

Takei, F. (1982). Immunogenetics *16*, 201.

Unkeless, J., Fleit, H., and Mellman, I. S. (1981). Adv. Immunol. *31*, 247.

Woodward, J. G., Orn, A., Harmon, R. C., Goodenow, R. S., Hood, L., and Frelinger, J. A. (1982). Proc. Natl. Acad. Sci. USA *79*, 3613.

29

Fc Receptors on Mononuclear and Polymorphonuclear Phagocytic Cells

Keith J. Dorrington* University of Toronto, Toronto, Ontario, Canada

I. INTRODUCTION

The primary purpose of this chapter is to review the current state of knowledge regarding the structure and function of Fc receptors (FcR) present on the surface membrane of two classes of effector cells active in the acute inflammatory response (Movat, 1978). We begin by introducing the cells in question.

The polymorphonuclear (PMN) leukocyte (sometimes referred to as a granulocyte or neutrophil) has a diameter of 10 to 15 μm, possesses a homogeneous nucleus, and its cytoplasm contains numerous granules. These granules or lysosomes have been shown to contain numerous enzymes, glycogen, lipid, a basic protein, and sulfated mycopolysaccharide. Other organelles are sparsely represented in the cytoplasm; the mature PMN leukocyte possesses no endoplasmic reticulum, a small Golgi apparatus, and a few small mitochondria.

PMN leukocytes arise in the bone marrow from a pluripotential stem cell which can differentiate into cells of the erythrocytic, granulocytic, thrombocytic, and monocytic series. Maturation proceeds through the following stages: myelolblast → promyelocyte → myelocyte → metamylocyte → PMN leukocyte. The mature PMN leukocyte has a short life span, with a half-life of nearly 7 hr. In humans, some 50 billion PMN leukocytes exist outside the bone marrow, in the blood and vascular system, with 50 to 100 times this number in reserve within the bone marrow. How the mobilization of these latter cells is controlled remains to be clearly defined.

The primary function of PMN leukocytes is phagocytosis, a process involving ingestion, neutralization, and whenever possible, destruction of the ingested particle. These

*Current affiliation: Connaught Research Institute, Willowdale, Canada

functions are accomplished by mechanisms involving intracellular lysosomal enzymes and other substances, the existence of which has been known since the pioneering work of Metchnikoff at the end of the nineteenth century. More than 40 discrete hydrolytic enzyme activities have been identified in the lysosomes of the PMN leukocyte (Movat, 1978).

Metchnikoff was the first to recognize the importance of mononuclear phagocytic cells in host defense and he referred to these cells as macrophages. The term *monocyte* is now used for the mononuclear phagocytic cells of the blood and macrophage refers to their counterparts in tissues. The differentiation of these cells, which also begins in the bone marrow, proceeds as follows: monoblast → promonocyte → monocyte → macrophage. The monocytes enter the circulation, where they have a half-life of up to 70 hr before migrating into tissues and becoming macrophages. The monocyte has a diameter of about 15 μm and undergoes a substantial increase in size (diameter up to 80 μm) upon differentiation into the macrophage. Unlike the PMN leukocytes, there is no reserve of mature monocytes in the bone marrow.

Mononuclear phagocytes are active cells. In vitro they attach to a variety of different substrates and move slowly with the extrusion of pseudopods. Their membranes move continuously, forming pinocytic* vacuoles. Ultrastructural analysis demonstrates moderate amounts of endoplasmic reticulum, a well-developed Golgi apparatus, mitochrondria, cytoplasmic granules, and numerous small vesicles. When stimulated, the macrophage undergoes a variety of ultrastructural changes, including further development of the Golgi complex and an increase in cytoplasmic granules. Such macrophages are referred to as being "activated" or even "angry."

Like the PMN leukocyte, the macrophage is a phagocytic cell, but unlike the former, whose appetite is restricted to bacteria, the macrophage will dispose of fungi, protozoa, viruses, erythrocytes, cell debris, and whole dead cells. The macrophage, however, is truly a cell for all seasons and plays a central role in a variety of effector mechanisms, including tumor surveillance, as well as in the induction of specific immunity (Nelson, 1976; Nathan et al., 1980). It is an active secretory cell producing more than 50 large and small molecules at last count (Nathan et al., 1980).

Both classes of cell depend on cell surface receptors to "sense" changes in their microenvironment prior to the expression of effector functions. One of these receptors, the Fc receptor (FcR), specifically recognizes a site on the Fc region of immunoglobulins and is intimately involved in phagocytosis, cytotoxic events, and certain secretory responses.

Before discussing Fc receptors in more detail, it is important to define what is implied by the term *receptor*. In the present context, FcR is defined as a distinct molecular entity present on the plasma membrane with the following properties.

1. *Saturability*. There are a finite number of receptors per cell, all of which can be occupied at high free-ligand concentrations.
2. *Specificity*. The receptor is capable of binding only certain classes or subclasses of immunoglobulin.
3. *High affinity*. The specific ligands are bound with high energy, implying an intimate stereochemical relationship to the binding site of the receptor.

**Pinocytosis*, literally "cell drinking," refers to the uptake of soluble molecules, whereas *phagocytosis*, or "cell eating," denotes the uptake of particles. Both activities are subsumed under the general term *endocytosis*.

4. *Induction of cellular response.* This property distinguishes a receptor from a binding or carrier protein (e.g., serum albumin).

Although there may be many molecular events separating productive receptor occupation from the response, it is this receptor-ligand association that initiates the response. However, as discussed below, the FcR shows only a low-to-moderate affinity for monomeric immunoglobulin and such interactions do not lead to the activation of cellular responses. Immunoglobulin oligomers are bound with higher affinity and are capable of cross-linking adjacent receptors, which may be the important event in the activation of such processes as phagocytosis. Such oligomers and higher polymers may be formed either nonspecifically by chemical cross-linking or heat aggregation, or specifically as soluble antibody-antigen complexes or sensitized cells or other particles.

II. DETECTION AND QUANTITATION OF Fc RECEPTORS

Similar methods may be used to establish the presence of FcRs on various cell types, including PMN leukocytes and mononuclear phagocytes, and to determine their specificity and other binding characteristics.

A. The Rosette Assay

Despite the fact that this assay is only semiquantitative at best, it has been widely used to detect FcRs and, in the inhibition mode, has proved useful in determining immunoglobulin class and subclass specificity. The indicator is a red cell that has been coated with immunoglobulin. Such coating may be achieved specifically using antibody specific for a red cell determinant or nonspecifically following chemical modification of the red cell surface (e.g., with chromic chloride or tannic acid). The former approach has been more widely used because it utilizes the natural antigen binding property of the immunoglobulin. Examples are anti-Rh_0 (D) antibodies specific for Rh-positive human erythrocytes, rabbit and anti-sheep erythrocyte, and rabbit anti-ox erythrocyte antibodies. More recently, monoclonal murine anti-sheep erythrocyte antibodies have been produced which have the advantage of class/subclass homogeneity which is not possible with polyclonal reagents.

When sensitized erythrocytes are mixed with FcR-bearing cells, the characteristic EA rosette is formed as shown in Fig. 1(A). The links between the peripheral erythrocytes and the central cell (in this case, a human peritoneal macrophage) are formed by the immunoglobulins and the FcRs. Figure 1 also shows a number of other consequences of FcR-immunoglobulin interaction. First, following rosette formation at 37°C, the bound red cells "cap" toward one pole of the macrophage [Fig. 1(B),(D)]. This phenomenon is inhibited at 4°C and in the presence of sodium azide (Romans et al., 1976). These observations suggest that the FcR is mobile in the plane of the membrane, a property observed for many cell surface molecules. If such capped cells are reexposed to sensitized erythrocytes, no further red cells are bound, suggesting that a critical number of receptors have been capped. Second, the bound erythrocytes lose their characteristic biconcave morphology and become spherocytic [Fig. 1(A),(B)] and, over time, are converted to hemolyzed ghosts [Fig. 1(C)]. Sometimes the red cells are phagocyted. Finally, dead macrophages, as judged by the uptake of trypan blue dye, retain the capability to form rosettes but apparently are unable to cap the bound erythrocytes or induce spherocytosis [Fig. 1(E)].

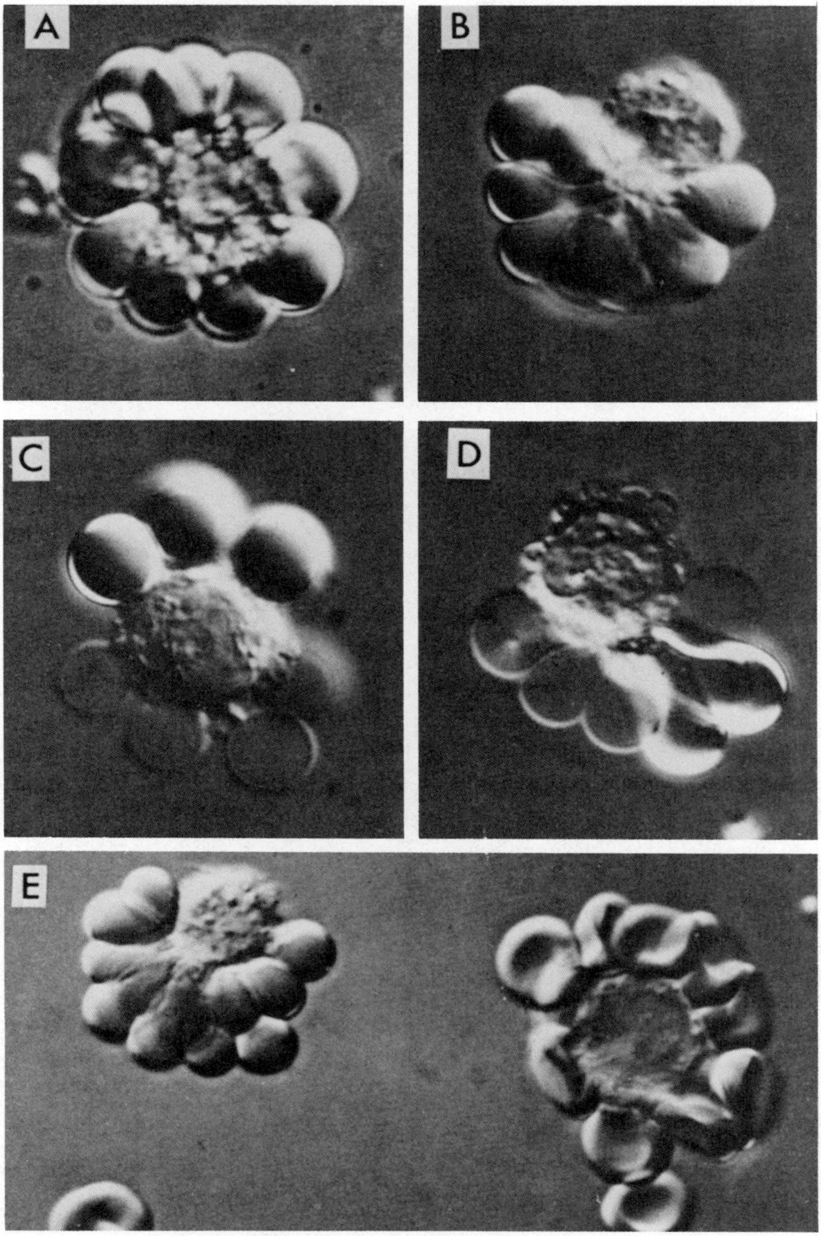

Figure 1 Interference contrast photomicrographs of adherent human blood monocyte EA [anti-Rh_0(D)] rosettes in serum. Nonrosetting EA appear as normal biconcave disks and serve as internal standards of size (ca. 7 μm) and morphology. (A) noncapped and (B) capped trypan blue-negative EA rosettes. Both cells have induced a tense spherocytosis in bound EA. The extreme cap in (B) shows morulalike rosetting EA tightly convergent on a single focus of cell membrane; (C, D) live noncapped and capped blood monocyte rosettes bearing EA both as spherocytes and as hemolyzed ghosts. In (D), capping has polarized to that region antipodal to the characteristic reniform nucleus; (E) adjacent live (left) and dead (right) human peritoneal macrophage EA rosettes in serum. The cell on the left has induced a spherocytic change in its capped EA. The cell on the right showed nuclear staining by trypan blue; its rosetting EA remained randomly distributed and of normal morphology.

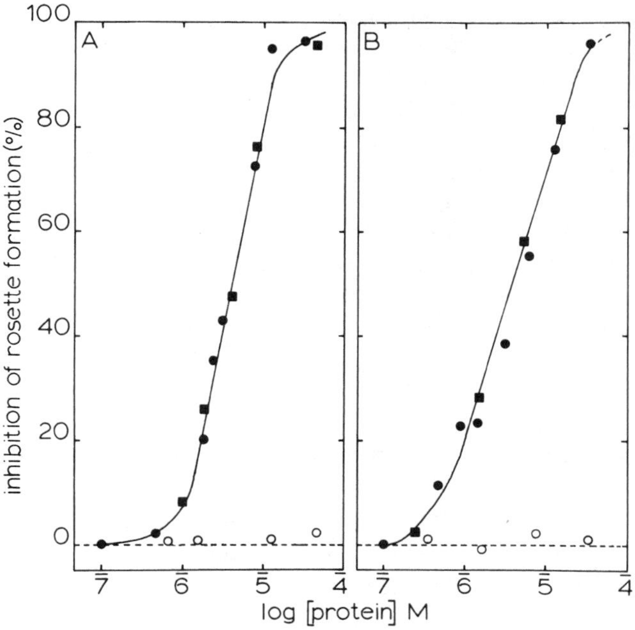

Figure 2 Typical results obtained in quantitative rosette-inhibition assays using anti-Rh_0-(D)-coated human Rh+ erythrocytes and either human monocytes (A) or human PMN leukocytes (B). Dose-dependent inhibition was obtained with either intact human IgG1 (●) or its Fc fragment (■). In contrast, Fab fragments were unable to interfere with rosette formation (○).

If increasing concentrations of soluble immunoglobulin are present when the sensitized erythrocytes are added to the FcR-bearing cells, rosette formation is decreased. Rosette-inhibition assays of this type have been widely used to determine the binding specificity of the FcR. The results of typical assays for human PMN leukocytes and macrophages are shown in Fig. 2.

B. Radioligand Binding Assays

Quantitative information concerning the number and intrinsic affinity of FcRs has been obtained by using radiolabeled immunoglobulins (usually with either ^{125}I or ^{131}I). Successful assays of this type must take several problems into account. Since equilibrium binding of ligand must be assumed for rigorous treatment of the binding data and because the intrinsic affinity between FcR and immunoglobulin is modest, it is necessary to separate bound from free ligand rapidly. Rapid washing techniques (e.g., Unkeless and Eisen, 1975) or separating the cells from free ligand by centrifugation through an oil-water interface are examples of successful solutions to this problem. A second severe problem relates to nonspecific binding of the ligand. Although such binding is not fully understood, it is frequently characterized as being of low affinity but high capacity. In other words, the cell surface possesses a large number of such binding sites, so large that it is experimentally difficult to saturate them, and these sites have low intrinsic affinity for the ligand. For very high affinity receptors, such as those for many hormones or that for IgE on basophils and mast cells, the low free concentration of ligand required to

saturate the receptors is insufficient to give significant nonspecific binding. In the case of most FcRs, however, the difference in their affinity compared to nonspecific sites is much smaller. The degree of nonspecific binding can be assessed by determining radioligand binding in the presence of high concentrations of unlabeled immunoglobulin. Nonspecific binding of up to 40% of the labeled ligand is not uncommon. With phagocytic cells, such as those of interest here, endocytosis must be minimized by the use of low temperature or metabolic inihibitors such as azide.

When appropriately controlled, radiolabeled-ligand binding assays can yield data that can be analyzed using the Scatchard equation, thus providing estimates of receptor number, intrinsic affinity, and heterogeneity. Much of the information discussed below has been obtained using such assays.

C. Ultrastructural Analysis

Electron microscopy has provided important information on the pathways of endocytosis following attachment of antibody-antigen complexes to putative FcRs at the cell surface. One approach in such studies has been to use soluble immune complexes in which the antigen is electron dense (e.g., ferritin). Another is to use antibody-coated particles, such as bacteria, which can be readily visualized in the electron microscope.

III. BINDING SPECIFICITY AND HETEROGENEITY OF Fc RECEPTORS

The term *Fc receptor* belies considerable complexity. We shall see that even for a single cell type, this term refers to a family of molecules, members of which exhibit different ligand binding specificities and appear to show structural polymorphism. Of the two cell types under discussion, the mononuclear phagocytic cells have received by far the greater attention. Many of the experimental data relate to the FcγRs specific for IgG and its subclasses, with particular attention having been paid to these receptors on murine macrophages. The latter emphasis derives largely from the existence of several stable continuously cultured mouse macrophage-like cell lines.

A. Number and Affinity of Fc Receptors

Quantitative information has been obtained using binding assays involving radiolabeled, monomeric IgG followed by Scatchard analysis of the data. The available information may be summarized as follows (Dorrington, 1977): macrophages possess approximately 1 million receptors which bind monomeric IgG with an intrinsic affinity (K_A) between 1×10^6 and 1×10^8 M^{-1}. Kinetic studies show that the ligand binds and dissociates rapidly. At the free concentration of IgG present in serum (e.g., human serum contains 12 mg/ml or 8×10^{-5} M IgG) most macrophage receptors will be occupied. Even for cells shown to possess subclass-specific receptors (see below), the free concentration of the relevant subclass is high enough to keep the receptors saturated. This reflects the fact, of course, that binding of monomeric IgG is not *functionally* important. The binding of polyvalent IgG ligands (e.g., soluble immune complexes or a sensitized bacterium) is greatly favored because of their ability to bind simultaneously to several receptors (see also Absolom et al. (1982), and Chap. 31). The avidity of such binding can reach very high levels and competes efficiently with IgG monomer.

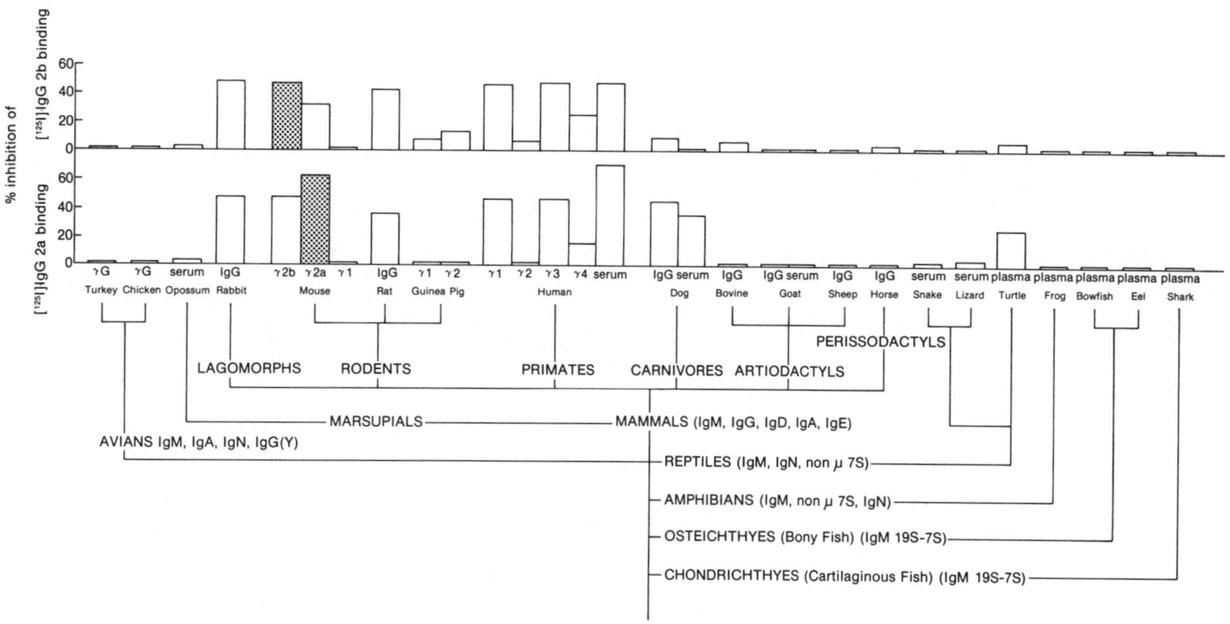

Figure 3 Species specificity of the IgG2a and IgG2b Fc receptors on $P388D_1$, a murine macrophagelike cell line. The ability of purified IgG or plasma from species representing the major phylogenetic families to inhibit the binding of radiolabeled murine IgG2a or IgG2b was assessed. A standard input of 50 μg was used for the IgG samples and 300 μg of the β-γ gfraction from each serum or plasma. (From Haeffner-Cavaillon et al., 1979.)

B. Class/Subclass Binding Specificity

Although almost all published studies indicate the presence of more than one class of Fcγ receptor on mouse macrophages, discrepancies in detail are apparent. These discrepancies may be due to the different types of assays used as well as the physical state of the ligand. For example, some results have been obtained with monomeric IgG, whereas others have been obtained using some form of polyvalent IgG as the ligand (i.e., antibody-antigen complexes, heat-aggregated or chemically cross-linked IgG and IgG coated onto erythrocytes). Since a full discussion of all the available information is not possible here, the reader is referred to the recent review by Unkeless et al. (1981) and the original papers for more details.

Unkeless (1977) showed that $P388D_1$, a murine macrophage cell line, possessed two classes of FcR: one binding IgG2a and the second binding IgG2b, aggregated IgG1, and rabbit IgG immune complexes. The IgG2a receptor was trypson sensitive, whereas the IgG2b receptor was not. Haeffner-Cavaillon et al. (1979) confirmed the existence of IgG2a and IgG2b receptors on $P388D_1$ but found extensive cross-reactivity. A third receptor specific for aggregated murine IgG3 has recently been described by Diamond and Yelton (1981).

A number of ingenious approaches have been used to explore the question of FcR heterogeneity.

1. Haeffner-Cavaillon et al. (1979) measured the ability of IgG from a variety of different species, to inhibit the binding of radiolabeled IgG2a and IgG2b to $P388D_1$ cells (Fig. 3). A marked species specificity was apparent. Bovine and equine IgG were not recognized by either receptor nor was IgG from the guinea pig, a close relative of the mouse. Human IgG subclass proteins showed the same binding hierarchy to $P388D_1$ receptors as they do for human monocytes and macrophages (i.e., IgG3 > IgG1 > IgG4 ≫ IgG2) and bound equally well to both receptors. Of particular interest were the findings that canine IgG and turtle immunoglobulins interacted only with the murine IgG2a receptor.
2. During a study of the biological effector functions of variant human IgG1 myeloma proteins, Klein et al. (1981) showed that these proteins could compete only with murine IgG2b for receptors on $P388D_1$. In addition to providing further evidence for more than one class of FcR on $P388D_1$, these results also suggest that topographically distinct regions of murine IgG2a and IgG2b Fc regions are involved in binding to their receptors (see below).
3. One advantage of having macrophage cell lines is that attempts can be made to select and expand mutant cells lacking one or more FcR (Unkeless et al., 1981). It has been possible to isolate mutants of J774 (another macrophage cell line) lacking IgG2b receptors but showing a normal capacity to bind IgG2a. Another J774 variant incapable of binding IgG3-sensitized erythrocytes has also been isolated.

Unkeless and his colleagues (Mellman and Unkeless, 1980; Unkeless et al., 1981) have produced a monoclonal antibody, referred to as 2.4G2 IgG, which recognizes an antigen on the trypsin-resistant IgG2b FcR and does not cross-react with the FcR specific for IgG2a on J774 and $P388D_1$ cells. This conclusion was based on several lines of evidence: (1) Fab fragments from 2.4G2 did not affect the binding of labeled monomeric IgG2a but did inhibit rosette formation with erythrocytes sensitized with IgG2b antibody; (2) trypsin treatment of macrophages did not affect the binding of labeled 2.4G2

Fab; and (3) J774 mutants lacking the IgG2b receptor did not bind 2.4G2 Fab. This monoclonal anti-FcR is discussed further in Sec. IV.

For other species that have IgG subclasses, it is reasonable to ask whether their macrophages possess more than one class of FcR. There is some evidence that guinea pig macrophages have separate receptors specific for IgG1 and IgG2 (Alexander et al., 1978). In contrast, human monocytes and macrophages appear to possess one class of FcγR capable of binding all four IgG subclasses, albeit with different affinities (order IgG3 > IgG1 > IgG2 ≫ IgG4; Huber et al., 1971). Human IgG3 and IgG1 also have the strongest tendency to adsorb aspecifically onto (usually nonpathogenic) bacteria and thus to promote their phagocytic ingestion by polymorphonuclear leukocytes (Absolom et al., 1982) and other phagocytic cells (see also Chap. 31).

Although the existence of discrete classes of FcγR on murine macrophages is supported by most of the available evidence, the biological significance of this heterogeneity is not readily apparent. There is no indication so far that the several receptors are linked to the expression of different effector mechanisms (Ralph et al., 1980). The murine IgG3 receptor appears to show absolute specificity for this subclass and it might be argued that this provides a mechanism for the clearance of antibody-antigen complexes containing an antibody subclass present in relatively low abundance. However, this argument is not particularly persuasive since a typical polyclonal antibody response would result in immune complexes (or opsonized particles) containing antibodies of different subclasses, thus ensuring their access to macrophage effector systems. A single class of receptor capable of recognizing all cytophilic IgG subclasses would seem to be a more parsimonious mechanism.

With regard to FcRs showing specificity for other classes of immunoglobulin, there is some evidence that these may exist. Melewicz and Spiegelberg (1980) have shown that a subpopulation of human monocytes possess receptors specific for IgE and that these recognized the Fc region of this class. Such FcϵRs were capable of mediating both phagocytosis and target-cell lysis. Subpopulations of both human PMN leukocytes and monocytes have been shown to carry Fc receptors specific for IgA (Fanger et al., 1980).

C. Location of the Binding Sites for Fc Receptors Within Fc

The Fc region is composed of four compact, globular domains, the relative disposition of which has been determined by x-ray crystallography (see Chap. 7). The two COOH-terminal Cγ3 domains form a tight noncovalent dimer across the twofold axis of the molecule. In contrast, the two Cγ2 domains do not interact and are separated by a solvent-filled channel. The carbohydrate prosthetic group covalently linked to each Cγ2 region is immobilized on the surface of the domain facing the solvent channel. The recognition that the heavy and light chains of IgG were folded into a series of domains led to the proposition that each domain carried the active site responsible for mediating a limited number of functions. This correspondence between structural and functional differentiation has proved to be an oversimplification, at least as far as binding to Fc receptors is concerned.

Several approaches have been used in attempts to locate FcR binding sites within Fc. First, fragments of Fc corresponding to the Cγ2 and Cγ3 regions have been tested for their ability to compete with Fc fragments (or IgG) for binding to FcR in a variety of assays. The second approach is to determine the functional consequences of structural abnormalities in genetic variants of IgG molecules. Third, one can evaluate the effects of binding third-party molecules to Fc if the binding sites for these ligands are known.

Experiments with murine macrophages failed to show any FcR binding activity in either Cγ2 or Cγ3 fragments (Haeffner-Cavaillon et al., 1979), whereas some activity toward human monocytes has been recovered in Cγ3 fragments of human IgG1 (Barnett-Foster et al., 1980). The FcγRs on human PMN leukocytes appear to require both Cγ2 and Cγ3 domains of Fc (Barnett-Foster et al., 1978). There is, in fact, a substantial body of literature providing conflicting conclusions regarding the important Fc domain for FcR binding. Recent studies have shed some light on these controversies.

Klein et al. (1981) showed that two human IgG1 variant proteins, which lacked the hinge region, could discriminate between IgG2a and IgG2b Fc receptors on $P388D_1$ cells; only the binding of IgG2b could be inhibited. These variant proteins show an unusually close physical association between their Fab and Fc regions, to the extent that portions of the Cγ2 domain surfaces would be inaccessible to large molecules. The Clq binding site, for example, which is known to be located on the Cγ2 region, is not expressed. In contrast, the binding of protein A of *Staphylococcus aureus*, which occurs at the junction of Cγ2 and Cγ3, is normal in these molecules. The results obtained with $P388D_1$ cells suggest that the IgG2a and IgG2b receptors recognize different regions on their ligands. The hinge-deleted proteins were also shown not to bind to the FcR on human PMN leukocytes, suggesting that this receptor recognizes a site on the Cγ2 domain. Diamond et al. (1979), using several deletion and recombinant murine IgG mutant molecules, have provided evidence that Cγ2 carries the site recognizing the IgG2b receptor.

Taken together, the available evidence suggests that different FcRs may recognize distinct sites on Fc, although a major role for Cγ2 is becoming apparent. The failure of isolated Cγ2 fragments to bind FcRs probably indicates that the nature conformation of this region depends on the presence of the neighboring Cγ3 domains.

IV. STRUCTURE OF Fc RECEPTORS

Progress in the isolation and characterization of Fcγ receptors has been slow (for a review, see Unkeless et al., 1981). A combination of factors has contributed to this: low abundance (ca. 10^6 receptors per cell or less than 1% of total membrane protein, relatively low intrinsic affinity ($K_A \sim 10^6$ to 10^8 M^{-1}); susceptibility to proteolysis, and limited solubility in the absence of detergents. The limited successes discussed below have paralleled the development of techniques for the handling of integral membrane proteins in general. The role played by intrinsic affinity in facilitating FcR isolation can be illustrated by reference to the receptor for IgE present on rat basophilic leukemia cells (Metzger, 1978). The FcεR has an affinity for monomeric IgE in excess of 10^{10} M^{-1}, a level comparable to many hormone-receptor interactions. Purification procedures based on this high specificity are efficient, whereas the four-orders-of-magnitude lower affinity of the FcγR leads to losses through dissociation and ambiguities due to nonspecific interactions between membrane proteins and the IgG ligands.

A well-designed protocol for the isolation of FcR should fulfill the following criteria. (1) the FcR would be obtained as a homogeneous protein with binding characteristics consistent with those observed in the membrane. (2) The yield of protein would be consistent with the known number of FcR on the cell surface. This assumes that a significant pool of receptors does not exist within the cell. (3) Yields of receptor from different cell types should be consistent with the known distribution of FcR. (4) Antibodies raised against the receptor should specifically precipitate directly or indirectly the FcR and block receptor activity. These goals have been achieved to varying degrees in the studies described below.

The most popular method used to isolate FcγRs has been affinity chromatography, in which detergent lysates have been passed over columns of IgG immobilized on a solid matrix. Membrane proteins are radiolabeled either extrinsically, using iodide and lactoperoxidase, or intrinsically, using labeled amino acid or sugar precursors. The membranes are solubilized using nonionic detergents (e.g., NP-40) prior to passage over the IgG column. Following extensive washing in detergent-containing buffers, bound radioactivity is eluted at either high or low pH. The eluted protein can then be analyzed by sodium dodecyl sulfate-polyacrylamide gel electrophoresis (SDS-PAGE) and by functional tests.

Table 1 summarizes the available information on the apparent molecular wieght (M_r) for FcγRs isolated from macrophages and PMN leukocytes using affinity techniques based upon FcR-IgG union. Even for such a fundamental property, a spectrum of values has been obtained.

Loube et al. (1978) isolated a 57,000-molecular weight protein from lysates of ^{125}I-labeled P388D$_1$ cells by chromatography over Sepharose columns carrying murine IgG2a or human IgG1 or Fc fragment (Fig. 4). Columns coupled with Fab or IgM did not bind this protein. The 57,000-molecular weight species appeared as a broad band on sodium dodecyl sulfate-polyacrylamide gel electrophoresis (SDS-PAGE), with evidence that more than one component was present. The isolated proteins accounted for less than 1% of the total radioactivity present in the membrane protein fraction, a level consistent with the abundance of the FcR on the cell surface. Comparable results were obtained with lysates prepared from cells labeled with radioactive amino acids and sugars, the latter providing evidence that the FcR was a glycoprotein. The FcR was apparently released from the cell surface and could be recovered from spent culture fluids. Shed material, from cells labeled with [^{35}S] methionine, gave two well-resolved polypeptides of 62,000 and 58,000 molecular weight which appeared to be related as judged by peptide mapping (Loube and Dorrington, 1980).

A variety of other studies using murine, rabbit, and human macrophages (see Unkeless et al., 1981) support the view that FcR, isolated by affinity chromatography, is composed of a family of polypeptides in the molecular weight range 50,000 to 70,000 which are not sensitive to reduction. Some studies suggest that receptors showing IgG-subclass specificity may be distinguished on the basis of their mass (Lane et al., 1980). Limited proteolysis and/or variable degrees of glycosylation may also contribute to the size polydispersity. In some instances retention of biological activity has been demonstrated for the isolated FcR.

In contrast to the consensus described above, at least with respect to molecular weight, some other studies have suggested that the FcR on a variety of cells is a 110,000 to 120,000-molecular weight polypeptide (e.g., Bourgois et al., 1977). The method of isolation used here was somewhat different in that immune complexes were incubated with detergent lysates of labeled cells. Extreme precautions were taken to limit proteolytic cleavage. If such cleavage did occur, the products remained linked by disulfide bonds and, upon reduction, yielded peptides having molecular weights of 90,000, 75,000, 46,000, and 23,000. It has been proposed that the FcR is composed of domains each having a mass near molecular weight 23,000. No binding activity has been demonstrated for the 120,000-molecular weight protein. The relationshp of this protein to the lower-molecular-weight FcRs is unclear, although it should be mentioned that Loube and Dorrington (1980) did sometimes recover a 120,000-protein which yielded the 60,000 species upon reduction.

Brief mention should be made here of some intriguing results which suggest that the trypsin-sensitive FcR specific for IgG2b on P388D$_1$ also carries phospholipase A$_2$

Table 1 Apparent Molecular Weights (M_r) Reported for Putative Fcγ Receptors from Normal Murine Macrophages, Macrophagelike Cell Lines, and Human PMN Leukocytes

Cells used	Method of isolation[a]	Ligand used[b]	$M_r \times 10^{-3}$ [c]
I. Macrophages			
Splenic macrophages	Ab-Ag	rab anti-SIII polysaccharide	120
$P388D_1$	AC	huIgG1, muIgG2a	57
$P388D_1$	AC	agg, huIgG	84
Peritoneal macrophages	AC	muIgG2a muIgG2b, IgG1	67 52
$P388D_1$, J744.2	AC	muIgG2a, IgG2b, IgG1, rabIgG	50–65
J774	MA	Fab of MA	47, 60
II. PMN leukocytes			
Peripheral blood	MA	Fab of MA	53, 66
Peripheral blood	AC	huIgG	33, 60

[a]Ab-Ag, antibody-antigen complexes; AC, affinity chromatography; MA, monoclonal antibody.
[b]hu, human; mu, murine; rab, rabbit; agg, aggregated.
[c]Apparent molecular weight determined by SDS-PAGE. Only major component is given.

activity. Suzuki et al. (1982) subjected detergent lysates of $P388D_1$ to two different affinity-chromatography systems: Sepharose coupled to either aggregated human IgG or a phosphatidylcholine (PC) derivative. Molecules bound by the IgG absorbant selectively recognized murine IgG2a but not IgG2b. In contrast, the PC binding protein selectively bound IgG2b and exhibited phospholipase A_2 activity. Phospholipase A_2 is involved in arachidonic acid production, which in turn is a precursor of prostaglandins. Since prostaglandins are potent modulators of cell function, these results suggest a mechanism whereby Fc receptors could be linked to effector mechanisms. In fact, the PC binding protein enzymic activity was augmented in the presence of aggregated IgG2b but not aggregated IgG2a.

Affinity chromatography has recently been used to isolated the FcR from human PMN leukocytes by Kulczycki et al. (1981). These workers recovered two polypeptides from detergent lysates of PMNs in an Fc-specific fashion. One had broad mobility on SDS-PAGE, giving an apparent molecular weight between 52,000 and 64,000 and the second gave a sharp band at 33,000. Some evidence suggested that the 33,000 protein was a proteolytic breakdown product of the larger species. Both proteins retained the ability to bind again to IgG-Sepharose. Human mononuclear cells yielded the same protein bands but also showed a polypeptide at molecular weight 43,000 which was not seen with PMNs.

The most promising recent developments in the area of FcR isolation have resulted from the application of the monoclonal antibody technology. Unkeless and his coworkers (Mellman and Unkeless, 1980; Unkeless et al., 1981) used the Fab fragment from their 2.4G2 monoclonal antibody (see above) coupled to Sepharose as an immuno-

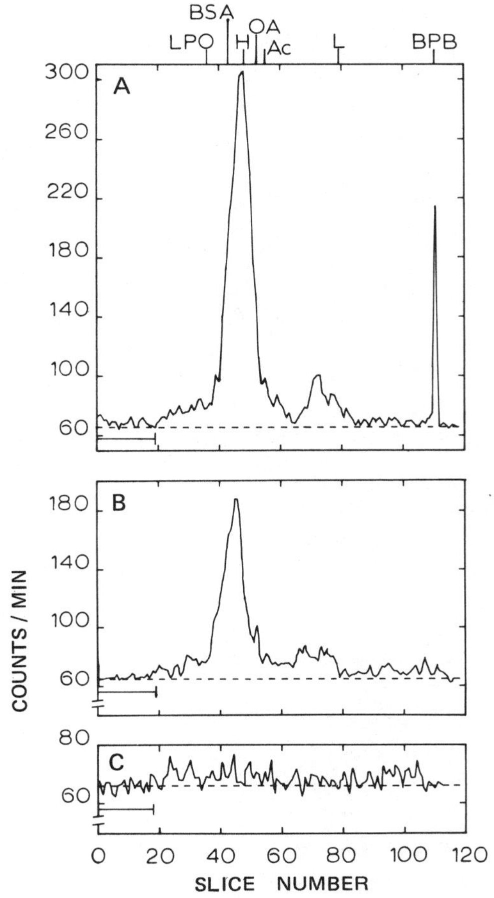

Figure 4 SDS-polyacrylamide gel electrophoresis of ^{125}I-labeled, detergent-solubilized cell surface proteins of P388D_1 cells eluted from affinity adsorbants bearing covalently coupled Fc fragments (A), human IgG1 (B), or Fab fragments (C). The apparent molecular weight of the major component bound to IgG and Fc was 57,000. Mobilities of various standard proteins are indicated by: LPO, lactoperoxidase; BSA, bovine serum albumin; H, heavy chain of IgG1; OA, ovalbumin; Ac, rabbit muscle action; L, immunoglobulin light chain; BPB, from Loube et al., 1978.

absorbant. The intrinsic affinity of the Fab fragment for the antigen was high and was not disrupted in the presence of either ionic or nonionic detergents. FcR was purified over 5000-fold in a single step in which detergent lysates of J774 macrophages were passed over anti-FcγR Fab-Sepharose and eluted at pH 11.5.

The 2.4G2 antigen showed two diffuse protein bands centered at molecular weight 60,000 and 47,000 upon SDS-PAGE. The presence of carbohydrate was demonstrated in several ways: the antigen bound to concanavalin A; and the isoelectric point became more alkaline following neuraminidase treatment and it was labeled by the galactose oxidase/[^{3}H] borohydride method. Unkeless et al. (1981) favor the possibility that both molecular weight species represent different molecular forms of the same gene product. Partial proteolytic cleavage and/or different extents of glycosylation could be

invoked to account for the differences. Clearly, both forms carry the same antigenic determinant. In addition, trypsinization of macrophages did not alter the number of 2.4G2 binding sites but did result in a higher yield of the lower-molecular-weight species.

When a variety of other murine FcR-bearing cells were examined with the 2.4G2 Fab, the two protein species were also recovered, albeit in different proportions (Fig. 5). This suggests that the antigen recognized by 2.4G2 is common to all murine FcRs. The antigen appears to be species specific, however, since it was not detectable on FcR-bearing cells from rat, guinea pig, rabbit, or humans.

The purified 2.4G2 antigen was shown to possess Fcγ binding activities in vitro except that its subclass specificity was less restricted. In the absence of detergent, the antigen formed large aggregates, as expected for an integral membrane protein. These polyvalent aggregates were able to agglutinate erythrocytes sensitized with monoclonal antierythrocyte antibodies of the IgG1, IgG2a, and IgG2b subclasses. The interaction with IgG2a was unexpected based on the specificity of the FcR in the membrane. Although there are a number of ways of accounting for this loss of specificity (Mellman and Unkeless, 1980), it would be of interest to incorporate the FcR into liposomes of different composition and assess the recovery of specificity. It should be noted that Schneider et al. (1981) isolated the FcR from the same cell line used by Unkeless's group, but using affinity chromatography, and were also unable to demonstrate subclass-specific binding.

Fleit et al. (1982) have recently described the production of a monoclonal antibody against the FcR of human PMN leukocytes. Unlike the situation described above, this monoclonal did not recognize an antigen common to FcRs on different cell types. The basic molecular characteristics of the PMN leukocyte FcR, such as its behavior on SDS-PAGE, were similar to those described for the murine FcR.

In summary, a consensus is developing regarding the basic molecular features of FcRs on mononuclear phagocytic cells and PMN leukocytes, although some of the discrepancies in the literature remain to be fully resolved. The use of monoclonal antibodies should facilitate the isolation of sufficient FcR to allow detailed biochemical characterization. Amino acid sequence data, obtained either by conventional microsequencing techniques or by DNA sequencing, will provide answers to several intriguing questions. How extensive are the structural homologies between FcRs on different cell types? Are FcRs similar to immunoglobulins, with a "variable" region responsible for ligand binding specificity and a "constant" region involved in membrane integration and linkage to effector mechanisms? Are these molecules structurally related to other membrane proteins involved in immune phenomena? Detailed structural studies will facilitate an understanding of the roles played by FcRs in the triggering of the effector functions of the cells that carry them on their membrane.

V. FUNCTIONAL SIGNIFICANCE OF Fc RECEPTORS

Fc receptors constitute an important link between the cellular and humoral aspects of immunity. The effector mechanisms exhibited by mononuclear cells and PMN leukocytes are not antigen specific, whereas antibodies appear as part of a specific, adaptive response to antigenic challenge. The ability of these cells to bind immune complexes via their Fc receptors provides an important mechanism for directing these effector pathways toward the specific needs of the host. It is convenient to consider the roles of Fc receptors under three general headings, although it should be recognized that these phenomena rarely proceed in isolation from one another.

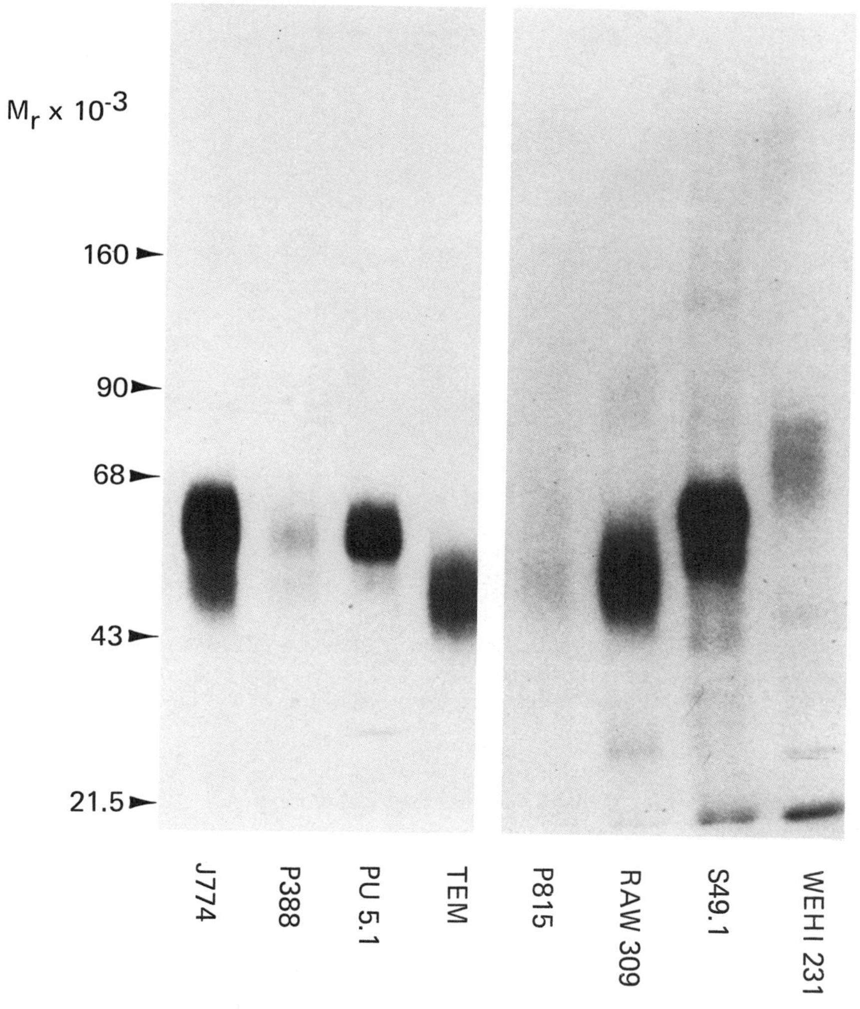

Figure 5 Antigens recognized by the 2.4G2 monoclonal antibody on different FcR-bearing cell types. From left to right the cell types are J774 (macrophage), P388 (null lymphocyte), PV 5.1 (null lymphocyte), thioglycollate-elicited macrophage (TEM), P815 (mastocytoma), RAQ 309 (macrophage), S49.1 (T lymphocyte), and WEHI 231 (B lymphocyte). (From Mellman and Unkeless, 1980.)

A. Receptor-Mediated Endocytosis

Phagocytosis is a characteristic property of both cell types* and immunoglobulin is not an obligatory requirement for this process to proceed. For example, latex beads and carbon particles may be ingested by a mechanism that is independent of the FcR. What does seem clear, however, is that the presence of antibody increases the efficiency of the

*Mononuclear phagocytes and PMNs are sometimes referred to as professional phagocytes since they make a full-time occupation of this function.

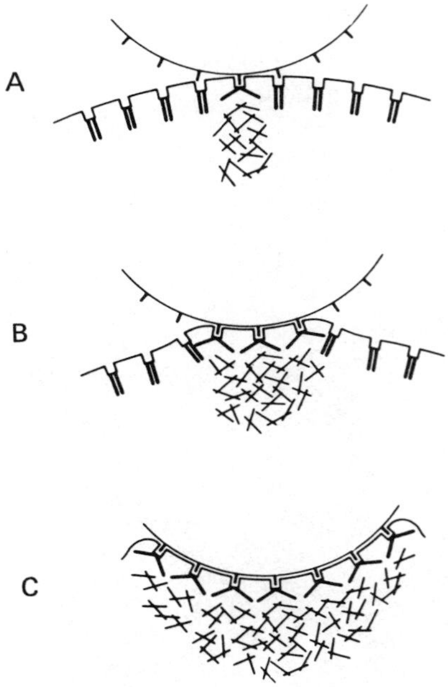

Figure 6 Model of possible phagocytic mechanism. ⊤ indicates IgG molecule bound to target cell membrane; Fc receptor in the inactive ( ) or activated ( ) state; , cytoplasmic contractile proteins. See the text for a discussion of the model. (From Griffin et al., 1976.)

process. Classical studies showed that a bacterial load was more effectively reduced in animals that had mounted a specific immune response when compared to those having no previous exposure. The most logical analysis of this enhancement suggests that the FcR functions to establish a close association between the membrane and the sensitized particle and maintaining this association for a critical time period.

The phagocytic process can be divided into two stages: (1) attachment of the particle to the cell surface, and (2) ingestion of the particle (see the excellent review by Silverstein et al., 1977). Attachment is generally temperature independent and does not require metabolic energy [in Fig. 1(E), for example, sensitized erythrocytes will bind to dead macrophages]. The ingestion phase, however, is markedly temperature dependent and requires active cellular metabolism.

Phagocytosis has been shown to be a local response to signals generated when antibody on the particle interacts with Fc receptors. For example, Fc receptors present on distant parts of the membrane are not affected. The process is also selective in that other receptors are not taken up with the ingested particle. A number of experiments have shown that a diffuse distribution of the immunoglobulin ligand on the particle is critical. It is believed that the movement of the membrane around the circumference of the particle involves increasing numbers of ligand-FcR interactions, sometimes referred to as the "zipper mechanism" of phagocytosis (Griffin et al., 1976).

The driving force of the zipper mechanism is not fully understood but probably involves the contractile elements of the cytoplasm, as shown in Fig. 6. The following represents a plausible model for the ingestion stage. First, IgG-FcR interaction generates

a signal, possibly the release of actin binding protein from the membrane, that triggers the association of actin filaments leading to the extension of pseudopods. These pseudopods form more ligand-FcR interactions and the aggregation of further contractile elements. This process continues until the leading edges of the membrane fuse and the particle is engulfed.

The formation of the phagocytic vacuole ("phagosome") is accompanied by an increase in oxygen consumption, a marked increase in H_2O_2 production, and enhanced glucose oxidation via the hexose monophosphate shut. The latter pathway leads to the formation of powerful oxidizing agents, such as superoxide and singlet oxygen, which constitute an important intracellular effector mechanism for the destruction of microorganisms (Nathan et al., 1980).

In the final stage of phagocytosis, the phagosome moves into the cytoplasm and quickly fuses with lysosomes, the contents of which are discharged into the phagosome. A variety of enzymes are present which effect the degradation of the endocytosed particle.

In the process described above, the FcR serves not only to bind the sensitized particle to the cell membrane but also triggers a specific cellular response. This fulfills the requirement of a true receptor outlined in Sec. I.

The mechanism of pinocytosis (e.g., the uptake of soluble immune complexes) is not as well defined as that of phagocytosis. There is evidence that antibody-antigen complexes containing at least three IgG molecules are rapidly taken up by phagocytic cells in vivo and in vitro. Such complexes may be taken up into both smooth endocytic vesicles and coated vesicles. The latter are membrane vesicles which carry a protein coat attached to their cytoplasmic side. The protein coat is composed largely of a single protein, clathrin. Coated vesicles are thought to play important roles in a variety of membrane-mediated intracellular events (Ockleford and White, 1980). Both types of vesicle probably fuse with lysosomes as an initial step in the degradation of the vesicle's contents.

B. Antibody-Dependent Cytotoxicity

It is established that macrophages, especially in the activated state, are capable of killing eukaryotic cells coated with antibody. The induction of spherocytosis in sensitized erythrocytes bound to macrophages via Fc receptors followed by lysis (Fig. 1) is a model for such cytotoxicity. A component of the antitumor activity exhibited by macrophages may be antibody dependent, although other mechanisms also exist (Nathan et al., 1980).

As in phagocytosis, the immunoglobulin-FcR interactions appears to serve two functions. First, these interactions serve to bring the membranes of the macrophage and the target cell into close apposition. Second, ligand-IgG interactions trigger the formation of the molecules responsible for inducing target cell damage. Although a variety of molecular mechanisms may result in such damage, it seems probable that the formation of strongly oxidizing species, arising from an oxidative burst in the effector cell, may be a major mechanism. Superoxide, hydrogen peroxide, and the hydroxyl radical each have the potential for inducing damage in the target cell membrane. Such species, however, are short-lived and the role of Fc receptors in maintaining close contact with the target cell must be important in promoting efficient killing. It should be recognized that cytotoxicity will frequently represent the extracellular phase of an overall process that includes phagocytosis.

Table 2 Secretory Products of Mononuclear Phagocytes

Enzymes	Reactive metabolites of oxygen
Lysozyme	Superoxide
Neutral proteases	Hydrogen peroxide
Plasminogen activator	Hydroxyl radical
Collagenase	Singlet oxygen (?)
Elastase	
Angiotensin convertase	Bioactive lipids
Acid hydrolases	
Proteases	Arachidonate metabolites
Lipases	Prostaglandin E_1
(deoxy) Ribonucleases	6-Ketoprostaglandin $F_{1\alpha}$ (from prostacyclin)
Phosphatases	Thromboxane
Glycosidases	Leukotriene
Sulfatases	Hydroxyeicosatetraeneoic acids (including slow-reacting substance of anaphylaxis)
Arginase	Platelet-activating factors
Complement components	Factor chemotactic for neutrophils
C1	Factors regulating synthesis of proteins by other cells
C4	Hepatocytes
C2	Serum amyloid A
C3	Haptoglobin
C5	Synovial-lining cells
Factor B	Collagenase
Factor D	Factors promoting replication of:
Properdin	Lymphocytes (lymphocyte-activating factors)
C3b inactivator	Myeloid precursors (colony-stimulating factors)
β1H	Erythroid precursors
Enzyme inhibitors	Fibroblasts
Plasmin inhibitors	Microvasculature
α_2 Macroglobulin	Factors inhibiting replication of:
Binding proteins	Lymphocytes
Transferrin	Tumor cells
Transcobalamin II	Viruses (interferon)
Fibronectin	*Listeria monocytogenes*
Nucleosides and metabolites	
Thymidine	
Uracil	
Uric acid	
Endogenous pyrogens	

Source: Nathan et al. (1980).

C. Secretion

Mononuclear phagocytes and, to a lesser extent, PMN leukocytes are secretory cells (Nathan et al., 1980). The repertoire of secreted molecules is both extensive and diverse (Table 2). Many of these molecules dramatically affect the microenvironment of the cells and play a central role in the development of and control of the inflammatory response. Although it has not been experimentally demonstrated for all secretory products, it seems probable that immune complexes trigger, or at least enhance, the rate of the secretion of many of these molecules. An exception is lysozyme secretion, which is a constitutive property of macrophages.

Consideration of the secretory products listed in Table 2 suggests that mononuclear phagocytes may modulate a variety of pathways relevant to inflammation. For example, plasmin, formed from plasminogen by the catalytic action of plasminogen activator, is involved in the blood clotting system, in the activation of the C1 and C3 complement components, and in the formation of kallikrein from prekallikrein. Two other neutral proteases, collagenase and elastase, can induce major changes in the architecture of connective tissue and blood vessel walls. Immune complexes are capable of enhancing and suppressing neutral protease secretion depending on whether or not they can be ingested. It is not possible here to analyze in detail the complex dynamics of secretion. Suffice it to say that the inflammatory response, which is an important homeostatic mechanism mounted to combat local injury, requires a complicated array of positive and negative feedback loops. In this way the extent of inflammation can be controlled.

It is important to appreciate that the cellular network underlying inflammation involves both mononuclear cells and PMN leukocytes (as well as B and T lymphocytes). Macrophages, for example, secrete endogenous pyrogen, a hormone that stimulates PMN leukocytes to release lactoferrin. Lactoferrin apparently mediates a complex series of responses, including the blocking of colony-stimulating factor release by macrophages.

The recognition of the diversity in the secretory potential of mononuclear cells and PMN leukocytes has revolutionized our understanding of the biological significance of these cells. It is also becoming clearer that Fc receptors play an important role in allowing these cells to sense changes in their environment.

ACKNOWLEDGMENTS

Work from the author's laboratory is supported by the Medical Research Council of Canada (Grant MT 4259). The helpful comments of Dr. D. G. Romans are gratefully acknowledged.

REFERENCES

Absolom, D. R., van Oss, C. J., and Neumann, A. W. (1982). J. Reticuloendothel. Soc. *31*, 59.

Alexander, M. D., Andrews, J. A., Leslie, R. G. Q., and Wood, N. J. (1978). Immunology *35*, 115.

Barnett-Foster, D. E., Dorrington, K. J., and Painter, R. H. (1978). J. Immunol. *120*, 1952.

Barnett-Foster, D. E., Dorrington, K. J., and Painter, R. H. (1980). J. Immunol. *124*, 2186.

Bourgois, A., Abney, E. R., and Parkhouse, R. M. E. (1977). Eur. J. Immunol. *7*, 691.

Diamond, B., and Yelton, D. E. (1981). J. Exp. Med. *153*, 514.

Diamond, B., Birshtein, B. K., and Scharff, M. D. (1979). J. Exp. Med. *150*, 721.

Dorrington, K. J. (1977). In *Immunology of Receptors* (B. Cinader, Ed.). Marcel Dekker, New York, p. 183.

Fanger, M. W., Shen, L., Pugh, J., and Bernier, G. M. (1980). Proc. Natl. Acad. Sci. USA *77*, 3640.

Fleit, H. B., Wright, S. D., and Unkeless, J. C. (1982). Proc. Natl. Acad. Sci. USA *79*, PAGE NO.

Griffin, F. M., Griffin, J. A., and Silverstein, S. C. (1976). J. Exp. Med. *144*, 788.

Haeffner-Cavaillon, N., Klein, M., and Dorrington, K. J. (1979). J. Immunol. *123*, 1905.

Huber, H., Douglas, S. D., Nusbacher, J., Kochwa, S., and Rosenfield, R. E. (1971). Nature (Lond.) *229*, 419.

Klein, M., Haeffner-Cavaillon, N., Isenman, D. E., Rivat, C., Navia, M. A., Davies, D. R., and Dorrington, K. J. (1981). Proc. Natl. Acad. Sci. USA *78*, 524.

Kulczycki, A., Solanki, L., and Cohen, L. (1981). J. Clin. Invest. *68*, 1558.

Lane, B. C., Kan-Mitchell, J., Mitchell, M. S., and Cooper, S. M. (1980). J. Exp. Med. *152*, 1147.

Loube, S. R., and Dorrington, K. J. (1980). J. Immunol. *125*, 970.

Loube, S. R., McNabb, T. C., and Dorrington, K. J. (1978). J. Immunol. *120*, 709.

Melewicz, F. M., and Spiegelberg, H. L. (1980). J. Immunol. *125*, 1026.

Mellman, I. S., and Unkeless, J. C. (1980). J. Exp. Med. *152*, 1048.

Metzger, H. (1978). Immunol. Rev. *41*, 186.

Movat, H. Z. (1978). In *Inflammation, Immunity and Hypersensitivity* (H. Z. Movat, Ed.). Harper & Row, Hagerstown, Md., p. 1.

Nathan, C. F., Murray, H. W., and Cohn, Z. A. (1980). N. Engl. J. Med. *303*, 622.

Nelson, D. S. (1976). *Immunobiology of the Macrophage*. Academic Press, New York.

Ockleford, C. D., and White, A. (1981). *Coated Vesicles*.

Romans, D. G., Pinteric, L., Falk, R. E., and Dorrington, K. J. (1976). J. Immunol. *116*, 1473.

Schneider, R. J., Atkinson, J. P., Krause, Y., and Kulczycki, A. (1981). J. Immunol. *126*, 735.

Silverstein, S. C., Steinman, R. M., and Cohn, Z. A. (1977). Annu. Rev. Biochem. *46*, 669.

Suzuki, T., Saito-Taki, T., Sadasivan, R., and Nitta, T. (1982). Proc. Natl. Acad. Sci. USA *79*, 591.

Unkeless, J. C. (1977). J. Exp. Med. *145*, 931.

Unkeless, J. C., and Eisen, H. N. (1975). J. Exp. Med. *142*, 1520.

Unkeless, J. C., Fleit, H., and Mellman, I. S. (1981). Adv. Immunol. *31*, 247.

30

Lymphokines

Takeshi Yoshida University of Connecticut Health Center, Farmington, Connecticut

I. INTRODUCTION

Immunologic reaction is generally divided into two major categories, antibody-mediated and cell-mediated immunity. The former represents the interaction of antigen with the specific antibody secreted by lymphocytes of B-cell lineage. This type of reactivity can be passively transferred to a nonimmune recipient with serum from an immune donor. Cell-mediated immunity, on the other hand, is expressed by the interaction of antigen with sensitized lymphocytes of T-cell lineage. Therefore, cell-mediated immunity cannot be passively transferred with immune serum, but can be transferred with viable T lymphocytes from an immune donor.

Lymphokines are defined as nonantibody, soluble mediators produced by lymphocytes. Historically, the production of mediators by antigen-stimulated lymphocytes was first demonstrated independently in 1966 by two groups of investigators: Bloom and Bennett, and David, in studies aimed at elucidation of the effector mechanisms of delayed-type hypersensitivity (DTH). It has become evident that lymphocyte-derived soluble factors mediate not only cutaneous DTH reactions but also various other types of cell-mediated immune reactions. The term *lymphokines* was introduced by Dumonde et al. (1969) to describe these mediators. In addition to the mediators of the effector arm of cell-mediated responses, lymphokines now include various molecules which mediate cell interactions involved in the afferent arm, or induction, of immune responses.

II. CLASSIFICATION OF LYMPHOKINES

A variety of in vitro biological assays for DTH responses have been described, and soluble factors mediating each of these bioassays have been named to represent currently more

Table 1 Classification of Lymphokines

Inflammatory (efferent) lymphokines	
Motility:	Migration inhibition factors
	Chemotactic factors
Activation:	Activating factors
Proliferation:	Mitogenic factors
	Cytotoxic factors
Differentiation:	Differentiation factors
Permeability:	Permeability factors
Immunoregulatory (afferent) lymphokines	
Helper function:	Interleukin 2
	Colony stimulating factors
Suppressor function:	SIRS
	Antigen-specific suppressors

than 100 lymphokine activities (Waksman, 1979). As discussed later, none of the lymphokines has been chemically purified, although some of them have reached a certain chemical homogeneity. At the present time, therefore, it is not possible to classify the lymphokines according to their chemical structures. Many of these factors share similar physicochemical characteristics. It is conceivable that several lymphokine activities may be expressed by the same molecular substance. It is useful, however, to categorize lymphokines into several different groups based on their target cells and functions. Lymphokintes can affect a great variety of cells, including granulocytes (polymorphonuclear cells), macrophages, lymphocytes, fibroblasts, endothelial cells, and various neoplastic tumor cells. Effects of lymphokines on target cells include almost every aspect of biological function. In general, lymphokines influence cell growth, cell motility, and activation. Such influences can be either suppressive or enhancing.

According to their major roles in different phases, namely, afferent or efferent arms of immunological responses, lymphokines may be divided into two major categories: (1) inflammatory (or efferent) lymphokines and (2) immunoregulatory (or afferent) lymphokines. This is illustrated in Table 1 with some examples of lymphokines in each category. Obviously, this division, although convenient, cannot be taken too rigidly since a few lymphokines are known to affect both the afferent and efferent limbs of immune responses. As discussed, for example, lymphocyte mitogenic factors (Interleukin II) not only influence afferent limbs of immune responses by controlling helper and suppressor lymphocyte functions but also increase the secondary production of inflammatory lymphokines by nonspecifically stimulating normal lymphocytes.

III. INFLAMMATORY (EFFERENT) LYMPHOKINES

Lymphokines were initially recognized as mediators of in vitro correlates or bioassays of delayed hypersensitivity. Therefore, the majority of lymphokines reported fall into this

category and some of the well-studied and representative examples are discussed in this section.

A. Lymphokines Affecting Cell Motility

Migration Inhibitory Factors

Since Rich and Lewis (1932) showed that tuberculin inhibited the in vitro cellular migration from spleen explants obtained from tuberculin-sensitive animals, many investigators have confirmed that such a migration-inhibitory phenomenon is a good in vitro correlate of DTH reactions. Using in vitro cellular migration from glass capillary tubes, David (1966) and Bloom and Bennett (1966) unequivocally demonstrated that the antigenic inhibition of cellular migration from such tubes was due to a soluble substance released from immune lymphocytes stimulated by specific antigen in vitro. This was the first lymphokine to be described and was named macrophage migration inhibition factor (MIF).

Macrophage migration inhibition factor. Although MIF was originally discovered in guinea pigs, most of the species capable of expressing cell-mediated immune responses have been found capable of producing MIF. Generally, MIF from all species appears to have similar physicochemical structures and the molecules seem to be active across the species, although it is well known that there are occasionally some strange species specificities [e.g., between guinea pigs and mice (Yoshida et al., 1973a)]. Physicochemically, MIF consists of protein or glycotprotein with molecular weights between 12,000 and 80,000. The difference in molecular weights is largely due to the heterogeneity of MIF molecules. For example, it was shown that guinea pig MIF activity is expressed by several molecular species, including average molecular weights of 12,000, 25,000, 45,000, 60,000, and 80,000, while human MIF seems to have average molecular weights of either 12,000 or 25,000. These heterogeneities may be due to polymerization of a smaller molecular species of MIF, although no direct evidence for such a contention has been reported. In addition to heterogeneity of molecular weights, guinea pig MIF can be divided into two different molecular species according to isoelectric focusing, one (pH 3-MIF) eluting at a pH between 3.2 and 4.4, the other (pH 5-MIF) between 5.2 and 5.5 (Remold and Mednis, 1977). A certain association exists between molecular sizes and difference in isoelectric points. A variety of physicochemical properties of guinea pig MIF are summarized in Table 3 (see also the review by Yoshida, 1979).

MIF is considered not to have antigen specificity, clearly distinguishing this type of lymphocyte product from antibody molecules (Yoshida et al., 1972). However, accumulating data seem to suggest that MIF produced by lymphocytes stimulated with an antigen can show a different molecular species from that produced by other immune lymphocytes stimulated with another specific antigen (see Sec. III). This indicates that the difference in antigenic nature may influence the nature of products generated by lymphocytes stimulated with such an antigen, although no direct and specific binding of antigen to MIF has been shown.

MIF seems to bind receptors on the macrophage cell surface. Thus MIF activity can be removed from lymphokine supernatants by incubation with macrophages. Trypsin treatment of macrophages temporarily destroys their responsiveness to MIF, but the effect is reversed after enzyme removal. Recent data indicate that the MIF receptor on macrophages may be an α-L-fucose-containing glycolipid (Liu et al., 1982).

The exact mechanism(s) for MIF to cause macrophage migration inhibition are not known. However, several possibilities with certain supporting data have been postulated: (1) MIF may interfere with some metabolic process necessary for macrophage motility, (2) MIF may have direct effects on cytoskeletal elements such as microtubules and microfilaments, and (3) MIF may cause surface-bound fibrin formation on macrophages and a network surrounding the cell, resulting in the restriction of movement.

MIF activity is occasionally associated with target cell aggregation and clumping, inhibition of macrophage spreading, and increased macrophage adherence and alterations of electrophoretic mobility. It is most likely that the factors reported to be responsible for those in vitro activities may be the same as MIF.

Leukocyte migration inhibition factor. Immune lymphocytes stimulated by specific antigen can also produce a factor which inhibits the migration of polymorphonuclear leukocytes (mainly neutrophils). This factor, called leukocyte migration inhibition factor (LIF), was shown to be distinct from MIF because LIF could not inhibit the migration of macrophages (Rocklin, 1974). Most of the studies on LIF have dealt with human LIF, without any systematic studies available on LIF from other species.

Physicochemically, LIF is a protein with a molecular weight between 60,000 and 70,000. Partially purified LIF preparations show charge heterogeneity, having two peaks of isoelectric points at pH 5 and pH 8.5. Unlike other lymphokines, LIF is known to possess enzyme activity. Thus various serine esterase inhibitors such as phenylmethylsulfonylfluoride (PMSF) and diisopropyl fluorophosphate (DFP) are able to abolish LIF activity. The esterase activity of LIF is arginine specific, and arginine esters with amides can protect LIF from destruction by PMSF. LIF seems to react with a specific receptor on polymorphonuclear cells. Human LIF activity can be blocked by n-acetylglucosamine as well as by α-L-fucose, which is an inhibitor for guinea pig and human MIF, although there are some controversial results reported about the effectiveness of α-L-fucose. On the other hand, some investigators consider that LIF activity may involve the formation and release of a secondary (or real) inhibitory factor due to the cleavage of precursor molecules by enzymatic activities of LIF.

Tumor cell migration inhibition factor. When various ascites-type tumor cells, instead of macrophages, are packed in capillary tubes, they can migrate in vitro in the same manner as macrophages. Such migration was found to be inhibited by lymphokine-containing culture supernatants. The finding of this factor, together with MAF activity, which is discussed in Sec. III.B, suggests that in addition to lymphotoxins (see Sec. III.C), noncytotoxic lymphokines may also play a role in host defense against neoplastic growth. The roles of lymphokines in tumor immunity have been discussed (D'Silva and Yoshida, 1981).

Migration inhibition of other cells. Using the capillary tube method or the agarose droplet method, migration of various other cell types has been examined. Recently, it was found that MIF-containing supernatants could inhibit the migration of cultured endothelial cells and of purified eosinophils. Whether these inhibitory factors are the same as MIF remains to be determined.

Migration Enhancing Factor

In conventional capillary tube MIF assays, enhanced macrophage migration is occasionally observed. Several investigators have described this activity and attributed it to a

lymphokine that is separable from MIF and macrophage chemotactic factor (MCF, see below) physicochemically. It is still unknown whether or not the same receptor sites on macrophages are responsible for binding both migration enhancing factor (MEF) and MIF.

Chemotactic Factors

In addition to the factors effecting random movement of the cells as described above, the activities of increasing unidirectional movement, chemotaxis, are detectable in lymphokine preparations. Chemotaxis is usually measured in vitro by quantitating the migration of target cells toward chemotactic agent across micropore filters of defined pore size. There are chemotactic lymphokines for a variety of cell types, as described below.

Macrophage chemotactic factor. Macrophage chemotactic factor (MCF) is known to be a molecule(s) distinct from other known chemotactic factors such as C3- and C5-derived factors and bacterial factors. MCF is heat stable (56°C for 30 min), sensitive to proteolytic enzymes, resistant to neuraminidase, and has an approximate molecular weight of 12,500. There are a few reports indicating the presence of different molecular species of MCF with a molecular weight between 35,000 and 55,000. Guinea pig MCF activity is inhibitable by incubation with α-L-fucose of α-L-rhamnose, suggesting that these monosaccharides are part of the MCF receptor site. Unlike MIF, there is no species specificity in MCF activity.

Chemotactic factors for granulocytes. Chemotactic lymphokines have been described for the three different types of granulocytes: neutrophils, eosinophils, and basophils. Neutrophil chemotactic factor (NCF) was the first to be described as a distinct lymphokine from MIF and MCF. However, detailed descriptions of the characteristics of NCF are not available. Basophil chemotactic factor (BCF) has been reported to have a molecular weight of approximately 12,500 and share similar physicochemical characteristics with MCF. Furthermore, BCF can be absorbed by macrophages, suggesting similarity, if not identity, between MCF and BCF.

Two different types of eosinophil chemotactic lymphokines are known. One of them is a precursor of eosinophil chemotactic factor (ECFp), which can be activated by its incubation with immune complex to generate ECF (Torisu et al., 1973). Only specific immune complexes containing the same antigen as that used to stimulate the lymphocyte cultures are effective in generating active ECF. This requirement for specific antigen makes this factor unique among other lymphokines, which, except for some antigen-specific regulatory factors (see below), do not have any antigen specificity. Another type of eosinophil chemotactic lymphokine, called eosinophil stimulator promoter (ESP), can affect the migration of eosinophils in the agarose droplet assay. ESP is sensitive to chymotrypsin but resistant to neuroaminidase treatment and has a molecular weight between 24,000 and 56,000. In view of clinical eosinophilia as well as eosinophil accumulation in certain types of immune reactions, further detailed studies seem to be warranted on the significance of these factors. The implication of these lymphokines as well as the role of eosinophils in immune responses was recently reviewed (Yoshida and Torisu, 1983).

Other chemotactic factors. Chemotactic activities for lymphocytes (LCF) and fibroblasts (FCF) can also be detected in lymphokine-containing culture supernatants. LCF has a molecular weight of approximately 22,000 and FCF of approximately 80,000. These factors, although they share many common physicochemical characteristics with MCF, are said to be distinct from MCF and/or NCF.

Table 2 Cellular Activation by Lymphokines

Enhancement of tumoricidal activity
Enhancement of bacteriocidal activity
Increased phagocytic activity
Increased production of biologically active molecules, complement components, collagenase, prostaglandins
Increased activity and release of lysosomal enzymes
Increased activity and release of plasminogen activator
Increased ruffled membrane movement
Increased adenylate cyclase activity
Activation of hexose monophosphate shunt
Increased protein synthesis

B. Lymphokines Affecting Cell Activation

This group of lymphokine activities includes those affecting specific cellular functions, such as phagocytosis, synthesis and release of secondary messengers and effector molecules such as various enzymes and mediators, cytoxic capability to kill target cells (e.g., tumor cells), and bacteriostatic or bacteriocidal ability. Different lymphokines seem to be responsible for activation of these functions in different cell types, as discussed below.

Macrophage Activating Factor

Macrophage activating factor (MAF) can "activate" macrophages to exhibit a number of biochemical, morphologic, and functional alterations compared to normal marcophages. The main alterations shown by activated macrophages are summarized in Table 2. Although enhancement of tumoricidal and bacteriocidal activities is biologically the most important alteration of macrophage functions, the procedures to assay these activities are usually quite cumbersome. Activation of macrophages to such a capable state is apparently a multistep process. For example, the activation process of macrophages for tumor cell killing requires at least two different signals (priming and triggering), which can also be modulated by factors such as endotoxin and serum other than MAF. A simple biochemical assay to measure the tumoricidal capacity of activated macrophages is yet to be discovered. In fact, some of the biochemical and functional alterations in the activated macrophage may be caused by mutually independent mechanisms which may be triggered by qualitatively as well as quantitatively different MAF.

MAF has almost exactly the same physicochemical characteristics as MIF, although several investigators have claimed the successful separation of MIF from MAF activity in lymphokine preparations (reviewed in Yoshida, 1979). Like MIF, MAF seems to have a partial species specificity. α-L-fucose is considered to be a part of the MAF receptor in the guinea pig as it is for the MIF receptor.

Osteoclast Activating Factor

Osteoclast activating factor (OAF) can cause induction of bone resorption nd localized bone loss. The in vitro assay for bone resorption uses the release of radioactivity from ^{45}Ca-labeled bone tissue in organ cultures after 4 to 6 days of incubation of the tissue with lymphokines. OAF is found in two molecular species, one of 18,000 and the other

between 1330 and 3500 molecular weight. The low-molecular-weight OAF is sensitive to trypsin and papain treatments, whereas the higher-molecular-weight species is not.

Similar to the activity of parathyroid hormone, OAF can increase the size and number of osteoclasts in bone explants and activate adenyl cyclase in bone cells. OAF, but not parathyroid hormone, causes enhanced synthesis of prostaglandin E (PGE) by explanted bones. This preceded production of PGE is necessary for the osteolytic activity of OAF. Since PGE secretion by monocytes is required for generation of OAF from lymphocytes, a positive feedback regulation seems to be operating between PGE and OAF levels.

Lymphokines Activating Other Cells

Fibroblast stimulating factor (FSF) is known to activate fibroblasts to synthesize and secrete a larger amount of collagens than that secreted by untreated fibroblasts. The relationship of this factor to fibroblast mitogenic factor (stimulation of DNA synthesis) is not yet clear. Granulocytes can also be activated by lymphokines, showing degranulation, lysosomal enzyme release, increased hydrogen peroxide production, and increased phagocytic ability. These effects of lymphokines on neutrophils have been well described, although the effect on basophils is not documented. It is, however, still unknown whether these activities are due to a distinct lymphokine or to the same molecule of the known lymphokines, such as chemotactic factors. In addition, the direct stimulatory effect of a lymphokine on endothelial cells has been shown by its in vivo capability to induce angiogenesis.

C. Lymphokines Affecting Cell Proliferation

There exists a group of lymphokines which either enhance or inhibit cell proliferation. It may not be appropriate to include some of the factors affecting proliferation of lymphocytes in these inflammatory or efferent lymphokines since their indirect effects on antibody production and generation of cytotoxic lymphocytes are far reaching into the afferent limb of immune responses. The implication of such factors, therefore, will also be discussed in Sec. IV.

Lymphocyte Mitogenic Factor

Lymphocyte mitogenic factor (LMF) can cause normal lymphocytes, including T cells, B cells, and thymocytes to undergo mitogenic divisions. Guinea pig LMF has heterologous molecular weights, one between 20,000 and 30,000, the other at 45,000. Human LMF has been reported to have a molecular weight between 15,000 and 55,000, while murine LMF was shown to have a molecular weight of 30,000 to 40,000. The molecular mechanisms by which LMF induces DNA synthesis is largely unknown. However, LMF may serve to amplify the immune response by activating other nonsensitized lymphocytes to proliferate and produce other lymphokines. Apparently, some of the molecules claimed to be LMF represent the same molecule as interleukin 2 (IL-2), which is discussed later. Future studies should clarify if any type of LMF is an independent entity from IL-2.

Mitogenic Factor of Other Cells

Macrophage growth factor or macrophage mitogenic factor (MMF) was reported to be present in lymphokine preparations. Biochemical properties are similar to those growth

factors obtained from fibrblast cultures. MMF has a molecular weight between 40,000 to 65,000 and a heterologous isoelectric point at pH 3.0 to 5.5 and pH 8.0. The identity of MMF to colony stimulating factor (CSF), discussed later, remains to be clarified.

No lymphokine activity has been reported to prliferate granulocytes. There are several preliminary reports available which indicate that endothelial cells and fibroblasts, both at a quiescent state, can be stimulated for increased DNA synthesis. However, no detailed characterization of these activities is available.

Inhibitor of DNA Synthesis

Activity of the inhibitor of DNA synthesis (IDS) is detected in vitro by its ability to suppress tritiated tymidine ([^{3}H]TdR) uptake in mitogen-stimulated lymphocytes. IDS activity is not species specific, but rather cell specific, in that it acts primarily on lymphocytes. IDS has no direct cytotoxic effects, even at high concentrations. The effect of IDS is restricted to the late G_1 phase of the cell cycle. Lymphocytes in proliferation express their receptors to IDS only during a limited portion of the cell cycle. Although the exact mechanism of IDS action is not clear, the activity may be mediated by increasing intracellular cyclic AMP levels and/or by inhibiting DNP polymerase. Rat IDS is a highly acidic glycoprotein with an isoelectric point of 2.7. In its monomeric form, rat IDS has a molecular weight of 20,000, although it usually exists as a tetramer. Human IDS appears to be a heat-sensitive protein with a molecular weight between 40,000 and 80,000.

Lymphotoxin

Unlike IDS (cytostatic factor), lymphotoxin (LT) is a cytotoxic lymphokine which causes direct target cell lysis. The cytolytic activity is usually assayed by measuring the release of [^{3}H]TdR from a monolayer of prelabeled fibroblast cell line L929 clutured 1 to 3 days in the presence of LT preparations. A variety of other target cells, such as HeLa, HEP 2, and P815 mastocytoma cells, can be used also. LT can cause various degrees of effects on target cells in vitro, ranging from reversible growth inhibition to cell lysis, based on the concentration of LT in the system.

The physicochemical characteristics of LT differ largely between species. For example, molecular weights of LT are for mouse, 90,000 to 150,000; for rat, 50,000 and 90,000 to 100,000; and for guinea pigs, 35,000 to 55,000. Human LT seems to be more heterogeneous. Thus "LT complex" has a molecular weight of more than 200,000, and three other forms, α-LT, β-LT, and γ-LT, have molecular weights between 70,000 and 90,000, 35,000 and 50,000, and 10,000 and 20,000, respectively. LT complex can combine immunoglobulinlike substances which possess antigen binding sites (reviewed in Granger et al., 1979). LT exerts its effect by binding to specific cell surface receptors and by causing an increased calcium influx followed by electrochemical imbalance, resulting in cell lysis.

Although LT apparently is not responsible for T-cell-mediated cytolytic reactions, the recent evidence indirectly suggests that LT-like molecules may play a role, intracellularly or in the vicinity of T cells, in direct T-cell-mediated killing mechanisms.

D. Lymphokines Affecting Cell Differentiation

Mouse lymphokine preparations, which had previously been shown to increase macrophage secretory, bacteriocidal, and tumoricidal activities, were shown to enhance the synthesis and expression of macrophage immune-associated (Ia) antigens (Steinman

et al., 1980). It is still disputable whether the expression of Ia antigens is really a good marker for macrophage differentiation. However, it is likely that lymphokines may affect not only proliferation of various target cells but also their differentiations. In view of various growth factor activities and colony stimulating factors (Sec. IV.A), future studies on this area of lymphokines appear to be promising.

E. Lymphokines Affecting Vascular Permeability and Other Lymphokines

In contrast to most lymphokine activities, the soluble mediators affecting vascular permeability are assayed in vivo. Thus the intradermal injection of lymphokine supernatants led to the evolution of skin reactions resembling DTH reactions. The activity responsible for this local inflammation was termed skin reactive factor (SRF). SRF may actually represent a combination of MIF and MCF, as well as factors that increase vascular permeability.

A lymphocyte-derived factor known as lymph node permeability factor (LNPF) was shown to affect vascular permeability. This was found in extracts of the lymph node, unlike other lymphokines, which are produced by stimulated lymphocytes. Vascular permeability factor (VPF) is a factor produced by mitogen stimulation of lymphocytes from patients with nephrotic syndrome. Occasionally, cells from normal individuals also produced VPF. Obviously, some kind of in vitro assay system for these factors may facilitate our understanding of this type of lymphokine. In this context, a lymphokine capable of stimulating endothelial cells (Sidkey and Auerbach, 1975) and a factor to inhibit endothelial cell migration (Cohen et al., 1982) may warrant further development of in vitro techniques relevant to vascular permeability.

Lymphokines have also been described to affect fibrin deposition and coagulation. Platelet aggregating factor (PAF) is capable of inducing the aggregation of platelets in vitro. Furthermore, the activities causing the decrease in clotting time, the acceleration of clot retraction, and coagulation of fibrinogen were all detectable in lymphokine supernatants.

Finally, but probably of great importance, a type of interferon (IFN) is found in lymphokine preparations. Biochemical purification of this type of interferon (γ type, or immune IFN) still has not been accomplished to the degree of other types of IFN. Biologically, however, various regulatory effects on immune responses by IFN_γ have been reported, in addition to antiviral activity.

IV. IMMUNOREGULATORY (AFFERENT) LYMPHOKINES

As mentioned earlier, lymphokines were originally defined as soluble mediators which involved the efferent arm (or effector mechanisms) of DTH reactions. It is now evident that lymphokines also participate in the afferent arm of the immune response (see Table 1). These molecules mediate cooperative cell interactions involved in antibody-mediated and cell-mediated immune responses. The effect can either enhance (helper factors) or inhibit (suppressor factors) the induction of an immune response. Helper and suppressor factors may be antigen specific or nonspecific, genetically restricted or nonrestricted. Some of the major factors are briefly discussed here.

A. Helper Factors

The requirement for T cells in the induction of antibody-mediated immune responses has been well documented in various experimental systems. Most of this helper function is mediated by soluble factors as discussed below.

Colony Stimulating Factor

Colony stimulating factor (CSF) is a protein with a molecular weight of 30,000. The target cells for CSF are monocytes or macrophages. Upon stimulation by CSF, these cells produce biologically active mediators such as interferon, prostaglandins, and interleukin 1 (IL-1). The production of IL-1 by CSF-activated macrophages may explain the augmentation by CSF of the PFC response of spleen cells. IL-1 is not discussed in this chapter since it belongs to the monokines rather than to the lymphokines. This was previously termed lymphocyte activating factor, and has a molecular weight in the range 12,000 to 18,000 (Aarden et al., 1979).

T-Cell Replacing Factor

The activity of T-cell replacing factor (TRF) in the primary immune response is usually assayed in vitro using the Mishell and Dutton system (1967) for the PFC response with murine spleen cells derived from athymic nu/nu mice. The target cell for TRF is the proliferating B cell, which is capable of receiving a maturational signal for antibody production and secretion. Thus the effect of TRF is not to enhance early B-cell triggering but to induce the ultimate differentiation of proliferating B cells into antibody-secreting plasma cells. TRF has heterogeneous molecular weights of 25,000, 30,000 to 35,000, and 45,000, but the biochemical nature of TRF is not well characterized.

Interleukin 2

In addition to its capacity to provide helper activity for the induction of PFC's in T-depleted cultures, interleukin 2 (IL-2) has been shown to support the continuous proliferation of factor-dependent T-cell clones (such as the murine CT6 cell line) in culture, to promote the proliferation of mitogen-treated thymocytes and to induce the generation of cytotoxic cells in thymocyte or nude mouse spleen cell cultures. These effects provide bioassays which can functionally distinguish IL-2 from other T-cell-derived helper factors. The target cell for all of the effects of IL-2 is the T cell. The mechanism of IL-2 enhancement of T-cell proliferation involves the direct binding of IL-2 to cell surface receptors which are expressed by target cells in response to mitogenic or antigenic stimulation.

Although murine IL-2 has a molecular weight of 30,000 by gel filtration and velocity sedimentation, the molecular weight of monomeric IL-2 is reported to be 16,000 to 21,000. IL-2 can be separated from CSF by phenyl Sepharose chromatography and from TRF by chromatography and isoelectric focusing. Murine IL-2 exhibits an isoelectric point heterogeneity between pH 3.8 and 5.0. A similar degree of charge heterogeneity has been reported for human IL-2 and is thought to be due to a variable degree of post-translational glycosylation of the factor. This also contributes to the apparent heterogeneity of molecular weight.

B. Suppressor Factors

In addition to the helper effects, T cells have been shown to exert suppressive effects which "down-regulate" the immune response. Since the initial studies of Horiuchi and Waksman in 1968, the influence of T-suppressor cells has been demonstrated in almost every aspect of cellular and humoral immunity. In the last decade, it has become clear that the mechanism of T-cell suppression involves the secretion of soluble mediators by activated T cells. These suppressor factors are both antigen specific and antigen nonspecific

and have been implicated in both humoral and cellular immune responses. A brief description of selected examples of these suppressor T-cell factors is provided below.

Soluble Immune Response Suppressor

Concanavalin A-activated murine spleen cells elaborate a soluble immune response suppressor factor (SIRS) that nonspecifically suppresses IgG and IgM anti-sheep erythrocyte PFX responses in vitro. This factor also suppressed PFC responses to the T-independent antigens dinitrophenyl-Ficoll and trinitrophenyl-lipopolysaccharide (LPS), as well as B-cell proliferative responses to LPS (Rich and Pierce, 1974). Physicochemically, SIRS is a glycoprotein with a molecular weight between 35,000 and 67,000. SIRS shows no strain specificity and is not adsorbed by anti-H2 alloantisera, immunoglobulin, or anti-immunoglobulin. SIRS can be distinguished from the antigen-specific suppressor T-cell factors described by Tada et al. (1975) in that it is not a product of the H2 complex and it does not bind antigen.

Antigen-Specific Suppressor Factors in Cell-Mediated Immunity

Lymph node cells from mice immunized with 2,4,6-trinitrobenzene sulfonic acid (TNBS) and later injected with 2-chloro-1,3,5-trinitrobenzene (PC1) elaborate in culture a suppressor factor which specifically suppresses the ability of PCl-sensitized cells to transfer contact sensitivity passively to an irradiated recipient. This factor is produced by T cells, has a molecular weight of about 50,000, and has an affinity for antigen. The PCl suppressor factor bears determinants coded by the I-J region of the *H-2* complex, and the I-J subregion of the major histocompatibility complex may specialize in coding for antigen-specific regulatory molecules.

In a similar system, a soluble factor suppresses contact sensitivity to 2,4-dinitrofluorobenzene (DNPB) by lymph node cells from DNFB-tolerized mice. Like the PCl suppressor factor, the DNFB suppressor factor is antigen and hapten specific and can be removed by anti-H-2 antiserum. The suppressive effect of this factor is genetically restricted to strains sharing H-2D- or H-2K-region haplotypes with the factor-producing strain.

V. CELLS THAT PRODUCE LYMPHOKINES

A. Lymphoid Cells

Most of the lymphoid tissues, including spleen, lymph nodes, peripheral blood leukocytes, thoracic duct cells, and peritoneal exudate cells from immune animals, are all good sources of lymphokines when stimulated by specific antigen. Thymus and bone marrow cells, probably because of a paucity of mature cells, are not effective. The cells from lymphoid tissues of normal animals can also be stimulated by nonspecific mitogens, such as concanavalin A or phytothemagglutinin, to generate lymphokines.

Based on the fact that cell-mediated immune reactions appear to be manifestations of T lymphocytes, it was initially assumed that T cells were the only source of lymphokines. In fact, only immune T cells can be activated by specific antigen to produce lymphokines; B cells cannot. However, B cells, if activated by nonspecific polyclonal mitogens such as endotoxin LPS, are able to produce lymphokines (Yoshida et al., 1973b). Obviously, T cells from normal animals can produce lymphokines when stimulated by

T-cell mitogens. Among subpopulations of murine T cells, only Ly-$1^+2^-3^-$ cells but not Ly-$1^-2^+3^+$ cells can produce inflammatory lymphokines. However, it was recently found that cytotoxic T-cell lines can produce IL-2. In addition, it is known that the mixed leukocyte reaction (MLR) results in significant production of lymphokines. Furthermore, a variety of lymphoid cell lines (To and B-cell lines) can spontaneously secrete various lymphokines, although any particular cell line may be able to produce only a limited set of lymphokines.

B. Nonlymphoid Cells

Nonlymphoid cells such as fibroblasts infected with viruses can elaborate substances which show the same in vitro activities as lymphokines. These factors are termed cytokines (reviewed in Ewan and Yoshia, 1979). Many nonlymphoid cell lines, including tumor cell lines, are known to produce cytokines. The mediators from these sources have physicochemical properties strikingly similar to those of lymphokines, with corresponding biological properties. Utilizing antilymphokine antibodies, cytokines have been shown to share antigenic determinants with lymphokines. These findings raise the interesting possibility that lymphokine production per se may be a general biological phenomenon and that what is unique to the immune system is the way in which lymphoid cells can be activated for such production by either mitogen or specific antigen.

VI. IN VIVO ROLES OF LYMPHOKINES

As discussed earlier, lymphokines are defined as mediators for various in vitro biological assays of DTH responses. However, obtaining evidence to support the contention that lymphokines mediate in vivo DTH reactions has been difficult, partly because of the impurity of the materials being dealt with. Three minimum conditions must be satisfied for lymphokines to be qualified as mediators of in vivo DTH reactions: (1) lymphokines should be recovered from the DTH reaction site, (2) lymphokines should induce qualitatively the same DTH reaction in normal animals as specific antigen does in immune animals, and (3) in vitro antagonists or potentiators of lymphokines should affect the expression of DTH appropriately. Four different types of in vivo DTH reactions will be discussed to see if lymphokines can fulfill these requirements as mediators of DTH. A part of the summarized data is shown in Table 3.

A. DTH Skin Reaction

Aqueous extracts from DTH skin lesions contain several lymphokine activities, including MCF and SRF, but no MIF is detectable. Conversely, exogenous lymphokines at a high concentration injected intradermally into normal guinea pigs incites a reaction similar to that of an active DTH response. This factor was discussed above in Sec. III.E as SRF. Finally, active DTH reactions in the immune guinea pig can be suppressed by intravenous as well as intraperitoneal injections of α-L-fucose, a known inhibitor of in vitro lymphokine activities. Furthermore, intravenous injection of antilymphokine antiserum can suppress the DTH reaction.

Based on these results, it appears reasonable to include lymphokines as mediators of DTH skin reactions, although it is still unknown which of the lymphokines is most responsible for the reaction. In view of various activities discussed above and the restricted set of lymphokine activities detectable in the partially purified SRF

Table 3 Lymphokines as Mediators of in Vivo DTH

DTH reactions	Lymphokines[a] recovered from lesions	Reactions reproduced by:
Cutnaeous reaction	MCF, SRF	Lks (MCF)
Macrophage disappearance reaction (MDR)	MIF, SRF	Lks (MIF)
Monocyte disappearance phenomenon	MIF, IFN, anergy factor	Lks (MIF)
Hypersensitivity granuloma formation	MIF, MCF, MFF, ESP, FSF, FCF	Lks (MAF)

[a]See the text for abbreviations of lymphokines.

preparations, it is likely that the reaction is mediated by a combination of lymphokines, such as MCF, MAF, and permeability factor.

B. Macrophage Disappearance Reaction

The microphage disappearance reaction (MDR) can be induced by injecting specific antigen into the peritoneal cavity of immune guinea pigs which have macrophage-rich inflammatory exudates due to previous glycogen injection. The absolute number of macrophages drops drastically within 6 hr. This reaction is an in vivo model of the DTH response. In such peritoneal exudates, MIF activity is detectable. Conversely, if exogenous MIF-rich supernatants are injected into normal guinea pigs bearing macrophage-rich peritoneal exudates, the number of macrophages decreases within a few hours. The active MDR can be inhibited by injection of α-L-fucose or antilymphokine antiserum as in the case of DTH skin reactions. These observations strongly indicate that lymphokines, probably MIF, mediate the MDR.

C. Monocyte Disappearance Phenomenon

In analogy to the MDR, the number of peripheral monocytes decreases within 6 hr when specific antigen is injected intravenously into immunized animals. This phenomenon has been known to be an in vivo model of the cell-mediated immunity. In this reaction, MIF activity is detectable in the serum at 6 to 12 hr after the antigenic challenge. Conversely, when exogenous MIF is injected intravenously into normal animals, the passive monocyte disappearance phenomenon is achieved. This phenomenon is inhibitable with the administration of α-L-fucose or antilymphokine antiserum.

D. Hypersensitivity Granuloma Formation

Previous studies on hypersensitivity granulomatous inflammation have established the fact that this is due to T-cell-mediated immune response. As one such model, pulmonary granuloma can be induced by injecting agarose beads coated with antigen into immunized guinea pigs. Several lymphokine activities are detectable in the extract of those lung granuloma lesions. These include MIF, MCF, macrophage fusion factor (MFF), fibroblast

stimulating factor (FSF), and fibroblast chemotactic factor (FCF). Conversely, agarose beads coated with exogenous lymphokines can induce granulomatous inflammation in lung when injected intravenously or intratracheally into normal guinea pigs (Yoshida, 1980). The treatment by α-L-fucose or antilymphokine antiserum is effective in reducing significantly the intensity of granulomatous inflammation.

E. Lymphokine Activities in Human Diseases

The increasing number of literature indicates that lymphokine activities can be detected in various tissues in many human diseases. Although this suggests a close relationship of such inflammatory lymphokines to the disease state, no direct evidence has been presented. However, it may be worth describing here some of these findings, with an expectation of the future discovery of a missing link between the detection of lymphokines and pathogenesis of the corresponding disease.

Serum MIF and LIF activities have been described in the majority of patients with lymphoma, chronic lymphocytic leukemia, Sezary syndrome, and other neoplastic diseases. In viral infections, serum MIF was detected in patients with viral hepatitis. A high incidence of positive MIF activity was also found in granulomatous diseases such as sarocoidosis and leprosy. In rheumatoid arthritis patients, MIF- and LT-like activities were recovered from synovial fluid. In multiple myeloma, OAF was detected in many patients, suggesting its role in bone destruction in this disease.

VII. REGULATION OF LYMPHOKINE PRODUCTION

A variety of direct and indirect evidence indicates that there exist both positive and negative feedback types of regulation in lymphokine production. It is possible that the imbalance of such regulation may result in excessive production or in a lack of lymphokine activities. For example, some of the lymphokines detected in patients may indicate excessive productivity. Following is a description of the state of the art as to our understanding of the regulation of lymphokine activities.

A. Positive Feedback Regulation

Lymphokines such as MAF activate macrophages to produce LAF or IL-1, with in turn recruits normal T lymphocytes to produce more lymphokines. Lymphokines stimulating nonlymphoid cells, such as fibroblasts and endothelial cells, may cause them to release cytokines mimicking lymphokines. These cytokines in their turn may activate normal lymphocytes to produce more lymphokines. All, or at least most of these pathways may be operating in vivo as a positive feedback regulatory loop.

B. Negative Feedback Regulation

Lymphokines such as MAF stimulate suppressor macrophages to produce prostaglandin E_1 and/or E_2. These endogenous factors can inhibit the production of lymphokines. This inhibition can be reversed by treatment with indomethacin or aspirin. Lymphokines induce and stimulate suppressor lymphocytes, which inhibit the production of antigen-specific lymphokine production. Suppressor lymphocytes seem to secrete soluble suppressor factors. These regulatory pathways have been reviewed recently (Yoshida and Cohen, 1982).

VIII. CONCLUSIONS

More than 15 years have passed since the discovery of lymphokines. The biology of lymphokines has experienced tremendous progress during this period. Nevertheless, crucial studies, such as of the in vivo roles of lymphokines, have been hampered because of a lack of purified materials. At the time of writing, however, we know that some lymphokines are already close to ultimate purity, and some monoclonal antibodies to these materials are available. Concerning other lymphokines which are not yet purified to such an extent, we know that the strategy for successful approaches appears to be established. Thus the next 5 to 10 years will be a period of great advances in lymphokinology, and more definitive knowledge on lymphokines should help us understand better the entire immunological system.

REFERENCES

Aarden, L. A., Brunner, T. K., Cerottini, J.-C. et al. (1979). J. Immunol. *123*, 2928.
Bloom, B. R., and Bennett, B. (1966). Science *153*, 180.
Cohen, M., Picciano, P. T., Douglas, W. J., et al., (1982). Science *215*, 301.
David, J. R. (1966). Proc. Natl. Acad. Sci. USA *56*, 72.
D'Silva, H. B., and Yoshida, T. (1981). In *Handbook of Cancer Immunology*, Vol. 6 (H. Waters, Ed.). Garland STPM Press, New York, p. 215.
Dumonde, D. C., Wolstencroft, R. A., Panayi, G. S., Matthew, M., Morley, J., and Howson, W. T. (1969). Nature (Lond.) *224*, 38.
Ewan, V., and Yoshida, T. (1979). In *Handbook of Inflammation*, Vol. 1: *Chemical Messengers of the Inflammatory Process* (J. C. Houck, Ed.). Elsevier/North-Holland, Amsterdam, p. 197.
Granger, G. A., Hiserrodt, J. C., and Ware, C. F. (1979). In *Biology of Lymphokines* (S. Cohen, E. Pick, and J. J. Oppenheim, Eds.). Academic Press, New York, p. 141.
Horiuchi, A., and Waksman, B. H. (1968). J. Immunol. *101*, 1322.
Liu, D. Y., Petschek, K. D., Remold, H. G., and David, J. R. (1982). J. Biol. Chem. *257*, 159.
Mishell, R. J., and Dutton, R. W. (1967). J. Exp. Med. *126*, 423.
Remold, H. G., and Mednis, A. (1977). J. Immunol. *118*, 2015.
Rich, A. R., and Lewis, M. R. (1932). Bull. Johns Hopkins Hosp. *50*, 115.
Rich, R. R., and Pierce, C. W. (1974). J. Immunol. *112*, 1360.
Rocklin, R. E. (1974). J. Immunol. *112*, 1461.
Sidky, Y. A., and Auerbach, R. (1975). J. Exp. Med. *141*, 1084.
Steinman, R. M., Noguira, N., Witmer, M., Tydings, J. D., and Mellman, I. S. (1980). J. Exp. Med. *152*, 1248.
Tada, T., Taniguchi, M., and Takemori, T. (1975). Transplant. Rev. *26*, 106.
Torisu, M., Yoshida, T., Cohen, S., and Ward, P. A. (1973). J. Immunol. *111*, 1450.
Waksman, B. H. (1979). In *Biology of Lymphokines* (S. Cohen, E. Pick, and J. J. Oppenheim, Eds.). Academic Press, New York, p. 585.
Yoshida, T. (1979). In *Biology of Lymphokines* (S. Cohen, E. Pick, and J. J. Oppenheim, Eds.). Academic Press, New York, p. 259.
Yoshida, T. (1980). In *Basic and Clinical Aspects of Granulomatous Diseases* (D. L. Boros and T. Yoshida, Eds.). Elsevier/North-Holland, New York, p. 81.
Yoshida, T., and Cohen, S. (1982). Fed. Proc. *41*, 2480.
Yoshida, T., and Torisu, M., Eds. (1983). *Immunobiology of the Eosinophil*. Elsevier/ North-Holland, New York.
Yoshida, T., Janeway, C. A., and Paul, W. E. (1972). J. Immunol. *109*, 201.
Yoshida, T., Nagai, T., and Hashimoto, T. (1973a). Lab. Invest. *29*, 329.
Yoshida, T., Sonozaki, H., and Cohen, S. (1973b). J. Exp. Med. *138*, 784.

31

Cell Adhesion and Phagocytic Engulfment

Darryl R. Absolom The Hospital for Sick Children, University of Toronto, Toronto, Ontario, Canada

Carel J. van Oss School of Medicine and School of Engineering and Applied Sciences, State University of New York at Buffalo, Buffalo, New York

I. INTRODUCTION

It can be shown that the fundamental process of phagocytocytic engulfment is closely linked to the general phenomenon of cell adhesiveness. Phagocytosis is the ingestion of particles by single cells and is phylogenetically the oldest and is still the most fundamental defense mechanism of the host body against foreign invaders such as bacteria. There are a few now classical observations by Davis and co-workers on the interaction between polymorphonuclear leukocytes (PMNs) and bacteria on a microscopic slide (Davis et al., 1973). Virulent bacteria, such as pneumococci, streptococci, klebsiella, and others, which are provided with capsules, escape phagocytosis by simply gliding away. Avirulent bacteria, however, which are lacking capsules, are readily engulfed. The fact that many microorganisms may be phagocytized by PMNs in the complete absence of antibody indicates that PMNs do not necessarily require very special receptors to ingest them.

Furthermore, primitive protozoa such as free-living amoeba feed on a variety of microorganisms and other materials and even avidly phagocytize polystyrene particles. Similarly, human platelets or granulocytes can be readily induced to ingest inert polymer particles for which it is highly unlikely that they have developed specific receptors. This also suggests that phagocytosis is a very primitive and general process. The primary requisite of phagocytosis is the encounter and adhesion of the particles and the phagocyte, followed by phagocytic ingestion. It is not the purpose of this chapter to consider the role of various biochemical interactions, leading to aspecific opsonization, or the role of fibronectin in such processes. Such biochemical specificities have been reviewed in detail elsewhere (Grinell, 1983). We wish to consider here the role of physicochemical interactions and focus on the fundamental mechanisms that promote or hinder the primary adhesion process and the engulfment that follows it.

The phagocytic process is comprised of a number of separate steps and activities: (1) the initial event, which may be described as the signal that initiates (2) the pursuit (i.e., the locomotion of the phagocyte toward the location of the foreign particle), and (3) the surface recognition of the particle or bacterium, followed by (4) the ingestion and (5) digestion of the foreign particle.

Cellular adhesion to large surfaces, as already mentioned, is based on the same mechanism as the phagocytic ingestion of small particles. Indeed, it is noteworthy that phagocytic cells are the very blood cells that are particularly prone to adhere to flat, relatively hydrophobic surfaces, while erythrocytes and lymphocytes have a much reduced tendency to adhere to such surfaces.

In both of these processes it is of fundamental importance to understand the circumstances under which either engulfment of (or initial adhesion to) a foreign particle, or nonengulfment (or nonadhesion) occurs, and the extent to which these processes can be manipulated. Also of interest is the extent of the specificity or nonspecificity of the surface recognition of particles or surfaces by cells is the engulfment/adhesion process. Since the particle or surface and the phagocyte have to be brought into contact with each other for phagocytosis or adhesion to occur, the thermodynamics of the free-energy changes caused by the creation or annihilation of interfaces, and the conditions under which such a contact may be expected to occur and to promote permanent adhesion, are discussed in this chapter. Considerable variation exists in the ability of cells to attach and adhere to surfaces upon contact. Factors that influence the extent of the interaction include the type of cell and its state of maturation, the nature of the surface of the particle, and the environmental conditions (e.g., ionic strength, protein composition, etc.).

II. THERMODYNAMICS OF PARTICLE ADHESION AND PARTICLE ENGULFMENT

Reactions between blood components and foreign particles or surfaces involve physical forces such as electrostatic interactions (e.g., electrostatic repulsion, plurivalent cationic bridging), van der Waals-Keesom forces, van der Waals-Debye forces, and most important, van der Waals-London or dispersion forces. Unbalanced surface forces of the foregoing types contribute to the free energy of the surface, and ultimately the decrease of the free energy of the process being considered (i.e., engulfment/adhesion) provides the driving force for the particle surface-cell interaction. This implies that the process will be favored if the free energy (F) for the system is decreased. If F increases, the process is thermodynamically unfavorable. The greater the extent of energy minimization, the more likely it will be that the process will occur.

A. Particle Adhesion

In order to make a quantitative prediction of the likelihood of the process occurring, it is necessary to "model" the process. Consider a particle P (e.g., a bacterium) initially suspended in a liquid L, adhering to a solid material S, which is also immersed in the same liquid (see Fig. 1). In the absence of gravity or any other external forces, as well as in the absence of any specific biochemical interactions (i.e., when the cell is considered simply as a deformable polymer particle), the change in the free energy ΔF^{adh} due to the process of cell adhesion is given by

$$\Delta F^{adh} = \gamma_{PS} - \gamma_{PL} - \gamma_{SL} \tag{1}$$

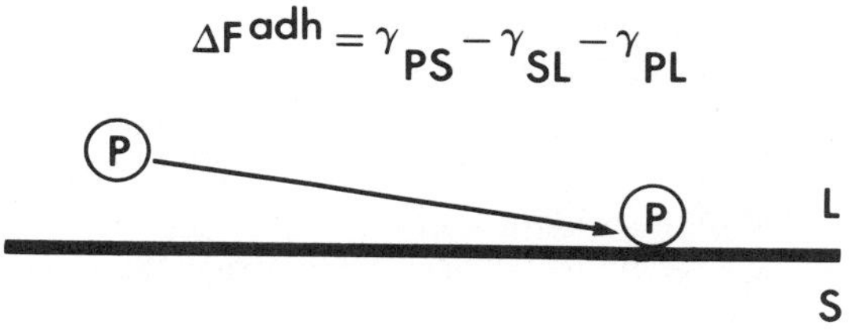

Figure 1 Schematic representation of the process of cell adhesion. P, cell, L, suspending liquid; S, substrate.

where γ_{PS} and γ_{PL} are the particle-solid and the particle-liquid interfacial tensions, respectively, and γ_{SL} is the solid-liquid interfacial tension. In this case, a cell-substrate interface is created and portions of the cell-liquid and substrate-liquid interfaces are destroyed. The free-energy difference is the maximum work that can be obtained from a process, here the adhesion of cells to a surface, and the interfacial tensions are, by definition, the work required to generate the interfacial areas under consideration.

B. Particle Engulfment

A similar model has been described for the purpose of phagocytic engulfment of small particles (see Fig. 2). The introduction of such models and their quantitative prediction [i.e., Eq. (1)] is not a panacea. Information with respect to the extent of the agreement or disagreement of the theoretical predictions of the model with the experimental data may provide important and useful insight into the *mechanisms* of the processes involved. For example, complete agreement between theory and experiment would suggest that the process being considered is entirely nonspecific in character; partial agreement, particularly the pattern in which the model breaks down, may provide valuable information on the more specific interactions in the system.

The success of the thermodynamic approach requires a quantitation of the various interfacial tensions. The solid-liquid interfacial tensions of solids have been studied using contact angle measurements and represent one of the oldest techniques for studying the nature of solid surfaces (Young, 1805). Yet, until the fairly recent contributions of Zisman and co-workers (Zisman, 1964), the interpretation and use of such measurements

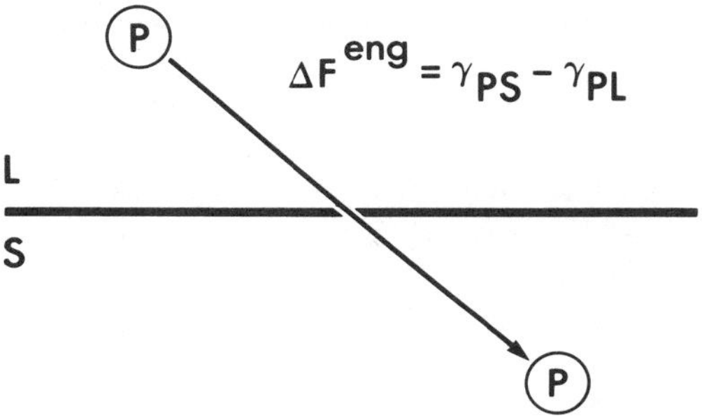

Figure 2 Schematic representation of the process of cell engulfment. P, particle; L, suspending liquid; S, cell membrane.

had not been studied. Furthermore, the method had not been previously applied to biological systems. With contact angle measurements with drops of saline water, an estimation of the surface tensions of phagocytic cells, as well as of various strains of bacteria, could be made (van Oss et al., 1975); and ΔF calculated. As mentioned above, a process is thermodynamically favored if the change in the free energy, ΔF, for the process is negative. The outcome of a process can be accurately predicted if ΔF is greater than the absolute value of 0.2 erg/cm^2 (van Oss et al., 1975).

It is clear from Eq. (1) that the surface tension of the suspending liquid can decisively influence the extent of both the phagocytic ingestion of small particles/bacteria and the adhesion of particles/cells to solid substrates. When the surface tension of the liquid, γ_{LV}, is greater than the surface tension of the adhering cells (or phagocytes), adhesion or engulfment becomes more pronounced with increasing hydrophobicity (decreasing surface tension) of the interacting surface (polymer substrate in the case of adhesion, bacteria in the case of engulfment). All types of blood cells appear to have surface tensions in the range 65 to 70 ergs/cm^2. As a general rule, in aqueous media such as Hanks' balanced salt solution (HBSS), which has a surface tension γ_{LV} of ~72.8 ergs/cm^2, the more hydrophobic the particle (bacteria), the larger the negative value of ΔF^{adh} and thus the stronger the adhesion. Table 1 compares the degree to which bacteria are ingested by normal human granulocytes with the contact angle measurements (van Oss et al., 1975). Adhesion of bacteria to phagocytes, while a minimum requirement for subsequent phagocytic ingestion, always should take place to a certain extent, at least in aqueous media, as there are to our knowledge few microorganisms even more hydrophilic than *Staphylococcus aureus*, s. Smith. Thus, in aqueous media, with a relatively high liquid surface tension, cell adhesion is most pronounced on the more hydrophobic surfaces.

In contrast, when cell adhesion and/or cell engulfment occurs in aqueous media of a lower surface tension than that of the cells themselves, adhesion to (and engulfment of) the more *hydrophilic* surface is more pronounced; that is, in low-surface-tension media, the hydrophilic bacteria are *more* readily engulfed by the phagocytic cells than are the hydrophobic bacteria (Absolom et al., 1982) and adhesion of cells is more extensive to the hydrophilic surfaces, contrary to the behavior in HBSS. This pattern of bacterial engulfment is contrary to what happens in vitro in aqueous media (van Oss et al., 1975) but also the opposite of what occurs in vivo (van Oss et al., 1975). Protein adsorption onto bacterial surfaces (and its consequences with respect to opsonization) is, however, not taken into account here, but is considered in Secs. II.C and III. Suffice it to say that the nonspecific adsorption of antibodies to the bacteria can qualitatively explain the anomalies pointed out above.

In conclusion, in vitro phagocytic ingestion of nonopsonized bacteria (B) largely parallels in vitro cell adhesion; both phenomena are dependent on the surface tension γ_{LV} of the liquid medium (L) as well as on the surface tensions γ_{PV} and γ_{BV} of the interacting surfaces (P standing for phagocyte and V for vapor). Both phenomena can be described by a relatively simple thermodynamic model in which the outcome of a process can be predicted from the net value and sign of the change of the free energy computed from the relevant interfacial energies.

Table 1 Comparison of the Extent of in Vitro Engulfment of Various Bacteria by Normal Human Granulocytes[a] in Aqueous Media[b]

Bacterium	Contact angle with drops of saline water (deg)	Phagocytic activity: N[c] ± SE.	ΔF^{adh} [d]
Staphylococcus aureus, s. Smith	16.5	0.2 ± 0.1	-0.31
Escherichia coli, t.0111	17.2	0.6 ± 0.2	-0.35
S. aureus	18.7	1.6 ± 0.3	-0.41
S. epidermidis	24.5	2.5 ± 0.4	-0.69
Listeria monocytogenes	26.5	3.1 ± 0.4	-0.84

[a]The contact angle of normal human PMNs is 18.0 ± 0.5°.
[b]For example, Hanks' balanced salt solution, γ_{LV} = 72.8 ergs/cm^2.
[c]Average number of bacteria ingested per PMN ± standard error.
[d]In ergs/cm^2 (1 erg/cm^2 = 1 millijoule/m^2).

C. Protein Adsorption and Opsonization

Although there are body fluids, serving as a liquid medium for phagocytosis, in which very little protein and only extremely small amounts of IgG are present (e.g., urine), most other body fluids contain significant concentrations of protein, including sizable amounts of albumin and IgG. The extent to which various proteins become physically adsorbed onto the bacterial surfaces must be taken into account in interpreting the

Table 2 Modes of Opsonization

	Heat-stable opsonization	Heat-labile opsonization
Aspecific opsonization		
Most pronounced with the most hydrophobic bacteria, which are therefore the least pathogenic[a]	Via aspecific adsorption of IgG1 and/or IgG3, through Fc γ receptors on phagocytes (PMNs as well as macrophages)	Via aspecific adsorption of IgG1 and/or IgG3; in the presence of $C\overline{1423}$; mediated by C3b through C3b receptors on phagocytes (PMNs as well as macrophages)
Specific opsonization		
Specific opsonization occurs only after specific antibodies have been formed especially against capsular antigens	Via specific binding of IgG1 and/or IgG3 through Fc γ receptors on phagocytes (PMNs as well as macrophages)	Via specific binding of IgG1, IgG3 and/or IgG2 and/or IgM, in the presence of $C\overline{1423}$; mediated by C3b through C3b receptors on phagocytes (PMNs as well as macrophages)

[a]That is, in the sense of being the most prone to phagocytic engulfment.

interfacial interactions occurring between bacteria and phagocytes under conditions approaching those prevailing in vivo. As each species of protein has a characteristic surface tension (van Oss et al., 1981), adsorption of the various proteins onto the surface of cells or bacteria will alter the surface properties of these particles, thereby affecting (increasing or decreasing) the extent of the engulfment process. Especially important is the physical adsorption of IgG by bacteria, as that, in most (but not all) cases, is tantamount to opsonization (see Table 2) (Absolom et al., 1982). This adsorption of IgG gives rise to altered bacterial surface properties but also gives rise to the possibility of specific biochemical interactions. Phagocytic cells possess membrane receptors for the Fc portion of the adsorbed IgG molecule. This specificity of interaction is discussed in more detail in Sec. III.

Protein adsorption onto various surfaces in many instances (van Oss et al., 1981) follows patterns similar to those of cell adhesion or engulfment, as described earlier. The same fundamental thermodynamic principles apply; in general, the more hydrophobic the bacterial species, the greater will be the extent of protein adsorption in aqueous media. The extent of protein adsorption onto the surface of any given bacterium also increases with increasing hydrophobicity of the protein species.

III. OPSONIZATION

Thus, the adsorption of certain proteins onto the surface of bacteria or other particles may influence the extent of phagocytic ingestion. Often, this adsorption greatly facilitates phagocytosis of the particles and these proteins are then referred to as *opsonins* (phagocytosis-promoting substances from the Greek *το οψον*: a relish, seasoning, or sauce) and the process of specific or aspecific adsorption of such proteins is referred to as *opsonization*.

A. Role of Specific and Aspecific IgG

The opsonic property of IgG, specific as well as nonspecific, is now well established (van Oss et al., 1975). It has been determined that the portion of the IgG molecule endowed with the opsonizing power is the Fc moiety (of IgG subclasses IgG1 and IgG3) while the Fab and $F(ab')_2$ fractions are devoid of this property.

Opsonization with IgG1 or IgG3 is accompanied by an increase in hydrophobicity of the bacterial surface, which gives rise to significantly increased phagocytic ingestion in vitro in aqueous media. Opsonization with IgG2 or IgG4 does not result in an increase in bacterial surface hydrophobicity, yet still promotes enhanced ingestion of the opsonized bacteria, at least in the case of IgG2. This strengthens the suspicion that there is more to the function of IgG than simply to increase the hydrophobicity of bacteria (or to trigger the fixation of complement), and that interactions between IgG (via its Fc moiety) and the IgG receptors on phagocytic cells play an important role in phagocytic ingestion (Absolom et al., 1982) (cf. Table 2).

IgG subclasses IgG1 and IgG3 are mainly implicated in opsonization, at least vis-à-vis monocytes and macrophages (Vernon-Roberts, 1972) and may also be the more important IgG subclasses as far as granulocytes are concerned. It is interesting to note that IgG1 and IgG3 also appear to be the IgG forms that are principally adsorbed onto polystyrene (latex) particles from whole human serum (Weening et al., 1979), which also points to their more pronounced hydrophobicity.

B. Role of Complement and IgM

The addition of complement to bacteria preopsonized with specific immunoglobulins (IgG as well as IgM) strongly enhances their phagocytic ingestion. Investigations using the consecutive addition of isolated complement (C) component to antibody-coated bacteria showed that opsonization took place at the bacteria-antibody-$C\overline{1423}$ stage (van Oss et al., 1975). Enhanced phagocytic engulfment of the complex was accompanied by a marked increase in its hydrophobicity. In vivo, however, C receptors (or more precisely C3b receptors) on the membranes of phagocytic cells are also bound to play the preponderant role in this type of opsonization. Addition of C5, C6, C7, C8, or C9 has no further effect on either hydrophobicity or phagocytic ingestion. Thus $C\overline{1423}$ (or more precisely C3b) is the principal "heat-labile" opsonin.

IgM alone, under conditions where C is inactive or absent, has no opsonic activity vis-à-vis granulocytes (van Oss et al., 1975). This agrees well with the pronounced hydrophilicity of IgM (van Oss et al., 1981). It is interesting to note that alveolar macrophages have receptors for IgG and C3b but not for IgM (Reynolds et al., 1975).

C. Other Opsonizing Agents

In principle a number of naturally occurring substances that can aspecifically adsorb to microorganisms may have an opsonizing effect. This may occur by two mechanisms: (1) the adsorbed substance may serve to increase the surface hydrophobicity of the bacteria, or (2) the phagocytic cell may have specific receptors for the opsonizing substance. The aspecific and specific opsonic roles of IgG and C have already been treated above. One other well-known opsonin is fibronectin (Yamada and Olden, 1978).

D. Role of IgA

IgA (as well as secretory IgA, sIgA) may be regarded as a true dysopsonin. It prevents the adherence of microorganisms to epithelial surfaces, as well as their ingestion by granulocytes and macrophages (Reynolds et al., 1978). IgA has a marked hydrophilicity, making it uniquely suited for preventing viruses from penetrating cells. The dysopsonic function of IgA and sIgA is especially important in preventing infection (particularly virus infections) in the upper respiratory and lower digestive tracts, that is, in those parts of the mammalian anatomy where exclusion and subsequent removal of microorganisms (e.g., by coughing) is feasible.

E. Opsonization in Vivo

In whole blood the more hydrophobic bacteria are more readily engulfed. This correlates well with the enhanced degree of adsorption of IgG from whole serum or plasma by the more hydrophobic bacteria (Absolom et al., 1982), leading to more Fc-membrane receptor and C3b receptor interactions.

F. Smallest Size of Immune Complexes Likely to Be Ingested

As mentioned earlier, antibody, in particular antibody of the IgG class, has a strong opsonizing power, even in the absence of complement. This opsonizing power is, to some extent, due to the increase in surface hydrophobicity contributed by the Fc portion of the adsorbed IgG. The following emerges: an antigenic particle (or molecule), by itself too hydrophilic to become phagocytized, is coated by antibody molecules (particularly of the IgG class) which are oriented with their (specific) Fab components toward the antigen and their Fc tails protruding outward, thus forming a more hydrophobic outer surface that facilitates phagocytosis. The protruding Fc tails mainly promote phagocytosis by binding to the Fc receptors present on the periphery of phagocytic cells. Thus the coating of particles with IgG molecules, with their Fc tails protruding outward, or the formation of antigen-antibody complexes with a sufficient number of Fc tails on their periphery, amply fulfil all the physical conditions necessary for phagocytic ingestion.

Nevertheless, these hydrophobic Fc tails do not appear to give rise to any engulfment of the myriads of single circulating IgG molecules. A minimum size appears to exist below which IgG molecules, or complexes comprising them, do not readily become phagocytized. Every freely mobile molecule or particle, regardless of its size, has a mean translational kinetic (or Brownian) energy equal to 1.5kT, where k is the molecular Boltzmann constant (= 1.38×10^{-16} erg/degree) and T the absolute temperature in degrees Kelvin. This energy is, at 37°C, 642×10^{-16} erg for all single molecules or particles. As mentioned above, a necessary condition for phagocytosis is that the particle (or IgG molecule, or antigen-antibody complex) adhere to the phagocyte. Thus an immune complex that in the course of its movement becomes attached to a phagocytic cell surface, with an energy of attachment less than 1.5kT, will be able to escape, owing to its kinetic energy of 1.5kT.

It is possible to estimate the minimum size of antigen-antibody complexes that will attach to the membrane with an energy greater than 1.5kT. According to that estimation, immune complexes comprising up to three Fc tails can escape attachment to granulocytes, and immune complexes comprising up to two Fc tails can escape attachment to macrophages (van Oss et al., 1975). These theoretical conclusions conform closely to the experimental

observations that all immune complexes comprising three or more IgG molecules are rapidly removed from the circulation in vivo (Mannik et al., 1971). More recent work further confirmed, in experimental immune complex disease in rabbits elicited with bovine serum albumin (BSA) as the antigen, that only immune complexes consisting of the antibody/antibody ratios of IgG_2-BSA or IgG_2-BSA_2 (molecular weight ca. 400,000 to 450,000) and IgG_3-BSA_2 or IgG_3-BSA_3 (molecular weight ca. 600,000 to 670,000) remain in the circulation (i.e., are not removed by phagocytosis) (Fagundus, 1980).

G. Bacterial Mechanisms for Eluding Opsonization and Ingestion

One of the principal means of protection against opsonization and ingestion is the bacterial capsule. This is a thick slimy layer of hydrophilic polysaccharide, protein, or glycoprotein (see Chap. 5), surrounding the bacterial cell. Aspecific opsonization through physical adsorption of IgG is made inoperative by the capsule, because the adsorption of IgG only takes place onto the bacterial cell wall, which lies well inside the capsule, so that no Fc can protrude (van Oss et al., 1975). At the same time aspecific ingestion is thwarted by the pronounced hydrophilicity of the capsular surface. Thus only those antibodies specifically formed to bind onto the capsule at its outer periphery are capable of opsonization (van Oss et al., 1975).

Another mechanism for thwarting opsonization (and thus phagocytic ingestion) is displayed by various strains of *Staphylococcus aureus*, which have protein A on their cell wall (see Chap. 15). Protein A binds IgG (IgG1, IgG2 and IgG4) at the Fc tail, so that only the Fab moieties protrude from the bacterial cell surface. This effectively prevents the Fc ends from contacting Fc receptors on phagocytic cells, while conferring a hydrophilic surface on the bacteria by means of the exposed (hydrophilic) Fab chains (van Oss et al., 1975; van Oss, 1978).

IV. "ACTIVATION" OF MACROPHAGES AND OTHER PHAGOCYTES

Macrophages "activated" by, for example, lymphokines such as migration inhibition factor (MIF) are characterized by enhanced spreading ability, phagocytic activity, microbicidal activity, hydrolytic enzyme production, respiration rate, and proliferation (van Oss et al., 1975). The biological characteristics of activated macrophages that are most suited to quantitative determination in vitro are macrophage migration inhibition and enhancement of macrophage adhesion. It has been shown that in vitro treatment with macrophage migration inhibition factor (MIF; see also Chap. 30) decreases the contact angle of guinea pig peritoneal macrophages, while concomitantly increasing the phagocytic activity (vis-à-vis sensitized erythrocytes) and adhesion capacity (i.e., decreased migration) of the macrophages, contrary to what would be expected from the thermodynamic predictions (van Oss et al., 1975). Ingestion of hydrophobic particles such as mycobacteria, polystyrene latex particles, or lipid droplets also leads to a decrease in contact angle of both human granulocytes and guinea pig macrophages, concomitant with an increase in phagocytic activity. Stimulation of macrophages with lipids or paraffins also leads to activation, in vitro as well as in vivo (van Oss et al., 1975). It has been suggested that the increase in the surface hydrophilicity of the activated phagocytes is accompanied by an associated decrease in membrane viscosity, which may well contribute to the observed increase in phagocytic activity (van Oss et al., 1982).

V. INFLUENCE OF ELECTROKINETIC SURFACE POTENTIAL AND OF CELL SHAPE

The electrokinetic or ζ potential [as measured by means of cell microelectrophoresis (Seaman and Brooks, 1979)] of human blood cells under physiological conditions is -11, -12, -16, and -18 mV for, respectively, platelets, PMNs, lymphocytes, and erythrocytes (van Oss et al., 1975). That surface potential suffices to prevent any of these cells from approaching one another more closely than to within about 100 Å, as long as they are reasonably smooth, and spherical or discoid shaped. However, as soon as cells have protruding pseudopodia, spicules, villi, or other projections, with a radius of curvature of $\leqslant$1000 Å, the diffuse electrostatic repulsion layer extending to the "secondary minimum" of 100 Å is easily pierced by such protrusions, so that contact to within a few Ångströms can occur, after which van der Waals attractions will prevail. Granulocytes as well as macrophages are characterized by their various protrusions of small radius of curvature (which with granulocytes tend to be pseudopodia, whereas with macrophages they are more ruffled in shape), so that their contact with (and subsequent adhesion to and engulfment of) bacterial and other particles generally ensues without being hindered by the negative charge of virtually all bacteria (van Oss et al., 1975; van Oss, 1978). This may account for the fact that attempts to correlate bacterial ζ potentials with the degree to which they become phagocytized have not been successful.

High levels of glucose cause a depression in the phagocytic activity of both granulocytes and of macrophages without, however, causing a change in the contact angle of either cell type (van Oss et al., 1975; van Oss, 1978). High levels of glucose also result in a decrease in cell adhesiveness. The cause of the decrease in phagocytic activity and cell adhesiveness at high levels of glucose (0.2 to 0.8%) appears to lie in the suppression of pseudopod formation and the increased tendency to assume a more spheroid shape under these conditions, favoring an electrostatic repulsion between the phagocytes and bacteria as well as between phagocytes and a surface. This phenomenon may well be one of the major underlying causes of the marked increased incidence of bacterial infections among insufficiently controlled diabetics.

When platelets become "sticky" under the influence of adenosine diphosphate (ADP), their surface hydrophobicity does not change but they develop marked pseudopod formation, which allows them to overcome the electrostatic repulsion of other negatively charged platelets (thereby permitting the onset of platelet aggregation) or other negatively charged surfaces (cells or tissues), permitting adhesion (van Oss et al., 1975; van Oss, 1978).

Hemagglutination (in the absence of any cross-linking agents such as dextran or polybrene) by blood group antibodies of the IgG class is feasible only in those instances where the antibodies cause a spiculation in the erythrocyte memranes, as with, for example, anti-A and anti-B (but not with anti-D) antibodies (van Oss et al., 1975, 1978). It is also interesting to note that viruses generally achieve contact with cells by means of spiky appendages which have a very small radius of curvature (van Oss et al., 1975; van Oss, 1978).

Thus spherical or otherwise smooth cells cannot normally approach one another closely enough for close-range interaction to occur, due to the electrostatic repulsion between the negatively charged electrical double layers surrounding the cells. Spicules or pseudopods with small radii of curvature (less than 1000 Å) can penetrate the electrical double layer of other cells or surfaces and thus establish contact and initiate adhesion and/or phagocytosis.

In addition to the foregoing electrokinetic repulsion effects, mitigated by surface roughness, one must not lose sight of the cross-linking role of plurivalent cations (e.g., calcium and magnesium) in mediating a process such as cell adhesion. This is generally achieved by the cation acting as a "bridging" agent between the two negatively charged interacting species. The extent of cell adhesion to various polymer surfaces can be significantly reduced by removing or chelating all the free Ca^{2+} and Mg^{2+} in the system (Absolom et al., 1980).

VI. METHODS FOR MEASURING THE SURFACE HYDROPHOBICITY OF CELLS, PARTICLES, AND PROTEINS

The main physicochemical properties of the outermost interface of bacteria and other particles and of phagocytic cells which determine the outcome of the interaction are surface hydrophobicity and charge (van Oss et al., 1975). The charge (or surface electrokinetic potential) of cells may be determined by means of microelectrophoresis under physiological in vitro conditions. This technique has been the subject of many reports (e.g., Seaman and Brooks, 1979; van Oss and Fike, 1979).

The surface hydrophobicity (or, in quantitative terms, surface tension) may be obtained by contact angle measurements (van Oss et al., 1975; Absolom et al., 1982). Two-phase partition methods, although widely used, have not yet yielded quantitative information of the surface tension of biological cells and particles (Albertsson, 1971). Typical results, using contact angle measurements, for some common biological entities are given in Table 3. Recently it was found that more correct results are obtained by measuring long-range (van der Waals) and short-range (hydrogen bond) surface tensions *separately*, by using different liquids (van Oss et al., 1986a).

VII. LONG-TERM PHENOMENA

Although the *initial* contact relationship of two confronting objects (e.g., a cell and a polymer surface) may be described and indeed predicted by the relatively simple thermodynamic principles outlined above, it should be remembered that cells are dynamic systems, and the influence of the interacting species on one another should also be considered. Some restricted and selected examples of cell-cell interactions are outlined below in an attempt to illustrate the diversity and complexity of the biophysical situation.

A. Development of Morphological Specialization Junctions

After the initial contact between cells has been established, attachment between the cells occurs. This is achieved through the development of desmosomes. The mechanism of desmosome formation is not well understood. It is generally agreed, however, that the desmosomes perform two functions: (1) to act as "permanent" adhesive sites between cells, and (2) to act as a communication channel between the cells. Desmosomes are thought to provide low-resistance junction sites where ion permeability of small ions is some 10^4 times greater than that at the cell surface (Weiss, 1972). In addition, these regions are permeable to molecules as large as 10^3 to 10^4 molecular weight. Although not conclusive, the strong possibility exists of information transfer via these junctions.

Table 3 Surface Tensions of Biological Entities (in ergs/cm^2, T = 22°C)

System	Contact angle (in degrees)
Granulocytes (Human)	69.1
Lymphocytes (Human)	70.1
Platelets (Porcine)	67.6
Bacteria	
E. coli	69.7
S. aureus	69.1
S. epidermidis	67.1
L. monocytogenes	66.3
Proteins	
Bovine serum albumin	70.2
Human serum albumin	70.3
Human immunoglobulin G	67.3
Human immunoglobulin M	69.4
Human $\alpha_2$macroglobulin	71.0
Human transferrin	66.8

B. Dynamic Nature of the Interaction of Cells

Adhering cells should not be considered as isolated units. The cells can establish a communication network, giving rise to a wide variety of activities, such as contact inhibition of movement and division, cell growth, cell segregation, and separation. Often these interactions manifest themselves only after the cells have attached themselves to a surface. For example, Stoker and co-workers (Stoker et al., 1968) have described how some cells maintained in suspension culture do not multiply, yet when attached to solid surfaces under otherwise identical conditions, they multiply vigorously. The dynamic interaction between adhering cells has been the subject of several excellent reviews (e.g., Curtis, 1981; Weiss, 1972) and will not be discussed further here.

REFERENCES

Absolom, D. R., van Oss, C. J., Zingg, W., and Neumann, A. W. (1982). J. Reticuloendothel. Soc. *31*, 59.

Albertsson, P.-A. (1971). In *Partition of Cells, Particles and Macromolecules*. Wiley-Interscience, New York.

Curtis, A. S. G. (1981). In *Lymphocyte Circulations: Experimental and Clinical Aspects*. Wiley, New York, Chap. 8.

Davis, B. D., Dulbecco, R., Eisen, H. N., Ginsberg, H. S., and Wood, W. B. (1973). *Microbiology*. Harper & Row, Hagerstown, Md.

Fagundus, A. M. (1980). Ph.D. dissertation, State University of New York, Buffalo.

Grinell, F. (1983). In *Biocompatible Polymers, Metals and Composites*. Wiley, New York, pp. 673–699.

Mannik, M., Arend, W. P., Hall, H. P., and Gilliland, B. C. (1971). J. Exp. Med., *133*, 713.

Reynolds, H. Y., Atkinson, J. P., Newball, H. H., and Frank, M. M. (1975). J. Immunol. *114*, 1813.

Reynolds, H. Y., Merrill, W. M., Amento, E. P., and Nagel, G. P. (1978). In *Secretory Immunity and Infection*, (J. R. McGhee, J. Mestecky, and J. L. Babb, Eds.). Plenum New York, pp. 533–564.

Seaman, G. V. F., and Brooks, D. E. (1979). In *Electrokinetic Separation Methods* (P. G. Righetti, C. J. van Oss, and J. W. van der Hoff, Eds.). Elsevier, Amsterdam, pp. 95–110.

Spelt, J. K., Absolom, D. R., van Oss, C. J., and Neumann, A. W. (1982). Cell Biophys. *4*, 113.

Stoker, M., O'Neil, C., Berryman, S., and Waxman, V. (1968). Int. J. Cancer *3*, 683.

van Oss, C. J. (1978). Annu. Rev. Microbiol. *32*, 19.

van Oss, C. J., and Fike, R. M. (1979). In *Electrokinetic Separation Methods* (P. G. Righetti, C. J. van Oss, and J. W. van der Hoff, Eds.). Elsevier, Amsterdam, pp. 111–120.

van Oss, C. J., Gillman, C. F., and Neumann, A. W. (1975). In *Phagocytic Engulfment and Cell Adhesiveness*. Marcel Dekker, New York.

van Oss, C. J., Mohn, J. F., and Cunningham, R. K. (1978). Vox Sang. *34*, 351.

van Oss, C. J., Absolom, D. R., Neumann, A. W., and Zingg, W. (1981). Biochim. Biophys. Acta *670*, 64.

van Oss, C. J. Absolom, D. R., and Neumann, A. W. (1983). In *The Reticuloendothelial System*, Vol. 4 (H. Friedman, M. R. Escobar , and S. M. Reichert, Eds.). Plenum, New York.

Vernon-Roberts, B. (1972). In *The Macrophage*. Cambridge University Press, Cambridge, p. 150.

Weening, R. S., Roos, D., and van Schalk, M. J. K. (1979). In *Inborn Errors of Immunity and Phagocytosis* (F. Gittler, J. W. T. Seakins, and R. A. Harkness, Eds.). University Park Press, Baltimore.

Weiss, L. (1972). In *The Chemistry of Biosurfaces*, Vol. 2 (M. L. Hair, Ed.). Marcel Dekker, New York, Chap. 8.

Yamada, K. M., and Olden, K. (1978). Nature (Lond.) *275*, 179.

Young, T. (1805). Philos. Trans. R. Soc. *95*, 65.

Zisman, W. (1964). Adv. Chem. Ser. *43*, 1.

Stoker, M., O'Neil, C., Berryman, S., and Waxman, V. (1968). Int. J. Cancer *3*, 683.

van Oss, C. J. (1978). Annu. Rev. Microbiol. *32*, 19.

van Oss, C. J., and Fike, R. M. (1979). In *Electrokinetic Separation Methods* (P. G. Righetti, C. J. van Oss, and J. W. van der Hoff, Eds.). Elsevier, Amsterdam, pp. 111–120.

van Oss, C. J., Gillman, C. F., and Neumann, A. W. (1975). In *Phagocytic Engulfment and Cell Adhesiveness*. Marcel Dekker, New York.

van Oss, C. J., Mohn, J. F., and Cunningham, R. K. (1978). Vox Sang. *34*, 351.

van Oss, C. J., Absolom, D. R., Neumann, A. W., and Zingg, W. (1981). Biochim. Biophys. Acta *670*, 64.

van Oss, C. J. Absolom, D. R., and Neumann, A. W. (1983). In *The Reticuloendothelial System*, Vol. 4 (H. Friedman, M. R. Escobar , and S. M. Reichert, Eds.). Plenum, New York.

Vernon-Roberts, B. (1972). In *The Macrophage*. Cambridge University Press, Cambridge, p. 150.

Ward, C. A. W., and Neumann, A. W. (1974). J. Colloid Interface Sci. *49*, 286.

Weening, R. S., Roos, D., and van Schalk, M. J. K. (1979). In *Inborn Errors of Immunity and Phagocytosis* (F. Gittler, J. W. T. Seakins, and R. A. Harkness, Eds.). University Park Press, Baltimore.

Weiss, L. (1972). In *The Chemistry of Biosurfaces*, Vol. 2 (M. L. Hair, Ed.). Marcel Dekker, New York, Chap. 8.

Yamada, K. M., and Olden, K. (1978). Nature (Lond.) *275*, 179.

Young, T. (1805). Philos. Trans. R. Soc. *95*, 65.

Zisman, W. (1964). Adv. Chem. Ser. *43*, 1.

32

Methods of Cell Separation

Nicholas Catsimpoolas Boston University School of Medicine, Boston, Massachusetts

I. INTRODUCTION

The cells of the immune system as found in biological fluids and tissue extracts represent a complex mixture of functionally distinct subpopulations. Therefore, in certain immunological studies, there exists a need to obtain isolated homogeneous populations of cells in order to define their functional role and to determine their biological and biochemical characteristics. Other areas where the availability of cells of a certain type is advantageous include transfusion, cell immunotherapy, and bone marrow transplantation. Furthermore, the production of hybridomas with specific biological functions can be facilitated if the starting cellular material is enriched with cells bearing the desirable markers.

Methods for cell separation (Catsimpoolas, 1977) vary from very simple manual laboratory techniques to the use of highly sophisticated computerized instruments depending on the isolation task at hand. The separation principles which are utilized depend either on the physical characteristics of the cells (e.g., density, electric charge, volume) or biological characteristics (e.g., specific receptors, markers, function). Often hybrid methods depending on both physical and biological properties are employed. An additional methodological distinction is that the cells can be separated in bulk as a population, or individually, one by one, as single cell units. The latter technique is known as *cell sorting*.

Some general restrictions and precautions have to be taken with any separation technique to avoid damaging the cells. Some of these requirements include maintenance of physiological osmolarity, the use of buffers with pH that does not damage the cells (preferably pH 7.4), use of nontoxic buffer and density media ingredients, employment of low temperatures to minimize metabolic activity, and the addition of osmotic expanders (e.g., sucrose, glucose) to maintain osmolarity in low-ionic-strength buffers.

Other important considerations are the avoidance of cell aggregation (clumping) and the preservation of cell viability during the separation process. In preparative procedures, the demonstration of biological function—after the separation has been accomplished—is often required.

II. BULK SEPARATIONS

A. Physical Methods

Definition

The cells of the immune system exhibit physical properties associated with the intact particle which may differ from one class of cell to another (e.g., between T and B lymphocytes, T lymphocytes of two different differentiation stages, antibody forming and memory B cells, etc.). Wherever these differences exist and are of sufficient magnitude to allow partial or complete separation, they can be utilized advantageously because the cells are not manipulated chemically, or altered by interactions with ligands or other cells as is the case wtih most biological methods. The physical properties most often utilized are those of cell volume, density, electric surface charge, and membrane affinity. The techniques that are used for the separation are based on sedimentation, electrokinetic, and two-phase partition processes.

Figure 1 DESAGA gravity sedimentation chamber equipped with a peristaltic pump for fraction collection from the top of the gradient.

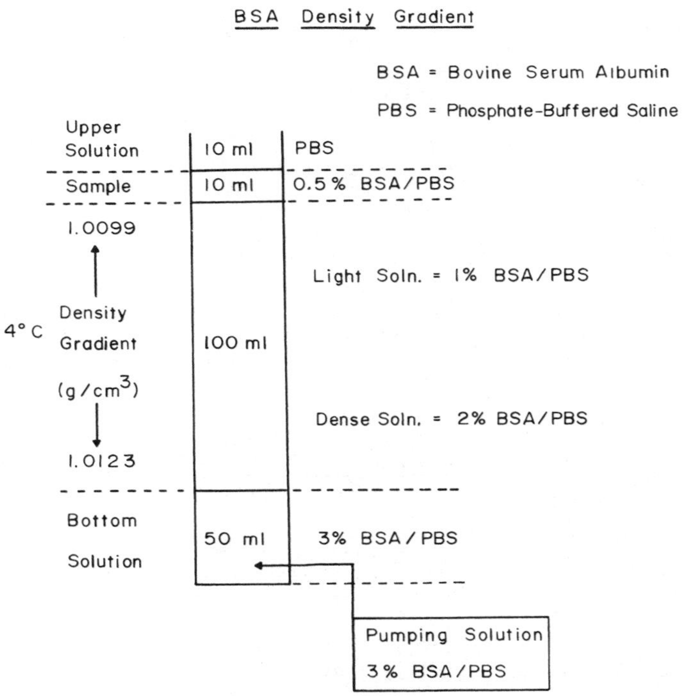

Figure 2 Schematic diagram of the arrangement of the solutions in a gravity sedimentation column. BSA, bovine serum albumin; PBS, phosphate buffered saline. The 3% BSA/PBS solution acts as a "cushion" for the density gradient, which is formed by a light (1%) and a dense (2%) BSA solution in PBS using a gradient maker. The cell particles are suspended in 0.5% BSA in PBS and layered on top of the gradient. An upper PBS solution is layered on top of the sample. At the end of the sedimentation experiment a 3% BSA in PBS solution is used to "chase" the gradient upward to collect fractions from the top of the column. (From Catsimpoolas and Griffith, 1977.)

Sedimentation Techniques

Gravity sedimentation. In this method, a shallow density gradient usually of bovine serum albumin (BSA) dissolved in phosphate buffered saline (PBS) is formed in a cylindrical column or chamber (Fig. 1) and the cells in suspension are layered on top of the gradient (Mel, 1963; Miller and Phillips, 1969; Tulp and Bont, 1979; Catsimpoolas et al., 1978). The BSA/PBS gradient covers the density range 1.0099 to 1.0123 g/cm^3 at 4°C (Fig. 2). The density of the gradient is thus much lower than the density of mammalian cells, which usually varies from 1.060 to 1.090 g/cm^3. Pulse loading of the cells on top of this gradient allows them to sediment toward the bottom of the column. The downward transport velocity of the cells is described by the Stokes equation:

$$v(x) = \frac{2}{q} \frac{r^2 g}{\eta(x)} [\rho_c - \rho(x)] \qquad (1)$$

where v(x) is the instantaneous velocity of cells at any position x in the column (cm/sec), g the acceleration of gravity (980.7 cm/sec^2), r the cell radius (cm), $\eta(x)$ the viscosity of

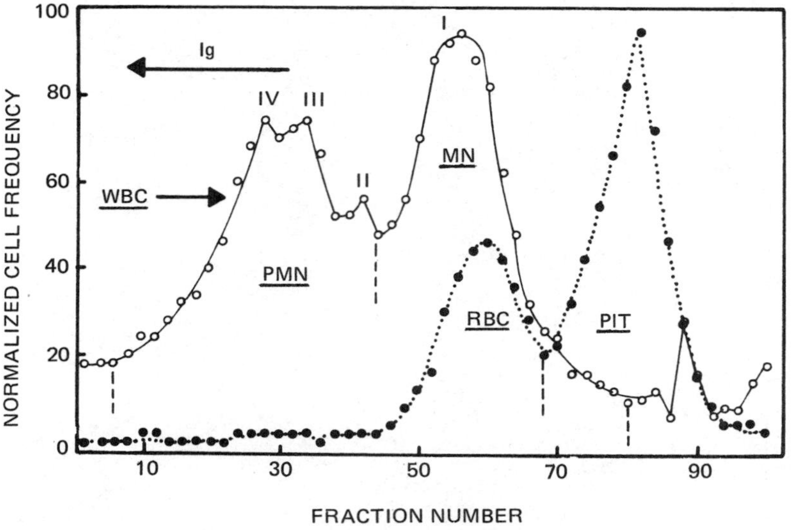

Figure 3 Velocity sedimentation at 1 g of human blood cells. Open circles (O) represent the distribution of white blood cells (WBC) and solid circles (●) represent the distribution of erythrocytes (RBC) and platelets (Plt). Each distribution has been normalized to the highest peak by computer. MN, mononuclear white cells; PMN, polymorphonuclear cells. (From Catsimpoolas et al., 1978.)

the medium at position x (poise), $\rho(x)$ the density of the medium at position x (g/cm^3), and ρ_c the cell density (g/cm^3). Assuming spherical shape, the cells sediment in a manner proportional to their $r^2\rho_c$. The volume V of spherical particles is proportional to r^3. Therefore, sedimentation depends on $V^{2/3}\rho_c$, which indicates that the combination of both volume and density plays a role in the fractionation of the cells. Other secondary parameters, such as deviation from spherical shape and surface "roughness," may contribute to the sedimentation effect by altering the frictional coefficient.

When the cells have been allowed to sediment for a sufficient period of time (i.e., 2 to 6 hr) to achieve separation, fractions are collected either from the bottom or the top of the column, and the cells in each fraction are counted and characterized by available methods. As an example, a typical fractionation of human blood cells by velocity sedimentation at 1 g is shown in Fig. 3. The open circles represent the distribution of the white cells [mononuclear (MN) plus polymorphonuclear (PMN) cells] and the solid circles the erythrocytes (RBC) and platelets (Plt). The curves were normalized by computer to the highest peak of each distribution. The dotted lines indicate that the fractions were highly enriched (90 to 100% populations of a particular type of cell were found). Thus the polymorphonuclear cells were distinctly separated from the mononuclear cells (lymphocytes and monocytes), erythrocytes, and platelets. However, separation of mononuclear cells from erythrocytes was not achieved. Highly enriched platelet fractions relatively free of other cells were obtained.

Zonal centrifugation. This technique, also known as velocity sedimentation in regular or isokinetic density gradients (Boone et al., 1968; Pretlow et al., 1969; Pretlow, 1971), utilizes the same principles as gravity sedimentation, except that the gravity force

is amplified by centrifugation of the sample. In the isokinetic variation, the density gradient has been designed and formed in such a manner as to provide constant sedimentation velocities. The most commonly used substance for the formation of the gradient is Ficoll (polysucrose, average MW = 400,000), although fetal calf serum and BSA have also been employed. The following equation describes the velocity sedimentation of cells:

$$v(x) = \frac{dx}{dt} = \frac{r^2(\rho_c - \rho_m)\omega^2 x}{18\eta} \tag{2}$$

where v is the velocity of the cell at time t and location x, x the distance of the cell in the gradient from the center of revolution, t the duration of centrifugation, r the cell radius, ρ_c the cell density, ρ_m the gradient density at position x, ω the angular velocity, and η the gradient viscosity at position x. Similarly to the gravity sedimentation technique, cells are separated according to both volume and density.

Isopychnic centrifugation. It may be seen from Eq. (2) that if the density of the gradient has sufficiently high values and, therefore, at some sedimentation distance x, the density of the cells equals the density of the gradient (i.e., $\rho_c = \rho_m$), the velocity becomes zero. Thus, at equilibrium, a band of cells is formed at that particular location which cannot be sedimented any further. Cell populations that have different densities are banded at different locations along the centrifuge tube. Thus the isopychnic centrifugation method separates cells according to their density only (Leif and Vinograd, 1964). The density of the cells can also be determined by measuring the density of the gradient fraction where the cells stopped migrating.

Materials that have been used for this type of separation include BSA, Ficoll, metrizamide, and colloidal silica. In simplified laboratory procedures (e.g., separation of peripheral lymphocytes from other blood cells) a two-step gradient is used instead of a continuous density variation. After centrifugation, the cells are separated by banding at the interface of the two density solutions, or sedimenting to the bottom of the tube (Böyum, 1968).

Centrifugal elutriation. In centrifugal elutriation (Lindahl, 1948) cells are separated by velocity sedimentation in a special centrifuge chamber which has the shape of a cone, the apex of which is in the direction of sedimentation. The cells are subjected to a calculated centrifugal force while suspended in a medium that flows continuously in a centripetal direction. Rapidly sedimenting cells encounter an increasingly rapid centripetal flow as they approach the apex of the cone, whereas the slow-sedimenting cells move centripetally and approach the base of the cone, where they are collected by leaving the chamber. The different velocities of the cells result from the summation of the centrifugal velocity and centripetal flow velocity (Sanderson et al., 1976).

The collection of the cells of known densities and radius can be predicted from the equation

$$r = \frac{3.6}{2\pi\omega}\left[\frac{\eta V_F}{2\pi(\rho_c - \rho_m)}\right]^{1/2} \frac{1}{x(R/L)^2(R-x)^2} \tag{3}$$

where r is the radius of the cell, ω the angular velocity, η the viscosity of the medium, V_F the volume of fluid passing through any cross section per unit time, ρ_c the cell density, ρ_m the medium density, x the distance from the center of revolution, R the greatest radius of the sedimentation chamber, Z the radial distance between the center of revolution and the apex of the chamber, and $L = Z - X_R$.

The advantage of centrifugal elutriation is that cells are separated continuously, and thus larger numbers can be prepared within a given time in comparison to other sedimentation techniques.

Electrophoresis

The charged particle. Electrophoresis is the physical process involved in the migration of electrically charged particles in an electric field provided by a direct-current potential. The net surface charge originates primarily by the ionization of chemical groups such as carboxyl, amino, phosphate, imidazolium, and so on, and secondarily by ion adsorption.

When an electrically charged particle is placed in a polar medium (e.g., composed of an aqueous solution of an electrolyte), ions of the same charge (coions) are repelled away from the surface and ions of opposite charge (counterions) are attracted. Therefore, near the surface there is an excess oc counterions over coions. This nonrandom spatial distribution of charges around the particle constitutes a second electrical layer despite the existing Brownian motion. The particle and its ionic environment are known as the *electric double layer* and many theories have been derived to describe its electrokinetic behavior (Shaw, 1969). Under certain assumptions Gouy and Chapman derived a simple expression to show that the potential generated in the diffuse layer decreases exponentially with distance from the charged surface:

$$\psi_x = \psi_0 \exp(-\kappa x) \tag{4}$$

where ψ_x is the potential at a distance x from the surface, ψ_0 the potential at the surface, and κ the Debye-Hückel constant. The distance from the surface at which the potential is decreased by one exponential factor is equal to $1/\kappa$, and it is customarily referred to as the thickness of the double layer. It is generally used as a measure of the extension of the double layer. The thickness of the layer depends on all the variables associated with the Debye-Hückel constant:

$$\kappa = \left(\frac{4\pi e^2}{\epsilon KT} \Sigma n_i z_i^2\right)^{1/2} \tag{5}$$

where e is the charge of the electron, ϵ the dielectric constant of the medium, K the Boltzmann constant, T the absolute temperature, n_i the number of ions per unit volume (L) of type i, and z_i the valency of charge of ion i. Thus the nature of the ions and their concentration, which determines the ionic strength of the solution, play an important role in the ionic cloud surrounding the particle if temperature is kept constant. For example, for a uni-univalent electrolyte the thickness of the double layer (i.e., $1/\kappa$) will be approximately 10 Å for a 0.1 M solution and 100 Å for a 0.001 M solution.

Behavior of the charged particle in an electric field. Electrophoresis of charged particles can be described as a directed movement of the "electrokinetic unit" in an electric field. The electrokinetic unit consists of the charged particle and a thin layer of bound fluid. The imaginery outer boundary of this layer corresponds to the surface of shear and represents the slippage plane between a particle and its suspending medium. This hypothetical plane is very close (i.e., few angstrom units) to the ionic groups of the particle surface. Its electrical potential is called the zeta potential (ζ) and its magnitude is somewhat smaller than the value of the cell surface potential (ψ_0). The zeta potential of particles can be estimated from the Helmholtz-Smoluchowski equation

$$\zeta = \frac{4\pi\eta u}{\epsilon} \qquad (6)$$

where ζ is the zeta potential (V), u the electrophoretic mobility (cm^2/V-sec), and η and ϵ are the viscosity (poise) and dielectric constant, respectively. The electrophoretic mobility value is obtained from measurement of the electrophoretic velocity v (cm/sec) and electric field strength E (V/cm):

$$u = \frac{v}{E} \qquad (7)$$

The electric field strength under constant current conditions can be estimated from

$$E = \frac{i}{qk} \qquad (8)$$

where E is the electric field strength (V/cm), i the current (A), k the specific conductance of the buffer (Ω^{-1}/cm), and q the cross-sectional area of the separation medium (cm^2). Thus the actual experimental measurements involve estimation of the electric field strength applied and the velocity of the particle from which the electrophoretic mobility (u) and zeta potential (ζ) can be derived. Furthermore, in consideration of the assumptions made by Gouy and Chapman on the diffuse double layer, one can estimate the surface charge density (σ_0) from

$$\sigma_0 = 2\left(\frac{NKT}{2000\pi}\right)^{1/2} (\epsilon I)^2 \sinh \frac{e\zeta}{2KT} \qquad (9)$$

where σ_0 is the surface charge density (stat. coulomb/cm^2), N is Avogadro's number, and I is the ionic strength. The other symbols are as defined above. The ionic strength is given by

$$I = \frac{1}{2} \Sigma c_i z_i^2 \qquad (10)$$

where c_i is the molar concentration of ion of type i, and z_i is as defined above.

Electrophoretic separations. Electrophoresis is performed in a number of types of apparatus operating under different principles (Catsimpoolas, 1980). Some are useful only as analytical devices, whereas others perform both analytical and preparative functions. Instruments for performing electrophoretic separations and/or measuring electrophoretic mobility can be classified into those which provide physical separation of the species and those which do not. Among the latter are the microscope/image analyzers (Bartels et al., 1981) and the laser light-scattering apparatus (Ware, 1974). Physical separation necessitates the use of some form of anticonvectional stabilization of the separated species, which is provided either by the particular mechanical design of the apparatus (capillary action, rotation) or by the use of density gradient media.

For preparative purposes the most commonly used techniques are free-flow and density gradient electrophoresis, which are described briefly below.

1. *Free-flow electrophoresis.* Continuous electrophoretic separation of cells can be achieved in the free-flow electrophoretic apparatus (Hannig, 1967), which consists of two plates which form a separation chamber with a gap of capillary dimensions (Fig. 4). Two electrode channels are placed at the sides of the chamber and are separated from the

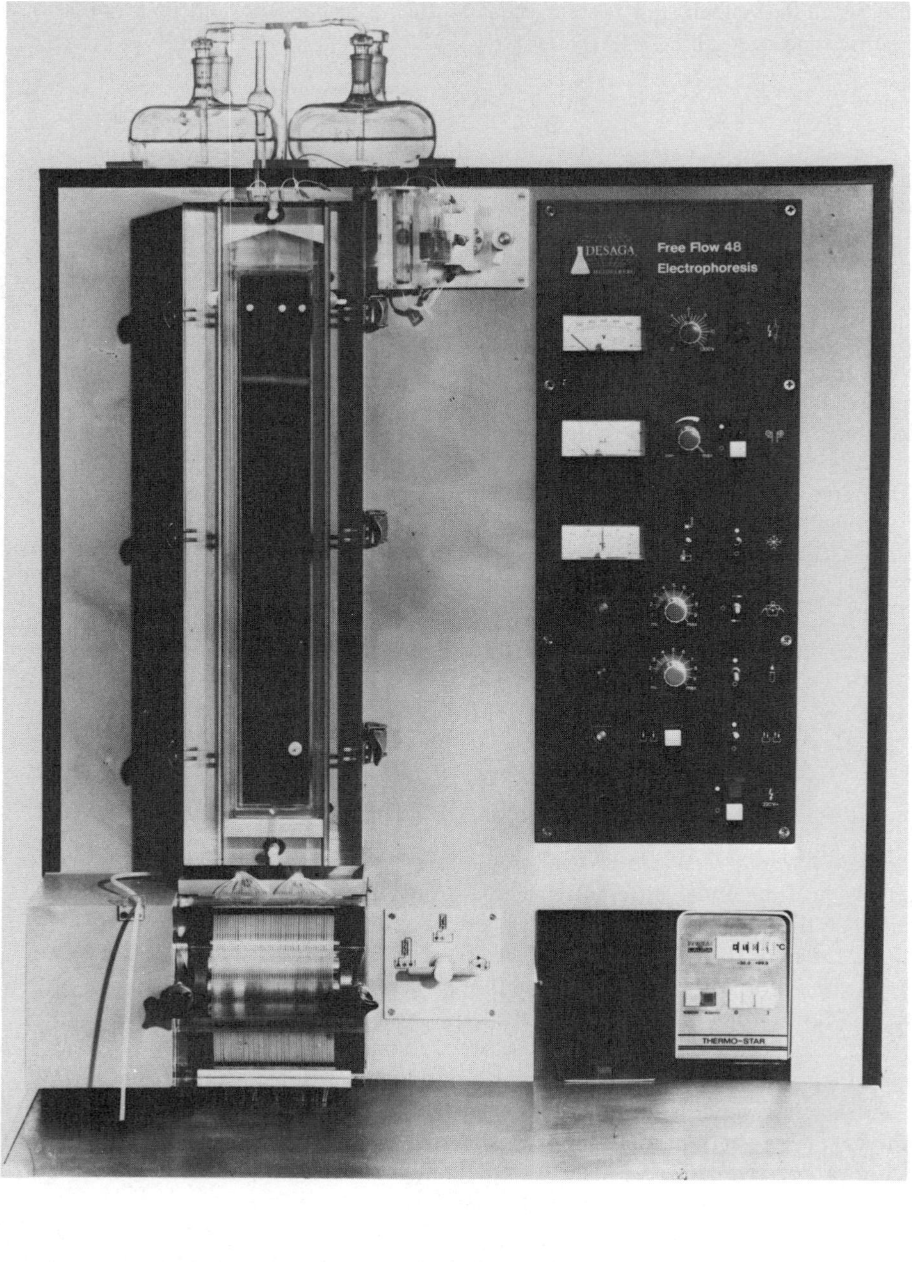

Figure 4 DESAGA free-flow electrophoresis apparatus.

separation compartment by membranes. By means of a multichannel peristaltic pump, buffer flows from the top to the bottom of the chamber and is collected in more than 30 tubes. Stabilization of the liquid curtain is achieved both by the capillary-thin separation gap of the electrophoresis chamber and by the laminar buffer flow. Effective cooling of the thin buffer curtain permits the application of high electrical field strengths, which shortens the time of analysis to a few minutes of residence. The cell sample is applied at the top of the apparatus as a continuous stream which is transferred by the buffer flow to the other end of the curtain, where it is collected. In the presence of the electric field (perpendicular to the buffer flow) the particles are deflected toward the positive electrode. The magnitude of deflection is a function of the electrophoretic velocity, buffer flow velocity, and applied electric field strength. Cells with different electrophoretic mobilities are deflected and separated in what appears to be continuous vertical streams exiting at different locations at the bottom of the chamber.

2. *Density gradient electrophoresis.* In the density gradient electrophoretic method (Boltz et al., 1973; Griffith et al., 1975) stabilization of the separated zones of cell populations is achieved by the formation of a density gradient using Ficoll or Ficoll/sucrose on which the sample is loaded (Figs. 5 and 6). The upper and lower parts of the gradient are in contact with the electrolyte in the electrode vessels. Usually, the density gradient is formed on top of a "cushion" dense solution of the same compound that was used for the gradient. Cooling is provided both by an external water jacket and an internal "cooling finger" of circulating 4°C water. Thus the cross section of the separation compartment has the form of an annulus cooled on both sides. After electrophoresis, fractions are collected by draining of the density gradient from the bottom of the column, or by pumping the gradient through a capillary tube in the center of the cooling finger. A typical separation of mouse spleen T and B lymphocytes using the density gradient technique is shown in Fig. 7. Ascending preparative cell electrophoresis in a D_2O/H_2O gradient has been described by Bronson and van Oss (1979) and has been applied to the separation of lymphocytes.

Partition in Two-Phase Polymer Systems

Two-phase liquid immiscible systems can be obtained by mixing two water-soluble polymers above certain critical concentrations (Albertsson, 1971) for cell separation, the most suitable compounds have been shown to be dextran (D) and polyethylene glycol (PEG). Since such systems are aqueous, they can be buffered and made isotonic. The partition of cells in the two phases (top/bottom) depends on the relative affinity of the membrane constituents to the polymers and salts in the phases and also to the cells' adsorption at the interface. Separation between two cell populations is achieved by appropriately manipulating these parameters to effect differences in the partition coefficients which are given by:

$$\bar{K} = \exp\left(\frac{\alpha + \gamma A}{KT}\right) \tag{11}$$

where $\bar{K}$ is the partition coefficient (i.e., concentration of cells in top/concentration of cells in bottom phase), α the difference in electrostatic free energy, γ the difference in interfacial free energy between the phases for the particles being partitioned, A the surface area of the particle, K the Boltzmann constant, and T the absolute temperature.

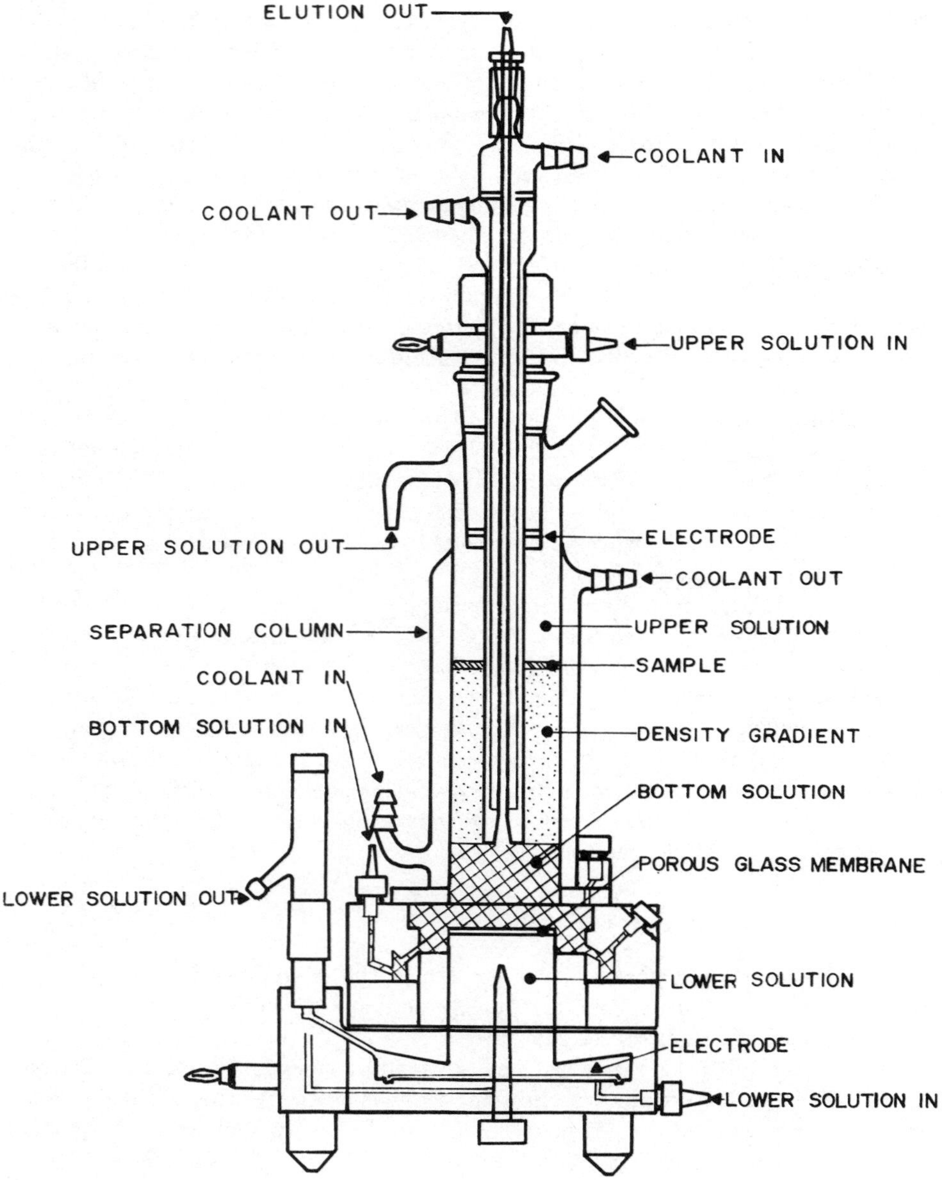

Figure 5 Schematic diagram of the Buchler Poly-Prep 200 density gradient electrophoresis apparatus. (From Catsimpoolas and Griffith, 1977.)

Although the partition coefficient of a soluble compound is defined as the ratio of the concentration in the top and bottom phases, the partition of cells takes place between one phase (e.g., top) and the interface (Fig. 8). Thus the partition is in practice expressed as the quantity of cells in the top phase.

The partition apparatus for cell separation consists of two circular plexiglass plates with 120 concentric cavities. The bottom plate remains stationary and the top plate

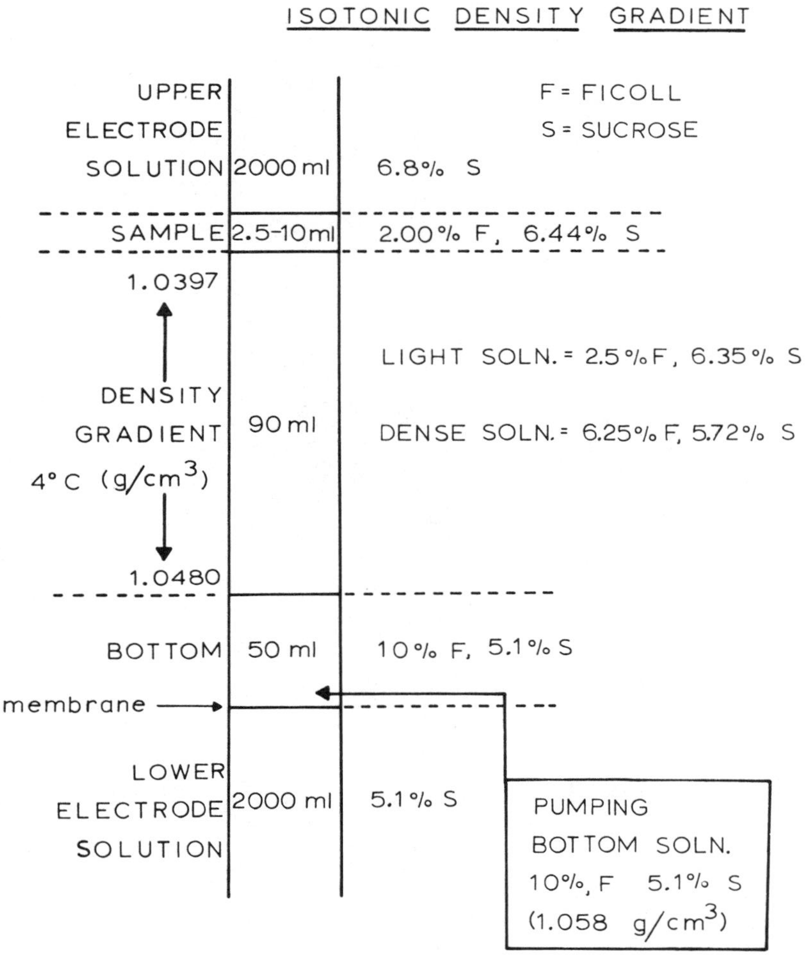

Figure 6 Schematic diagram of the arrangement of solutions in the preparative density gradient electrophoresis column shown in Fig. 5. (From Catsimpoolas and Griffith, 1977.)

rotates by means of a motor. There is a hole through the top plate at each cavity for loading the phases and the cells. All the cavities are loaded with the bottom and top phases except for cavities No. 0 to 5, which receive the bottom phase and the cells suspended in the upper phase. In addition, the volumes of the phases in each cavity are predetermined to assure that the interface is treated as part of the bottom phase. Operation of the apparatus consists of (1) rotatory shaking for 25 sec, during which top and bottom phases are mixed; (2) a settling cycle for 5 min, during which top and bottom phases separate; and (3) a transfer during which the top plate rotates clockwise through 3° (3° x 120 cavities = 360°), lining up each top cavity with the next cavity in the bottom plate. This cycle (1 to 3) is repeated for a preset number of transfers (e.g., 60). With each transfer the cells in the top phase are carried forward, where they are reextracted with fresh bottom phase while those cells at the interface are reextracted with fresh top phase. At

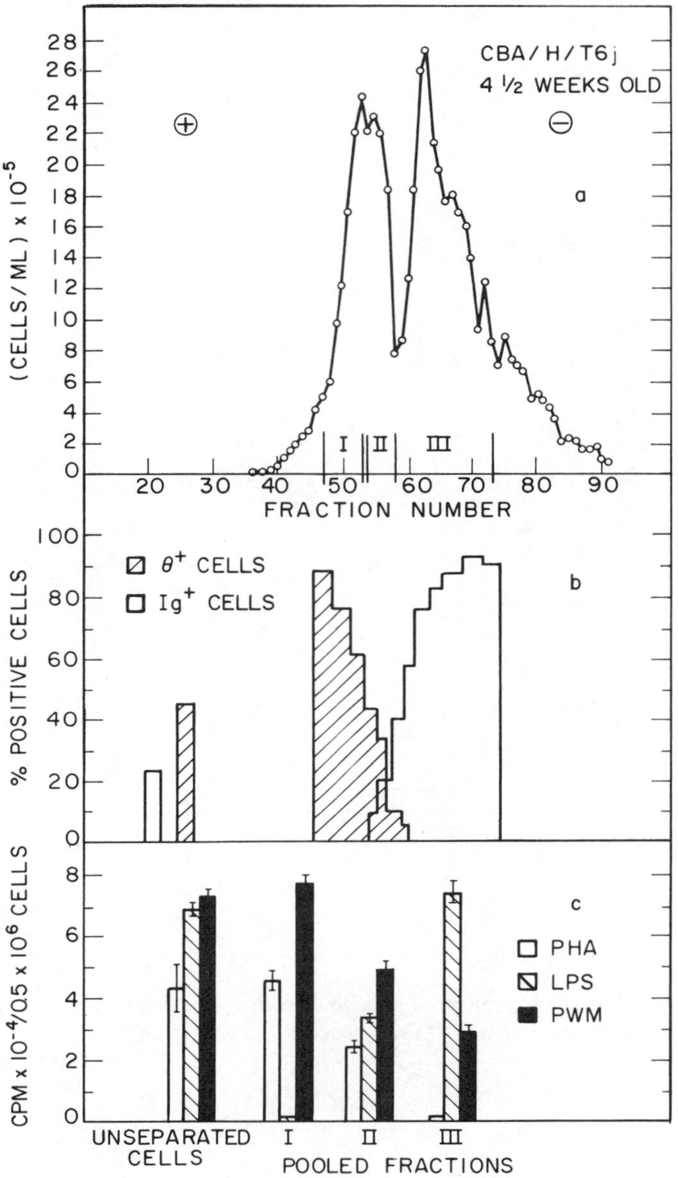

Figure 7 Density gradient electrophoretic distribution profile of CBA/H/TGJ mouse spleen lymphocytes from a 4.5-week-old animal and identification of θ and IgG-positive cells as well as cells stimulated by mitogens (PHA, LPS, PWM). (From Platsoucas and Catsimpoolas, 1979.)

the end of the partition experiment the cells in each cavity are collected, counted, and further characterized.

The phases can be modified by adding salts that have different affinities for the two phases, or using substituted polymer ligands (e.g., diethylaminoethyldextran, trimethylaminopolyethylene glycol, etc.).

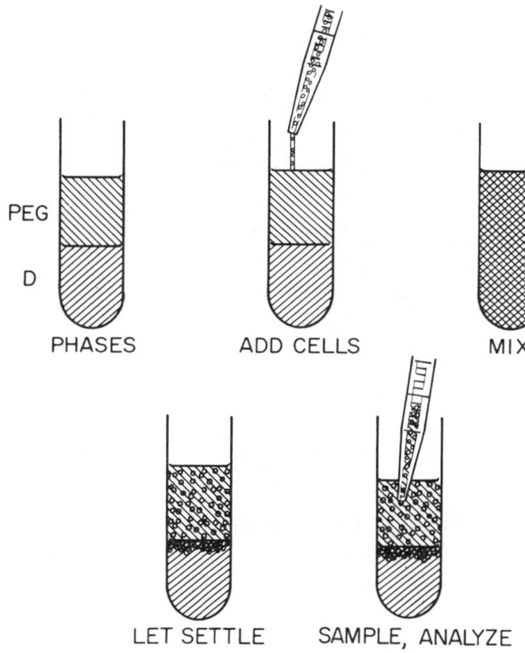

Figure 8 Diagrammatic presentation of the partition procedure with cells. PEG denotes the polyethylene glycol-rich phase and D the dextran-rich phase. A known quantity of cells is added to the phases, which are then mixed and permitted to settle for a certain period of time. At the end of this time, an aliquot is withdrawn from the top phase and the cell quantity is determined. Partition is expressed as the quantity of cells in the top phase. (From Walter, 1977.)

B. Biological Methods

The presence of various specific receptors on the surface of cells provide the basis for their separation by bioaffinity methods. These are clonal or class receptors, differentiation or histocompatibility antigens, and lectin receptors. The techniques that are used for isolation (or elimination) of specific cell populations are rosetting, affinity chromatography, adherence, magnetic attachment, and negative selection (elimination) either by lysing the cells with an antiserum to a surface antigen and complement, or by killing them by incorporation of a high-specific-activity radioisotype during the proliferation stage.

Rosetting

Lymphocytes, upon reaction with native heterologous erythrocytes (RBCs) (e.g., sheep, mouse, monkey) or with specific antibody-coated erythrocytes, form cellular complexes known as rosettes. These structures consist of a single lymphocyte surrounded by a number of attached erythrocytes. Examples of such interactions include the spontaneous rosette formation of human T lymphocytes with sheep erythrocytes, the rosetting of cells bearing Fc receptors by erythrocytes coated with IgG or IgM heterologous anti-RBC antibody, interaction of T_μ and $T\gamma$ lymphocytes with ox red blood cells coated with rabbit IgM or IgG anti-ox RBCs, and other.

Rosetting from nonrosetting lymphocytes can be separated by centrifugation of the mixture on a Ficoll-Hypaque cushion with a density of 1.077 g/cm^3 (Böyum, 1968). This separation is possible because of the high density of the erythrocytes (1.12 g/cm^3) and large size of the rosette structures (Wahl et al., 1976). Briefly, the cells are layered carefully on top of the Ficoll-Hypaque density cushion and centrifuged for 20 to 30 min at 400*g*. Rosettes and free RBCs are collected from the bottom of the cushion. Nonrosetting lymphocytes remain at the interface and are removed with a Pasteur pipette. Free or lymphocyte-attached red blood cells are removed by hypotonic lysis with Tris or $KHCO_3$ buffered ammonium chloride. An alternative way to recover the lymphocytes is dissociation of the rosettes by incubation at 37°C and mechanical disruption. The red blood cells are subsequently removed by centrifugation on Ficoll-Hypaque.

Affinity Chromatography and Immunoadsorbents

The affinity chromatography methods utilize covalently bound molecules, usually specific antibodies and lectins, to a solid-phase support material such as Sephadex, polyacrylamide gel beads, glass beads, and other (Wigzell and Anderson, 1969). The mixture of the cells to be separated is applied to the column and the nonretained cells are eluted. The solid-phase material binds a specific population of cells by means of their cell surface antigens, receptors, or other markers. The bound cells can also be released by materials that compete strongly for the binding site of the support matrix.

The affinity methods have been used for the separation or depletion of various lymphocyte subpopulations, including B lymphocytes, complement receptor bearing cells, specific haptens or antigen binding lymphocytes, Fc receptor-bearing cells, cytotoxic T lymphocytes, and others.

A variation of the affinity technique utilizes polystyrene plastic dishes for surface adsorption of antigens, antibodies, and lectins by adherence instead of covalent binding. After incubation with the cell mixture of interest, the nonadhering cells are removed by gentle stirring and washing.

Elimination by Lysis and "Suicide"

Incubation of lymphocytes with antiserum specific for one of their surface markers causes lysis of the cells in the presence of complement (Boyse et al., 1964; Raff, 1970). Thus the method can be used for the elimination of specific cell populations such as T lymphocytes by anti-θ serum, B lymphocytes by anti-immunoglobulin serum, Ly-1+ or Ly-2,3+ cells by treatment with specific anti-Ly serum, and others.

An additional method for elimination of specific lymphocytes is the use of a high-activity radioisotope which kills the cells after binding, or incorporation. Radioiodinated antigens have been used for this purpose as well as incorporation of the thymidine analog 5-bromodeoxyuridine (BUdR) into the DNA of dividing cells and inactivation of the latter by exposure to visible light ("suicide" technique).

Miscellaneous Techniques

Adherence to nylon fiber columns (Greaves and Brown, 1974) has been used as a simple method for preparing reasonable quantities of T-cell-enriched lymphocytic suspensions of greater than 85% purity, but with the limitation of poor yields (about 40 to 50% of T cells are lost). Adherence to other materials (e.g., glass surfaces, plastic, Sephadex G-10)

has also been used to remove monocytes and variable percentages of lymphocytes, particularly B cells. From heparinized whole blood, nylon fibers selectively adsorb the granulocytes (Greenwalt et al., 1962). After washing with saline the granulocytes can be eluted in a pure and undamaged state, with phosphate buffered saline containing 10% dimethyl sulfoxide (or 2.5% ethanol) and 0.1% Na_2 EDTA (Absolom et al., 1981).

Magnetic separation of monocytes and phagocytes by injection of carbonyl iron (Wofsy et al., 1971) or magnetic microspheres (Kronick et al., 1978) is often used for their selective removal from cell suspensions by attraction to a magnet.

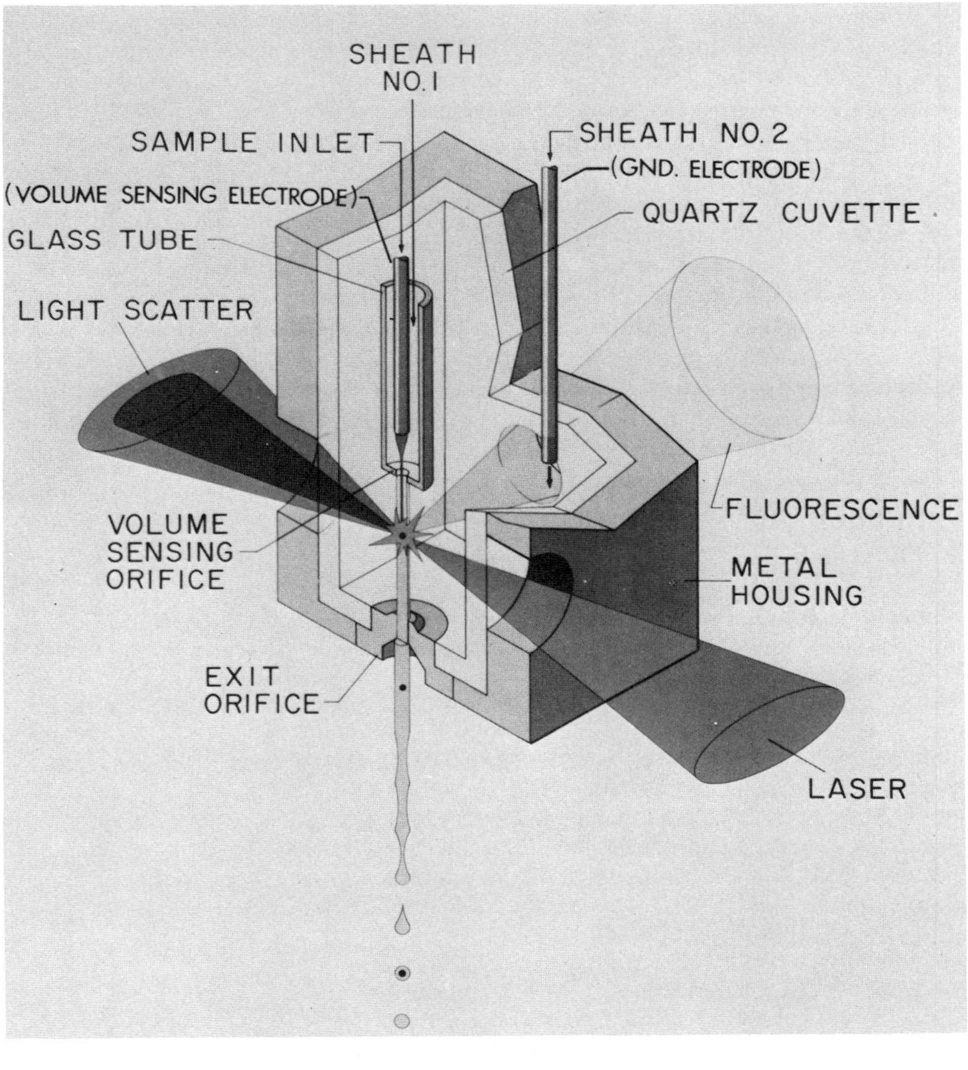

Figure 9 Sectional illustration of the flow chamber of a cell sorter, showing sample and sheath fluid inlet tubes, cell stream, cell volume-sensing orifice, laser illumination, and liquid jet from the exit orifice. (From Steinkamp, 1977.)

III. SINGLE-CELL SEPARATIONS

A. Flow Sorting

Over the past 15 years gradual progress in the development of high-technology-based instrumentation systems has made possible the isolation of individual cells (sorting) on the başis of multiple parameters (Melamed et al., 1979). Fluorescent staining techniques for cells based on monoclonal antibodies in conjunction with quantitative volume measurements are the most frequently used two-parameter separation systems. The cells are sorted by the instrument individually and at high speed (i.e., up to 10^5 cells per minute).

Typically, cells stained with fluorescent-labeled antibodies bound to specific surface markers are suspended in a carrier liquid (e.g., phosphate buffered saline) and are introduced into a flow chamber. The cell suspension is sufficiently dilute to allow cells to pass one by one through an electronic volume sensor and to intersect an argon laser beam of excitation light which makes the labeled cells fluoresce (Fig. 9). The emitted light due to fluorescence is detected by a photomultiplier and quantitated by a photometer. Only cells that contain bound fluorescent-labeled antibody emit light upon excitation. After this measurement, the stream carrying the cells emerges into the air as a liquid jet. A piezoelectric transducer is used to produce liquid droplets mechanically. Statistically, 1 droplet out of 40 contains a cell. At this time, if the measured (as above) volume-fluorescence parameters fall within a preset (by the operator) range, an electronic time relay is activated that triggers a droplet-charging pulse as the cell reaches the droplet formation point. As a result, a group of droplets, one of which contains the cell, is charged and subsequently is deflected by a static electric field (5000 V/cm) into a collection vessel. Cells that do not meet the present criterion do not trigger a droplet-charging pulse and therefore are not collected.

REFERENCES

Absolom, D. R., van Oss, C. J., and Neumann, A. W. (1981). Transfusion *21*, 663.

Albertsson, P. Å. (1971). *Partition of Cell Particles and Macromolecules.* Wiley-Interscience, New York.

Bartels, P. H., Olson, G. B., Bartels, H. G., Brooks, D. E., and Seaman, G. V. F. (1981). Cell Biophys. *3*, 373.

Boltz, R. C., Todd, P., Streibel, M. J., and Louie, M. K. (1973). Prep. Biochem. *3*, 383.

Boone, C. W., Harell, G. S., and Bond, H. E. (1968). J. Cell Biol. *36*, 369.

Boyse, E. A., Old, L. J., and Chomonlinkov, I. (1964). Methods Med. Res. *10*, 39.

Böyum, A. (1968). Scand. J. Clin. Lab. Invest. *21*(Suppl. 97), 51.

Bronson, P. M., and van Oss, C. J. (1979). Prep. Biochem. *9*, 61.

Catsimpoolas, N., Ed. (1977). *Methods of Cell Separation*, Vol. 1, 1977; Vol. 2, 1979; Vol. 3, 1980. Plenum, New York.

Catsimpoolas, N. (1980). Electrophoresis *1*, 73.

Catsimpoolas, N., and Griffith, A. L. (1977). In *Methods of Cell Separation*, Vol. 1 (N. Catsimpoolas, Ed.). Plenum, New York, p. 1.

Catsimpoolas, N., Griffith, A. L., Skrabut, E. M., and Valeri, C. R. (1978). Anal. Biochem. *87*, 243.

Greaves, M. F., and Brown, G. (1974). J. Immunol. *112*, 420.

Greenwalt, T. J., Gajewski, M., and McKenna (1962). Transfusion *2*, 221.

Griffith, A. L., Catsimpoolas, N., and Wortis, H. H. (1975). Life Sci. *16*, 1693.

Hannig, K. (1967). In *Electrophoresis: Theory, Methods, and Applications*, Vol. 2 (M. Bier, Ed.). Academic Press, New York, p. 423.
Kronick, P. L., Campbell, G. LeM., and Joseph, K. (1978). Science *200*, 1074.
Leif, R. C., and Vinograd, J. (1964). Proc. Natl. Sci. USA *51*, 520.
Lindahl, P. E. (1948). Nature (Lond.) *161*, 448.
Mel, H. (1963). Nature (Lond.) *200*, 423.
Melamed, M. R., Mullaney, P. F., and Mendelsohn, M. D., Eds. (1979). *Flow Cytometry and Sorting*. Wiley, New York.
Miller, R. G., and Phillips, R. A. (1969). J. Cell. Physiol. *73*, 191.
Platsoucas, C., and Catsimpoolas, N. (1979). Cell Biophys. *1*, 161.
Pretlow, T. G. (1971). Anal. Biochem. *41*, 248.
Pretlow, T. G., Boone, C. W., Shrager, R. I., and Weiss, G. H. (1969). Anal. Biochem. *29*, 230.
Raff, M. C. (1970). Nature (Lond.) *226*, 1257.
Sanderson, R. J., Bird, K. E., Palmer, N. F., and Brennan, J. (1976). Anal. Biochem. *71*, 615.
Shaw, D. J. (1969). *Electrophoresis*. Academic Press, New York.
Steinkamp, J. A. (1977). In *Methods of Cell Separation*, Vol. 1 (N. Catsimpoolas, Ed.). Plenum, New York, p. 251.
Tulp, A., and Bont, W. S. (1975). Anal. Biochem. *67*, 11.
Wahl, S. M., Rosenstreich, D. L., and Oppenheim, J. J. (1976). In *In Vitro Methods in Cell-Mediated and Tumor Immunity* (R. R. Bloom, and J. R. David, Eds.). Academic Press, New York, p. 231.
Walter, H. (1977). In *Methods of Cell Separation*, Vol. 1 (N. Catsimpoolas, Ed.). Plenum, New York, p. 307.
Ware, B. R. (1974). Adv. Colloid Interface Sci. *4*, 1.
Wigzell, H., and Andersson, B. (1969). J. Exp. Med. *129*, 23.
Wofsy, L., Kimura, J., and Truffa-Bachi, P. (1971). J. Immunol. *107*, 725.

Appendix

Histocompatibility Antigens, Lymphocyte Interactions, and the Immune Response: Toward a Solution of the Immunological Puzzle

Christopher J. Krco Mayo Clinic, Rochester, Minnesota

Among the major goals of immunologic research are the unraveling, identification, and characterization of the components that comprise an optimal immune response. Through the years the immunological subspecialities have become increasingly more diversified and at times seemingly disjointed from one another. Of course this is only an illusion since ultimately all parts will intertwince into a unified theory of immunologic structure and function. Several chapters in this textbook summarize the current theories and knowledge of several immunologic subspecialities. A cursory examination of the titles may not immediately reveal their interrelationships but they all indeed are interdependent, and collectively they contribute a significant piece to the solution of the immunologic puzzle.

In "Genetic Regulation of Immune Response" by Long and David (Chap. 25), an historical review of the concept of genetic control of immunity is presented and the role of the mouse as an indispensable model for immunologic analyses is described. Since the turn of the century the mouse has served as an immunogenetic yardstick by which one measures and monitors phylogenetically the evolution and diversification of immune response genes in mammals and other species. By challenging mice with batteries of cell-surface native and synthetic proteins as antigenic probes, fundamental concepts of immunology have been formulated. The data from such experimentation have resulted in the identification of a gene cluster, the major histocompatibility complex, (MHC) that confers and regulates immunological responses.

The murine MHC has been divided into four regions (*H-2K, H-2D, I*, and *S*). The *H-2D* genes encode products that are called class I histocompatibility antigens and which act as targets for graft rejection. The *I* region genes encode immune response gene products (identified as class II products) that confer and regulate immune responsiveness. The *S* region encodes complement components. By comparison, the genetic organzation of the human major histocompatibility complex (called the HLA complex) is remarkably

similar. Knowledge concerning the HLA complex is summarized by Juji and Kano in "The HLA System" (Chap. 26). The HLA complex consists of three genes (HLA-*A*, HLA-*B*, and HLA-*C*) that determine histocompatibility (class I) gene products. The HLA-*D*(*DR*) region encodes the human equivalent of class II immune response gene products. Thus, with allowances for differences in the number and sequence of genes, both the human and mouse MHC encode both class I and class II gene products that function in graft rejection and regulation of immune responses. The role of HLA-*D*(HLA-*DR*) in disease susceptibility is outlined by Juji and Kano and correlates with the role of mouse *I* region genes in conferring susceptibility/resistance in mouse strains to disease-provoking challenges. Immunochemistry of the HLA class I and class II antigens is also summarized by Juji and Kano. The HLA-*A*, HLA-*B*, and HLA-*C* transplantation antigens are approximately 45,000 in m.w. and are noncovalently associated with a 12,000 m.w. component called β_2-microglobulin. The class II [HLA-*D*(HLA-*DR*)] encoded gene products are composed of two subunits of 28,000 and 32,000 m.w. which are noncovalently associated with the cell surface.

One of the problems in immunochemical analyses of gene products of the HLA complex is the lack of genetically defined cells and antisera necessary for fine genetic analyses. Once again the mouse system has permitted the gathering of detailed immunochemical data concerning the class I and class II gene products. That information, summarized by Zaleski (Chap. 27) in "Immunochemistry of Murine Major Histocompatibility Antigens," demonstrated the power of mouse immunogenetics in providing a data bank for comparing and contrasting the genetic organizations of mammalian MHCs. The mouse *H-2K* and *H-2D* gene products are also of m.w. 45,000 and also are noncovalently associated on the cell surface with the 12,000 m.w. β_2-microglobulin. The immune response associated gene products (class II, or Ia antigens) also have two subunits of 28,000 and 32,000 m.w. In addition, the structure of carbohydrate residues on the class I antigens is known and is discussed by Zaleski. As usual, groundbreaking data using mice have revealed some possible functions of the histocompatibility and immune response gene products. This is due in part to the availability of well-defined mutants. By correlating structure/function relationships it is becoming possible to identify the functions and mechanism mode of action of these gene products. One well-established finding has been the observation that the class II gene products regulate intercellular communication among lymphocytes.

Just how intricate the pathways of intercellular communication can be is clearly outlined in Janeway's "Lymphocyte Interactions" (Chap. 24). In this overview Janeway defines the functions of lymphocytes and the pathways by which these functions are performed and regulated. At least eight lymphocyte interaction pathways are identified. Many of the pathways involve the interactions of distinct specialized subpopulations of lymphocytes. In order to monitor the function of a given lymphocyte subpopulation unique cell surface markers (differentiation antigens) are used. The differentiation antigens themselves are exceedingly complex. Knowledge of the serology, chemistry, and genetics of these antigens is crucial for understanding the immunobiology of lymphocytes.

The complexity of differentiation antigens and the role they play in immunoregulation is outlined by Skelly, Potter, and Ahmed in "Lymphocyte Antigens and Receptors" (Chap. 28). They describe the known differentiation markers on murine and human T and B cells, and the chemistry and methods of detection are summarized. In addition, the roles of lectins, hormones, and lymphokine receptors in lymphocyte immunobiology are discussed. One common component of many transplantation antigens is the 12,000 m.w. β_2-microglobulin. The chemistry of this important molecule is summarized by Lillehoj and Poulik in "β_2-Microglobulins" (Chap. 9).

Author Index

Numbers indicate that an author's work is referred to. Italic numbers give the page on which the complete reference is listed.

Subject Index